D1383049

Non-Uniform Random Variate Generation

Luc Devroye

Non-Uniform Random Variate Generation

Springer-Verlag
New York Berlin Heidelberg Tokyo

Luc Devroye
School of Computer Science
McGill University
Montreal   H3A 2K6
Canada

AMS Classifications: 62-H12, 62-G05, 68-K05, 90-A99, 90-B99

Library of Congress Cataloging in Publication Data
Devroye, Luc.
   Non-uniform random variate generation.
   Bibliography: p.
   Includes index.
   1. Random variables.   I. Title.
QA274.D48   1986   519.2   86-3783

Printed and bound by R.R. Donnelley & Sons, Harrisonburg, Virginia.
Printed in the United States of America.

9 8 7 6 5 4 3 2 1

ISBN 0-387-96305-7 Springer-Verlag New York Berlin Heidelberg Tokyo
ISBN 3-540-96305-7 Springer-Verlag Berlin Heidelberg New York Tokyo

PREFACE

This text is about one small field on the crossroads of statistics, operations research and computer science. Statisticians need random number generators to test and compare estimators before using them in real life. In operations research, random numbers are a key component in large scale simulations. Computer scientists need randomness in program testing, game playing and comparisons of algorithms.

The applications are wide and varied. Yet all depend upon the same computer generated random numbers. Usually, the randomness demanded by an application has some built-in structure: typically, one needs more than just a sequence of independent random bits or independent uniform [0,1] random variables. Some users need random variables with unusual densities, or random combinatorial objects with specific properties, or random geometric objects, or random processes with well defined dependence structures. This is precisely the subject area of the book, the study of non-uniform random variates.

The plot evolves around the expected complexity of random variate generation algorithms. We set up an idealized computational model (without overdoing it), we introduce the notion of uniformly bounded expected complexity, and we study upper and lower bounds for computational complexity. In short, a touch of computer science is added to the field. To keep everything abstract, no timings or computer programs are included.

This was a labor of love. George Marsaglia created CS690, a course on random number generation at the School of Computer Science of McGill University. The text grew from course notes for CS690, which I have taught every fall since 1977. A few ingenious pre-1977 papers on the subject (by Ahrens, Dieter, Marsaglia, Chambers, Mallows, Stuck and others) provided the early stimulus. Bruce Schmeiser's superb survey talks at various ORSA/TIMS and Winter Simulation meetings convinced me that there was enough structure in the field to warrant a separate book. This belief was reinforced when Ben Fox asked me to read a preprint of his book with Bratley and Schrage. During the preparation of the text, Ben's critical feedback was invaluable. There are many others whom I would like to thank for helping me in my understanding and suggesting interesting problems. I am particularly grateful to Richard Brent, Jo Ahrens, Uli Dieter, Brian Ripley, and to my ex-students Wendy Tse, Colleen Yuen and Amir

Naderlsamanl. For stlmull of another nature during the past few months, I thank my wlfe Bea, my chlldren Natasha and Blrglt, my Burger Klng mates Jeanne Yuen and Kent Chow, my sukebe frlends ln Toronto and Montreal, and the supreme sukebe, Bashekku Shubataru. Wlthout the flnanclal support of NSERC, the research leadlng to thls work would have been lmposslble. The text was typed (wlth one flnger) and edlted on LISA's Offlce System before lt was sent on to the School's VAX for troff-lng and laser typesettlng.

TABLE OF CONTENTS

XI. MULTIVARIATE DISTRIBUTIONS   554

Chapter One
INTRODUCTION

1. GENERAL OUTLINE.

Random number generation has intrigued scientists for a few decades, and a lot of effort has been spent on the creation of randomness on a deterministic (non-random) machine, that is, on the design of computer algorithms that are able to produce "random" sequences of integers. This is a difficult task. Such algorithms are called generators, and all generators have flaws because all of them construct the n-th number in the sequence in function of the $n-1$ numbers preceding it, initialized with a nonrandom seed. Numerous quantities have been invented over the years that measure just how "random" a sequence is, and most well-known generators have been subjected to rigorous statistical testing. However, for every generator, it is always possible to find a statistical test of a (possibly odd) property to make the generator flunk. The mathematical tools that are needed to design and analyze these generators are largely number theoretic and combinatorial. These tools differ drastically from those needed when we want to generate sequences of integers with certain non-uniform distributions, given that a perfect uniform random number generator is available. The reader should be aware that we provide him with only half the story (the second half). The assumption that a perfect uniform random number generator is available is now quite unrealistic, but, with time, it should become less so. Having made the assumption, we can build quite a powerful theory of non-uniform random variate generation.

The existence of a perfect uniform random number generator is not all that is assumed. Statisticians are usually more interested in continuous random variables than in discrete random variables. Since computers are finite memory machines, they cannot store real numbers, let alone generate random variables with a given density. This led us to the following assumptions:

Assumption 1.   Our computer can store and manipulate real numbers.

Assumption 2.   There exists a perfect uniform [0,1] random variate generator, i.e. a generator capable of producing a sequence $U_1, U_2, ...$ of independent random variables with a uniform distribution on [0,1].

The generator of assumption 2 is our fundamental building block. The sequence of U_i 's can be intelligently manipulated to give us random variables with specified distributions in R^d, d -dimensional Euclidean space. Occasionally, we mention the effect that the finite word-length of the computer has on the manipulated sequence. With the two assumptions given above, we demand that the random variables obtained by combining the U_i 's have the exact distribution that was asked. Algorithms or generators with this property is called exact. Exact algorithms approach reality if we use extended precision arithmetic (some languages allow users to work with integers of virtually unlimited length by linking words together in a linked list). Inexact algorithms, which are usually algorithms that are based upon a mathematical approximation of sorts, are forever excluded, because neither extended precision arithmetic nor improvements in the basic random number generator make them more exact.

A random variate generation algorithm is a program that halts with probability one and exits with a real number X. This X is called a **random variate**. Because of our assumptions, we can treat random variates as if they were random variables! Note also that if we can produce one random variate X, then we are able to produce a sequence $X_1, X_2, ...$ of independent random variates distributed as X (this follows from assumption 2). This facilitates our task a lot: rather than having to concentrate on infinite sequences, we just need to look at the properties of single random variates.

Simple, easy-to-understand algorithms will survive longer, all other things being roughly equal. Unfortunately, such algorithms are usually slower than their more sophisticated counterparts. The notion of time itself is of course relative. For theoretical purposes, it is necessary to equate time with the number of "fundamental" operations performed before the algorithm halts. This leads to our third assumption:

Assumption 3. The fundamental operations in our computer include addition, multiplication, division, compare, truncate, move, generate a uniform random variate, exp, log, square root, arc tan, sin and cos. (This implies that each of these operations takes one unit of time regardless of the size of the operand(s). Also, the outcomes of the operations are real numbers.)

The complexity of an algorithm, denoted by C, is the time required by the algorithm to produce one random variate. In many cases, C itself is a random variable since it is a function of $U_1, U_2, ...$ We note here that we are mainly interested in generating independent sequences of random variables. The average complexity per random variate in a sequence of length n is

$$\frac{1}{n} \sum_{i=1}^{n} C_i$$

where C_i is the complexity for the i -th random variate. By the strong law of large numbers, we know that this average tends with probability one to the expected complexity, $E(C)$. There are examples of algorithms with infinite expected complexity, but for which the probability that C exceeds a certain small constant is extremely small. These should not be a priori discarded.

We have now set the stage for the book. Our program is ambitious. In the remainder of this chapter, we introduce our notation, and define some distributions. By carefully selecting sections and exercises from the book, teachers could use it to introduce their students to the fundamental properties of distributions and random variables. Chapters II and III are crucial to the rest of the book: here, the principles of inversion, rejection, and composition are explained in all their generality. Less universal methods of random variate generation are developed in chapter IV. All of these techniques are then applied to generate random variates with specific univariate distributions. These include small families of densities (such as the normal, gamma or stable densities), small families of discrete distributions (such as the binomial and Poisson distributions), and families of distributions that are too large to be described by a finite number of parameters (such as all unimodal densities or all densities with decreasing hazard rate). The corresponding chapters are IX, X and VII. We devote chapter XI to multivariate random variate generation, and chapter VI to random process generation. In these chapters, we want to create dependence in a very specific way. This effort is continued in chapters XII and XIII on the generation of random subsets and the generation of random combinatorial objects such as random trees, random permutations and random partitions.

We do not touch upon the applications of random variate generation in Monte Carlo methods for solving various problems (see e.g. Rubinstein,1981): these problems include stochastic optimization, Monte Carlo integration, solving linear equations, deciding whether a large number is prime, etcetera. We will spend an entire section, however, on the important topic of discrete event simulation, driven by the beauty of some data structures used to make the simulation more efficient. As usual, we will not describe what happens inside some simulation languages, but merely give timeless principles and some analysis. Some of this is done in chapter XIV.

There are a few other chapters with specialized topics: the usefulness of order statistics is pointed out in chapter V. Shortcuts in simulation are highlighted in chapter XVI, and the important table methods are given special treatment in a chapter of their own (VIII). The reader will note that not a single experimental result is reported, and not one computer is explicitly named. The issue of programming in assembler language versus a high level language is not even touched (even though we think that assembler language implementations of many algorithms are essential). All of this is done to insure the universality of the text. Hopefully, the text will be as interesting in 1995 as in 1985 by not dwelling upon the shortcomings of today's computers. In fact, the emphasis is plainly upon complexity, the number of operations (instructions) needed to carry out certain tasks. Thus, chapter XV could very well be the most important chapter in the book for the future of the subject: here computers are treated as bit manipulating machines. This approach allows us to deduce lower bounds for the time needed to generate random variates with certain distributions.

We have taught some of the material at McGill University's School of Computer Science. For a graduate course on the subject for computer scientists, we recommend the material with a combinatorial and algorithmic flavor. One could cover, not necessarily in the order given, parts of chapters I and II, all of chapter III, sections V.2 and V.3, selected examples from chapter X, all of chapters XII, XIII and XV, and section XIV.5. In addition, one could add chapter VIII. We usually cover I.1-3, II.1-2, II.3.1-2, II.3.6, II.4.1-2, III, V.1-3, V.4.1-4, VI.1, VIII.2-3, XII.1-2, XII.3.1, XII.4-5, XIII.1, XIII.2.1, XIII.3.3, XIII.4-5, and XIV.5.

In a statistics department, the needs are very different. A good sequence would be chapters II, III, V, VI, VII.2.1-3, selected examples from chapters IX,X, and chapter XII. In fact, this book can be used to introduce some of these students to the famous distributions in statistics, because the generators demand that we understand the connections between many distributions, that we know useful representations of distributions, and that we are well aware of the shape of densities and distribution functions. Some designs require that we disassemble some distributions, break densities up into parts, find tight inequalities for density functions.

The attentive reader notices very quickly that inequalities are ubiquitous. They are required to obtain efficient algorithms of all kinds. They are also useful in the analysis of the complexity. When we can make a point with inequalities, we will do so. A subset of the book could be used as the basis of a fun reading course on the development and use of inequalities: use parts of chapter I as needed, cover sections II.2, II.3, II.4.1, II.5.1, brush through chapter III, cover sections IV.5-7, include nearly all of chapter VII, and move on to sections VIII.1-2, IX.1.1-2, IX.3.1-3, IX.4, IX.6, X.1-4, XIV.3-4.

This book is intended for students in operations research, statistics and computer science, and for researchers interested in random variate generation. There is didactical material for the former group, and there are advanced technical sections for the latter group. The intended audience has to a large extent dictated the layout of the book. The introduction to probability theory in chapter I is not sufficient for the book. It is mainly intended to make the reader familiar with our notation, and to aid the students who will read the simpler sections of the book. A first year graduate level course in probability theory and mathematical statistics should be ample preparation for the entire book. But pure statisticians should be warned that we use quite a few ideas and "tricks" from the rich field of data structures and algorithms in computer science. Our short PASCAL programs can be read with only passing familiarity with the language.

Nonuniform random variate generation has been covered in numerous books. See for example Jansson (1966), Knuth (1969), Newman and Odell (1971), Yakowitz (1977), Fishman (1978), Kennedy and Gentle (1980), Rubinstein (1981), Payne (1982), Law and Kelton (1982), Bratley, Fox and Schrage (1983), Morgan (1984) and Banks and Carson (1984). In addition, there are quite a few survey articles (Zelen and Severo (1972), McGrath and Irving (1973), Patil, Boswell and Friday (1975), Marsaglia (1976), Schmeiser (1980), Devroye (1981), Ripley (1983) and Deak (1984)) and bibliographies (Sowey (1972), Nance and Overstreet (1972), Sowey (1978), Deak and Bene (1979), Sahai (1979)).

2. ABOUT OUR NOTATION.

In this section, we will briefly introduce the reader to the different formats that are possible for specifying a distribution, and to some of the most important densities in mathematical statistics.

2.1. Definitions.

A random variable X has a density f on the real line if for any Borel set A,

$$P(X \in A) = \int_A f(x)\, dx.$$

In other words, the probability that X belongs to A is equal to the area under the graph of f. The distribution function F of X is defined by

$$F(x) = P(X \leq x) = \int_{-\infty}^{x} f(y)\, dy\ , \qquad (x \in R)\ .$$

We have $F'(x) = f(x)$ for almost all x. The mean value of X is

$$E(X) = \int x\ f(x)\ dx\ ,$$

provided that this integral exists. The r-th moment of X is defined by $E(X^r)$. If the second moment of X is finite, then its variance is defined by

$$Var(X) = E((X - E(X))^2) = E(X^2) - E^2(X)\ .$$

A mode of X, if it exists, is a point at which f attains its maximal value. If g is an arbitrary Borel measurable function and X has density f, then $E(g(X)) = \int g(x)\ f(x)\ dx$. A p-th quantile of a distribution, for $p \in (0,1)$, is any point x for which $F(x) = p$. The 0.5 quantile is also called the median. It is known that for nonnegative X,

$$E(X) = \int_0^{\infty} P(X \geq x)\, dx\ .$$

A distribution is completely specified when its distribution function is given. We recall that any nondecreasing function F, right-continuous, with limits 0 and 1 as $x \to -\infty$ and $x \to \infty$ respectively, is always the distribution function of some random variable. The distribution of a random variable is also completely known when the characteristic function

$$\phi(t) = E(e^{itX})\ , t \in R\ ,$$

is given. For more details on the properties of distribution functions and characteristic functions, we refer to standard texts in probability such as Chow and Teicher (1978).

A random vector in R^d has a distribution function

$$F(x_1, \ldots, x_d) = P(X_1 \leq x_1, \ldots, X_d \leq x_d) .$$

The random vector $(X_1, \ldots, X_d)$ has a density $f(x_1, \ldots, x_d)$ if and only if for all Borel sets A of R^d,

$$P((X_1, \ldots, X_d) \in A) = \int_A \int f(x_1, \ldots, x_d) \; dx_1 \cdots dx_d .$$

The characteristic function of this random variable is

$$\phi(t_1, \ldots, t_d) = E(e^{it_1 X_1 + \cdots + it_d X_d}) \qquad ((t_1, \ldots, t_d) \in R^d) .$$

The X_i's are called marginal random variables. The marginal distribution function of X_1 is

$$F_1(x) = F(x, \infty, \ldots, \infty) \qquad (x \in R) .$$

Its marginal characteristic function is

$$\phi_1(t) = \phi(t, 0, \ldots, 0) , \qquad (t \in R) .$$

Another important notion is that of independence. Two random variables X_1 and X_2 are independent if and only if for all Borel sets A and B,

$$P(X_1 \in A, X_2 \in B) = P(X_1 \in A) P(X_2 \in B) .$$

Thus, if F is the distribution function of (X_1, X_2), then X_1 and X_2 are independent if and only if

$$F(x_1, x_2) = F_1(x_1) F_2(x_2) , \quad \text{all } (x_1, x_2) \in R^2 ,$$

for some functions F_1 and F_2. Similarly, if (X_1, X_2) has a density f, then X_1 and X_2 are independent if and only if this density can be written as the product of two marginal densities. Finally, X_1 and X_2 are independent if and only if for all bounded Borel measurable functions g_1 and g_2:

$$E(g_1(X_1)g_2(X_2)) = E(g_1(X_1)) E(g_2(X_2)) .$$

In particular, the characteristic function of two independent random variables is the product of their characteristic functions:

$$\phi(t_1, t_2) = E(e^{it_1 X_1} e^{it_2 X_2}) = E(e^{it_1 X_1}) E(e^{it_2 X_2}) = \phi_1(t_1) \phi_2(t_2) .$$

All the previous observations can be extended without trouble towards d random variables $X_1, \ldots, X_d$.

2.2. A few important univariate densities.

In the table shown below, several important densities are listed. Most of them have one or two parameters. From a random variate generation point of view, several of these parameters are unimportant. For example, if X is a random variable with a distribution having three parameters, a, b, c, and when $kX + l$ has a distribution with parameters $ka + l, kb, c$, then b is called a scale parameter, and a is called a translation parameter. The shape of the distribution is only determined by the parameter c: since c is invariant to changes in scale and to translations, it is called a shape parameter. For example, the normal distribution has no shape parameter, and the gamma distribution has one shape parameter.

Some univariate densities.				
$f(x)$	$E(X)$	$Var(X)$	$\text{Mode}(X)$	$F(x)$
Normal(μ,σ^2) $\dfrac{1}{\sigma\sqrt{2\pi}}e^{-\frac{(x-\mu)^2}{2\sigma^2}}$	μ	σ^2	μ	$\int_{-\infty}^{x} f(y)\ dy$
Gamma(a,b) $\dfrac{1}{\Gamma(a)b^a}x^{a-1}e^{-\frac{x}{b}}$ $(x>0)$	ab	ab^2	$(a-1)b$	$\int_{-\infty}^{x} f(y)\ dy$
Exponential(λ) $\lambda e^{-\lambda x}$ $(x>0)$	$\dfrac{1}{\lambda}$	$\dfrac{1}{\lambda^2}$	0	$1-e^{-\lambda x}$
Cauchy(σ) $\dfrac{\sigma}{\pi(x^2+\sigma^2)}$	does not exist	does not exist	0	$\dfrac{1}{2}+\dfrac{1}{\pi}\arctan(\dfrac{x}{\sigma})$
Pareto(a,b) $\dfrac{ab^a}{x^{a+1}}$ $(x>b)$	$\dfrac{ab}{a-1}$ $(a>1)$	$\dfrac{ab^2}{(a-2)(a-1)^2}$ $(a>2)$	b	$1-\dfrac{b^a}{x}$
Beta(a,b) $\dfrac{\Gamma(a+b)}{\Gamma(a)\Gamma(b)}x^{a-1}(1-x)^{b-1}$ $(x\in[0,1])$	$\dfrac{a}{a+b}$	$\dfrac{ab}{(a+b)^2(a+b+1)}$	$\dfrac{a-1}{a+b-2}$ $(a,b>1)$	$\int_{-\infty}^{x} f(y)\ dy$

A variety of shapes can be found in this table. For example, the beta family of densities on [0,1] has two shape parameters, and the shapes vary from standard unimodal forms to J-shapes and U-shapes. For a comprehensive description of most parametric families of densities, we refer to the two volumes by Johnson and Kotz (1970). When we refer to normal random variables, we mean normal random variables with parameters 0 and 1. Similarly, exponential random variables are exponential (1) random variables. The uniform [0,1] density is the density which puts its mass uniformly over the interval [0,1]:

$$f(x) = I_{[0,1]}(x) \qquad (x\in R).$$

Here I is the indicator function of a set. Finally, when we mention the gamma (a) density, we mean the gamma $(a,1)$ density.

The strategy in this book is to build from simple cases: simple random variables and distributions are random variables and distributions that can easily be generated on a computer. The context usually dictates which random variables are meant. For example, the uniform [0,1] distribution is simple, and so are the exponential and normal distributions in most circumstances. At the other end of the scale we have the difficult random variables and distributions. Most of this book is about the generation of random variates with difficult distributions. To clarify the presentation, it is convenient to use the same capital letters for all simple random variables. We will use N, E and U for normal, exponential and uniform [0,1] random variables. The notations G and B are often used for gamma and beta random variables. For random variables in general, we will reserve the symbols X, Y, W, Z, V.

3. ASSESSMENT OF RANDOM VARIATE GENERATORS.

One of the most difficult problems in random variate generation is the choice of an appropriate generator. Factors that play an important role in this choice include:

1. Speed.
2. Set-up (initialization) time.
3. Length of the compiled code.
4. Machine independence, portability.
5. Range of the set of applications.
6. Simplicity and readability.

Of these factors, the last one is perhaps the most neglected in the literature. Users are more likely to work with programs they can understand. Five line programs are easily typed in, and the likelihood of making errors is drastically reduced. Even packaged generators can have subtle bugs in their conception or implementation. It is nearly impossible to certify that programs with dozens, let alone hundreds, of lines of code are correct. So, we will often spend more time on simple algorithms than on sophisticated ultra-fast ones.

Subprograms for random variate generation can be divided into three groups: (1) subprograms with no variable parameters, such as subprograms for the normal (0,1) density; (2) subprograms with a finite number of variable parameters (these are typically for parametric classes of densities such as the class of all beta densities); (3) subprograms that accept names of other subprograms as arguments, and can be applied for a wide class of distributions (the description of this class is of course not dependent upon parameters).

3.1. Distributions with no variable parameters.

A frequently used subprogram for distributions with no variable parameters should be chosen very carefully: usually, speed is very important, while the length of the compiled code is less crucial. Clearly, the initialization time is zero, and in some cases it is worthwhile to write the programs in machine language. This is commonly done for distributions such as the normal distribution and the exponential distribution.

For infrequently used subprograms, it is probably not worth to spend a lot of time developing a fast algorithm. Rather, a simple expedient method will often do. In many cases, the portability of a program is the determining factor: can we use the program in different installations under different circumstances? Portable programs have to be written in a machine-independent language. Furthermore, they should only use standard library subprograms and be compiler-independent. Optimizing compilers often lead to unsuspected problems. Programs should follow the universal conventions for giving names to variables, and be protected against input error. The calling program should not be told to use special statements (such as the COMMON statement in FORTRAN). Finally, the subprogram itself is not assumed to perform unasked tasks (such as printing messages), and all conventions for subprogram linkage must be followed.

Assume now that we have narrowed the competition down to a few programs, all equally understandable and portable. The programs take expected time t_i per random variate where i refers to the i-th program ($1 \leq i \leq K$). Also, they require s_i bytes of storage. Among these programs, the j-th program is said to be **inadmissible** if there exists an i such that $t_j \geq t_i$ and $s_j \geq s_i$ (with at least one of these inequalities strict). If no such i exists, then the j-th program is admissible. If we measure the cost of the i-th program by some function $\psi(t_i, s_i)$, increasing in both its arguments, then it is obvious that the best program is an admissible program.

3.2. Parametric families.

The new ingredient for multi-parameter families is the set-up time, that is, the time spent computing constants that depend only upon the parameters of the distribution. We are often in one of two situations:

Case 1.   The subprogram is called very often for fixed values of the parameters. The set-up time is unimportant, and one can only gain by initializing as many constants as possible.

Case 2.   The parameters of the distribution change often between calls of the subprogram. The total time per variate is definitely influenced by the set-up time.

An example.

The admissibility of a method now depends upon the set-up time as well, as is seen from this example. Stadlober (1981) gave the following table of expected times per variate (in microseconds) and size of the program (in words) for several algorithms for the t distribution:

Algorithm:	TD	TROU	T3T
t a=3.5	65	66	78
t a=5	70	67	81
t a=10	75	68	84
t a=50	78	69	88
t a=1000	79	70	89
s	255	100	83
u	12	190	0

Here t stands for the expected time, a for the parameter of the distribution, s for the size of the compiled code, and u for the set-up time. TD, TROU and T3T refer to three algorithms in the literature. For any algorithm and any a, the expected time per random variate is $t + \lambda u$ where $\lambda \in [0,1]$ is the fraction of the variates that required a set-up. The most important cases are $\lambda = 0$ (one set-up in a large sample for fixed a) and $\lambda = 1$ (parameter changes at every call). Also, $1/\lambda$ is about equal to the waiting time between set-ups. Clearly, one algorithm dominates another timewise if $t + \lambda u$ considered as a function of λ never exceeds the corresponding function for the other algorithm. One can do this for each a, and this leads to quite a complicated situation. Usually, one should either randomize the entries of t over various values of a. Alternatively, one can compare on the basis of $t_{max} = \max_a t$. In our example, the values would be 79, 70 and 89 respectively. It is easy to check that $t_{max} + \lambda u$ is minimal for TROU when $0 \leq \lambda \leq 9/178$, for TD when $9/178 \leq \lambda \leq 5/6$, and for T3T when $5/6 \leq \lambda \leq 1$. Thus, there are no inadmissible methods if we want to include all values of λ. For fixed values of λ however, we have a given ranking of the $t_{max} + \lambda u$ values and the discussion of the inadmissibility in terms of $t_{max} + \lambda u$ and s is as for the distributions without parameters. Thus, TD is inadmissible in this sense for $\lambda > 5/6$ or $\lambda < 9/178$, and TROU is inadmissible for $\lambda > 1/10$. ■

Speed versus size.

It is a general rule in computer science that speed can be reduced by using longer more sophisticated programs. Fast programs are seldom short, and short programs are likely to be slow. But it is also true that long programs are often not elegant and more error-prone. Short smooth programs survive longer and are understood by a larger audience. This bias towards short programs will be apparent in chapters IV, IX and X where we must make certain recommendations to the general readership. ∎

4. OPERATIONS ON RANDOM VARIABLES.

In this section we briefly indicate how densities and distribution functions change when random variables are combined or operated upon in certain ways. This will allow us to generate new random variables from old ones. We are specially interested in operations on simple random variables (from a random variate generation point of view) such as uniform [0,1] random variables. The actual applications of these operations in random variate generation are not discussed in this introductory chapter. Most of this material is well-known to students in statistics, and the chapter could be skipped without loss of continuity by most readers. For a unified and detailed treatment of operations on random variables, we refer to Springer(1979).

4.1. Transformations.

Transformations of random variables are easily taken care of by the following device:

Theorem 4.1.

Let X have distribution function F, and let $h : R \rightarrow B$ be a strictly increasing function where B is either R or a proper subset of R. Then $h(X)$ is a random variable with distribution function $F(h^{-1}(x))$.

If F has density f and h^{-1} is absolutely continuous, then $h(X)$ has density

$$(h^{-1})'(x) \ f(h^{-1}(x)), \quad \text{for almost all } x .$$

Proof of Theorem 4.1.

Observe first that for arbitrary x,

$$P(h(X) \leq x) = P(X \leq h^{-1}(x)) = F(h^{-1}(x)).$$

This is thus the distribution function of $h(X)$. If this distribution function is absolutely continuous in x, then we know (Chow and Teicher (1978)) that $h(X)$ has a density that is almost everywhere equal to the derivative of the distribution function. This is the case for example when both F and h^{-1} are absolutely continuous, and the formal derivative is the one shown in the statement of the Theorem. ∎

Example 4.1. Linear transformations.

If F is the distribution function of a random variable X, then $aX + b$ has distribution function $F((x - b)/a)$ when $a > 0$. The corresponding densities, if they exist, are $f(x)$ and $\frac{1}{a} f(\frac{x-b}{a})$. Verify that when X is gamma (a, b) distributed, then cX is gamma (a, cb), all $c > 0$. ∎

Example 4.2. The exponential distribution.

When X has distribution function F and $\lambda > 0$ is a real number, then $-\frac{1}{\lambda} \log X$ has distribution function $1 - F(e^{-\lambda x})$, which can be verified directly:

$$P(-\frac{1}{\lambda} \log X \leq x) = P(X \geq e^{-\lambda x}) = 1 - F(e^{-\lambda x}) \quad (x > 0).$$

In particular, if X is uniform $[0,1]$, then $-\frac{1}{\lambda} \log X$ is exponential (λ). Vice versa, when X is exponential (λ), then $e^{-\lambda X}$ is uniform $[0,1]$. ∎

Example 4.3. Power transformations.

When X has distribution function F and density f, then X^p ($p>0$ is a real number, and the power is defined as a sign-preserving transformation) has distribution function $F(x^{\frac{1}{p}})$ and density

$$\frac{1}{p}x^{\frac{1}{p}-1}f(x^{\frac{1}{p}}).\blacksquare$$

Example 4.4. Non-monotone transformations.

Non-monotone transformations are best handled by computing the distribution function first from general principles. To illustrate this, let us consider a random variable X with distribution function F and density f. Then, the random variable X^2 has distribution function

$$P(X^2\leq x) = P(\mid X\mid \leq\sqrt{x}) = F(\sqrt{x})-F(-\sqrt{x}) \qquad (x>0)$$

and density

$$\frac{1}{\sqrt{x}}\frac{f(\sqrt{x})+f(-\sqrt{x})}{2}.$$

In particular, when X is normal $(0,1)$, then X^2 is gamma distributed, as can be seen from the form of the density

$$\frac{1}{\sqrt{x}}\frac{1}{2\sqrt{2\pi}}(e^{-\frac{x}{2}}+e^{-\frac{x}{2}}) = \frac{1}{\sqrt{2\pi}}x^{-\frac{1}{2}}e^{-\frac{x}{2}} \qquad (x\geq 0).$$

The latter density is known as the chi-square density with one degree of freedom (in shorthand: χ_1^2). $\blacksquare$

Example 4.5. A parametric form for the density.

Let X have density f and let h be as in Theorem 4.1. Then, putting $x=h(u)$ and $y=f(u)/h'(u)$, where y stands for the value of the density of $h(X)$ at x, and y and x are related through the parameter u, we verify by elimination of u that

$$y = f(h^{-1}(x)) / h'(h^{-1}(x)).$$

This is equal to $f\,(h^{-1}(x))h^{-1\prime}(x\,)$, which was to be shown. Thus, the parametric representation in terms of u given above is correct, and will give us a plot of the density versus x. This is particularly useful when the inverse of h is difficult to obtain in closed analytical form. For example, when X is uniform [0,1], then for $a\,,b>0$, $aX+bX^3$ has a density with parametric representation

$$x\ =\ au+bu^3\,,\ y\ =\ \frac{1}{a+3bu^2}\qquad(0\leq au+bu^3\leq1)\,.$$

By elimination of u, we obtain a simple formula of x in terms of y:

$$x\ =\ \sqrt{\frac{1}{y}-a}\ \sqrt{\frac{1}{3b}(\frac{2a}{3}+\frac{1}{3y})}\,.$$

The plot of y versus x has the following general form: it vanishes outside [0,1], and decreases monotonically on this interval from $y=\dfrac{1}{a}$ at $x=0$ to a nonzero value at $x=1$. Furthermore, $\dfrac{\partial y}{\partial x}$ at $u=0$ (i.e. at $x=0$), is 0, so that the shape of the density resembles that of a piece of the normal density near 0. ■

Let us now look at functions of several random variables. We can obtain many distributions as relatively uncomplicated functions of simple random variables. Many cases can be handled by the following d-dimensional generalization of Theorem 4.1:

Theorem 4.2.

Let X have a continuous density f on R^d and let $h:R^d\rightarrow R^d$ be a one-to-one and onto mapping to T, the image of S, the support set of f, under h. Thus, the inverse of the transformation $Y=h(X)$ exists: $X=h^{-1}(Y)=g(Y)$. If we write $y=(y_1,\ldots,y_d)$ and $g=(g_1,\ldots,g_d)$, then if the partial derivatives

$$g_{ij}=\frac{\partial g_i}{\partial y_j}$$

exist and are continuous on T, Y has density

$$f\,(g\,(y))\ |\,J\,|\qquad(y\in T)\,,$$

where J is the Jacobian of the transformation and is defined as the determinant of the matrix

$$\begin{bmatrix} g_{11} & \cdots & g_{1d} \\ \cdots & \cdots & \cdots \\ \cdots & \cdots & \cdots \\ g_{d1} & \cdots & g_{dd} \end{bmatrix}.$$

Example 4.6. The t distribution.

We will show here that when X is normal $(0,1)$ and Y is independent of X and gamma $(\frac{a}{2},2)$ distributed (this is called the chi-square distribution with a degrees of freedom), then

$$Z = X / \sqrt{\frac{Y}{a}}$$

is t distributed with a degrees of freedom, that is, Z has density

$$\frac{\Gamma(\frac{a+1}{2})}{\Gamma(\frac{a}{2})\sqrt{\pi a}} \frac{1}{(1+\frac{z^2}{a})^{\frac{a+1}{2}}} \qquad (z \in R).$$

What one does in a situation like this is "invent" a 2-dimensional vector random variable (for example, (Z,W)) that is a function of (X,Y), one of whose component random variables is Z. The obvious choice in our example is

$$Z = X / \sqrt{\frac{Y}{a}}$$
$$W = Y$$

The inverse transformation is determined by $X = Z\sqrt{\frac{W}{a}}$, $Y = W$. This inverse transformation has a Jacobian $\sqrt{\frac{w}{a}}$ where we use x,y,z,w for the running values that correspond to the random variables X,Y,Z,W. Thus, the density of (Z,W) is

$$c \; e^{-\frac{wz^2}{2a}} \; w^{\frac{a}{2}-1} \; e^{-\frac{w}{2}} \sqrt{\frac{w}{a}}$$

where

$$c = \frac{1}{\Gamma(\frac{a}{2})2^{\frac{a}{2}}\sqrt{2\pi}} \qquad (w>0, z \in R)$$

is a normalization constant. From a joint density, we obtain a marginal density by taking the integral with respect to the non-involved variables (in this case with respect to dw). In w, we have for fixed z a gamma $(\frac{a+1}{2}, \frac{2}{1+z^2/a})$ density times $\frac{c}{\sqrt{a}}$. After integration with respect to dw, we obtain

$$\frac{c}{\sqrt{a}}\Gamma(\alpha)\beta^{\alpha}$$

where α and β are the parameters of the gamma density given above. This is precisely what we needed to show. ∎

4.2. Mixtures.

Discrete mixtures.

Let Y be a positive integer valued random variable, and, given that $Y = i$, let X have density f_i. Then the (unconditional) density of X is

$$\sum_{i=1}^{\infty} P(Y=i) f_i(x) .$$

This device can be used to cut a given density f up into simpler pieces f_i that can be handled quite easily. Often, the number of terms in the mixture is finite. For example, if f is a piecewise linear density with a finite number of break-points, then it can always be decomposed (rewritten) as a finite mixture of uniform and triangular densities.

Continuous mixtures.

Let Y have density g on R, and given that $Y = y$, let X have density f_y (thus, y can be considered as a parameter of the density of X), then the density f of X is given by

$$f(x) = \int f_y(x) g(y) \ dy .$$

As an example, we consider a mixture of exponential densities with parameter Y itself exponentially distributed with parameter 1. Then X has density

$$f(x) = \int y e^{-yx} e^{-y} \ dy$$

$$= \int y e^{-\frac{y}{(x+1)^{-1}}} \ dy$$

$$= \frac{1}{(x+1)^2} \qquad (x > 0) .$$

Since the parameter of the exponential distribution is the inverse of the scale parameter, we see without work that when E_1, E_2 are independent exponential random variables, then E_1/E_2 has density $1/(x+1)^2$ on $[0,\infty)$.

Mixtures of uniform densities.

If we consider a mixture of uniform $[0,y]$ densities where y is the mixture parameter, then we obtain a density that is nonincreasing on $[0,\infty)$. The random variables X thus obtained are distributed as the product UY of a uniform $[0,1]$ random variable U and an arbitrary (mixture) random variable Y. These distributions will be of great interest to us since U is the fundamental random variable in random variate generation.

4.3. Order statistics.

If $U_1, \ldots, U_n$ are iid uniform $[0,1]$ random variables, then the order statistics for this sample are $U_{(1)}, \ldots, U_{(n)}$, where

$$U_{(1)} \le U_{(2)} \le \cdots \le U_{(n)}$$

and $U_{(1)}, \ldots, U_{(n)}$ is a permutation of $U_1, \ldots, U_n$. We know that $(U_1, \ldots, U_n)$ is uniformly distributed in the unit cube $[0,1]^n$. Thus, $(U_{(1)}, \ldots, U_{(n)})$ is uniformly distributed in the simplex S_n:

$$S_n = \{(x_1, \ldots, x_n) : 0 < x_1 < x_2 < \cdots < x_n < 1\} .$$

Theorem 4.3.

The joint density of $(U_{(1)}, \ldots, U_{(n)})$ is

$$n! I_{S_n}(x_1, \ldots, x_n) .$$

The i-th order statistic $U_{(i)}$ has the beta density with parameters i and $n-i+1$, i.e. its density is

$$\frac{\Gamma(n+1)}{\Gamma(i)\Gamma(n-i+1)} x^{i-1}(1-x)^{n-i} \qquad (x \in [0,1]) .$$

Proof of Theorem 4.3.

The first part is shown by a projection argument: there are $n!$ points in $[0,1]^n$ that map to a given point in S_n when we order them. This can be formalized as follows. Let A be an arbitrary Borel set contained in S_n. Writing $x_{(1)} < \cdots < x_{(n)}$ for the ordered permutation of $x_1, \ldots, x_n$, we have

$$\int_A dx_1 \cdots dx_n$$

$$= \sum_{\sigma} \int_{A,(x_1 = x_{\sigma(1)}, \ldots, x_n = x_{\sigma(n)})} dx_1 \cdots dx_n$$

$$(\sigma = \sigma(1), \ldots, \sigma(n) \text{ is a permutation of } 1, \ldots, n)$$

$$= \sum_{\sigma} \int_A dx_{(1)} \cdots dx_{(n)}$$

$$= \int_A n! \, dx_{(1)} \cdots dx_{(n)} .$$

The first part of the Theorem follows by the arbitrariness of A. For the second part, we choose x in $[0,1]$, and compute the marginal density of $U_{(i)}$ at x by integrating the density with respect to all variables x_j, $j \neq i$. This yields

$$n! \int_0^{x_2} \cdots \int_0^{x_1} \int_x^1 \int_{x_{n-1}}^1 dx_n \cdots dx_{i+1} dx_{i-1} \cdots dx_1 .$$

This gives the beta density with parameters i and $n-i+1$. ■

Of particular importance will be the distribution of $\max(U_1, \ldots, U_n)$: the distribution function is easily obtained by a direct argument because

$$P(\max(U_1, \ldots, U_n) \leq x) \qquad (x \in [0,1])$$
$$= P(U_1 \leq x) \cdots P(U_n \leq x)$$
$$= x^n$$
$$= P(U_1 \leq x^n)$$
$$= P(U_1^{\frac{1}{n}} \leq x) .$$

Thus, the distribution function is x^n on $[0,1]$, and the density is nx^{n-1} on $[0,1]$. We have also shown that $\max(U_1, \ldots, U_n)$ is distributed as $U_1^{1/n}$.

Another important order statistic is the median. The median of $U_1, \ldots, U_{2n+1}$ is $U_{(n)}$. We have seen in Theorem 4.3 that the density is

$$\frac{(2n+1)!}{n!^2} (x(1-x))^n \qquad (x \in [0,1]) .$$

Example 4.7.

If $U_{(1)}, U_{(2)}, U_{(3)}$ are the order statistics of three independent uniform $[0,1]$ random variables, then their densities on $[0,1]$ are respectively,

$$3(1-x)^2 ,$$
$$6x(1-x)$$

and

$$3x^2 . ■$$

The generalizations of the previous results to other distributions are straight-forward. If $X_1, \ldots, X_n$ are iid random variables with density f and distribution function F, then the maximum has distribution function F^n. From Theorem 4.3, we can also conclude that the i-th order statistic $X_{(i)}$ has density

$$\frac{n!}{(i-1)!(n-i)!} F(x)^{i-1}(1-F(x))^{n-i} f(x) .$$

4.4. Convolutions. Sums of independent random variables.

The distribution of the sum S_n of n random variables $X_1, \ldots, X_n$ is usually derived by one of two tools, convolution integrals or characteristic functions. In this section, we will write f_i, F_i, ϕ_i for the density, distribution function and characteristic function of X_i, and we will use the notation f, F, ϕ for the corresponding functions for the sum S_n. In the convolution method, we argue as follows:

$$F(x) = P(X_1 + \cdots + X_n \leq x)$$
$$= \int \prod_{i<n} f_i(y_i) F_n(x - y_1 - \cdots - y_{n-1}) \prod_{i<n} dy_i .$$

Also,

$$f(x) = \int \prod_{i<n} f_i(y_i) f_n(x - y_1 - \cdots - y_{n-1}) \prod_{i<n} dy_i .$$

Except in the simplest cases, these convolution integrals are difficult to compute. In many instances, it is more convenient to derive the distribution of S_n by finding its characteristic function. By the independence of the X_i's, we have

$$\phi(t) = E(e^{it(X_1 + \cdots + X_n)})$$
$$= \prod_{j=1}^{n} E(e^{itX_j})$$
$$= \prod_{j=1}^{n} \phi_j(t) .$$

If the X_i's are iid, then $\phi = \phi_1^n$.

Example 4.8. Sums of normal random variables.

First, we show that the characteristic function of a normal $(0,1)$ random variable is $e^{-t^2/2}$. To see this, note that it can be computed as follows for $t \in R$:

$$\int \frac{1}{\sqrt{2\pi}} e^{ity - y^2/2} dy$$

$$= e^{-t^2/2} \int \frac{1}{\sqrt{2\pi}} e^{-(y-it)^2/2} \, dy$$

$$= e^{-t^2/2} \, .$$

From the definition of the characteristic function we see that if X has characteristic function $\phi(t)$, then $aX + b$ has characteristic function $e^{ibt} \phi(at)$. Thus, a normal (μ, σ^2) random variable has characteristic function

$$e^{it\mu} \phi(\sigma t) \, .$$

If X_i is normal $(\mu_i, \sigma_i^{\,2})$, then S_n has characteristic function

$$\prod_{j=1}^{n} e^{it\mu_j} e^{-\sigma_j^{\,2}t^2/2}$$

$$= e^{it\sum_{j=1}^{n}\mu_j} e^{-\sum_{j=1}^{n}\sigma_j^{\,2}t^2/2} \, ,$$

which is the characteristic function of a normal random variable with parameters $\sum \mu_j$ and $\sum \sigma_j^{\,2}$. ∎

Example 4.9. Sums of gamma random variables.

In this example too, it is convenient to first obtain the characteristic function of a gamma (a, b) random variable. It can be computed as follows:

$$\int_0^\infty \frac{y^{a-1} e^{-y/b}}{\Gamma(a) b^a} e^{ity} \, dy \qquad \text{(by definition)}$$

$$= \int_0^\infty \frac{y^{a-1} e^{-y(1-itb)/b}}{\Gamma(a) b^a} \, dy$$

$$= \int_0^\infty \frac{z^{a-1} e^{-z/b}}{(1-itb)^a \Gamma(a) b^a} \, dz \qquad \text{(use } z = y(1-itb) \text{)}$$

$$= \frac{1}{(1-itb)^a} \, .$$

Thus, if $X_1, \ldots, X_n$ are independent gamma random variables with parameters a_i and b, then the sum S_n is gamma with parameters $\sum a_i$ and b. ∎

It is perhaps worth to mention that when the X_i's are iid random variables, then S_n, properly normalized, is nearly normally distributed when n grows large.

If the distribution of X_1 has mean μ and variance $\sigma^2 > 0$, then $(S_n - n\mu)/(\sigma\sqrt{n})$ tends in distribution to a normal $(0,1)$ random variable, i.e,

$$\lim_{n \to \infty} P(\frac{S_n - n\mu}{\sigma\sqrt{n}} \le x) = \frac{1}{\sqrt{2\pi}} e^{-x^2/2}, \quad \text{all } x .$$

This is called the central limit theorem (Chow and Teicher, 1978). This will be exploited further on in the design of algorithms for families of distributions that are closed under additions, such as the gamma or Poisson families. If the variance is not finite, then the limit law is no longer normal. See for example exercise 4.17, where an example is found of such non-normal attraction.

4.5. Sums of independent uniform random variables.

In this section we consider the distribution of

$$\sum_{i=1}^{n} a_i U_i$$

where the a_i's are positive constants and the U_i's are independent uniform $[0,1]$ random variables. We start with the main result of this section.

Theorem 4.4.

The distribution function of $\sum_{i=1}^{n} a_i U_i$ (where $a_i > 0$, all i, and the U_i's are independent uniform $[0,1]$ random variables) is given by

$$F(x) = \frac{1}{a_1 a_2 \cdots a_n \, n!} (x_+^n - \sum_i (x - a_i)_+^n + \sum_{i \ne j} (x - a_i - a_j)_+^n - \cdots) .$$

Here $(.)_+$ is the positive part of $(.)$. The density is obtained by taking the derivative with respect to x.

Proof of Theorem 4.4.

Consider the simplex S formed by the origin and the vertices on the n coordinate axes at distances $x/a_1, \ldots, x/a_n$, where $x > 0$ is the point at which we want to calculate $F(x)$. Let us define the sets B_i as

$$B_i = [0,\infty)^{i-1} \times (1,\infty) \times [0,\infty)^{n-i}$$

where $1 \le i \le n$. Note now that the first quadrant minus the unit cube $[0,1]^n$ can be decomposed by the inclusion/exclusion principle as follows:

$$[0,\infty)^n - [0,1]^n$$
$$= \sum_i B_i - \sum_{i \ne j} B_i \cap B_j + \cdots .$$

Now, since $F(x) = \text{area}(S \cap [0,1]^n) = \text{area}(S) - \text{area}(S \cap ([0,\infty)^n - [0,1]^n))$, we obtain

$$F(x) = \text{area}(S) - \sum_i \text{area}(S \cap B_i) + \sum_{i \neq j} \text{area}(S \cap B_i \cap B_j) - \cdots .$$

This is all we need, because for any subset J of $1, \ldots, n$, we have

$$\text{area}(S \cap_{i \in J} B_i) = \frac{(x - \sum_{i \in J} a_i)_+^n}{a_1 \cdots a_n \, n!} .$$

This concludes the proof of Theorem 4.4. ■

It is instructive to do the proof of Theorem 4.4 for the special case $n = 2$, and to draw the simplex and the various sets used in the geometric proof. For the important case $a_1 = a_2 = \cdots = a_n = 1$, the distribution function is

$$F(x) = \frac{1}{n!} \left(x_+^n - \binom{n}{1}(x-1)_+^n + \binom{n}{2}(x-2)_+^n - \cdots \right) .$$

In particular, for $n = 2$, obtaining the density by taking the derivative of the distribution function, we have

$$f(x) = x_+ - 2(x-1)_+ + (x-2)_+$$

$$= \begin{cases} 0 & \text{if } x < 0 \\ x & \text{if } 0 \leq x \leq 1 \\ 2-x & \text{if } 1 \leq x \leq 2 \\ 0 & \text{if } 2 < x \end{cases} .$$

In other words, the density has the shape of an isosceles triangle. In general, the density of $U_1 + U_2 + \cdots + U_n$ consists of pieces of polynomials of degree $n-1$ with breakpoints at the integers. The form approaches that of the normal density as $n \to \infty$.

4.6. Exercises.

1. If h is strictly monotone, h' exists and is continuous, g is a given density, and X is a random variable with density $h'(x)g(h(x))$, then $h(X)$ has density g. (This is the inverse of Theorem 4.1.)

2. If X has density $1/(x^2\sqrt{\pi\log x})$ $(x \geq 1)$, then $\sqrt{2\log X}$ is distributed as the absolute value of a normal random variable. (Use exercise 1.)

3. If X is a gamma $(\frac{1}{2}, 1)$ random variable, i.e. X has density $e^{-x}/\sqrt{\pi x}$ $(x > 0)$, then $\sqrt{2X}$ is distributed as the absolute value of a normal random variable. (Use exercise 1.)

4. Let A be a $d \times d$ matrix with nonzero determinant. Let $Y = AX$ where both X and Y are R^d-valued random vectors. If X has density f, then Y has density

    $$f(A^{-1}y) \mid \det A^{-1} \mid \qquad (y \in R^d).$$

    Thus, if X has a uniform density on a set B of R^d, then Y is uniformly distributed on a set C of R^d. Also, determine C from B and A.

5. If Y is gamma $(a, 1)$ and X is exponential (Y), then the density of X is

    $$f(x) = \frac{a}{(x+1)^{a+1}} \qquad (x \geq 0).$$

6. A random variable is said to have the F distribution with a and b degrees of freedom when its density is

    $$f(x) = \frac{cx^{\frac{a}{2}-1}}{(1+\frac{ax}{b})^{\frac{a+b}{2}}}, \qquad (x > 0).$$

    Here, c is the constant $\Gamma(\frac{a+b}{2})(\frac{a}{b})^{\frac{a}{2}}/\Gamma(\frac{a}{2})\Gamma(\frac{b}{2})$. Show that when X and Y are independent chi-square random variables with parameters a and b respectively, then $(\frac{X}{a})/(\frac{Y}{b})$ is $F(a,b)$. Show also that when X is $F(a,b)$, then $\frac{1}{X}$ is $F(b,a)$. Show finally that when X is t-distributed with a degrees of freedom, X^2 is $F(1,a)$. Draw the curves of the densities of $F(2,2)$ and $F(3,1)$ random variables.

7. When N_1 and N_2 are independent normal random variables, the random variables $N_1^2 + N_2^2$ and N_1/N_2 are independent.

8. Let f be the triangular density defined by

    $$f(x) = \begin{cases} 1 - \dfrac{x}{2} & \text{if } 0 \leq x \leq 2 \\ 0 & \text{elsewhere} \end{cases}.$$

When U_1 and U_2 are independent uniform $[0,1]$ random variables, then the following random variables all have density f :

$2 \min (U_1, U_2)$;

$2 \mid U_1 + U_2 - 1 \mid$;

$2(1 - \sqrt{U_1})$.

9. Show that the density of the product $\prod_{i=1}^{n} U_i$ of n iid uniform $[0,1]$ random variables is

$$f(x) = \begin{cases} \dfrac{1}{(n-1)!} \log(\dfrac{1}{x})^{n-1} & 0 \le x \le 1 \\ 0 & \text{elsewhere} \end{cases}$$

10. When X is gamma $(a, 1)$, then $1/X$ has density

$$f(x) = (\dfrac{1}{x})^{a+1} \dfrac{e^{-\frac{1}{x}}}{\Gamma(a)} \qquad (x \ge 0) .$$

11. Let $Y = \prod_{i=1}^{k} X_i$ where $X_1, \ldots, X_k$ are iid random variables each distributed as the maximum of n iid uniform $[0,1]$ random variables. Then Y has density

$$f(x) = \dfrac{n^k}{\Gamma(k)} x^{n-1} (-\log(x))^{k-1} \qquad (0 \le x \le 1) .$$

(Rider, 1955; Rahman, 1964).

12. Let $X_1, \ldots, X_n$ be iid uniform $[-1,1]$ random variables, and let Y be equal to $(\min(X_1, \ldots, X_n) + \max(X_1, \ldots, X_n))/2$. Show that Y has density

$$f(x) = \dfrac{n}{2}(1 - \mid x \mid)^{n-1} \qquad (\mid x \mid \le 1) ,$$

and variance $\dfrac{2}{(n+1)(n+2)}$ (Neyman and Pearson, 1928; Carlton, 1946).

13. We say that the power distribution with parameter $a > -1$ is the distribution corresponding to the density

$$f(x) = (a+1)x^a \qquad (0 < x < 1) .$$

If $X_1, \ldots$ are iid random variables having the power distribution with parameter a, then show that

A. X_1/X_2 has density

$$\begin{cases} \dfrac{a+1}{2} x^a & 0 < x < 1 \\ \dfrac{a+1}{2} x^{-(a+2)} & 1 \le x \end{cases}$$

B. $\prod_{i=1}^{n} X_i$ has density

$$\frac{(a+1)^n}{\Gamma(n)} x^a \left(\log \frac{1}{x}\right)^{n-1} \qquad (0 < x < 1) .$$

(Springer, 1979, p. 161).

14. The ratio G_a / G_b of two independent gamma random variables with parameters $(a,1)$ and $(b,1)$ respectively has density

$$\frac{1}{B(a,b)} \frac{x^{a-1}}{(1+x)^{a+b}} \qquad (x > 0) .$$

Here $B(a,b)$ is the standard abbreviation for the constant in the beta integral, i.e. $B(a,b) = \Gamma(a)\Gamma(b)/\Gamma(a+b)$. This is called the beta density of the second kind. Furthermore, $G_a/(G_a+G_b)$ has the beta density with parameters a and b.

15. Let $U_1, \ldots, U_4$ be iid uniform $[0,1]$ random variables. Show that $(U_1+U_2)/(U_3+U_4)$ has density

$$\begin{cases} \dfrac{7x}{6} & 0 < x < \dfrac{1}{2} \\[2ex] \dfrac{8}{3} - \dfrac{3x}{2} - \dfrac{2}{3x^2} + \dfrac{1}{6x^3} & \dfrac{1}{2} \le x < 1 \\[2ex] -\dfrac{2}{3} + \dfrac{x}{6} + \dfrac{8}{3x^2} - \dfrac{3}{2x^3} & 1 \le x < 2 \\[2ex] \dfrac{7}{6x^3} & 2 \le x \end{cases}$$

16. Show that $N_1 N_2 + N_3 N_4$ has the Laplace density (i.e., $\frac{1}{2} e^{-|x|}$), whenever the N_i's are iid normal random variables (Mantel, 1973).

17. Show that the characteristic function of a Cauchy random variable is $e^{-|t|}$. Using this, prove that when $X_1, \ldots, X_n$ are iid Cauchy random variables, then $\dfrac{1}{n} \sum_{i=1}^{n} X_i$ is again Cauchy distributed, i.e. the average is distributed as X_1.

18. Use the convolution method to obtain the densities of U_1+U_2 and $U_1+U_2+U_3$ where the U_i's are iid uniform $[-1,1]$ random variables.

19. In the oldest FORTRAN subroutine libraries, normal random variates were generated as

$$X_n = \frac{1}{\sqrt{n/3}} \sum_{j=1}^{n} \left(U_j - \frac{1}{2}\right)$$

where the U_j's are iid uniform $[0,1]$ random variates. Usually n was equal to 12. This generator is of course inaccurate. Verify however that the mean and variance of such random variables are correct. Bolshev (1959) later proposed the corrected random variate

$$Y = X_5 - \frac{3X_5 - X_5{}^3}{100} .$$

Define a notion of closeness between densities, and verify that Y is closer to a normal random variable than X_5.

20. Let $U_1, \ldots, U_n, V_1, \ldots, V_m$ be iid uniform [0,1] random variables. Define $X = \max(U_1, \ldots, U_n)$, $Y = \max(V_1, \ldots, V_m)$. Then X/Y has density

$$f(x) = \begin{cases} cx^{n-1} & , 0 \leq x \leq 1 \\ \dfrac{c}{x^{m+1}} & , x \geq 1 \end{cases}$$

where $c = \dfrac{nm}{n+m}$ (Murty, 1955).

21. Show that if $X \leq Y \leq Z$ are the order statistics for three iid normal random variables, then

$$\frac{\min(Z-Y, Y-X)}{Z-X}$$

has density

$$f(x) = \frac{3\sqrt{3}}{\pi(1-x+x^2)} , \qquad (0 \leq x \leq \tfrac{1}{2}) .$$

See e.g. Lieblein (1952).

Chapter Two
GENERAL PRINCIPLES
IN RANDOM VARIATE GENERATION

1. INTRODUCTION.

In this chapter we introduce the reader to the fundamental principles in non-uniform random variate generation. This chapter is a must for the serious reader. On its own it can be used as part of a course in simulation.

These basic principles apply often, but not always, to both continuous and discrete random variables. For a structured development it is perhaps best to develop the material according to the guiding principle rather than according to the type of random variable involved. The reader is also cautioned that we do not make any recommendations at this point about generators for various distributions. All the examples found in this chapter are of a didactical nature, and the most important families of distributions will be studied in chapters IX,X,XI in more detail.

2. THE INVERSION METHOD.

2.1. The inversion principle.

The inversion method is based upon the following property:

Theorem 2.1.

Let F be a continuous distribution function on R with inverse F^{-1} defined by

$$F^{-1}(u) = \inf \{x : F(x) = u, 0 < u < 1\} .$$

If U is a uniform [0,1] random variable, then $F^{-1}(U)$ has distribution function F. Also, if X has distribution function F, then $F(X)$ is uniformly distributed on [0,1].

Proof of Theorem 2.1.

The first statement follows after noting that for all $x \in R$,

$$P(F^{-1}(U) \le x) = P(\inf \{y : F(y) = U\} \le x)$$
$$= P(U \le F(x)) = F(x) .$$

The second statement follows from the fact that for all $0 < u < 1$,

$$P(F(X) \le u) = P(X \le F^{-1}(u))$$
$$= F(F^{-1}(u)) = u \quad\blacksquare .$$

Theorem 2.1 can be used to generate random variates with an arbitrary continuous distribution function F provided that F^{-1} is explicitly known. The faster the inverse can be computed, the faster we can compute X from a given uniform [0,1] random variate U. Formally, we have

The inversion method

Generate a uniform [0,1] random variate U.
RETURN $X \leftarrow F^{-1}(U)$

In the next table, we give a few important examples. Often, the formulas for

$F^{-1}(U)$ can be simplified, by noting for example that $1-U$ is distributed as U.

Density $f(x)$	$F(x)$	$X=F^{-1}(U)$	Simplified form
Exponential(λ) $\lambda e^{-\lambda x}$, $x \geq 0$	$1-e^{-\lambda x}$	$-\dfrac{1}{\lambda}\log(1-U)$	$-\dfrac{1}{\lambda}\log(U)$
Cauchy(σ) $\dfrac{\sigma}{\pi(x^2+\sigma^2)}$	$\dfrac{1}{2}+\dfrac{1}{\pi}\arctan(\dfrac{x}{\sigma})$	$\sigma\tan(\pi(U-\dfrac{1}{2}))$	$\sigma\tan(\pi U)$
Rayleigh(σ) $\dfrac{x}{\sigma}e^{-\frac{x^2}{2\sigma^2}}$, $x \geq 0$	$1-e^{-\frac{x^2}{2\sigma^2}}$	$\sigma\sqrt{-\log(1-U)}$	$\sigma\sqrt{-\log(U)}$
Triangular on$(0,a)$ $\dfrac{2}{a}(1-\dfrac{x}{a})$, $0 \leq x \leq a$	$\dfrac{2}{a}(x-\dfrac{x^2}{2a})$	$a(1-\sqrt{1-U})$	$a(1-\sqrt{U})$
Tail of Rayleigh $xe^{\frac{a^2-x^2}{2}}$, $x \geq a > 0$	$1-e^{\frac{a^2-x^2}{2}}$	$\sqrt{a^2-2\log(1-U)}$	$\sqrt{a^2-2\log U}$
Pareto(a,b) $\dfrac{ab^a}{x^{a+1}}$, $x \geq b > 0$	$1-(\dfrac{b}{x})^a$	$\dfrac{b}{(1-U)^{1/a}}$	$\dfrac{b}{U^{1/a}}$

There are many areas in random variate generation where the inversion method is of particular importance. We cite four examples:

Example 2.1. Generating correlated random variates.

When two random variates X and Y are needed with distribution functions F and G respectively, then these can be obtained as $F^{-1}(U)$ and $G^{-1}(V)$ where U and V are uniform [0,1] random variates. If U and V are dependent, then so are $F^{-1}(U)$ and $G^{-1}(V)$. Maximal correlation is achieved by using $V=U$, and maximal negative correlation is obtained by setting $V=-U$. While other methods may be available for generating X and/or Y individually, few methods allow the flexibility of controlling the correlation as described here. In variance reduction, negatively correlated random variates are very useful (see e.g. Hammersley and Handscomb, 1964, or Bratley, Fox and Schrage, 1984). ∎

Example 2.2. Generating maxima.

To generate $X = \max(X_1, \ldots, X_n)$, where the X_i's are iid random variates with distribution function F, we could:

(i) Generate $X_1, \ldots, X_n$, and take the maximum.

(ii) Generate a uniform $[0,1]$ random variate U and find the solution X of $F^n(X) = U$.

(iii) Generate V, a random variate distributed as the maximum of n iid uniform $[0,1]$ random variates, and find the solution X of $F(X) = V$.

Thus, the elegant solutions (ii) and (iii) involve inversion. ∎

Example 2.3. Generating all order statistics.

A sample $X_{(1)}, \ldots, X_{(n)}$ of order statistics of a sequence $X_1, \ldots, X_n$ of iid random variables with distribution function F can be obtained as $F^{-1}(U_{(1)}), \ldots, F^{-1}(U_{(n)})$, where the $U_{(i)}$'s are the order statistics of a uniform sample. As we will see further on, this is often more efficient than generating the X_i sample and sorting it. ∎

Example 2.4. A general purpose generator.

The inversion method is the only truly universal method: if all we can do is compute $F(x)$ for all x, and we have enough (i.e., infinite) time on our hands, then we can generate random variates with distribution function F. All the other methods described in this book require additional information in one form or another. ∎

2.2. Inversion by numerical solution of F(X)=U.

The inversion method is exact when an explicit form of F^{-1} is known. In other cases, we must solve the equation $F(X) = U$ numerically, and this requires an infinite amount of time when F is continuous. Any stopping rule that we use with the numerical method leads necessarily to an inexact algorithm. In this section we will briefly describe a few numerical inversion algorithms and stopping rules. Despite the fact that the algorithms are inexact, there are situations in which we are virtually forced to use numerical inversion, and it is important to compare different inversion algorithms from various points of view.

In what follows, X is the (unknown, but exact) solution of $F(X) = U$, and $X*$ is the value returned by the numerical inversion algorithm. A stopping rule which insists that $|X*-X| < \delta$ for some small $\delta > 0$ is not realistic because for large values of X, this would probably imply that the number of significant digits is greater than the built-in limit dictated by the wordsize of the computer. A second choice for our stopping rule would by $|F(X*)-F(X)| < \epsilon$, where $\epsilon > 0$ is a small number. Since all F values are in the range [0,1], we do not face the above-mentioned problem any more, were it not for the fact that small variations in X can lead to large variations in $F(X)$-values. Thus, it is possible that even the smallest realizable increment in X yields a change in $F(X)$ that exceeds the given constant ϵ. A third possibility for our stopping rule would be $|X*-X| < \delta |X|$ where the value of δ is determined by the wordsize of the computer. While this addresses the problem of relative accuracy correctly, it will lead to more accuracy than is orinarily required for values of X near 0. Thus, no stopping rule seems universally recommendable. If we know that X takes values in [-1,1], then the rule $|X*-X| < \delta$ seems both practical and amenable to theoretical analysis. Let us first see what we could do when the support of F falls outside [-1,1].

Let $h : R \rightarrow (-1,1)$ be a strictly monotone continuous transformation. Assume now that we obtain $X*$ by the following method:

Let $Y*$ be the numerical solution of $F(h^{-1}(y)) = U$, where U is a uniform [0,1] random variable and $Y*$ is such that it is within δ of the exact solution Y of the given equation.
$X* \leftarrow h^{-1}(Y*)$

Here we used the fact that Y has distribution function $F(h^{-1}(y))$, $|y| \leq 1$. Let us now look at what happens to the accuracy of the solution. A variation of dy on the value of y leads to variation of $h^{-1\prime}(y) \, dx = h^{-1\prime}(h(x)) \, dx$ on the corresponding value of x. The expected variation thus is about equal to $V\delta$ where

$$V = E(h^{-1\prime}(h(X))) = E(\frac{1}{h'(X)}).$$

Unfortunately, the best transformation h, i.e. the one that minimizes V, depends upon the distribution of X. We can give the reader some insight in how to choose h by an example. Consider for example the class if transformations

$$h(x) = \frac{x-m}{s + \mid x-m \mid},$$

where $s > 0$ and $m \in R$ are constants. Thus, we have $h^{-1}(y) = m + sy/(1 - \mid y \mid)$, and

$$V = E(\frac{1}{s}(s + \mid X-m \mid)^2) = s + 2E(\mid X-m \mid) + \frac{1}{s}E((X-m)^2).$$

For symmetric random variables X, this expression is minimized by setting $m = 0$ and $s = \sqrt{Var(X)}$. For asymmetric X, the minimization problem is very difficult. The next best thing we could do is minimize a good upper bound for V, such as the one provided by applying the Cauchy-Schwarz inequality,

$$V \le s + 2\sqrt{E(X-m)^2)} + \frac{1}{s}E((X-m)^2).$$

This upper bound is minimal when

$$m = E(X), \; s = \sqrt{Var(X)}.$$

The upper bound for V then becomes $4\sqrt{Var(X)}$. This approach requires either exact values or good approximations for m and s. We refer to Exercise 1 for a detailed comparison of the average accuracy of this method with that of the direct solution of $F(X) = U$ given that the same stopping rule is used.

We will discuss three popular numerical inversion algorithms for $F(X) = U$:

The bisection method

Find an initial interval $[a, b]$ to which the solution belongs.
REPEAT
         $X \leftarrow (a+b)/2$
         IF $F(X) \le U$
                THEN $a \leftarrow X$
                ELSE $b \leftarrow X$
UNTIL $b - a \le 2\delta$
RETURN X

The secant method (regula falsi method)

Find an interval $[a,b]$ to which the solution belongs.
REPEAT

$$X \leftarrow a + (b-a)\frac{U-F(a)}{F(b)-F(a)}$$

   IF $F(X) \leq U$

        THEN $a \leftarrow X$

        ELSE $b \leftarrow X$

UNTIL $b-a \leq \delta$
RETURN X

The Newton-Raphson method

Choose an initial guess X.
REPEAT

$$X \leftarrow X - \frac{(F(X)-U)}{f(X)}$$

UNTIL stopping rule is satisfied. (Note: f is the density corresponding to F.)
RETURN X.

In the first two methods, we need an initial interval $[a,b]$ known to contain the solution. If the user knows functions G and H such that $G(x) \geq F(x) \geq H(x)$ for all x, then we could start with $[a,b] = [G^{-1}(U), H^{-1}(U)]$. In particular, if the support of F is known, then we can set $[a,b]$ equal to it. Because it is important to have reasonably small intervals, any a priori information should be used to select $[a,b]$. For example, if F has variance σ^2 and is symmetric about 0, then by Cantelli's extension of Chebyshev's inequality,

$$F(x) \geq \frac{x^2}{x^2+\sigma^2} \qquad (x>0) .$$

This suggests that when $U > \frac{1}{2}$, we take

$$[a,b] = [0, \sigma\sqrt{\frac{U}{1-U}}] .$$

When $U \leq \frac{1}{2}$, we argue by symmetry. Thus, information about moments and quantiles of F can be valuable for initial guesswork. For the Newton-Raphson method, we can often take an arbitrary point such as 0 as our initial guess.

The actual choice of an algorithm depends upon many factors such as

> (i) Guaranteed convergence.
> (ii) Speed of convergence.
> (iii) A priori information.
> (iv) Knowledge of the density f .

If f is not explicitly known, then the Newton-Raphson method should be avoided because the approximation of $f(x)$ by $\frac{1}{\delta}(F(x+\delta)-F(x))$ is rather inaccurate because of cancelation errors.

Only the bisection method is guaranteed to converge in all cases. If $F(X)=U$ has a unique solution, then the secant method converges too. By "convergence" we mean of course that the returned variable $X*$ would approach the exact solution X if we would let the number of iterations tend to ∞. The Newton-Raphson method converges when F is convex or concave. Often, the density f is unimodal with peak at m . Then, clearly, F is convex on $(-\infty,m\,]$, and concave on $[m,\infty)$, and the Newton-Raphson method started at m converges.

Let us consider the speed of convergence now. For the bisection method started at $[a\,,b\,] = [g_1(U),g_2(U)]$ (where g_1,g_2 are given functions), we need N iterations if and only if

$$2^{N-1} < g_2(U)-g_1(U) \leq 2^N .$$

The solution of this is

$$N = 1+ \left\lfloor \log_+((g_2(U)-g_1(U))/\delta) \right\rfloor ,$$

where $\log_+$ is the positive part of the logarithm with base 2. From this expression, we retain that $E(N)$ can be infinite for some long-tailed distributions. If the solution is known to belong to $[-1,1]$, then we have deterministically,

$$N \leq 1+\log_+(\frac{1}{\delta}) .$$

And in all cases in which $E(N) < \infty$, we have as $\delta \downarrow 0$, $E(N) \sim \log(\frac{1}{\delta})$. Essentially, adding one bit of accuracy to the solution is equivalent to adding one iteration. As an example, let us take $\delta = 10^{-7}$, which corresponds to the standard choice for problems with solutions in $[-1,1]$ when a 32-bit computer is used. The value of N in that case is in the neighborhood of 24, and this is often inacceptable.

The secant and Newton-Raphson methods are both faster, albeit less robust, than the bisection method. For a good discussion of the convergence and rate of convergence of the given methods, we refer to Ostrowski (1973). Let us merely state one of the results for $E(N)$, the quantity of interest to us, where N is the number of iterations needed to get to within δ of the solution (note that this is impossible to verify when an algorithm is running !). Also, let F be the distribution function corresponding to a unimodal density with absolutely bounded derivative f'. The Newton-Raphson method started at the mode converges, and for some number N_0 depending only upon F (but possibly ∞) we have

$$E(N) \leq N_0 + \mathrm{loglog}(\frac{1}{\delta})$$

where all logarithms are base 2. For the secant method, a similar statement can be made but the base should be replaced by the golden ratio, $\frac{1}{2}(1+\sqrt{5})$. In both cases, the influence of δ on the average number of iterations is practically nil, and the asymptotic expression for $E(N)$ is smaller than in the bisection method (when $\delta\downarrow 0$). Obviously, the secant and Newton-Raphson methods are not universally faster than the bisection method. For ways of accelerating these methods, see for example Ostrowski (1973, Appendix I, Appendix G).

2.3. Explicit approximations.

When F^{-1} is not explicitly known, it can sometimes be well approximated by another explicitly known function $g(U)$. In iterative methods, the stopping rule usually takes care of the accuracy problem. Now, by resorting to a one-step procedure, we squarely put the burden of verifying the accuracy of the solution on the shoulders of the theoretician. Also, we should define once again what we mean by accuracy (see Devroye (1982) for a critical discussion of various definitions). Iterative methods can be notoriously slow, but this is a small price to pay for their conciseness, simplicity, flexibility and accuracy. The four main limitations of the direct approximation method are:

(i)   The approximation is valid for a given F: to use it when F changes frequently during the simulation experiment would probably require extraordinary set-up times.

(ii) The function g must be stored. For example, g is often a ratio of two polynomials, in which case all the coefficients must be put in a long table.

(iii) The accuracy of the approximation is fixed. If a better accuracy is needed, the entire function g must be replaced. This happens for example when one switches to a computer with a larger wordsize. In other words, future computer upgrades will be expensive.

(iv) Certain functions cannot be approximated very well by standard approximation techniques, except possibly by inacceptably complicated functions. Also, approximations are difficult to develop for multiparameter families of functions.

How one actually goes about designing approximations g will not be explained here. For example, we could start from a very rough approximation of F^{-1}, and then explicitly compute the function that corresponds to one or two or a fixed number of Newton-Raphson iterations. This is not systematic enough in general. A spline method was developed in Kohrt (1980) and Ahrens and Kohrt (1981). In the general literature, one can find many examples of approximations by ratios of polynomials. For example, for the inverse of the normal distribution function, Odeh and Evans (1974) suggest

$$g(u) = \sqrt{-2\log(u)} + \frac{A(\sqrt{-2\log(u)})}{B(\sqrt{-2\log(u)})} , \frac{1}{2} \geq u \geq 10^{-20} ,$$

where $A(x) = \sum_{i=0}^{4} a_i x^i$, and $B(x) = \sum_{i=0}^{4} b_i x^i$, and the coefficients are as shown in the table below:

i	a_i	b_i
0	-0.322232431088	0.0993484626060
1	-1.0	0.588581570495
2	-0.342242088547	0.531103462366
3	-0.0204231210245	0.103537752850
4	-0.0000453642210148	0.0038560700634

For u in the range $[\frac{1}{2}, 1-10^{-20}]$, we take $-g(1-u)$, and for u in the two tiny left-over intervals near 0 and 1, the approximation should not be used. Rougher approximations can be found in Hastings (1955) and Bailey (1981). Bailey's approximation requires fewer constants and is very fast. The approximation of Beasley and Springer (1977) is also very fast, although not as accurate as the Odeh-Evans approximation given here. Similar methods exist for the inversion of beta and gamma distribution functions.

2.4. Exercises.

1. Most stopping rules for the numerical iterative solution of $F(X) = U$ are of the type $b - a \leq \delta$ where $[a, b]$ is an interval containing the solution X, and $\delta > 0$ is a small number. These algorithms may never halt if for some u, there is an interval of solutions of $F(X) = u$ (this applies especially to the secant method). Let A be the set of all u for which we have for some $x < y$, $F(x) = F(y) = u$. Show that $P(U \in A) = 0$, i.e. the probability of ending up in an infinite loop is zero. Thus, we can safely lift the restriction imposed throughout this section that $F(X) = u$ has one solution for all u.

2. Show that the secant method converges if $F(X) = U$ has one solution for the given value of U.

3. Show that if $F(0) = 0$ and F is concave on $[0, \infty)$, then the Newton-Raphson method started at 0 converges.

4. **Student's t distribution with 3 degrees of freedom.**

Consider the density

$$f(x) = \frac{2}{\pi(1+x^2)^2} \ ,$$

and the corresponding distribution function

$$F(x) = \frac{1}{2} + \frac{1}{\pi}(\text{arc tan } x + \frac{x}{1+x^2}) \ .$$

These functions define the t distribution with 3 degrees of freedom. Elsewhere we will see very efficient methods for generating random variates from this distribution. Nevertheless, because F^{-1} is not known explicitly (except perhaps as an infinite series), this distribution can be used to illustrate many points made in the text. Note first that the distribution is symmetric about 0. Prove first that

$$\frac{1}{2} + \frac{1}{\pi}\text{arc tan } x \ \le F(x) \le \frac{1}{2} + \frac{2}{\pi}\text{arc tan } x \quad (x \ge 0) \ .$$

Thus, for $U \ge \frac{1}{2}$, the solution of $F(X)=U$ lies in the interval

$$[\tan(\frac{\pi}{2}(U-\frac{1}{2})),\tan(\pi(U-\frac{1}{2}))] \ .$$

Using this interval as a starting interval, compare and time the bisection method, the secant method and the Newton-Raphson method (in the latter method, start at 0 and keep iterating until X does not change in value any further). Finally, assume that we have an efficient Cauchy random variate generator at our disposal. Recalling that a Cauchy random variable C is distributed as $\tan(\pi(U-\frac{1}{2}))$, show that we can generate X by solving the equation

$$\text{arc tan } X + \frac{X}{1+X^2} = \text{arc tan } C \ ,$$

and by starting with initial interval

$$[\sqrt{\frac{\sqrt{1+C^2}-1}{\sqrt{1+C^2}+1}},C]$$

when $C > 0$ (use symmetry in the other case). Prove that this is a valid method.

5. Develop a general purpose random variate generator which is based upon inversion by the Newton-Raphson method, and assumes only that F and the corresponding density f can be computed at all points, and that f is unimodal. Verify that your method is convergent. Allow the user to specify a mode if this information is available.

6. Write general purpose generators for the bisection and secant methods in which the user specifies an initial interval $[g_1(U), g_2(U)]$.

7. Discuss how you would solve $F(X) = U$ for X by the bisection method if no initial interval is available. In a first stage, you could look for an interval $[a, b]$ which contains the solution X. In a second stage, you proceed by ordinary bisection until the interval's length drops below δ. Show that regardless of how you organize the original search (this could be by looking at adjacent intervals of equal length, or adjacent intervals with geometrically increasing lengths, or adjacent intervals growing as $2, 2^2, 2^{2^2}, ...$), the expected time taken by the entire algorithm is ∞ whenever $E(\log_+ |X|) = \infty$. Show that for extrapolatory search, it is not a bad strategy to double the interval sizes. Finally, exhibit a distribution for which the given expected search time is ∞. (Note that for such distributions, the expected number of bits needed to represent the integer portion is infinite.)

8. **An exponential class of distributions.** Consider the distribution function $F(x) = 1 - e^{-A_n(x)}$ where $A_n(x) = \sum_{i=1}^{n} a_i x^i$ for $x \geq 0$ and $A_n(x) = 0$ for $x < 0$. Assume that all coefficients a_i are nonnegative and that $a_1 > 0$. If U is a uniform $[0,1]$ random variate, and E is an exponential random variate, then it is easy to see that the solution of $1 - e^{-A_n(X)} = U$ is distributed as the solution of $A_n(X) = E$. The basic Newton-Raphson step for the solution of the second equation is

$$X \leftarrow X - \frac{A_n(X) - E}{A_n'(X)} .$$

Since $a_1 > 0$ and A_n is convex, any starting point $X \geq 0$ will yield a convergent sequence of values. We can thus start at $X = 0$ or at $X = E/a_1$ (which is the first value obtained in the Newton-Raphson sequence started at 0). Compare this algorithm with the algorithm in which X is generated as

$$\min_{1 \leq i \leq n} (\frac{E_i}{a_i})^{\frac{1}{i}}$$

where $E_1, \ldots, E_n$ are iid exponential random variates.

9. **Adaptive inversion.** Consider the situation in which we need to generate a sequence of n iid random variables with continuous distribution function F by the method of inversion. The generated couples $(X_1, U_1), ...$ are stored ($X_1 = F^{-1}(U_1)$ and U_1 is uniform $[0,1]$). Define an algorithm based upon a dynamic hash table for the U_i's in which the table is used to find a good starting interval for inversion. Implement, and compare this adaptive method with memoryless algorithms (Yuen, 1981).

10. **Truncated distributions.** Let X be a random variable with distribution function F. Define the truncated random variable Y by its distribution

function

$$G(x) = \begin{cases} 0 & x < a \\ \dfrac{F(x)-F(a)}{F(b)-F(a)} & a \le x \le b \\ 1 & x > b \end{cases}.$$

Here $-\infty \le a < b \le \infty$. Show that Y can be generated as $F^{-1}(F(a)+U(F(b)-F(a)))$ where U is a uniform $[0,1]$ random variate.

11. Find a monotonically decreasing density f on $[0,\infty)$ such that the Newton-Raphson procedure started at $X=0$ needs N steps to get within δ of the solution of $F(X)=U$ where N is a random variable with mean $E(N)=\infty$ for all $\delta > 0$.

12. **The logistic distribution.** A random variable X is said to have the logistic distribution with parameters $a \in R$ and $b > 0$ when

$$F(x) = \frac{1}{1+e^{-\frac{x-a}{b}}}.$$

It is obvious that a is a translation parameter and that b is a scale parameter. The standardized logistic distribution has $a=0, b=1$. The density is

$$f(x) = \frac{e^{-x}}{(1+e^{-x})^2} = F(x)(1-F(x)).$$

The logistic density is symmetric about 0 and resembles in several respects the normal density. Show the following:

A. When U is uniformly distributed on $[0,1]$, then $X=\log(\dfrac{U}{1-U})$ has the standard logistic distribution.

B. $\dfrac{U}{1-U}$ is distributed as the ratio of two iid exponential random variables.

C. We say that a random variable Z has the extremal value distribution with parameter a when $F(x)=e^{-ae^{-x}}$. If X is distributed as Z with parameter Y where Y is exponentially distributed, then X has the standard logistic distribution.

D. $E(X^2)=\dfrac{\pi^2}{3}$, and $E(X^4)=\dfrac{7\pi^4}{15}$.

E. If X_1, X_2 are independent extremal value distributed random variables with the same parameter a, then X_1-X_2 has a logistic distribution.

3. THE REJECTION METHOD.

3.1. Definition.

The rejection method is based upon the following fundamental property of densities:

Theorem 3.1.

Let X be a random vector with density f on R^d, and let U be an independent uniform $[0,1]$ random variable. Then $(X, cUf(X))$ is uniformly distributed on $A = \{(x, u) : x \in R^d, 0 \leq u \leq cf(x)\}$, where $c > 0$ is an arbitrary constant. Vice versa, if (X, U) is a random vector in R^{d+1} uniformly distributed on A, then X has density f on R^d.

Proof of Theorem 3.1.

For the first statement, take a Borel set $B \subseteq A$, and let B_x be the section of B at x, i.e. $B_x = \{u : (x, u) \in B\}$. By Tonelli's theorem,

$$P((X, cUf(X)) \in B) = \int \int_{B_x} \frac{1}{cf(x)} \, du \; f(x) \, dx = \frac{1}{c} \int_B du \; dx \; .$$

Since the area of A is c, we have shown the first part of the Theorem. The second part follows if we can show that for all Borel sets B of R^d, $P(X \in B) = \int_B f(x) \, dx$ (recall the definition of a density). But

$$P(X \in B) = P((X, U) \in B_1 = \{(x, u) : x \in B, 0 \leq u \leq cf(x)\})$$

$$= \frac{\int \int_{B_1} du \; dx}{\int \int_A du \; dx} = \frac{1}{c} \int_B cf(x) \, dx = \int_B f(x) \, dx \; ,$$

which was to be shown. ∎

Theorem 3.2.

Let $X_1, X_2, ...$ be a sequence of iid random vectors taking values in R^d, and let $A \subseteq R^d$ be a Borel set such that $P(X_1 \in A) = p > 0$. Let Y be the first X_i taking values in A. Then Y has a distribution that is determined by

$$P(Y \in B) = \frac{P(X_1 \in A \cap B)}{p} , \ B \text{ Borel set of } R^d .$$

In particular, if X_1 is uniformly distributed in A_0 where $A_0 \supseteq A$, then Y is uniformly distributed in A.

Proof of Theorem 3.2.

For arbitrary Borel sets B, we observe that

$$P(Y \in B) = \sum_{i=1}^{\infty} P(X_1 \notin A, \ldots, X_{i-1} \notin A, X_i \in B \cap A)$$

$$= \sum_{i=1}^{\infty} (1-p)^{i-1} P(X_1 \in A \cap B)$$

$$= \frac{1}{1-(1-p)} P(X_1 \in A \cap B),$$

which was to be shown. If X_1 is uniformly distributed in A_0, then

$$P(Y \in B) = \frac{P(X_1 \in A \cap B)}{P(X_1 \in A)} = \frac{\int\limits_{A_0 AB} dx}{\int\limits_{A_0} dx} \cdot \frac{\int\limits_{A_0} dx}{\int\limits_{AA_0} dx} = \frac{\int\limits_{AB} dx}{\int\limits_{A} dx} .$$

This concludes the proof of Theorem 3.2. ∎

The basic version of the rejection algorithm assumes the existence of a density g and the knowledge of a constant $c \geq 1$ such that

$$f(x) \leq cg(x) \quad \text{(all } x) .$$

Random variates with density f on R^d can be obtained as follows:

The rejection method

REPEAT

    Generate two independent random variates X (with density g on R^d) and U (uniformly distributed on $[0,1]$).

    Set $T \leftarrow c \dfrac{g(X)}{f(X)}$.

UNTIL $UT \leq 1$

RETURN X

By Theorem 3.1, $(X, cUg(X))$ (where X and U are as explained in the first line of the REPEAT loop) is uniformly distributed under the curve of cg in R^{d+1}. By Theorem 3.2, we conclude that the random variate $(X, cUg(X))$ generated by this algorithm (i.e. at time of exit) is uniformly distributed under the curve of f. By the second part of Theorem 3.1, we can then conclude that its d-dimensional projection X must have density f.

The three things we need before we can apply the rejection algorithm are (i) a dominating density g; (ii) a simple method for generating random variates with density g; and (iii) knowledge of c. Often, (i) and (iii) can be satisfied by a priori inspection of the analytical form of f. Basically, g must have heavier tails and sharper infinite peaks than f. In some situations, we can determine cg for entire classes of densities f. The dominating curves cg should always be picked with care: not only do we need a simple generator for g (requirement (ii)), but we must make sure that the computation of $\dfrac{g(X)}{f(X)}$ is simple. Finally, cg must be such that the algorithm is efficient.

Let N be the number of iterations in the algorithm, i.e. the number of pairs (X, U) required before the algorithm halts. We have

$$P(N=i) = (1-p)^{i-1}p \ ; \ P(N \geq i) = (1-p)^{i-1} \quad (i \geq 1),$$

where

$$p = P(f(X) \geq cUg(X)) = \int P(U \leq \frac{f(x)}{cg(x)}) \, dx$$

$$= \int \frac{f(x)}{cg(x)} g(x) \, dx = \frac{1}{c} \int f(x) \, dx = \frac{1}{c} \ .$$

Thus, $E(N) = \dfrac{1}{p} = c$, $E(N^2) = \dfrac{2}{p^2} - \dfrac{1}{p}$ and $Var(N) = \dfrac{1-p}{p^2} = c^2 - c$. In other words, $E(N)$ is one over the probability of accepting X. From this we conclude that we should keep c as small as possible. Note that the distribution of N is geometric with parameter $p = \dfrac{1}{c}$. This is good, because the probabilities

$P(N=i)$ decrease monotonically, and at an exponential rate (note that $P(N>i)=(1-p)^i \leq e^{-pi}$).

The rejection method has an almost unlimited potential. We have given up the principle that one uniform [0,1] random variate yields one variate X (as in the inversion method), but what we receive in return is a powerful, simple and exact algorithm.

Example 3.1. Bounded densities of compact support.

Let $C_{M,a,b}$ be the class of all densities on $[a,b]$ bounded by M. Any such density is clearly bounded by M. Thus, the rejection algorithm can be used with uniform dominating density $g(x)=(b-a)^{-1}$ $(a \leq x \leq b)$, and the constant c becomes $M(b-a)$. Formally, we have

The rejection method for $C_{M,a,b}$

REPEAT
      Generate two independent uniform [0,1] random variates U and V.
      Set $X \leftarrow a+(b-a)V$.
UNTIL $UM \leq f(X)$
RETURN X ∎

The reader should be warned here that this algorithm can be horribly inefficient, and that the choice of a constant dominating curve should be avoided except in a few cases.

3.2. Development of good rejection algorithms.

Generally speaking, g is chosen from a class of easy densities. This class includes the uniform density, triangular densities, and most densities that can be generated quickly by the inversion method. The situation usually dictates which densities are considered as "easy". There are two major techniques for determining c and g in the inequality $f \leq cg$: one could first study the form of f and apply one of many analytical devices for obtaining inequalities. Many of these are illustrated throughout this book (collecting them in a special chapter would have forced us to duplicate too much material). While this approach gives often quick results (see Example 3.2 below), it is ad hoc, and depends a lot on the mathematical background and insight of the designer. In a second approach, which is also illustrated in this section, one starts with a family of dominating densities g and chooses the density within that class for which c is smallest. This approach is more structured but could sometimes lead to difficult optimization problems.

Example 3.2. A normal generator by rejection from the Laplace density.

Let f be the normal density. Obtaining an upper bound for f boils down to obtaining a lower bound for $\dfrac{x^2}{2}$. But we have of course

$$\frac{1}{2}(\,|\,x\,|-1)^2 = \frac{x^2}{2} + \frac{1}{2} - |\,x\,| \geq 0 \ .$$

Thus,

$$\frac{1}{\sqrt{2\pi}} e^{-\frac{x^2}{2}} \leq \frac{1}{\sqrt{2\pi}} e^{\frac{1}{2} - |\,x\,|} = cg(x) \ ,$$

where $g(x) = \dfrac{1}{2} e^{-|\,x\,|}$ is the Laplace density, and $c = \sqrt{\dfrac{2e}{\pi}}$ is the rejection constant. This suggests the following algorithm:

A normal generator by the rejection method

REPEAT

    Generate an exponential random variate X and two independent uniform [0,1] random variates U and V. If $U < \dfrac{1}{2}$, set $X \leftarrow -X$ (X is now distributed as a Laplace random variate).

UNTIL $V \dfrac{1}{\sqrt{2\pi}} e^{\frac{1}{2} - |\,X\,|} \leq \dfrac{1}{\sqrt{2\pi}} e^{-\frac{X^2}{2}}$.

RETURN X

The condition in the UNTIL statement can be cleaned up. The constant $\dfrac{1}{\sqrt{2\pi}}$ cancels out on left and right hand sides. It is also better to take logarithms on both sides. Finally, we can move the sign change to the RETURN statement because there is no need for a sign change of a random variate that will be rejected. The random variate U can also be avoided by the trick implemented in the algorithm given below.

A normal generator by rejection from the Laplace density

REPEAT

> Generate an exponential random variate X and an independent uniform $[-1,1]$ random variate V.

UNTIL $(X-1)^2 \leq -2\log(\mid V \mid)$

RETURN $X \leftarrow X$ sign (V) ∎

For given densities f and g, the rejection constant c should be at least equal to

$$\sup_x \frac{f(x)}{g(x)} .$$

We cannot loose anything by setting c equal to this supremum, because this insures us that the curves of f and cg touch each other somewhere. Instead of letting g be determined by some inequality which we happen to come across as in Example 3.2, it is often wiser to take the best g_θ in a family of densities parametrized by θ. Here θ should be thought of as a subset of R^k (in which case we say that there are k parameters). Define the optimal rejection constant by

$$c_\theta = \sup_x \frac{f(x)}{g_\theta(x)} .$$

The optimal θ is that for which c_θ is minimal, i.e. for which c_θ is closest to 1.

We will now illustrate this optimization process by an example. For the sake of argument, we take once again the normal density f. The family of dominating densities is the Cauchy family with scale parameter θ:

$$g_\theta(x) = \frac{\theta}{\pi} \frac{1}{\theta^2+x^2} .$$

There is no need to consider a translation parameter as well because both f and the Cauchy densities are unimodal with peak at 0. Let us first compute the optimal rejection constant c_θ. We will prove that

$$c_\theta = \begin{cases} \dfrac{\sqrt{2\pi}}{e\,\theta} e^{\frac{\theta^2}{2}} & ,\theta < \sqrt{2} \\[3mm] \theta\sqrt{\dfrac{\pi}{2}} & ,\theta \geq \sqrt{2} \end{cases} .$$

We argue as follows: f / g_θ is maximal when $\log(f / g_\theta)$ is maximal. Setting the derivative with respect to x of $\log(f / g_\theta)$ equal to 0 yields the equation

$$-x + \frac{2x}{\theta^2 + x^2} = 0 .$$

This gives the values $x = 0$ and $x = \pm\sqrt{2-\theta^2}$ (the latter case can only happen when $\theta^2 \le 2$). At $x = 0$, f / g_θ takes the value $\theta\sqrt{\dfrac{\pi}{2}}$. At $x = \pm\sqrt{2-\theta^2}$, f / g_θ takes the value $\dfrac{\sqrt{2\pi}}{e\,\theta} e^{\frac{\theta^2}{2}}$. It is easy to see that for $\theta < \sqrt{2}$, the maximum of f / g_θ is attained at $x = \pm\sqrt{2-\theta^2}$ and the minimum at $x = 0$. For $\theta \ge \sqrt{2}$, the maximum is attained at $x = 0$. This concludes the verification of the expression for c_θ.

The remainder of the optimization is simple. The function c_θ has only one minimum, at $\theta = 1$. The minimal value is $c_1 = \sqrt{\dfrac{2\pi}{e}}$. With this value, the condition of acceptance $U c_\theta g_\theta(X) \le f(X)$ can be rewritten as

$$U\sqrt{\frac{2\pi}{e}}\frac{1}{\pi}\frac{1}{1+X^2} \le \frac{1}{\sqrt{2\pi}} e^{-\frac{X^2}{2}} ,$$

or as

$$U \le (1+X^2)\frac{\sqrt{e}}{2} e^{-\frac{X^2}{2}} .$$

A normal generator by rejection from the Cauchy density

[SET-UP]
$\alpha \leftarrow \dfrac{\sqrt{e}}{2}$
[GENERATOR]
REPEAT
        Generate two independent uniform [0,1] random variates U and V.
        Set $X \leftarrow \tan(\pi V)$, $S \leftarrow X^2$ (X is now Cauchy distributed).
UNTIL $U \le \alpha(1+S)e^{-\frac{S}{2}}$
RETURN X

The algorithm derived here, though it has a rejection constant near 1.4 is no match for most normal generators developed further on. The reason for this is that we need fairly expensive Cauchy random variates, plus the evaluation of exp in the acceptance step.

3.3. Generalizations of the rejection method.

Some generalizations of the rejection method are important enough to warrant special treatment in this key chapter. The first generalization concerns the following case:

$$f(x) = c \ g(x) \ \psi(x) \ ,$$

where the function ψ is [0,1]-valued, g is an easy density and c is a normalization constant at least equal to 1. The rejection algorithm for this case can be rewritten as follows:

The rejection method

REPEAT

      Generate independent random variates X, U where X has density g and U is uniformly distributed on [0,1].

UNTIL $U \le \psi(x)$

RETURN X

Vaduva (1977) observed that for special forms of ψ, there is another way of proceeding. This occurs when $\psi = 1-\Psi$ where Ψ is a distribution function of an easy density.

Vaduva's generalization of the rejection method

REPEAT

      Generate two independent random variates X, Y, where X has density g and Y has distribution function Ψ.

UNTIL $X \le Y$

RETURN X

For $\psi = \Psi$, we need to replace $X \le Y$ in the acceptance step by $X \ge Y$.

Theorem 3.3.

Vaduva's rejection method produces a random variate X with density $f = cg(1-\Psi)$, and the rejection constant (the expected number of iterations) is c.

Proof of Theorem 3.3.

We prove this by showing that Vaduva's algorithm is entirely equivalent to the original rejection algorithm. Note that the condition of acceptance, $X \leq Y$ is with probability one satisfied if and only if $1-\Psi(X) \geq 1-\Psi(Y)$. But by the probability integral transform, we know that $1-\Psi(Y)$ is distributed as U, a uniform [0,1] random variable. Thus, we need only verify whether $U \leq 1-\Psi(X)$, which yields the original acceptance condition given at the beginning of this section. ■

The choice between generating U and computing $1-\Psi(X)$ on the one hand (the original rejection algorithm) and generating Y with distribution function Ψ on the other hand (Vaduva's method) depends mainly upon the relative speeds of computing a distribution function and generating a random variate with that distribution.

Example 3.3.

Consider the density

$$f(x) = c \ (ax^{a-1}) \ e^{-x} \ , 0 < x \leq 1 \ ,$$

where $a > 0$ is a parameter and c is a normalization constant. This density is part of the gamma (a) density, written here in a form convenient to us. The dominating density is $g(x) = ax^{a-1}$, and the function ψ is e^{-x}. Random variates with density g can be obtained quite easily by inversion (take $V^{\frac{1}{a}}$ where V is a uniform [0,1] random variate). In this case, the ordinary rejection algorithm would be

REPEAT

      Generate two iid uniform [0,1] random variates U, V, and set $X \leftarrow V^{\frac{1}{a}}$.

UNTIL $U \leq e^{-X}$

RETURN X

Vaduva's modification essentially consists in generating X and an exponential random variate E until $E \geq X$. It is faster if we can generate E faster than we can compute e^{-X} (this is sometimes the case). Of course, in this simple example, we could have deduced Vaduva's modification by taking the logarithm of the acceptance condition and noting that E is distributed as $-\log(U)$. ■

We now proceed with another generalization found in Devroye (1984):

Theorem 3.4.

Assume that a density f on R^d can be decomposed as follows:

$$f(x) = \int g(y,x) h(y,x) \, dy \ ,$$

where $\int dy$ is an integral in R^k, $g(y,x)$ is a density in y for all x, and there exists a function $H(x)$ such that $0 \leq h(y,x) \leq H(x)$ for all y, and $H/\int H$ is an easy density. Then the following algorithm produces a random variate with density f, and takes N iterations where N is geometrically distributed with parameter $\dfrac{1}{\int H}$ (and thus $E(N) = \int H$).

   Generalized rejection method

   REPEAT
   　　Generate X with density $H/\int H$ (on R^d).
   　　Generate Y with density $g(y,X), y \in R^k$ (X is fixed).
   　　Generate a uniform [0,1] random variate U.
   UNTIL $UH(X) \leq h(Y,X)$
   RETURN X

Proof of Theorem 3.4.

We will prove that this Theorem follows directly from Theorem 3.2. Let us define the new random vector $W_1 = (X,Y,U)$ where W_1 refers to the triple generated in the REPEAT loop. Then, if A is the set of values $w_1 = (x,y,u)$ for which $uH(x) \leq h(y,x)$, we have for all Borel sets B in the space of w_1,

$$P(W \in B) = \frac{P(W_1 \in A \cap B)}{p}$$

where $p = P(W_1 \in A)$ and W refers to the value of W_1 upon exit. Take $B = (-\infty, x] \times R^k \times [0,1]$, and conclude that

$$P(X \text{(returned)} \leq x) = \frac{1}{p} P(X \leq x, UH(X) \leq h(Y,X)) \ .$$

$$= \int \int_{-\infty}^{x} g(y,z) \, \frac{h(y,z)}{H(z)} \, \frac{H(z)}{\int H} \, dz \, dy$$

$$= \frac{1}{p \int H} \int_{-\infty}^{z} f(z) \, dz \ .$$

We note first that by setting $x = \infty$, $p = \frac{1}{\int H}$. But then, clearly, the variate produced by the algorithm has density f as required. ■

3.4. Wald's equation.

We will rather often be asked to evaluate the expected value of

$$\sum_{i=1}^{N} \psi(W_i) \ ,$$

where W_i is the collection of all random variables used in the i-th iteration of the rejection algorithm, ψ is some function, and N is the number of iterations of the rejection method. The random variable N is known as a stopping rule because the probabilities $P(N=n)$ are equal to the probabilities that $W_1, \ldots, W_n$ belong to some set B_n. The interesting fact is that, regardless of which stopping rule is used (i.e., whether we use the one suggested in the rejection method or not), as long as the W_i's are iid random variables, the following remains true:

Theorem 3.5. (Wald's equation.)

Assume that $W_1, \ldots$ are iid R^d-valued random variables, and that ψ is an arbitrary nonnegative Borel measurable function on R^d. Then, for all stopping rules N,

$$E(\sum_{i=1}^{N} \psi(W_i)) = E(N) \, E(\psi(W_1)) \ .$$

Proof of Theorem 3.5.

To simplify the notation we write $Z_i = \psi(W_i)$ and note that the Z_i's are iid nonnegative random variables. The proof given here is standard (see e.g. Chow and Teicher (1978, pp. 137-138)), but will be given in its entirety. We start by noting that Z_i and $I_{[N < i]}$ are independent for all i. Thus, so are Z_i and $I_{[N \geq i]}$. We will assume that $E(Z_1) < \infty$ and $E(N) < \infty$. It is easy to verify that the chain of equalities given below remains valid when one or both of these expectations is ∞.

$$E(\sum_{i=1}^{N} Z_i) = E(\sum_{i=1}^{\infty} Z_i I_{[N \geq i]})$$

$$= \sum_{i=1}^{\infty} E\left(Z_i I_{[N \geq i]}\right)$$

$$= \sum_{i=1}^{\infty} E\left(Z_i\right) P\left(N \geq i\right)$$

$$= E\left(Z_1\right) \sum_{i=1}^{\infty} P\left(N \geq i\right)$$

$$= E\left(Z_1\right) E\left(N\right).$$

The exchange of the expectation and infinite sum is allowed by the monotone convergence theorem: just note that for any sequence of nonnegative random variables $Y_1, \ldots,$ $\sum_{i=1}^{n} E\left(Y_i\right) = E\left(\sum_{i=1}^{n} Y_i\right) \rightarrow E\left(\sum_{i=1}^{\infty} Y_i\right).$ ■

It should be noted that for the rejection method, we have a special case for which a shorter proof can be given because our stopping rule N is an instantaneous stopping rule: we define a number of decisions D_i , all 0 or 1 valued and dependent upon W_i only: $D_1 = 0$ indicates that we "reject" based upon W_1, etcetera. A 1 denotes acceptance. Thus, N is equal to n if and only if $D_n = 1$ and $D_i = 0$ for all $i < n$. Now,

$$E\left(\sum_{i=1}^{N} \psi(W_i)\right)$$

$$= E\left(\sum_{i < N} \psi(W_i)\right) + E\left(\psi(W_N)\right)$$

$$= E\left(N-1\right) E\left(\psi(W_1) \mid D_1 = 0\right) + E\left(\psi(W_1) \mid D_1 = 1\right)$$

$$= \left(\frac{1}{P\left(D_1 = 1\right)} - 1\right) \frac{E\left(\psi(W_1) I_{D_1 = 0}\right)}{P\left(D_1 = 0\right)} + \frac{E\left(\psi(W_1) I_{D_1 = 1}\right)}{P\left(D_1 = 1\right)}$$

$$= \frac{E\left(\psi(W_1)\right)}{P\left(D_1 = 1\right)},$$

which proves this special case of Theorem 3.5.

3.5. Letac's lower bound.

In a profound but little publicized paper, Letac (1975) asks which distributions can be obtained for $X = U_N$ where N is a stopping time and $U_1, U_2, \ldots$ is an iid sequence of uniform [0,1] random variables. He shows among other things that all densities on [0,1] can be obtained in this manner. In exercise 3.14, one universal stopping time will be described. It does not coincide with Letac's universal stopping rule, but will do for didactical purposes.

More importantly, Letac has obtained lower bounds on the performance of any algorithm of this type. His main result is:

Theorem 3.6. (Letac's lower bound)

Assume that $X = U_N$ has density f on [0,1], where N and the U_i's are as defined above. For any such stopping rule N (i.e., for any algorithm), we have

$$E(N) \geq ||f||_\infty,$$

where $||.||_\infty$ is the essential supremum of f.

Proof of Theorem 3.6.

By the independence of the events $[N \geq n]$ and $[U_n \in B]$ (which was also used in the proof of Wald's equation), we have

$$P(N \geq n, U_n \in B) = P(N \geq n)P(U_1 \in B).$$

But,

$$P(X \in B) = \sum_{n=1}^{\infty} P(N = n, U_n \in B)$$

$$\leq \sum_{n=1}^{\infty} P(N \geq n, U_n \in B)$$

$$= \sum_{n=1}^{\infty} P(N \geq n) P(U_1 \in B)$$

$$= E(N)P(U_1 \in B).$$

Thus, for all Borel sets B,

$$E(N) \geq \frac{P(X \in B)}{P(U_1 \in B)}.$$

If we take the supremum of the right-hand-side over all B, then we obtain $||f||_\infty$. ∎

There are quite a few algorithms that fall into this category. In particular, if we use rejection with a constant dominating curve on [0,1], then we use N uniform random variates where for continuous f ,

$$E(N) \geq \sup_x f(x) .$$

We have seen that in the rejection algorithm, we come within a factor of 2 of this lower bound. If the U_i's have density g on the real line, then we can construct stopping times for all densities f that are absolutely continuous with respect to g , and the lower bound reads

$$E(N) \geq \left|\left|\frac{f}{g}\right|\right|_\infty .$$

For continuous $\frac{f}{g}$, the lower bound is equal to $\sup\frac{f}{g}$ of course. Again, with the rejection method with g as dominating density, we come within a factor of 2 of the lower bound.

There is another class of algorithms that fits the description given here, notably the Forsythe-von Neumann algorithms, which will be presented in section IV.2.

3.6. The squeeze principle.

In the rejection method based on the inequality $f \leq cg$, we need to compute the ratio $\frac{f}{g}$ N times where N is the number of iterations. In most cases, this is a slow operation because f is presumably not a simple function of its argument (for otherwise, we would know how to generate random variates from f by other means). In fact, sometimes f is not known explicitly: in this book, we will encounter cases in which it is the integral of another function or the solution of a nonlinear equation. In all these situations, we should try to avoid the computation of $\frac{f}{g}$ either entirely, or at least most of the time. For principles leading to the total avoidance of the computation, we refer to the more advanced chapter IV. Here we will briefly discuss the squeeze principle (a term introduced by George Marsaglia (1977)) designed to avoid the computation of the ratio with high probability. One should in fact try to find functions h_1 and h_2 that are easy to evaluate and have the property that

$$h_1(x) \leq f(x) \leq h_2(x) .$$

Then, we have:

The squeeze method

REPEAT

      Generate a uniform [0,1] random variate U.

      Generate a random variate X with density g.

      Set $W \leftarrow Ucg(X)$.

      Accept $\leftarrow [W \leq h_1(X)]$.

      IF NOT Accept

            THEN IF $W \leq h_2(X)$ THEN Accept $\leftarrow [W \leq f(X)]$.

UNTIL Accept

RETURN X

In this algorithm, we introduced the boolean variable "Accept" to streamline the exit from the REPEAT loop. Such boolean variables come in handy whenever a program must remain structured and readable. In the algorithm, we count on the fact that "Accept" gets its value most of the time from the comparison between W and $h_1(X)$, which from now on will be called a quick acceptance step. In the remaining cases, we use a quick rejection step ($W > h_2(X)$), and in the rare cases that W is sandwiched between $h_1(X)$ and $h_2(X)$, we resort to the expensive comparison of W with $f(X)$ to set the value of "Accept".

The validity of the algorithm is not jeopardized by dropping the quick acceptance and quick rejection steps. In that case, we simply have the statement Accept$\leftarrow [W \leq f(X)]$, and obtain the standard rejection algorithm. In many cases, the quick rejection step is omitted since it has the smallest effect on the efficiency. Note also that it is not necessary that $h_1 \geq 0$ or $h_2 \leq cg$, although nothing will be gained by considering violations of these boundary conditions.

We note that N is as in the rejection algorithm, and thus, $E(N) = c$. The gain will be in the number of computations N_f of f, the dominating factor in the time complexity. The computation of $E(N_f)$ demonstrates the usefulness of Wald's equation once again. Indeed, we have

$$N_f = \sum_{i=1}^{N} I_{[h_1(X_i) < W_i < h_2(X_i)]} \, ,$$

where W_i is the W obtained in the i-th iteration, and X_i is the X used in the i-th iteration. To this sum, we can apply Wald's equation, and thus,

$$E(N_f) = E(N) \, P(h_1(X_1) < W_1 < h_2(X_1))$$
$$= c \int g(x) \frac{h_2(x) - h_1(x)}{cg(x)} \, dx$$
$$= \int (h_2(x) - h_1(x)) \, dx \, .$$

Here we used the fact that we have proper sandwiching, i.e. $0 \leq h_1 \leq f \leq h_2 \leq cg$. If $h_1 \equiv 0$ and $h_2 \equiv cg$ (i.e., we have no squeezing), then we obtain the result $E(N_f) = c$ for the rejection method. With only a quick acceptance step (i.e. $h_2 = cg$), we have $E(N_f) = c - \int h_1$. When $h_1 \geq 0$ and/or $h_2 \leq cg$ are violated, equality in the expression for $E(N_f)$ should be replaced by inequality (exercise 3.13).

Inequalities via Taylor's series expansion.

A good source of inequalities for functions f in terms of simpler functions is provided by Taylor's series expansion. If f has n continuous derivatives (denoted by $f^{(1)}, \ldots, f^{(n)}$), then it is known that

$$f(x) = f(0) + \frac{x}{1!} f^{(1)}(0) + \cdots + \frac{x^{n-1}}{n-1!} f^{(n-1)}(0) + \frac{x^n}{n!} f^{(n)}(\xi) ,$$

where ξ is a number in the interval $[0, x]$ (or $[x, 0]$, depending upon the sign of x). From this, by inspection of the last term, one can obtain inequalities which are polynomials, and thus prime candidates for h_1 and h_2. For example, we have

$$e^{-x} = 1 - x + \frac{x^2}{2!} - \cdots + (-1)^{n-1} \frac{x^{n-1}}{n-1!} + (-1)^n \frac{x^n}{n!} e^{-\xi} .$$

From this, we see that for $x \geq 0$, e^{-x} is sandwiched between consecutive terms of the well-known expansion

$$e^{-x} = \sum_{i=0}^{\infty} (-1)^i \frac{x^i}{i!} .$$

In particular,

$$1 - x \leq e^{-x} \leq 1 - x + \frac{x^2}{2} \qquad (x \geq 0) . \blacksquare$$

Example 3.4. The normal density.

For the normal density f, we have developed an algorithm based upon rejection from the Cauchy density in Example 3.2. We used the inequality $f \leq cg$ where $c = \sqrt{\dfrac{2\pi}{e}}$ and $g(x) = \dfrac{1}{\pi(1+x^2)}$. For h_1 and h_2 we should look for simple functions of x. Applying the Taylor series technique described above, we see that

$$1 - \frac{x^2}{2} \leq \sqrt{2\pi} f(x) \leq 1 - \frac{x^2}{2} + \frac{x^4}{8} .$$

Using the lower bound for h_1, we can now accelerate our normal random variate generator somewhat:

Normal variate generator by rejection and squeezing

REPEAT

Generate a uniform [0,1] random variate U.

Generate a Cauchy random variate X.

Set $W \leftarrow \dfrac{2U}{\sqrt{e}\ (1+X^2)}$.   (Note: $W \leftarrow cUg(X)\sqrt{2\pi}$.)

Accept $\leftarrow [W \leq 1 - \dfrac{X^2}{2}]$.

IF NOT Accept THEN Accept $\leftarrow [W \leq e^{-\dfrac{X^2}{2}}]$.

UNTIL Accept

RETURN X

This algorithm can be improved in many directions. We have already got rid of the annoying normalization constant $\sqrt{2\pi}$. For $|X| > \sqrt{2}$, the quick acceptance step is useless in view of $h_1(X) < 0$. Some further savings in computer time result if we work with $Y \leftarrow \dfrac{1}{2}X^2$ throughout. The expected number of computations of f is

$$c - \int h_1 = \sqrt{\frac{2\pi}{e}} - \frac{1}{\sqrt{2\pi}} \int\limits_{|x| \leq \sqrt{2}} (1 - \frac{x^2}{2})\ dx = \sqrt{\frac{2\pi}{e}} - \frac{4}{3\sqrt{\pi}} \cdot \blacksquare$$

Example 3.5. Proportional squeezing.

It is sometimes advantageous to sandwich f between two functions of the same form as in

$$bg \leq f \leq cg\ ,$$

where g is an easy density (as in the rejection method), and b is a positive constant. When b and c are close to 1, such a proportional squeeze can be very useful. For example, random variates can be generated as follows:

The proportional squeeze method

REPEAT

    Generate a uniform $[0,1]$ random variate U.

    Generate a random variate X with density g.

    Accept $\leftarrow [U \leq \frac{b}{c}]$.

    IF NOT Accept THEN Accept $\leftarrow [U \leq \frac{f(X)}{cg(X)}]$.

UNTIL Accept

RETURN X

Here the expected number of computations of f is quite simply $c-b$. The main area of application of this method is in the development of universally applicable algorithms in which the real line is first partitioned into many intervals. On each interval, we have a nearly constant or nearly linear piece of density. For this piece, proportional squeezing with dominating density of the form $g(x) = a_0 + a_1 x$ can usually be applied (see exercises 3.10 and 3.11 below). ∎

Example 3.6. Squeezing based upon an absolute deviation inequality.

Assume that a density f is close to another density h in the following sense:

$$| f - h | \leq g \ .$$

Here g is another function, typically with small integral. Here we could implement the rejection method with as dominating curve $g + h$, and apply a squeeze step based upon $f \geq h - g$. After some simplifications, this leads to the following algorithm:

REPEAT

Generate a random variate X with density proportional to $h+g$, and a uniform $[0,1]$ random variate U.

Accept $\leftarrow [\dfrac{g(X)}{h(X)} \leq \dfrac{1-U}{1+U}]$

IF NOT Accept THEN Accept $\leftarrow [U(g(X)+h(X)) \leq f(X)]$

UNTIL Accept

RETURN X

This algorithm has rejection constant $1+\int g$, and the expected number of evaluations of f is at most $2\int g$. Algorithms of this type are mainly used when g has very small integral. One instance is when the starting absolute deviation inequality is known from the study of limit theorems in mathematical statistics. For example, when f is the gamma (n) density normalized to have zero mean and unit variance, it is known that f tends to the normal density as $n \rightarrow \infty$. This convergence is studied in more detail in local central limit theorems (see e.g. Petrov (1975)). One of the by-products of this theory is an inequality of the form needed by us, where g is a function depending upon n, with integral decreasing at the rate $1/\sqrt{n}$ as $n \rightarrow \infty$. The rejection algorithm would thus have improved performance as $n \rightarrow \infty$. What is intriguing here is that this sort of inequality is not limited to the gamma density, but applies to densities of sums of iid random variables satisfying certain regularity conditions. In one sweep, one could thus design general algorithms for this class of densities. See also sections XIV.3.3 and XIV.4. ∎

3.7. Recycling random variates.

In this section we have emphasized the expected number of iterations in our algorithms. Sometimes we have looked at the number of function evaluations. But by and large we have steered clear of making statements about the expected number of uniform random variates needed before an algorithm halts. One of the reasons is that we can always recycle unused parts of the uniform random variate. The recycling principle is harmless for our infinite precision model, but should be used with extreme care in standard single precision arithmetic on computers.

For the rejection method, based upon the inequality $f \leq cg$ where g is the dominating density, and c is a constant, we note that given a random variate X

with density g and an independent uniform $[0,1]$ random variate U, the halting rule is $Ucg(X)/f(X) \leq 1$. Given that we halt, then $Ucg(X)/f(X)$ is again uniform on $[0,1]$. If we reject, then

$$\frac{\dfrac{Ucg(X)}{f(X)} - 1}{\dfrac{g(X)}{f(X)} - 1}$$

is again uniformly distributed on $[0,1]$. These recycled uniforms can be used either in the generation of the next random variate (if more than one random variate is needed), or in the next iteration of the rejection algorithm. Thus, in theory, the cost of uniform $[0,1]$ random variates becomes negligible: it is one if only one random variate must be generated, and it remains one even if n random variates are needed. The following algorithm incorporates these ideas:

Rejection algorithm with recycling of one uniform random variate

Generate a uniform $[0,1]$ random variate U.
REPEAT
    REPEAT
        Generate a random variate X with density g.
        $T \leftarrow \dfrac{cg(X)}{f(X)}$, $V \leftarrow UT$
        $U \leftarrow \dfrac{V-1}{T-1}$ (prepare for recycling)
        UNTIL $U \leq 0$ (equivalent to $V \leq 1$)
        RETURN X (X has density f)
        $U \leftarrow V$ (recycle)
UNTIL False (this is an infinite loop; add stopping rule)

In this example, we merely want to make a point about our idealized model. Recycling can be (and usually is) dangerous on finite-precision computers. When f is close to cg, as in most good rejection algorithms, the upper portion of U (i.e. $(V-1)/(T-1)$ in the notation of the algorithm) should not be recycled since $T-1$ is close to 0. The bottom part is more useful, but this is at the expense of less readable algorithms. All programs should be set up as follows: a uniform random variate should be provided upon input, and the output consists of the returned random variate and another uniform random variate. The input and output random variates are dependent, but it should be stressed that the returned random variate X and the recycled uniform random variate are independent! Another argument against recycling is that it requires a few multiplications and/or divisions. Typically, the time taken by these operations is longer than the time needed to generate one good uniform $[0,1]$ random variate. For all these reasons, we do not pursue the recycling principle any further.

3.8. Exercises.

1. Let f and g be easy densities for which we have subprograms for computing $f(x)$ and $g(x)$ at all $x \in R^d$. These densities can be combined into other densities in several manners, e.g.

    $$h = c \max(f, g)$$
    $$h = c \min(f, g)$$
    $$h = c \sqrt{fg}$$
    $$h = c f^{\alpha} g^{1-\alpha}$$

    where c is a normalization constant (different in each case) and $\alpha \in [0,1]$ is a constant. How would you generate random variates with density h ? Give the expected time complexity (expected number of iterations, comparisons, etc.).

2. Decompose the density $h(x) = \dfrac{2}{\pi} \sqrt{1-x^2}$ on $[-1,1]$ as follows:

    $$h(x) = c \sqrt{f(x) g(x)}$$

    where $c = \dfrac{2}{\pi} \sqrt{\dfrac{8}{3}}$, $f(x) = \dfrac{3}{4}(1-x^2)$ and $g(x) = \dfrac{1}{2}$, and $|x| \leq 1$. Thus, h is in one of the forms specified in exercise 3.1. Give a complete algorithm and analysis for generating random variates with density h by the general method of exercise 3.1.

3. The algorithm

    > REPEAT
    >> Generate X with density g.
    >>
    >> Generate an exponential random variate E.
    >
    > UNTIL $h(X) \leq E$
    > RETURN X

    when used with a nonnegative function h produces a random variate X with density

    $$c g(x) e^{-h(x)},$$

    where c is a normalization constant. Show this.

4. How does c, the rejection constant, change with n (i.e., what is its rate of increase as $n \to \infty$) when the rejection method is used on the beta (n,n) density and the dominating density g is the uniform density on $[0,1]$?

5. Lux (1979) has generalized the rejection method as follows. Let g be a given density, and let F be a given distribution function. Furthermore, assume

that r is a fixed positive-valued monotonically decreasing function on $[0,\infty)$. Then a random variate X with density

$$f(x) = g(x) \int_{-\infty}^{r(x)} \left[\frac{1}{\int_0^{r^{-1}(y)} g(z)\, dz} \right] dF(y) \qquad (x > 0):$$

Lux's algorithm

REPEAT

      Generate a random variate X with density g.

      Generate a random variate Y with distribution function F.

UNTIL $Y \le r(X)$

RETURN X

Also, the probability of acceptance of a random couple (X,Y) in Lux's algorithm is $\int_0^{\infty} F(r(x))\, g(x)\, dx$.

6. The following density on $[0,\infty)$ has both an infinite peak at 0 and a heavy tail:

$$f(x) = \frac{2}{(1+x)\sqrt{x^2+2x}} \qquad (x > 0).$$

Consider as a possible candidate for a dominating curve $c_\theta\, g_\theta$ where

$$c_\theta\, g_\theta(x) = \begin{cases} \dfrac{2}{\pi\sqrt{2x}} & ,0 \le x \le \theta \\[2mm] \dfrac{2}{\pi x^2} & ,x > \theta \end{cases},$$

where c_θ is a constant depending upon θ only and $\theta > 0$ is a design parameter. Prove first that indeed $f \le c_\theta\, g_\theta$. Then show that c_θ is minimal for $\theta = 2^{1/3}$ and takes the value $\dfrac{6}{\pi 2^{\frac{1}{3}}}$. Give also a description of the entire rejection algorithm together with the values for the expected number of iterations, comparisons, square root operations, multiplications/divisions, and assignment statements. Repeat the same exercise when the dominating density is the density of the random variable $\theta U^2/V$ where $\theta > 0$ is a parameter and U and V are two iid uniform $[0,1]$ random variates. Prove that in this case too we obtain the same rejection constant $\dfrac{6}{\pi 2^{\frac{1}{3}}}$.

7. **Optimal rejection algorithms for the normal density.** Assume that normal random variates are generated by rejection from a density g_θ where θ is a design parameter. Depending upon the class of g_θ's that is considered, we may obtain different optimal rejection constants. Complete the following table:

$g_\theta(x)$	Optimal θ	Optimal rejection constant c
Cauchy (θ): $\dfrac{\theta}{\pi(\theta^2+x^2)}$	1	$\sqrt{\dfrac{2\pi}{e}}$
Laplace (θ): $\dfrac{\theta}{2}e^{-\theta\mid x\mid}$	1	$\sqrt{\dfrac{2e}{\pi}}$
Logistic (θ): $\dfrac{\theta e^{-\theta x}}{(1+e^{-\theta x})^2}$	?	?
$\min(\dfrac{1}{4\theta},\dfrac{\theta}{4x^2})$	?	?

8. **Sibuya's modified rejection method.** Sibuya (1962) noted that the number of uniform random variates in the rejection algorithm can be reduced to one by repeated use of the same uniform random variate. His algorithm for generating a random variate with density f (known not to exceed cg for an easy density g) is:

> Generate a uniform [0,1] random variate U.
> REPEAT
>   Generate a random variate X with density g .
> UNTIL $cg(X)U \leq f(X)$
> RETURN X

Show the following:

(i) The algorithm is valid if and only if $c =$ ess sup $(f(X)/g(X))$.

(ii) If N is the number of X's needed in Sibuya's algorithm, and $N*$ is the number of X's needed in the original rejection algorithm, then

$$E(N) \geq E(N*)$$

and

$$P(N \geq i) \geq P(N* \geq i) \quad \text{(all } i\text{)} .$$

(Hint: use Jensen's inequality.) We conclude from (ii) that Sibuya's method is worse than the rejection method in terms of number of required iterations.

(iii) We can have $P(N=\infty)>0$ (just take $g =f$, $c >1$). We can also have $P(N=\infty)=0, E(N)=\infty$ (just take $f(x)=2(1-x)$ on [0,1], $c =2$

and $g(x)=1$ on $[0,1]$). Give a necessary and sufficient condition for $P(N=\infty)=0$, and show that this requires that c is chosen optimally. See also Greenwood (1976).

9. There exists a second moment analog of Wald's equation which you should try to prove. Let $W_1, \ldots,$ and $\psi \geq 0$ be as in Theorem 3.5. Assume further that $\psi(W_1)$ has mean μ and variance $\sigma^2 < \infty$. Then, for any stopping rule N with $E(N) < \infty$,

$$ E\left(\left(\sum_{i=1}^{N}(W_i-\mu)\right)^2\right) = \sigma^2 E(N) . $$

See for example Chow and Teicher (1978, pp. 139).

10. Assume that we use proportional squeezing for a density f on $[0,1]$ which is known to be between $2b(1-x)$ and $2c(1-x)$ where $0 \leq b \leq 1 \leq c < \infty$. Then, we need in every iteration a uniform random variate U and a triangular random variate X (which in turn can be obtained as $\min(U_1,U_2)$ where U_1, U_2 are also uniform $[0,1]$ random variates). Prove that if $U_{(1)} \leq U_{(2)}$ are the order statistics of U_1, U_2, then

$$ (U_{(1)}, \frac{U_{(2)}-U_{(1)}}{1-U_{(1)}}) $$

is distributed as (X,U). Thus, using this device, we can "save" one uniform random variate per iteration. Write out the details of the corresponding proportional squeeze algorithm.

11. Assume that the density f has support on $[0,1]$ and that we know that it is Lipschitz with constant C, i.e.

$$ |f(y)-f(x)| \leq C|x-y| \qquad (x,y \in R) . $$

Clearly, we have $f(0)=f(1)=0$. Give an efficient algorithm for generating a random variate with density f which is based upon an n-part equi-spaced partition of $[0,1]$ and the use of the proportional squeeze method for nearly linear densities (see previous exercise) for generating random variates from the n individual pieces. Your algorithm should be asymptotically efficient, i.e. it should have $E(N_f)=o(1)$ as $n \to \infty$ where N_f is the number of computations of f.

12. **Random variates with density** $f(x)=c(1-x^2)^a$ ($|x| \leq 1$). The family of densities treated in this exercise coincides with the family of symmetric beta densities properly translated and rescaled. For example, when the parameter a is integer, f is the density of the median of $2a+1$ iid uniform $[-1,1]$ random variates. It is also the density of the marginal distribution of a random vector uniformly distributed on the surface of the unit sphere in R^d where d and a are related by $a=\dfrac{d-3}{2}$. For the latter reason, we will use it later as an important tool in the generation of random vectors that are uniformly distributed on such spheres. The parameter a must be greater than -1. We have

$$c = \frac{\Gamma(a + \frac{3}{2})}{\sqrt{\pi}\Gamma(a+1)} \ ,$$

and the inequalities

$$ce^{-\frac{ax^2}{1-x^2}} \leq f(x) \leq ce^{-ax^2} \qquad (|x| \leq 1).$$

The following rejection algorithm with squeezing can be used:

Translated symmetric beta generator by rejection and squeezing

REPEAT
    REPEAT
        Generate a normal random variate X.
        Generate an exponential random variate E.
    UNTIL $Y \leq 1$
    $X \leftarrow \dfrac{X}{\sqrt{2a}}$, $Y \leftarrow X^2$
    Accept $\leftarrow [1 - Y(1 + \frac{a}{E}Y) \geq 0]$.
    IF NOT Accept THEN Accept $\leftarrow [aY + E + a\log(1-Y) \geq 0]$.
    UNTIL Accept
    RETURN X

A.   Verify that the algorithm is valid.

B.   The expected number of normal/exponential pairs needed is
$\dfrac{\Gamma(a + \frac{3}{2})}{\sqrt{a}\ \Gamma(a+1)}$. Selected values are

$a = 1$	$\frac{3}{4}\sqrt{\pi}$	1.329340...
$a = 2$	$\frac{15}{16}\sqrt{\frac{\pi}{2}}$	1.174982...
$a = 3$	$\frac{105}{96}\sqrt{\frac{\pi}{3}}$	1.119263...

Show that this number tends to 1 as $a \to \infty$ and to ∞ as $a \downarrow 0$.

C.   From part B we conclude that it is better to take care of the case $0 \leq a \leq 1$ separately, by bounding as follows: $c(1-x^2) \leq f(x) \leq c$. The expected number of iterations becomes $2c$, which takes the values $\frac{3}{2}$ at $a = 1$ and 1 at $a = 0$. Does this number vary monotonically with a? How does $E(N_f)$ vary with a?

D. Write a generator which works for all $a > -1$. (This requires yet another solution for a in the range $(-1,0)$.)

E. Random variates from f can also obtained in other ways. Show that all of the following recipes are valid:

(i) $S \sqrt{B}$ where B is beta$(\frac{1}{2}, a+1)$ and S is a random sign.

(ii) $S \sqrt{\dfrac{Y}{Y+Z}}$ where Y, Z are independent gamma$(\frac{1}{2},1)$ and gamma$(a+1,1)$ random variates, and S is a random sign.

(iii) $2B-1$ where B is a beta$(a+1, a+1)$ random variate.

13. Consider the squeeze algorithm of section 3.6 which uses the inequality $f \le cg$ for the rejection-based generator, and the inequalities $h_1 \le f \le h_2$ for the quick acceptance and rejection steps. Even if h_1 is not necessarily positive, and h_2 is not necessarily smaller than cg, show that we always have

$$E(N_f) = \int (\min(h_2, cg) - \max(h_1, 0)) \le \int (h_2 - h_1) ,$$

where N_f is the number of evaluations of f.

14. **A universal generator a la Letac.** Let f be **any** density on $[0,1]$, and assume that the cumulative mass function $M(t) = \int_{f \ge t} f(x)\, dx$ is known. Consider the following algorithm:

```
Generate a random integer Z  where P(Z=i)=M(i)-M(i+1).
REPEAT
        Generate (X,V) uniformly in [0,1]²
UNTIL Z+V≤f(X)
RETURN X
```

Show that the algorithm is valid (relate it to the rejection method). Relate the expected number of X's generated before halting to $||f||_\infty$, the essential supremum of f. Among other things, conclude that the expected time is ∞ for every unbounded density. Compare the expected number of X's with Letac's lower bound. Show also that if inversion by sequential search is used for generating Z, then the expected number of iterations in the search before halting is finite if and only if $\int f^2 < \infty$. A final note: usually, one does not have a cumulative mass function for an arbitrary density f.

4. DECOMPOSITION AS DISCRETE MIXTURES.

4.1. Definition.

If our target density f can be decomposed into a discrete mixture

$$f(x) = \sum_{i=1}^{\infty} p_i f_i(x)$$

where the f_i's are given densities and the p_i's form a probability vector (i.e., $p_i \geq 0$ for all i and $\sum_i p_i = 1$), then random variates can be obtained as follows:

The composition method.

Generate a random integer Z with probability vector $p_1, \ldots, p_i, \ldots$ (i.e. $P(Z=i)=p_i$).
Generate a random variate X with density f_Z.
RETURN X

This algorithm is incomplete, because it does not specify just how Z and X are generated. Every time we use the general form of the algorithm, we will say that the composition method is used.

We will show in this section how the decomposition method can be applied in the design of good generators, but we will not at this stage address the problem of the generation of the discrete random variate Z. Rather, we are interested in the decomposition itself. It should be noted however that in many, if not most, practical situations, we have a finite mixture with K components.

4.2. Decomposition into simple components.

Very often, we will decompose the graph of f into a bunch of very simple structures such as rectangles and triangles, mainly because random variates with rectangular-shaped or triangular-shaped densities are so easy to generate (by linear combinations of one or two uniform [0,1] random variates). This decomposition is finite if f is piecewise linear with a finite number of pieces (this forces f to have compact support). In general, one will decompose f as follows:

$$f(x) = \sum_{i=1}^{K-2} p_i f_i(x) + p_{K-1} f_{K-1}(x) + p_K f_K(x)$$

where f_K is a tall density (it is zero on a central interval $[a, b]$), p_K is usually very small, and all other f_i's vanish outside the central interval $[a, b]$. The structure of $f_1, \ldots, f_{K-2}$ is simple, e.g. rectangular. After having picked the rectangles in such a way that the corresponding p_i's add up to nearly 1, we collect the leftover piece in $p_{K-1} f_{K-1}$. This last piece is often strangely shaped, and random variates from it are generated by the rejection method. The point is that p_{K-1} and p_K are so small that we do not have to generate random variates with this density very often. Most of the time, i.e. with probability $p_1 + \cdots + p_{K-2}$, it suffices to generate one or two uniform $[0,1]$ random variates and to shift or rescale them. This technique will be called the **jigsaw puzzle method**, a term coined by Marsaglia. The careful decomposition requires some refined analysis, and is usually only worth the trouble for frequently used fixed densities such as the normal density. We refer to the section on normal variate generation for several applications of this sort of decomposition. Occasionally, it can be applied to families of distributions (such as the beta and gamma families), but the problem is that the decomposition itself is a function of the parameter(s) of the family. This will be illustrated for the beta family (see section IX.4).

To give the readers a flavor of the sort of work that is involved, we will try to decompose the normal density into a rectangle and one residual piece: the rectangle will be called $p_1 f_1(x)$, and the residual piece $p_2 f_2(x)$. It is clear that p_1 should be as large as possible. But since $p_1 f_1(x) \leq f(x)$, the largest p_1 must satisfy

$$p_1 \leq \inf_x \frac{f(x)}{f_1(x)} .$$

Thus, with $f_1(x) = \frac{1}{2}\theta$, $\mid x \mid \leq \theta$ where θ is the width of the centered rectangle, we see that at best we can set

$$p_1 = \inf_{\mid x \mid \leq \theta} \frac{2\theta e^{-\frac{x^2}{2}}}{\sqrt{2\pi}} = 2\frac{\theta}{\sqrt{2\pi}} e^{-\frac{\theta^2}{2}} .$$

The function p_1 is maximal (as a function of θ) when $\theta = 1$, and the corresponding value is $\sqrt{\dfrac{2}{\pi e}}$. Of course, this weight is not close to 1, and the present decomposition seems hardly useful. The work involved when we decompose in terms of several rectangles and triangles is basically not different from the short analysis done here.

4.3. Partitions into intervals.

Many algorithms are based on the following principle: partition the real line into intervals $A_1, \ldots, A_K$, and decompose f as

$$f(x) = \sum_{i=1}^{K} f(x) I_{A_i}(x) .$$

If we can generate random variates from the restricted densities $f I_{A_i}/p_i$ (where $p_i = \int_{A_i} f$), then the decomposition method is applicable. The advantages offered by partitions into intervals cannot be denied: the decomposition is so simple that it can be mechanized and used for huge classes of densities (in that case, there are usually very many intervals); troublespots on the real line such as infinite tails or unbounded peaks can be conveniently isolated; and most importantly, the decomposition is easily understood by the general user.

In some cases, random variates from the component densities are generated by the rejection method based on the inequalities

$$f(x) \leq h_i(x) \ , x \in A_i \ , 1 \leq i \leq K \ .$$

Here the h_i's are given dominating curves. There are two subtly different methods for generating random variates with density f, given below. One of these needs the constants $p_i = \int_{A_i} f$, and the other one requires the constants $q_i = \int_{A_i} h_i$. Note that the q_i's are nearly always known because the h_i's are chosen by the user. The p_i's are usually known when the distribution function for F is easy to compute at arbitrary points.

The composition method.

Generate a discrete random variate Z with probability vector $p_1, \ldots, p_K$ on $\{1, \ldots, K\}$.
REPEAT
       Generate a random variate X with density h_i/q_i on A_i.
       Generate an independent uniform [0,1] random variate U.
UNTIL $U h_i(X) \leq f(X)$
RETURN X

The modified composition method.

REPEAT

Generate a discrete random variate Z with probability vector proportional to $q_1, \ldots, q_K$ on $\{1, \ldots, K\}$.

Generate a random variate X with density h_i / q_i on A_i.

Generate a uniform $[0,1]$ random variate U.

UNTIL $U h_i(X) \leq f(X)$

RETURN X

In the second algorithm we use the rejection method with as dominating curve $h_1 I_{A_1} + \cdots + h_K I_{A_K}$, and use the composition method for random variates from the dominating density. In contrast, the first algorithm uses true decomposition. After having selected a component with the correct probability we then use the rejection method. A brief comparison of both algorithms is in order here. This can be done in terms of four quantities: N_Z, N_U, N_h and N_{h_i}, where N is the number of random variates required of the type specified by the index with the understanding that N_h refers to $\sum_{i=1}^{K} h_i$, i.e. it is the total number of random variates needed from any one of the K dominating densities.

Theorem 4.1.

Let $q = \sum_{i=1}^{K} q_i$, and let N be the number of iterations in the second algorithm. For the second algorithm we have $N_U = N_Z = N_h = N$, and N is geometrically distributed with parameter $\dfrac{1}{q}$. In particular,

$$E(N) = q \; ; \; E(N^2) = 2q^2 - q \; .$$

For the first algorithm, we have $N_Z = 1$. Also, $N_U = N_h$ satisfy

$$E(N_U) = q \; ; \; E(N_U^2) = \sum_{i=1}^{K} \frac{2q_i^2}{p_i} - q \geq 2q^2 - q \; .$$

Finally, for both algorithms, $E(N_{h_i}) = q_i$.

Proof of Theorem 4.1.

The statement for the second algorithm is obvious when we note that the rejection constant is equal to the area q under the dominating curve (the sum of the h_i's in this case). For the first algorithm, we observe that given the value of Z, N_U is geometrically distributed with parameter p_Z / q_Z. From the properties of the geometric distribution, we then conclude the following:

$$E(N_U) = \sum_{i=1}^{K} p_i \left(\frac{q_i}{p_i} \right) = \sum_{i=1}^{K} q_i = q \ ,$$

$$E(N_U{}^2) = \sum_{i=1}^{K} p_i \left(\frac{2}{\left(\frac{p_i}{q_i} \right)^2} - \frac{1}{\frac{p_i}{q_i}} \right) = \sum_{i=1}^{K} 2p_i \left(\frac{q_i}{p_i} \right)^2 - q \ .$$

To show that the last expression is always greater or equal to $2q^2 - q$ we use the Cauchy-Schwarz inequality:

$$\sum_{i=1}^{K} 2p_i \left(\frac{q_i}{p_i} \right)^2 - q \geq 2 \left(\sum_{i=1}^{K} p_i \frac{q_i}{p_i} \right)^2 \left(\sum_{i=1}^{K} p_i \times 1 \right)^2 - q = 2q^2 - q \ .$$

Finally, we consider $E(N_{h_i})$. For the first algorithm, its expected value is $p_i \left(\frac{q_i}{p_i} \right) = q_i$. For the second algorithm, we employ Wald's equality after noting that $N_{h_i} = \sum_{j=1}^{N} I_{[\text{piece } h_i \text{ is used in the } j\text{-th iteration}]}$. Thus, the expected value is $E(N)P$ (piece h_i is used in the first iteration), which is equal to $q \left(\frac{q_i}{q} \right) = q_i$. ∎

In standard circumstances, q is close to 1, and discrete random variate generators are ultra efficient. Thus, N_Z is not a great factor. For all the other quantities involved in the comparison, the expected values are equal. But when we examine the higher moments of the distributions, we notice a striking difference, because the second method has in all cases a smaller second moment. In fact, the difference can be substantial when for some i, the ratio q_i / p_i is large. If we take $q_i = p_i$ for $i \geq 2$ and $q_1 = q - (1 - p_1)$, then for the first method,

$$E(N_U{}^2) = \frac{2(q - 1 + p_1)^2}{p_1} + 2(1 - p_1) - q = (2q^2 - q) + 2(q - 1)^2 \left(\frac{1}{p_1} - 1 \right) \ .$$

The difference between the two second moments in this example is $2(q - 1)^2 \left(\frac{1}{p_1} - 1 \right)$. Thus, isolating a small probability piece in the decomposition method and using a poor rejection rate for that particular piece is dangerous. In such situations, one is better off using a global rejection method as suggested in the second algorithm.

4.4. The waiting time method for asymmetric mixtures.

In large simulations, one needs iid random variates $X_1, \ldots, X_n, \ldots$. If these random variates are generated by the composition method, then for every random variate generated we need one discrete random variate Z for selecting a component. When f is decomposed into a main component $p_1 f_1$ (p_1 is close to 1) and a small component $p_2 f_2$, then most of these selections will choose the first component. In those cases, it is useful to generate the times of occurrence of selection of the second component instead. If the second component is selected at times $T_1, T_2, \ldots$, then it is not difficult to see that $T_1, T_2 - T_1, \ldots$ are iid geometric random variables with parameter p_2, i.e.

$$P(T_1 = i) = (1 - p_2)^{i-1} p_2 \quad (i \geq 1) .$$

A random variate T_1 can be generated as $\left\lceil -\dfrac{E}{\log(p_2)} \right\rceil$ where E is an exponential random variate. Of course, we need to keep track of these times as we go along, occasionally generating a new time. These times need to be stored locally in subprograms for otherwise we need to pass them as parameters. In some cases, the overhead associated with passing an extra parameter is comparable to the time needed to generate a uniform random variate. Thus, one should carefully look at how the large simulation can be organized before using the geometric waiting times.

4.5. Polynomial densities on $[0,1]$.

In this section, we consider densities of the form

$$f(x) = \sum_{i=0}^{K} c_i x^i \quad (0 \leq x \leq 1) ,$$

where the c_i's are constants and K is a positive integer. Densities with polynomial forms are important further on as building blocks for constructing piecewise polynomial approximations of more general densities. If K is 0 or 1, we have the uniform and triangular densities, and random variate generation is no problem. There is also no problem when the c_i's are all nonnegative. To see this, we observe that the distribution function F is a mixture of the form

$$F(x) = \sum_{i=1}^{K+1} \left(\frac{c_{i-1}}{i} \right) x^i$$

where of course $\sum_{i=1}^{K+1} \dfrac{c_{i-1}}{i} = 1$. Since x^i is the distribution function of the maximum of i iid uniform $[0,1]$ random variables, we can proceed as follows:

Generate a discrete random variate Z where $P(Z=i)=\dfrac{c_{i-1}}{i}$, $1\leq i\leq K+1$.

RETURN X where X is generated as $\max(U_1, \ldots, U_Z)$ and the U_i's are iid uniform $[0,1]$ random variates.

We have a nontrivial problem on our hands when one or more of the c_i's are negative. The solution given here is due to Ahrens and Dieter (1974), and can be applied whenever $c_0+\sum\limits_{i\,:\,c_i\,<0} c_i \geq 0$. They decompose f as follows: let A be the collection of integers in $\{0, \ldots, K\}$ for which $c_i \geq 0$, and let B the collection of indices in $\{0, \ldots, K\}$ for which $c_i < 0$. Then, we have

$$f(x) = \sum_{i=0}^{K} c_i x^i$$

$$= p_0+\sum_{i\in A}\frac{c_i}{i+1}((i+1)x^i)+\sum_{i\in B}(-\frac{ic_i}{i+1})(\frac{i+1}{i}(1-x^i)) \qquad (0\leq x\leq 1)\,,$$

where $p_0=c_0+\sum\limits_{i\in B}c_i$ (which is ≥ 0 by assumption). If we set p_i equal to $c_i/(i+1)$ for $i\in A$,$i\geq 1$, and to $-ic_i/(i+1)$ for $i\in B$, then $p_0,p_1, \ldots, p_K$ is a probability vector, and we have thus decomposed f as a finite mixture. Let us briefly mention how random variate generation for the component densities can be done.

Lemma 4.1.

Let $U_1,U_2,...$ be iid uniform $[0,1]$ random variables.

A. For $a>1$, $U_1^{\frac{1}{a}}U_2$ has density

$$\frac{a}{a-1}(1-x^{a-1}) \qquad (0\leq x\leq 1)\,.$$

B. Let L be the index of the first U_i not equal to $\max(U_1, \ldots, U_n)$ for $n\geq 2$. Then U_L has density

$$\frac{n}{n-1}(1-x^{n-1}) \qquad (0\leq x\leq 1)\,.$$

C. The density of $\max(U_1, \ldots, U_n)$ is nx^{n-1} $(0\leq x\leq 1)$.

Proof of Lemma 4.1.

Part C is trivially true. Part A is a good exercise on transformations of random variables. Part B has a particularly elegant short proof. The density of a randomly chosen U_i is 1 (all densities are understood to be on [0,1]). Thus, when f is the density of U_L, we must have

$$\frac{n-1}{n} f(x) + \frac{1}{n} n x^{n-1} = 1 \ .$$

This uses the fact that with probability $\frac{1}{n}$, the randomly chosen U_i is the maximal U_i, and that with the complimentary probability, the randomly chosen U_i is distributed as U_L. ∎

We are now in a position to give more details of the polynomial density algorithm of Ahrens and Dieter.

Polynomial density algorithm of Ahrens and Dieter

[SET-UP]

Compute the probability vector $p_0, p_1, \ldots, p_K$ from $c_0, \ldots, c_K$ according to the formulas given above. For each $i \in \{0,1, \ldots, K\}$, store the membership of i ($i \in A$ if $c_i \geq 0$ and $i \in B$ otherwise).

[GENERATOR]

Generate a discrete random variate Z with probability vector $p_0, p_1, \ldots, p_K$.

IF $Z \in A$

THEN RETURN $X \leftarrow U^{\frac{1}{Z+1}}$ (or $X \leftarrow \max(U_1, \ldots, U_{Z+1})$ where $U, U_1, \ldots$ are iid uniform [0,1] random variates).

ELSE RETURN $X \leftarrow U_1^{\frac{1}{Z+1}} U_2$ (or $X \leftarrow U_L$ where L is the U_i with the lowest index not equal to $\max(U_1, \ldots, U_{Z+1})$).

4.6. Mixtures with negative coefficients.

Assume that the density $f(x)$ can be written as

$$f(x) = \sum_{i=1}^{\infty} p_i f_i(x),$$

where the f_i's are densities, but the p_i's are real numbers summing to one. A general algorithm for these densities was given by Bignami and de Matteis (1971). It uses the fact that if p_i is decomposed into its positive and negative parts, $p_i = p_{i+} - p_{i-}$, then

$$f(x) \leq g(x) = \sum_{i=1}^{\infty} p_{i+} f_i(x).$$

Then, the following rejection algorithm can be used:

Negative mixture algorithm of Bignami and de Matteis

REPEAT

Generate a random variate X with density $\sum_{i=1}^{\infty} p_{i+} f_i / \sum_{i=1}^{\infty} p_{i+}$.

Generate a uniform [0,1] random variate U.

UNTIL $U \sum_{i=1}^{\infty} p_{i+} f_i(X) \leq \sum_{i=1}^{\infty} p_i f_i(X)$

RETURN X

The rejection constant here is $\int g = \sum_{i=1}^{\infty} p_{i+}$. The algorithm is thus not valid when this constant is ∞. One should observe that for this algorithm, the rejection constant is probably not a good measure of the expected time taken by it. This is due to the fact that the time needed to verify the acceptance condition can be very large. For finite mixtures, or mixtures that are such that for every x, only a finite number of $f_i(x)$'s are nonzero, we are in good shape. In all cases, it is often possible to accept or reject after having computed just a few terms in the series, provided that we have good analytical estimates of the tail sums of the series. Since this is the main idea of the series method of section IV.5, it will not be pursued here any further.

Example 4.1.

The   density $f(x) = \frac{3}{4}(1-x^2)$, $|x| \leq 1$,   can   be   written   as

$f(x) = \frac{6}{4}(\frac{1}{2} I_{[-1,1]}(x)) - \frac{2}{4}(\frac{x^2}{6} I_{[-1,1]}(x))$. The   algorithm   given   above   is   then entirely equivalent to ordinary rejection from a uniform density, which in this case has a rejection constant of $\dfrac{3}{2}$:

REPEAT
    Generate a uniform $[-1,1]$ random variate X.
    Generate a uniform $[0,1]$ random variate U.
UNTIL $U \leq 1 - X^2$
RETURN X ■

5. THE ACCEPTANCE-COMPLEMENT METHOD.

5.1. Definition.

Let f be a given density on R^d which can be decomposed into a sum of two nonnegative functions:

$$f(x) = f_1(x) + f_2(x) .$$

Assume furthermore that there exists an easy density g such that $f_1 \leq g$. Then the following algorithm can be used to generate a random variate X with density f :

The acceptance-complement method

Generate a random variate X with density g.
Generate a uniform $[0,1]$ random variate U.
IF $U > \dfrac{f_1(X)}{g(X)}$
    THEN Generate a random variate X with density $\dfrac{f_2}{p}$ where $p = \int f_2$.
RETURN X

This, the acceptance-complement method, was first proposed by Kronmal and Peterson (1981,1984). It generalizes the composition method as can be seen if we take $f_1 = f I_A$, $g = f_1 / \int f_1$ and $f_2 = f I_{A^c}$ where A is an arbitrary set of R^d

and A^c is its complement. It is competitive if three conditions are met:

(i) g is an easy density.

(ii) f_2/p is an easy density when p is not small (when p is small, this does not matter much).

(iii) f_1/g is not difficult to evaluate.

As with the composition method, the algorithm given above is more a principle than a detailed recipe. When we compare it with the rejection method, we notice that instead of one design variable (a dominating density) we find two design variables, f_2 and g. Moreover, there is no rejection involved at all, although very often, it turns out that a random variate from $\dfrac{f_2}{p}$ is generated by the rejection method.

Let us first show that this method is valid. For this purpose, we need only show that for all Borel sets $B \subseteq R^d$, the random variate generated by the algorithm (which will be denoted here by X) satisfies $P(X \in B) = \int_B F(x)\, dx$. To avoid confusion with too many X's, we will use Y for the random variate with density g. Thus,

$$
\begin{aligned}
P(X \in B) &= P(Y \in B, U \le \frac{f_1(Y)}{g(Y)}) + P(U > \frac{f_1(Y)}{g(Y)}) \frac{\int_B f_2(x)\, dx}{p} \\
&= \int_B g(x) \frac{f_1(x)}{g(x)}\, dx + (1 - \int g(x) \frac{f_1(x)}{g(x)}\, dx) \frac{\int_B f_2(x)\, dx}{p} \\
&= \int_B f_1(x)\, dx + \int_B f_2(x)\, dx \\
&= \int_B f(x)\, dx .
\end{aligned}
$$

In general, we gain if we can RETURN the first X generated in the algorithm. Thus, it seems that we should try to maximize its probability of acceptance,

$$
P(U \le \frac{f_1(Y)}{g(Y)}) = \int f_1 = 1 - p
$$

subject of course to the constraint $f_1 \le g$ where g is an easy density. Thus, good algorithms have g "almost" equal to f.

There is a visual explanation of the method related to that of the rejection method. What is important here is that the areas under the graphs of $g - f_1$ and f_2 are equal. In the next section, we will give a simplified version of the acceptance-complement algorithm developed independently by Ahrens and Dieter (1981,1983). Examples and details are given in the remaining sections and in some of the exercises.

5.2. Simple acceptance-complement methods.

Ahrens and Dieter (1981,1983) and Deak (1981) considered the special case defined by an arbitrary density g on R^d and the following decomposition:

$$f(x) = f_1(x) + f_2(x);$$
$$f_1(x) = \min(f(x), g(x)) \qquad (\text{note}: f_1 \leq g);$$
$$f_2(x) = (f(x) - g(x))_+.$$

We can now rewrite the acceptance-complement algorithm quite simply as follows:

Simple acceptance-complement method of Ahrens and Dieter

Generate a random variate X with density g.

Generate a uniform [0,1] random variate U.

IF $U > \dfrac{f(X)}{g(X)}$

      THEN Generate a random variate X with density $(f-g)_+/p$ where $p = \int\limits_{f>g} (f-g)$.

RETURN X

Deak (1981) calls this the **economical method**. Usually, g is an easy density close to f. It should be obvious that generation from the leftover density $(f-g)_+/p$ can be problematic. If there is some freedom in the design (i.e. in the choice of g), we should try to minimize p. This simple acceptance-complement method has been used for generating gamma and t variates (see Ahrens and Dieter (1981,1983) and Stadlober (1981) respectively). One of the main technical obstacles encountered (and overcome) by these authors was the determination of the set on which $f(x) > g(x)$. If we have two densities that are very close, we must first verify where they cross. Often this leads to complicated equations whose solutions can only be determined numerically. These problems can be sidestepped by exploiting the added flexibility of the general acceptance-complement method.

5.3. Acceleration by avoiding the ratio computation.

The time-consuming ratio evaluation $\dfrac{f_1}{g}$ in the acceptance condition can be avoided some of the time if we know two easy-to-compute functions h and $h*$ with the property that

$$h(x) \leq \frac{f_1(x)}{g(x)} \leq h*(x) .$$

The IF step in the acceptance-complement algorithm can be replaced in those cases by

Squeeze step in acceptance-complement method

IF $U > h(X)$

> THEN IF $U \geq h*(X)$

>> THEN Generate a random variate X with density $\dfrac{f_2}{p}$ where $p = \int f_2$.

>> ELSE IF $U > \dfrac{f_1(X)}{g(X)}$

>>> THEN Generate a random variate X with density $\dfrac{f_2}{p}$ where $p = \int f_2$.

RETURN X

A similar but more spectacular acceleration is possible for the Ahrens-Dieter algorithm if one can quickly determine whether a point belongs to A, where A is a subset of $f > g$. In particular, one will find that the set on which $f > g$ often is an interval, in which case this acceleration is easy to apply.

Accelerated version of the Ahrens-Dieter algorithm

Generate a random variate X with density g.

IF $X \notin A$

> THEN

>> Generate a uniform [0,1] random variate U.

>> IF $U > \dfrac{f(X)}{g(X)}$

>>> THEN Generate a random variate X with density $(f - g)_+/p$.

RETURN X

With probability $P(X \in A)$, no uniform random variate is generated. Thus, what one should try to do is to choose g such that $P(X \in A)$ is maximal. This in turn suggests choosing g such that

$$\int_{f \geq g} g$$

is large.

5.4. An example: nearly flat densities on $[0,1]$.

We will develop a universal generator for all densities f on $[-1,1]$ which satisfy the following property: $\sup_x f(x) - \inf_x f(x) \leq \frac{1}{2}$. Because we always have $0 \leq \inf_x f(x) \leq \frac{1}{2} \leq \sup_x f(x)$, we see that $\sup_x f(x) \leq 1$. We will apply the acceptance-complement method here with as simple a decomposition as possible, for example

$$g(x) = \frac{1}{2} \qquad (\mid x \mid \leq 1) ;$$

$$f_1(x) = f(x) - (f_{max} - \frac{1}{2}) \qquad (f_{max} = \sup_x f(x)) ;$$

$$f_2(x) = f_{max} - \frac{1}{2} \qquad (\mid x \mid \leq 1) .$$

The condition imposed on the class of densities follows from the fact that we must ask that f_1 be nonnegative. The algorithm now becomes:

Acceptance-complement method for nearly flat densities

Generate a uniform $[-1,1]$ random variate X.
Generate a uniform $[0,1]$ random variate U.
IF $U > 2(f(X) - f_{max} + \frac{1}{2})$
       THEN Generate a uniform $[-1,1]$ random variate X.
RETURN X

To this, we could add a squeeze step, because we can exit whenever $U \leq 2(\inf_x f(x) - f_{max} + \frac{1}{2})$, and the probability of this fast exit increases with the "flatness" of f. It is 1 when f is the uniform density.

A comparison with the rejection method is in order here. First we observe that because we picked g and f_2 both uniform, we need only uniform random variates. The number N of such uniform random variates used up in the algorithm is either 2 or 3. We have

$$E(N) = 2 + 1 \times P(U > 2(f(X) - f_{max} + \frac{1}{2}) ,$$

where X stands for a uniform $[-1,1]$ random variate. Thus,

$$E(N) = 2 + \int_{-1}^{1} \frac{1}{2} 2(f_{max} - f(x)) \, dx$$

$$= 2 + 2f_{max} - 1 = 1 + 2f_{max} .$$

In addition, if no squeeze step is used, we require exactly one computation of f per variate. The obvious rejection algorithm for this example is

Rejection algorithm for nearly flat densities

REPEAT
>   Generate a uniform $[-1,1]$ random variate X.
>   Generate a uniform $[0,1]$ random variate U.

UNTIL $Uf_{max} \leq f(X)$

RETURN X

Here too we could insert a squeeze step $(Uf_{max} \leq \inf_{x} f(x))$. Without it, the expected number of uniform random variates needed is 2 times the expected number of iterations, i.e. $4f_{max}$. In addition, the expected number of computations of f is $2f_{max}$. On both counts, this is strictly worse than the acceptance-complement method.

We have thus established that for some fairly general classes of densities, we have a strict improvement over the rejection algorithm. The universality of the algorithms depends upon the knowledge of the infimum and supremum of f. This is satisfied for example if we know that f is symmetric unimodal in which case the infimum is $f(1)$ and the supremum is $f(0)$.

The algorithm given above can be applied to the main body of most symmetric unimodal densities such as the normal and Cauchy densities. For the truncated Cauchy density

$$f(x) = \frac{2}{\pi(1+x^2)} \qquad (|x| \leq 1) ,$$

our conditions are satisfied because $f_{max} = \frac{2}{\pi}$ and the infimum of f is $\frac{1}{\pi}$, the difference being smaller than $\frac{1}{2}$. In this case, the expected number of uniform random variates needed is $1 + \frac{4}{\pi}$. Next, note that if we can generate a random variate X with density f, then a standard Cauchy random variate can be obtained by exploiting the property that the random variate Y defined by

$$Y = \begin{cases} X & \text{with probability } \dfrac{1}{2} \\ \dfrac{1}{X} & \text{with probability } \dfrac{1}{2} \end{cases}$$

is Cauchy distributed. For this, we need an extra coin flip. Usually, extra coin flips are generated by borrowing a random bit from U. For example, in the universal algorithm shown above, we could have started from a uniform $[-1,1]$ random variate U, and used $|U|$ in the acceptance condition. Since $\text{sign}(U)$ is independent of $|U|$, $\text{sign}(U)$ can be used to replace X by $\dfrac{1}{X}$, so that the returned random variate has the standard Cauchy density. The Cauchy generator thus obtained was first developed by Kronmal and Peterson (1981).

We were forced by technical considerations to limit the densities somewhat. The rejection method can be used on all bounded densities with compact support. This typifies the situation in general. In the acceptance-complement method, once we choose the general form of g and f_2, we loose in terms of universality. For example, if both f_2 and g are constant on $[-1,1]$, then $f = f_1 + f_2 \le g + f_2 \le 1$. Thus, no density f with a peak higher than 1 can be treated by the method. If universality is a prime concern, then the rejection method has little competition.

5.5. Exercises.

1. Kronmal and Peterson (1981) developed yet another Cauchy generator based upon the acceptance-complement method. It is based upon the following decomposition of the truncated Cauchy density f (see text for the definition) into $f_1 + f_2$:

    $$f_1(x) = f(x) - \frac{1}{\pi}(1 - |x|) \quad (|x| \le 1);$$

    $$f_2(x) = \frac{1}{\pi}(1 - |x|) \quad (|x| \le 1);$$

    $$g(x) = \frac{1}{2} \quad (|x| \le 1).$$

    We have:

A Cauchy generator of Kronmal and Peterson

Generate iid uniform $[-1,1]$ random variates X and U.

IF $|U| > \dfrac{2}{\pi}$

       THEN IF $|U| \leq 0.7225$

                   THEN IF $|U| > \dfrac{4}{\pi(1+X^2)} - \dfrac{2}{\pi}(1-|X|)$

                         THEN Generate iid uniform $[-1,1]$ random variates X,U.
$X \leftarrow |X| - |U|$.

                         ELSE Generate a uniform $[-1,1]$ random variate U.
$X \leftarrow |X| - |U|$.

IF $U \leq 0$

       THEN RETURN X

       ELSE RETURN $\dfrac{1}{X}$

The first two IF's are not required for the algorithm to be correct: they correspond to squeeze steps. Verify that the algorithm generates standard Cauchy random variates. Prove also that the acceleration steps are valid. The constant 0.7225 is but an approximation of an irrational number, which should be determined.

Chapter Three
DISCRETE RANDOM VARIATES

1. INTRODUCTION.

A discrete random variable is a random variable taking only values on the nonnegative integers. In probability theoritical texts, a discrete random variable is a random variable which takes with probability one values in a given countable set of points. Since there is a one-to-one correspondence between any countable set and the nonnegative integers, it is clear that we need not consider the general case. In most cases of interest to the practitioner, this one-to-one correspondence is obvious. For example, for the countable set $1, \frac{1}{2}, \frac{1}{4}, \frac{1}{8}, ...$, the mapping is trivial.

The distribution of a discrete random variable X is determined by the probability vector $p_0, p_1, ...$:

$$P(X=i) = p_i \quad (i=0,1,2,...) .$$

The probability vector can be given to us in several ways, such as

A. A table of values $p_0, p_1, \ldots, p_K$. Note that here it is necessary that X can only take finitely many values.

B. An analytical expression such as $p_i = 2^{-i}$ $(i \geq 1)$. This is the standard form in statistical applications, and most popular distributions such as the binomial, Poisson and hypergeometric distributions are given in this form.

C. A subprogram which allows us to compute p_i for each i. This is the "black box" model.

D. Indirectly.. For example, the **generating function**

$$m(s) = \sum_{i=0}^{\infty} p_i s^i \quad (s \in R)$$

can be given. Sometimes, a recursive equation allowing us to compute p_i from $p_j, j < i$, is given.

In cases B, C and D, we should also distinguish between methods for the generation of X when X has a fixed distribution, and methods that should be applicable when X belongs to a certain family of integer-valued random variables.

The methods that will be described below apply usually to only one or two of the cases listed above. Some of these are based on principles that are equally applicable to continuous random variate generation like inversion, composition and rejection. Other principles are unique to discrete random variate generation like the alias principle and the method of guide tables. In any case, this chapter goes hand in hand with chapter II. Very often, the best generator for a certain density uses a clever combination of discrete random variate generation principles and standard methods for continuous random variates. The actual discussion of such combinations is deferred until chapter VIII.

When we give examples in this chapter, we will refer to well-known discrete distributions. At this point, it is instructive to summarize some of these distributions.

Name of distribution	Parameters	$P(X=i)$	Range for i
Poisson(λ)	$\lambda > 0$	$\dfrac{e^{-\lambda}\lambda^i}{i!}$	$i \geq 0$
Binomial(n,p)	$n \geq 1; 0 \leq p \leq 1$	$\dbinom{n}{i} p^i (1-p)^{n-i}$	$0 \leq i \leq n$
Negative binomial(n,p)	$n \geq 1; p > 0$	$\dbinom{n+i-1}{i} p^i (1+p)^{n+i}$	$i \geq 0$
Logarithmic series(θ)	$0 < \theta < 1$	$\dfrac{\theta^i}{-\log(1-\theta)\,i}$	$i \geq 1$
Geometric(p)	$0 < p < 1$	$p(1-p)^{i-1}$	$i \geq 1$

We refer the reader to Johnson and Kotz (1969, 1982) or Ord (1972) for a survey of the properties of the most frequently used discrete distributions in statistics. For surveys of generators, see Schmeiser (1983), Ahrens and Kohrt (1981) or Ripley (1983).

Some of the methods described below are extremely fast: this is usually the case for well-designed table methods, and for the alias method or its variant, the alias-urn method. The method of guide tables is also very fast. Only finite-valued discrete random variates can be generated by table methods because tables must be set-up beforehand. In dynamic situations, or when distributions are infinite-tailed, slower methods such as the inversion method can be used.

2. THE INVERSION METHOD.

2.1. Introduction.

In the **inversion method**, we generate one uniform $[0,1]$ random variate U and obtain X by a monotone transformation of U which is such that $P(X=i)=p_i$. If we define X by

$$F(X-1) = \sum_{i<X} p_i < U \leq \sum_{i \leq X} p_i = F(X),$$

then it is clear that $P(X=i)=F(i)-F(i-1)=p_i$. This is comparable to the inversion method for continuous random variates. The solution of the inequality shown above is uniquely defined with probability one. An exact solution of the inversion inequalities can always be obtained in finite time, and the inversion method can thus truly be called universal. Note that for continuous distributions, we could not invert in finite time except in special cases.

There are several possible techniques for solving the inversion inequalities. We start with the simplest and most universal one, i.e. a method which is applicable to all discrete distributions.

Inversion by sequential search

Generate a uniform $[0,1]$ random variate U.
Set $X \leftarrow 0$, $S \leftarrow p_0$.
WHILE $U > S$ DO
    $X \leftarrow X+1; S \leftarrow S + p_X$
RETURN X

Note that S is adjusted as we increase X in the **sequential search algorithm.** This method applies to the "black box" model, and it can handle infinite tails. The time taken by the algorithm is a random variable N, which can be equated in first approximation with the number of comparisons in the WHILE condition. But

$$P(N=i) = P(X=i-1) = p_{i-1} \quad (i \geq 1).$$

Thus, $E(N)=E(X)+1$. In other words, the tail of the distribution of X determines the expected time taken by the algorithm. This is an uncomfortable situation in view of the fact that $E(X)$ can possibly be ∞. There are other more practical objections: p_i must be computed many times, and the consecutive additions $S \leftarrow S + p_X$ may lead to inadmissible accumulated errors. For these reasons, the sequential search algorithm is only recommended as a last resort. In the remainder of this section, we will describe various methods for improving the sequential search algorithm. In particular cases, the computation of p_X can be avoided altogether as can be seen from the following example.

Example 2.1. Poisson random variates by sequential search.

We can quickly verify that for the Poisson (λ) distribution,

$$p_{i+1} = \frac{\lambda}{i+1} p_i \ , \ p_0 = e^{-\lambda} \ .$$

Thus, the sequential search algorithm can be simplified somewhat by recursively computing the values of p_i during the search:

Poisson generator using sequential search

Generate a uniform [0,1] random variate U.

Set $X \leftarrow 0$,$P \leftarrow e^{-\lambda}$,$S \leftarrow P$.

WHILE $U > S$ DO

$\quad\quad X \leftarrow X+1$,$P \leftarrow \dfrac{\lambda P}{X}$,$S \leftarrow S+P$.

RETURN X

We should note here that the expected number of comparisons is equal to $E(X+1) = \lambda+1$. ■

A slight improvement in which the variable S is not needed was suggested by Kemp(1981). Note however that this forces us to destroy U:

Inversion by sequential search (Kemp, 1981)

Generate a uniform [0,1] random variate U.

$X \leftarrow 0$

WHILE $U > p_X$ DO

$\quad\quad U \leftarrow U - p_X$

$\quad\quad X \leftarrow X+1$

RETURN X

2.2. Inversion by truncation of a continuous random variate.

If we know a continuous distribution function G on $[0,\infty)$ with the property that G agrees with F on the integers, i.e.

$$G(i+1) = F(i) \quad (i=0,1,\dots), \ G(0) = 0,$$

then we could use the following algorithm for generating a random variate X with distribution function F :

Inversion by truncation of a continuous random variate

Generate a uniform $[0,1]$ random variate U.

RETURN $X \leftarrow \left\lfloor G^{-1}(U) \right\rfloor$

This method is extremely fast if G^{-1} is explicitly known. That it is correct follows from the observation that for all $i \geq 0$,

$$P(X \leq i) = P(G^{-1}(U) < i+1) = P(U < G(i+1)) = G(i+1) = F(i).$$

The task of finding a G such that $G(i+1)-G(i)=p_i$, all i, is often very simple, as we illustrate below with some examples.

Example 2.2. The geometric distribution.

When $G(x)=1-e^{-\lambda x}$, $x \geq 0$, we have

$$G(i+1)-G(i) = e^{-\lambda i} - e^{-\lambda(i+1)}$$
$$= e^{-\lambda i}(1-e^{-\lambda})$$
$$= (1-q)q^i \quad (i \geq 0),$$

where $q = e^{-\lambda}$. From this, we conclude that

$$\left\lfloor -\frac{1}{\lambda}\log U \right\rfloor$$

is geometrically distributed with parameter $e^{-\lambda}$. Equivalently, $\left\lfloor \dfrac{\log U}{\log(1-p)} \right\rfloor$ is

geometrically distributed with parameter p. Equivalently, $\left\lfloor -\dfrac{E}{\log(1-p)} \right\rfloor$ is

geometrically distributed with the same parameter, when E is an exponential random variate. ∎

Example 2.3. A family of monotone distributions.

Consider $G(x)=1-x^{-b}$, $x \geq 1$, $G(1)=0$, $b > 0$.   We   see   that
$G(i+1)-G(i)=i^{-b}-(i+1)^{-b}$. Thus a random variate X with probability vector

$$p_i = \frac{1}{i^b} - \frac{1}{(i+1)^b} \qquad (i \geq 1)$$

can be generated as $\left\lfloor U^{-\frac{1}{b}} \right\rfloor$. In particular, $\left\lfloor \frac{1}{U} \right\rfloor$ has probability vector

$$p_i = \frac{1}{i(i+1)} \qquad (i \geq 1) . \blacksquare$$

Example 2.4. Uniformly distributed discrete random variates.

A discrete random variable is said to be uniformly distributed on $\{1,2, \ldots , K\}$ when $p_i = \frac{1}{K}$ for all $1 \leq i \leq K$. Since $p_i = G(i+1)-G(i)$ where $G(x)=\frac{x-1}{K}$, $1 \leq x \leq K+1$, we see that $X \leftarrow \lfloor 1+KU \rfloor$ is uniformly distributed on the integers 1 through K . $\blacksquare$

2.3. Comparison-based inversions.

The sequential search algorithm uses comparisons only (between U and certain functions of the p_j's). It was convenient to compare U first with p_0, then with p_0+p_1 and so forth, but this is not by any means an optimal strategy. In this section we will highlight some other strategies that are based upon comparisons only. Some of these require that the probability vector be finite.

For example, if we were allowed to permute the integers first and then perform sequential search, then we would be best off if we permuted the integers in such a way that $p_0 \geq p_1 \geq p_2 \geq \cdots$. This is a consequence of the fact that the number of comparisons is equal to $1+X$ where X is the random variate generated. Reorganizations of the search that result from this will usually not preserve the monotonicity between U and X. Nevertheless, we will keep using the term inversion.

The improvements in expected time by reorganizations of sequential search can sometimes be dramatic. This is the case in particular when we have peaked distributions with a peak that is far removed from the origin. A case in point is the binomial distribution which has a mode at $\lfloor np \rfloor$ where n and p are the parameters of the binomial distribution. Here one could first verify whether $U \leq F(\lfloor np \rfloor)$, and then perform a sequential search "up" or "down" depending upon the outcome of the comparison. For fixed p, the expected number of comparisons grows as $\sqrt{n}$ instead of as n as can easily be checked. Of course, we have to compute either directly or in a set-up step, the value of F at $\lfloor np \rfloor$. A similar improvement can be implemented for the Poisson distribution. Interestingly, in this simple case, we do preserve the monotonicity of the transformation.

Other reorganizations are possible by using ideas borrowed from computer science. We will replace linear search (i.e., sequential search) by tree search. For good performance, the search trees must be set up in advance. And of course, we will only be able to handle a finite number of probabilities in our probability vector.

One can construct a binary search tree for generating X. Here each node in the tree is either a leaf (terminal node), or an internal node, in which case it has two children, a left child and a right child. Furthermore, each internal node has associated with it a real number, and each leaf contains one value, an integer between 0 and K. For a given tree, we obtain X from a uniform $[0,1]$ random variate U in the following manner:

Inversion by binary search

Generate a uniform $[0,1]$ random variate U.
Ptr ← Root of tree (Ptr points to a node).
WHILE Ptr $\neq$ Leaf DO
    IF Value (Ptr) $> U$
        THEN Ptr ← Leftchild (Ptr)
        ELSE Ptr ← Rightchild (Ptr).
RETURN X ← Value (Ptr)

Here, we travel down the tree, taking left and right turns according to the comparisons between U and the real numbers stored in the nodes, until we reach a leaf. These real numbers must be chosen in such a way that the leafs are reached with the correct probabilities. There is no particular reason for choosing $K+1$ leaves, one for each possible outcome of X, except perhaps economy of storage. Having fixed the shape of the tree and defined the leaves, we are left with the task of determining the real numbers for the K internal nodes. The real number for a given internal node should be equal to the probabilities of all the leaves encountered before the node in an inorder traversal. At the root, we turn left with the correct probability, and by induction, it is obvious that we keep on doing so when we travel to a leaf. Of course, we have quite a few possibilities where the shape of the tree is concerned. We could make a complete tree, i.e. a tree where all levels are full except perhaps the lowest level (which is filled from left to right). Complete trees with $2K+1$ nodes have

$$L = 1 + \left\lfloor \log_2(2K+1) \right\rfloor$$

levels, and thus the search takes at most L comparisons. In linear search, the worst case is always $\Omega(K)$, whereas now we have $L \sim \log_2 K$. The data structure that can be used for the inversion is as follows: define an array of $2K+1$ records. The last $K+1$ records correspond to the leaves (record $K+i$ corresponds to integer $i-1$). The first K records are internal nodes. The j-th record has as children records $2j$ and $2j+1$, and as father $\left\lfloor \dfrac{j}{2} \right\rfloor$. Thus, the root of the tree is record 1, its children are records 2 and 3, etcetera. This gives us a complete binary tree structure. We need only store one value in each record, and this can be done for the entire tree in time $O(K)$ by noting that we need only do an inorder traversal and keep track of the cumulative probability of the leaves visited when a node is encountered. Using a stack traversal, and notation similar to that of Aho, Hopcroft and Ullman (1982), we can do it as follows:

Set-up of the binary search tree

(BST[1] ,..., BST[2K+1] is our array of values. To save space, we can store the probabilities $p_0, \ldots, p_K$ in BST[K+1] ,..., BST[2K+1].)

(S is an auxiliary stack of integers.)

MAKENULL(S) (create an empty stack).

Ptr←1, PUSH(Ptr,S) (start at the root).

$P \leftarrow 0$ (set cumulative probability to zero).

REPEAT

    IF Ptr$\leq K$

        THEN PUSH(Ptr,S), Ptr←2 Ptr

        ELSE

            $P \leftarrow P + $ BST[Ptr]

            Ptr←TOP(S), POP(S)

            BST[Ptr] $\leftarrow P$

            Ptr←2 Ptr+1

UNTIL EMPTY (S)

The binary search tree method described above is not optimal with respect to the expected number of comparisons required to reach a decision. For a fixed binary search tree, this number is equal to $\sum\limits_{i=0}^{K} p_i D_i$ where D_i is the depth of the i-th leaf (the depth of the root is one, and the depth of a node is the number of nodes encountered on the path from that node to the root). A binary search tree is optimal when the expected number of comparisons is minimal. We now define **Huffman's tree** (Huffman, 1952, Zimmerman, 1959), and show that it is optimal.

The two smallest probability leaves should be furthest away from the root, for if they are not, then we can always swap one or both of them with other nodes at a deeper level, and obtain a smaller value for $\sum p_i D_i$. Because internal nodes have two children, we can always make these leaves children of the same internal node. But if the indices of these nodes are j and k, then we have

$$\sum_{i=0}^{K} p_i D_i = \sum_{i \neq j,k} p_i D_i + (p_j + p_k)D^* + (p_j + p_k) \, .$$

Here D^* is the depth of the internal father node. We see that minimizing the right-hand-side of this expression reduces to a problem with K instead of $K+1$ nodes, one of these nodes being the new internal node with probability $p_j + p_k$ associated with it. Thus, we can now construct the entire (Huffman) tree. Perhaps a small example is informative here.

Example 2.5.

Consider the probabilities

p_0	0.11
p_1	0.30
p_2	0.25
p_3	0.21
p_4	0.13

We note that we should join nodes 0 and 4 first and form an internal node of cumulative weight 0.24. Then, this node and node 3 should be joined into a supernode of weight 0.45. Next, nodes 1 and 2 are made children of the same internal node of weight 0.55, and the two leftover internal nodes finally become children of the root. ∎

For a data structure, we can no longer use a complete binary tree, but we can make use of the array implementation in which entries 1 through K denote internal nodes, and entries $K+1$ through $2K+1$ define leaves. For leaves, the entries are the given probabilities, and for the internal nodes, they are the threshold values as defined for general binary search trees. Since the shape of the tree must also be determined, we are forced to add for entries 1 through K two fields, a leftchildpointer and a rightchildpointer. For the sake of simplicity, we use BST[.] for the threshold values and probabilities, and Left[.], Right[.] for the pointer fields. The tree can be constructed in time $O(K \log K)$ by the Hu-Tucker algorithm (Hu,Tucker, 1971):

Construction of the Huffman tree

Create a heap H with elements $(K+1, p_0), \ldots, (2K+1, p_K)$ and order defined by the keys p_i (the smallest key is at the top of the heap). (For the definition of a heap, we refer to Aho, Hopcroft and Ullman (1982)). Note that this operation can be done in $O(K)$ time.

FOR i:=1 TO K DO

        Take top element (j, p) off the heap H and fix the heap.

        Take top element (k, q) off the heap H and fix the heap.

        Left$[i] \leftarrow j$, Right$[i] \leftarrow k$.

        Insert $(i, p+q)$ in the heap H.

Compute the array BST by an inorder traversal of the tree. (This is analogous to the traversal seen earlier, except that for travel down the tree, we must make use of the fields Left[.] and Right[.] instead of the positional trick that in a complete binary tree the index of the leftchild is twice that of the father. The time taken by this portion is $O(K)$.)

The entire set-up takes time $O(K \log K)$ in view of the fact that insertion and deletion-off-the-top are $O(\log K)$ operations for heaps.

It is worth pointing out that for families of discrete distributions, the extra cost of setting up a binary search tree is often inacceptable.

We close this section by showing that for most distributions the expected number of comparisons $(E(C))$ with the Huffman binary search tree is much less than with the complete binary search tree. To understand why this is possible, consider for example the simple distribution with probability vector $\frac{1}{2}, \frac{1}{4}, \ldots, \frac{1}{2^K}, \frac{1}{2^K}$. It is trivial to see that the Huffman tree here has a linear shape: we can define it recursively by putting the largest probability in the right child of the root, and putting the Huffman tree for the leftover probabilities in the left subtree of the root. Clearly, the expected number of comparisons is $(\frac{1}{2})2 + (\frac{1}{4})3 + (\frac{1}{8})4 + \cdots$. For any K, this is less than 3, and as $K \to \infty$, the value 3 is approached. In fact, this finite bound also applies to the extended Huffman tree for the probability vector $\frac{1}{2^i}$ $(i \geq 1)$. Similar asymmetric trees are obtained for all distributions for which $E(e^{tX}) < \infty$ for some $t > 0$: these are distributions with roughly speaking exponentially or subexponentially decreasing tail probabilities. The relationship between the tail of the distribution and $E(C)$ is clarified in Theorem 2.1.

Theorem 2.1.

Let $p_1, p_2,...$ be an arbitrary probability vector. Then it is possible to construct a binary search tree (including the Huffman tree) for which

$$E(C) \leq 1 + 4 \left\lceil \log_2(1 + E(X)) \right\rceil ,$$

where X is the discrete random variate generated by using the binary search tree for inversion.

Proof of Theorem 2.1.

The tree that will be considered here is as follows: choose first an integer $k \geq 1$. We put leaves at levels $k+1, 2k+1, 3k+1,...$ only. At level $k+1$, we have 2^k slots, and all but one is filled from left to right. The extra slot is used as a root for a similar tree with $2^k - 1$ leaves at level $2k+1$. Thus, C is equal to:

$$k+1 \qquad \text{with probability } \sum_{i=1}^{2^k-1} p_i$$

$$2k+1 \qquad \text{with probability } \sum_{i=2^k}^{2(^k-1)} p_i .$$

. . .

Taking expected values gives

$$E(C) = 1 + k \sum_{j=1}^{\infty} j \sum_{i=(j-1)(2^k-1)+1}^{j(2^k-1)} p_i$$

$$= 1 + k \sum_{i=1}^{\infty} p_i \sum_{\frac{i}{2^k-1} \leq j \leq 1 + \frac{i-1}{2^k-1}} j$$

$$\leq 1 + k \sum_{i=1}^{\infty} p_i \left(1 + \frac{i-1}{2^k-1}\right)\left(2 - \frac{1}{2^k-1}\right)$$

$$\leq 1 + 2k \sum_{i=1}^{\infty} p_i \left(1 + \frac{i}{2^k-1}\right)$$

$$\leq 1 + 2k + \frac{2k}{2^k-1} \sum_{i=1}^{\infty} i p_i$$

$$= 1 + 2k + \frac{2k}{2^k-1} E(X) .$$

If we take $k = \left\lceil \log_2(1 + E(X)) \right\rceil$, then $2^k - 1 \geq E(X)$, and thus,

$$E(C) \leq 1 + 2 \left\lceil \log_2(1 + E(X)) \right\rceil \left(1 + \frac{E(X)}{E(X)}\right) = 1 + 4 \left\lceil \log_2(1 + E(X)) \right\rceil .$$

This concludes the proof of Theorem 2.1. ∎

We have shown two things in this theorem. First, of all, we have exhibited a particular binary search tree with design constant $k \geq 1$ (k is an integer) for which

$$E(C) \leq 1+2k+\frac{2k}{2^k-1}\ E(X)\ .$$

Next, we have shown that the value of $E(C)$ for the Huffman tree does not exceed the upper bound given in the statement of the theorem by manipulating the value of k and noting that the Huffman tree is optimal. Whether in practice we can use the construction successfully depends upon whether we have a fair idea of the value of $E(X)$, because the optimal k depends upon this value. The upper bound of the theorem grows logarithmically in $E(X)$. In contrast, the expected number of comparisons for inversion by sequential search grows linearly with $E(X)$. It goes without saying that if the p_i's are not in decreasing order, then we can permute them to order them. If in the construction we fill empty slots by borrowing from the ordered vector $p_{(1)}, p_{(2)}, ...,$ then the inequality remains valid if we replace $E(X)$ by $\sum_{i=1}^{\infty} i p_{(i)}$. We should also note that Theorem 2.1 is useless for distributions with $E(X) = \infty$. In those situations, there are other possible constructions. The binary tree that we construct has once again leaves at levels $k+1, 2k+1, ...,$ but now, we define the leaf positions as follows: at level $k+1$, put one leaf, and define 2^k-1 roots of subtrees, and recurse. This means that at level $2k+1$ we find 2^k-1 leaves. We associate the p_i's with leaves in the order that they are encountered in this construction, and we keep on going until K leaves are accommodated.

Theorem 2.2.

For the binary search tree constructed above with fixed design constant $k \geq 1$, we have

$$E(C) \leq 1+kp_1+\frac{2k}{\log(2^k-1)}E(\log X)$$

and, for $k=2$,

$$E(C) \leq 1+2p_1+\frac{4}{\log 3}E(\log X) \leq 3+\frac{4}{\log 3}E(\log X)\ ,$$

where X is a random variate with the probability vector $p_1, \ldots, p_K$ that is used in the construction of the binary search tree, and C is the number of comparisons in the inversion method.

Proof of Theorem 2.2.

Let us define $m = 2^k - 1$ to simplify the notation. It is clear that

$$
C = \begin{cases}
k+1 & \text{with probability } p_1 \\
2k+1 & \text{with probability } p_2 + \cdots + p_{m+1} \\
3k+1 & \text{with probability } p_{m+2} + \cdots + p_{m^2+m+1} \\
\cdots
\end{cases}
$$

In such expressions, we assume that $p_i = 0$ for $i > K$. The construction also works for infinite-tailed distributions, so that we do not need K any further. Now,

$$
E(C) \leq 1 + kp_1 + k \sum_{j=2}^{\infty} j \sum_{i=1+1+\cdots+m^{j-2}}^{1+\cdots+m^{j-1}} p_i
$$

$$
= 1 + kp_1 + k \sum_{i=2}^{\infty} p_i \sum_{1+1+\cdots+m^{j-2} \leq i \leq 1+\cdots+m^{j-1}} j
$$

$$
\leq 1 + kp_1 + k \sum_{i=2}^{\infty} p_i \sum_{m^{j-2} < i \leq m^j} j
$$

$$
\leq 1 + kp_1 + k \sum_{i=2}^{\infty} \left(2 \frac{\log i}{\log m}\right)
$$

$$
= 1 + kp_1 + \frac{2k}{\log m} \sum_{i=2}^{\infty} p_i \log i
$$

$$
= 1 + kp_1 + \frac{2k}{\log m} E(\log X) .
$$

This proves the first inequality of the theorem. The remainder follows without work. ■

The bounds of Theorem 2.2 grow as $E(\log X)$ and not as $\log(E(X))$. The difference is that $E(\log X) \leq \log(E(X))$ (by Jensen's inequality), and that for long-tailed distributions, the former expression can be finite while the second expression is ∞.

2.4. The method of guide tables.

We have seen that inversion can be based upon sequential search, ordinary binary search or modified binary search. All these techniques are comparison-based. Computer scientists have known for a long time that hashing methods are ultra fast for searching data structures provided that the elements are evenly distributed over the range of values of interest. This speed is bought by the exploitation of the truncation operation.

Chen and Asau (1974) first suggested the use of hashing techniques to handle the inversion. To insure a good expected time, they introduced an ingenious trick, which we shall describe here. Their method has come to be known as the **method of guide tables**. Again, we have a monotone relationship between X, the generated random variate, and U, the uniform [0,1] random variate which is inverted.

We assume that a probability vector $p_0, p_1, \ldots, p_K$ is given. The cumulative probabilities are defined as

$$q_i = \sum_{j=0}^{i} p_j \qquad (0 \leq i \leq K) \ .$$

If we were to throw a dart (in this case U) at the segment [0,1], which is partitioned into $K+1$ intervals $[0,q_0), [q_0,q_1), \ldots, [q_{K-1},1]$, then it would come to rest in the interval $[q_{i-1},q_i)$ with probability $q_i - q_{i-1} = p_i$. This is another way of rephrasing the inversion principle of course. It is another matter to find the interval to which U belongs. This can be done by standard binary search in the array of q_i's (this corresponds roughly to the complete binary search tree algorithm). If we are to exploit truncation however, then we somehow have to consider equi-spaced intervals, such as $[\frac{i-1}{K+1}, \frac{i}{K+1})$, $1 \leq i \leq K+1$. The method of guide tables helps the search by storing in each of the $K+1$ intervals a "guide table value" g_i where

$$g_i = \max_{q_j < \frac{i}{K+1}} j \ .$$

This helps the inversion tremendously:

Method of guide tables

Generate a uniform [0,1] random variate U.

Set $X \leftarrow \lfloor (K+1)U+1 \rfloor$ (this is the truncation).

Set $X \leftarrow g_X + 1$ (guide table look-up).

WHILE $q_{X-1} > U$ DO $X \leftarrow X - 1$.

RETURN X

It is easy to determine the validity of this algorithm. Note also that no expensive computations are involved.

Theorem 2.3.

The expected number of comparisons (of q_{X-1} and U) in the method of guide tables is always bounded from above by 2.

Proof of Theorem 2.3.

Observe that the number of comparisons C is not greater than the number of q_i values in the interval X (the returned random variate) plus one. But since all intervals are equi-spaced, we have

$$E(C) \leq 1 + \frac{1}{K+1} \sum_{i=0}^{K} (\text{number of values of } q_j \text{ in interval } i)$$

$$\leq 1+1 = 2 \; . \blacksquare$$

Theorem 2.3 is very important because it guarantees a uniformly good performance for all distributions as long as we make sure that the number of intervals and the number of possible values of the discrete random variable are equal.

This inversion method too requires a set-up step. The table of values $g_1, g_2, \ldots, g_{K+1}$ can be found in time $O(K)$:

Set-up of guide table

FOR $i := 1$ TO $K + 1$ DO $g_i \leftarrow 0$.
$S \leftarrow 0$.
FOR $i := 0$ TO K DO
    $S \leftarrow S + p_i$ (S is now q_i).
    $j \leftarrow \lfloor S(K + 1) + 1 \rfloor$. (Determine interval for q_i.)
    $g_j \leftarrow i$.
FOR $i := 2$ TO $K + 1$ DO $g_i \leftarrow \max(g_{i-1}, g_i)$.

There is a trade-off between expected number of comparisons and the size of the guide table. It is easy to see that if we have a guide table of $\alpha(K + 1)$ elements for some $\alpha > 0$, then we have

$$E(C) \leq 1 + \frac{1}{\alpha}.$$

If speed is extremely important, one should not hesitate to set α equal to 5 or 10. Of all the inversion methods discussed so far, the method of guide tables shows clearly the greatest potential in terms of speed. This is confirmed in Ahrens and Kohrt(1981).

2.5. Inversion by correction.

It is sometimes possible to find another distribution function G that is close to the distribution function F of the random variable X. Here G is the distribution function of another discrete random variable, Y. It is assumed that G is an easy distribution. In that case, it is possible to generate X by first generating Y and then applying a small correction. It should be stressed that the fact that G is close to F does not imply that the probabilities $G(i) - G(i-1)$ are close to the probabilities $F(i) - F(i-1)$. Thus, other methods that are based upon the closeness of these probabilities, such as the rejection method, are not necessarily applicable. We are simply using G to obtain an initial estimate of X.

Inversion by correction; direct version

Generate a uniform $[0,1]$ random variate U.

Set $X \leftarrow G^{-1}(U)$ (i.e. X is an integer such that $G(X-1) < U \leq G(X)$. This usually means that X is obtained by truncation of a continuous random variable.)

IF $U \leq F(X)$

        THEN WHILE $U \leq F(X-1)$ DO $X \leftarrow X-1$.

        ELSE WHILE $U > F(X+1)$ DO $X \leftarrow X+1$.

RETURN X

We can measure the time taken by this algorithm in terms of the number of F-computations. We have:

Theorem 2.4.

    The number of computations C of F in the inversion algorithm shown above is

$$2 + \mid Y - X \mid$$

where X, Y are defined by

$$F(X-1) < U \leq F(X) , \ G(Y-1) < U \leq G(Y) .$$

It is clear that $E(C) = 2 + E(\mid Y-X \mid)$ where Y, X are as defined in the theorem. Note that Y and X are dependent random variables in this definition. We observe that in the algorithm, we use inversion by sequential search and start this search from the initial guess Y. The correction is $\mid Y-X \mid$.

There is one important special case, occurring when F and G are stochastically ordered, for example, when $F \leq G$. Then one computation of F can be saved by noting that we can use the following implementation.

Inversion by correction; $F \leq G$

Generate a uniform [0,1] random variate U. Set $X \leftarrow G^{-1}(U)$.
WHILE $U > F(X)$ DO $X \leftarrow X+1$.
RETURN X

What is saved here is the comparison needed to decide whether we should search up or down. Since in the notation of Theorem 2.4, $Y \leq X$, we see that

$$E(C) = 1 + E(X-Y).$$

When $E(X)$ and $E(Y)$ are finite, this can be written as $1 + E(X) - E(Y)$. In any case, we have

$$E(C) = 1 + \sum_i |F(i) - G(i)|.$$

To see this, use the fact that $E(X) = \sum_i (1 - F(i))$ and $E(Y) = \sum_i (1 - G(i))$. When $F \geq G$, we have a symmetric development of course.

In some cases, a random variate with distribution function G can more easily be obtained by methods other than inversion. Because we still need a uniform [0,1] random variate, it is necessary to cook up such a random variate from the previous one. Thus, the initial pair of random variates (U, X) can be generated indirectly:

Inversion by correction; indirect version

Generate a random variate X with distribution function G.

Generate an independent uniform [0,1] random variate V, and set $U \leftarrow G(X-1) + V(G(X) - G(X-1))$.
IF $U \leq F(X)$
      THEN WHILE $U \leq F(X-1)$ DO $X \leftarrow X-1$.
      ELSE WHILE $U > F(X+1)$ DO $X \leftarrow X+1$.
RETURN X

It is easy to verify that the direct and indirect versions are equivalent because the joint distributions of the starting pair (U, X) are identical. Note that in both cases, we have the same monotone relation between the generated X and the random variate U, even though in the indirect version, an auxiliary uniform [0,1] random variate V is needed.

Example 2.6.

Consider

$$F(i) = 1 - \frac{1+a}{i^p + ai} \qquad (i \geq 1),$$

where $a > 0$ and $p > 1$ are given constants. Explicit inversion of F is not feasible except perhaps in special cases such as $p = 2$ or $p = 3$. If sequential search is used started at 0, then the expected number of F computations is

$$1 + \sum_{i=1}^{\infty} (1 - F(i)) = 1 + \sum_{i=1}^{\infty} \frac{1+a}{i^p + ai} \geq 1 + \sum_{i=1}^{\infty} \frac{1}{i^p} .$$

Assume next that we use inversion by correction, and that as easy distribution function we take $G(i) = 1 - \frac{1}{i^p}$, $i \geq 1$. First, we have stochastic ordering because $F \leq G$. Note first that $G^{-1}(U)$ (the inverse being defined as in Theorem 2.4) is equal to $\left\lceil 1 + U^{-\frac{1}{p}} \right\rceil$. Furthermore, the expected number of computations of F is

$$1 + \sum_{i=1}^{\infty} G(i) - F(i) = 1 + \sum_{i=2}^{\infty} \frac{ai^p - ai}{i^p(i^p + ai)} \leq 1 + \sum_{i=2}^{\infty} \frac{a}{i^p} .$$

Thus, the improvement in terms of expected number of computations of F is at least $1 + (1-a) \sum_{i=2}^{\infty} \frac{1}{i^p}$, and this can be considerable when a is small. ■

2.6. Exercises.

1. Give a one-line generator (based upon inversion via truncation of a continuous random variate) for generating a random variate X with distribution

    $$P(X = i) = \frac{i}{\dfrac{n(n+1)}{2}} \qquad (1 \leq i \leq n) .$$

2. By empirical measurement, the following discrete cumulative distribution function was obtained by Nigel Horspool when studying operating systems:

    $$F(i) = \min(1 , 0.114 \log(1 + \frac{i}{0.731}) - 0.069) \qquad (i \geq 1) .$$

Give a one-line generator for this distribution which uses truncation of a continuous random variate.

3.   Give one-line generators based upon inversion by truncation of a continuous random variate for the following probability distributions on the positive integers:

p_n
$\dfrac{a}{(n+a)(n+a+1)}$ $(a>0)$
$\dfrac{1}{2^n}$
$\dfrac{3n^2+3n+1}{n^3(n+1)^3}$
$\dfrac{2n+1}{n^2(n+1)^2}$
$\dfrac{1}{\sqrt{n}(n+1)(\sqrt{n}+\sqrt{n+1})}$

3. TABLE LOOK-UP METHODS.

3.1. The table look-up principle.

We can generate a random variate X very quickly if all probabilities p_i are rational numbers with common denominator M. It suffices to note that the sum of the numerators is also M. Thus, if we were to construct an array A of size M with Mp_0 entries 0, Mp_1 entries 1, and so forth, then a uniformly picked element of this array would yield a random variate with the given probability vector $p_0, p_1, ...$ Formally we have:

Table look-up method

[SET-UP]

Given the probability vector $(p_0 = \dfrac{k_0}{M}, p_1 = \dfrac{k_1}{M}, ...)$, where the k_i's and M are nonnegative integers, we define a table $A = (A[0], \ldots, A[M-1])$ where k_i entries are i, $i \geq 0$.
[GENERATOR]
Generate a uniform $[0,1]$ random variate U.
RETURN $A[\lfloor MU \rfloor]$

The beauty of this technique is that it takes a constant time. Its disadvantages include its limitation (probabilities are rarely rational numbers) and its large table size (M can be phenomenally big).

We will give two important examples to illustrate its use.

Example 3.1. Simulating dice.

We are asked to generate the sum of n independently thrown unbiased dice. This can be done naively by using $X_1+X_2+\cdots+X_n$ where the X_i's are iid uniform $\{1,2,\ldots,6\}$ random variates. The time for this algorithm grows as n. Usually, n will be small, so that this is not a major drawback. We could also proceed as follows: first we set up a table $A[0],\ldots,A[M-1]$ of size $M=6^n$ where each entry corresponds to one of the 6^n possible outcomes of the n throws (for example, the first entry corresponds to $1,1,1,1,\ldots,1$, the second entry to $2,1,1,1,\ldots,1$, etcetera). The entries themselves are the sums. Then $A[\lfloor MU \rfloor]$ has the correct distribution when U is a uniform $[0,1]$ random variate. Note that the time is $O(1)$, but that the space requirements now grow exponentially in n. Interestingly, we have one uniform random variate per random variate that is generated. And if we wish to implement the inversion method, the only thing that we need to do is to sort the array according to increasing values. We have thus bought time and paid with space. It should be noted though that in this case the space requirements are so outrageous that we are practically limited to $n \leq 5$. Also, the set-up is only admissible if very many iid sums are needed in the simulation. ■

Example 3.2. The histogram method.

Statisticians often construct histograms by counting frequencies of events of a certain type. Let events $0,1,\ldots,K$ have associated with them frequencies $k_0,k_1,\ldots,k_K$. A question sometimes asked is to generate a new event with the probabilities defined by the histogram, i.e. the probability of event i should be $\dfrac{k_i}{M}$ where $M=\sum_{i=0}^{K} k_i$. In this case, we are usually given the original events in table form $A[0],\ldots,A[M-1]$, and it is obvious that the table method can be applied here without set-up. We will refer to this special case as the **histogram method**. Note that for Example 3.1, we could also construct a histogram, but it differs in that a table must be set up. ■

Assume next that we wish to generate the number of heads in n perfect coin tosses. It is known that this number is binomially distributed with parameters n and $\dfrac{1}{2}$. By the method of Example 3.1, we can use a table look-up method with table of size 2^n, so for $n \leq 10$, this is entirely reasonable. Unfortunately, when the coin is not perfect and the probability of heads is an irrational number p, the table look-up method cannot be used.

3.2. Multiple table look-ups.

The table look-up method has a geometric interpretation. When the table size is M, then we can think of the algorithm in terms of the selection of one of M equi-spaced intervals of $[0,1]$ by finding the interval to which a uniform $[0,1]$ random variate U belongs. Each interval has an integer associated with it, which should be returned.

One of the problems highlighted in the previous section is the table size. One should also recognize that there normally are many identical table entries. These duplicates can be grouped together to reduce the table size. Assume for example that there are k_i entries with value i where $i \geq 0$ and $\sum_{i \geq 0} k_i = M$. Then, if $M = M_0 M_1$ for two integers M_0, M_1, we can set up an auxiliary table $B[0], \ldots, B[M_0-1]$ where each $B[i]$ points to a block of M_1 entries in the true table $A[0], \ldots, A[M-1]$. If this block is such that all values are identical, then it is not necessary to store the block. If we think geometrically again, then this corresponds to defining a partition of $[0,1]$ into M_0 intervals. The original partition of M intervals is finer, and the boundaries are aligned because M is a multiple of M_0. If for the i-th big interval, all M_1 values of $A[j]$ are identical, then we can store that value directly in $B[i]$ thereby saving M_1-1 entries in the A table. By rearranging the A table, it should be possible to repeat this for many large intervals. For the few large intervals covering small intervals with non-identical values for A, we do store a placeholder such as $*$. In this manner, we have built a three-level tree. The root has M_0 children with values $B[i]$. When $B[i]$ is an integer, then i is a terminal node. When $B[i] = *$, we have an internal node. Internal nodes have in turn M_1 children, each carrying a value $A[j]$. It is obvious that this process can be extended to any number of levels. This structure is known as a trie (Fredkin, 1960) or an extendible hash structure (Fagin, Nievergelt, Pippenger and Strong, 1979). If all internal nodes have precisely two children, then we obtain in effect the binary search tree structure of section III.2. Since we want to get as much as possible from the truncation operation, it is obvious that the fan-out should be larger than 2 in all cases.

Consider for example a table for look-up with 1000 entries defined for the following probability vector:

Probability		Number of entries in table A
p_1	0.005	5
p_2	0.123	123
p_3	0.240	240
p_4	0.355	355
p_5	0.277	277

Suppose now that we set up an auxiliary table B which will allow us to refer to sections of size 100 in the table A. Here we could set

B[0]	2
B[1]	3
B[2]	3
B[3]	4
B[4]	4
B[5]	4
B[6]	5
B[7]	5
B[8]	*
B[9]	*

The interpretation is that if $B[i]=j$ then j appears 100 times in table A, and if $B[i]=*$ then we must consult a block of 100 entries of A which are not all identical. Thus, if $B[8]$ or $B[9]$ are chosen, then we need to consult $A[800], \ldots, A[999]$, where we make sure that there are 5 "1"'s, 23 "2"'s, 40 "3"'s, 55 "4"'s and 77 "5"'s. Note however that we need no longer store $A[0], \ldots, A[799]$! Thus, our space requirements are reduced from 1000 words to 210 words.

After having set-up the tables $B[0], \ldots, B[9]$ and $A[800], \ldots, A[999]$, we can generate X as follows:

Example of a multiple table look-up

Generate a uniform [0,1] random variate U.
Set $X \leftarrow B[\lfloor 10U \rfloor]$.
IF $X \neq *$
  THEN RETURN X
  ELSE RETURN $A[\lfloor 1000U \rfloor]$

Here we have exploited the fact that the same U can be reused for obtaining a random entry from the table A. Notice also that in 80% of the cases, we need not access A at all. Thus, the auxiliary table does not cost us too much timewise. Finally, observe that the condition $X \neq *$ can be replaced by $X > 7$, and that therefore B [8] and B [9] need not be stored.

What we have described here forms the essence of **Marsaglia's table look-up method** (Marsaglia, 1963; see also Norman and Cannon, 1972). We can of course do a lot of fine-tuning. For example, the table A [800], . . . , A [999] can in turn be replaced by an auxiliary table C grouping now only 10 entries, which could be picked as follows:

C[80]	2
C[81]	2
C[82]	3
C[83]	3
C[84]	3
C[85]	3
C[86]	4
C[87]	4
C[88]	4
C[89]	4
C[90]	4
C[91]	5
C[92]	5
C[93]	5
C[94]	5
C[95]	5
C[96]	5
C[97]	5
C[98]	*
C[99]	*

Given that $B[i]=*$ for our value of U, we can in 90% of the cases return $C[\lfloor 100U \rfloor]$. Only if once more an entry $*$ is seen do we have to access the table A [980], . . . , A [999] at position $\lfloor 1000U \rfloor$. The numbering in our arrays is convenient for accessing elements for our representation, i.e. $B[i]$ stands for $C[10i]$, . . . , $C[10i+9]$, or for $A[100i]$, . . . , $A[100i+99]$. Some high level languages do not permit the use of subranges of the integers as indices. It is also convenient to combine A,B and C into one big array. All of this requires additional work during the set-up stage.

We observe that in the multilevel table look-up we must group identical entries in the original table, and this forces us to introduce a nonmonotone relationship between U and X.

The method described here can be extended towards the case where all p_i's are multiples of either 10^{-7} or 2^{-32}. In these cases, the p_i's are usually approximations of real numbers truncated by the wordsize of the computer.

4. THE ALIAS METHOD.

4.1. Definition.

Walker (1974, 1977) proposed an ingenious method for generating a random variate X with probability vector $p_0, p_1, \ldots, p_{K-1}$ which requires a table of size $O(K)$ and has a worst-case time that is independent of the probability vector and K. His method is based upon the following property:

Theorem 4.1.

Every probability vector $p_0, p_1, \ldots, p_{K-1}$ can be expressed as an equiprobable mixture of K two-point distributions.

Proof of Theorem 4.1.

We have to show that there are K pairs of integers $(i_0, j_0), \ldots, (i_{K-1}, j_{K-1})$ and K probabilities $q_0, \ldots, q_{K-1}$ such that

$$p_i = \frac{1}{K} \sum_{l=0}^{K-1} (q_l I_{[i_l = i]} + (1 - q_l) I_{[j_l = i]}) \qquad (0 \leq i < K) .$$

This can be shown by induction. It is obviously true when $K = 1$. Assuming that it is true for $K < n$, we can show that it is true for $K = n$ as follows. Choose the minimal p_i. Since it is at most equal to $\frac{1}{K}$, we can take i_0 equal to the index of this minimum, and set q_0 equal to $K p_{i_0}$. Then choose the index j_0 which corresponds to the largest p_i. This defines our first pair in the equiprobable mixture. Note that we used the fact that $\frac{(1-q_0)}{K} \leq p_{j_0}$ because $\frac{1}{K} \leq p_{j_0}$. The other $K-1$ pairs in the equiprobable mixture have to be constructed from the leftover probabilities

$$p_0, \ldots, p_{i_0} - p_{i_0}, \ldots, p_{j_0} - \frac{1}{K}(1-q_0), \ldots, p_{K-1}$$

which, after deletion of the i_0-th entry, is easily seen to be a vector of $K-1$ non-negative numbers summing to $\frac{K-1}{K}$. But for such a vector, an equiprobable mixture of $K-1$ two-point distributions can be found by our induction hypothesis. ∎

To turn this theorem into profit, we have two tasks ahead of us: first we need to actually construct the equiprobable mixture (this is a set-up problem), and then we need to generate a random variate X. The latter problem is easy to solve. Theorem 4.1 tells us that it suffices to throw a dart at the unit square in the plane and to read off the index of the region in which the dart has landed.

The unit square is of course partitioned into regions by cutting the x-axis up into K equi-spaced intervals which define slabs in the plane. These slabs are then cut into two pieces by the threshold values q_l. If

$$p_i = \frac{1}{K} \sum_{l=0}^{K-1} (q_l I_{[i_l=i]} + (1-q_l)I_{[j_l=i]}) \qquad (0 \le i < K) \, ,$$

then we can proceed as follows:

The alias method

Generate a uniform [0,1] random variate U. Set $X \leftarrow \lfloor KU \rfloor$. Generate a uniform [0,1] random variate V.

IF $V < q_X$

      THEN RETURN i_X

      ELSE RETURN j_X

Here one uniform random variate is used to select one component in the equiprobable mixture, and one uniform random variate is used to decide which part in the two-point distribution should be selected. This unsophisticated version of the alias method thus requires precisely two uniform random variates and two table look-ups per random variate generated. Also, three tables of size K are needed.

We observe that one uniform random variate can be saved by noting that for a uniform [0,1] random variable U, the random variables $X = \lfloor KU \rfloor$ and $V = KU - X$ are independent: X is uniformly distributed on $0, \ldots , K-1$, and the latter is again uniform [0,1]. This trick is not recommended for large K because it relies on the randomness of the lower-order digits of the uniform random number generator. With our idealized model of course, this does not matter.

One of the arrays of size K can be saved too by noting that we can always insure that $i_0, \ldots , i_{K-1}$ is a permutation of $0, \ldots , K-1$. This is one of the duties of the set-up algorithm of course. If a set-up gives us such a permuted table of i-values, then it should be noted that we can in time $O(K)$ reorder the structure such that $i_l = l$, for all l. The set-up algorithm given below will directly compute the tables j and q in time $O(K)$ and is due to Kronmal and Peterson (1979, 1980):

Set-up of tables for alias method

Greater $\leftarrow \emptyset$, Smaller $\leftarrow \emptyset$ (Greater and Smaller are sets of integers.)
FOR $l := 0$ TO $K-1$ DO
    $q_l \leftarrow Kp_l$.
    IF $q_l < 1$
        THEN Smaller $\leftarrow$ Smaller $+\{l\}$.
        ELSE Greater $\leftarrow$ Greater $+\{l\}$.
WHILE NOT EMPTY (Smaller) DO
    Choose $k \in$ Greater $, l \in$ Smaller $[q_l$ is finalized].
    Set $j_l \leftarrow k$ $[j_l$ is finalized].
    $q_k \leftarrow q_k - (1 - q_l)$.
    IF $q_k < 1$ THEN Greater $\leftarrow$ Greater - $\{k\}$, Smaller $\leftarrow$ Smaller $+\{k\}$.
    Smaller $\leftarrow$ Smaller $-\{l\}$.

The sets Greater and Smaller can be implemented in many ways. If we can do it
in such a way that the operations "grab one element", "is set empty ?", "delete
one element" and "add one element" can be done in constant time, then the algorithm given above takes time $O(K)$. This can always be insured if linked lists
are used. But since the cardinalities sum to K at all times, we can organize it by
using an ordinary array in which the top part is occupied by Smaller and the bottom part by Greater. The alias algorithm based upon the two tables computed
above reads:

Alias method with two tables

Generate a random integer X uniformly distributed on $0, \ldots, K-1$.
Generate a uniform [0,1] random variate V.
IF $V \leq q_X$
    THEN RETURN X
    ELSE RETURN j_X

Thus, per random variate, we have either 1 or 2 table accesses. The expected
number of table accesses is

$$1 + \frac{1}{K} \sum_{l=0}^{K-1} (1 - q_l) .$$

The alias method can further be improved by minimizing this expression, but this won't be pursued any further here. The main reason for not doing so is that there exists a simple generalization of the alias method, called the alias-urn method, which is designed to reduce the expected number of table accesses. Because of its importance, we will describe it in a separate section.

4.2. The alias-urn method.

Peterson and Kronmal (1982) suggested a generalization of the alias method in the following manner: think of the probability vector $p_0, p_1, \ldots, p_{K-1}$ as a special case of a probability vector with $K* \geq K$ components where $p_i = 0$ for all $i \geq K$. Everything that was said in the previous section remains valid for this case. In particular, if we use the linear set-up algorithm for the tables q and j, then it should be noted that $q_l > 0$ for at most K values of l. At least for all $l > K-1$ we must have $q_l = 0$. For these values of l, one table access is necessary:

The alias-urn method

Generate a random integer X uniformly distributed on $0, \ldots, K*-1$.

IF $X \geq K$

        THEN RETURN j_X

        ELSE

                Generate a uniform [0,1] random variate V.

            IF $V \leq q_X$

                THEN RETURN X

                ELSE RETURN j_X

Per random variate, we require either one or two table look-ups. It is easy to see that the expected number of table look-ups (not counting q_X) is

$$\frac{K*-K}{K*} + \frac{1}{K*}\sum_{l=0}^{K-1}(1-q_l) \leq 1 \ .$$

The upper bound of 1 may somehow seem like magic, but one should remember that instead of one comparison, we now have either one or two comparisons, the expected value being

$$1+\frac{K}{K*} \ .$$

Thus, as $K*$ becomes large compared to K, the expected number of comparisons and the expected number of table accesses both tend to one, as for the urn method. In this light, the method can be considered as an urn method with slight fine-tuning. We are paying for this luxury in terms of space, since we need to store $K*+K$ values: $j_0, \ldots, j_{K*-1}, q_0, \ldots, q_{K-1}$. Finally, note that the comparison $X \geq K$ takes much less time than the comparison $V \leq q_X$.

4.3. Geometrical puzzles.

We have seen the geometrical interpretation of the alias method: throw a dart at random and uniformly on the unit square of R^2 properly partitioned into $2K$ rectangles, and return the index that is associated with the rectangle that is hit. The indices, or aliases, are stored in a table, and so are the definitions of the rectangles. The power of the alias method is due to the fact that we can take K identical slabs of height 1 and base $\dfrac{1}{K}$ and then split each slab into two rectangles. It should be obvious that there are an unlimited number of ways in which the unit square can be cut up conveniently. In general, if the components are $A_1, \ldots, A_M$, and the aliases are $j_1, \ldots, j_M$, then the algorithm

General alias algorithm

Generate a random variate (X, Y) uniformly distributed in $[0,1]^2$.
Determine the index Z in $1, \ldots, M$ such that $(X, Y) \in A_Z$.
RETURN j_Z

produces a random variate which takes the value k with probability

$$\sum_{l \,:\, j_l = k} \text{area}(A_l) .$$

Let us illustrate this with an example. Let the probabilities for consecutive integers $1, 2, \ldots$ be $c, \dfrac{c}{2}, \dfrac{c}{2}, \dfrac{c}{4}, \dfrac{c}{4}, \dfrac{c}{4}, \dfrac{c}{4}, \ldots, \dfrac{c}{2^n}$, where n is a positive integer, and $c = \dfrac{1}{n+1}$ is a normalization constant. It is clear that we can group the values in groups of size $1, 2, 4, \ldots, 2^n$, and the probability weights of the groups are all equal to c. This suggests that we should partition the square first into $n+1$ equal vertical slabs of height 1 and base $\dfrac{1}{n+1}$. Then, the i-th slab should be further subdivided into 2^i equal rectangles to distinguish between different integers in the groups. The algorithm then becomes:

Generate a random variate X with a uniform distribution on $\{0,1, \ldots , n \}$.
Generate a random variate Y with a uniform distribution on $2^X , \ldots , 2^{X+1}-1$.
RETURN Y.

In this simple example, it is possible to combine the uniform variate generation and membership determination into one. Also, no table is needed.

Consider next the probability vector

$$p_i = \frac{2}{n+1}(1-\frac{i}{n}) \quad (0 \le i \le n) .$$

Now, we can partition the unit square into $n(n+1)$ equal rectangles and assign aliases as in the matrix shown below:

$$\begin{vmatrix} 0 & 0 & 0 & 0 & 0 \\ 0 & 1 & 1 & 1 & 1 \\ 0 & 1 & 2 & 2 & 2 \\ 0 & 1 & 2 & 3 & 3 \\ 0 & 1 & 2 & 3 & 4 \\ 0 & 1 & 2 & 3 & 4 \end{vmatrix} .$$

We can verify first that the probabilities are correct. Then, it is easily seen that the alias method applied here requires no table either. Both examples illustrate the virtually unlimited possibilities of the alias method.

4.4. Exercises.

1. Give a simple linear time algorithm for sorting a table of records $R_1, \ldots , R_n$ if it is known that the vector of key values used for sorting is a permutation of $1, \ldots , n$.

2. Show that there exists a one-line FORTRAN or PASCAL language generator for random variates with probability vector $p_i = \frac{2}{n+1}(1-\frac{i}{n})$, $0 \le i \le n$ (Duncan McCallum).

3. Combine the rejection and geometric puzzle method for generating random variates with probability vector $p_i = \frac{c}{i}$, $1 \le i \le K$, where c is a normalization constant. The method should take expected time bounded uniformly over K . Hint: note that the vector $c, \frac{c}{2}, \frac{c}{2}, \frac{c}{4}, \frac{c}{4}, \frac{c}{4}, \frac{c}{4}, \ldots$ dominates the

given probability vector.

4.   Repeat the previous exercise for the two-parameter class of probability vectors $p_i = \dfrac{c}{i^M}$, $1 \leq i \leq K$ where M is a positive integer.

5. OTHER GENERAL PRINCIPLES.

5.1. The rejection method.

The rejection principle remains of course valid for discrete distributions. If the probability vector p_i , $i \geq 0$, is such that

$$p_i \leq cq_i \qquad (i \geq 0) ,$$

where $c \geq 1$ is the rejection constant and q_i , $i \geq 0$, is an easy probability vector, then the following algorithm is valid:

The rejection method

REPEAT
    Generate a uniform $[0,1]$ random variate U.
    GENERATE a random variate X with discrete distribution determined by q_i , $i \geq 0$.
UNTIL $Ucq_X \leq p_X$
RETURN X

We recall that the number of iterations is geometrically distributed with parameter $\dfrac{1}{c}$ (and thus mean c). Also, in each iteration, we need to compute $\dfrac{p_X}{cq_X}$. In view of the ultra fast methods described in the previous sections for finite-valued random variates, it seems that the rejection method is mainly applicable in one of two situations:

A.   The distribution has an infinite tail.

B.   The distribution changes frequently (so that we do not have the time to set up long tables every time).

Often, the body of a distribution can be taken care of by the guide table, alias or alias-urn methods, and the tail (which carries small probability anyway) is dealt with by the rejection method.

Example 5.1.

Consider the probability vector

$$p_i = \frac{6}{\pi^2 i^2} \qquad (i \geq 1) .$$

Sequential search for this distribution is undesirable because the expected number of comparisons would be $1 + \sum_{i=1}^{\infty} i p_i = \infty$. With the easy probability vector

$$q_i = \frac{1}{i(i+1)} \qquad (i \geq 1) ,$$

we can apply the rejection method. The best possible rejection constant is

$$c = \sup_{i \geq 1} \frac{p_i}{q_i} = \frac{6}{\pi^2} \sup_{i \geq 1} \frac{i+1}{i} = \frac{12}{\pi^2} .$$

Since $\left\lfloor \dfrac{1}{U} \right\rfloor$ has probability vector q (where U is a uniform [0,1] random variable), we can proceed as follows:

REPEAT

  Generate iid uniform [0,1] random variates U, V. Set $X \leftarrow \left\lfloor \dfrac{1}{U} \right\rfloor$.

UNTIL $2VX \leq X + 1$
RETURN X ∎

Example 5.2. Monotone distributions.

When the probability vector $p_1, \ldots, p_n$ is nonincreasing, then it is obvious that $p_i \leq \dfrac{1}{i}$ for all i. Thus, the following rejection algorithm is valid:

REPEAT

    Generate a random variate X with probability vector proportional to $1, \dfrac{1}{2}, \ldots, \dfrac{1}{n}$.

    Generate a uniform $[0,1]$ random variate U.

UNTIL $U \leq X p_X$

RETURN X

The expected number of iterations is $\displaystyle\sum_{i=1}^{n} \frac{1}{i} \leq 1 + \log(n)$. For example, a binomial (n,p) random variate can be generated by this method in expected time $O(\log(n))$ provided that the probabilities can be computed in time $O(1)$ (this assumes that the logarithm of the factorial can be computed in constant time). For the dominating distribution, see for example exercise III.4.3. ∎

Example 5.3. The hybrid rejection method.

As in example 5.1, random variates with the dominating probability vector are usually obtained by truncation of a continuous random variate. Thus, it seems important to discuss very briefly how we can apply a hybrid rejection method based on the following inequality:

$$p_i \leq c g(x) \qquad (\text{all } x \in [i, i+1) \, , \ i \geq 0) \, .$$

Here $c \geq 1$ is the rejection constant, and g is an easy density on $[0, \infty)$. Note that p can be extended to a density f in the obvious manner , i.e. $f(x) = p_i$, all $x \in [i, i+1)$. Thus, random variates with probability vector p can be generated as follows:

Hybrid rejection algorithm

REPEAT
    Generate a random variate Y with density g. Set $X \leftarrow \lfloor Y \rfloor$.
    Generate a uniform $[0,1]$ random variate U.
UNTIL $U c g (Y) \leq p_X$
RETURN X ∎

5.2. The composition and acceptance-complement methods.

It goes without saying that the entire discussion of the composition and acceptance-complement methods for continuous random variates can be repeated for discrete random variates.

5.3. Exercises.

1. Develop a rejection algorithm for the generation of an integer-valued random variate X where

    $$P(X=i) = \frac{c}{2i-1} - \frac{c}{2i} \qquad (i=1,2,...)$$

    and $c = \frac{1}{2\log 2}$ is a normalization constant. Analyze the efficiency of your algorithm. Note: the series $1 - \frac{1}{2} + \frac{1}{3} - \frac{1}{4} + \frac{1}{5} - \cdots$ converges to $\log 2$. Therefore, the terms considered in pairs and divided by $\log 2$ can be considered as probabilities defining a probability vector.

2. Consider the family of probability vectors $\frac{c(a)}{(a+i)^2}$, $i \geq 1$, where $a \geq 0$ is a parameter and $c(a) > 0$ is a normalization constant. Develop the best possible rejection algorithm that is based upon truncation of random variables with distribution function

    $$F(x) = 1 - \frac{a+1}{a+x} \qquad (x > 1).$$

    Find the probability of acceptance, and show that it is at least equal to $\frac{a}{a+2}$. Show that the infimum of the probability of acceptance over

$a \in [0,\infty)$ is nonzero.

3. **The discrete normal distribution.** A random variable X has the discrete normal distribution with parameter $\sigma > 0$ when

$$P(X=i) = ce^{-\frac{(|i|+\frac{1}{2})^2}{2\sigma^2}} \qquad (i \text{ integer}).$$

Here $c > 0$ is a normalization constant. Show first that

$$c = \frac{1}{\sigma}\left(\frac{1}{\sqrt{2\pi}}+o(1)\right)$$

as $\sigma \to \infty$. Show then that X can be generated by the following rejection algorithm:

> REPEAT
>> Generate a normal random variate Y, and let X be the closest integer to Y, i.e. $X \leftarrow \text{round}(Y)$. Set $Z \leftarrow |X|+\frac{1}{2}$.
>>
>> Generate a uniform $[0,1]$ random variate U.
>
> UNTIL $-2\sigma^2\log(U) \geq Z^2-Y^2$
> RETURN X

Note that $-\log(U)$ can be replaced by an exponential random variate. Show that the probability of rejection does not exceed $\frac{2}{\sigma}\sqrt{\frac{2}{\pi}}$. In other words, the algorithm is very efficient when σ is large.

Chapter Four
SPECIALIZED ALGORITHMS

1. INTRODUCTION.

1.1. Motivation for the chapter.

The main techniques for random variate generation were developed in chapters II and III. These will be supplemented in this chapter with a host of other techniques: these include historically important methods (such as the Forsythe-von Neumann method), methods based upon specific properties of the uniform distribution (such as the polar method for the normal density), methods for densities that are given as convergent series (the series method) and methods that have proven particularly successful for many distributions (such as the ratio-of-uniforms method).

To start off, we insert a section of exercises requiring techniques of chapters II and III.

1.2. Exercises.

1. Give one or more reasonably efficient methods for the generation of random variates from the following densities (which should be plotted too to gain some insight):

Density	Range for x	Range for the parameter(s)
$\left(\pi\log(\frac{1}{x})\right)^{-\frac{1}{2}}$	$0<x<1$	
$2\sqrt{\frac{1}{\pi}\log(\frac{1}{x})}$	$0<x<1$	
$\frac{4}{\pi^2 x}\log(\frac{1+x}{1-x})$	$0<x<1$	
$\frac{8}{\pi^2(1-x^2)}\log(\frac{1}{x})$	$0<x<1$	
$\frac{2e^{2a}}{\sqrt{2\pi}}e^{-x^2-\frac{a^2}{x^2}}$	$x>0$	$a>0$
$\frac{4x^2}{\sqrt{\pi}}e^{-x^2}$	$x\geq 0$	
$\sqrt{\frac{\theta\pi}{x}}\,e^{-\theta x}$	$x>0$	$\theta>0$

2. Write short and fast programs for generating random variates with the densities given in the table below. In the programs, use only uniform [0,1] and/or uniform [−1,1] random variates.

Density	Range for x	Range for the parameter(s)
$\frac{n}{n-1}(1-x^{n-1})$	$0\leq x\leq 1$	$n\geq 2,\ n$ integer
$\frac{1}{2x^4}e^{-\frac{1}{x}}$	$x>0$	
$\frac{2}{e^{\pi x}+e^{-\pi x}}$	$x\in R$	
$\frac{4\log(2x-1)}{\pi^2(x-1)x}$	$x>1$	

3. Write one-line generators (i.e., assignment statements) for generating random variates with densities as described below. You can use log,exp,cos,atan,max,min and functions that generate uniform [0,1] and normal random variates.

Density	Range of x	Range of the parameter(s)
$\frac{(-\log x)^n}{n!}$	$0<x<1$	n positive integer
$\frac{1}{2}e^{-|x|}$	$x\in R$	
$\frac{2}{2^{\frac{n}{2}}\Gamma(\frac{n}{2})}x^{n-1}e^{-\frac{x^2}{2}}$	$x>0$	n positive integer
$\frac{1}{2+e^x+e^{-x}}$	$x\in R$	
$a-(2a-2)x$	$0\leq x\leq 1$	$1\leq a\leq 2$

In number 2 we recognize the Laplace density. Number 4 is the logistic density.

4. Show how one can generate a random variate of one's choice having a density f on $[0,\infty)$ with the property that $\lim_{x\downarrow 0} f(x)=\infty$, $f(x)>0$ for all x.

5. Give random variate generators for the following simple densities:

Density	Range for x
$\dfrac{6}{\pi^2}\dfrac{x}{e^x-1}$	$x>0$
$\dfrac{12}{\pi^2}\dfrac{x}{e^x+1}$	$x>0$
$\dfrac{6}{\pi^2}\dfrac{\log(\frac{1}{x})}{1-x}$	$0<x<1$
$\dfrac{12}{\pi^2}\dfrac{\log(1+x)}{x}$	$0<x<1$
$\dfrac{\arctan(x)}{Gx}$	$0<x<1$
$\dfrac{\log(\frac{1}{x})}{G(1+x^2)}$	$0<x<1$
$\dfrac{2\tan(x)}{\pi x}$	$x\geq 0$
$\dfrac{2}{\pi}(\dfrac{\sin(x)}{x})^2$	$x\geq 0$

Here G is Catalan's constant $(0.9159655941772190...\,)$.

6. Find a direct method (i.e., one not involving rejection of any kind) for generating random variates with distribution function $F(x)=1-e^{-ax-bx^2-cx^3}$ $(x\geq 0)$, where $a,b,c>0$ are parameters.

7. Someone shows you the rejection algorithm given below. Find the density of the generated random variate. Find the dominating density used in the rejection method, and determine the rejection constant.

REPEAT

    Generate iid uniform $[0,1]$ random variates U_1,U_2,U_3.

UNTIL $U_3(1+U_1U_2)\leq 1$

RETURN $X \leftarrow -\log(U_1U_2)$

8. Find a simple function of two iid uniform $[0,1]$ random variates which has distribution function $F(x)=1-\dfrac{\log(1+x)}{x}$ $(x>0)$. This distribution function is important in the theory of records (see e.g. Shorrock, 1972).

9. Give simple rejection algorithms with good rejection constants for generating discrete random variates with distributions determined as follows:

p_n	Range for n
$\dfrac{4}{\pi}\arctan(\dfrac{1}{2n^2})$	$n \geq 1$
$\dfrac{8}{\pi}\dfrac{1}{(4n+1)(4n+3)}$	$n \geq 0$
$\dfrac{8}{\pi^2}\dfrac{1}{(2n+1)^2}$	$n \geq 0$
$\dfrac{4}{\pi}\arctan(\dfrac{1}{n^2+n+1})$	$n \geq 1$

10. **The hypoexponential distribution.** Give a uniformly fast generator for the family of hypoexponential densities given by

$$f(x) = \frac{\lambda\mu}{\mu-\lambda}(e^{-\lambda x} - e^{-\mu x}) \quad (x > 0),$$

where $\mu > \lambda > 0$ are the parameters of the distribution.

2. THE FORSYTHE-VON NEUMANN METHOD.

2.1. Description of the method.

In 1951, von Neumann presented an ingenious method for generating exponential random variates which requires only comparisons and a perfect uniform [0,1] random variate generator. The exponential distribution is entirely obtained by manipulating the outcomes of the comparisons. Forsythe (1972) later generalized the technique to other distributions, albeit at the expense of simplicity since the method requires more than just comparisons. The method was then applied with a great deal of success in normal random variate generation (Ahrens and Dieter, 1973; Brent, 1974) and even in beta and gamma generators (Atkinson and Pearce, 1976). Unfortunately, in the last decade, most of the algorithms based on the Forsythe-von Neumann method have been surpassed by other algorithms partially due to the discovery of the alias and acceptance-complement methods. The method is expensive in terms of uniform [0,1] random variates unless special "tricks" are used to reduce the number. In addition, for general distributions, there is a tedious set-up step which makes the algorithm virtually inaccessible to the average user.

Just how comparisons can be manipulated to create exponentially distributed random variables is clear from the following Theorem.

Theorem 2.1.

Let $X_1, X_2, \ldots$ be iid random variables with distribution function F. Then:

(1) $\quad P(x \geq X_1 \geq \cdots \geq X_{k-1} < X_k) = \dfrac{F(x)^{k-1}}{(k-1)!} - \dfrac{F(x)^k}{k!}$ \quad (all x).

(11) If the random variable K is determined by the condition $x \geq X_1 \geq \cdots \geq X_{K-1} < X_K$, then $P(K \text{ odd}) = e^{-F(x)}$, all x.

(111) If Y has distribution function G and is independent of the X_i's, and if K is defined by the condition $Y \geq X_1 \geq \cdots \geq X_{K-1} < X_K$, then

$$P(Y \leq x \mid K \text{ odd}) = \frac{\displaystyle\int_{-\infty}^{x} e^{-F(y)} \, dG(y)}{\displaystyle\int_{-\infty}^{+\infty} e^{-F(y)} \, dG(y)} \qquad \text{(all } x\text{)} .$$

Proof of Theorem 2.1.

For fixed x,

$$P(x \geq X_1 \geq \cdots \geq X_k) = \frac{1}{k!} P(\max_{i \leq k} X_i \leq x) = \frac{F(x)^k}{k!} .$$

Thus,

$$P(x \geq X_1 \geq \cdots \geq X_{k-1} < X_k)$$
$$= P(x \geq X_1 \geq \cdots \geq X_{k-1}) - P(x \geq X_1 \geq \cdots \geq X_k)$$
$$= \frac{F(x)^{k-1}}{(k-1)!} - \frac{F(x)^k}{k!} .$$

Also,

$$P(K \text{ odd}) = (1 - \frac{F(x)}{1!}) + (\frac{F(x)^2}{2!} - \frac{F(x)^3}{3!}) + \cdots = e^{-F(x)} .$$

Part (111) of the theorem finally follows from the following equalities:

$$P(Y \leq x, K \text{ odd}) = \int_{-\infty}^{x} P(K \text{ odd} \mid Y = y) \, dG(y) = \int_{-\infty}^{x} e^{-F(y)} \, dG(y) ,$$

$$P(K \text{ odd}) = \int_{-\infty}^{+\infty} e^{-F(y)} \, dG(y) . \blacksquare$$

We can now describe Forsythe's method (Forsythe, 1972) for densities f which can be written as follows:

$$f(x) = cg(x)e^{-F(x)} ,$$

where g is a density, $0 \le F(x) \le 1$ is some function (not necessarily a distribution function), and c is a normalization constant.

Forsythe's method

REPEAT
    Generate a random variate X with density g.
    $W \leftarrow F(X)$
    $K \leftarrow 1$
    Stop $\leftarrow$ False (Stop is an auxiliary variable for getting out of the next loop.)
    REPEAT
        Generate a uniform [0,1] random variate U.
        IF $U > W$
            THEN Stop $\leftarrow$ True
            ELSE $W \leftarrow U, K \leftarrow K+1$
    UNTIL Stop
UNTIL K odd
RETURN X

We will first verify with the help of Theorem 2.1 that this algorithm is valid. First, for fixed $X = x$, we have for the first iteration of the outer loop,

$$P(K \text{ odd}) = e^{-F(x)} .$$

Thus, at the end of the first iteration,

$$P(X \le x, K \text{ odd}) = \int_{-\infty}^{x} e^{-F(y)} g(y) \, dy .$$

Arguing as in the proof of the properties of the rejection method, we deduce that:

(1)   The returned random variate X satisfies

$$P(X \le x) = \int_{-\infty}^{x} ce^{-F(y)} g(y) \, dy .$$

Thus, it has density $ce^{-F(x)} g(x)$.

(11)   The expected number of outer loops executed before halting is $\dfrac{1}{p}$ where p is the probability of exit, i.e. $p = P(K \text{ odd}) = \int_{-\infty}^{+\infty} e^{-F(y)} g(y) \, dy .$

(iii) In any single iteration,

$$E(K) = \int \left[1(1 - \frac{F(x)}{1!}) + 2(\frac{F(x)}{1!} - \frac{F(x)^2}{2!}) + \cdots \right] g(x) \, dx$$

$$= \int \left[1 + \frac{F(x)}{1!} + \frac{F(x)^2}{2!} + \cdots \right] g(x) \, dx$$

$$= \int e^{F(x)} g(x) \, dx \ .$$

(iv) If N is the total number of uniform $[0,1]$ random variates required, then (by Wald's equation)

$$E(N) = \frac{1 + E(K)}{p} = \frac{1 + \int e^{F(x)} g(x) \, dx}{\int e^{-F(x)} g(x) \, dx} \ .$$

In addition to the N uniform random variates, we also need on the average $\frac{1}{p}$ random variates with density g. It should be mentioned though that g is often uniform on $[0,1]$ so that this causes no major drawbacks. In that case, the total expected number of uniform random variates needed is at least equal to $||f||_\infty$ (this follows from Letac's lower bound). From (iv) above, we deduce that

$$2 \le E(N) \le \frac{1 + e}{\frac{1}{e}} = e + e^2 \ .$$

Observe that Forsythe's method does not require any exponentiation. There are of course about $\frac{1}{p}$ evaluations of F. If we were to use the rejection method with as dominating density g, then p would be exactly the same as here. Per iteration, we would also need a g-distributed random variate, one uniform random variate, and one computation of e^{-F}. In a nutshell, we have replaced the latter evaluation by a (usually) cheaper evaluation of F and some additional uniform random variates. If exponential random variates are cheap, then we can in the rejection method replace the e^{-F} evaluation by an evaluation of F if we replace also the uniform random variate by the exponential random variate. In such situations, it seems very unlikely that Forsythe's method will be faster.

One of the disadvantages of the algorithm shown above is that F must take values in $[0,1]$, yet many common densities such as the exponential and normal densities when put in a form useful for Forsythe's method, have unbounded F such as $F(x) = x$ or $F(x) = \frac{x^2}{2}$. To get around this, the real line must be broken up into pieces, and each piece treated separately. This will be documented further on. It should be pointed out however that the rejection method for $f = ce^{-F} g$ puts no restrictions on the size of F.

2.2. Von Neumann's exponential random variate generator.

A basic property of the exponential distribution is given in Lemma 2.1:

Lemma 2.1.

An exponential random variable E is distributed as $(Z-1)\mu + Y$ where Z, Y are independent random variables and $\mu > 0$ is an arbitrary positive number: Z is geometrically distributed with

$$P(Z=i) = \int_{(i-1)\mu}^{i\mu} e^{-x} \, dx = e^{-(i-1)\mu} - e^{-i\mu} \qquad (i \geq 1),$$

and Y is a truncated exponential random variable with density

$$f(x) = \frac{e^{-x}}{1 - e^{-\mu}} \qquad (0 \leq x \leq \mu).$$

Proof of Lemma 2.1.

Straightforward. ∎

If we choose $\mu = 1$, then Forsythe's method can be used directly for the generation of Y. Since in this case $F(x) = x$, nothing but uniform random variates are required:

von Neumann's exponential random variate generator

REPEAT

      Generate a uniform [0,1] random variate Y. Set $W \leftarrow Y$.

      $K \leftarrow 1$

      Stop $\leftarrow$ False

      REPEAT

            Generate a uniform [0,1] random variate U.

            IF $U > W$

                  THEN Stop $\leftarrow$ True

                  ELSE $W \leftarrow U, K \leftarrow K+1$

      UNTIL Stop

UNTIL K odd

Generate a geometric random variate Z with $P(Z=i)=(1-\frac{1}{e})(\frac{1}{e})^{i-1}$ $(i \geq 1)$.

RETURN $X \leftarrow (Z-1)+Y$

The remarkable fact is that this method requires only comparisons, uniform random variates and a counter. A quick analysis shows that $p = P(K \text{ odd}) = \int_0^1 e^{-x}\ dx = 1-\frac{1}{e}$. Thus, the expected number of uniform random variates needed is

$$E(N) = \frac{1+\int_0^1 e^x\ dx}{\int_0^1 e^{-x}\ dx} = \frac{e^2}{e-1}.$$

This is a high bottom line. Von Neumann has noted that to generate Z, we need not carry out a new experiment. It suffices to count the number of executions of the outer loop: this is geometrically distributed with the correct parameter, and turns out to be independent of Y.

2.3. Monahan's generalization.

Monahan (1979) generalized the Forsythe-von Neumann method for generating random variates X with distribution function

$$F(x) = \frac{H(-G(x))}{H(-1)}$$

where

$$H(x) = \sum_{n=1}^{\infty} a_n x^n \ ,$$

$1 = a_1 \geq a_2 \geq \cdots \geq 0$ is a given sequence of constants, and G is a given distribution function.

Theorem 2.2. (Monahan, 1979)

The following algorithm generates a random variate X with distribution function F :

Monahan's algorithm

REPEAT
    Generate a random variate X with distribution function G .
    $K \leftarrow 1$
    Stop $\leftarrow$ False
    REPEAT
        Generate a random variate U with distribution function G .
        Generate a uniform [0,1] random variate V .
        IF $U \leq X$ AND $V \leq \dfrac{a_{K+1}}{a_K}$
            THEN $K \leftarrow K+1$
            ELSE Stop $\leftarrow$ True
    UNTIL Stop
UNTIL K odd
RETURN X

The expected number of random variates with distribution function G is

$$\frac{1+H(1)}{-H(-1)} \ .$$

Proof of Theorem 2.2.

We define the event A_n by $[X = \max(X, U_1, \ldots, U_n), Z_1 = \cdots = Z_n = 1]$, where the U_i's refer to the random variates U generated in the inner loop, and the Z_i's are Bernoulli random variables equal to consecutive values of $I_{[V \leq \frac{a_{i+1}}{a_i}]}$.

Thus,

$$P(X \leq x, A_n) = a_n G(x)^n ,$$

$$P(X \leq x, A_n, A_{n+1}{}^c) = a_n G(x)^n - a_{n+1} G(x)^{n+1} .$$

We will call the probability that X is accepted p_0. Then

$$p_0 = P(K \text{ odd}) = \sum_{n=1}^{\infty} a_n (-1)^{n+1} = H(-1) .$$

Thus, the returned X has distribution function

$$F(x) = P(X \leq x) = \frac{\sum\limits_{n=1}^{\infty} a_n G(x)^n (-1)^{n+1}}{p_0} = \frac{H(-G(x))}{H(-1)} .$$

The expected number of G-distributed random variates needed is $E(N)$ where

$$E(N) = \frac{1}{p_0} \sum_{n=1}^{\infty} (n+1) P(A_n A_{n+1}{}^c)$$

$$= \sum_{n=1}^{\infty} (n+1) \frac{a_n - a_{n+1}}{p_0}$$

$$= \frac{1 + \sum\limits_{n=1}^{\infty} a_n}{p_0}$$

$$= \frac{1 + H(1)}{-H(-1)} . \blacksquare$$

Example 2.1.

Consider the distribution function

$$F(x) = 1 - \cos(\frac{\pi x}{2}) \quad (0 \leq x \leq 1) .$$

To put this in the form of Theorem 2.2, we choose another distribution function, $G(x) = x^2$ $(0 \leq x \leq 1)$, and note that

$$F(x) = \frac{H(-G(x))}{H(-1)}$$

where

$$H(x) = x + \frac{\pi^2}{48} x^2 + \frac{\pi^4}{5760} x^3 + \cdots + \frac{\pi^{2i-2}}{2^{2i-3}(2i)!} x^i + \cdots .$$

One can easily show that $p_0 = H(-1) = \dfrac{8}{\pi^2}$, while $E(N)$ is approximately 2.74. Also, all the conditions of Theorem 2.2 are satisfied. Random variates with this distribution function can of course be obtained by the inversion method too, as $\dfrac{2}{\pi} \arccos(U)$ where U is a uniform [0,1] random variate. Monahan's algorithm avoids of course any evaluation of a transcendental function. The complete algorithm can be summarized as follows, after we have noted that

$$\frac{a_{n+1}}{a_n} = \left(\frac{\pi}{2}\right)^2 \frac{1}{(2n+2)(2n+1)} :$$

REPEAT
    Generate $X \leftarrow \max(U_1, U_2)$ where U_1, U_2 are iid uniform [0,1] random variates.
    $K \leftarrow 1$
    Stop $\leftarrow$ False
    REPEAT
        Generate U, distributed as X.
        Generate a uniform [0,1] random variate V.
        IF $U \leq X$ AND $V \leq \dfrac{\left(\dfrac{\pi}{2}\right)^2}{4K^2 + 6K + 2}$
           THEN $K \leftarrow K + 1$
           ELSE Stop $\leftarrow$ True
      UNTIL Stop
    UNTIL K odd
    RETURN X ∎

2.4. An example: Vaduva's gamma generator.

We will apply the Forsythe-von Neumann method to develop a gamma generator when the parameter a is in $(0,1]$. Vaduva (1977) suggests handling the part of the gamma density on $[0,1]$ separately. This part is

$$f(x) = c(ax^{a-1})e^{-x} \qquad (0 < x \le 1),$$

where c is a normalization constant. This is in the form $cg(x)e^{-F(x)}$ for a density g and a $[0,1]$-valued function F. Random variates with density $g(x) = ax^{a-1}$ can be generated as $U^{\frac{1}{a}}$ where U is a uniform $[0,1]$ random variate. Thus, we can proceed as follows:

Vaduva's generator for the left part of the gamma density

REPEAT

> Generate a uniform $[0,1]$ random variate U. Set $X \leftarrow U^{\frac{1}{a}}$.
>
> $W \leftarrow X$
>
> $K \leftarrow 1$
>
> Stop $\leftarrow$ False
>
> REPEAT
>
>> Generate a uniform $[0,1]$ random variate U.
>>
>> IF $U > W$
>>
>>> THEN Stop $\leftarrow$ True
>>>
>>> ELSE $W \leftarrow U, K \leftarrow K+1$
>>
>> UNTIL Stop
>
> UNTIL K odd
>
> RETURN X

Let N be the number of uniform $[0,1]$ random variates required by this method. Then, as we have seen,

$$E(N) = \frac{1 + \int_0^1 ax^{a-1}e^x \, dx}{\int_0^1 ax^{a-1}e^{-x} \, dx}.$$

Lemma 2.2.

For Vaduva's partial gamma generator shown above, we have

$$2 \leq E(N) \leq (2+a(e-1))e^{\frac{a}{a+1}} \leq \sqrt{e}\,(e+1),$$

and

$$\lim_{a \downarrow 0} E(N) = 2 .$$

Proof of Lemma 2.2.

First, we have

$$1 = \int_0^1 ax^{a-1}\,dx \geq \int_0^1 ax^{a-1}e^{-x}\,dx$$

$$= E(e^{-Y}) \quad (\text{where } Y \text{ is a random variable with density } ax^{a-1})$$

$$\geq e^{-E(Y)} \quad (\text{by Jensen's inequality})$$

$$= e^{\frac{-a}{a+1}}.$$

Also,

$$1 \leq \int_0^1 ax^{a-1}e^x\,dx$$

$$= 1 + \frac{a}{a+1} + \frac{a}{2!(a+2)} + \cdots \quad (\text{by expansion of } e^x)$$

$$\leq 1 + a\left(1 + \frac{1}{2!} + \frac{1}{3!} + \cdots\right)$$

$$= 1 + a(e-1).$$

Putting all of this together gives us the first inequality. Note that the supremum of the upper bound for $E(N)$ is obtained for $a=1$. Also, the limit as $a \downarrow 0$ follows from the inequality. ∎

What is important here is that the expected time taken by the algorithm remains uniformly bounded in a. We have also established that the algorithm seems most efficient when a is near 0. Nevertheless, the algorithm seems less efficient than the rejection method with dominating density g developed in Example II.3.3. There the rejection constant was

$$c = \frac{1}{\displaystyle\int_0^1 ax^{a-1}e^{-x}\,dx}$$

which is known to lie between 1 and $e^{\frac{a}{a+1}}$. Purely on the basis of expected number of uniform random variates required, we see that the rejection method has $2 \leq E(N) \leq 2e^{\frac{a}{a+1}} \leq 2\sqrt{e}$. This is better than for Forsythe's method for all values of a. See also exercise 2.2.

2.5. Exercises.

1. Apply Monahan's theorem to the exponential distribution where
    $H(x)=e^x-1$, $G(x)=x$, $0<x<1$, and $F(x)=\dfrac{(1-e^{-x})}{1-\dfrac{1}{e}}$. Prove that
    $p_0=1-\dfrac{1}{e}$ and that $E(N)=\dfrac{e}{e-1}$ (Monahan, 1979).

2. We can use decomposition to generate gamma random variates with parameter $a \leq 1$. The restriction of the gamma density to [0,1] is dealt with in the text. For the gamma density restricted to $[1,\infty)$ rejection can be used based upon the dominating density $g(x)=e^{1-x}$ $(x \geq 1)$. Show that this leads to the following algorithm:

    > **REPEAT**
    > > Generate an exponential random variate E. Set $X \leftarrow 1+E$.
    > >
    > > Generate a uniform [0,1] random variate U. Set $Y \leftarrow U^{-\frac{1}{1-a}}$.
    > **UNTIL** $X \leq Y$
    > **RETURN** X

    Show that the expected number of iterations is $\dfrac{1}{\displaystyle\int_1^\infty e^{1-x} x^{a-1} \, dx}$, and that this varies monotonically from 1 (for $a=1$) to $\dfrac{1}{\displaystyle\int_1^\infty \dfrac{e^{1-x}}{x} \, dx}$ (as $a \downarrow 0$).

3. Complicated densities are often cut up into pieces, and each piece is treated separately. This usually yields problems of the following type: $f(x)=ce^{-F(x)}$ $(a \leq x \leq b)$, where $0 \leq F(x) \leq F* \leq 1$, and $F*$ is usually much smaller than 1. This is another way of putting that f varies very little on $[a,b]$. Show that the expected number of uniform random variates

needed in Forsythe's algorithm does not exceed $e^{F*}+e^{2F*}$. In other words, this approaches 2 very quickly as $F* \downarrow 0$.

3. ALMOST-EXACT INVERSION.

3.1. Definition.

A random variate with absolutely continuous distribution function F can be generated as $F^{-1}(U)$ where U is a uniform [0,1] random variate. Often, F^{-1} is not feasible to compute, but can be well approximated by an easy-to-compute strictly increasing absolutely continuous function ψ. Of course, $\psi(U)$ does not have the desired distribution unless $\psi=F^{-1}$. But it is true that $\psi(Y)$ has distribution function F where Y is a random variate with a nearly uniform density. The density h of Y is given by

$$h(y) = f(\psi(y))\psi'(y),$$

where f is the density corresponding to F. The almost-exact inversion method can be summarized as follows:

Almost-exact inversion

Generate a random variate Y with density h.
RETURN $\psi(Y)$

The point is that we gain if two conditions are satisfied: (i) ψ is easy to compute; (ii) random variates with density h are easy to generate. But because we can choose ψ from among wide classes of transformations, it should be obvious that this freedom can be exploited to make generation with density h easier. Marsaglia (1977, 1980, 1984) has made the almost-exact inversion method into an art. His contributions are best explained in a series of examples and exercises, including generators for the gamma and t distributions.

Just how one measures the goodness of a certain transformation ψ depends upon how one wants to generate Y. For example, if straightforward rejection from a uniform density is used, then the smallness of the rejection constant

$$c = \sup_{y} h(y)$$

would be a good measure. On the other hand, if h is treated via the mixture method and h is decomposed as

$$h(y) = pI_{[0,1]}(y) + (1-p)r(y),$$

then the probability p is a good measure, since the residual density r is normally difficult. A value close to 1 is highly desirable here. Note that in any case,

$$p \leq \inf_{y \in [0,1]} h(y).$$

Thus, ψ will often be chosen so as to minimize c or to maximize p, depending upon the generator for h.

All of the above can be repeated if we take a convenient non-uniform distribution as our starting point. In particular, the normal density seems a useful choice when the target densities are the gamma or t densities. This generalization too will be discussed in this section.

3.2. Monotone densities on $[0,\infty)$.

Nonincreasing densities f on the positive real line have sometimes a shape that is similar to that of $\dfrac{\theta}{(1+\theta x)^2}$ where $\theta > 0$ is a parameter. Since this is the density of the distribution function $\dfrac{\theta x}{1+\theta x}$, we could look at transformations ψ defined by

$$\psi(y) = \frac{y}{\theta(1-y)}.$$

In this case, h becomes:

$$h(y) = f\left(\frac{y}{\theta(1-y)}\right) \frac{1}{\theta(1-y)^2} \qquad (0 \leq y \leq 1).$$

For example, for the exponential density, we obtain

$$h(y) = e^{-\frac{y}{\theta(1-y)}} \frac{1}{\theta(1-y)^2} \qquad (0 \leq y \leq 1).$$

Assume that we use rejection from the uniform density for generation of random variates with density h. This suggests that we should try to minimize $\sup h$. By elementary computations, one can see that h is maximal for $1-y = \dfrac{1}{2\theta}$, and that the maximal value is

$$4\theta e^{\frac{1}{\theta}-2},$$

which is minimal for $\theta = 1$. The minimal value is $\dfrac{4}{e} = 1.4715177....$ The rejection algorithm for h requires the evaluation of an exponent in every iteration, and is therefore not competitive. For this reason, the composition approach is much more likely to produce good results.

3.3. Polya's approximation for the normal distribution.

In this section, we will illustrate the composition approach. The example is due to Marsaglia (1984). For the inverse F^{-1} of the absolute normal distribution function F, Polya (1949) suggested the approximation

$$\psi(y) = \sqrt{-\theta\log(1-y^2)} \quad (0\leq y \leq 1) ,$$

where he took $\theta = \dfrac{\pi}{2}$. Let us keep θ free for the time being. For this transformation, the density $h(y)$ of Y is

$$h(y) = \frac{1}{\sqrt{2\pi}} \frac{\theta y (1-y^2)^{\frac{\theta}{2}-1}}{\sqrt{-\theta\log(1-y^2)}} \quad (0\leq y \leq 1) .$$

Let us now choose θ so that $\inf\limits_{[0,1]} h(y)$ is maximal. This occurs for $\theta\approx 1.553$ (which is close to but not equal to Polya's constant, because our criterion for closeness is different). The corresponding value p of the infimum is about 0.985. Thus, random variates with density h can be generated as shown in the next algorithm:

Normal generator based on Polya's approximation

Generate a uniform [0,1] random variate U.

IF $U \leq p$ (p is about 0.985 for the optimal choice of θ)

      THEN RETURN $\psi(\dfrac{U}{p})$ (where $\psi(y)=\sqrt{-\theta\log(1-y^2)}$)

      ELSE

            Generate a random variate Y with residual density $\dfrac{(h(y)-p)}{(1-p)}$ $(0\leq y \leq 1)$.

            RETURN $\psi(Y)$

The details, such as a generator for the residual density, are delegated to exercise 3.5. It is worth pointing out however that the uniform random variate U is used in the selection of a mixture density and in the returned variate $\psi(\dfrac{U}{p})$. For this reason, it is "almost" true that we have one normal random variate per uniform random variate.

3.4. Approximations by simple functions of normal random variates.

In analogy with the development for the uniform distribution, we can look at other common distributions such as the normal distribution. The question now is to find an easy to compute function ψ such that $\psi(Y)$ has the desired density, where now Y is nearly normally distributed. In fact, Y should have density h given in the introduction:

$$h(y) = f(\psi(y))\psi'(y) \quad (y \in R).$$

Usually, the purpose is to maximize p in the decomposition

$$h(y) = p(\frac{1}{\sqrt{2\pi}} e^{-\frac{y^2}{2}}) + (1-p)r(y)$$

where r is a residual density. Then, the following algorithm suggested by Marsaglia (1984) can be used:

Marsaglia's almost-exact inversion algorithm

Generate a uniform [0,1] random variate U.

IF $U \leq p$

      THEN Generate a normal random variate Y.

      ELSE Generate a random variate Y with residual density r.

RETURN $\psi(Y)$

For the selection of ψ, one can either look at large classes of simple functions or scan the literature for transformations. For popular distributions, the latter route is often surprisingly efficient. Let us illustrate this for the gamma (a) density. In the table shown below, several choices for ψ are given that transform normal random variates in nearly gamma random variates (and hopefully nearly normal random variates into exact gamma random variates).

Method	$\psi(y)$	Reference
	$a + y\sqrt{a}$	Central limit theorem
Freeman-Tukey	$\dfrac{(y + \sqrt{4a})^2}{4}$	Freeman and Tukey (1950)
Fisher	$\dfrac{(y + \sqrt{4a-1})^2}{4}$	
Wilson-Hilferty	$a(\dfrac{y}{\sqrt{9a}} + 1 - \dfrac{1}{9a})^3$	Wilson and Hilferty (1931)
Marsaglia	$a - \dfrac{1}{3} + py\sqrt{a} + \dfrac{y^2}{3}$, $p = 1 - \dfrac{0.16}{a}$	Marsaglia (1984)

In this table we omitted on purpose more complicated and often better approximations such as those of Cornish-Fisher, Severo-Zelen and Pelzer-Pratt. For a comparative study and a bibliography of such approximations, the reader should consult Narula and Li (1977). Bolshev (1959, 1963) gives a good account of how one can obtain normalizing transformations in general. Note that our table contains only simple polynomial transformations. For example, Marsaglia's quadratic transformation is such that

$$h(y) = p(\frac{1}{\sqrt{2\pi}}e^{-\frac{y^2}{2}})+(1-p)r(y),$$

where $p = 1 - \frac{0.16}{a}$. For example, when $a = 16$, we have $p = 0.99$. See exercise 3.1 for more information.

The Wilson-Hilferty transformation was first used by Greenwood (1974) and later by Marsaglia (1977). We first verify that h now is

$$h(y) = cz^{3a-1}e^{-az^3} \quad (z = \frac{y}{\sqrt{9a}}+1-\frac{1}{9a} \geq 0),$$

where c is a normalization constant. The algorithm now becomes:

Gamma generator based upon the Wilson-Hilferty approximation

Generate a random variate Y with density h.
RETURN $X \leftarrow \psi(Y) = a(\frac{Y}{\sqrt{9a}}+1-\frac{1}{9a})^3$

Generation from h is done now by rejection from a normal density. The details require careful analysis, and it is worthwhile to do this once. The normal density used for the rejection differs slightly from that used by Marsaglia (1977). The story is told in terms of inequalities. We have

---

Lemma 3.1.

Assume that $a > \frac{1}{3}$. Define $z = \frac{y}{\sqrt{9a}}+1-\frac{1}{9a}$, and $z_0 = (\frac{3a-1}{3a})^{\frac{1}{3}}$. Define the density $h(y) = cz^{3a-1}e^{-az^3}, z \geq 0$ (note: this is a density in y, not in z), where c is a normalization constant. Then, the following inequality is valid for $z \geq 0$:

$$\frac{z^{3a-1}e^{-az^3}}{z_0^{3a-1}e^{-az_0^3}} \leq e^{-\frac{(z-z_0)^2}{2\sigma^2}},$$

where $\sigma^2 = \dfrac{1}{9a(1-\frac{1}{3a})^{\frac{1}{3}}}$.

Proof of Lemma 3.1.

The proof is based upon the Taylor series expansion. We will write $e^{g(z)}$ instead of $h(y)$ for notational convenience. Thus,

$$g(z) = -az^3 + (3a-1)\log z + \log c \ .$$

This function is majorized by a quadratic polynomial in z for this will give us a normal dominating density. In such situations, it helps to expand the function about a point z_0. This point should be picked in such a way that it corresponds to the peak of g because doing so will eliminate the linear term in Taylor's series expansion. Note that

$$g'(z) = -3az^2 + \frac{3a-1}{z} \ ,$$

$$g''(z) = -6az - \frac{3a-1}{z^2} \ ,$$

$$g'''(z) = -6a + \frac{6a-2}{z^3} \ .$$

We see that $g'(z)=0$ for $z=z_0$. Thus, by Taylor's series expansion,

$$g(z) = g(z_0) + \frac{1}{2}(z-z_0)^2 g''(\xi) \ ,$$

where ξ is in the interval $[z, z_0]$ (or $[z_0, z]$). We obtain our result if we can show that

$$\sup_{\xi \geq 0} g''(\xi) \leq -\frac{1}{\sigma^2} \ .$$

But when we look at g''', we notice that it is zero for $z = \left(\frac{3a-1}{3a}\right)^{\frac{1}{3}}$. It is not difficult to verify that for this value, g'' attains a maximum on the positive half of the real line. Thus,

$$\sup_{\xi \geq 0} g''(\xi) \leq -9a \left(1 - \frac{1}{3a}\right)^{\frac{1}{3}} \ .$$

This concludes the proof of Lemma 3.1. ∎

The first version of the rejection algorithm is given below.

First version of the Wilson-Hilferty based gamma generator

[SET-UP]

Set $\sigma^2 \leftarrow \dfrac{1}{9a\,(1-\dfrac{1}{3a})^{\frac{1}{3}}}$, $z_0 = (\dfrac{3a-1}{3a})^{\frac{1}{3}}$.

[GENERATOR]

REPEAT

      Generate a normal random variate N and a uniform $[0,1]$ random variate U.

      Set $Z \leftarrow z_0 + \sigma N$

UNTIL $Z \geq 0$ AND $U e^{-\frac{(Z-z_0)^2}{2\sigma^2}} \leq (\dfrac{Z}{z_0})^{3a-1} e^{-a(Z^3-z_0^3)}$

RETURN $X \leftarrow aZ^3$

Note that we have used here the fact that $z = \dfrac{y}{\sqrt{9a}} + 1 - \dfrac{1}{9a}$. There are two things left to the designer. First, we need to check how efficient the algorithm is. This in effect boils down to verifying what the rejection constant is. Then, we need to streamline the algorithm. This can be done in several ways. For example, the acceptance condition can be replaced by

UNTIL $Z \geq 0$ AND $-E - \dfrac{(Z-z_0)^2}{2\sigma^2} \leq (3a-1)\log(\dfrac{Z}{z_0}) - a\,(Z^3-z_0^3)$

where E is an exponential random variate. Also, $\dfrac{(Z-z_0)^2}{2\sigma^2}$ is nothing but $\dfrac{N^2}{2}$. Additionally, we could add a squeeze step by using sharp inequalities for the logarithm. Note that $\dfrac{Z}{z_0} = 1 + \dfrac{\sigma N}{z_0}$, so that for large values of a, Z is close to z_0 which in turn is close to 1. Thus, inequalities for the logarithm should be sharp near 1. Such inequalities are given for example in the next Lemma.

Lemma 3.2.

Let $x \in [0,1)$. Then the following series expansion is valid:

$$\log(1-x) = -x - \frac{1}{2}x^2 - \frac{1}{3}x^3 - \cdots .$$

Thus, for $k \geq 1$,

$$-\sum_{i=1}^{k} \frac{1}{i}x^i \geq \log(1-x) \geq -\sum_{i<k} \frac{1}{i}x^i - \frac{1}{k}\frac{x^k}{1-x} .$$

Furthermore, for $x \leq 0$, and k odd,

$$-\sum_{i=1}^{k+1} \frac{1}{i}x^i \leq \log(1-x) \leq -\sum_{i=1}^{k} \frac{1}{i}x^i .$$

Proof of Lemma 3.2.

We note that in all cases,

$$-\log(1-x) = \sum_{i=1}^{k} \frac{1}{i}x^i + \frac{x^k}{k(1-\xi)^k}$$

where ξ is between 0 and x. The bounds are obtained by looking at the k-th term in the sums. Consider first $0 \leq \xi \leq x < 1$. Then, the k-th term is at least equal to $\frac{x^k}{k}$. If $x \leq \xi \leq 0$ and k is odd, then the same is true. If however k is even, then the k-th term is majorized by $\frac{x^k}{k}$.

We also note that for $0 \leq x < 1$,

$$-\log(1-x) = x + \frac{1}{2}x^2 + \cdots \leq x + \cdots + \frac{1}{k}x^k(1+x+x^2+x^3+\cdots)$$

$$= \sum_{i=1}^{k} \frac{1}{k}\frac{x^i}{1-x} . \blacksquare$$

Let us return now to the algorithm, and use these inequalities to avoid computing the logarithm most of the time by introducing a quick acceptance step.

Second version of the Wilson-Hilferty based gamma generator

[SET-UP]

Set $\sigma^2 \leftarrow \dfrac{1}{9a\left(1-\dfrac{1}{3a}\right)^{\frac{1}{3}}}$, $z_0 = \left(\dfrac{3a-1}{3a}\right)^{\frac{1}{3}}$, $z_1 \leftarrow a - \dfrac{1}{3}$.

[GENERATOR]

REPEAT

  Generate a normal random variate N and an exponential random variate E.

  Set $Z \leftarrow z_0 + \sigma N$ (auxiliary variate)

  Set $X \leftarrow aZ^3$ (variate to be returned)

  $W \leftarrow \dfrac{\sigma N}{Z}$ (note that $W = 1 - \dfrac{z_0}{Z}$)

  Set $S \leftarrow -E - \dfrac{N^2}{2} + (X - z_1)$

  Accept $\leftarrow [S \le (3a-1)(W + \dfrac{1}{2}W^2 + \dfrac{1}{3}W^3)]$ AND $[Z \ge 0]$

  IF NOT Accept

      THEN Accept $\leftarrow [S \le -(3a-1)\log(1-W)]$ AND $[Z \ge 0]$

UNTIL Accept

RETURN X

In this second version, we have implemented most of the suggested improvements. The algorithm is only applicable for $a > \dfrac{1}{3}$ and differs slightly from the algorithms proposed in Greenwood (1974) and Marsaglia (1977). Obvious things such as the observation that $(W + \dfrac{1}{2}W^2 + \dfrac{1}{3}W^3)$ should be evaluated by Horner's rule, are not usually shown in our algorithms. There are two quantities that should be analyzed:

(1)   The expected number of iterations before halting.

(11)   The expected number of computations of the logarithm in the acceptance step (a comparison with (1) will show us how efficient the squeeze step is).

Lemma 3.3.

The expected number of iterations of the algorithm given above (or its rejection constant) is

$$(\frac{\sqrt{a}\, a^{a-1}}{\Gamma(a)})(\frac{3a-1}{3a})^{a-\frac{1}{2}}e^{-a+\frac{1}{3}}\sqrt{2\pi}\ .$$

For $a \geq \frac{1}{2}$, this is less than $e^{\frac{1}{6a}}$. It tends to 1 as $a \to \infty$ and to ∞ as $a \downarrow \frac{1}{3}$.

Proof of Lemma 3.3.

The area under the dominating curve for h is

$$\int_{-\infty}^{\infty} h(z_0)e^{-\frac{(z-z_0)^2}{2\sigma^2}}\, dy$$

where we recall that $z = \frac{y}{\sqrt{9a}}+1-\frac{1}{9a}$, $z_0 = (\frac{3a-1}{3a})^{\frac{1}{3}}$. Since $dy = \sqrt{9a}\ dz$, we see that this equals

$$h(z_0)\sqrt{2\pi}\sqrt{9a}\ \sigma$$

$$= cz_0^{3a-1}e^{-az_0^3}\sqrt{2\pi}\frac{1}{(1-\frac{1}{3a})^{\frac{1}{6}}}$$

$$= (\frac{\sqrt{a}\, a^{a-1}}{\Gamma(a)})(\frac{3a-1}{3a})^{a-\frac{1}{3}}e^{-a+\frac{1}{3}}\sqrt{2\pi}(\frac{3a}{3a-1})^{\frac{1}{6}}\ .$$

Here we used the fact that the normalization constant c in the definition of h is $\frac{\sqrt{a}\, a^{a-1}}{\Gamma(a)}$, which is verified by noting that

$$\int_{z\geq 0} z^{3a-1}e^{-az^3}\, dy = \frac{\Gamma(a)}{\sqrt{a}\, a^{a-1}}\ .$$

The remainder of the proof is based upon simple facts about the Γ function: for example, the function stays bounded away from 0 on $[0,\infty)$. Also, for $a > 0$,

$$\Gamma(a) = (\frac{a}{e})^a\sqrt{\frac{2\pi}{a}}e^{\frac{\theta}{12a}}\ ,$$

where $0 \leq \theta \leq 1$. We will also need the elementary exponential inequalities

$$e^{-px} \geq (1-x)^p \geq e^{-\frac{px}{1-x}} \quad (p \geq 0, 0 \leq x \leq 1)\ .$$

Using this in our expression for the rejection constant gives an upper bound

$$\frac{\sqrt{a}\, a^{a-1} e^{a} \sqrt{2\pi a}\, e^{-a+\frac{1}{3}}}{a^{a}\sqrt{2\pi}} \left(\frac{3a-1}{3a}\right)^{a-\frac{1}{2}}$$

$$= e^{\frac{1}{3}} \left(1-\frac{1}{3a}\right)^{a-\frac{1}{2}}$$

$$\leq e^{\frac{1}{3}-(a-\frac{1}{2})(3a)^{-1}}$$

$$= e^{\frac{1}{6a}},$$

which is $1+\dfrac{1}{6a}+O\left(\dfrac{1}{a^{2}}\right)$ as $a\to\infty$. ■

From Lemma 3.3, we conclude that the algorithm is not uniformly fast for $a\in(\frac{1}{3},\infty)$. On the other hand, since the rejection constant is $1+\dfrac{1}{6a}+O\left(\dfrac{1}{a^{2}}\right)$ as $a\to\infty$, it should be very efficient for large values of a. Because of this good fit, it does not pay to introduce a quick rejection step. The quick acceptance step on the other hand is very effective, since asymptotically, the expected number of computations of a logarithm is $o(1)$ (exercise 3.1). In fact, this example is one of the most beautiful applications of the effective use of the squeeze principle.

3.5. Exercises.

1.   Consider the Wilson-Hilferty based gamma generator developed in the text. Prove that the expected number of logarithm calls is $o(1)$ as $a\to\infty$.

2.   For the same generator, give all the details of the proof that the expected number of iterations tends to ∞ as $a\downarrow\dfrac{1}{3}$.

3.   For Marsaglia's quadratic gamma-normal transformation, develop the entire comparison-based algorithm. Prove the validity of his claims about the value of p as a function of a. Develop a fixed residual density generator based upon rejection for

$$r*(x) = \sup_{a\geq a_0} r(x).$$

Here a_0 is a real number. This helps because it avoids setting up constants each time. See Marsaglia (1984) for graphs of the residual densities r.

4.   **Student's t-distribution.** Consider the t-density

$$f(x) = \frac{1}{\sqrt{\pi a}}\frac{\Gamma(\frac{a+1}{2})}{\Gamma(\frac{a}{2})}\frac{1}{(1+\frac{x^2}{a})^{\frac{a+1}{2}}}.$$

Find the best constant p if f is to be decomposed into a mixture of a normal and a residual density (p is the weight of the normal density). Repeat the same thing for $h(y)$ if we use almost-exact inversion with transformation

$$\psi(y) = y + \frac{y + y^3}{4a}.$$

Compare both values of p as a function of a. (This transformation was suggested by Marsaglia (1984).)

5.   Work out all the details of the normal generator based on Polya's approximation.

6.   Bolshev (1959, 1963) suggests the following transformations which are supposed to produce nearly normally distributed random variables based upon sums of iid uniform [0,1] random variates. If X_n is $\sqrt{\dfrac{3}{n}} \sum\limits_{i=1}^{n} U_i$ where the U_i's are iid uniform [0,1] random variates, then

$$Y_n = X_n - \frac{1}{20n}(3X_n - X_n^3)$$

and

$$Z_n = X_n - \frac{41}{13440n^2}(X_n^5 - 10X_n^3 + 15X_n)$$

are nearly normally distributed. Use this to generate normal random variates. Take $n = 1,2,3$.

7.   Show that the rejection constant of Lemma 3.3 is at most $\left(\dfrac{e^2}{3a-1}\right)^{\frac{1}{6}}$ when $\dfrac{1}{3} < a \leq \dfrac{1}{2}$.

8.   For the gamma density, the quadratic transformations lead to very simple rejection algorithms. As an example, take $s = a - \dfrac{1}{2}, t = \sqrt{\dfrac{s}{2}}$. Prove the following:

A.   The density of $X = s\left(\sqrt{\dfrac{Z}{s}} - 1\right)$ (where Z is gamma (a) distributed) is

$$f(x) = c\left(1 + \frac{x}{s}\right)^{2a-1} e^{-2x} e^{-\frac{x^2}{s}} \qquad (x \geq -s)$$

where $c = 2s^{a-1}e^{-s^2}/\Gamma(a)$.

B.   We have

$$f(x) \leq ce^{-\frac{x^2}{s}}.$$

C.   If this inequality is used to generate random variates with density f, then the rejection constant, $c\sqrt{\pi s}$, is $\sqrt{\dfrac{2\pi}{e}}$ at $a = 1$, and tends to

$\sqrt{2}$ as $a \uparrow \infty$. Prove also that for all values $a > \dfrac{1}{2}$, the rejection constant is bounded from above by $\sqrt{2}e^{\frac{1}{4a}}$.

D. The raw almost-exact inversion algorithm is:

Almost-exact inversion algorithm for gamma variates

REPEAT

        Generate a normal random variate N and an exponential random variate E.

        $X \leftarrow tN$

UNTIL $X \geq s$ AND $E - 2X + 2s \log(1 + \dfrac{X}{s}) \geq 0$

RETURN $s(1 + \dfrac{X}{s})^2$

E. Introduce quick acceptance and rejection steps in the algorithm that are so accurate that the expected number of evaluations of the logarithm is $o(1)$ as $a \uparrow \infty$. Prove the claim.

Remark: for a very efficient implementation based upon another quadratic transformation, see Ahrens and Dieter (1982).

4. MANY-TO-ONE TRANSFORMATIONS.

4.1. The principle.

Sometimes it is possible to exploit some distributional properties of random variables. Assume for example that $\psi(X)$ has an easy density h, where X has density f. When ψ is a one-to-one transformation, X can then be generated as $\psi^{-1}(Y)$ where Y is a random variate with the easy density h. A point in case is the inversion method of course where the easy density is the uniform density. There are important examples in which the transformation ψ is many-to-one, so that the inverse is not uniquely defined. In that case, if there are k solutions $X_1, \ldots, X_k$ of the equation $\psi(X) = Y$, it suffices to choose among the X_i's. The probabilities however depend upon Y. The usefulness of this approach was first realized by Michael, Schucany and Haas (1976), who gave a comprehensive description and discussion of the method. They were motivated by a simple fast algorithm for the inverse gaussian family based upon this approach.

By far the most important case is $k = 2$, which is the one that we shall deal with here. Several important examples are developed in subsections.

Assume that there exists a point t such that ψ' is of one sign on $(-\infty, t)$ and on (t, ∞). For example, if $\psi(x) = x^2$, then $\psi'(x) = 2x$ is nonpositive on $(-\infty, 0)$ and nonnegative on $(0, \infty)$, so that we can take $t = 0$. We will use the notation

$$x = l(y) , \ x = r(y)$$

for the two solutions of $y = \psi(x)$: here, l is the solution in $(-\infty, t)$, and r is the solution in (t, ∞). If ψ satisfies the conditions of Theorem I.4.1 on each interval, and X has density f, then $\psi(X)$ has density

$$h(y) = |l'(y)| f(l(y)) + |r'(y)| f(r(y)) .$$

This is quickly verified by computing the distribution function of $\psi(X)$ and then taking the derivative. Vice versa, given a random variate Y with density h, we can obtain a random variate X with density f by choosing $X = l(Y)$ with probability

$$\frac{|l'(Y)| f(l(Y))}{h(Y)} ,$$

and choosing $X = r(Y)$ otherwise. Note that $|l'(y)| = 1/|\psi'(l(y))|$. This, the method of Michael, Schucany and Haas (1976), can be summarized as follows:

Inversion of a many-to-one transformation

Generate a random variate Y with density h.

Generate a uniform $[0,1]$ random variate U.

Set $X_1 \leftarrow l(Y) , X_2 \leftarrow r(Y)$

IF $U \leq \dfrac{1}{1 + \dfrac{f(X_2)}{f(X_1)} \left| \dfrac{\psi'(X_1)}{\psi'(X_2)} \right|}$

    THEN RETURN $X \leftarrow X_1$
    ELSE RETURN $X \leftarrow X_2$

It will be clear from the examples that in many cases the expression in the selection step takes a simple form.

4.2. The absolute value transformation.

The transformation $y = |x-t|$ for fixed t satisfies the conditions of the previous section. Here we have $l(y)=t-y$, $r(y)=t+y$. Since $|\psi'|$ remains constant, the decision is extremely simple. Thus, we have

---

Generate a random variate Y with density $h(y)=f(t-y)+f(t+y)$.

Generate a uniform $[0,1]$ random variate U.

IF $U \leq \dfrac{f(t-Y)}{f(t-Y)+f(t+Y)}$

THEN RETURN $X \leftarrow t-Y$

ELSE RETURN $X \leftarrow t+Y$

If f is symmetric about t, then the decisions $t-Y$ and $t+Y$ are equally likely. Another interesting case occurs when h is the uniform density. For example, consider the density

$$f(x) = \frac{1+\cos x}{\pi} \quad (0 \leq x \leq \pi) .$$

Then, taking $t = \dfrac{\pi}{2}$, we see that

$$h(y) = f(t-y)+f(t+y) = \frac{2}{\pi} \quad (0 \leq y \leq \frac{\pi}{2}) .$$

Thus, we can generate random variates with this density as follows:

Generate two iid uniform $[0,1]$ random variates U,V.

Set $Y \leftarrow \dfrac{\pi V}{2}$.

IF $U \leq \dfrac{1+\cos Y}{2}$

THEN RETURN $X \leftarrow Y$

ELSE RETURN $X \leftarrow \pi-Y$

Here we have made use of additional symmetry in the problem. It should be noted that the evaluation of the cos can be avoided altogether by application of the series method (see section 5.4).

4.3. The inverse gaussian distribution.

Michael, Schucany and Haas (1976) have successfully applied the many-to-one transformation method to the **inverse gaussian distribution**. Before we proceed with the details of their algorithm, it is necessary to give a short introductory tour of the distribution (see Folks and Chhikara (1978) for a survey).

A random variable $X \geq 0$ with density

$$ f(x) = \sqrt{\frac{\lambda}{2\pi x^3}} \, e^{-\frac{\lambda(x-\mu)^2}{2\mu^2 x}} \qquad (x \geq 0) $$

is said to have the inverse gaussian distribution with parameters $\mu > 0$ and $\lambda > 0$. We will say that a random variate X is $I(\mu,\lambda)$. Sometimes, the distribution is also called Wald's distribution, or the first passage time distribution of Brownian motion with positive drift.

The densities are unimodal and have the appearance of gamma densities. The mode is at

$$ \mu \left(\sqrt{1 + \frac{9\mu^2}{4\lambda^2}} - \frac{3\mu}{2\lambda} \right) . $$

The densities are very flat near the origin and have exponential tails. For this reason, all positive and negative moments exist. For example, $E(X^{-a}) = E(X^{a+1})/\mu^{2a+1}$, all $a \in R$. The mean is μ and the variance is $\frac{\mu^3}{\lambda}$. The main distributional property is captured in the following Lemma:

Lemma 4.1. (Shuster, 1968)

When X is $I(\mu,\lambda)$, then

$$ \frac{\lambda(X-\mu)^2}{\mu^2 X} $$

is distributed as the square of a normal random variable , i.e. it is chi-square with one degree of freedom.

Proof of Lemma 4.1.

Straightforward. ∎

Based upon Lemma 4.1, we can apply a many-to-one transformation

$$ \psi(x) = \frac{\lambda(x-\mu)^2}{\mu^2 x} . $$

Here, the inverse has two solutions, one on each side of μ. The solutions of $\psi(X)=Y$ are

$$X_1 = \mu + \frac{\mu^2 Y}{2\lambda} - \frac{\mu}{2\lambda}\sqrt{4\mu\lambda Y + \mu^2 Y^2}$$

$$X_2 = \frac{\mu^2}{X_1}$$

One can verify that

$$\frac{f(X_2)}{f(X_1)} = \left(\frac{X_1}{\mu}\right)^3 ,$$

$$\frac{\psi'(X_1)}{\psi'(X_2)} = -\left(\frac{\mu}{X_1}\right)^2 .$$

Thus, X_1 should be selected with probability $\dfrac{\mu}{\mu + X_1}$. This leads to the following algorithm:

Inverse gaussian distribution generator of Michael, Schucany and Haas

Generate a normal random variate N.

Set $Y \leftarrow N^2$

Set $X_1 \leftarrow \mu + \dfrac{\mu^2 Y}{2\lambda} - \dfrac{\mu}{2\lambda}\sqrt{4\mu\lambda Y + \mu^2 Y^2}$

Generate a uniform $[0,1]$ random variate U.

IF $U \leq \dfrac{\mu}{\mu + X_1}$

        THEN RETURN $X \leftarrow X_1$

        ELSE RETURN $X \leftarrow \dfrac{\mu^2}{X_1}$

This algorithm was later rediscovered by Padgett (1978). The time-consuming components of the algorithm are the square root and the normal random variate generation. There are a few shortcuts: a few multiplications can be saved if we replace Y by μY at the outset, for example. There are several exercises about the inverse gaussian distribution following this sub-section.

4.4. Exercises.

1. **First passage time distribution of drift-free Brownian motion.** Show that as $\mu \to \infty$ while λ remains fixed, the $I(\mu,\lambda)$ density tends to the density

$$f(x) = \sqrt{\frac{\lambda}{2\pi x^3}} e^{-\frac{\lambda}{2x}} \qquad (x \geq 0) ,$$

which is the one-sided stable density with exponent $\frac{1}{2}$, or the density for the first passage time of drift-free Brownian motion. Show that this is the density of the inverse of a gamma $(\frac{1}{2}, \frac{2}{\lambda})$ random variable (Wasan and Roy, 1967). This is equivalent to showing that it is the density of $\frac{\lambda}{N^2}$ where N is a normal random variable.

2. This is a further exercise about the properties of the inverse gaussian distribution. Show the following:

    (i)   If X is $I(\mu,\lambda)$, then cX is $I(c\mu, c\lambda)$.

    (ii) The characteristic function of X is $e^{\frac{\lambda}{\mu}(1 - \sqrt{1 - \frac{2i\mu^2 t}{\lambda}})}$.

    (iii) If X_i , $1 \leq i \leq n$, are independent $I(\mu_i, c\mu_i^2)$ random variables, then $\sum_{i=1}^{n} X_i$ is $I(\sum \mu_i, c(\sum \mu_i)^2)$. Thus, if the X_i' s are iid $I(\mu,\lambda)$, then $\sum X_i$ is $I(n\mu, n^2\lambda)$.

    (iv) Show that when N_1 , N_2 are independent normal random variables with variances σ_1^2 and σ_2^2, then $\dfrac{N_1 N_2}{\sqrt{N_1^2 + N_2^2}}$ is normal with variance σ_3^2 determined by the relation $\dfrac{1}{\sigma_3} = \dfrac{1}{\sigma_1} + \dfrac{1}{\sigma_2}$.

    (v)   The distribution function of X is

$$F(x) = \Phi(\sqrt{\frac{\lambda}{x}}(\frac{x}{\mu}-1)) + e^{\frac{2\lambda}{\mu}} \Phi(-\sqrt{\frac{\lambda}{x}}(1+\frac{x}{\mu})) ,$$

where Φ is the standard normal distribution function (Zigangirov, 1962).

5. THE SERIES METHOD.

5.1. Description.

In this section, we consider the problem of the computer generation of a random variable X with density f where f can be approximated from above and below by sequences of functions f_n and g_n. In particular, we assume that:

(1) $\lim\limits_{n \to \infty} f_n = f$;

    $\lim\limits_{n \to \infty} g_n = f$.

(11) $f_n \leq f \leq g_n$.

(111) $f \leq ch$ for some constant $c \geq 1$ and some easy
       density h .

The sequences f_n and g_n should be easy to evaluate, while the dominating density h should be easy to sample from. Note that f_n need not be positive, and that g_n need not be integrable. This setting is common: often f is only known as a series, as in the case of the Kolmogorov-Smirnov distribution or the stable distributions, so that random variate generation has to be based upon this series. But even if f is explicitly known, it can often be expanded in a fast converging series such as in the case of a normal or exponential density. The series method described below actually avoids the exact evaluation of f all the time. It can be thought of as a rejection method with an infinite number of acceptance and rejection conditions for squeezing. Nearly everything in this section was first developed in Devroye (1980).

The series method

REPEAT
       Generate a random variate X with density h .
       Generate a uniform [0,1] random variate U .
       $W \leftarrow Uch(X)$
       $n \leftarrow 0$
       REPEAT
              $n \leftarrow n + 1$
              IF $W \leq f_n(X)$ THEN RETURN X
       UNTIL $W > g_n(X)$
UNTIL False

The fact that the outer loop in this algorithm is an infinite loop does not matter, because with probability one we will exit in the inner loop (in view of $f_n \to f$, $g_n \to f$). We have here a true rejection algorithm because we exit when $W \leq Uch(X)$. Thus, the expected number of outer loops is c , and the choice of the dominating density h is important. Notice however that the time should be measured in terms of the number of f_n and g_n evaluations. Such analysis will be given further on. While in many cases, the convergence to f is so fast that the expected number of f_n evaluations is barely larger than c, it is true that there are examples in which this expected number is ∞. It is also worth observing that the squeeze steps are essential here for the correctness of the algorithm. They actually form the algorithm.

In the remainder of this section, we will give three important special cases of approximating series. The series method and its variants will be illustrated with the aid of the exponential, Raab-Green and Kolmogorov-Smirnov distributions further on.

Assume first that f can be written as a convergent series

$$f(x) = \sum_{n=1}^{\infty} s_n(x) \le ch(x)$$

where

$$\left| \sum_{i=n+1}^{\infty} s_i(x) \right| \le R_{n+1}(x)$$

is a known estimate of the remainder, and h is a given density. In this special instance, we can rewrite the series method in the following form:

The convergent series method

REPEAT
    Generate a random variate X with density h.
    Generate a uniform $[0,1]$ random variate U.
    $W \leftarrow Uch(X)$
    $S \leftarrow 0$
    $n \leftarrow 0$
    REPEAT
        $n \leftarrow n+1$
        $S \leftarrow S + s_n(X)$
    UNTIL $|S-W| > R_{n+1}(X)$
UNTIL $S \le W$
RETURN X

Assume next that f can be written as an alternating series
$$f(x) = ch(x)(1 - a_1(x) + a_2(x) - a_3(x) + \cdots)$$

where a_n is a sequence of functions satisfying the condition that $a_n(x) \downarrow 0$ as $n \to \infty$, for all x, c is a constant, and h is an easy density. Then, the series method can be written as follows:

The alternating series method

REPEAT

    Generate a random variate X with density h.

    Generate a uniform $[0,c]$ random variate U.

    $n \leftarrow 0, W \leftarrow 0$

    REPEAT

        $n \leftarrow n + 1$

        $W \leftarrow W + a_n(X)$

        IF $U \geq W$ THEN RETURN X

        $n \leftarrow n + 1$

        $W \leftarrow W - a_n(X)$

    UNTIL $U < W$

UNTIL False

This algorithm is valid because f is bounded from above and below by two converging sequences:

$$1 + \sum_{j=1}^{k} (-1)^j \, a_j(x) \leq \frac{f(x)}{ch(x)} \leq 1 + \sum_{j=1}^{k+1} (-1)^j \, a_j(x) \, , \, k \text{ odd }.$$

That this is indeed a valid inequality follows from the monotonicity of the terms (consider the terms pairwise). As in the ordinary series method, f is never fully computed. In addition, h is never evaluated either.

A second important special case occurs when

$$f(x) = ch(x) \, e^{-a_1(x) + a_2(x) - \cdots}$$

where c, h, a_n are as for the alternating series method. Then, the alternating series method is equivalent to:

The alternating series method; exponential version

REPEAT
    Generate a random variate X with density h .
    Generate an exponential random variate E .
    $n \leftarrow 0, W \leftarrow 0$
    REPEAT
        $n \leftarrow n + 1$
        $W \leftarrow W + a_n(X)$
        IF $E \geq W$ THEN RETURN X
        $n \leftarrow n + 1$
        $W \leftarrow W - a_n(X)$
    UNTIL $E < W$
UNTIL False

5.2. Analysis of the alternating series algorithm.

For the four versions of the series method defined above, we know that the expected number of iterations is equal to the rejection constant, c . In addition, there is a hidden contribution to the time complexity due to the fact that the inner loop, needed to decide whether $Uch(X) \leq f(X)$, requires a random number of computations of a_n . The computations of a_n are assumed to take a constant time independent of n - if they do not, just modify the analysis given in this section slightly. In all the examples that will follow, the a_n computations take a constant time.

In Theorem 5.1, we will give a precise answer for the alternating series method.

Theorem 5.1.

Consider the alternating series method for a density f decomposed as follows:

$$f(x) = ch(x)(1 - a_1(x) + a_2(x) - \cdots),$$

where $c \geq 1$ is a normalization constant, h is a density, and $a_0 \equiv 1 \geq a_1 \geq a_2 \geq \cdots \geq 0$. Let N be the total number of computations of a factor a_n before the algorithm halts. Then,

$$E(N) = c \int_0^\infty [\sum_{i=0}^\infty a_i(x)] h(x) \, dx \ .$$

Proof of Theorem 5.1.

By Wald's equation, $E(N)$ is equal to c times the expected number of a_n computations in the first iteration. In the first iteration, we fix $X = x$ with density h. Then, dropping the dependence on x, we see that for the odd terms a_n, we require

1	with probability $1 - a_1$
2	with probability $a_1 - a_2$
3	with probability $a_2 - a_3$
4	with probability $a_3 - a_4$
...	

computations of a_n. The expected value of this is

$$\sum_{i=1}^\infty i(a_{i-1} - a_i) = \sum_{i=0}^\infty a_i \ .$$

Collecting these results gives us Theorem 5.1. ∎

Theorem 5.1 shows that the expected time complexity is equal to the oscillation in the series. Fast converging series lead to fast algorithms.

5.3. Analysis of the convergent series algorithm.

As in the previous section, we will let N be the number of computations of terms s_n before the algorithm halts. We have:

Theorem 5.2.

For the convergent series algorithm of section 5.1,

$$E(N) \leq 2\int(\sum_{n=1}^{\infty} R_n(x)) \, dx \quad .$$

Proof of Theorem 5.2.

By Wald's equation, $E(N)$ is equal to c times the expected number of s_n computations in the first global iteration. If we fix X with density h, then if N is the number of s_n computations in the first iteration alone,

$$P(N > n \mid X) \leq \frac{2R_{n+1}(X)}{ch(X)} \quad .$$

Thus,

$$E(N \mid X) = \sum_{n=0}^{\infty} P(N > n \mid X)$$

$$\leq \sum_{n=0}^{\infty} \frac{2R_{n+1}(X)}{ch(X)} \quad .$$

Hence, turning to the overall number of s_n computations,

$$E(N) \leq c \sum_{n=1}^{\infty} \int h(x) \frac{2R_n(x)}{ch(x)} \, dx$$

$$= 2\int(\sum_{n=1}^{\infty} R_n(x)) \, dx \quad . \blacksquare$$

It is important to note that a series converging at the rate $\frac{1}{n}$ or slower cannot yield finite expected time. Luckily, many important series, such as those of all the remaining subsections on the series method converge at an exponential rather than a polynomial rate. In view of Theorem 5.2, this virtually insures the finiteness of their expected time. It is still necessary however to verify whether the expected time statements are not upset in an indirect way through the dependence of $R_n(x)$ upon x: for example, the bound of Theorem 5.2 is infinite when $\int R_n(x) \, dx = \infty$ for some n .

5.4. The exponential distribution.

It is known that for all odd k and all $x > 0$,

$$\sum_{j=0}^{k-1} (-1)^j \frac{x^j}{j!} \geq e^{-x} \geq \sum_{j=0}^{k} (-1)^j \frac{x^j}{j!} .$$

We will apply the alternating series method to the truncated exponential density

$$f(x) = \frac{e^{-x}}{1-e^{-\mu}} \quad (0 \leq x \leq \mu) ,$$

where $1 \geq \mu > 0$ is the truncation point. As dominating curve, we can use the uniform density (called h) on $[0,\mu]$. Thus, in the decomposition needed for the alternating series method, we use

$$c = \frac{\mu}{1-e^{-\mu}} ,$$

$$h(x) = \frac{1}{\mu} I_{[0,\mu]}(x) ,$$

$$a_n(x) = \frac{x^n}{n!} .$$

The monotonicity of the a_n's is insured when $|x| \leq 1$. This forces us to choose $\mu \leq 1$. The expected number of a_n computations is

$$E(N) = c \int_0^\mu \sum_{j=0}^\infty \frac{x^j}{j!} \frac{1}{\mu} \, dx$$

$$= c \, \frac{e^\mu - 1}{\mu}$$

$$= \frac{e^\mu - 1}{1-e^{-\mu}} .$$

For example, for $\mu = 1$, the value e is obtained. But interestingly, $E(N) \downarrow 1$ as $\mu \downarrow 0$. The truncated exponential density is important, because standard exponential random variates can be obtained by adding an independent properly scaled geometric random variate (see for example section IV.2.2 on the Forsythe-von Neumann method or section IX.2 about exponential random variates). The algorithm for the truncated exponential density is given below:

A truncated exponential generator via the alternating series method

REPEAT

Generate a uniform $[0,\mu]$ random variate X.

Generate a uniform $[0,1]$ random variate U.

$n \leftarrow 0, W \leftarrow 0, V \leftarrow 1$ (V is used to facilitate evaluation of consecutive terms in the alternating series.)

REPEAT

$n \leftarrow n + 1$

$V \leftarrow \dfrac{VX}{n}$

$W \leftarrow W + V$

IF $U \geq W$ THEN RETURN X

$n \leftarrow n + 1$

$V \leftarrow \dfrac{VX}{n}$

$W \leftarrow W - V$

UNTIL $U < W$

UNTIL False

The alternating series method based upon Taylor's series is not applicable to the exponential distribution on $[0,\infty)$ because of the impossibility of finding a dominating density h based upon this series. In the exercise section, the ordinary series method is applied with a family of dominating densities, but the squeezing is still based upon the Taylor series for the exponential density.

5.5. The Raab-Green distribution.

The density

$$f(x) = \frac{1+\cos(x)}{2\pi} \qquad (\mid x \mid \leq \pi)$$

$$= \frac{1}{\pi}(1 - \frac{1}{2}\frac{x^2}{2!} + \frac{1}{2}\frac{x^4}{4!} - \cdots)$$

was suggested by Raab and Green (1961) as an approximation for the normal density. The series expansion is very similar to that of the exponential function. Again, we are in a position to apply the alternating series method, but now with $h(x) = \dfrac{1}{2\pi}$ ($\mid x \mid \leq \pi$), $c = 2$ and $a_n(x) = \dfrac{1}{2}\dfrac{x^{2n}}{2n!}$. It is easy to verify that $a_n \downarrow 0$ as $n \rightarrow \infty$ for all x in the range:

$$\frac{a_{n+1}(x)}{a_n(x)} = \frac{x^2}{(2n+2)(2n+1)} \le \frac{\pi^2}{12} \quad (n \ge 1).$$

Note however that a_1 is not smaller than 1, which was a condition necessary for the application of Theorem 5.1. Nevertheless, the alternating series method remains formally valid, and we have:

A Raab-Green density generator via the alternating series method

REPEAT
  Generate a uniform $[-\pi,\pi]$ random variate X.
  Generate a uniform $[0,1]$ random variate U.
  $n \leftarrow 0, W \leftarrow 0, V \leftarrow 1$ (V is used to facilitate evaluation of consecutive terms in the alternating series.)
  REPEAT
      $n \leftarrow n+1$
      $V \leftarrow \dfrac{VX^2}{(2n)(2n-1)}$
      $W \leftarrow W+V$
      IF $U \ge W$ THEN RETURN X
      $n \leftarrow n+1$
      $V \leftarrow \dfrac{VX^2}{(2n)(2n-1)}$
      $W \leftarrow W-V$
  UNTIL $U < W$
UNTIL False

The drawback with this algorithm is that c, the rejection constant, is 2. But this can be avoided by the use of a many-to-one transformation described in section IV.4. The principle is this: if (X,U) is uniformly distributed in $[-\frac{\pi}{2},\frac{\pi}{2}] \times [0,2]$, then we can exit with X when $U \le 1+\cos(X)$ and with $\pi \operatorname{sign} X - X$ otherwise, thereby avoiding rejections altogether. With this improvement, we obtain:

An improved Raab-Green density generator based on the alternating series method

Generate a uniform $[-\frac{\pi}{2},\frac{\pi}{2}]$ random variate X.

Generate a uniform $[0,1]$ random variate U.

$n \leftarrow 0, W \leftarrow 0, V \leftarrow 1$ (V is used to facilitate evaluation of consecutive terms in the alternating series.)

REPEAT

$\quad n \leftarrow n+1$

$\quad V \leftarrow \dfrac{VX^2}{(2n)(2n-1)}$

$\quad W \leftarrow W+V$

$\quad$ IF $U \geq W$ THEN RETURN X

$\quad n \leftarrow n+1$

$\quad V \leftarrow \dfrac{VX^2}{(2n)(2n-1)}$

$\quad W \leftarrow W-V$

$\quad$ IF $U \leq W$ THEN RETURN π sign$X-X$

UNTIL False

This algorithm improves over the algorithm of section IV.4 for the same distribution in which the cos was evaluated once per random variate. We won't give a detailed time analysis here. It is perhaps worth noting that the probability that the UNTIL step is reached, i.e. the probability that one iteration is completed, is about 2.54%. This can be seen as follows: If $N*$ is the number of completed iterations, then

$$P(N*>i) = \frac{2}{\pi} \int_0^{\frac{\pi}{2}} \frac{1}{2} \frac{x^{4i}}{(4i)!} \, dx = \frac{1}{\pi} \frac{(\frac{\pi}{2})^{4i+1}}{(4i+1)!} \ .$$

and thus

$$E(N*) = \sum_{i=0}^{\infty} \frac{1}{\pi} \frac{(\frac{\pi}{2})^{4i+1}}{(4i+1)!} \ .$$

In particular, $P(N*>1)=\dfrac{\pi^4}{3840}\approx 0.0254$. Also, $E(N*)$ is about equal to $1+2P(N*>1)\approx 1.0254$ because $P(N*>2)$ is extremely small.

5.6. The Kolmogorov-Smirnov distribution.

The **Kolmogorov-Smirnov distribution function**

$$F(x) = \sum_{n=-\infty}^{\infty} (-1)^n e^{-2n^2 x^2} \quad (x \geq 0)$$

appears as the limit distribution of the Kolmogorov-Smirnov test statistic (Kolmogorov (1933); Smirnov (1939); Feller (1948)). No simple procedure for inverting F is known, hence the inversion method is likely to be slow. Also, both the distribution function and the corresponding density are only known as infinite series. Thus, exact evaluation of these functions is not possible in finite time. Yet, by using the series method, we can generate random variates with this distribution extremely efficiently. This illustrates once more that generating random variates is simpler than computing a distribution function.

First, it is necessary to obtain convenient series expansions for the density. Taking the derivative of F, we obtain the density

$$f(x) = 8 \sum_{n=1}^{\infty} (-1)^{n+1} n^2 x e^{-2n^2 x^2} \quad (x \geq 0) ,$$

which is in the format of the alternating series method if we take

$$ch(x) = 8xe^{-2x^2} ,$$

$$a_n(x) = (n+1)^2 e^{-2x^2((n+1)^2-1)} \quad (n \geq 0) .$$

There is another series for F and f which can be obtained from the first series by the theory of theta functions (see e.g. Whittaker and Watson , 1927):

$$F(x) = \frac{\sqrt{2\pi}}{x} \sum_{n=1}^{\infty} e^{-\frac{(2n-1)^2 \pi^2}{8x^2}} \quad (x > 0) ;$$

$$f(x) = \frac{\sqrt{2\pi}}{x} \sum_{n=1}^{\infty} [\frac{(2n-1)^2 \pi^2}{4x^3} - \frac{1}{x}] e^{-\frac{(2n-1)^2 \pi^2}{8x^2}} \quad (x > 0) .$$

Again, we have the format needed for the alternating series method, but now with

$$ch(x) = \frac{\sqrt{2\pi}\pi^2}{4x^4} e^{-\frac{\pi^2}{8x^2}} \quad (x > 0) ,$$

$$a_n(x) = \begin{cases} \dfrac{4x^2}{\pi^2} e^{-\frac{(n^2-1)\pi^2}{8x^2}} & (n \text{ odd} , x > 0) \\ \\ (n+1)^2 e^{-\frac{((n+1)^2-1)\pi^2}{8x^2}} & (n \text{ even} , x > 0) \end{cases}$$

We will refer to this series expansion as the second series expansion. In order for the alternating series method to be applicable, we must verify that the a_n's satisfy the monotonicity condition. This is done in Lemma 5.1:

Lemma 5.1.

The terms a_n in the first series expansion are monotone $\downarrow$ for $x > \sqrt{\dfrac{1}{3}}$.
For the second series expansion, they are monotone $\downarrow$ when $x < \dfrac{\pi}{2}$.

Proof of Lemma 5.1.

In the first series expansion, we have

$$\log(\frac{a_{n-1}(x)}{a_n(x)}) = -2\log(1+\frac{1}{n})+2(2n+1)x^2$$

$$\geq -\frac{2}{n}+2(2n+1)x^2 \geq -2+6x^2 > 0 \ .$$

For the second series expansion, when n is even,

$$\frac{a_n(x)}{a_{n+1}(x)} = \frac{(n+1)^2\pi^2}{4x^2} \geq \frac{\pi^2}{4x^2} > 1 \ .$$

Also,

$$\log(\frac{a_{n-1}(x)}{a_n(x)}) = -\log(\frac{(n+1)^2\pi^2}{4x^2})+\frac{n\pi^2}{2x^2} = y-2\log(n+1)-\log(\frac{y}{2})$$

where $y = \dfrac{\pi^2}{2x^2}$. The last expression is increasing in y for $y \geq 2$ and all $n \geq 2$.
Thus, it is not smaller than $2n-2\log(n+1) \geq 0$. ∎

We now give the algorithm of Devroye (1980). It uses the mixture method because one series by itself does not yield easily identifiable upper and lower bounds for f on the entire real line. We are fortunate that the monotonicity conditions are satisfied on $(\sqrt{\dfrac{1}{3}},\infty)$ and on $(0,\dfrac{\pi}{2})$ for the two series respectively. Had these intervals been disjoint, then we would have been forced to look for yet another approximation. We define the breakpoint for the mixture method by $t \in (\sqrt{\dfrac{1}{3}},\dfrac{\pi}{2})$. The value 0.75 is suggested. Define also $p = F(t)$.

Generate a uniform [0,1] random variate U.

IF $U < p$

      THEN RETURN a random variate X with density $\dfrac{f}{p}, 0 < x < t$.

      ELSE RETURN a random variate X with density $\dfrac{f}{1-p}, t < x$.

For generation in the two intervals, the two series expansions are used. Another constant needed in the algorithm is $t' = \dfrac{\pi^2}{8t^2}$. We have:

Generator for the leftmost interval

REPEAT

    REPEAT

        Generate two iid exponential random variates, E_0, E_1.

$$E_0 \leftarrow \frac{E_0}{1 - \dfrac{1}{2t'}}$$

$$E_1 \leftarrow 2E_1$$

$$G \leftarrow t' + E_0$$

        Accept $\leftarrow [(E_0)^2 \leq t' E_1 (G + t')]$

        IF NOT Accept

$$\text{THEN Accept} \leftarrow [\frac{G}{t'} - 1 - \log(\frac{G}{t'}) \leq E_1]$$

    UNTIL Accept

$$X \leftarrow \frac{\pi}{\sqrt{8G}}$$

$$W \leftarrow 0$$

$$Z \leftarrow \frac{1}{2G}$$

$$P \leftarrow e^{-G}$$

$$n \leftarrow 1$$

$$Q \leftarrow 1$$

    Generate a uniform [0,1] random variate U.

    REPEAT

$$W \leftarrow W + ZQ$$

        IF $U \geq W$ THEN RETURN X

$$n \leftarrow n + 2$$

$$Q \leq P^{n^2 - 1}$$

$$W \leftarrow W - n^2 Q$$

    UNTIL $U < W$

UNTIL False

Generator for the rightmost interval

REPEAT
  Generate an exponential random variate E.
  Generate a uniform $[0,1]$ random variate U.
  $X \leftarrow \sqrt{t^2 + \dfrac{E}{2}}$
  $W \leftarrow 0$
  $n \leftarrow 1$
  $Z \leftarrow e^{-2X^2}$
  REPEAT
    $n \leftarrow n + 1$
    $W \leftarrow W + n^2 Z^{n^2 - 1}$
    IF $U \geq W$ THEN RETURN X
    $n \leftarrow n + 1$
    $W \leftarrow W - n^2 Z^{n^2 - 1}$
  UNTIL $U \leq W$
UNTIL False

The algorithms are both straightforward applications of the alternating series method, but perhaps a few words of explanation are in order regarding the algorithms used for the dominating densities. This is done in two lemmas.

Lemma 5.2.

  The random variable $\sqrt{t^2 + \dfrac{E}{2}}$ (where E is an exponential random variable and $t > 0$) has density

$$cxe^{-2x^2} \quad (x \geq t),$$

where $c > 0$ is a normalization constant.

Proof of Lemma 5.2.

  Verify that the distribution function of the random variable is $1 - e^{-2(x^2 - t^2)}$ ($x \geq t$). Taking the derivative of this distribution function yields the desired result. ∎

Lemma 5.3.

If G is a random variable with truncated gamma $(\frac{3}{2})$ density $c\sqrt{y}\,e^{-y}$ $(y \geq t' = \frac{\pi^2}{8t^2})$, then $\frac{\pi}{\sqrt{8G}}$ has density

$$\frac{c}{x^4} e^{-\frac{\pi^2}{8x^2}} \quad (0 < x \leq t),$$

where the c's stand for (possibly different) normalization constants, and $t > 0$ is a constant. A truncated gamma $(\frac{3}{2})$ random variate can be generated by the algorithm:

Truncated gamma generator

REPEAT
    Generate two iid exponential random variates, E_0, E_1.

$$E_0 \leftarrow \frac{E_0}{1 - \frac{1}{2t'}}$$

$E_1 \leftarrow 2E_1$
$G \leftarrow t' + E_0$
Accept $\leftarrow [(E_0)^2 \leq t' E_1 (G + t')]$
    IF NOT Accept
        THEN Accept $\leftarrow [\frac{G}{t'} - 1 - \log(\frac{G}{t'}) \leq E_1]$
UNTIL Accept
RETURN G

Proof of Lemma 5.3.

The Jacobian of the transformation $y = \frac{\pi^2}{8x^2}$ is $\frac{4\pi}{(8y)^{\frac{3}{2}}}$. This gives the distributional result without further work if we argue backwards. The validity of the rejection algorithm with squeezing requires a little work. First, we start from the inequality

$$y \leq e^{\frac{y}{t'}} \frac{t'}{e} \quad (y \geq t'),$$

which can be obtained by maximizing $ye^{\frac{-y}{t'}}$ in the said interval. Thus,

$$\sqrt{y}\,e^{-y} \le \sqrt{\frac{t'}{e}}\,e^{-(1-\frac{1}{2t'})y} \qquad (y \ge t') \,.$$

The upper bound is proportional to the density of $t' + \dfrac{E}{1-\dfrac{1}{2t'}}$ where E is an

exponential random variate. This random variate is called G in the algorithm. Thus, if U is a uniform random variate, we can proceed by generating couples G, U until

$$e^{\frac{G}{2t'}}\sqrt{\frac{t'}{e}}\,U \le \sqrt{G} \,.$$

This condition is equivalent to

$$\frac{G}{t'}-1-\log(\frac{G}{t'}) \le 2E_1$$

where E_1 is another exponential random variable. A squeeze step can be added by noting that $\log(1+u) \ge \dfrac{2u}{2+u}$ $(u \ge 0)$ (exercise 5.1). ■

All the previous algorithms can now be collected into one long (but fast) algorithm. For generalities on good generators for the tail of the gamma density, we refer to the section on gamma variate generation. In the implementation of Devroye (1980), two further squeeze steps were added. For the rightmost interval, we can return X when $U \ge 4e^{-6t^2}$ (which is a constant). For the leftmost interval, the same can be done when $U \ge \dfrac{4t^2}{\pi^2}$. For $t = 0.75$, we have $p \approx 0.373$, and the quick acceptance probabilities are respectively ≈ 0.86 and ≈ 0.77 for the latter squeeze steps.

Related distributions.

The **empirical distribution function** $F_n(x)$ for a sample $X_1, \ldots, X_n$ of iid random variables is defined by

$$F_n(x) = \sum_{i=1}^{n} \frac{1}{n} I_{[X_i \le x]}$$

where I is the indicator function. If X_i has distribution function $F(x)$, then the following goodness-of-fit statistics have been proposed by various authors:

(1)   The   asymmetrical   Kolmogorov-Smirnov   statistics
$K_n^+ = \sqrt{n}\,\sup\,(F_n - F)$, $K_n^- = \sqrt{n}\,\sup\,(F - F_n)$.

(ll)   The Kolmogorov-Smirnov statistic $K_n = \max(K_n{}^+, K_n{}^-)$.

(lll) Kuiper's statistic $V_n = K_n{}^+ + K_n{}^-$.

(lv)   von Mises' statistic $W_n{}^2 = n \int (F_n - F)^2 dF$.

(v)   Watson's statistic $U_n = n \int (F_n - F - (\int (F_n - F) dF))^2 dF$.

(vl)   The Anderson-Darling statistic $A_n{}^2 = n \int \dfrac{(F_n - F)^2}{F(1-F)} dF$.

For surveys of the properties and applications of these and other statistics, see
Darling (1955), Barton and Mallows (1965), and Sahler (1968). The limit random
variables (as $n \to \infty$) are denoted with the subscripts ∞. The limit distributions
have characteristic functions that are infinite products of characteristic functions
of gamma distributed random variables except in the case of A_∞. From this, we
note several relations between the limit distributions. First, $2K_\infty{}^{+2}$ and $2K_\infty{}^{-2}$
are exponentially distributed (Smirnov, 1939; Feller, 1948). K_∞ has the
Kolmogorov-Smirnov distribution function discussed in this section (Kolmogorov,
1933; Smirnov, 1939; Feller, 1948). Interestingly, V_∞ is distributed as the sum of
two independent random variables distributed as K_∞ (Kuiper, 1960). Also, as
shown by Watson (1961, 1962), U_∞ is distributed as $\dfrac{1}{\pi}\sqrt{K_\infty}$. Thus, generation
for all these limit distributions poses no problems. Unfortunately, the same can-
not be said for A_∞ (Anderson and Darling, 1952) and W_∞ (Smirnov, 1937;
Anderson and Darling, 1952).

5.7. Exercises.

1. Prove   the   following   inequality   needed   in   Lemma   5.3:
    $\log(1+u) \geq \dfrac{2u}{2+u}$ $(u > 0)$.

2. **The exponential distribution.** For the exponential density, choose a
    dominating density h from the family of densities

    $$\frac{na^n}{(x+a)^{n+1}} \quad (x > 0) ,$$

    where $n \geq 1$ and $a > 0$ are design parameters. Show the following:

    (l) h is the density of $a(U^{-\frac{1}{n}} - 1)$ where U is a uniform [0,1] random vari-
           able. It is also the density of $a(\max^{-1}(U_1, \ldots, U_n) - 1)$ where the U_i's
           are iid uniform [0,1] random variables.

    (ll)   Show that the rejection constant is $c = (\dfrac{n+1}{e})^{n+1} \dfrac{e^a a^{-n}}{n}$, and show
           that this is minimal when $a = n$.

    (lv)   Show that with $a = n$, we have $c = \dfrac{1}{e}(1 + \dfrac{1}{n})^{n+1} \to 1$ as $n \to \infty$.

(v) Give the series method based upon rejection from h (where $a = n$ and $n \geq 1$ is an integer). Use quick acceptance and rejection steps based upon the Taylor series expansion.

(vi) Show that the expected time of the algorithm is ∞ when $n = 1$ (this shows the danger inherent in the use of the series method). Show also that the expected time is finite when $n \geq 2$.

(Devroye, 1980)

3. Apply the series method for the normal density truncated to $[-a, a]$ with rejection from a uniform density. Since the expected number of iterations is

$$\frac{2a}{\sqrt{2\pi}(F(a) - F(-a))}$$

where F is the normal distribution function, we see that it is important that a be small. How would you handle the tails of the distribution ? How would you choose a for the combined algorithm ?

4. In the study of spectral phenomena, the following densities are important:

(i) $f_1(x) = \dfrac{1}{\pi}\left(\dfrac{\sin(x)}{x}\right)^2$ (the Fejer-de la Vallee Poussin density);

(ii) $f_2(x) = \dfrac{3}{\pi}\left(\dfrac{\sin(x)}{x}\right)^4$ (the Jackson-de la Vallee Poussin density) .

These densities have oscillating tails. Using the fact that

$$\frac{\sin(x)}{x} = 1 - \frac{x^2}{3!} + \frac{x^4}{5!} - \cdots \ ,$$

and that $\dfrac{\sin(x)}{x}$ falls between consecutive partial sums in this series, derive a good series algorithm for random variate generation for f_1 and f_2. Compare the expected time complexity with that of the obvious rejection algorithms.

5. **The normal distribution.** Consider the series method for the normal density based upon the dominating density $h(x) = \min(a, \dfrac{1}{16ax^2})$ where $a > 0$ is a parameter. Show the following:

(i) If (U, V) are iid uniform $[-1,1]$ random variates, then $\dfrac{V}{4aU}$ has density h .

(ii) Show that

$$e^{-\frac{x^2}{2}} \leq \max(\frac{1}{a}, \frac{32a}{e})h(x)$$

and deduce that the best constant a is $\sqrt{\dfrac{e}{32}}$.

(iii) Prove that the following algorithm is valid:

Normal generator via the series method

REPEAT

Generate two iid uniform $[-1,1]$ random variates V_1, V_2 and a uniform $[0,1]$ random variate U.

$$X \leftarrow \sqrt{\frac{2}{e}\frac{V_1}{V_2}}$$

IF $|X| \leq \sqrt{\frac{2}{e}}$

THEN $W \leftarrow \sqrt{\frac{32}{e}} U - 1$

ELSE $W \leftarrow \dfrac{U}{\sqrt{8e} X^2} - 1$

$n \leftarrow 0, Y \leftarrow \dfrac{X^2}{2}, P \leftarrow -1$

REPEAT

$n \leftarrow n + 1$

$P \leftarrow -\dfrac{PY}{n}$

$W \leftarrow W + P$

IF $W \leq 0$ THEN RETURN X

$n \leftarrow n + 1$

$P \leftarrow -\dfrac{PY}{n}$

$W \leftarrow W + P$

UNTIL $W > 0$

UNTIL False

(iv) Show that in this algorithm, the expected number of iterations is $\dfrac{4}{\sqrt{\pi e}}$.

(An iteration is defined as a check of the UNTIL False statement or a permanent return.)

6.   Erdos and Kac (1946) encountered the following distribution function on $[0, \infty)$:

$$F(x) = \frac{4}{\pi} \sum_{j=0}^{\infty} (-1)^j \frac{1}{2j+1} e^{-(2j+1)^2 \pi^2 / (8x^2)} \qquad (x > 0) .$$

This shows some resemblance to the Kolmogorov-Smirnov distribution function. Apply the series method to obtain an efficient algorithm for generating random variates with this distribution function. Furthermore, show the identity

$$F(x) = \sum_{j=-\infty}^{\infty} (-1)^j \left(\Phi((2j+1)x) - \Phi((2j-1)x) \right) ,$$

where Φ is the normal distribution function (Grenander and Rosenblatt, 1953), which can be of some help in the development of your algorithm.

6. REPRESENTATIONS OF DENSITIES AS INTEGRALS.

6.1. Introduction.

For most densities, one usually first tries the inversion, rejection and mixture methods. When either an ultra fast generator or an ultra universal algorithm is needed, we might consider looking at some other methods. But before we go through this trouble, we should verify whether we do not already have a generator for the density without knowing it. This occurs when there exists a special distributional property that we do not know about, which would provide a vital link to other better known distributions. Thus, it is important to be able to decide which distributional properties we can or should look for. Luckily, there are some general rules that just require knowledge of the shape of the density. For example, by Khinchine's theorem (given in this section), we know that a random variable with a unimodal density can be written as the product of a uniform random variable and another random variable, which turns out to be quite simple in some cases. Khinchine's theorem follows from the representation of the unimodal density as an integral. Other representations as integrals will be discussed too. These include a representation that will be useful for generating stable random variates, and a representation for random variables possessing a Polya type characteristic function. There are some general theorems about such representations which will also be discussed. It should be mentioned though that this section has no direct link with random variate generation, since only probabilistic properties are exploited to obtain a convenient reduction to simpler problems. We also need quite a lot of information about the density in question. Thus, were it not for the fact that several key reductions will follow for important densities, we would not have included this section in the book. Also, representing a density as an integral really boils down to defining a continuous mixture. The only novelty here is that we will actually show how to track down and invent useful mixtures for random variate generation.

6.2. Khinchine's and related theorems.

By far the most important class of densities is the class of unimodal densities. Thus, it is useful to have some integral representations for such densities. Formally, a distribution is called convex on a set A of the real line if for all $x, y \in A$,

$$F(\lambda x + (1-\lambda)y) \leq \lambda F(x) + (1-\lambda)F(y) \quad (0 \leq \lambda \leq 1).$$

It is concave if the inequality is reversed. It is **unimodal** if it is convex on $(-\infty,0]$ and concave on $[0,\infty)$, and in that case the point 0 is called a mode of the distribution. The rationale for this definition becomes obvious when translated to the density (if it exists). We will not consider other possible locations for the mode to keep the notation simple.

Theorem 6.1. Khinchine's theorem.

A random variable X is unimodal if and only if X is distributed as UY where U,Y are independent random variables: U is uniformly distributed on $[0,1]$ and Y is another random variable not necessarily possessing a density. If Y has distribution function G on $[0,\infty)$, then UY has distribution function

$$F(x) = \int_0^1 G\left(\frac{x}{u}\right) du \ .$$

Proof of Theorem 6.1.

We refer to Feller (1971, p. 158) for the only if part. For the if part we observe that $P(UY \leq x \mid U=u) = \dfrac{G(x/u)}{u}$, and thus, integrating over $[0,1]$ with respect to du gives us the result. ∎

To handle the corollaries of Khinchine's theorem correctly, we need to recall the definition of an absolutely continuous function f on an interval $[a,b]$: for all $\epsilon > 0$, there exists a $\delta > 0$ such that for all nonoverlapping intervals $(x_i,y_i),1 \leq i \leq n$, and all integers n,

$$\sum_{i=1}^n |x_i - y_i| < \delta$$

implies

$$\sum_{i=1}^n |f(x_i) - f(y_i)| < \epsilon \ .$$

When f is absolutely continuous on $[a,b]$, its derivative f' is defined almost everywhere on $[a,b]$. Also, it is the indefinite integral of its derivative:

$$f(x) - f(a) = \int_a^x f'(u) \, du \qquad (a \leq x \leq b) \ .$$

See for example Royden (1968). Thus, Lipschitz functions are absolutely continuous. And if f is a density on $[0,\infty)$ with distribution function F, then F is absolutely continuous,

$$F(x) = \int_0^x f(u) \, du \ ,$$

and

$$F'(x) = f(x) \text{ almost everywhere .}$$

A density f is called monotone on $[0,\infty)$ (or, in short, monotone) when f is nonincreasing on $[0,\infty)$ and f vanishes on $(-\infty,0)$. However, it is possible that $\lim_{x \downarrow 0} f(x)=\infty$.

Theorem 6.2.

Let X be a random variable with a monotone density f. Then

$$\lim_{x \to \infty} xf(x) = \lim_{x \downarrow 0} xf(x) = 0 .$$

If f is absolutely continuous on all closed intervals of $(0,\infty)$, then f' exists almost everywhere,

$$f(x) = -\int_x^\infty f'(u) \, du ,$$

and X is distributed as UY where U is a uniform $[0,1]$ random variable, and Y is independent of U and has density

$$g(x) = -xf'(x) \quad (x > 0) .$$

Proof of Theorem 6.2.

Assume that $\limsup_{x \to \infty} xf(x) \geq 2a > 0$. Then there exists a subsequence $x_1 < x_2 < \cdots$ such that $x_{i+1} \geq 2x_i$ and $x_i f(x_i) \geq a > 0$ for all i. But

$$1 = \int_0^\infty f(x) \, dx \geq \sum_{i=1}^\infty (x_{i+1}-x_i)f(x_{i+1}) \geq \sum_{i=1}^\infty \frac{1}{2}x_{i+1}f(x_{i+1}) = \infty ,$$

which is a contradiction. Thus, $\lim_{x \to \infty} xf(x)=0$.

Assume next that $\limsup_{x \downarrow 0} xf(x) \geq 2a > 0$. Then we can find $x_1 > x_2 > \cdots$ such that $x_{i+1} \leq \dfrac{x_i}{2}$ and $x_i f(x_i) \geq a > 0$ for all i. Again, a contradiction is obtained:

$$1 = \int_0^\infty f(x) \, dx \geq \sum_{i=1}^\infty (x_i-x_{i+1})f(x_i) \geq \sum_{i=1}^\infty \frac{1}{2}x_i f(x_i) = \infty .$$

Thus, $\lim_{x \downarrow 0} xf(x)=0$. This brings us to the last part of the Theorem. The first two statements are trivially true by the properties of absolutely continuous functions. Next we show that g is a density. Clearly, $f' \leq 0$ almost everywhere. Also, xf is absolutely continuous on all closed intervals of $(0,\infty)$. Thus, for $0 < a < b < \infty$, we have

$$bf(b)-af(a) = \int_a^b f(x)\ dx + \int_a^b xf'(x)\ dx\ .$$

By the first part of this Theorem, the left-hand-side of this equation tends to 0 as $a \downarrow 0, b \to \infty$. By the monotone convergence theorem, the right-hand side tends to $1 + \int_0^\infty xf'(x)\,dx$, which proves that g is indeed a density. Finally, if Y has density g, then UY has density

$$\int_x^\infty \frac{g(u)}{u}\ du = -\int_x^\infty f'(u)\ du = f(x)\ .$$

This proves the last part of the Theorem. ∎

The extra condition on f in Theorem 6.2 is needed because some monotone densities have $f'=0$ almost everywhere (think of staircase functions). The extra condition in Theorem 6.2 not present in Khinchine's theorem essentially guarantees that the mixing Y variable has a density too. In general, Y needs to have distribution function

$$1 - xf(x) - \int_x^\infty f(u)\ du \qquad (x > 0)\ .$$

(exercise 6.9). We also note that Theorem 6.2 has an obvious extension to unimodal densities.

For monotone f that are absolutely continuous on all closed intervals of $(0,\infty)$, the following generator is thus valid:

Generator for monotone densities based on Khinchine's theorem

Generate a uniform $[0,1]$ random variate U.
Generate a random variate Y with density $g(x) = -xf'(x), x > 0$.
RETURN $X \leftarrow UY$

Example 6.1. The exponential power distribution (EPD).

Subbotin (1923) introduced the following symmetric unimodal densities:

$$f(x) = (2\Gamma(1+\frac{1}{\tau}))^{-1} e^{-|x|^\tau},$$

where $\tau > 0$ is a parameter. This class contains the normal ($\tau = 2$) and Laplace ($\tau = 1$) densities, and has the uniform density as a limit ($\tau \to \infty$). By Theorem 6.2, and the symmetry in f, it is easily seen that

$$X \leftarrow VY^{\frac{1}{\tau}}$$

has the given density where V is uniformly distributed on $[-1,1]$ and Y is gamma$(1 + \frac{1}{\tau}, 1)$ distributed. In particular, a normal random variate can be obtained as $V\sqrt{2Y}$ where Y is gamma $(\frac{3}{2})$ distributed, and a Laplace random variate can be obtained as $V(E_1 + E_2)$ where E_1, E_2 are iid exponential random variates. Note also that X can be generated as $SY^{1/\tau}$ where Y is gamma $(\frac{1}{\tau})$ distributed. For direct generation from the EPD distribution by rejection, we refer to Johnson (1979). ∎

Example 6.2. The Johnson-Tietjen-Beckman family of densities.

Another still more flexible family of symmetric unimodal densities was proposed by Johnson, Tietjen and Beckman (1980):

$$f(x) = \frac{1}{2\Gamma(\alpha)} \int_{\frac{1}{x^\tau}}^{\infty} u^{\alpha - \tau - 1} e^{-u} \, du \ ,$$

where $\alpha > 0$ and $\tau > 0$ are shape parameters. An infinite peak at 0 is obtained whenever $\alpha \leq \tau$. The EPD distribution is obtained for $\alpha = \tau + 1$, and another distribution derived by Johnson and Johnson (1978) is obtained for $\tau = \frac{1}{2}$. By Theorem 6.2 and the symmetry in f, we observe that the random variable

$$X \leftarrow VY^\tau$$

has density f whenever V is uniformly distributed on $[-1,1]$ and Y is gamma (α) distributed. For the special case $\tau = 1$, the gamma-integral distribution is obtained which is discussed in exercise 6.1. ∎

Example 6.3. Simple relations between densities.

In the table below, a variety of distributional results are given that can help for the generation of some of them.

Density of Y	Density of UY (U is uniform on [0,1])
Exponential	Exponential-integral ($\int_x^{\infty} \dfrac{e^{-u}}{u}\, du$)
Gamma (2)	Exponential
Beta($2,b$)	Beta($1,b+1$)
Rayleigh ($xe^{-x^2/2}$)	$\int_x^{\infty} e^{-u^2/2}\, du$
Uniform [0,1]	$-\log(x)$
$(1+a)x^a$ ($x \in [0,1]$) ($a>0$)	$\dfrac{a+1}{a}(1-x^a)$
Maxwell ($\dfrac{x^2}{\sqrt{2\pi}}e^{-\frac{x^2}{2}}$)	Normal

■

There are a few other representation theorems in the spirit of Khinchine's theorem. For particular forms, one could consult Lux (1978) and Mikhailov (1965). For the stable distribution discussed in this section, we will need:

Theorem 6.3.

Let U be a uniform [0,1] random variable, let E be an exponential random variable, and let $g:[0,1] \to [0,\infty)$ be a given function. Then $\dfrac{E}{g(U)}$ has distribution function

$$F(x) = 1 - \int_0^1 e^{-xg(u)}\, du$$

and density

$$f(x) = \int_0^1 g(u)e^{-xg(u)}\, du \ .$$

Proof of Theorem 6.3.

For $x > 0$,

$$P\left(\frac{E}{g(U)} > x\right) = P(E > xg(U)) = E(e^{-xg(U)}) = \int_0^1 e^{-xg(u)} \, du \ .$$

The derivative with respect to x is $-f(x)$ where f is defined above. ∎

Finally, we mention a useful theorem of Mikhailov's about convolutions with exponential random variables:

Theorem 6.4. (Mikhailov, 1965)

If Y has density f and E is an exponential random variable independent of Y, then $E + Y$ has density

$$h(x) = \int_0^\infty e^{-u} f(x+u) \, du = \int_{-\infty}^x f(u) e^{u-x} \, du \ .$$

Furthermore, if g is an absolutely continuous density on $[0,\infty)$ with $g(0) = 0$ and $g + g' \geq 0$, then $X \leftarrow E + Y$ has density g where now Y has density $g + g'$, and E is still exponentially distributed.

Proof of Theorem 6.4.

The first statement is trivial. For part two, we note that $g + g'$ is indeed a density since $g + g' \geq 0$ and $\int_0^\infty (g + g') = 1$. (This follows from the fact that g is absolutely continuous and has $g(0) = 0$.) But then, by partial integration, X has density

$$\int_{-\infty}^x (h(u) + h'(u)) e^{u-x} \, du = h(x) \ . \ \blacksquare$$

6.3. The inverse-of-f method for monotone densities.

Assume that f is monotone on $[0,\infty)$ and continuous, and that its inverse f^{-1} can be computed relatively easily. Since f^{-1} itself is a monotone density, we can use the following method for generating a random variate with density f :

The inverse-of-f method for monotone densities

Generate a random variate Y with density f^{-1}.
Generate a uniform $[0,1]$ random variate U.
RETURN $X \leftarrow Uf^{-1}(Y)$

The correctness of the algorithm follows from the fact that (Y,X) is uniformly distributed under the curve of f^{-1}, and thus that (X,Y) is uniformly distributed under the curve of f .

Example 6.4.

If Y is exponentially distributed, then Ue^{-Y} has density $-\log(x)$ $(0 < x \leq 1)$ where U is uniformly distributed on $[0,1]$. But by the well-known connection between exponential and uniform distributions, we see that the product of two iid uniform $[0,1]$ random variables has density $-\log(x)$ $(0 < x \leq 1)$. ∎

Example 6.5.

If Y has density

$$f^{-1}(y) = (\log(\frac{2}{\pi y^2}))^{\frac{1}{2}} \quad (0 \leq y \leq \sqrt{\frac{\pi}{2}}) ,$$

and U is uniformly distributed on $[0,1]$, then $X \leftarrow Uf^{-1}(Y)$ has the halfnormal distribution. ∎

6.4. Convex densities.

The more we know about a density, the easier it is to generate random variates with this density. There are for example a multitude of tools available for monotone densities, ranging from very specific methods based upon Khinchine's theorem to black box or universal methods. In this section we look at an even smaller class of densities, the convex densities. We will consider the class C_+ of convex densities on $[0,\infty)$, and the class C of densities that are convex on $[0,\infty)$ and on $(-\infty,0)$. Thus, C_+ is a subclass of the monotone densities dealt with in the previous section.

Convex densities are absolutely continuous on all closed subintervals of $(0,\infty)$, and possess monotone right and left derivatives everywhere that are equal except possibly on a countable set. If the second derivative f'' exists, then f is convex if $f''\geq 0$. We will give one useful representation for convex densities.

Theorem 6.5. (Mixture of triangles)

For every $f \in C_+$, we have

$$f(x) = \int\limits_0^\infty \frac{2}{u}(1-\frac{x}{u})_+ \, dF(u) \,,$$

where F is a distribution function with $F(0)=0$ defined by:

$$F(u) = 1+\frac{u^2}{2} f'(u)-(uf(u)+\int\limits_u^\infty f) \quad (u>0) \,,$$

where f' is the right-hand derivative of f (which exists on $[0,\infty)$). If F is absolutely continuous, then it has density

$$g(u) = \frac{1}{2} uf''(u) \quad (u>0) \,.$$

Proof of Theorem 6.5.

We have to show first that if V,Y are independent random variables, where V has a triangular density $2(1-x)_+$ and Y has distribution function F, then $X \leftarrow VY$ has density f. But for $x>0$,

$$\int\limits_x^\infty f = \int\limits_x^\infty (1-\frac{x}{u})^2 \, dF(u)$$

$$= \int\limits_x^\infty dF(u)-2x\int\limits_x^\infty \frac{dF(u)}{u}+x^2\int\limits_x^\infty \frac{dF(u)}{u^2} \,,$$

$$f(x) = \int\limits_x^\infty \frac{2}{u}(1-\frac{x}{u})dF(u) = 2\int\limits_x^\infty \frac{dF(u)}{u}-2x\int\limits_x^\infty \frac{dF(u)}{u^2} \,,$$

and

$$-f'(x) = 2 \int\limits_x^\infty \frac{dF(u)}{u^2} \ .$$

In our case, it can be verified that the interchange of integrals and derivatives is allowed. Substitute the value of f' in the right-hand sides of the equalities for f and $\int\limits_x^\infty f$. Then check that

$$xf(x) + \int\limits_x^\infty f(u) \ du = \int\limits_x^\infty dF(u) + \frac{x^2}{2} f'(x)$$

and this gives us the first result. If F is absolutely continuous, then taking the derivative gives its density, $\frac{x^2}{2} f''(x)$. ■

This theorem states that for $f \in C_+$, we can use the following algorithm:

Generator for convex densities

Generate a triangular random variate V (this can be done as $\min(U_1, U_2)$ where the U_i's are iid uniform $[0,1]$ random variates).

Generate a random variate Y with distribution function $F(u) = 1 + \frac{u^2}{2} f'(u) - (uf(u) + \int\limits_u^\infty f)$ $(u > 0)$. (If F is absolutely continuous, then Y has density $\frac{x^2}{2} f''(x)$.)

RETURN $X \leftarrow VY$

6.5. Recursive methods based upon representations.

Representations of densities as integrals lead sometimes to properties of the following kind: assume that three random variables X, Y, Z have densities f, g, h which are related by the decomposition

$$g(x) = ph(x) + (1-p)f(x) \ .$$

Assume that X is distributed as $\psi(Y, U)$ for some function ψ and a uniform $[0,1]$ random variable U independent of Y (this is always the case). Then, we have with probability p, $X \approx \psi(Z,U)$ and with probability $1-p$, $X \approx \psi(\psi(Y',U'),U)$ where (Y',U') is another pair distributed as (Y,U). (The notation $\approx$ is ued for "is distributed as".) This process can be repeated until we reach a substitution by Z. We assume that Z has an easy density h. Notice that we never need to actually generate from g! Formally, we have , starting with Z:

Recursive generator

Generate a random variate Z with density h, and a uniform [0,1] random variate U.
$X \leftarrow \psi(Z,U)$
REPEAT
      Generate a uniform [0,1]random variate V.
      IF $V \leq p$
            THEN RETURN X
            ELSE
                  Generate a uniform [0,1] random variate U.
                  $X \leftarrow \psi(X,U)$
UNTIL False

The expected number of iterations in the REPEAT loop is $\frac{1}{p}$ because the number of V-variates needed is geometrically distributed with parameter p. This algorithm can be fine-tuned in most applications by discovering how uniform variates can be re-used.

Let us illustrate how this can help us. We know that for the gamma density with parameter $a \in (0,1)$,

$$f(x) = \frac{x^{a-1}e^{-x}}{\Gamma(a)} \quad (x > 0) :$$

$$g(x) = -xf'(x) = ah(x)+(1-a)f(x) ,$$

where h is the gamma $(a+1)$ density. This is a convenient decomposition since the parameter of h is greater than one. Also, we know that a gamma (a) random variate is distributed as UY where U is a uniform [0,1] random variate and Y has density $-xf'(x)$ (apply Theorem 6.2). Recall that we have seen several fast gamma generators for $a \geq 1$ but none that was uniformly fast over all a. The previous recursive algorithm would boil down to generating X as

$$Z \prod_{i=1}^{L} U_i$$

where Z is gamma $(a+1)$ distributed, L is geometric with parameter a, and the U_i's are iid uniform [0,1] random variates. Note that this in turn is distributed as Ze^{-G_L} where G_L is a gamma (L) random variate. But the density of G_L is

$$\sum_{i=1}^{\infty} a\,(1-a\,)^{i-1}\frac{x^{i-1}e^{-x}}{(i-1)!}=e^{-ax}\qquad(x>0)\ .$$

Thus, we have shown that the following generator is valid:

A gamma generator for a < 1

Generate a gamma $(a+1)$ random variate Z .

Generate an exponential random variate E .

RETURN $X \leftarrow Ze^{-\frac{E}{a}}$

The recursive algorithm does not require exponentiation, but the expected number of iterations before halting is $\frac{1}{a}$, and this is not uniformly bounded over $(0,1)$. The algorithm based upon the decomposition as $Ze^{-\frac{E}{a}}$ on the other hand is uniformly fast.

Example 6.6. Stuart's theorem.

Without knowing it, we have proved a special case of a theorem of Stuart's (Stuart, 1962): If Z is gamma (a) distributed, and Y is beta $(b,a-b)$ distributed and independent of Z, then $ZY,Z(1-Y)$ are independent gamma (b) and gamma$(a-b)$ random variables. If we put $b=1$, and formally replace a by $a+1$ then it is clear that $ZU^{\frac{1}{a}}$ is gamma (a) distributed, where U is a uniform $[0,1]$ random variable. ∎

There are other simple examples. The von Neumann exponential generator is also based upon a recursive relationship. It is true that an exponential random variate E is with probability $1-\frac{1}{e}$ distributed as a truncated exponential random variate (on $[0,1]$) , and that E is with probability $\frac{1}{e}$ distributed as $1+E$. This recursive rule leads precisely to the exponential generator of section IV.2.

6.6. A representation for the stable distribution.

The standardized stable distribution is best defined in terms of its characteristic function ϕ:

$$
\log\phi(t) = \begin{cases} -\mid t\mid^{\alpha}e^{-i\frac{\pi}{2}\bar{\alpha}\delta\,\mathrm{sgn}(t)} & (\alpha\neq1) \\[2mm] -\mid t\mid(1+i\;\delta\frac{2}{\pi}\mathrm{sgn}(t)\log(\mid t\mid)) & (\alpha=1) \end{cases}
$$

Here $\delta\in[-1,1]$ and $\alpha\in(0,2]$ are the shape parameters of the stable distribution, and $\bar{\alpha}$ is defined by $\min(\alpha,2-\alpha)$. We omit the location and scale parameters in this standard form. To save space, we will say that X is stable(α,δ) when it has the above mentioned characteristic function. This form of the characteristic function is due to Zolotarev (1959). By far the most important subclass is the class of symmetric stable distributions which have $\delta=0$: their characteristic function is simply

$$
\phi(t) = e^{-\mid t\mid^{\alpha}}\,.
$$

Despite the simplicity of this characteristic function, it is quite difficult to obtain useful expressions for the corresponding density except perhaps in the special cases $\alpha=2$ (the normal density) and $\alpha=1$ (the Cauchy density). Thus, it would be convenient if we could generate stable random variates without having to compute the density or distribution function at any point. There are two useful representations that will enable us to apply Theorem 6.4 with a slight modification. These will be given below.

Theorem 6.6. (Ibragimov and Chernin, 1959; Kanter, 1975)

For $\alpha<1$, the density of a stable$(\alpha,1)$ random variable can be written as

$$
f(x) = \frac{\alpha x^{\frac{1}{\alpha-1}}}{(1-\alpha)\pi}\int_{0}^{\pi} g(u)e^{-g(u)x^{\frac{\alpha}{\alpha-1}}}\,du\;,
$$

where

$$
g(u) = (\frac{\sin(\alpha u)}{\sin(u)})^{\frac{1}{1-\alpha}}\;\frac{\sin((1-\alpha)u)}{\sin(\alpha u)}\,.
$$

When U is uniformly distributed on $[0,1]$ and E is independent of U and exponentially distributed, then

$$
(\frac{g(\pi U)}{E})^{\frac{1-\alpha}{\alpha}}
$$

is stable$(\alpha,1)$ distributed.

Proof of Theorem 6.6.

For the first statement, we refer to Ibragimov and Chernin (1959). The latter statement is an observation of Kanter's (1975) which is quite easily verified by computing the distribution function of $\left(\dfrac{g(\pi U)}{E}\right)^{\frac{1-\alpha}{\alpha}}$, and noting that it is equal to

$$\frac{1}{\pi}\int_{0}^{\pi} e^{-g(u)x^{\frac{\alpha}{\alpha-1}}}\, du \ .$$

Taking the derivative gives us the density f . ∎

The second part of the proof uses a slight extension of Theorem 6.4. This representation allows us to generate stable(α,1) random variates quite easily - in most computer languages, one line of computer code will suffice! There are two problems however. First, we are stuck with the evaluation of several trigonometric functions and of two powers. We will see some methods of generating stable random variates that do not require such costly operations, but they are much more complicated. Our second problem is that Theorem 6.6 does not cover the case $\delta\neq1$. But this is easily taken care of by the following Lemma for which we refer to Feller (1971):

Lemma 6.1.

A. If X and Y are iid stable(α,1), then $Z\leftarrow pX-qY$ is stable(α,δ) where

$$p^{\alpha} = \sin(\frac{\pi\overline{\alpha}(1+\delta)}{2})/\sin(\pi\overline{\alpha}) \ ,$$

$$q^{\alpha} = \sin(\frac{\pi\overline{\alpha}(1-\delta)}{2})/\sin(\pi\overline{\alpha}) \ .$$

B. If X is stable($\dfrac{\alpha}{2}$,1) and N is independent of X and normally distributed, then $N\sqrt{2X}$ is stable(α,0), all $\alpha\in(0,2]$.

Using this Lemma and Theorem 6.6, we see that we can generate all stable random variates with either $\alpha<1$ or $\delta=0$. To fill the void, Chambers, Mallows and Stuck (1976) proposed to use a representation of Zolotarev's (1966):

Theorem 6.7. (Zolotarev, 1966; Chambers, Mallows and Stuck,1976)

Let E be an exponential random variable, and let U be a uniform $[-\frac{\pi}{2},\frac{\pi}{2}]$ random variable independent of E. Let further $\gamma = -\frac{\pi\delta\overline{\alpha}}{2\alpha}$. Then, for $\alpha\neq1$,

$$X \leftarrow \frac{\sin(\alpha(U-\gamma))}{(\cos U)^{\frac{1}{\alpha}}} \left(\frac{\cos(U-\alpha(U-\gamma))}{E}\right)^{\frac{1-\alpha}{\alpha}}$$

is stable(α,δ) distributed. Also,

$$X \leftarrow \frac{2}{\pi}\left((\frac{\pi}{2}+\delta U)\tan(U)-\delta\log(\frac{\pi E \cos(U)}{\pi+2\delta U})\right)$$

is stable(1,δ) distributed.

We leave the determination of the integral representation of f to the reader. It is noteworthy that Theorem 6.7 is a true extension of Theorem 6.6 (just note that for $\alpha<1,\delta=1$, we obtain $\gamma=-\frac{\pi}{2}$. There are three special cases worth noting:

(1) A stable(2,0) random variate can be generated as $\sqrt{E}\frac{\sin(2U)}{\cos(U)} = 2\sqrt{E}\sin(U)$. This is the well-known Box-Muller representation of $\sqrt{2}$ times a normal random variate (see section V.4).

(ii) A stable(1,0) random variate can be obtained as $\tan(U)$, which yields the inversion method for generating Cauchy random variates.

(iii) A stable($\frac{1}{2}$,1) random variate can be obtained as

$$\frac{1}{4E\sin^2(\frac{U}{2}-\frac{\pi}{4})} ,$$

which is distributed in turn as

$$\frac{1}{4E\cos^2(U)} ,$$

which is in turn distributed as $\frac{1}{2N^2}$ where N is normally distributed.

6.7. Densities with Polya type characteristic functions.

This section is added because it illustrates that representations offer unexpected help in many ways. It is frustrating to come across a distribution with a very simple characteristic function in one's research, and not be able to generate random variates with this characteristic function, at least not without a lot of work. But we do know of course how to generate random variates with some characteristic functions such as normal, uniform and exponential random variates. Thus, if we can find a representation of the characteristic function ϕ in terms of one of these simpler characteristic functions, then there is hope of generating random variates with characteristic function ϕ. By this process, we can take care of quite a few characteristic functions, even some for which the density is not known in a simple analytic form. This will be illustrated now for the class of Polya characteristic functions, i.e. real even continuous functions ϕ with $\phi(0)=1$, $\lim_{t\to\infty} \phi(t)=0$, convex on $(0,\infty)$. This class is important both from a practical point of view (it contains many important distributions) and from a didactical point of view. The examples that we will consider in this subsection are listed in the table below.

Characteristic function $\phi(t)$	Name
$e^{-|t|^{\alpha}}, 0<\alpha\leq 1$	Symmetric stable distribution
$\dfrac{1}{1+|t|^{\alpha}}, 0<\alpha\leq 1$	Linnik's distribution
$(1-|t|)^{\alpha}, |t|\leq 1, \alpha\geq 1$	
$1-|t|^{\alpha}, |t|\leq 1, 0<\alpha\leq 1$	

The second entry in this table is the characteristic function of a unimodal density for $\alpha\in(0,2]$ (Linnik (1962), Lukacs (1970, pp. 96-97)), yet no simple form for the density is known. We are now ready for the representation.

Theorem 6.8. (Girault, 1954; Dugue and Girault, 1955)

Every Polya characteristic function ϕ can be decomposed as follows:

$$\phi(t) = \int_0^{\infty}(1-|\frac{t}{s}|)_+ \, dF(s) \quad (t>0),$$

$$\phi(t) = -\phi(-t) \quad (t<0),$$

where F is a distribution function with $F(0)=0$ and defined by

$$F(s) = 1-\phi(s)+s\,\phi'(s) \quad (s>0).$$

Here ϕ' is the right-hand derivative of ϕ (which exists everywhere). If F is absolutely continuous, then it has density

$$g(s) = s\,\phi''(s) \quad (s>0).$$

From this, it is a minor step to conclude:

Theorem 6.9. (Devroye, 1984)

If ϕ is a Polya characteristic function, then $X \leftarrow \dfrac{Y}{Z}$ has this characteristic function when Y, Z are independent random variables: Z has the distribution function F of Theorem 6.8, and Y has the Fejer-de la Vallee Poussin (or: FVP) density

$$\frac{1}{2\pi} \left(\frac{\sin(\frac{x}{2})}{\frac{x}{2}} \right)^2 .$$

Theorem 6.9 uses Theorem 6.8 and the fact that the FVP density has characteristic function $(1-|t|)_+$. There are but two things left to do now: first, we need to obtain a fast FVP generator because it is used for all Polya type distributions. Second, it is important to demonstrate that the distribution function F in the various examples is often quite simple and easy to handle.

Remark 6.1. A generator for the Fejer-de la Vallee Poussin density.

Notice that if X has density

$$\frac{1}{\pi} \left(\frac{\sin(x)}{x} \right)^2 ,$$

then $2X$ has the FVP density. In view of the oscillating behavior of this density, it is best to proceed by the rejection method or the series method We note first that $\sin(x)$ is bounded from above and below by consecutive terms in the series expansion

$$\sin(x) = x - \frac{1}{3!} x^3 + \frac{1}{5!} x^5 - \cdots ,$$

and that it s bounded in absolute value by 1. Thus, the density f of X is bounded as follows:

$$f(x) \leq \frac{4}{\pi} h(x) ,$$

where $h(x) = \min(\frac{1}{4}, \frac{1}{4x^2})$, which is the density of V^B, where V is a uniform $[-1,1]$ random variable, and B is ± 1 with equal probability. The rejection constant of $\dfrac{4}{\pi}$ in this inequality is usually quite acceptable. Thus, we have:

FVP generator based upon rejection

REPEAT
        Generate iid uniform $[-1,1]$ random variates U,X.
        IF $U<0$
            THEN

$$X \leftarrow \frac{1}{X}$$

            Accept $\leftarrow [\,|U| \leq \sin^2(X)]$
        ELSE Accept $\leftarrow [\,|U| X^2 \leq \sin^2(X)]$
        UNTIL Accept
        RETURN $2X$

The expected time can be reduced by the judicious use of squeeze steps. First, if $|X|$ is outside the range $[0,\dfrac{\pi}{2}]$, it can always be reduced to a value within that range (as far as the value of $\sin^2(X)$ is concerned). Then there are two cases:

(i)   If $|X| \leq \dfrac{\pi}{4}$, we can use

$$X - \frac{X^3}{6} \leq \sin(X) \leq X \ .$$

(ii) If $|X| \in (\dfrac{\pi}{4},\dfrac{\pi}{2}]$, then we can use the fact that $\sin(X) = \cos(\dfrac{\pi}{2}-X) = \cos(Y)$, where Y now is in the range of (i). The following inequalities will be helpful:

$$1 - \frac{Y^2}{2} \leq \sin(X) \leq 1 - \frac{Y^2}{2} + \frac{Y^4}{24} \ . \ \blacksquare$$

Example 6.7. The symmetric stable distribution.

In Theorem 6.9, Z has density g given by

$$g(s) = (\alpha^2 s^{2\alpha-1} + \alpha(1-\alpha)s^{\alpha-1})e^{-s} \qquad (s>0) \ .$$

But we note that Z^α has density

$$\alpha(se^{-s}) + (1-\alpha)(e^{-s}) \qquad (s>0) \ ,$$

which is a mixture of a gamma (2) and an exponential density. Thus, Z is distributed as

$$(E_1 + E_2 I_{[U<\alpha]})^{\frac{1}{\alpha}}$$

where E_1, E_2 and U are independent random variables: E_1 and E_2 have an exponential density, and U is uniformly distributed on $[0,1]$. It is also worth observing that if we use $U_1,...$ for iid uniform $[0,1]$ random variables, then Z is distributed as

$$(E_1 + \max(E_2 + \log(\alpha),0))^{\frac{1}{\alpha}}$$

and as

$$\log^{\frac{1}{\alpha}}(\max(\frac{\alpha}{U_1 U_2},\frac{1}{U_1})) \cdot \blacksquare$$

Example 6.8. Linnik's distribution

We verify that Z in Theorem 6.9 has density g given by

$$g(s) = ((\alpha^2+\alpha)s^{2\alpha-1}+(\alpha-\alpha^2)s^{\alpha-1})(1+s^\alpha)^{-3} \quad (s>0) .$$

It is perhaps easier to work with the density of Z^α:

$$\frac{s(\alpha+1)+(1-\alpha)}{(1+s)^3} \quad (s>0) .$$

The latter density has distribution function $1-\dfrac{1+\alpha}{1+s}+\dfrac{\alpha}{(1+s)^2}$, and this is easy to invert. Thus, a random variate Z can be generated as

$$(\frac{\alpha+1-\sqrt{(\alpha+1)^2-4\alpha U}}{2U}-1)^{\frac{1}{\alpha}} ,$$

where U is a uniform $[0,1]$ random variate. If speed is extremely important, the square root can be avoided if we use the rejection method for the density of Z^α, with dominating density $(1+s)^{-2}$, which is the density of $\dfrac{1}{U}-1$. A little work shows that Z can be generated as follows:

REPEAT

    Generate iid uniform [0,1] random variates U,V.

$$X \leftarrow \frac{1}{U}-1$$

    UNTIL $2\alpha U \leq V$ (Now, X is distributed as Z^{α}.)

RETURN $X^{\frac{1}{\alpha}}$

The expected number of iterations is $1+\alpha$. ■

Example 6.9. Other examples.

Assume that $\phi(t) = (1-|t|)_{+}^{\alpha}$ for $\alpha > 1$. Then $\phi(s)-s\,\phi'(s)$ is absolutely continuous. Thus, the random variable Z of Theorem 6.9 has beta $(2,\alpha-1)$ density $g(s)=\alpha(\alpha-1)s(1-s)^{\alpha-2}$ $(0 \leq s \leq 1)$.

There are situations in which the distribution function F of Theorems 6.8 and 6.9 is not absolutely continuous. To illustrate this, take $\phi(t)=(1-|t|^{\alpha})_{+}$, and note that $F(s) = (1-\alpha)s^{\alpha}$ $(0 \leq s \leq 1)$. Also, $F(1)=1$. Thus, F has an atom of weight α at 1, and it has an absolutely continuous part of weight $1-\alpha$ with support on $(0,1)$. The absolutely continuous part has density $\alpha s^{\alpha-1}$ $(0 \leq s \leq 1)$, which is the density of $U^{\frac{1}{\alpha}}$ where U is uniform on $[0,1]$. Thus,

$$Z = \begin{cases} 1 & \text{with probability } \alpha \\ U^{\frac{1}{\alpha}} & \text{with probability } 1-\alpha \end{cases}.$$

Here we can use the standard trick of recuperating part of the uniform [0,1] random variate used to make the "with probability α" choice. ■

6.8. Exercises.

1. **The gamma-integral distribution.** We say that X is GI(a) (has the gamma-integral distribution with parameter $a > 0$) when its density is

$$f(x) = \int_x^\infty \frac{u^{a-2} e^{-u}}{\Gamma(a)} \, du \quad (x > 0).$$

This distribution has a few remarkable properties: it decreases monotonically on $[0, \infty)$. It has an infinite peak at 0 when $a \leq 1$. At $a = 1$, we obtain the exponential-integral density. When $a > 1$, we have $f(0) = \dfrac{1}{a-1}$. For $a = 2$, the exponential density is obtained. When $a > 2$, there is a point of inflection at $a - 2$, and $f'(0) = 0$. For $a = 3$, the distribution is very close to the normal distribution. In this exercise we are mainly interested in random variate generation. Show the following:

A. X can be generated as UY where U is uniformly distributed on $[0,1]$ and Y is gamma (a) distributed.

B. When a is integer, X is distributed as G_Z where Z is uniformly distributed on $1, \ldots, a-1$, and G_Z is a gamma (Z) random variate. Note that X is distributed as $-\log(U_1 \cdots U_Z)$ where the U_i's are iid uniform $[0,1]$ random variates. Hint: use induction on a.

C. As $a \to \infty$, $\dfrac{X}{a}$ tends in distribution to the uniform $[0,1]$ density.

D   Compute all moments of the GI(a) distribution. (Hint: use Khinchine's theorem.)

2. The density of the energy spectrum of fission neutrons is

$$f(x) = \frac{1}{\sqrt{\pi a b}} e^{-(a+x)b} \sinh\left(\frac{2\sqrt{ax}}{b}\right) \quad (x > 0),$$

where $a, b > 0$ are parameters. Recall that $\sinh(x) = \dfrac{1}{2}(e^x - e^{-x})$. Apply Theorem 6.4 for designing a generator for this distribution(Mikhailov, 1965).

3. How would you compute $f(x)$ with seven digits of accuracy for the exponential-integral density of Example 6.3? Prove also that for the same distribution, $F(x) = (1 - e^{-x}) + x f(x)$ where F is the distribution function.

4. If U, V are iid uniform $[0,1]$ random variables, then for $0 < a < 1$, $UV^{\frac{1}{1-a}}$ has density $x^{-a} - 1$ $(0 < x < 1)$.

5. In the next three exercises, we consider the following class of monotone densities on $[0,1]$:

$$f(x) = \frac{\Gamma(a+b+1)}{\Gamma(a+1)\Gamma(b+1)} (1 - x^{\frac{1}{b}})^a \quad (0 \leq x \leq 1),$$

where $a, b > 0$ are parameters. The coefficient will be called B. The mode of the density occurs at $x = 0$, and $f(0) = B$. Show the following:

A. f is convex if and only if $a,b \geq 1$. It is concave if and only if $a,b \leq 1$.

B. Y^b has density f , where Y is beta $(b,a+1)$ distributed.

C. $(\dfrac{Y}{Y+Z})^b$ has density f where Y is gamma (b) distributed, and Z is gamma $(a+1)$ distributed and independent of Y .

6.   This is a continuation of exercise 5 for the special case $b=1$. The density is $f(x)=(a+1)(1-x)^a$ $(0 \leq x \leq 1)$. From the previous exercise we recall that a random variate with this distribution can be obtained as $1-U^{\frac{1}{a+1}}$ and as $\dfrac{E}{E+G_{a+1}}$ where U is a uniform [0,1] random variate, E is an exponential random variate, and G_{a+1} is a gamma $(a+1)$ random variate independent of E . Both these methods require costly operations. The following rejection algorithms are usually faster:

Rejection method #1, recommended for a > 1

REPEAT
 REPEAT
  Generate two iid exponential random variates, E_1,E_2.
  $X \leftarrow \dfrac{E_1}{a}$
 UNTIL $X \leq 1$
 Accept $\leftarrow [E_2(1-X)-aX^2 \geq 0]$
 IF NOT Accept THEN Accept $\leftarrow [aX+E_2+a\log(1-X) \geq 0]$
UNTIL Accept
RETURN X

Rejection method #2, recommended for a < 1

REPEAT
 Generate two iid uniform [0,1] random variates, U,X .
UNTIL $U \leq (1-X)^a$
RETURN X

Show that the rejection algorithms are valid. Show furthermore that the expected number of iterations is $\dfrac{a+1}{a}$ and $a+1$ respectively. (Thus, a uniformly fast algorithm can be obtained by using the first method for $a \geq 1$

and the second method for $a < 1$.)

7. Continuation of exercise 5 for $b = \dfrac{1}{2}$. The density we are considering here can be written as follows:

$$f(x) = B(1-x^2)^a \qquad (0 \le x \le 1).$$

(Here $B = \dfrac{2}{\sqrt{\pi}} \dfrac{\Gamma(a + \frac{3}{2})}{\Gamma(a+1)}$.) From exercise 5 we recall that a random variate with this density can be generated as $\dfrac{N}{\sqrt{N^2 + 2G_{a+1}}}$ where N is a normal random variate, and G_{a+1} is a gamma $(a+1)$ random variate independent of N.

A. Show that we can also use $|2Y-1|$ where Y is beta $(a+1, a+1)$ distributed.

B. Show that if we keep generating iid uniform $[0,1]$ random variates U, X until $U \le (1-X^2)^a$, then X has density f, the expected number of iterations is B, and B increases monotonically from 1 $(a=0)$ to ∞ $(a \to \infty)$.

C. Show that the following rejection algorithm is valid and has rejection constant $\dfrac{\Gamma(a + \frac{3}{2})}{\sqrt{a}\, \Gamma(a+1)}$ (which tends monotonically to 1 as $a \to \infty$):

Rejection from a normal density

REPEAT

Generate independent normal and exponential random variates N, E.

$X \leftarrow \dfrac{|N|}{\sqrt{2a}}$, $Y \leftarrow X^2$

Accept $\leftarrow [Y \le 1] \text{AND} [1 - Y(1 + \frac{aY}{E}) \ge 0]$

IF NOT Accept THEN Accept $\leftarrow [Y \le 1]$ AND $[aY + E + a \log(1-Y) \ge 0]$

UNTIL Accept

RETURN X

Hint: use the inequalities $-\dfrac{x}{1-x} \le \log(1-x) \le -x$ $(0 < x < 1)$.

8. **The exponential power distribution.** Show that if S is a random sign, and $G_{\frac{1}{\tau}}$ is a gamma $(\frac{1}{\tau})$ random variate, then $S(G_{\frac{1}{\tau}})^{\frac{1}{\tau}}$ has the exponential

power distribution with parameter τ, that is, its density is of the form $ce^{-|x|^\tau}$ where c is a normalization constant.

9. Extend Theorem 6.2 by showing that for all monotone densities, it suffices to take Y with distribution function

$$F(x) = 1 - \int_x^\infty f(u)\,du - xf(x) \quad (x \in R).$$

10. Extend Theorem 6.5 to all convex densities in C.

11. **The Pareto distribution.** Let E, Y be independent random variables, where E is exponentially distributed, and Y has density g on $[0,\infty)$. Give an integral form for the density and distribution function of $X = E/Y$. Random variables of this type are called exponential scale mixtures. Show that when Y is gamma (a), then $1 + E/Y$ is Pareto with parameter a, i.e. $1 + E/Y$ has density a/x^{a+1} $(x > 1)$ (see e.g. Harris, 1968).

12. Develop a uniformly fast generator for the family of densities

$$f(x) = C_n \left(\frac{\sin(x)}{x} \right)^n,$$

where $n \geq 1$ is an integer parameter, and C_n is a constant depending upon n only.

7. THE RATIO-OF-UNIFORMS METHOD.

7.1. Introduction.

The rejection method has one big drawback: densities with infinite tails have to be handled with care; often, tails have to be cut off and treated separately. In many cases, this can be avoided if the ratio-of-uniforms method is used. This method is particularly well suited for bell-shaped densities with tails that decrease at least as fast as x^{-2}. The ratio-of-uniforms method was first proposed by Kinderman and Monahan (1977), and later applied to a variety of distributions such as the t distribution (Kinderman and Monahan,1979) and the gamma distribution (Cheng and Feast, 1979).

Because the resulting algorithms are short and often fast, and because we have yet another beautiful illustration of the rejection and squeeze principles, we will devote quite a bit of space to this method. The treatment will be systematic and simple: we are not looking for the most general form of algorithm but for one that is easy to understand.

We begin with

Theorem 7.1. (Kinderman and Monahan, 1977)

Let $A = \{(u, v) : 0 \leq u \leq \sqrt{f(\frac{v}{u})}\}$ where $f \geq 0$ is an integrable function. If (U, V) is a random vector uniformly distributed over A, then $\frac{V}{U}$ has density $\frac{1}{c} f$ where $c = \int f = 2$ area (A).

Proof of Theorem 7.1.

Define (X, Y) by $X = U, Y = \frac{V}{U}$. The Jacobian of the transformation $u = x, v = xy$ is x. The density of (U, V) is $I_A(u, v)/(c/2)$. Thus, the density of (X, Y) is x times $I_A(x, yx)/(c/2) = xI_{[0, f(y)]}(x)/(c/2)$. The density of $Y = \frac{V}{U}$ is the marginal density computed as

$$\int_0^{\sqrt{y}} \frac{x}{(c/2)} dx = \frac{f(y)}{c} . \blacksquare$$

But we already know how to generate uniformly distributed random vectors: it suffices to enclose the area A by a simple set such as a rectangle, in which we know how to generate uniform random vectors, and to apply the rejection principle. Thus, it is important to verify what A looks like in general. First, A is a subset of $[0, \infty) \times R$. It is symmetric about the u-axis if f is symmetric about 0. It vanishes in the negative v-quadrant when f is the density of a nonnegative random variable. But what interests us more than anything else are conditions insuring that $A \subseteq [0, b) \times [a_-, a_+]$ for some finite constants $b \geq 0, a_- \leq 0, a_+ \geq 0$. It helps to note that the boundary of A can be found parametrically by $\{(u(x), v(x)) : x \in R\}$ where

$$u(x) = \sqrt{f(x)},$$
$$v(x) = x\sqrt{f(x)}.$$

Thus, A can be enclosed in a rectangle if and only if

(i) $f(x)$ is bounded;
(ii) $x^2 f(x)$ is bounded.

Basically, this includes all bounded densities with subquadratic tails, such as the normal, gamma, beta, t and exponential densities. From now on, the enclosing rectangle will be called $B = [0, b) \times [a_-, a_+]$. For the sake of simplicity, we will only treat densities satisfying (i) and (ii) in this section.

The ratio-of-uniforms method

[SET-UP]

Compute b, a_-, a_+ for an enclosing rectangle. Note that
$b \geq \sup \sqrt{f(x)}, a_- \leq \inf x \sqrt{f(x)}, a_+ \geq \sup x \sqrt{f(x)}$.

[GENERATOR]

REPEAT

    Generate U uniformly on $[0, b]$, and V uniformly on $[a_-, a_+]$.

$$X \leftarrow \frac{V}{U}$$

UNTIL $U^2 \leq f(X)$

RETURN X

By Theorem II.3.2, (U, V) is uniformly distributed in A. Thus, the algorithm is valid, i.e. X has density proportional to the function f. We can also replace f by cf for any constant c. This allows us to eliminate all annoying normalization constants. In any case, the expected number of iterations is

$$\frac{b(a_+ - a_-)}{\text{area } A} = \frac{2b(a_+ - a_-)}{\int\limits_{-\infty}^{\infty} f(x) \, dx}.$$

This will be called the rejection constant. Good densities are densities in which A fills up most of its enclosing rectangle. As we will see from the examples, this is usually the case when f puts most of its mass near zero and has monotonically decreasing tails. Roughly speaking, most bell-shaped f are acceptable candidates.

The acceptance condition $U^2 \leq f(X)$ cannot be simplified by using logarithmic transformations as we sometimes did in the rejection method - this is because U is explicitly needed in the definition of X. The next best thing is to make sure that we can avoid computing f most of the time. This can be done by introducing one or more quick acceptance and quick rejection steps. Typically, the algorithm takes the following form.

The ratio-of-uniforms method with two-sided squeezing

[SET-UP]

Compute b, a_-, a_+ for an enclosing rectangle. Note that
$b \geq \sup \sqrt{f(x)}, a_- \leq \inf x \sqrt{f(x)}, a_+ \geq \sup x \sqrt{f(x)}$.

[GENERATOR]

REPEAT

    Generate U uniformly on $[0,b]$, and V uniformly on $[a_-, a_+]$.

    $X \leftarrow \dfrac{V}{U}$

    IF [Quick acceptance condition]

        THEN Accept ← True

        ELSE IF [Quick rejection condition]

            THEN Accept ← False

            ELSE Accept ← [Acceptance condition ($U^2 \leq f(X)$)]

UNTIL Accept

RETURN X

In the next sub-section, we will give various quick acceptance and quick rejection conditions for the distributions listed in this introduction, and analyze the performance for these examples.

7.2. Several examples.

We will need various inequalities in the design of squeeze steps. The following Lemma can be useful in this respect.

Lemma 7.1.

(1) $-x \geq \log(1-x) \geq -\dfrac{x}{1-x}$ $(0 \leq x < 1)$.

(11) $-x - \dfrac{x^2}{2} \geq \log(1-x)$

$\qquad \geq -x - \dfrac{x^2}{2(1-x)}$ $(0 \leq x < 1)$.

(111) $\log(x) \leq x - 1$ $(x > 0)$.

(1v) $x - \dfrac{x^2}{2} \leq \log(1+x)$

$\qquad \leq x - \dfrac{x^2}{2} + \dfrac{x^3}{3} \leq x$ $(0 < x < 1)$.

(v) $\dfrac{2x + 3x^2}{2(1+x)^2} \leq \log(1+x)$

$\qquad \leq \dfrac{2x + 3x^2 + x^3}{2(1+x)^2}$

$\qquad = x - \dfrac{x^2}{2(1+x)}$ $(x \geq 0)$.

(v1)   Reverse the inequalities in (v) when $-1 < x \leq 0$.

Proof of Lemma 7.1.

Parts (1) through (1v) were obtained in Lemma IV.3.2. By the Taylor series for $g(x) = (1+x)\log(1+x)$, we see that

$$g(x) = g(0) + xg'(0) + \dfrac{x^2}{2}g''(\xi)$$

for   some ξ between   0   and x .   But
$g(0) = 0, g'(x) = \log(1+x) - 1, g'(0) = 1, g''(x) = \dfrac{1}{1+x}$. Thus, for $x > 0$,

$$x + \dfrac{x^2}{2(1+x)} \leq g(x) \leq x + \dfrac{x^2}{2} .$$

This proves (v) and (v1). ∎

For various densities, we list quick acceptance and rejection conditions in terms of u, v, x. When used in the algorithm, these running variables should be replaced by the random variates U, V, X of course. Other useful quantities such as the rejection constant and values for b, a_-, a_+ are listed too.

Example 7.1. The normal density.

All of the above is summarized in the table given below:

$f(x)$	$e^{-\frac{x^2}{2}} \quad (x \in R)$
$b = \sup \sqrt{f(x)}$	1
$a_+ = \sup x \sqrt{f(x)}, a_- = \inf x \sqrt{f(x)}$	$\sqrt{\dfrac{2}{e}}, -\sqrt{\dfrac{2}{e}}$
area (A)	$\sqrt{\dfrac{\pi}{2}}$
Rejection constant	$\dfrac{4}{\sqrt{\pi e}}$
Acceptance condition	$x^2 \leq -4\log u$
Quick acceptance condition	$x^2 \leq 4(-cu+1+\log c) \quad (c>0)$
	$x^2 \leq 4-4u$
	$x^2 \leq 6-8u+2u^2$
	$x^2 \leq \dfrac{44}{6}-12u+6u^2-\dfrac{4}{3}u^3$
Quick rejection condition	$x^2 > 4(\dfrac{c}{u}-1-\log c) \quad (c>0)$
	$x^2 \geq \dfrac{4}{u}-4$
	$x^2 \geq \dfrac{2}{u}-2u$

The table is nearly self-explanatory. The quick acceptance and rejection conditions were obtained from the acceptance condition and Lemma 7.1. Most of these are rather straightforward. The fastest experimental results were obtained with the third entries in both lists. It is worth pointing out that the first quick acceptance and rejection conditions are valid for all constants $c>0$ introduced in the conditions, by using inequalities for $\log(uc)$ given in Lemma 7.1. The parameter c should be chosen so that the area under the quick acceptance curve is maximal, and the area under the quick rejection curve is minimal. ∎

Example 7.2. The exponential density.

In analogy with the normal density, we present the following table.

$f(x)$	e^{-x} $(x \in R)$
$b = \sup \sqrt{f(x)}$	1
$a_+ = \sup x \sqrt{f(x)}, a_- = \inf x \sqrt{f(x)}$	$\dfrac{2}{e}, 0$
area (A)	$\dfrac{2}{e}$
Rejection constant	$\dfrac{4}{e}$
Acceptance condition	$x \leq -2\log u$
Quick acceptance condition	$x \leq 2(1-u)$
Quick rejection condition	$x \geq \dfrac{2}{u} - 2$
	$x \geq \dfrac{2}{eu} - \dfrac{(u - \frac{1}{e})^2}{u}$

It is insightful to draw A and to construct simple quick acceptance and rejection conditions by examining the shape of A. Since A is convex, several linear functions could be useful. ∎

Example 7.3. The t distribution.

The ratio-of-uniforms method has led to some of the fastest known algorithms for the t distribution. In this section, we omit, as we can, the normalization constant of the t density with parameter a, which is

$$\frac{\Gamma(\frac{a+1}{2})}{\sqrt{\pi a}\ \Gamma(\frac{a}{2})} .$$

Since for large values of a, the t density is close to the normal density, we would expect that the performance of the algorithm would be similar too. This is indeed the case. For example, as $a \to \infty$, the rejection constant tends to $\dfrac{4}{\sqrt{\pi e}}$, which is the value for the normal density.

$f(x)$	$\dfrac{1}{(1+\dfrac{x^2}{a})^{\frac{a+1}{2}}}$ $(x\in R)$
$b=\sup\sqrt{f(x)}$	1
$a_+=\sup x\sqrt{f(x)}, a_-=\inf x\sqrt{f(x)}$	$\dfrac{\sqrt{2a}\,(a-1)^{\frac{a-1}{4}}}{(a+1)^{\frac{a+1}{4}}},\ \dfrac{\sqrt{2a}\,(a-1)^{\frac{a-1}{4}}}{(a+1)^{\frac{a+1}{4}}}$
area (A)	$2\dfrac{\sqrt{2a}\,(a-1)^{\frac{a-1}{4}}}{(a+1)^{\frac{a+1}{4}}}$
Rejection constant	$4\dfrac{\sqrt{2a}\,(a-1)^{\frac{a-1}{4}}}{(a+1)^{\frac{a+1}{4}}}\dfrac{\Gamma(\frac{a+1}{2})}{\sqrt{\pi a}\,\Gamma(\frac{a}{2})}$
Acceptance condition	$x^2\le a(u^{-\frac{4}{a+1}}-1)$
Quick acceptance condition	$x^2\le 5-4u(1+\frac{1}{a})^{\frac{a+1}{4}}$
Quick rejection condition	$x^2\ge -3+\dfrac{4}{u}(1+\frac{1}{a})^{-\frac{a+1}{4}}$ (only valid for $a\ge3$)

We observe that the ratio-of-uniforms method can only be useful when $a\ge1$ for otherwise A would be unbounded. The quick acceptance and rejection steps follow from inequalities obtained by Kinderman and Monahan (1979). The corresponding algorithm is known in the literature as algorithm TROU: one can show that the expected number of iterations is uniformly bounded over $a\ge1$, and that it varies from $\dfrac{4}{\pi}$ at $a=1$ to $\dfrac{4}{\sqrt{\pi e}}$ as $a\to\infty$.

There are two important special cases. For the Cauchy density $(a=1)$, the acceptance condition is $u^2\le\dfrac{1}{1+x^2}$, or, put differently, $u^2+v^2\le1$. Thus, we obtain the result that if (U,V) is uniformly distributed in the unit circle, then $\dfrac{V}{U}$ is Cauchy distributed. Without squeeze steps, we have:

A Cauchy generator based upon the ratio-of-uniforms method

REPEAT
    Generate iid uniform $[-1,1]$ random variates U,V.
UNTIL $U^2+V^2\le1$
RETURN $X\leftarrow\dfrac{V}{U}$

For the t density with 3 degrees of freedom $(a = 3)$,

$$\frac{2}{\pi\sqrt{3}} \frac{1}{(1+\frac{x^2}{3})^2} \, ,$$

the acceptance condition is $\frac{x^2}{3} \leq \frac{1}{u}-1$, or $v^2 \leq 3u(1-u)$. Thus, once again, the acceptance region A is ellipsoidal. The unadorned ratio-of-uniforms algorithm is:

t3 generator based upon ratio-of-uniforms method

REPEAT
>   Generate U uniformly on $[0,1]$.
>   Generate V uniformly on $[-\frac{\sqrt{3}}{2}, \frac{\sqrt{3}}{2}]$.
>
> UNTIL $V^2 \leq 3U(1-U)$
>
> RETURN $X \leftarrow \frac{V}{U}$

This is equivalent to

t3 generator based upon ratio-of-uniforms method

REPEAT
>   Generate iid uniform $[-1,1]$ random variates U, V.
>
> UNTIL $U^2 + V^2 \leq 1$
>
> RETURN $X \leftarrow \sqrt{3} \frac{V}{1+U}$

Both the Cauchy and $t3$ generators have obviously rejection constants of $\frac{4}{\pi}$, and should be accelerated by the judicious use of quick acceptance and rejection conditions that are linear in their arguments. ■

Example 7.4. The gamma density.

In this example, we consider the centered gamma (a) density with mode at the origin,

$$f(x) = c\frac{e^{a-1}}{(a-1)^{a-1}}(x+a-1)^{a-1}e^{-(x+a-1)} \qquad (x+a-1\geq 0) .$$

Here c is a normalization constant equal to $\dfrac{(a-1)^{a-1}}{e^{a-1}\Gamma(a)}$ which will be dropped. The table with facts is given below. Notice that the expected number of iterations is $\dfrac{4}{e}$ at $a=1$, and $\dfrac{4}{\sqrt{\pi e}}$ as $a\to\infty$, just as for the t density.

$f(x)$	$\frac{e^{a-1}}{(a-1)^{a-1}}(x+a-1)^{a-1}e^{-(x+a-1)}$ $\quad(x+a-1\geq 0)$
$b=\sup\sqrt{f(x)}$	1
$a_+=\sup x\sqrt{f(x)}, a_-=\inf x\sqrt{f(x)}$	$z_+\sqrt{f(z_+)}$ where $z_+=1+\sqrt{2a-1}$, $z_-\sqrt{f(z_-)}$ where $z_-=1-\sqrt{2a-1}$
area (A)	a_+-a_-
Rejection constant	$2c(a_+-a_-)$
Acceptance condition	$u \leq (\frac{e(x+a-1)}{a-1})^{\frac{a-1}{2}}e^{-\frac{x+a-1}{2}}$
	$2\log u + x \leq (a-1)\log(1+\frac{x}{a-1})$
Quick acceptance condition	$(x+a-1)^2(-2u^2+8u-6) \leq -x^2(2x+a-1)\;(x\geq 0)$
	$(x+a-1)(-2u^2+8u-6) \leq -x^2\;(x\leq 0)$
Quick rejection condition	$(x+a-1)(2u^2-2) \geq -ux^2\;(x\geq 0)$
	$(a-1)(2u^2-2) \geq -ux^2\;(x\leq 0)$

We leave the verification of the inequalities implicit in the quick acceptance and rejection steps to the readers. All one needs here is Lemma 7.1. Timings with this algorithm have shown that good speeds are obtained for a greater than 5. The algorithm is uniformly fast for $a\in[1,\infty)$. The ratio-of-uniforms algorithms of Cheng and Feast (1979), Robertson and Walls (1980) and Kinderman and Monahan (1979) are different in conception. ■

7.3. Exercises.

1. For the quick acceptance and rejection conditions for Student's t distribution, the following inequality due to Kinderman and Monahan (1979) was used:

$$5-4(1+\frac{1}{a})^{\frac{a+1}{4}}u \leq a(u^{-\frac{4}{a+1}}-1) \leq -3+\frac{4(1+\frac{1}{a})^{-\frac{a+1}{4}}}{u} \qquad (u\geq 0) .$$

The upper bound is only valid for $a \geq 3$. Show this. Hint: first show that the middle expression $g(u)$ is convex in u. Thus,

$$g(u) \geq g(z) + (u-z)g'(z) .$$

Here z is to be picked later. Show that the area under the quick acceptance curve is maximal when $z = (1 + \frac{1}{a})^{-\frac{a+1}{4}}$, and substitute this value. For the lower bound, show that $g(u)$ as a function of $\frac{1}{u}$ is concave, and argue similarly.

2. Barbu (1982) has pointed out that when (U,V) is uniformly distributed in $A = \{(u,v) : 0 \leq u \leq f(u+v)\}$, then $U+V$ has a density which is proportional to f. Similarly, if in the definition of A, we replace $f(u+v)$ by $(f(\frac{v}{\sqrt{u}}))^{\frac{2}{3}}$, then $\frac{V}{\sqrt{U}}$ has a density which is proportional to f. Show this.

3. Prove the following property. Let X have density f and define $Y = \sqrt{f(X)}\max(U_1, U_2)$ where U_1, U_2 are iid uniform [0,1] random variables. Define also $U = Y, V = XY$. Then (U,V) is uniformly distributed in $A = \{(u,v) : 0 \leq u \leq \sqrt{f(\frac{v}{u})}\}$. Note that this can be useful for rejection in the (u,v) plane when rectangular rejection is not feasible.

4. In this exercise, we study sufficient conditions for convergence of performances. Assume that f_n is a sequence of densities converging in some sense to a density f as $n \to \infty$. Let b_n, a_{+n}, a_{-n} be the defining constants for the enclosing rectangles in the ratio-of-uniforms method. Let b, a_+, a_- be the constants for f. Show that the rejection constants converge, i.e.

$$\lim_{n \to \infty} b_n(a_{+n} - a_{-n}) = b(a_+ - a_-)$$

when

$$\sup_x \left| \frac{f_n(x)}{f(x)} - 1 \right| = o(1) ,$$

or when

$$\sup_x x^2 |f_n(x) - f(x)| = o(1) .$$

5. Give an example of a bounded density on $[0,\infty)$ for which the region A is unbounded in the v-direction, i.e. $b = \infty$.

6. Let f be a mixture of nonoverlapping uniform densities of varying widths and heights. Draw the region A.

7. From general principles (such as exercise 4), prove that the rejection constant for the t distribution tends to the rejection constant for the normal density as $a \to \infty$.

8.   Prove that all the quick acceptance and rejection inequalities used for the
      gamma density are valid.

Chapter Five
UNIFORM AND EXPONENTIAL SPACINGS

1. MOTIVATION.

The goal of this book is to demonstrate that random variates with various distributions can be obtained by cleverly manipulating iid uniform [0,1] random variates. As we will see in this chapter, normal, exponential, beta, gamma and t distributed random variates can be obtained by manipulation of the order statistics or spacings defined by samples of iid uniform [0,1] random variates. For example, the celebrated polar method or Box-Muller method for normal random variates will be derived in this manner (Box and Muller, 1958).

There is a strong relationship between these spacings and radially symmetric distributions in R^d, so that with a little extra effort we will be able to handle the problem of generating uniform random variates in and on the unit sphere of R^d.

The polar method can also be considered as a special case of a more general method, the method of deconvolution. Because of this close relationship it will also be presented in this chapter.

We start with the fundamental properties of uniform order statistics and uniform spacings. This material is well-known and can be found in many books on probability theory and mathematical statistics. It is collected here for the convenience of the readers. In the other sections, we will develop various algorithms for univariate and multivariate distributions. Because order statistics and spacings involve sorting random variates, we will have a short section on fast expected time sorting methods. Just as chapter IV, this chapter is highly specialized, and can be skipped too. Nevertheless, it is recommended for new students in the fields of simulation and mathematical statistics.

2. PROPERTIES OF UNIFORM AND EXPONENTIAL SPACINGS.

2.1. Uniform spacings.

Let $U_1, \ldots, U_n$ be iid uniform $[0,1]$ random variables with order statistics $U_{(1)} \leq U_{(2)} \leq \cdots \leq U_{(n)}$. The statistics S_i defined by

$$S_i = U_{(i)} - U_{(i-1)} \qquad (1 \leq i \leq n+1)$$

where by convention $U_{(0)} = 0$, $U_{(n+1)} = 1$, are called the uniform spacings for this sample.

Theorem 2.1.

$(S_1, \ldots, S_n)$ is uniformly distributed over the simplex

$$A_n = \{(x_1, \ldots, x_n) : x_i \geq 0, \sum_{i=1}^{n} x_i \leq 1\}.$$

Proof of Theorem 2.1.

We know that $U_{(1)}, \ldots, U_{(n)}$ is uniformly distributed over the simplex

$$B_n = \{(x_1, \ldots, x_n) : 0 \leq x_1 \leq \cdots \leq x_n \leq 1\}.$$

The transformation

$$s_1 = u_1$$
$$s_2 = u_2 - u_1$$
$$\cdots$$
$$s_n = u_n - u_{n-1}$$

has as inverse

$$u_1 = s_1$$
$$u_2 = s_1 + s_2$$
$$\cdots$$
$$u_n = s_1 + s_2 + \cdots + s_n$$

and the Jacobian of the transformation, i.e. the determinant of the matrix formed by $\dfrac{\partial s_i}{\partial u_j}$ is 1. This shows that the density of $S_1, \ldots, S_n$ is uniformly distributed on the set A_n. ∎

Proofs of this sort can often be obtained without the cumbersome transformations. For example, when X has the uniform density on a set $A \subseteq R^d$, and B is a linear nonsingular transformation: $R^d \to R^d$, then $Y = BX$ is uniformly distributed on BA as can be seen from the following argument: for all Borel sets $C \subseteq R^d$,

$$P(Y \in C) = P(BX \in C) = P(X \in B^{-1}C)$$

$$= \frac{\int\limits_{(B^{-1}C) \cap A} dx}{\int\limits_A dx} = \frac{\int\limits_{C \cap (BA)} dx}{\int\limits_{BA} dx} .$$

Theorem 2.2.

$S_1, \ldots, S_{n+1}$ is distributed as

$$\frac{E_1}{\sum\limits_{i=1}^{n+1} E_i}, \ldots, \frac{E_{n+1}}{\sum\limits_{i=1}^{n+1} E_i}$$

where $E_1, \ldots, E_{n+1}$ is a sequence of iid exponential random variables. Furthermore, if G_{n+1} is independent of $(S_1, \ldots, S_{n+1})$ and is gamma $(n+1)$ distributed, then

$$S_1 G_{n+1}, \ldots, S_{n+1} G_{n+1}$$

is distributed as $E_1, E_2, \ldots, E_{n+1}$.

The proof of Theorem 2.2 is based upon Lemma 2.1:

Lemma 2.1.

For any sequence of nonnegative numbers $x_1, \ldots, x_{n+1}$, we have

$$P(S_1 > x_1, \ldots, S_{n+1} > x_{n+1}) = \left(1 - \sum_{i=1}^{n+1} x_i \right)_+^n .$$

Proof of Lemma 2.1.

Assume without loss of generality that $\sum\limits_{i=1}^{n+1} x_i \leq 1$ (for otherwise the lemma is obviously true). We use Theorem 2.1. In the notation of Theorem 2.1, we start from the fact that $S_1, \ldots, S_n$ is uniformly distributed in A_n. Thus, our probability is equal to

$$P(S_1 > x_1, \ldots, S_n > x_n, 1 - \sum_{i=1}^n S_i > x_{n+1}) .$$

This is the probability of a set A_n^* which is a simplex just as A_n except that its top is not at $(0, 0, \ldots, 0)$ but rather at $(x_1, \ldots, x_n)$, and that its sides are not of length 1 but rather of length $1 - \sum_{i=1}^{n+1} x_i$. For uniform distributions, probabilities can be calculated as ratios of areas. In this case, we have

$$\frac{\int_{A_n^*} dx}{\int_{A_n} dx} = (1 - \sum_{i=1}^{n+1} x_i)^n \quad . \blacksquare$$

Proof of Theorem 2.2.

Part one. Let $G = G_{n+1}$ be the random variable $\sum_{i=1}^{n+1} E_i$. Note that we need only show that

$$\frac{E_1}{G}, \ldots, \frac{E_n}{G}$$

is uniformly distributed in A_n. The last component $\frac{E_{n+1}}{G}$ is taken care of by noting that it equals 1 minus the sum of the first n components. Let us use the symbols e_i, y, x_i for the running variables corresponding to $E_i, G, \frac{E_i}{G}$. We first compute the joint density of $E_1, \ldots, E_n, G$:

$$f(e_1, \ldots, e_n, y) = \prod_{i=1}^{n} e^{-e_i} \; e^{-(y - e_1 - \cdots - e_n)} = e^{-y} \; ,$$

valid when $e_i \geq 0$, all i, and $y \geq \sum_{i=1}^{n} e_i$. Here we used the fact that the joint density is the product of the density of the first n variables and the density of G given $E_1 = e_1, \ldots, E_n = e_n$. Next, by a simple transformation of variables, it is easily seen that the joint density of $\frac{E_1}{G}, \ldots, \frac{E_n}{G}, G$ is

$$y^n f(x_1 y, \ldots, x_n y, y) = y^n e^{-y} \quad (x_i y \geq 0, \; \sum_{i=1}^{n} x_i y \leq y) \; .$$

This is easily obtained by the transformation $\{x_1 = \frac{e_1}{y}, \ldots, x_n = \frac{e_n}{y}, y = y\}$. Finally, the marginal density of $\frac{E_1}{G}, \ldots, \frac{E_n}{G}$ is obtained by integrating the last density with respect to dy, which gives us

$$\int_0^\infty y^n e^{-y} \; dy \; I_{A_n}(x_1, \ldots x_n) = n! \; I_{A_n}(x_1, \ldots, x_n) \; .$$

This concludes the proof of part one.

Part two. Assume that $x_1 \geq 0, \ldots, x_{n+1} \geq 0$. By Lemma 2.1, we have

$$P(GS_1 > x_1, \ldots, GS_{n+1} > x_{n+1})$$

$$= \int_0^\infty P\left(S_1 > \frac{x_1}{y}, \ldots, S_{n+1} > \frac{x_{n+1}}{y} \mid G = y\right) \frac{y^n e^{-y}}{n!} dy$$

$$= \int_{y: \sum \frac{x_i}{y} \leq 1} \left(1 - \sum_{i=1}^{n+1} \frac{x_i}{y}\right)^n \frac{y^n e^{-y}}{n!} dy$$

$$= \int_c^\infty (y - c)^n \frac{e^{-y}}{n!} dy \quad \text{(where } c = \sum_{i=1}^{n+1} x_i)$$

$$= e^{-c}$$

$$= \prod_{i=1}^{n+1} e^{-x_i} \cdot \blacksquare$$

A myriad of results follow from Theorem 2.2. For example, if $U, U_1, \ldots, U_n$ are iid uniform $[0,1]$ random variables, E is an exponential random variable, and G_n is a gamma (n) random variable, then the following random variables have identical distributions:

$$\min(U_1, \ldots, U_n)$$

$$1 - U^{\frac{1}{n}}$$

$$1 - e^{-\frac{E}{n}}$$

$$\frac{E}{E + G_n} \quad (E, G_n \text{ are independent })$$

$$\left(\frac{E}{n}\right) - \frac{1}{2!}\left(\frac{E}{n}\right)^2 + \frac{1}{3!}\left(\frac{E}{n}\right)^3 - \cdots .$$

It is also easy to show that $\dfrac{\max(U_1, \ldots, U_n)}{\min(U_1, \ldots, U_n)}$ is distributed as $1 + \dfrac{G_{n-1}}{E}$, that $\max(U_1, \ldots, U_n) - \min(U_1, \ldots, U_n)$ is distributed as $1 - S_1 - S_{n+1}$ (i.e. as $\dfrac{G_{n-1}}{G_{n-1} + G_2}$), and that $U_{(k)}$ is distributed as $\dfrac{G_k}{G_k + G_{n+1-k}}$ where G_k and G_{n+1-k} are independent. Since we already know from section I.4 that $U_{(k)}$ is beta $(k, n+1-k)$ distributed, we have thus obtained a well-known relationship between the gamma and beta distributions.

2.2. Exponential spacings.

In this section, $E_{(1)} \leq E_{(2)} \leq \cdots \leq E_{(n)}$ are the order statistics corresponding to a sequence of iid exponential random variables $E_1, E_2, \ldots, E_n$.

Theorem 2.3. (Sukhatme, 1937)

If we define $E_{(0)} = 0$, then the normalized exponential spacings

$$(n-i+1)(E_{(i)} - E_{(i-1)}), \ 1 \leq i \leq n,$$

are iid exponential random variables. Also,

$$\frac{E_1}{n}, \ \frac{E_1}{n} + \frac{E_2}{n-1}, \ \ldots, \ \frac{E_1}{n} + \cdots + \frac{E_n}{1}$$

are distributed as $E_{(1)}, \ldots, E_{(n)}$.

Proof of Theorem 2.3.

The second statement follows from the first statement: it suffices to call the random variables of the first statement $E_1, E_2, \ldots, E_n$ and to note that

$$E_{(1)} = \frac{E_1}{n},$$

$$E_{(2)} = E_{(1)} + \frac{E_2}{n-1},$$

$$\cdots$$

$$E_{(n)} = E_{(n-1)} + \frac{E_n}{1}.$$

To prove the first statement, we note that the joint density of $E_{(1)}, \ldots, E_{(n)}$ is

$$n! \, e^{-\sum\limits_{i=1}^{n} x_i} \quad (0 \leq x_1 \leq x_2 \leq \cdots \leq x_n < \infty)$$

$$= n! \, e^{-\sum\limits_{i=1}^{n}(n-i+1)(x_i - x_{i-1})} \quad (0 \leq x_1 \leq x_2 \leq \cdots \leq x_n < \infty).$$

Define now $Y_i = (n-i+1)(E_{(i)} - E_{(i-1)})$, $y_i = (n-i+1)(x_i - x_{i-1})$. Thus, we have

$$x_1 = \frac{y_1}{n},$$

$$x_2 = \frac{y_1}{n} + \frac{y_2}{n-1},$$

$$\cdots$$

$$x_n = \frac{y_1}{n} + \cdots + \frac{y_n}{1}.$$

The determinant of the matrix formed by $\dfrac{\partial x_i}{\partial y_j}$ is $\dfrac{1}{n!}$. Thus, $Y_1, \ldots, Y_n$ has density

$$e^{-\sum_{i=1}^{n} y_i} \qquad (y_i \geq 0 \text{ ,all } i) \text{ ,}$$

which was to be shown. ■

Theorem 2.3 has an important corollary: in a sample of two iid exponential random variates, $E_{(2)}-E_{(1)}$ is again exponentially distributed. This is basically due to the memoryless property of the exponential distribution: given that $E \geq x$, $E-x$ is again exponentially distributed. In fact, if we show the memoryless property (this is easy), and if we show that the minimum of n iid exponential random variables is distributed as $\dfrac{E}{n}$ (this is easy too), then we can prove Theorem 2.3 by induction.

Theorem 2.4. (Malmquist, 1950)

Let $0 \leq U_{(1)} \leq \cdots \leq U_{(n)} \leq 1$ be the order statistics of $U_1, U_2, \ldots, U_n$, a sequence of iid uniform $[0,1]$ random variables. Then , if $U_{(n+1)}=1$,

A. $\{(\dfrac{U_{(i)}}{U_{(i+1)}})^i , 1 \leq i \leq n\}$ is distributed as $U_1, \ldots, U_n$.

B. $U_n^{\frac{1}{n}}, U_n^{\frac{1}{n}} U_{n-1}^{\frac{1}{n-1}}, \ldots, U_n^{\frac{1}{n}} \cdots U_1^{\frac{1}{1}}$ is distributed as $U_{(n)}, \ldots, U_{(1)}$.

Proof of Theorem 2.4.

In Theorem 2.3, replace U_i by e^{-E_i} and $U_{(i)}$ by $e^{-E_{(n-i+1)}}$. Then, in the notation of Theorems 2.3 and 2.4 we see that the following sequences are identically distributed:

$$(\frac{U_{(i)}}{U_{(i+1)}})^i , 1 \leq i \leq n \text{ ,}$$

$$(e^{-E_{(n-i+1)}+E_{(n-i)}})^i , 1 \leq i \leq n \text{ ,}$$

$$e^{-E_i} , 1 \leq i \leq n \text{ ,}$$

$$U_i , 1 \leq i \leq n \text{ .}$$

This proves part A. Part B follows without work from part A. ■

2.3. Exercises.

1. Give an alternative proof of Theorem 2.3 based upon the memoryless property of the exponential distribution (see suggestion following the proof of that theorem).

2. Prove that in a sample of n iid uniform $[0,1]$ random variates, the maximum minus the minimum (i.e., the range) is distributed as

$$U^{\frac{1}{n}} V^{\frac{1}{n-1}}$$

where U, V are iid uniform $[0,1]$ random variates.

3. Show that the minimum spacing in a uniform sample of size n is distributed as $\dfrac{1}{n+1}(1-U^{\frac{1}{n}})$ where U itself is uniformly distributed on $[0,1]$.

4. Prove or disprove: $\dfrac{U}{U+V}$ is uniformly distributed on $[0,1]$ when U, V are iid uniform $[0,1]$ random variables.

5. Prove Whitworth's formula: if S_i , $1 \le i \le n+1$ are uniform spacings, then

$$P(\max_i S_i \ge x) = \binom{n}{1}(1-x)_+ - \binom{n}{2}(1-2x)_+^2 + \cdots .$$

(Whitworth, 1897)

6. Let E_1, E_2, E_3 be iid exponential random variables. Show that the following random variables are independent: $\dfrac{E_1}{E_1+E_2}$, $\dfrac{(E_1+E_2)}{E_1+E_2+E_3}$, $E_1+E_2+E_3$. Furthermore, show that their densities are the uniform $[0,1]$ density, the triangular density on $[0,1]$ and the gamma (3) density, respectively.

3. GENERATION OF ORDERED SAMPLES.

The first application that one thinks of when presented with Theorem 2.2 is a method for generating the order statistics $U_{(1)} \le \cdots \le U_{(n)}$ directly. By this we mean that it is not necessary to generate $U_1, \ldots, U_n$ and then apply some sorting method.

In this section we will describe several problems which require such ordered samples. We will not be concerned here with the problem of the generation of one order statistic such as the maximum or the median.

3.1. Generating uniform [0,1] order statistics.

The previous sections all suggest methods for generating uniform [0,1] order statistics:

A. Sorting

Generate iid uniform [0,1] random variates $U_1, \ldots, U_n$.
Obtain $U_{(1)}, \ldots, U_{(n)}$ by sorting the U_i's.

B. Via uniform spacings

Generate iid exponential random variates $E_1, \ldots, E_{n+1}$, and compute their sum G.
$U_{(0)} \leftarrow 0$
FOR $j := 1$ TO n DO
$$U_{(j)} \leftarrow U_{(j-1)} + \frac{E_j}{G}$$

C. Via exponential spacings

$U_{(n+1)} \leftarrow 1$
FOR $j := n$ DOWNTO 1 DO
Generate a uniform [0,1] random variate U.
$$U_{(j)} \leftarrow U^{\frac{1}{j}} U_{(j+1)}$$

Algorithm A is the naive approach. Sorting methods usually found in computer libraries are comparison-based. This means that information is moved around in tables based upon pairwise comparisons of elements only. It is known (see e.g. Knuth (1973) or Baase (1978)) that the worst-case and expected times taken by these algorithms are $\Omega(n \log n)$. Heapsort and mergesort have worst-case times that are $O(n \log n)$. Quicksort has expected time $O(n \log n)$, but worst-case time both $O(n^2)$ and $\Omega(n^2)$. For details, any standard textbook on data structures can be consulted (see e.g. Aho, Hopcroft and Ullman , 1983). What is different in the present case is that the U_i's are uniformly distributed on [0,1]. Thus, we can hope to take advantage of truncation. As we will see in the next section, we can bucket sort the U_i's in expected time $O(n)$.

Algorithms B and C are $O(n)$ algorithms in the worst-case. But only method C is a one-pass method. But because method C requires the computation of a power in each iteration, it is usually slower than either A or B. Storagewise, method A is least efficient since additional storage proportional to n is needed. Nevertheless, for large n, method A with bucket sorting is recommended. This is due to the accumulation of round-off errors in algorithms B and C.

Algorithms B and C were developed in a series of papers by Lurie and Hartley (1972), Schucany (1972) and Lurie and Mason (1973). Experimental comparisons can be found in Rabinowitz and Berenson (1974), Gerontides and Smith (1982), and Bentley and Saxe (1980). Ramberg and Tadikamalla (1978) consider the case of the generation of $U_{(k)}, U_{(k+1)}, \ldots, U_{(m)}$ where $1 \leq k \leq m \leq n$. This requires generating one of the extremes $U_{(k)}$ or $U_{(m)}$, after which a sequential method similar to algorithms B or C can be used, so that the total time is proportional to $m - k + 1$.

3.2. Bucket sorting. Bucket searching.

We start with the description of a data structure and an algorithm for sorting n [0,1] valued elements $X_1, \ldots, X_n$.

Bucket sorting

[SET-UP]

We need two auxiliary tables of size n called Top and Next. Top $[i]$ gives the index of the top element in bucket i (i.e. $[\frac{i-1}{n}, \frac{i}{n})$). A value of 0 indicates an empty bucket. Next $[i]$ gives the index of the next element in the same bucket as X_i. If there is no next element, its value is 0.

FOR $i := 1$ TO n DO Next $[i] \leftarrow 0$

FOR $i := 0$ TO $n - 1$ DO Top $[i] \leftarrow 0$

FOR $i := 1$ TO n DO

    Bucket $\leftarrow \left\lfloor nX_i \right\rfloor$

    Next $[i] \leftarrow$ Top [Bucket]

    Top [Bucket] $\leftarrow i$

[SORTING]

Sort all elements within the buckets by ordinary bubble sort or selection sort, and concatenate the nonempty buckets.

The set-up step takes time proportional to n in all cases. The sort step is where we notice a difference between distributions. If each bucket contains one element, then this step too takes time proportional to n. If all elements on the other hand fall in the same bucket, then the time taken grows as n^2 since selection sort for that one bucket takes time proportional to n^2. Thus, for our analysis, we will have to make some assumptions about the X_i's. We will assume that the X_i's are iid with density f on $[0,1]$. In Theorem 3.1 we show that the expected time is $O(n)$ for nearly all densities f.

Theorem 3.1. (Devroye and Klincsek, 1981)

The bucket sort given above takes expected time $O(n)$ if and only if
$$\int f^2(x)\, dx < \infty.$$

Proof of Theorem 3.1.

Assume that the buckets receive $N_0, \ldots, N_{n-1}$ points. It is clear that each N_i is binomially distributed with parameters n and p_i where

$$p_i = \int_{\frac{i}{n}}^{\frac{i+1}{n}} f(x)\, dx .$$

By the properties of selection sort, we know that there exist finite positive constants c_1, c_2, such that the time T_n taken by the algorithm satisfies:

$$c_1 \le \frac{T_n}{n + \sum_{i=0}^{n-1} N_i^2} \le c_2 .$$

By Jensen's inequality for convex functions, we have

$$\sum_{i=0}^{n-1} E(N_i^2) = \sum_{i=0}^{n-1} \left(np_i(1-p_i) + n^2 p_i^2 \right)$$

$$\le \sum_{i=0}^{n-1} np_i + \sum_{i=0}^{n-1} \left(n \int_{\frac{i}{n}}^{\frac{i+1}{n}} f(x)\, dx \right)^2$$

$$\le n + \sum_{i=0}^{n-1} n \int_{\frac{i}{n}}^{\frac{i+1}{n}} f^2(x)\, dx$$

$$= n \left(1 + \int_0^1 f^2(x)\, dx \right) .$$

This proves one implication. The other implication requires some finer tools, especially if we want to avoid imposing smoothness conditions on f. The key measure theoretical result needed is the Lebesgue density theorem, which (phrased in a form suitable to us) states among other things that for any density f on R, we have

$$\lim_{n \to \infty} n \int_{x - \frac{1}{n}}^{x + \frac{1}{n}} |f(y) - f(x)| \, dy = 0 \quad \text{(for almost all } x \text{)} .$$

Consult for example Wheeden and Zygmund (1977).

If we define the density

$$f_n(x) = p_i \quad (0 \le \frac{i}{n} \le x < \frac{i+1}{n} \le 1) ,$$

then it is clear that

$$|f_n(x) - f(x)| \le \int_{\frac{i}{n}}^{\frac{i+1}{n}} |f(y) - f(x)| \, dy \quad (\frac{i}{n} \le x < \frac{i+1}{n})$$

$$\le n \int_{x - \frac{1}{n}}^{x + \frac{1}{n}} |f(y) - f(x)| \, dy ,$$

and this tends to 0 for for almost all x. Thus, by Fatou's lemma,

$$\lim \inf \int_0^1 f_n^2(x) \, dx \ge \int_0^1 \lim \inf f_n^2(x) \, dx = \int_0^1 f^2(x) \, dx .$$

But

$$\frac{1}{n} \sum_{i=0}^{n-1} E(N_i^2) \ge \sum_{i=0}^{n-1} np_i^2 = \sum_{i=0}^{n-1} \int_{\frac{i}{n}}^{\frac{i+1}{n}} f_n^2(x) \, dx = \int_0^1 f_n^2(x) \, dx .$$

Thus, $\int f^2 = \infty$ implies $\lim \inf \dfrac{T_n}{n} = \infty$. $\blacksquare$

In selection sort, the number of comparisons of two elements is $(n-1) + (n-2) + \cdots + 1 = \dfrac{n(n-1)}{2}$. Thus, the total number of comparisons needed in bucket sort is, in the notation of the proof of Theorem 3.1,

$$\sum_{i=0}^{n-1} \frac{N_i(N_i - 1)}{2} .$$

The expected number of comparisons is thus

$$\sum_{i=0}^{n-1} \frac{1}{2}(n^2 p_i{}^2 + n p_i(1-p_i) - n p_i)$$

$$= \frac{n(n-1)}{2} \sum_{i=0}^{n-1} p_i{}^2$$

$$\leq \frac{n-1}{2} \int_0^1 f^2(x)\ dx \ .$$

This upper bound is, not unexpectedly, minimized for the uniform density on [0,1], in which case we obtain the upper bound $\frac{n-1}{2}$. In other words, the expected number of comparisons is less than the total number of elements ! This is of course due to the fact that most of the sorting is done in the set-up step.

If selection sort is replaced by an $O(n \log n)$ expected time comparison-based sorting algorithm (such as quicksort, mergesort or heapsort), then Theorem 3.1 remains valid provided that the condition $\int f^2 < \infty$ is replaced by

$$\int_0^\infty f(x) \log_+ f(x)\ dx \ < \ \infty \ .$$

See Devroye and Klincsek (1981). The problem with extra space can be alleviated to some extent by clever programming tricks. These tend to slow down the algorithm and won't be discussed here.

Let us now turn to searching. The problem can be formulated as follows. [0,1]-valued data $X_1, \ldots, X_n$ are given. We assume that this is an iid sequence with common density f. Let T_n be the time taken to determine whether X_Z is in the structure where Z is a random integer taken from $\{1, \ldots, n\}$ independent of the X_i's. This is called the successful search time. The time T_n^* taken to determine whether X_{n+1} (a random variable distributed as X_1 but independent of the data sequence) is in the structure is called the unsuccessful search time. If we store the elements in an array, then linear (or sequential search) yields expected search times that are proportional to n. If we use binary search and the array is sorted, then it is proportional to $\log(n)$. Assume now that we use the bucket data structure, and that the elements within buckets are not sorted. Then, with linear search within the buckets, the expected number of comparisons of elements for successful search, given $N_0, \ldots, N_{n-1}$, is

$$\sum_{i=0}^{n-1} \frac{N_i}{n} \frac{N_i+1}{2} \ .$$

For unsuccessful search, we have

$$\sum_{i=0}^{n-1} \frac{N_i}{n} N_i \ .$$

Arguing now as in Theorem 3.1, we have:

Theorem 3.2.

When searching a bucket structure we have $E(T_n) = O(1)$ if and only if $\int f^2 < \infty$. Also, $E(T_n^*) = O(1)$ of and only if $\int f^2 < \infty$.

3.3. Generating exponential order statistics.

To generate a sorted sample of exponential random variables, there are two algorithms paralleling algorithms A and C for the uniform distribution.

A. Bucket sorting

Generate iid exponential random variates $E_1, \ldots, E_n$.

Obtain $E_{(1)} \leq \cdots \leq E_{(n)}$ by bucket sorting.

C. Via exponential spacings

$E_{(0)} \leftarrow 0$

FOR $i := 1$ TO n DO

    Generate an exponential random variate E.

    $E_{(i)} \leftarrow E_{(i-1)} + \dfrac{E}{n-i+1}$

Method C uses the memoryless property of the exponential distribution. It takes time $O(n)$. Careless bucket sorting applied to algorithm A could lead to a superlinear time algorithm. For example, this would be the case if we were to divide the interval $[0, \max E_i]$ up into n equi-sized intervals. This can of course be avoided if we first generate $U_{(1)} \leq \cdots \leq U_{(n)}$ for a uniform sample in expected time $O(n)$, and then return $-\log U_{(n)}, \ldots, -\log U_{(1)}$. Another possibility is to construct the bucket structure for $E_i \bmod 1$, $1 \leq i \leq n$, i.e. for the fractional parts only, and to sort these elements. Since the fractional parts have a bounded density,

$$\frac{e^{-x} I_{[0,1]}(x)}{1 - \dfrac{1}{e}} \ ,$$

we know from Theorem 3.1 that a sorted array can be obtained in expected time $O(n)$. But this sorted array has many sorted sub-arrays. In one extra pass, we can untangle it provided that we have kept track of the unused integer parts of the data, $\lfloor E_i \rfloor$. Concatenation of the many sub-arrays requires another pass, but we still have linear behavior.

3.4. Generating order statistics with distribution function F.

The order statistics $X_{(1)} \leq \cdots \leq X_{(n)}$ that correspond to $X_1, \ldots, X_n$, a sequence of iid random variables with absolutely continuous distribution function F on R^1 can be generated as

$$F^{-1}(U_{(1)}), \ldots, F^{-1}(U_{(n)})$$

or as

$$F^{-1}(1 - e^{-E_{(1)}}), \ldots, F^{-1}(1 - e^{-E_{(n)}})$$

starting from uniform or exponential order statistics. The exponential order statistics method based on C (see previous section) was proposed by Newby (1979). In general, the choice of one method over the other one largely depends upon the form of F. For example, for the Weibull distribution function

$$F(x) = 1 - e^{-(\frac{x}{b})^a} \qquad (x \geq 0)$$

we have $F^{-1}(u) = b(-\log(1-u))^{\frac{1}{a}}$ and $F^{-1}(1 - e^{-u}) = b u^{\frac{1}{a}}$, so that the exponential order statistics method seems better suited.

In many cases, it is much faster to just sort $X_1, \ldots, X_n$ so that the costly inversions can be avoided. If bucket sorting is used, one should make sure that the expected time is $O(n)$. This can be done for example by transforming the data in a monotone manner for the purpose of sorting to [0,1] and insuring that the density f of the transformed data has a small value for $\int f^2$. Transformations that one might consider should be simple, e.g. $\dfrac{x}{a+x}$ is useful for transforming nonnegative data. The parameter $a > 0$ is a design parameter which should be picked such that the density after transformation has the smallest possible value for $\int f^2$.

The so-called grouping method studied by Rabonowitz and Berenson (1974) and Gerontides and Smith (1982) is a hybrid of the inversion method and the bucket sorting method. The support of the distribution is partitioned into k intervals, each having equal probability. Then one keeps for each interval a linked list. Intervals are selected with equal probability, and within each interval, random points are generated directly. In a final pass, all linked lists are sorted and concatenated. The sorting and concatenating take linear expected time when

$k = n$, because the interval cardinalities are as for the bucket method in case of a uniform distribution. There are two major differences with the bucket sorting method: first of all, the determination of the intervals requires $k-1$ implicit inversions of the distribution function. This is only worthwhile when it can be done in a set-up step and very many ordered samples are needed for the same distribution and the same n (recall that k is best taken proportional to n). Secondly, we have to be able to generate random variates with a distribution restricted to these intervals. Candidates for this include the rejection method. For monotone densities or unimodal densities and large n, the rejection constant will be close to one for most intervals if rejection from uniform densities is used.

But perhaps most promising of all is the **rejection method** itself for generating an ordered sample. Assume that our density f is dominated by cg where g is another density, and $c > 1$ is the rejection constant. Then, exploiting properties of points uniformly distributed under f, we can proceed as follows:

Rejection method for generating an ordered sample

[NOTE: n is the size of the ordered sample; $m > n$ is an integer picked by the user. Its recommended value is $\left\lfloor nc + \sqrt{nc(c-1)\log\left(\dfrac{cn}{2\pi(c-1)}\right)} \right\rfloor$.]

REPEAT

        Generate an ordered sample $X_1, \ldots, X_m$ with density g.

        Generate m iid uniform [0,1] random variates $U_1, \ldots, U_m$.

        Delete all X_i's for which $U_i > cg(X_i)/f(X_i)$.

UNTIL the edited (but ordered) sample has $N \geq n$ elements

Delete another $N-n$ randomly selected X_i's from this sample, and return the edited sample.

The main loop of the algorithm, when successful, gives an ordered sample of random size $N \geq n$. This sample is further edited by one of the well-known methods of selecting a random (uniform) sample of size $N-n$ from a set of size n (see chapter XII). The expected time taken by the latter procedure is $E(N-n \mid N \geq n)$ times a constant not depending upon N or n. The expected time taken by the global algorithm is $m / P(N \geq n) + E(N-n \mid N \geq n)$ if constants are omitted, and a uniform ordered sample with density g can be generated in linear expected time.

Theorem 3.3.

Let m, n, N, f, c, g keep their meaning of the rejection algorithm defined above. Then, if $m \geq cn$ and $m = O(n)$, the algorithm takes expected time $O(n)$. If in addition $m - cn = o(n)$ and $(m - cn)/\sqrt{n} \to \infty$, then

$$T_n = \frac{m}{P(N \geq n)} + E(N - n \mid N \geq n) \sim cn$$

as $n \to \infty$.

Proof of Theorem 3.3.

In order to analyze the success probability, we need some result about the closeness between the binomial and normal distributions. First of all, since N is binomial $(m, \frac{1}{c})$, we know from the central limit theorem that as $m \to \infty$,

$$P(N < n) \sim \Phi\left(\frac{n - \frac{m}{c}}{\sqrt{m\frac{1}{c}(1 - \frac{1}{c})}}\right)$$

where Φ is the normal distribution function. If $m \geq cn$ at all times, then we see that $P(N < n)$ stays bounded away from 1, and oscillates asymptotically between 0 and 1/2. It can have a limit. If $(m - cn)/\sqrt{n} \to \infty$, then we see that $P(N < n) \to 0$.

We note that $E(N - n \mid N \geq n) = E((N - n)_+)/P(N \geq n)$. Since $N - n \leq m - n$, we see that $T_n \leq (2m - n)/P(N \geq n)$. The bound is $O(n)$ when $m = O(n)$ and $P(N \geq n)$ is bounded away from zero. Also, $T_n \sim cn$ when $P(N < n) \to 0$ and $m \sim cn$. ∎

Remark 3.1. Optimal choice of m.

The best possible value for T_n is cn because we cannot hope to accept n points with large enough probability of success unless the original sample is at least of size cn. It is fortunate that we need not take m much larger than cn. Detailed computations are needed to obtain the following recommendation for m: take m close to

$$nc + \sqrt{nc(c-1)\log\left(\frac{cn}{2\pi(c-1)}\right)}.$$

With this choice, T_n is $cn + O(\sqrt{n \log(n)})$. See exercise 3.7 for guidance with the derivation. ∎

3.5. Generating exponential random variates in batches.

By Theorem 2.2, iid exponential random variates $E_1, \ldots, E_n$ can be generated as follows:

Exponential random variate generator

Generate an ordered sample $U_{(1)} \leq \cdots \leq U_{(n-1)}$ of uniform $[0,1]$ random variates.
Generate a gamma (n) random variate G_n.
RETURN $(G_n U_{(1)}, G_n (U_{(2)} - U_{(1)}), \ldots, G_n (1 - U_{(n-1)}))$.

Thus, one gamma variate (which we are able to generate in expected time $O(1)$) and a sorted uniform sample of size $n-1$ are all that is needed to obtain an iid sequence of n exponential random variates. Thus, the contribution of the gamma generator to the total time is asymptotically negligible. Also, the sorting can be done extremely quickly by bucket sort if we have a large number of buckets (exercise 3.1), so that for good implementations of bucket sorting, a super-efficient exponential random variate generator can be obtained. Note however that by taking differences of numbers that are close to each other, we loose some accuracy. For very large n, this method is not recommended.

One special case is worth mentioning here: UG_2 and $(1-U)G_2$ are iid exponential random variates.

3.6. Exercises.

1. In bucket sorting, assume that instead of n buckets, we take kn buckets where $k \geq 1$ is an integer. Analyze how the expected time is affected by the choice of k. Note that there is a time component for the set-up which increases as kn. The time component due to selection sort within the buckets is a decreasing function of k and f. Determine the asymptotically optimal value of k as a function of $\int f^2$ and of the relative weights of the two time components.

2. Prove the claim that if an $O(n \log n)$ expected time comparison-based sorting algorithm is used within buckets, then $\int_0^1 f \log_+ f < \infty$ implies that the

expected time is $O(n)$.

3. Show that $\int f \log_+ f < \infty$ implies $\int f^2 < \infty$ for any density f. Give an example of a density f on $[0,1]$ for which $\int f \log_+ f < \infty$, yet $\int f^2 = \infty$. Give also an example for which $\int f \log_+ f = \infty$.

4. The randomness in the time taken by bucket sorting and bucket searching can be appropriately measured by $\sum_{i=0}^{n-1} N_i^2$, a quantity that we shall call T_n. It is often good to know that T_n does not become very large with high probability. For example, we may wish to obtain good upper bounds for $P(T_n > E(T_n) + \alpha)$, where $\alpha > 0$ is a constant. For example, obtain bounds that decrease exponentially fast in n for all bounded densities on $[0,1]$ and all $\alpha > 0$. Hint: use an exponential version of Chebyshev's inequality and a Poissonization trick for the sample size.

5. Give an $O(n)$ expected time generator for the maximal uniform spacing in a sample of size n. Then give an $O(1)$ expected time generator for the same problem.

6. If a density f can be decomposed as $pf_1 + (1-p)f_2$ where f_1, f_2 are densities and $p \in [0,1]$ is a constant, then an ordered sample $X_{(1)} \leq \cdots \leq X_{(n)}$ of f can be generated as follows:

Generate a binomial (n, p) random variate N.

Generate the order statistics $Y_{(1)} \leq \cdots \leq Y_{(N)}$ and $Z_{(1)} \leq \cdots \leq Z_{(n-N)}$ for densities f_1 and f_2 respectively.

Merge the sorted tables into a sorted table $X_{(1)} \leq \cdots \leq X_{(n)}$.

The acceleration is due to the fact that the method based upon inversion of F is sometimes simple for f_1 and f_2 but not for f; and that n coin flips needed for selection in the mixture are avoided. Of course, we need a binomial random variate. Here is the question: based upon this decomposition method, derive an efficient algorithm for generating an ordered sample from any monotone density on $[0,\infty)$.

7. This is about the optimal choice for m in Theorem 3.3 (the rejection method for generating an ordered sample). The purpose is to find an m such that for that choice of m, $T_n - cn \sim \inf_m (T_n - cn)$ as $n \to \infty$. Proceed as follows: first show that it suffices to consider only those m for which $T_n \sim cn$. This implies that $E((N-n)_+) = o(m-cn)$, $P(N < n) \to 0$, and $(m-cn)/\sqrt{n} \to \infty$. Then deduce that for the optimal m,

$$T_n = cn\left(1 + (1 + o(1))\left(\frac{m-cn}{cn} + P(N < n)\right)\right).$$

Clearly, $m \sim cn$, and $(m-cn)/cn$ is a term which decreases much slower than $1/\sqrt{n}$. By the Berry-Esseen theorem (Chow and Teicher (1978, p. 299) or Petrov (1975)), find a constant C depending upon c only such that

$$\left| P(N < n) - \Phi\left(\frac{n - \dfrac{m}{c}}{\sqrt{\dfrac{m}{c}\left(1 - \dfrac{1}{c}\right)}} \right) \right| \le \frac{C}{\sqrt{n}} .$$

Conclude that it suffices to find the m which minimizes

$$(m - cn)/(cn) + \Phi\left(\frac{n - \dfrac{m}{c}}{\sqrt{\dfrac{m}{c}\left(1 - \dfrac{1}{c}\right)}} \right) . \text{ Next, using the fact that as } u \to \infty,$$

$$1 - \Phi(u) \sim \frac{1}{u\sqrt{2\pi}} e^{-\frac{u^2}{2}} ,$$

reduce the problem to that of minimizing

$$\rho\sqrt{\frac{c-1}{cn}} + \frac{1}{\rho\sqrt{2\pi}} e^{-\frac{\rho^2}{2}} ,$$

where $m - cn = \rho\sqrt{c(c-1)n}$ for some $\rho \to \infty$, $\rho = o(\sqrt{n})$. Approximate asymptotic minimization of this yields

$$\rho = \sqrt{\log\left[\frac{cn}{2\pi(c-1)} \right]} .$$

Finally, verify that for the corresponding value for m, the minimal value of T_n is asymptotically obtained (in the " $\sim$ " sense).

4. THE POLAR METHOD.

4.1. Radially symmetric distributions.

Here we will explain about the intimate connection between order statistics and random vectors with radially symmetric distributions in R^d. This connection will provide us with a wealth of algorithms for random variate generation. Most importantly, we will obtain the time-honored Box-Muller method for the normal distribution.

A random vector $X = (X_1, \ldots, X_d)$ in R^d is **radially symmetric** if AX is distributed as X for all orthonormal $d \times d$ matrices A. It is strictly radially symmetric if also $P(X = 0) = 0$. Noting that AX corresponds to a rotated version of X, radial symmetry is thus nothing else but invariance under rotations of the coordinate axes. We write C_d for the unit sphere in R^d. X is uniformly distributed on C_d when X is radially symmetric and $||X||=1$ with probability one. Here $||.||$ is the standard L_2 norm. Sometimes, a radially symmetric random vector has a density f, and then necessarily it is of the form

$$f(x_1, \ldots, x_d) = g(||x||) \quad (x=(x_1, \ldots, x_d)\in R^d)$$

for some function g. This function g on $[0,\infty)$ is such that

$$\int_0^\infty dV_d r^{d-1} g(r)\, dr = 1,$$

where

$$V_d = \frac{\pi^{\frac{d}{2}}}{\Gamma(\frac{d}{2}+1)}$$

is the volume of the unit sphere C_d. We say that g defines or determines the radial density. Elliptical radial symmetry is not be treated in this early chapter, nor do we specifically address the problem of multivariate random variate generation. For a bibliography on radial symmetry, see Chmielewski (1981). For the fundamental properties of radial distributions not given below, see for example Kelker (1970).

Theorem 4.1. (Uniform distributions on the unit sphere.)

1. If X is strictly radially symmetric, then $\dfrac{X}{||X||}$ is uniformly distributed on C_d.

2. If X is uniformly distributed on C_d, then $(X_1^2, \ldots, X_d^2)$ is distributed as $(\dfrac{Y_1}{S}, \ldots, \dfrac{Y_d}{S})$, where $Y_1, \ldots, Y_d$ are independent gamma $(\dfrac{1}{2})$ random variables with sum S.

3. If X is uniformly distributed on C_d, then X_1^2 is beta $(\dfrac{1}{2}, \dfrac{d-1}{2})$. Equivalently, X_1^2 is distributed as $\dfrac{Y}{Y+Z}$ where Y, Z are independent gamma $(\dfrac{1}{2})$ and gamma $(\dfrac{d-1}{2})$ random variables. Furthermore, X_1 has density

$$\frac{\Gamma(\frac{d}{2})}{\Gamma(\frac{1}{2})\Gamma(\frac{d-1}{2})}(1-x^2)^{\frac{d-3}{2}} \qquad (\,|\,x\,|\, \le 1)\,.$$

Proof of Theorem 4.1.

For all orthogonal $d \times d$ matrices A, $\dfrac{AX}{||X||}$ is distributed as $\dfrac{AX}{||AX||}$, which in turn is distributed as $\dfrac{X}{||X||}$ because X is strictly radially symmetric. Since $||\dfrac{X}{||X||}|| = \dfrac{||X||}{||X||} = 1$, statement 1 follows.

To prove statement 2, we define the iid normal random variables $N_1, \ldots, N_d$, and note that $N = (N_1, \ldots, N_d)$ is radially symmetric with density determined by

$$g(r) = \frac{1}{(2\pi)^{\frac{d}{2}}} e^{-\frac{r^2}{2}} \qquad (r \ge 0)\,.$$

Thus, by part 1, the vector with components $\dfrac{N_i}{||N||}$ is uniformly distributed on C_d. But since N_i^2 is gamma $(\dfrac{1}{2}, 2)$, we deduce that the random vector with components $\dfrac{N_i^2}{||N||^2}$ is distributed as a random vector with components $\dfrac{2Y_i}{2S}$. This proves statement 2.

The first part of statement 3 follows easily from statement 2 and known properties of the beta and gamma distributions. The beta $(\frac{1}{2}, \frac{d-1}{2})$ density is

$$c\frac{(1-x)^{\frac{d-3}{2}}}{\sqrt{x}} \quad (0 < x < 1),$$

where $c = \dfrac{\Gamma(\frac{d}{2})}{\Gamma(\frac{1}{2})\Gamma(\frac{d-1}{2})}$. Putting $Y = \sqrt{X}$, we see that Y has density

$$c\,(1-y^2)^{\frac{d-3}{2}}\frac{1}{y}2y \quad (0 < y < 1),$$

when X is beta $(\frac{1}{2}, \frac{d-1}{2})$ distributed. This proves statement 3. ■

Theorem 4.2. (The normal distribution.)

If $N_1, \ldots, N_d$ are iid normal random variables, then $(N_1, \ldots, N_d)$ is radially symmetric with density defined by

$$g(r) = \frac{1}{(2\pi)^{\frac{d}{2}}} e^{-\frac{r^2}{2}} \quad (r \geq 0).$$

Furthermore, if $(X_1, \ldots, X_d)$ is strictly radially symmetric and the X_i's are independent, then the X_i's are iid normal random variables with nonzero variance.

Proof of Theorem 4.2.

The first part was shown in Theorem 4.1. The second part is proved for example in Kelker (1970). ■

Theorem 4.3. (Radial transformations.)

1. If X is strictly radially symmetric in R^d with defining function g, then $R = ||X||$ has density

$$dV_d \, r^{d-1} g(r) \quad (r \geq 0).$$

2. If X is uniformly distributed on C_d, and R is independent of X and has the density given above, then RX is strictly radially symmetric in R^d with defining function g.

3. If X is radially symmetric in R^d with defining function g, and if R is a random variable on $[0,\infty)$ with density h, independent of X, then RX is radially symmetric with defining function

$$g*(r) = \int\limits_0^\infty \frac{h(u)}{u^d} g\left(\frac{r}{u}\right) du \; .$$

Proof of Theorem 4.3.

For statement 1, we need the fact that the surface of C_d has $d-1$-dimensional volume dV_d. By a simple polar transformation,

$$P(R \leq r) = \int\limits_{||x|| \leq r} g(||x||) \, dx = \int\limits_{y \leq r} dV_d \, y^{d-1} g(y) \, dy \quad (r \geq 0).$$

This proves statement 1.

RX is radially symmetric because for all orthogonal $d \times d$ matrices A, $A(RX)$ is distributed as $R(AX)$ and thus as RX. But such distributions are uniquely determined by the distribution of $||RX|| = R||X|| = R$, and thus, statement follows from statement 1.

Consider finally part 3. Clearly, RX is radially symmetric. Given R, $R||X||$ has density

$$\frac{1}{R} dV_d \left(\frac{r}{R}\right)^{d-1} g\left(\frac{r}{R}\right) \quad (r \geq 0).$$

Thus, the density of $||X||$ is the expected value of the latter expression with respect to R, which is seen to be $g*$. ∎

Let us briefly discuss these three theorems. Consider first the marginal distributions of random vectors that are uniformly distributed on C_d:

d	Density of X_1 (on $[-1,1]$)	Name of density
2	$\dfrac{1}{\pi\sqrt{1-x^2}}$	Arc sine density
3	$\dfrac{1}{2}$	Uniform $[-1,1]$ density
4	$\dfrac{2}{\pi}\sqrt{1-x^2}$	
5	$\dfrac{3}{4}(1-x^2)$	
6	$\dfrac{8}{3\pi}(1-x^2)^{\frac{3}{2}}$	

Since all radially symmetric random vectors are distributed as the product of a uniform random vector on C_d and an independent random variable R, it follows that the first component X_1 is distributed as R times a random variable with densities as given in the table above or in part 3 of Theorem 4.1. Thus, for $d \geq 2$, X_1 has a marginal density whenever X is strictly radially symmetric. By Khinchine's theorem, we note that for $d \geq 3$, the density of X_1 is unimodal.

Theorem 4.2 states that radially symmetric distributions are virtually useless if they are to be used as tools for generating independent random variates $X_1, \ldots, X_n$ unless the X_i's are normally distributed. In the next section, we will clarify the special role played by the normal distribution.

4.2. Generating random vectors uniformly distributed on C$_d$.

The following two algorithms can be used to generate random variates with a uniform distribution on C_d:

Via normal random variates

Generate iid normal random variates, $N_1, \ldots, N_d$, and compute $S \leftarrow \sqrt{N_1^2 + \cdots + N_d^2}$.
RETURN $(\dfrac{N_1}{S}, \ldots, \dfrac{N_d}{S})$.

Via rejection from the enclosing hypercube

REPEAT

Generate iid uniform [−1,1] random variates $X_1, \ldots, X_d$, and compute $S \leftarrow X_1^2 + \cdots + X_d^2$.

UNTIL $S \leq 1$

$S \leftarrow \sqrt{S}$

RETURN $(\dfrac{X_1}{S}, \ldots, \dfrac{X_d}{S})$

In addition, we could also make good use of a property of Theorem 4.1. Assume that d is even and that a d-vector X is uniformly distributed on C_d. Then,

$$(X_1^2 + X_2^2, \ldots, X_{d-1}^2 + X_d^2)$$

is distributed as

$$(\dfrac{E_1}{S}, \ldots, \dfrac{E_{\frac{d}{2}}}{S})$$

where the E_i's are iid exponential random variables and $S = E_1 + \cdots + E_{\frac{d}{2}}$. Furthermore, given $X_1^2 + X_2^2 = r^2$, $(\dfrac{X_1}{r}, \dfrac{X_2}{r})$ is uniformly distributed on C_2. This leads to the following algorithm:

Via uniform spacings

Generate iid uniform [0,1] random variates $U_1, \ldots, U_{\frac{d}{2}-1}$.

Sort the uniform variates (preferably by bucket sorting), and compute the spacings $S_1, \ldots, S_{\frac{d}{2}}$.

Generate independent pairs $(V_1, V_2), \ldots, (V_{d-1}, V_d)$, all uniformly distributed on C_2.

RETURN $(V_1\sqrt{S_1}, V_2\sqrt{S_1}, V_3\sqrt{S_2}, V_4\sqrt{S_2}, \ldots, V_{d-1}\sqrt{\dfrac{S_d}{2}}, V_d\sqrt{\dfrac{S_d}{2}})$.

The normal and spacings methods take expected time $O(d)$, while the rejection method takes time increasing faster than exponentially with d. By Stirling's formula, we observe that the expected number of iterations in the rejection method is

$$\frac{2^d}{V_d} = \frac{2^d \, \Gamma(\frac{d}{2+1})}{\pi^{\frac{d}{2}}} \sim (\frac{2d}{\pi e})^{\frac{d}{2}} \sqrt{\pi d} \quad ,$$

which increases very rapidly to ∞. Some values for the expected number of iterations are given in the table below.

d	Expected number of iterations
1	1
2	$\frac{4}{\pi} \approx 1.27$
3	$\frac{6}{\pi} \approx 1.91$
4	$\frac{32}{\pi^2} \approx 3.24$
5	$\frac{60}{\pi^2} \approx 6.06$
6	$\frac{384}{\pi^3} \approx 12.3$
7	$\frac{840}{\pi^3} \approx 27.0$
8	$\frac{6144}{\pi^4} \approx 62.7$
10	$\frac{122880}{\pi^5} \approx 399$

The rejection method is not recommended except perhaps for $d \leq 5$. The normal and spacings methods differ in the type of operations that are needed: the normal method requires d normal random variates plus one square root, whereas the spacings method requires one bucket sort , $\frac{d}{2}$ square roots and $\frac{d}{2} - 1$ uniform random variates. The spacings method is based upon the assumption that a very fast method is available for generating random vectors with a uniform distribution on C_2. Since we work with spacings, it is also possible that some accuracy is lost for large values of d. For all these reasons, it seems unlikely that the spacings method will be competitive with the normal method. For theoretical and experimental comparisons, we refer the reader to Deak (1979) and Rubinstein (1982). For another derivation of the spacings method, see for example Sibuya (1962), Tashiro (1977), and Guralnik, Zemach and Warnock (1985).

4.3. Generating points uniformly in and on C_2.

We say that a random vector is uniformly distributed in C_d when it is radially symmetric with defining function $g(r) = \dfrac{1}{V_d}$ $(0 \leq r \leq 1)$. For $d = 2$, such random vectors can be conveniently generated by the rejection method:

Rejection method

REPEAT
        Generate two iid uniform $[-1,1]$ random variates U_1, U_2.
UNTIL $U_1{}^2 + U_2{}^2 \leq 1$
RETURN (U_1, U_2)

On the average, $\dfrac{4}{\pi}$ pairs of uniform random variates are needed before we exit. For each pair, two multiplications are required as well. Some speed-up is possible by squeezing:

Rejection method with squeezing

REPEAT
        Generate two iid uniform $[-1,1]$ random variates U_1, U_2 , and compute
        $Z \leftarrow |U_1| + |U_2|$.
        Accept $\leftarrow [Z \leq 1]$
        IF NOT Accept THEN IF $Z \geq \sqrt{2}$
                THEN Accept $\leftarrow [U_1{}^2 + U_2{}^2 \leq 1]$
UNTIL Accept
RETURN (U_1, U_2)

In the squeeze step, we avoid the two multiplications precisely 50% of the time.

The second, slightly more difficult problem is that of the generation of a point uniformly distributed on C_2. For example, if (X_1, X_2) is strictly radially symmetric (this is the case when the components are iid normal random variables, or when the random vector is uniformly distributed in C_2), then it suffices to take $(\dfrac{X_1}{S}, \dfrac{X_2}{S})$ where $S = \sqrt{X_1{}^2 + X_2{}^2}$. At first sight, it seems that the costly square root is unavoidable. That this is not so follows from the following key theorem:

Theorem 4.4.

If (X_1, X_2) is uniformly distributed in C_2, and $S = \sqrt{X_1{}^2 + X_2{}^2}$, then:

1. S and $(\dfrac{X_1}{S}, \dfrac{X_2}{S})$ are independent.

2. S^2 is uniformly distributed on $[0,1]$.

3. $\dfrac{X_2}{X_1}$ is Cauchy distributed.

4. $(\dfrac{X_1}{S}, \dfrac{X_2}{S})$ is uniformly distributed on C_2.

5. When U is uniform $[0,1]$, then $(\cos(2\pi U), \sin(2\pi U))$ is uniformly distributed on C_2.

6. $(\dfrac{X_1{}^2 - X_2{}^2}{S^2}, \dfrac{2X_1 X_2}{S^2})$ is uniformly distributed on C_2.

Proof of Theorem 4.4.

Properties 1,3 and 4 are valid for all strictly radially symmetric random vectors (X_1, X_2). Properties 1 and 4 follow directly from Theorem 4.3. From Theorem 4.1, we recall that S has density $dV_d\, r^{d-1} = 2r$ $(0 \le r \le 1)$. Thus, S^2 is uniformly distributed on $[0,1]$. This proves property 2. Property 5 is trivially true, and will be used to prove properties 3 and 6. From 5, we know that $\dfrac{X_2}{X_1}$ is distributed as $\tan(2\pi U)$, and thus as $\tan(\pi U)$, which in turn is Cauchy distributed (property 3). Finally, in view of

$$\cos(4\pi U) = \cos^2(2\pi U) - \sin^2(2\pi U),$$
$$\sin(4\pi U) = 2\sin(2\pi U)\cos(2\pi U),$$

we see that $(\dfrac{X_1{}^2 - X_2{}^2}{S^2}, \dfrac{2X_1 X_2}{S^2})$ is uniformly distributed on C_2, because it is distributed as $(\cos(4\pi U), \sin(4\pi U))$. This concludes the proof of Theorem 4.4. ∎

Thus, for the generation of a random vector uniformly distributed on C_2, the following algorithm is fast:

REPEAT
> Generate iid uniform $[-1,1]$ random variates X_1, X_2.
> Set $Y_1 \leftarrow X_1^2, Y_2 \leftarrow X_2^2, S \leftarrow Y_1 + Y_2$.

UNTIL $S \leq 1$

RETURN $(\dfrac{Y_1 - Y_2}{S}, \dfrac{2X_1 X_2}{S})$

4.4. Generating normal random variates in batches.

We begin with the description of the polar method for generating d iid normal random variates:

Polar method for normal random variates

Generate X uniformly on C_d.

Generate a random variate R with density $dV_d \, r^{d-1} e^{-\frac{r^2}{2}}$ $(r \geq 0)$. (R is distributed as $\sqrt{2G}$ where G is gamma $(\dfrac{d}{2})$ distributed.)

RETURN RX

In particular, for $d = 2$, two independent normal random variates can be obtained by either one of the following methods:

$\sqrt{2E} \; (\dfrac{X_1}{S}, \dfrac{X_2}{S})$
$\sqrt{2E} \; (\cos(2\pi U), \sin(2\pi U))$
$\sqrt{2E} \; (\dfrac{X_1^2 - X_2^2}{X_1^2 + X_2^2}, \dfrac{2X_1 X_2}{X_1^2 + X_2^2})$
$\sqrt{-4\log(S)}(\dfrac{X_1}{S}, \dfrac{X_2}{S})$

Here (X_1, X_2) is uniformly distributed in C_2, $S = \sqrt{X_1^2 + X_2^2}$, U is uniformly distributed on $[0,1]$ and E is exponentially distributed. Also, E is independent of the other random variables. The validity of these methods follows from Theorems 4.2, 4.3 and 4.4. The second formula is the well-known Box-Muller method

(1958). Method 4, proposed by Marsaglia, is similar to method 1, but uses the observation that S^2 is a uniform $[0,1]$ random variate independent of $(\frac{X_1}{S},\frac{X_2}{S})$ (see Theorem 4.4), and thus that $-2\log(S)$ is exponentially distributed. If the exponential random variate in E is obtained by inversion of a uniform random variate, then it cannot be competitive with method 4. Method 3, published by Bell (1968), is based upon property 6 of Theorem 4.4, and effectively avoids the computation of the square root in the definition of S. In all cases, it is recommended that (X_1,X_2) be obtained by rejection from the enclosing square (with an accelerating squeeze step perhaps). A closing remark about the square roots. Methods 1 and 4 can always be implemented with just one (not two) square roots, if we compute, respectively,

$$\sqrt{\frac{2E}{S^2}}$$

and

$$\sqrt{\frac{-2\log(S^2)}{S^2}} \, .$$

In one of the exercises, we will investigate the polar method with the next higher convenient choice for d, $d=4$. We could also make d very large , in the range $100 \cdots 300$, and use the spacings method of section 4.2 for generating X with a uniform distribution on C_d (the normal method is excluded since we want to generate normal random variates). A gamma $(\frac{d}{2})$ random variate can be generated by one of the fast methods described elsewhere in this book.

4.5. Generating radially symmetric random vectors.

Theorem 4.3 suggests the following method for generating radially symmetric random vectors in R^d with defining function g :

Generate a random vector X uniformly distributed on C_d .
Generate a random variate R with density $dV_d r^{d-1}g(r)$ $(r \geq 0)$.
RETURN RX

Since we already know how to generate random variates with a uniform distribution on C_d, we are just left with a univariate generation problem. But in the multiplication with R, most of the information in X is lost. For example, to insure that X is on C_d, the rejection method generates X uniformly in C_d and divides then by $||X||$. But when we multiply the result with R, this division by $||X||$ seems somehow wasteful. Johnson and Ramberg (1977) observed that it is sometimes better to start from a random vector with a uniform distribution in C_d:

The Johnson-Ramberg method for generating radially symmetric random vectors

Generate a random vector X uniformly in C_d (preferably by rejection from the enclosing hypercube).

Generate a random variate R with density $-V_d r^d g'(r)$ $(r \geq 0)$, where g is the defining function of the radially symmetric distribution.

RETURN RX

This method only works when $-V_d r^d g'(r)$ is indeed a density in r on $[0,\infty)$. A sufficient condition for this is that g is continuously differentiable on $(0,\infty)$, $g'(r) < 0$ $(r > 0)$, and $r^d g(r) \to 0$ as $r \downarrow 0$ and $r \uparrow \infty$.

Example 4.1. The multivariate Pearson II density.

Consider the multivariate Pearson II density with parameter $a \geq 1$, defined by

$$g(r) = c (1-r^2)^{a-1} \quad (0 \leq r \leq 1),$$

where

$$c = \frac{\Gamma(a + \frac{d}{2})}{\pi^{\frac{d}{2}} \Gamma(a)}.$$

The density of R in the standard algorithm is the density of $\sqrt{B}$ where B is a beta $(\frac{d}{2}, a)$ random variable:

$$g(r) = cdV_d r^{d-1}(1-r^2)^{a-1} \quad (0 \leq r \leq 1).$$

For $d = 2$, R can thus be generated as $\sqrt{1 - U^{\frac{1}{a}}}$ where U is a uniform $[0,1]$ random variate. We note further that in this case, very little is gained by using the Johnson-Ramberg method since R must have density

$$g(r) = 2cV_d r^{d+1}(a-1)(1-r^2)^{a-2} \quad (0 \leq r \leq 1).$$

This is the density of the square root of a beta $(\frac{d}{2}+1, a-1)$ random variable. ■

Example 4.2. The multivariate Pearson VII density.

The multivariate Pearson VII density with parameter $a > \frac{d}{2}$ is defined by the function

$$g(r) = \frac{c}{(1+r^2)^a} \; ,$$

where

$$c = \frac{\Gamma(a)}{\pi^{\frac{d}{2}} \Gamma(a-\frac{d}{2})} \; .$$

The densities of R for the standard and Johnson-Ramberg methods are respectively,

$$\frac{cd V_d \, r^{d-1}}{(1+r^2)^a}$$

and

$$\frac{2c V_d \, r^{d+1} a}{(1+r^2)^{a+1}} \; .$$

In both cases, we can generate random R as $\sqrt{\dfrac{B}{1-B}}$ where B is beta $(\frac{d}{2}, a-\frac{d}{2})$ in the former case, and beta $(\frac{d}{2}+1, a-\frac{d}{2})$ in the latter case. Note here that for the special choice $a = \frac{d+1}{2}$, the multivariate Cauchy density is obtained. ■

Example 4.3.

The multivariate radially symmetric distribution determined by

$$g(r) = \frac{1}{V_d (1+r^d)^2}$$

leads to a density for R given by

$$\frac{dr^{d-1}}{(1+r^d)^2} \, .$$

This is the density of $(\frac{U}{1-U})^{\frac{1}{d}}$ where U is a uniform [0,1] random variable.

4.6. The deconvolution method.

Assume that we know how to generate Z, a random variable which is distributed as the sum $X+Y$ of two iid random variables X,Y with density f. We can then generate the pair X,Y by looking at the conditional density of X given the value of Z. The following algorithm can be used:

The deconvolution method

Generate a random variate Z with the density $h(z) = \int f(x) f(z-x)\, dx$.

Generate X with density $\dfrac{f(x) f(Z-x)}{h(Z)}$.

RETURN $(X, Z-X)$

First, we notice that h is indeed the density of the sum of two iid random variables with density f. Also, given Z, X has density $\dfrac{f(x) f(Z-x)}{h(Z)}$. Thus, the algorithm is valid.

To illustrate this, recall that if X,Y are iid gamma $(\frac{1}{2})$, then $X+Y$ is exponentially distributed. In this example, we have therefore,

$$f(x) = \frac{1}{\sqrt{\pi x}} e^{-x} \quad (x \geq 0) \, ,$$

$$h(z) = e^{-z} \quad (z \geq 0) \, .$$

Furthermore, the density $\dfrac{f(x) f(Z-x)}{h(Z)}$ can be written as

$$\frac{1}{\pi \sqrt{x(Z-x)}} \quad (x \in (0,Z)) \, ,$$

which is the arc sine density. Thus, applying the deconvolution method shows the following: if E is an exponential random variable, and W is a random variable with the standard arc sine density

$$\frac{1}{\pi \sqrt{x(1-x)}} \quad (x \in (0,1)) \, ,$$

then $(EW, E(1-W))$ is distributed as a pair of iid gamma $(\frac{1}{2})$ random variables.
But this leads precisely to the polar method because the following pairs of random variables are identically distributed:

(N_1, N_2) (two iid normal random variables);

$(\sqrt{2EW}, \sqrt{2E(1-W)})$;

$(\sqrt{2E} \cos(2\pi U), \sqrt{2E} \sin(2\pi U))$.

Here U is a uniform $[0,1]$ random variable. The equivalence of the first two pairs is based upon the fact that a normal random variable is distributed as the square root of 2 times a gamma $(\frac{1}{2})$ random variable. The equivalence of the first and the third pair was established in Theorem 4.4. As a side product, we observe that W is distributed as $\cos^2(2\pi U)$, i.e. as $\dfrac{X_1^2}{X_1^2 + X_2^2}$ where (X_1, X_2) is uniformly distributed in C_2.

4.7. Exercises.

1. Write one-line random variate generators for the normal, Cauchy and arc sine distributions.

2. If N_1, N_2 are iid normal random variables, then $\dfrac{N_1}{N_2}$ is Cauchy distributed, $N_1^2 + N_2^2$ is exponentially distributed, and $\sqrt{N_1^2 + N_2^2}$ has the Rayleigh distribution (the Rayleigh density is $xe^{-\frac{x^2}{2}}$ ($x \geq 0$)).

3. Show the following. If X is uniformly distributed on C_d and R is independent of X and generated as $\max(U_1, \ldots, U_d)$ where the U_i's are iid uniform $[0,1]$ random variates, then RX is uniformly distributed in C_d.

4. Show that if X is uniformly distributed on C_d, then $Y/||Y||$ is uniformly distributed on C_k where $k \leq d$ and $Y = (X_1, \ldots, X_k)$.

5. Prove by a geometrical argument that if (X_1, X_2, X_3) is uniformly distributed on C_3, then X_1, X_2 and X_3 are uniform $[-1,1]$ random variables.

6. If X is radially symmetric with defining function g, then its first component, X_1, has density

$$\frac{2\pi^{\frac{d-1}{2}}}{\Gamma(\frac{d-1}{2})} \int_r^\infty u (u^2 - r^2)^{\frac{d-3}{2}} g(u) \, du \quad (r \geq 0).$$

7. Show that two independent gamma $(\frac{1}{2})$ random variates can be generated

as $(-S \log(U_2), -(1-S)\log(U_2))$, where $S = \sin^2(2\pi U_1)$ and U_1, U_2 are independent uniform [0,1] random variates.

8. Consider the pair of random variables defined by

$$(\sqrt{2E} \, \frac{2S}{1+S}, \sqrt{2E} \, \frac{1-S}{1+S})$$

where E is an exponential random variable, and $S \leftarrow \tan^2(\pi U)$ for a uniform [0,1] random variate U. Prove that the pair is a pair of iid absolute normal random variables.

9. Show that when (X_1, X_2, X_3, X_4) is uniformly distributed on C_4, then (X_1, X_2) is uniformly distributed in C_2.

10. Show that both $\dfrac{N}{\sqrt{N^2+2E}}$ and $\sqrt{\dfrac{G}{G+E}}$ are uniformly distributed on [0,1] when N, E and G are independent normal, exponential and gamma $(\frac{1}{2})$ random variables, respectively.

11. **Generating uniform random vectors on C_4.** Show why the following algorithm is valid for generating random vectors uniformly on C_4:

> Generate two iid random vectors uniformly in C_2, $(X_1, X_2), (X_3, X_4)$ (this is best done by rejection).
> $S \leftarrow X_1^2 + X_2^2, W \leftarrow X_3^2 + X_4^2$
> RETURN $(X_1, X_2, X_3 \sqrt{\dfrac{1-S}{W}}, X_4 \sqrt{\dfrac{1-S}{W}})$

(Marsaglia, 1972).

12. **Generating random vectors uniformly on C_3.** Prove all the starred statements in this exercise. To obtain a random vector with a uniform distribution on C_3 by rejection from $[-1,1]^3$ requires on the average $\dfrac{18}{\pi} = 5.73...$ uniform $[-1,1]$ random variates, and one square root per random vector. The square root can be avoided by an observation due to Cook (1957): If (X_1, X_2, X_3, X_4) is uniformly distributed on C_4, then

$$\frac{1}{X_1^2 + X_2^2 + X_3^2 + X_4^2} \left(2(X_2 X_4 + X_1 X_3), 2(X_3 X_4 - X_1 X_2), X_1^2 - X_2^2 - X_3^2 + X_4^2\right)$$

is uniformly distributed on C_3 (*). Unfortunately, if a random vector with a uniform distribution on C_4 is obtained by rejection from the enclosing hypercube, then the expected number of uniform random variates needed is $4(\dfrac{32}{\pi^2}) \approx 13$. Thus, both methods are quite expensive. Using Theorem 4.4 and exercises 4 and 5, one can show (∗) that

$$\left(\frac{X_1\sqrt{1-Z^2}}{\sqrt{S}},\frac{X_2\sqrt{1-Z^2}}{\sqrt{S}},Z\right)$$

is uniformly distributed on C_3 when (X_1,X_2) is uniformly distributed in C_2, $S=X_1{}^2+X_2{}^2$, and Z is independent of $(\frac{X_1}{\sqrt{S}},\frac{X_2}{\sqrt{S}})$ and uniformly distributed on $[-1,1]$. But $2S-1$ itself is a candidate for Z (∗). Replacing Z by $2S-1$, we conclude that

$$(2X_1\sqrt{1-S},2X_2\sqrt{1-S},2S-1)$$

is uniformly distributed on C_3 (this method was suggested by Marsaglia (1972)). If the random vector (X_1,X_2) is obtained by rejection from $[-1,1]^2$, the expected number of uniform $[-1,1]$ random variates needed per three-dimensional random vector is $\frac{8}{\pi}\approx 2.55$ (∗).

13. **The polar methods for normal random variates, d=4.** Random vectors uniformly distributed on C_4 can be obtained quite efficiently by Marsaglia's method described in exercise 11. To apply the polar method for normal random variates, we need an independent random variate R distributed as $\sqrt{2(E_1+E_2)}$ where E_1,E_2 are independent exponential random variates. Such an R can be generated in a number of ways:

(i) As $\sqrt{2(E_1+E_2)}$.

(ii) As $\sqrt{-2\log(U_1U_2)}$ where U_1,U_2 are independent uniform $[0,1]$ random variates.

(iii) As $\sqrt{-2\log(\overline{WU_2})}$ where U_2 is as in (ii) and W is an independent random variate as in exercise 11.

Why is method (iii) valid ? Compare the three methods experimentally. Compare also with the polar method for $d=2$.

14. Implement the polar method for normal random variates when d is large. Generate random vectors on C_d by the spacings method when you do so. Plot the average time per random variate versus d.

15. **The spacings method for uniform random vectors on C_d when d is odd.** Show the validity of the following method for generating a uniform random vector on C_d:

Generate $\dfrac{d-1}{2}-1$ iid uniform [0,1] random variates.

Obtain the spacings $S_1, \ldots, S_{\frac{d-1}{2}}$ by bucket sorting the uniform random variates.

Generate independent gamma $(\dfrac{d-1}{2})$ and gamma $(\dfrac{1}{2})$ random variates G,H.

$R \leftarrow \sqrt{\dfrac{G}{G+H}}$, $R* \leftarrow \sqrt{1-R^2} = \sqrt{\dfrac{H}{H+G}}$

Generate iid random vectors $(V_1,V_2), \ldots, (V_{d-2},V_{d-1})$ uniformly on C_2.

RETURN $(RV_1\sqrt{S_1}, RV_2\sqrt{S_1}, RV_3\sqrt{S_2}, \ldots, RV_{d-1}\sqrt{S_{\frac{d-1}{2}}}, R*)$.

16. Let X be a random vector uniformly distributed on C_{d-1}. Then the random vector Y generated by the following procedure is uniformly distributed on C_d :

Generate independent gamma $(\dfrac{d-1}{2})$ and gamma $(\dfrac{1}{2})$ random variates G,H.

$R \leftarrow \sqrt{\dfrac{G}{G+H}}$

RETURN $Y \leftarrow (RX, \pm\sqrt{1-R^2})$ where $\pm$ is a random sign.

Show this. Notice that this method allows one to generate Y inductively by starting with $d=1$ or $d=2$. For $d=1$, X is merely ± 1. For $d=2$, R is distributed as $\sin(\dfrac{\pi U}{2})$. For $d=3$, R is distributed as $\sqrt{1-U^2}$ where U is a uniform [0,1] random variable. To implement this procedure, a fast gamma generator is required (Hicks and Wheeling, 1959; see also Rubinstein, 1982).

17. In a simulation it is required at one point to obtain a random vector (X,Y) uniformly distributed over a star on R^2. A star S_a with parameter $a>0$ is defined by four curves, one in each quadrant and centered at the origin. For example, the curve in the positive quadrant is a piece of a closed line satisfying the equation

$$| 1-x |^a + | 1-y |^a = 1 .$$

The three other curves are defined by symmetry about all the axes and about the origin. For $a = \frac{1}{2}$, we obtain the circle, for $a = 1$, we obtain a diamond, and for $a = 2$, we obtain the complement of the union of four circles. Give an algorithm for generating a point uniformly distributed in S_a, where the expected time is uniformly bounded over a.

18. **The Johnson-Ramberg method for normal random variates.** Two methods for generating normal random variates in batches may be competitive with the ordinary polar method because they avoid square roots. Both are based upon the Johnson-Ramberg technique:

Generate X uniformly in C_2 by rejection from $[-1,1]^2$.

Generate R, which is distributed as $\sqrt{2G}$ where G is a gamma $(\frac{3}{2})$ random

variable. (Note that R has density $\frac{r^2}{2} e^{-\frac{r^2}{2}}$.)

RETURN RX

Generate X uniformly in C_3 by rejection from $[-1,1]^3$.

Generate R, where R is distributed as $\sqrt{2G}$ and G is a gamma (2) random

variable. (Note that R has density $(\frac{r}{\sqrt{2}})^3 \Gamma^{-1}(\frac{5}{2}) e^{-\frac{r^2}{2}}$.)

RETURN RX

These methods can only be competitive if fast direct methods for generating R are available. Develop such methods.

19. Extend the entire theory towards other norms, i.e. C_d is now defined as the collection of all points for which the p-th norm is less than or equal to one. Here $p > 0$ is a parameter. Reprove all theorems. Note that the role of the normal density is now inherited by the density

$$f(x) = c e^{-|x|^p},$$

where $c > 0$ is a normalization constant. Determine this constant. Show that a random variate with this density can be obtained as $X^{\frac{1}{p}}$ where X is gamma $(\frac{1}{p})$ distributed. Find a formula for the probability of acceptance when random variates with a uniform distribution in C_d are obtained by rejection from $[-1,1]^2$. (To check your result, the answer for $d=2$ is $\Gamma^2(\frac{1}{p})\Gamma^{-1}(\frac{2}{p})$ (Beyer, 1968, p. 630).) Discuss various methods for generating random vectors uniformly distributed on C_d, and deduce the marginal density of such random vectors .

Chapter Six
THE POISSON PROCESS

1. THE POISSON PROCESS.

1.1. Introduction.

One of the most important processes occurring in nature is the Poisson point process. It is therefore important to understand how such processes can be simulated. The methods of simulation vary with the type of Poisson point process, i.e. with the space in which the process occurs, and with the homogeneity or non-homogeneity of the process. We will not be concerned with the genesis of the Poisson point process, or with important applications in various areas. To make this material come alive, the reader is urged to read the relevant sections in Feller (1965) and Cinlar (1975) for the basic theory, and some sections in Trivedi (1982) for computer science applications.

In a first step, we will define the homogeneous Poisson process on $[0,\infty)$: the process is entirely determined by a collection of random events occurring at certain random times $0 < T_1 < T_2 < \cdots$. These events can correspond to a variety of things, such as bank robberies, births of quintuplets and accidents involving Montreal taxi cabs. If $N(t_1,t_2)$ is the number of events occurring in the time interval (t_1,t_2), then the following two conditions are often satisfied:

(i) For disjoint intervals $(t_1,t_2),(t_3,t_4), \ldots,$ the random variables $N(t_1,t_2),N(t_3,t_4),\ldots$ are independent.

(ii) $N(t_1,t_2)$ is distributed as $N(0,t_2-t_1)$, i.e. the distribution of the number of events in a certain time interval just depends upon the length of the interval.

The amazing fact is that these two conditions imply that all random variables $N(t_1,t_2)$ are Poisson distributed, and that there exists a constant $\lambda \geq 0$ such that $N(t,t+a)$ is Poisson λa for all $t \geq 0$, $a > 0$. See e.g. Feller (1965). Thus, the Poisson distribution occurs very naturally.

The previous concept can be generalized to R^d. Let A be a subset of R^d, and let N be a random variable taking only integer values. Let $X_1, \ldots, X_N$ be a sequence of random vectors taking values in A. Then we say that the X_i's define a uniform (or: homogeneous) Poisson process on A if

(A) For any finite collection of finite-volume nonoverlapping subsets of A, say $A_1, \ldots, A_k$, the random variables $N(A_1), \ldots, N(A_k)$ are independent.

(B) For any Borel subset $B \subseteq A$, the distribution of $N(B)$ depends upon $Vol(B)$ only.

Again, these assumptions imply that all $N(B)$'s are Poisson distributed with parameter $\lambda \, Vol(B)$ for some $\lambda \geq 0$. λ will be called the rate, or rate parameter, of the homogeneous Poisson process on A. Examples of such processes in multidimensional Euclidean space include bacteria on a Petri plate and locations of murders in Houston.

Theorem 1.1.

  Let $B \subseteq A$ be fixed sets from R^d, and let $0 < Vol(B) < \infty$. Then:

(i) If $X_1, X_2, \ldots$ determines a uniform Poisson process on A with parameter λ, then for any parition $B_1, \ldots, B_k$ of B, we have that $N(B_1), \ldots, N(B_k)$ are independent Poisson distributed with parameters $\lambda \, Vol(B_i)$.

(ii) Let N be Poisson distributed with parameter $\lambda \, Vol(B)$, and let $X_1, \ldots, X_N$ be the first N random vectors from an iid sequence of random vectors uniformly distributed on B. For any partition $B_1, \ldots, B_k$ of B, the sequence $N(B_1), \ldots, N(B_k)$ is sequence of independent Poisson random variables with parameters $\lambda \, Vol(B_1), \ldots, \lambda \, Vol(B_k)$. In other words, $X_1, \ldots, X_N$ determines a uniform Poisson process on B with rate parameter λ.

Proof of Theorem 1.1.

  We will only show part (ii). Assume that $Vol(B) = 1$ and that B is partitioned into two sets, A_1, A_2 with respective volumes p and $q = 1 - p$. For any two integers $i, j \geq 0$ with $i + j = k$, we have

$$P(N(A_1) = i, N(A_2) = j)$$

$$= P(N(B) = k) P(N(A_1) = i, N(A_2) = j \mid N(B) = k)$$

$$= (e^{-\lambda} \frac{\lambda^k}{k!}) \binom{k}{i} p^i q^j$$

$$= (e^{-\lambda p} \frac{\lambda p^i}{i!})(e^{-\lambda q} \frac{\lambda q^j}{j!}),$$

and therefore, $N(A_1)$ and $N(A_2)$ are independent Poisson random variables as claimed. This argument can be extended towards all finite partitions and all positive values for $Vol(B)$. ∎

1.2. Simulation of homogeneous Poisson processes.

If we have to simulate a uniform Poisson process on a set $A \subseteq R^d$, then we need to generate a number of random vectors $X_i \in A$. This can be done as follows (by Theorem 1.1):

Homogeneous Poisson process generator

Generate a Poisson random variate N with parameter $\lambda \, Vol(A)$.

Generate iid random vectors $X_1, \ldots, X_N$ uniformly distributed on A.

RETURN $X_1, \ldots, X_N$

To generate N it is virtually useless to use an $O(1)$ expected time algorithm because in the remainder of the algorithm, at least time $\Omega(N)$ is spent. Thus, it is recommended that if the algorithm is used, the Poisson random variate be generated by a very simple algorithm (with expected time typically growing as λ). For specific sets A, other methods can be used which do not require the explicit generation of a Poisson random variate. There are three cases that we will use to illustrate this:

(i) A is $[0,\infty)$.
(ii) A is a circle.
(iii) A is a rectangle.

To do so, we need an interesting connection between Poisson processes and the exponential distribution.

Theorem 1.2.

Let $0 < T_1 < T_2 < \cdots$ be a uniform Poisson process on $[0,\infty)$ with rate parameter $\lambda > 0$. Then

$$\lambda(T_1-0), \lambda(T_2-T_1), \lambda(T_3-T_2), \ldots$$

are distributed as iid exponential random variables.

Proof of Theorem 1.2.

For any $k \geq 0$ and any $x > 0$,

$$P(T_{k+1} > T_k + x \mid T_k) = P(T_{k+1} \notin [T_k, T_k + x] \mid T_k)$$
$$= P(N_{[0,x]} = 0)$$
$$= e^{-\lambda x} \frac{\lambda^0}{0!}$$
$$= e^{-\lambda x} \ .$$

Thus, given T_k, $T_{k+1} - T_k$ is exponential with parameter λ. Generalizing this argument to obtain the claimed independence as well, we see that for any finite k, and any sequence of nonnegative numbers $x_0, x_1, \ldots$,

$$P(T_{k+1} - T_k > x_k, T_k - T_{k-1} > x_{k-1}, \ldots, T_2 - T_1 > x_1, T_1 - 0 > x_0)$$
$$= P(N_{(T_k, T_k + x_k)} = 0, \ldots, N_{(0, x_0)} = 0)$$
$$= P(N_{(0, x_0 + x_1 + \cdots + x_k)} = 0)$$
$$= e^{-\lambda \sum_{i=0}^{k} x_i}$$
$$= \prod_{i=0}^{k} e^{-\lambda x_i} .$$

This concludes the proof of Theorem 1.2. ∎

Theorem 1.2 suggests the following method for simulating a uniform Poisson process on $A = [0, \infty)$:

Uniform Poisson process generator on the real line: the exponential spacings method

$T \leftarrow 0$ (auxiliary variable used for updating the "time")
$k \leftarrow 0$ (initialize the event counter)
REPEAT
      Generate an exponential random variate E.
      $k \leftarrow k + 1$
      $T \leftarrow T + \dfrac{E}{\lambda}$
      $T_k \leftarrow T$
UNTIL False (this is an infinite loop; a stopping rule can be added if desired).

This algorithm is easy to implement because no Poisson random variates are needed. For other simple sets A, there exist trivial generalizations of Theorem 1.2. For example, when A is $[0, t] \times [0, 1]$ where possibly $t = \infty$, $0 < T_1 < T_2 < \cdots$ is a uniform Poisson process with rate λ on $[0, t]$, and $U_1, U_2, \ldots$ is a sequence of iid uniform $[0, 1]$ random variables, then

$$(T_1, U_1), (T_2, U_2), \ldots$$

determines a uniform Poisson process with rate λ on A .

Example 1.1. A uniform Poisson process on the unit circle.

If the set A is the circle with unit radius, then the various properties of uniform Poisson processes can be used to come up with several methods of generation (these can be extended to d dimensional spheres). Assume that λ is the desired rate. First, we could simply generate a Poisson $\lambda\pi$ random variate N, and then return a sequence of N iid random vectors uniformly distributed in the unit circle. If we apply the order statistics method suggested by Theorem 1.2, then the Poisson random variate is implicitly obtained. For example, by switching to polar coordinates (R, θ), we note that for a uniform Poisson process, R and θ are independent, and that a randomly chosen R has density $2r$ $(0 \leq r \leq 1)$ and that a randomly chosen θ is uniformly distributed on $[0, 2\pi]$. Thus, we could proceed as follows: generate a uniform Poisson process $0 < \theta_1 < \theta_2 < \cdots < \theta_N$ with rate parameter $\dfrac{\lambda}{2\pi}$ on $[0, 2\pi]$ by the exponential spacings method. Exit with

$$(\theta_1, R_1), \ldots, (\theta_N, R_N)$$

where the R_i's are iid random variates with density $2r$ $(0 \leq r \leq 1)$ which can be generated individually as the maxima of two independent uniform $[0,1]$ random variates. There is no special reason for applying the exponential spacings method to the angles. We could have picked the radii as well. Unfortunately, the ordered radii do not form a one-dimensional uniform Poisson process on $[0,1]$. They do form a nonhomogeneous Poisson process however, and the generation of such processes will be clarified in the next subsection. ∎

1.3. Nonhomogeneous Poisson processes.

There are situations in which events occur at "random times" but some times are more likely than others. This is the case for arrivals in intensive care units, for job submissions in a computer centre and for injuries to NFL players. A very good model for these cases is the nonhomogeneous Poisson process model, defined here for the sake of convenience on $[0, \infty)$. This is the most important case because "time" is usually the running variable.

A nonhomogeneous Poisson process on $[0, \infty)$ is determined by a rate function $\lambda(t) \geq 0$ $(t \geq 0)$, which can be considered as a density of sorts, with the difference that $\int_0^\infty \lambda(t) \, dt$ is not necessarily 1 (usually, it is ∞). The process is defined by the following property: for all finite collections of disjoint intervals $A_1, \ldots, A_k$, the numbers of events happening in these intervals $(N_1, \ldots, N_k)$ are independent Poisson random variables with parameters

$$\int_{A_i} \lambda(t) \ dt \qquad (1 \le i \le k) \ .$$

Let us now review how such processes can be simulated. By simulation, we understand that the times of occurrences of events $0 < T_1 < T_2 < \cdots$ are to be given in increasing order. The major work on simulation of nonhomogeneous Poisson processes is Lewis and Shedler (1979). This entire section is a reworked version of their paper. It is interesting to observe that the general principles of continuous random variate generation can be extended: we will see that there are analogs of the inversion, rejection and composition methods.

The role of the distribution function will be taken over by the integrated rate function

$$\Lambda(t) = \int_0^t \lambda(u) \ du \ .$$

We begin by noting that given $T_n = t$, $T_{n+1} - T_n$ has distribution function

$$F(x) = 1 - e^{-(\Lambda(t+x) - \Lambda(t))} \qquad (x \ge 0)$$

provided that $\lim_{t \to \infty} \Lambda(t) = \infty$ (i.e, $\int_0^\infty \lambda(t) \ dt = \infty$). This follows from the fact that

$$F(x) = P(T_{n+1} - T_n > x \mid T_n = t)$$
$$= P(N(t, t+x) = 0 \mid T_n = t)$$
$$= e^{-(\Lambda(t+x) - \Lambda(t))} \qquad (x \ge 0) \ .$$

Thus, T_{n+1} is distributed as $T_n + F^{-1}(U)$ where U is a uniform [0,1] random variate. Interestingly, writing U as $1 - e^{-E}$ (where E denotes an exponential random variable), we see that T_{n+1} is also distributed as $\Lambda^{-1}(E + \Lambda(T_n))$. In other words, we need to invert Λ. Formally, we have (see also Cinlar (1975) or Bratley, Fox and Schrage, 1983):

Algorithm based on inversion of the integrated rate function

$T \leftarrow 0$ (T will be an auxiliary variable)
$k \leftarrow 0$ (k is a counter)
REPEAT
      Generate an exponential random variate E.
      $k \leftarrow k + 1$
      $T \leftarrow T + \Lambda^{-1}(E + \Lambda(T))$
      $T_k \leftarrow T$
UNTIL False

Example 1.2. Homogeneous Poisson process.

For the special case $\lambda(t) = \lambda$, $\Lambda(t) = \lambda t$, it is easily seen that in the algorithm given above, the step $T \leftarrow T + \Lambda^{-1}(E + \Lambda(T))$ reduces to $T \leftarrow T + \frac{E}{\lambda}$. Thus, we obtain the exponential spacings method again. ■

Example 1.3.

To model morning pre-rush hour traffic, we can sometimes take $\lambda(t) = t$, which gives $\Lambda(t) = \frac{t^2}{2}$. The step $T \leftarrow T + \Lambda^{-1}(E + \Lambda(T))$ now needs to be replaced by

$$T \leftarrow \sqrt{T^2 + 2E} \quad . ■$$

If the rate function can be split into a sum of rate functions, as in

$$\lambda(t) = \sum_{i=1}^{n} \lambda_i(t)$$

and if $0 < T_{i1} < T_{i2} < \cdots$, $1 \le i \le n$ are independent realizations of the individual nonhomogeneous Poisson processes, then the merged ordered sequences form a realization of the nonhomogeneous Poisson process with rate function $\lambda(t)$. This corresponds to the **composition method**, but the difference now is that we need realizations of all component processes. The decomposition can be used when there is a natural decomposition dictated by the analytical form of $\lambda(t)$. Because the basic operation in merging the processes is to take the minimal value from the n processes, it could be advantageous for large n to store the times in a heap containing n elements. We summarize:

The composition method

Generate $T_{11}, \ldots, T_{n1}$ for the n Poisson processes, and store these values together with the indices of the corresponding processes in a table.

$T \leftarrow 0$ (T is the running time)

$k \leftarrow 0$ REPEAT

Find the minimal element (say, T_{ij}) in the table and delete it.

$k \leftarrow k + 1$

$T_k \leftarrow T_{ij}$

Generate the value $T_{i\ j+1}$ and insert it into the table.

UNTIL False

The third general principle is that of **thinning** (Lewis and Shedler, 1979). Similar to what we did in the rejection method, we assume the existence of an easy dominating rate function $\mu(t)$:

$$\lambda(t) \leq \mu(t), \text{ all } t.$$

Then the idea is to generate a homogeneous Poisson process on the part of the positive halfplane between 0 and $\mu(t)$, then to consider the homogeneous Poisson process under λ, and finally to exit with the x-components of the events in this process. This requires a theorem similar to that preceding the rejection method.

Theorem 1.3.

 Let $\lambda(t) \geq 0$ be a rate function on $[0,\infty)$, and let A be the set of all (x,y) with $x \geq 0, 0 \leq y \leq \lambda(x)$. The following is true:

(i) If $(X_1, Y_1),...$ (with ordered X_i's) is a homogeneous Poisson process with unit rate on A, then $0 < X_1 < X_2 < \cdots$ is a nonhomogeneous Poisson process with rate function $\lambda(t)$.

(ii) If $0 < X_1 < X_2 < \cdots$ is a nonhomogeneous Poisson process with rate function $\lambda(t)$, and $U_1, U_2,...$ are iid uniform $[0,1]$ random variables, then $(X_1, U_1\lambda(X_1)), (X_2, U_2\lambda(X_2)),...$ is a homogeneous Poisson process with unit rate on A.

(iii) If $B \subseteq A$, and $(X_1, Y_1),...$ (with ordered X_i's) is a homogeneous Poisson process with unit rate on A, then the subset of points (X_i, Y_i) belonging to B forms a homogeneous Poisson process with unit rate function on B.

Proof of Theorem 1.3. We verify that for nonoverlapping intervals $A_1, \ldots, A_k$, the number of X_i's falling in the intervals (which we shall denote by $N(A_1), \ldots, N(A_k)$), satisfy:

$$P(N(A_1) = i_1, \ldots, N(A_k) = i_k)$$
$$= P(N(\overline{A}_1) = i_1, \ldots, N(\overline{A}_k) = i_k)$$
$$= \prod_{j=1}^{k} \frac{\left(\int_{A_i} \lambda(t)dt\right)^{i_j}}{i_j!} e^{-\int_{A_i} \lambda(t)dt},$$

where $\overline{A}_i$ refers to the intersection of the infinite slice with vertical projection A_i with A. This concludes the proof of part (i).

 To show (ii), we can use Theorem 1.1: it suffices to show that for all finite sets $\overline{A}_1$, the number of random vectors N falling in $\overline{A}_1$ is Poisson distributed with parameter $Vol(\overline{A}_1)$, and that every random vector in this set is uniformly distributed in it. The distribution of N is indeed Poisson with the given parameter because the X_i sequence determines a nonhomogeneous Poisson process with the correct rate function. Also, by Theorem II.3.1, a random vector $(X, U\lambda(X))$ is uniformly distributed in $\overline{A}_1$ if U is uniformly distributed on $[0,1]$ and X is a random vector with density proportional to $\lambda(x)$ restricted to A_1. Thus, it suffices to show that if an X is picked at random from among the X_i's in A_1, then X is a random vector with density proportional to $\lambda(x)$ restricted to A_1. Let B be a Borel set contained in A_1, and let us write λ_B and λ_{A_1} for the integrals of λ over B and A_1 respectively. Thus,

$$P(X \in B \mid X \in A_1) = P(X \in B \mid N(A_1) = 1)$$
$$= \frac{P(N(B) = 1, N(A_1 - B) = 0)}{P(N(A_1) = 1)}$$

$$= \frac{\lambda_B \, e^{-\lambda_B} \, e^{-(\lambda_{A_1} - \lambda_B)}}{\lambda_{A_1} e^{-\lambda_{A_1}}}$$

$$= \frac{\lambda_B}{\lambda_{A_1}}$$

$$= \frac{1}{\lambda_{A_1}} \int_B \lambda(x) \, dx \ ,$$

which was to be shown.

Part 3 follows from Theorem 1.1 on homogeneous Poisson processes without further work. ∎

Consider now the thinning algorithm of Lewis and Shedler (1979):

The thinning method (Lewis and Shedler)

$T \leftarrow 0$
$k \leftarrow 0$
REPEAT
      Generate Z, the first event in a nonhomogeneous Poisson process with rate function μ occurring after T. Set $T \leftarrow Z$.
      Generate a uniform [0,1] random variate U.
      IF $U \leq \dfrac{\lambda(Z)}{\mu(Z)}$
            THEN $k \leftarrow k+1$, $X_k \leftarrow T$
UNTIL False

The sequence of X_k's thus generated is claimed to determine a nonhomogeneous Poisson process with rate function λ. Notice that we have taken a nonhomogeneous Poisson process $0 < Y_1 < Y_2 < \cdots$ with rate function μ and eliminated some points. As we know, $(Y_1, U_1 \mu(Y_1)), \ldots$ is a homogeneous Poisson process with unit rate under the curve of μ if $U_1, U_2, \ldots$ are iid uniform [0,1] random variates (Theorem 1.3). Thus, the subsequence falling under the curve of λ determines a homogeneous Poisson process with unit rate under that curve (part (iii) of the same theorem). Finally, taking the x-coordinates only of that subsequence gives a nonhomogeneous Poisson process with rate function λ.

The nonhomogeneous Poisson process with rate function μ is usually obtained by the inversion method.

Example 1.4. Cyclic rate functions.

The following example is also due to Lewis and Shedler (1979): consider a cyclic rate function

$$\lambda(t) = \lambda(1+\cos(t))$$

with as obvious choice for dominating rate function $\mu(t)=2\lambda$. We have

$T \leftarrow 0$

$k \leftarrow 0$

REPEAT

> Generate an exponential random variate E.
>
> $T \leftarrow T + \dfrac{E}{2\lambda}$
>
> Generate a uniform $[0,1]$ random variate U.
>
> IF $U \leq \dfrac{1+\cos(T)}{2}$
>
> > THEN $k \leftarrow k+1$, $X_k \leftarrow T$

UNTIL False

It goes without saying that the squeeze principle can be used here to help avoiding the cosine computation most of the time. ∎

A final word about the efficiency of the algorithm when used for generating a nonhomogeneous Poisson process on a set $[0,t]$. The expected number of events needed from the dominating process is $\int_0^t \mu(u)\, du$, whereas the expected number of random variates returned is $\int_0^t \lambda(u)\, du$. The ratio of the expected values can be considered as a fair measure of the efficiency, comparable in spirit to the rejection constant in the standard rejection method. Note that we cannot use the expected value of the ratio because that would in general be ∞ in view of the positive probability ($e^{-\int_0^t \lambda(u)\, du}$) of returning no variates.

1.4. Global methods for nonhomogeneous Poisson process simulation.

Nonhomogeneous Poisson processes on $[0,\infty)$ can always be obtained from homogeneous Poisson processes on $[0,\infty)$ by the following property (see e.g. Cinlar (1975, pp. 98-99)):

Theorem 1.4.

If $0<T_1<T_2< \cdots$ is a homogeneous Poisson process with unit rate on $[0,\infty)$, and if Λ is an integrated rate function, then

$$0<\Lambda^{-1}(T_1)<\Lambda^{-1}(T_2)< \cdots$$

determines a nonhomogeneous Poisson process with integrated rate function Λ.

Proof of Theorem 1.4.

We have implicitly shown this in the previous section. Let $i_1,...$ be integers, let k be an integer, and let $N(A)$ be the number of points in a set $A \subseteq [0,\infty)$. Then, $N(A)$ is equal to $N*(\Lambda(A))$ where $N*$ refers to the homogeneous Poisson process, and $\Lambda(A)$ is the set A transformed under Λ. Thus, if $A_1, \ldots, A_k$ are disjoint sets, it is easily seen that $N(A_1), \ldots, N(A_k)$ are distributed as $N*(\Lambda(A_1)), \ldots, N*(\Lambda(A_k))$, which is a sequence of independent Poisson random variables with parameters equal to the Lebesgue measures of the sets $\Lambda(A_i)$, i.e. $\int_{A_i} \lambda(t) \, dt$ where λ is the a.e. derivative of Λ. This shows that the transformed process is a nonhomogeneous Poisson process with integrated rate function Λ. ∎

We observe that if $\int_0^\infty \lambda(t) \, dt <\infty$, then the function Λ^{-1} is not defined for very large arguments. In that case, the T_i's with values exceeding $\int_0^\infty \lambda(t) \, dt$ should be ignored. We conclude thus that only a finite number of events occur in such cases. No matter how large the finite value of the integral is, there is always a positive probability of not having any event at all.

Let us apply this theorem to the simulation restricted to a finite interval $[0,t_0]$. This is equivalent to the infinite interval case provided that $\lambda(t)$ is replaced by

$$\begin{cases} \lambda(t) & (0\leq t \leq t_0) \\ 0 & (t >t_0) \end{cases}$$

Thus, it suffices to use $\Lambda^{-1}(T_1),...$ for all T_i's not exceeding $\Lambda(t_0)$. The inversion of Λ is sometimes not practical. The next property can be used to avoid it, provided that we have fast methods for generating order statistics with non-uniform densities (see e.g. chapter V). The straightforward proof of its validity is left to the reader (see e.g. Cox and Lewis, 1966, chapter 2).

Theorem 1.5.

Let N be a Poisson random variate with parameter $\Lambda(t_0)$. Let $0 < T_1 < T_2 < \cdots < T_N$ be order statistics corresponding to the distribution function

$$\frac{\Lambda(t)}{\Lambda(t_0)} \qquad (0 \leq t \leq t_0) \,,$$

then this subsequence determines a nonhomogeneous Poisson process on $[0, t_0]$ with integrated rate function Λ.

Both Theorem 1.4 and Theorem 1.5 lead to global methods, i.e. methods in which a nonhomogeneous Poisson process can be obtained from another process, usually in a separate pass of the data. The methods of the previous section, in contrast, are sequential: the event times of the process are generated directly from left to right. Since the one-pass sequential approach allows optional stopping and restarting anywhere in the process, it is definitely of more practical value. In some applications, there is also a considerable savings in storage because no intermediate (or auxiliary) process needs to be stored. Finally, some global methods require the computation of Λ^{-1}, whereas the thinning method does not. This is an important consideration when Λ is difficult to compute.

For more examples, and additional details, we refer to the exercises and the other sections in this chapter. Readers who do not specialize in random process generation will probably not gain very much from reading the other sections in this chapter.

1.5. Exercises.

1. When $\int_0^\infty \lambda(t)\,dt < \infty$, the inversion and thinning methods for nonhomogeneous Poisson process generation need modifying. Show how.

2. Let N be the total number of events (points) in a nonhomogeneous Poisson process on the positive real line with rate function $\lambda(t)$. Show that there are only two possible situations:

$$P(N < \infty) = 1 \quad (\int_0^\infty \lambda(t)\, dt < \infty)$$

$$P(N < \infty) = 0 \quad (\int_0^\infty \lambda(t)\, dt = \infty).$$

3. The following rate function is given to you: $\lambda(t)$ is piecewise constant with breakpoints at $a, 2a, 3a, 4a, \ldots$, where for $t \in [ia, (i+1)a)$, $\lambda(t) = \lambda_i$, $i = 0,1,2,\ldots$. Generalize the exponential spacings method for generating a nonhomogeneous Poisson process with this rate function. Hint: do not use transformations of exponential random variates when you cross breakpoints, but rely on the memoryless property of the exponential distribution.

4. We are interested in the generation of a nonhomogeneous Poisson process with log-linear rate function

$$\lambda(t) = c_0 e^{c_0 + ct} \quad (t \geq 0) .$$

where $c_0 > 0$, $c \in R$. There are two important situations: when $c < 0$, the process dies out and only a finite number of events occurs. The process corresponds to an exponential population explosion however when $c > 0$. Generate such a process by the inversion-of-Λ method.

5. This is a continuation of the previous exercise related to a method of Lewis and Shedler (1976) for simulating non-homogeneous Poisson processes with log-linear rate function. Show that if N is a Poisson $(-\dfrac{c_0}{c})$ random variable, and $E_1, E_2, \ldots$ is a sequence of iid exponential random variates, then, assuming $c < 0$,

$$-\frac{E_i}{c(N-i+1)} \quad (1 \leq i \leq N)$$

are distributed as the gaps between events in a nonhomogeneous Poisson process with rate function $\lambda(t) = c_0 e^{c_0 + ct}$ $(t \geq 0)$ on $[0, \infty)$. Give the algorithm that exploits this property. Note that this implies that the expected number of events in such a process is $-\dfrac{c_0}{c} < \infty$. For the case $c > 0$, show how by flipping the time axis around, you can reduce the problem to that of the case $c < 0$ provided that one is only interested in simulation on a finite time interval.

6. Give an algorithm for generating random variates with a log-quadratic rate function. Hint: consider several cases as in the previous two exercises (Lewis and Shedler, 1979).

2.   GENERATION OF RANDOM VARIATES WITH A GIVEN HAZARD RATE.

2.1. Hazard rate. Connection with Poisson processes.

In this section we consider the problem of the computer generation of random variables with a given hazard rate h on $[0,\infty)$. If X is a random variable with density f and distribution function F, then the **hazard rate** h and **cumulative hazard rate** H are inter-related as follows:

$$h(x) = \frac{f(x)}{1-F(x)} \; ;$$

$$H(x) = \int_0^x h(y)\, dy = -\log(1-F(x)) \; ;$$

$$F(x) = 1-e^{-H(x)} \; ;$$

$$f(x) = h(x)e^{-H(x)} \; .$$

The hazard rate plays a crucial role in reliability studies (Barlow and Proschan, 1965) and in all situations involving lifetime distributions. Note that $\int_0^\infty h(y)\, dy = \infty$ and thus $\lim_{x \to \infty} H(x) = \infty$. The key distribution now is the exponential: it has constant hazard rate of value 1. Roughly speaking, hazard rates tending to 0 correspond to densities with larger-than-exponential tails, and diverging hazard rates are for densities with smaller-than-exponential tails. For compact support distributions, we have $\lim_{x \uparrow c} H(x) = \infty$ for some finite c (corresponding to the rightmost point in the support). Sometimes, h or H is given, and not f or F. In particular, when only h is given, f cannot be computed exactly because we would first need to compute H by numerical integration. Thus, there is a need for methods which allow us to generate random variates with a given hazard rate h. Fortunately, such random variates are intimately connected to Poisson point processes.

Theorem 2.1.

Let $0 < T_1 < T_2 < \cdots$ be a nonhomogeneous Poisson process with rate function h (and thus integrated rate function H). Then T_1 is a random variable with hazard rate h.   ◢

Proof of Theorem 2.1.

Note that for $x > 0$,

$$P(T_1 \leq x) = 1-P(\text{no event times in } [0,x])$$

$$= 1-e^{-\int_0^x h(t)\, dt}$$

$$= 1 - e^{-H(x)} \, ,$$

which was to be shown. ∎

This connection helps us understand the algorithms of this section. We will discuss the inversion, composition and thinning methods. For special sub-classes of hazard rate functions, there are universally applicable (black box) methods that are worth reporting. In particular for **DHR distributions** (distributions with decreasing hazard rate), the method of dynamic thinning will be introduced and analyzed (Devroye, 1985). Other classes, such as the class of **IHR distributions** (distributions with increasing hazard rate), are dealt with indirectly in the text and exercises.

2.2. The inversion method.

For generating a random variate with cumulative hazard rate H, it suffices to invert an exponential random variate:

Inversion method

Generate an exponential random variate E.
RETURN $X \leftarrow H^{-1}(E)$

If H^{-1} is not explicitly known, then we are forced to solve $H(X) = E$ for X by some iterative method. Here the discussion of the standard inversion method for distribution functions applies again.

We can easily verify that the algorithm is valid, either by using the connection with Poisson processes given in Theorem 2.1, or directly: for $x > 0$ observe that if H is strictly increasing, then

$$P(H^{-1}(E) \leq x) = P(E \leq H(x)) = 1 - e^{-H(x)} = F(x) \, .$$

When H is not strictly increasing, then the chain of inequalities remains valid for any consistent definition of H^{-1}.

This method is difficult to attribute to one person. It was mentioned in the works of Cinlar (1975), Kaminsky and Rumpf (1977), Lewis and Shedler (1979) and Gaver (1979). In the table below, a list of examples is given. Basically, this list contains distributions with an easily invertible distribution function because

$F(x) = 1 - e^{-H(x)}$.

$f(x)$	$h(x)$	$H(x)$	$H^{-1}(E)$
$ax^{a-1}e^{-x^a}$ $(a > 0)$(Weibull)	ax^{a-1}	x^a	$E^{\frac{1}{a}}$
$\dfrac{a}{(1+x)^{a+1}}$ (Pareto)	$\dfrac{a}{1+x}$	$a\log(1+x)$	$e^{\frac{E}{a}} - 1$
ax^{a-1} $(a > 0, x \le 1)$ (power function)	$\dfrac{ax^{a-1}}{1-x^a}$	$-\log(1-x^a)$	$(1-e^{-E})^{\frac{1}{a}}$

2.3. The composition method.

When $h = h_1 + \cdots + h_n$ where the h_i's are in turn hazard rates, then we can use Theorem 2.1 directly and use the fact that it suffices to consider the minimum of n random variables $X_1, \ldots, X_n$ with the individual hazard rates h_i. When the individual cumulative hazard rates are H_i, then this can be shown directly: for $x > 0$,

$$P(\min(X_1, \ldots, X_n) \ge x) = \prod_{i=1}^{n} e^{-H_i(x)} = e^{-H(x)}.$$

If the decomposition is such that for some h_i we have $\int_0^\infty h_i(t)\, dt < \infty$, then the method is still applicable if we switch to nonhomogeneous Poisson processes.

Composition method

$X \leftarrow \infty$

FOR $i = 1$ TO n DO

> Generate Z distributed as the first event time in a nonhomogeneneous Poisson process with rate function h_i (ignore this if there are no events in the process; if $\int_0^\infty h_i = \infty$, then Z has hazard rate h_i).
>
> IF $Z < X$ THEN $X \leftarrow Z$

RETURN X

Usually, the composition method is slow because we have to deal with all the individual hazard rates. There are shortcuts to speed things up a bit. For example, after we have looked at the first component and set X equal to the random variate with hazard rate h_1, it suffices to consider the nonhomogeneous Poisson processes restricted to $[0, X]$. The point is that if X is small, then the probability of observing one or more event times in this interval is also small. Thus, often a

quick check suffices to avoid random variate generation for the remaining nonhomogeneous Poisson processes. To illustrate this, decompose h as follows:

$$h(x) = h_1(x) + h_2(x)$$

where h_1 is a hazard rate which puts its mass near the origin. The function h_2 is
nonnegative, but does not have to be a hazard rate. It can be considered as a
small adjustment, h_1 being the main (easy) component. Then the following algorithm can be used:

Composition method with quick acceptance

Generate a random variate X with hazard rate h_1.
Generate an exponential random variate E.
IF $E \leq H_2(X)$ (H_2 is the cumulative hazard rate for h_2)
      THEN RETURN $X \leftarrow H_2^{-1}(E)$
      ELSE RETURN X

Something can be gained if we replace $X \leftarrow H_2^{-1}(E)$ by a step in which we return
a random variate X distributed with hazard rate

$$h_2(x) \frac{1 - F_2(x)}{F_2(X) - F_2(x)}$$

which can be done by methods that do not involve inversion. The expected
number of times that we need to use the second (time-consuming) step in the
algorithm is the probability that $E \leq H_2(X)$ where X has hazard rate h_1:

$$P(E \leq H_2(X)) = \int\limits_0^\infty h_1(y) e^{-H_1(y)} (1 - e^{-H_2(y)}) \, dy$$

$$= 1 - \int\limits_0^\infty h_1(y) e^{-H(y)} \, dy$$

$$= 1 - \int\limits_0^\infty (h(y) - h_2(y)) e^{-H(y)} \, dy$$

$$= \int\limits_0^\infty h_2(y) e^{-H(y)} \, dy$$

$$= \int\limits_0^\infty (\frac{h_2(y)}{h(y)}) f(y) \, dy$$

where f is the density corresponding to f. From the last expression we conclude
that it is important to keep $\frac{h_2}{h}$ small.

2.4. The thinning method.

Combining the theorem about thinning Poisson processes (Theorem 1.4) with Theorem 2.1 shows that the following algorithm produces a random variate with hazard rate h, provided that we can generate a nonhomogeneous Poisson point process with rate function g where

$$h(x) \leq g(x) \quad \text{(all } x\text{)} .$$

Thinning method (Lewis and Shedler, 1979)

$X \leftarrow 0$

REPEAT

    Generate a random variate Δ with hazard rate $g(X+x)$ $(x \geq 0)$ (equivalently, generate the first occurrence in a nonhomogeneous Poisson point process with the same rate function).

    Generate a uniform $[0,1]$ random variate U.

    $X \leftarrow X + \Delta$

UNTIL $Ug(X) \leq h(X)$

RETURN X

This algorithm is most efficient when g is very simple. In particular, constant dominating rate functions $g = g_0$ are practical, because Δ can be obtained as $\dfrac{E}{g_0}$ where E is an exponential random variate. We will now see what the expected complexity is for this algorithm. It is annoying that the distribution of the number of iterations (which we shall call N) depends very heavily on h and g. Recall, in comparison, that for the rejection method, the distribution is always geometric. For the thinning method, we might even have $E(N) = \infty$, so that it is absolutely essential to clarify just how $E(N)$ depends upon h and g. The following theorem is due to Devroye (1985):

Theorem 2.2. (Analysis of the thinning method.)

Let f and F be the density and distribution function corresponding to a hazard rate h. Let $g \geq h$ be another hazard rate having cumulative hazard rate G. Then the expected number of iterations in the thinning algorithm given above is

$$E(N) = \int\limits_0^\infty g(x)(1-F(x)) \, dx = \int\limits_0^\infty f(x)G(x) \, dx .$$

Proof of Theorem 2.2.

Let us call the X variates in subsequent iterations X_i, where $i = 1, 2, \ldots$. Similarly, the uniform $[0,1]$ random variates used in the algorithm have also subscripts referring to the iteration, as in $U_1, U_2, \ldots$. In Theorem 1.4 we have shown that $(X_1, U_1 g(X_1)), (X_2, U_2 g(X_2)), \ldots$ if continued at infinitum form a homogeneous Poisson process with unit rate on the area bounded by the x-axis and the curve g. The only thing that we introduce in the thinning method is a stopping rule. We condition now on X, the random variate returned in the algorithm. Notice that N is 1 plus the number of event times in a nonhomogeneous Poisson process with rate function $g - h$ restricted to $[0, X)$. Thus, conditioned on X, $N-1$ is Poisson distributed with parameter $\int_0^X (g - h)$. This observation uses the properties of Theorem 1.4 connecting homogeneous Poisson processes in the plane with nonhomogeneous Poisson processes on the line.

It is a simple matter to compute $E(N)$:

$$E(N) = 1 + E\left(\int_0^X (g - h)\right) = 1 + E\left(\int_0^X g\right) - E(H(X))$$

$$= E\left(\int_0^X g\right)$$

$$= \int_0^\infty f\, G$$

$$= \int_0^\infty g\, (1 - F)\,.$$

Here we used the fact that $H(X)$ is exponentially distributed, and, in the last step, partial integration. ■

Theorem 2.2 establishes a connection between $E(N)$ and the size of the tail of X. For example, when $g = c$ is a constant, then

$$E(N) = cE(X)\,.$$

Not unexpectedly, the value of $E(N)$ is scale-invariant: it depends only upon the shapes of h and g. When g increases, as for example in

$$g(x) = \sum_{i=0}^n c_i\, x^i\,,$$

then $E(N)$ depends upon more than just the first moment:

$$E(N) = \sum_0^n \frac{c_i}{i+1} E(X^{i+1})\,.$$

There are plenty of examples for which $E(N)=\infty$ even when $g(x)=1$ for all x. Consider for example $h(x)=\dfrac{1}{x+1}$, which corresponds to the long-tailed density $f(x)=\dfrac{1}{(x+1)^2}$. Generally speaking, $E(N)$ is small when g and h are close. For example, we have the following helpful inequalities:

Theorem 2.3.

The thinning algorithm satisfies

$$E(N) \le \sup_{x>0} \frac{g(x)}{h(x)} \; ;$$

$$E(N) \le \sup_{x>0} \frac{1-F(x)}{1-F*(x)} \; ,$$

where $F, F*$ are the distribution functions for h and g respectively.

Proof of Theorem 2.3.

The first inequality follows from

$$E(N) = \int_0^\infty \frac{g(x)}{h(x)} f(x)\, dx \; ,$$

and the second inequality is a consequence of

$$E(N) = \int_0^\infty f*(x) \frac{1-F(x)}{1-F*(x)}\, dx \; ,$$

where $f*$ is the density corresponding to g. ∎

There are examples in which g and h appear to be far apart $(\lim_{x\uparrow\infty} \frac{g(x)}{h(x)} = \infty)$, yet $E(N)<\infty$: consider for example $h(x)=\dfrac{1}{x+1}$, $g(x)=\dfrac{1}{(x+1)^a}$, $0<a\le 1$. The explanation is that g and h should be close to each other near the origin and that the difference does not matter too much in low density regions such as the tails.

The expression for $E(N)$ can be manipulated to choose the best dominating hazard rate g from a parametrized class of hazard rates. This will not be explored any further.

2.5. DHR distributions. Dynamic thinning.

In this section we will try to obtain a black box generator for DHR distributions, i.e. a generator which does not require a priori explicit knowledge of the form of h. The method that will be given in this section is the method of **dynamic thinning**. This principle in itself is also useful for other distributions and for the nonhomogeneous Poisson process on the real line. The algorithm resembles the thinning algorithm, but the dominating hazard rate is dynamic, i.e. it varies during the execution of the algorithm.

The DHR distributions form a sub-class of the monotone densities because $f = h e^{-H}$, $h \downarrow$ and $H \uparrow$. It contains the Pareto distribution with parameter $a > 0$:

$$h(x) = \frac{a}{x+1} ,$$

the Weibull distribution with parameter $a \leq 1$ and the gamma distribution with parameter $a \leq 1$. The peak of the density is at 0, with value $f(0) = h(0)$. This value can of course be ∞ as for the gamma (a) density with $0 < a < 1$. The class has some desirable properties, for example, it is closed under convex combinations (see exercises), which means that mixtures of DHR distributions are again DHR.

The inversion method is based upon the fact that the solution X of $H(X) = E$ where E is exponentially distributed, has cumulative hazard rate H. But for DHR distributions, H is concave (its derivative h is nonincreasing). Thus, Newton-Raphson iterations started at 0 converge whenever $h(0) < \infty$:

Inversion method for DHR distributions

$X \leftarrow 0$

REPEAT

$$X \leftarrow X + \frac{E - H(X)}{h(X)}$$

UNTIL False

In practical applications, an appropriate stopping rule must be added. An exact solution usually requires infinite time (this is not the case if h is piecewise constant !). The thinning method, if it is to be used in black box mode, can only use the constant dominating hazard rate $g = h(0)$, in which case the expected number of iterations becomes

$$h(0)E(X) .$$

We recall however that DHR distributions have heavier-than-exponential tails. Thus, the fact that $E(N)$, the expected number of iterations, is proportional to $E(X)$ could be a serious drawback. The two prototype examples that we will consider throughout this section are the exponential density $(E(N) = h(0)E(X) = 1)$ and the Pareto (a) density

$$f(x) = \frac{a}{(1+x)^{a+1}},$$

for which $h(x) = \frac{a}{1+x}, H(x) = a \log(1+x), h(0) = a$, and, if $a > 1$, $E(X) = \frac{1}{a-1}$.
Thus,

$$E(N) = \begin{cases} \infty & 0 < a \leq 1 \\ \dfrac{a}{a-1} & a > 1 \end{cases}.$$

We are now ready to present the dynamic thinning algorithm:

Dynamic thinning algorithm for DHR distributions

$X \leftarrow 0$
REPEAT
    $\lambda \leftarrow h(X)$
    Generate an exponential random variate E and a uniform $[0,1]$ random variate U.
    $X \leftarrow X + \dfrac{E}{\lambda}$
UNTIL $\lambda U \leq h(X)$
RETURN X

The method uses thinning with a constant but continuously adjusted dominating hazard rate λ. When h decreases as X grows, so will λ. This forces the probability of acceptance up. The complexity can again be measured in terms of the number of iterations before halting, N. Note that the number of evaluations of h is $1+N$ (and not $2N$ as one might conclude from the algorithm shown above, because some values can be recuperated by introducing auxiliary variables). If $\lambda \leftarrow h(X)$ is taken out of the loop, and replaced at the top by $\lambda \leftarrow h(0)$, we obtain the standard thinning algorithm. While both algorithms do not require any knowledge about h except that h is DHR, a reduction in N is hoped for when dynamic thinning is used. In Devroye (1985), various useful upper bounds for $E(N)$ are obtained. Some of these are given in the next subsection and in the exercise section. The value of $E(N)$ is always less than or equal that of the thinning method. For example, for the Pareto (a) distribution, we obtain

$$E(N) = \frac{1}{\displaystyle\int_0^\infty e^{-z}\left(1+\frac{z}{a}\right)^{-1} dz},$$

which is finite for all $a > 0$. In fact, we have the following chain of inequalities showing the improvement over standard thinning:

$$E(N) = \cfrac{1}{\int\limits_0^\infty e^{-z}\left(1+\cfrac{z}{a}\right)^{-1} dz}$$

$$\leq \frac{a+1}{a} \quad \text{(use Jensen's inequality; note: } \cfrac{1}{\left(1+\cfrac{z}{a}\right)} \text{ is convex in } z \text{)}$$

$$< \frac{a}{a-1} \quad \text{(for all } a > 1\text{)}$$

$$= \mu = h(0)E(X).$$

For example, at $a = 1$, we have $E(N) \leq 2$ whereas $h(0)E(X) = \infty$.

2.6. Analysis of the dynamic thinning algorithm.

Throughout this section, we will use the following notation:

$$\mu = h(0)E(X),$$

$$\beta = \sup_{x \geq 0} \int\limits_0^\infty e^{-yh(x)}(h(x) - h(x+y)) \, dy,$$

$$\gamma = \sup_{x \geq 0} \frac{h(x)}{h\left(x + \cfrac{1}{h(x)}\right)},$$

$$\xi = E(\log_+(h(0)X))$$

where μ, β, γ and ξ are various quantities that will appear in the upper bounds for $E(N)$ given in this subsection. Note that ξ is the logarithmic moment of $h(0)X$, for which we have, by Jensen's inequality,

$$\xi \leq E(\log(h(0)X+1)) \leq \log(\mu+1) \leq \mu,$$

so that ξ is always finite when μ is finite. Obtaining an upper bound of the form $O(\xi)$ is, needless to say, strong proof that dynamic thinning is a drastic improvement over standard thinning. This is the goal of this subsection. Before we proceed with the technicalities, it is perhaps helpful to collect all the results.

Theorem 2.4.

The expected number of iterations in the dynamic thinning algorithm applied to a DHR distribution with bounded h does not exceed any of the following quantities:

A. μ;

B. $\dfrac{1}{1-\beta}$;

C. $\dfrac{e}{e-1}\gamma$;

D. γ (when h is also convex);

E. $(8\mu)^{\frac{1}{2}}+4(8\mu)^{\frac{1}{4}}$;

F. $\inf\limits_{p>2}\left(4p+2+\dfrac{2\xi}{\log(p-1)}\right)$;

G. $O\left(\dfrac{\xi}{\log(\xi)}\right)$ as $\xi\to\infty$;

H. $O\left(\dfrac{\log(\mu)}{\log\log(\mu)}\right)$ as $\mu\to\infty$.

Part A states that we have an improvement over standard thinning. Inequalities B and D are sharp: for example, for the exponential distribution, we have $\beta=0$, $\gamma=1$, which leads to $E(N)\le1$. Inequality B is also sharp for the Pareto family defined above. One can easily verify that

$$\beta = \int_0^\infty e^{-\frac{ya}{1+z}}\left(\frac{a}{1+x}-\frac{a}{1+x+y}\right)dy$$

$$= \int_0^\infty e^{-z}\left[1-\frac{1}{1+\frac{z}{a}}\right]dz$$

where we used the transformation $z=\dfrac{ay}{1+x}$. By carefully checking the induction argument used in the proof of Theorem 2.4, we see that for any $i\ge0$, $P(N>i)=\beta^i$ and thus that

$$E(N)= \frac{1}{\int_0^\infty e^{-z}\left(1+\frac{z}{a}\right)^{-1}dz}$$

$$= \frac{1}{1-\beta}\ .$$

Bounds C and D are never better than bound B, but often γ is easier to compute than β. For the Pareto family, we obtain via D,

$$E(N) \leq \gamma = \frac{a+1}{a} ,$$

a result that can be obtained from B via Jensen's inequality too. Inequalities E-H relate the size of the tail of X to $E(N)$, and give us more insight into the behavior of the algorithm. Of these, inequality H is perhaps the easiest to understand: $E(N)$ cannot grow faster than the logarithm of μ. Unfortunately, when $\mu = \infty$, it is of little help. In those cases, the logarithmic moment ξ is often finite. For example, this is the case for all members of the Pareto family. We will now prove Theorem 2.4. It requires a few Lemmas and other technical facts. Yet the proofs are instructive to those wishing to learn how to apply embedding techniques and well-known inequalities in the analysis of algorithms. Non-technical readers should most certainly not read beyond this point.

Proof of Theorem 2.4.

Part A. This part uses embedding. Consider the sequence of random vectors $(Y_1, h(0)U_1), (Y_2, h(0)U_2), \ldots$ where the U_i's are iid uniform $[0,1]$ random variables, and $0 = Y_0 < Y_1 < Y_2 < \cdots$ are defined by the relations:

$$Y_{i+1} = Y_i + \frac{E_{i+1}}{h(0)}$$

where $E_1, E_2, \ldots$ are iid exponential random variates. This is the sequence considered in standard thinning, where we stop when for the first time $h(0)U_i \leq h(Y_i)$. We recall from Theorem 2.3 that in that case $E(N) = \mu = E(h(0)X)$. Let us use starred random variables for the subsequence satisfying $h(0)U_i \leq h(Y_{i-1})$. Observe first that this sequence is distributed as the sequence of random vectors used in dynamic thinning. Then, part A follows without work because we still stop when the first random vector falling below the curve of h is encountered.

Part B. Let the E_i's be as before. The sequence $Y_0 < Y_1 < \cdots$ used in dynamic thinning satisfies: $Y_0 = 0$, and

$$Y_{i+1} = Y_i + \frac{E_{i+1}}{h(Y_i)} .$$

Note that this is the sequence of possible candidates for the returned random variate X in the algorithm. The index i refers to the iteration. Taking the stopping rule into account, we have for $i \geq 1$,

$$P(N > i \mid Y_0, \ldots, Y_i) = \prod_{j=1}^{i} (1 - \frac{h(Y_j)}{h(Y_{j-1})}) .$$

Thus, for $i \geq 2$,

$$P(N > i \mid Y_0, \ldots, Y_{i-1})$$

$$= \prod_{j=1}^{i-1} (1 - \frac{h(Y_j)}{h(Y_{j-1})}) \int_0^\infty e^{-yh(Y_{i-1})}(h(Y_{i-1}) - h(Y_{i-1} + y)) \, dy$$

$$\leq \beta \prod_{j=1}^{i-1} (1 - \frac{h(Y_j)}{h(Y_{j-1})}),$$

and we obtain, by a simple induction argument on i, that

$$P(N > i) \leq \beta^i \qquad (i \geq 0).$$

Thus,

$$E(N) = \sum_{i=0}^\infty P(N > i) \leq \frac{1}{1-\beta}.$$

Part C. Part C is obtained from B by bounding β from above. Fix x and $c > 0$. Then

$$\int_0^\infty e^{-yh(x)}(h(x) - h(x+y)) \, dy$$

$$\leq \int_{y > \frac{c}{h(x)}} e^{-yh(x)}(h(x) - h(x+y)) \, dy + \int_0^{\frac{c}{h(x)}} e^{-yh(x)}(h(x) - h(x+y)) \, dy$$

$$\leq \int_c^\infty e^{-z} \, dz + \int_0^{\frac{c}{h(x)}} e^{-yh(x)}(h(x) - h(x + \frac{c}{h(x)})) \, dy$$

$$= e^{-c} + (1 - e^{-c}) \left[\frac{h(x + \frac{c}{h(x)})}{h(x)} \right]$$

$$= 1 - (1 - e^{-c}) \frac{h(x + \frac{c}{h(x)})}{h(x)}.$$

Inequality C follows after taking $c = 1$.

Part D. Inequality D follows by applying Jensen's inequality to an intermediate expression in the preceding chain of inequalities:

$$\int_0^\infty e^{-yh(x)}(h(x) - h(x+y)) \, dy$$

$$= \int_0^\infty e^{-yh(x)}h(x)(1 - \frac{h(x+y)}{h(x)}) \, dy$$

$$\leq 1 - \frac{1}{h(x)} h \left[x + \int_0^\infty e^{-yh(x)}h(x)y \, dy \right]$$

$$= 1 - \frac{1}{h(x)} h(x + \frac{1}{h(x)}) .$$

Lemma 2.1, needed for parts E-H. We will show that for $x \geq 0$, $p > 2$, and integer m in $\{0, 1, \ldots, n\}$,

$$P(N > n) \leq P(X > x) + \frac{h(0)x}{p^{n-m}} + (1 - \frac{1}{p})^m \qquad (n > 0) .$$

Define the E_i and Y_i sequences as in the proof of part B, and let $U_1, U_2, \ldots$ be a sequence of iid uniform $[0,1]$ random variables. Note that the random variate X returned by the algorithm is Y_N where N is the first index i for which $U_i h(Y_{i-1}) \leq h(Y_i)$. Define N_1, N_2 by:

$$N_1 = \sum_{i=1}^{n} I_{[h(Y_i) \leq \frac{1}{p} h(Y_{i-1})]} ,$$

$$N_2 = \sum_{i=1}^{n} I_{[h(Y_i) > \frac{1}{p} h(Y_{i-1})]} .$$

Then we can write the following:

$$[N > n] \subseteq [X > x] \cup [X \leq x, N_1 \geq n - m, N > n] \cup [N_2 \geq m, N > n] .$$

Now,

$$P(X \leq x, N_1 \geq n - m, N > n) \leq P(E_1 \leq \frac{xh(0)}{p^{n-m}}) \leq \frac{xh(0)}{p^{n-m}}$$

and

$$P(N_2 \geq m, N > n) \leq P(N > n \mid N_2 \geq m) \leq (1 - \frac{1}{p})^m .$$

This concludes the proof of the Lemma.

Part E. Consider Lemma 2.1, and take $x = x_n$ random, independent of X and uniformly distributed on $[\frac{n}{Ch(0)}, \frac{n+1}{Ch(0)}]$ where $C > 0$ is a constant to be chosen further on. Take $m = m_n = \lfloor \frac{n}{2} \rfloor$, and take p constant and independent of n.

We will apply the formula

$$E(N) = \sum_{n=0}^{\infty} P(N > n)$$

and use Lemma 2.1, averaged over x_n. This yields an upper bound consisting of three terms:

(1)

$$\sum_{n=0}^{\infty} P(X > x_n) = \sum_{n=0}^{\infty} \int_{n}^{n+1} P(Ch(0)X > t) \, dt$$

$$= \int_0^\infty P\,(Ch\,(0)X > t)\,\,dt\ =\ E\,(Ch\,(0)X)\ =\ C\,\mu\ .$$

(11)

$$\sum_{n=0}^\infty (1-\frac{1}{p})^{m_n}\ =\ 1+2\sum_{j=1}^\infty (1-\frac{1}{p})^{j}\ =\ 1+\frac{2(1-\frac{1}{p})}{\frac{1}{p}}\ =\ 2p-1\ .$$

(111)

$$\frac{1}{C}\sum_{n=0}^\infty (\int_n^{n+1} t\ dt\,)p^{-(n-m_n)}\ =\ \frac{1}{C}\sum_{n=0}^\infty \frac{2n+1}{2}p^{-(n-m_n)}$$

$$=\ \frac{2}{C}\sum_{n=0}^\infty (2n+1)p^{-n}$$

$$=\ \frac{2}{C}(\frac{1}{1-\frac{1}{p}}+2\sum_{n=0}^\infty np^{-n}\,)$$

$$=\ \frac{2}{C}(\frac{1}{1-\frac{1}{p}}+\frac{2}{p}\frac{1}{(1-\frac{1}{p})^2})$$

$$=\ \frac{2}{C}\frac{1+\frac{1}{p}}{(1-\frac{1}{p})^2}\ .$$

These estimates are substituted in

$$E\,(N)\ \le\ 1+\sum_{n=1}^\infty (P\,(X > x_n\,)+\frac{h\,(0)x_n}{p^{n-m_n}}+(1-\frac{1}{p})^{m_n}\ .$$

This gives the upper bound

$$E\,(N)\ \le\ 1+C\,\mu-P\,(X > x_0)+2(p-1)+\frac{2}{C}(\frac{p\,(p+1)}{(p-1)^2}-\frac{1}{4})\ .$$

Since $h\,(0)X$ is stochastically greater than an exponential random variate, we have

$$P\,(X > x_0)\ =\ \int_0^1 P\,(Ch\,(0)X > t)\,\,dt\ \ge\ \int_0^1 e^{-\frac{t}{C}}\,\,dt$$

$$=\ C\int_0^{\frac{1}{C}} e^{-z}\,\,dz\ =\ C\,(1-e^{-\frac{1}{C}})\ \ge\ 1-\frac{1}{2C}\ .$$

Thus,

$$E(N) \leq C\mu + 2(p-1) + \frac{2}{C}\frac{p(p+1)}{(p-1)^2} .$$

The optimal choice for C is

$$C = \sqrt{\frac{2p(p+1)}{\mu(p-1)^2}} ,$$

which, after substitution, gives

$$E(N) \leq 2(p-1) + \sqrt{8\mu}\sqrt{\frac{p(p+1)}{(p-1)^2}}$$

$$< 2(p-1) + \sqrt{8\mu}\frac{p+1}{p-1}$$

$$= 2(p-1) + \frac{2\sqrt{8\mu}}{p-1} + \sqrt{8\mu} .$$

The right-hand-side is minimal for $p-1 = (8\mu)^{\frac{1}{4}}$, and this choice gives inequality E.

Part F. In Lemma 2.1, replace n by $2j$, and sum over j. Set $m_{2j} = j$, $p_{2j} = p > 2$, and $h(0)x_{2j} = (p-1)^j$. Since for any random variable Z,

$$\sum_{j=0}^{\infty} P(Z > j) \leq 1 + \int_0^{\infty} P(Z > t)\, dt = 1 + E(Z_+) ,$$

we see that

$$E(N) \leq 2\sum_{j=0}^{\infty} P(N > 2j)$$

$$\leq 2\sum_{j=0}^{\infty} (P(h(0)X > (p-1)^j) + 2(1-\frac{1}{p})^j)$$

$$= 2\sum_{j=0}^{\infty} P(\frac{\log_+(h(0)X)}{\log(p-1)} > j) + 4p$$

$$\leq 2E(\frac{\log_+(h(0)X)}{\log(p-1)}) + 4p + 2 .$$

Part G. Inequality G follows from inequality F for the following choice of p:

$$p = 2 + \frac{\xi}{2\log^2(1+\xi)} .$$

This value was obtained as follows: inequality F is sharpest when p is picked as the solution of $(p-1)\log^2(p-1) = \frac{\xi}{2}$. But because we want $p > 2$, and because we want a good p for large values of ξ, it is good to obtain a rough solution by functional iteration, and then adding 2 to this to make sure that the restrictions on p are satisfied. Resubstitution yields:

$$E(N) \leq 10 + \frac{2\xi}{\log^2(1+\xi)} + \frac{2\xi}{\log(1 + \frac{\xi}{2\log^2(1+\xi)})} ,$$

which is $O\left(\dfrac{\xi}{\log(\xi)}\right)$ as $\xi \to \infty$.

Part H. Use the bound of part G, and the fact that $\xi \leq \log(1+\mu)$. In fact, we have shown that

$$E(N) \leq (2+o(1))\frac{\log(\mu)}{\log\log(\mu)}$$

as $\mu \to \infty$. ∎

2.7. Exercises.

1. Sketch the hazard rate for the halfnormal density for $x > 0$. Determine whether it is monotone, and show that $\lim\limits_{x \uparrow \infty} \dfrac{h(x)}{x} = 1$.

2. Give an efficient algorithm for the generation of random variates from the left tail of the extreme value distribution truncated at $c < 0$ (the extreme value distribution function before truncation is $e^{-e^{-x}}$). Hint: when E is exponentially distributed, then $\dfrac{1}{b}\log(1+bEe^{-a})$ has hazard rate $h(x) = e^{a+bx}$ for $x > 0$, $b > 0$.

3. Show that when H is a cumulative hazard rate on $[0,\infty)$, then $\dfrac{H(x)}{x}$ is a hazard rate on $[0,\infty)$. Assume now that random variates with cumulative hazard rate H are easy to generate. How would you generate random variates with hazard rate $\dfrac{H(x)}{x}$?

4. Prove that $\dfrac{1}{x}$ cannot be a hazard rate on $[0,\infty)$.

5. Construct a hazard rate on $[0,\infty)$, continuous at all points except at $c > 0$, having the additional properties that $h(x) > 0$ for all $x > 0$, and that $\lim\limits_{x \uparrow c} h(x) = \lim\limits_{x \downarrow c} h(x) = \infty$.

6. In this exercise, we consider a tight fit for the thinning method: $M = \int(g-h) < \infty$. Show first that

$$E(N) \leq 1 + \int_0^\infty (g-h) .$$

    Prove also that the probability that N is larger than Me decreases very rapidly to 0, by establishing the inequality

$$P(N \geq i) \leq e^{-M}\left(\frac{eM}{i}\right)^i \quad (i \geq M) .$$

    To do this, start with $P(N \geq i) \leq e^{-ti}E(e^{tN})$ where $t \geq 0$ is arbitrary (this is Jensen's inequality). Evaluate the expected value, bound this value by introducing M, and optimize with respect to t.

7. Consider the family of hazard rates $h_b(x) = \dfrac{x}{1+bx}$ $(x > 0)$, where $b > 0$ is a parameter. Discuss random variate generation for this family. The average time needed per random variate should remain uniformly bounded over b.

8. Give an algorithm for the generation of random variates with hazard rate $h_b(x) = b + x$ $(x > 0)$ where $b \geq 0$ is a parameter. Inversion of an exponential random variate requires the evaluation of a square root, which is considered a slow operation. Can you think of a potentially faster method?

9. Develop a thinning algorithm for the family of gamma densities with parameter $a \geq 1$ which takes expected time uniformly bounded over a.

10. The hazard rate has infinite peaks at all locations at which the density has infinite peaks, plus possibly an extra infinite peak at ∞. Construct a monotone density f which is such that it oscillates infinitely often in the following extreme sense:

    $$\lim_{x \uparrow \infty} \sup\ h(x) = \infty\ ;$$

    $$\lim_{x \uparrow \infty} \inf\ h(x) = 0\ .$$

    Notice that h is neither DHR nor IHR.

11. If X is a random variate with hazard rate h, and ψ is a suitable smooth monotone transformation, give a formula for the hazard rate of $\psi(X)$ and conditions under which your formula is valid. See Gaver (1979) for several examples of such transformations.

12. Show that a mixture of DHR distributions is again a DHR distribution (Barlow, Marshall and Proschan, 1963).

13. Show that for any DHR random variable X, $\mu = E(h(0)X) \geq 1$.

14. Construct a DHR distribution for which the logarithmic moment $\xi = E(\log_+(h(0)X)) = \infty$.

15. For the Pareto family (density $f(x) = \dfrac{a}{(1+x)^{a+1}}$, $x > 0$), find the rate of increase of ξ, the logarithmic moment, as $a \downarrow 0$ (the answer should be of the form: $\xi \sim$ simple expression involving a).

16. Develop a black box method for DHR distributions with $h(0) = \infty$.

17. Let the hazard rate h be piecewise constant with breakpoints at $0 = x_0 < x_1 < x_2 < \cdots$ and values h_i on $(x_{i-1}, x_i]$, $i \geq 1$. Assume that these numbers are given in an infinite table. Describe the inversion algorithm. Determine the expected number of iterations as a function of the x_i's and the h_i's.

18. Show that for the dynamic thinning method for DHR distributions, $E(N) \leq 4 + \sqrt{24\mu}$, where $\mu = E(h(0)X)$ (Devroye, 1985).

19. This exercise is concerned with an improvement over inequalities $F - H$ in Theorem 2.4. Define the random variable $Y = \log_+(h(0)X)$, and the quantity

    $$\chi = E\left(\frac{Y}{\log(1+Y)}\right).$$

A. Show that $\chi < \infty$ implies $\xi < \infty$ (try to do this by establishing an inequality).

B. Show by example that there exists a density f for which $\chi < \infty$, yet $\xi = \infty$.

C. Find positive constants $a > 0, b > 0$ such that for the dynamic thinning method, $E(N) \leq a + b\chi$. Hint: in Lemma 2.1, choose

$$p = p_n = \frac{n}{\log^3(n+1)}, \qquad m = m_n = \left\lfloor \frac{n}{\log(n+1)} \right\rfloor,$$

$$x = x_n = \frac{1}{h(0)} e^{n \log n - 4n \log\log(n+1)}; \text{ and use } E(N) \leq n_0 + \sum_{n=n_0}^{\infty} P(N > n)$$

for an appropriate n_0 (Devroye, 1985).

3. GENERATING RANDOM VARIATES WITH A GIVEN DISCRETE HAZARD RATE.

3.1. Introduction.

Assume that we wish to generate a random variate with a given probability vector $p_1, p_2, \ldots$, and that the **discrete hazard rate function** $h_n, n = 1, 2, \ldots$ is given, where

$$h_n = \frac{p_n}{q_n},$$

$$q_n = \sum_{i=n}^{\infty} p_i.$$

One verifies quickly that

$$p_n = h_n \prod_{i < n} (1 - h_i).$$

In some applications, the original probability vector of p_n's has a more complicated form than the discrete hazard rate function.

The general methods for random variate generation in the continuous case have natural extensions here. As we will see, the role of the exponential distribution is inherited by the geometric distribution. In different sections, we will briefly touch upon various techniques, while examples will be drawn from the classes of logarithmic series distributions and negative binomial distributions. In general, if we have finite-valued random variables that remain fixed throughout the simulation, table methods should be used. Thus, it seems appropriate to draw all the examples from classes of distributions with unbounded support.

3.2. The sequential test method.

The following method will be called the **sequential test method**. Although it is conceptually very simple, it seems to have been formally proposed for the first time by Shanthikumar (1983,1985).

Sequential test method

$X \leftarrow 0$
REPEAT
      Generate a uniform [0,1] random variate U.
      $X \leftarrow X + 1$
UNTIL $U \leq h_X$
RETURN X

The validity of this method follows directly from the fact that all h_n's are numbers in [0,1], and that

$$p_n = h_n \prod_{i<n} (1-h_i) .$$

It is obvious that the number of iterations needed here is equal to X. The strength of this method is that it is universally applicable, and that it can be used in the black box mode. When it is compared with the inversion method for discrete random variates, one should observe that in both cases the expected number of iterations is $E(X)$, but that in the inversion method, only one uniform random variate is needed, versus one uniform random variate per iteration in the sequential test method. If h_n is computed in $O(1)$ time and p_n is computed as the product of n factors involving $h_1, \ldots, h_n$, then the expected time of the inversion method grows as $E(X^2)$. Fortunately, there is a simple recursive formula for p_n :

$$p_{n+1} = p_n (\frac{h_{n+1}}{h_n})(1-h_n) .$$

Thus, if the p_n's are computed recursively in this manner, the inversion method takes expected time proportional to $E(X)$, and the performance should be comparable to that of the sequential test method.

3.3. Hazard rates bounded away from 1.

Consider the class of discrete hazard rates h_n with supremum $\rho < 1$. This class will be called the class $H(\rho)$. For such hazard rates, the sequential test method can be accelerated by observing that we can jump ahead more than 1 in each iteration. To see this, assume that X is geometrically distributed with parameter p:

$$P(X = n) = p(1-p)^{n-1} \quad (n \geq 1).$$

Then X has hazard rate $h_n = p$. But in that case the sequential test method counts the number of iid uniform $[0,1]$ random variates generated until for the first time a number smaller than p is obtained. This is of course known to be geometrically distributed with parameter p. In this special case, the individual uniform random variates can be avoided, because we can generate X directly by inversion of a uniform random variate U as

$$X \leftarrow \left\lceil \frac{-\log U}{-\log(1-p)} \right\rceil$$

or as $X \leftarrow \left\lceil \dfrac{E}{-\log(1-p)} \right\rceil$, where E is an exponential random variate. For the limit case $p = 1$, we have $X = 1$ with probability one. The smaller p, the more dramatic the improvement. For non-geometric distributions, it is possible to give an algorithm which parallels to some extent the thinning algorithm.

Thinning method for discrete distributions

NOTE: This algorithm is valid for hazard rates in $H(\rho)$ where $\rho \in (0,1]$ is a given number.
$X \leftarrow 0$
REPEAT
      Generate iid uniform $[0,1]$ random variates U, V.
$$X \leftarrow X + \left\lceil \frac{\log U}{\log(1-\rho)} \right\rceil$$
UNTIL $V \leq \dfrac{h_X}{\rho}$
RETURN X

This algorithm is due to Shanthikumar (1983,1985). We have to show that it is valid, and verify what the expected time complexity is.

> **Theorem 3.1. (Shanthikumar, 1983,1985)**
>
> The discrete thinning method generates a random variate with discrete hazard rate h_n .

Proof of Theorem 3.1.

Let $G_1, G_2, \ldots$ be the sequence of iid geometric (p) random variates used in the discrete thinning method. Let X be the returned random variate. Thus, $X = G_1 + \cdots + G_N$ where N is the number of iterations. Let us define the partial sums $S_n = \sum_{i=1}^{n} G_i$. Thus, $X = S_N$. We compute the probability $P(S_N = n)$ from the following formula:

$$P(X = n, N = k+1, S_1 = n_1, \ldots, S_r = n_k)$$
$$= h_n \, \rho^k \, (1-\rho)^{n-1-k} \prod_{i=1}^{k} (1 - \frac{h_{n_i}}{\rho}) \quad (k \leq n-1) \ .$$

This can be seen by just computing individual probabilities of independent events. To obtain $P(X = n)$, it suffices to sum over all possible values of k and n_i . We note now that the following multinomial expansion is valid:

$$\prod_{i=1}^{n-1} (\rho(1 - \frac{h_i}{\rho}) + 1 - \rho)$$
$$= \sum_{k=0}^{n-1} \rho^k \, (1-\rho)^{n-1-k} \, (\sum_{1 \leq n_1 < n_2 < \cdots \leq n_k \leq n-1} \prod_{i=1}^{k} (1 - \frac{h_{n_i}}{\rho})) \ .$$

Thus,

$$P(X = n) = \rho \frac{h_n}{\rho} \prod_{i=1}^{n-1} (\rho(1 - \frac{h_i}{\rho}) + 1 - \rho)$$
$$= h_n \prod_{i=1}^{n-1} (1 - h_i) \quad (n = 1, 2, \ldots) \ ,$$

which was to be shown. ∎

If we use the algorithm with $\rho = 1$ (which is always allowed), then the sequential test algorithm is obtained. For some distributions, we are forced into this situation. For example, when X has compact support, with $p_n > 0, p_{n+i} = 0$ for some n and all $i \geq 1$, then $h_n = 1$. In any case, we have

Theorem 3.2.

For the discrete thinning algorithm, the expected number of iterations $E(N)$ can be computed as follows:

$$E(N) = \rho E(X).$$

Proof of Theorem 3.2.

We observe that in the notation of the proof of the previous theorem,

$$X = \sum_{i=1}^{N} G_i,$$

so that by Wald's equation,

$$E(X) = E(N)E(G_1) = E(N)\rho,$$

which was to be shown. ∎

Example 3.1. The logarithmic series distribution.

For the logarithmic series distribution defined by

$$P(X=n) = \frac{1}{-\log(1-\theta)} \frac{\theta^n}{n} \quad (n \geq 1),$$

where $\theta \in (0,1)$ is a parameter, we observe that h_n is not easy to compute (thus, some preprocessing seems necessary for this distribution). However, several key properties of h_n can be obtained with little difficulty:

(i) $\dfrac{1}{h_n} = 1 + \dfrac{n\theta}{n+1} + \dfrac{n\theta^2}{n+2} + \cdots$;

(ii) $h_n \downarrow 1-\theta$ as $n \to \infty$;

(iii) $\rho = \sup\limits_n h_n = h_1 = \dfrac{\theta}{-\log(1-\theta)}$.

Thus, while the sequential method has $E(N)=E(X)=\dfrac{\rho}{1-\theta}$, the discrete thinning method satisfies

$$E(N) = \rho E(X) = \frac{\rho^2}{1-\theta}.$$

Since $\rho \to 0$ as $\theta \to 1$, we see that the improvement in performance can be dramatic. Unfortunately, even with the thinning method, we do not obtain an algorithm that is uniformly fast over all values of θ:

$$\sup_{\theta \in (0,1)} E(N) = \infty \ . \ \blacksquare$$

3.4. Discrete dynamic thinning.

Shanthikumar (1983,1985) has also observed that for distributions with decreasing discrete hazard rate (also referred to below as DHR distributions) that the value of ρ can be dynamically modified to increase the jumps for the geometric random variates, and thus increase the performance. The formal algorithm is given below.

Dynamic thinning method for discrete DHR distributions

$X \leftarrow 0$

REPEAT

　　Generate iid uniform [0,1] random variates U, V.

　　$\rho \leftarrow h_{X+1}$

　　$X \leftarrow X + \left\lceil \dfrac{\log U}{\log(1-\rho)} \right\rceil$

UNTIL $V \leq \dfrac{h_X}{\rho}$

RETURN X

The validity of this algorithm follows by a short recursive argument:

$P(X > n \mid X > n-1,$ the last partial sum of geometric variates

less than n takes the value k)

$= (1-h_k) + h_k (1 - \dfrac{h_n}{h_k})$

$= 1 - h_n \ .$

Thus, because this does not depend upon k,

$P(X > n) = (1-h_n) P(X > n-1)$

$= \displaystyle\prod_{i=1}^{n} (1-h_i) \ .$

3.5. Exercises.

1. Prove the following for the logarithmic series distribution with parameter $\theta \in (0,1)$:

   (1) $\dfrac{1}{h_n} = 1 + \dfrac{n\,\theta}{n+1} + \dfrac{n\,\theta^2}{n+2} + \cdots$,

   (11) $h_n \downarrow 1-\theta$ as $n \to \infty$,

   (111) $E(X) = \dfrac{\theta}{-\log(1-\theta)} \dfrac{1}{1-\theta}$.

2. Assume that discrete dynamic thinning is used for a DHR distribution. Obtain good upper bounds for $E(N)$ in terms of the size of the tail of the distribution. Show also that for the logarithmic series distribution the value of $E(N)$ is not uniformly bounded in $\theta \in (0,1)$, the parameter of the distribution.

3. Show that in the discrete thinning algorithm, quick acceptance and rejection steps can be introduced that would effectively reduce the expected number of evaluations of h_n. Compute the expected number of such evaluations for two squeezing sequences.

4. A continuation of exercise 3. For the logarithmic series distribution with parameter θ, show that

   $$1-\theta \le h_n \le \dfrac{n\,(1-\theta)+1}{n+1} \quad (n \ge 1) .$$

   Show that if these bounds are used for squeeze steps in the discrete dynamic thinning method, then the expected number of evaluations of h_n is $o\,(1)$ as $\theta \uparrow 1$. (The inequalities are due to Shanthikumar (1983,1985).)

5. **The negative binomial distribution.** A random variable Y has the negative binomial distribution with parameters (k,p) where $k \ge 1, p \in (0,1)$ if

   $$P(Y=n) = \binom{n-1}{k-1} p^k (1-p)^{n-k} \quad (n \ge k) .$$

   Then, the normalized random variable $X = Y - k + 1$ has a distribution on all positive integers. For this random variable X, show that $h_n \uparrow p$ as $n \uparrow \infty$. (Hint: the relationship

   $$h_{n+1} = (\dfrac{n+k-1}{n})(\dfrac{h_n}{1-h_n})(1-p) \quad (n \ge 1)$$

   is helpful.) Show that in the sequential test algorithm, $E(N) = \dfrac{k}{p}$, while in the discrete thinning algorithm (with $\rho = p$), we have $E(N) = k$. Compare this algorithm with the algorithm based upon the observation that Y is distributed as the sum of k iid geometric (p) random variates. Finally, show the squeeze type inequalities

   $$1 - \dfrac{n+k-1}{n}(1-p) \le h_n \le p \quad (n \ge 1) .$$

6. Example 3.1 for the logarithmic series distribution and the previous exercise
    for the negative binomial distribution require the computation of h_n. This
    can be done by setting up a table up to some large value. If the parameters
    of the distributions change very often, this is not feasible. Show that we can
    compute the sequence of values recursively during the generation process by

$$h_1 = p_1 ;$$

$$h_{n+1} = \frac{p_{n+1}}{p_n} \frac{h_n}{1-h_n} .$$

Chapter Seven
UNIVERSAL METHODS

1. BLACK BOX PHILOSOPHY.

In the next two chapters we will apply the tools of the previous chapters in the design of algorithms that are applicable to large families of distributions. Described in terms of a common property, such as the family of all unimodal densities with mode at 0, these families are generally speaking nonparametric in nature. A method that is applicable to such a large family is called a **universal method**. For example, the rejection method can be used for all bounded densities on [0,1], and is thus a universal method. But to actually apply the rejection method correctly and efficiently would require knowledge of the supremum of the density. This value cannot be estimated in a finite amount of time unless we have more information about the density in question, usually in the form of an explicit analytic definition. Universal methods which do not require anything beyond what is given in the definition of the family are called black box methods.

Consider for example all discrete distributions on the positive integers. Assume only that for each i we can evaluate p_i (consider this evaluation as being performed by a black box). Then the sequential inversion method (section III.2) can be used to generate a random variate with this distribution, and can thus be called a black box method for this family. The inversion method for distributions with a continuous distribution function is not a black box method because finite time generation is only possible in special cases (e.g., the distribution function is piecewise linear).

The larger the family for which we design a black box method, the less we should expect from the algorithm timewise: a case in point is the sequential inversion method for discrete random variates. The undeniable advantage of having a few black box methods in one's computer library is that one can always fall back on these when everything else fails. Comparative timings with algorithms specially designed for particular distributions are not fair.

In chapters IX and X we will mainly be concerned with fast algorithms for parametric families that are widely used by the statistical community. In this chapter too, we will be concerned with speed, but it is by no means the driving force. Because continuous distributions are more difficult to handle in general, we will only focus on families with densities. In section 2, we present a case study for the class of log-concave densities, to wet the appetite. Since the whole story in black box methods is told in terms of inequalities when the rejection method is involved, it is important to show how standard probability theoretical inequalities can aid in the design of black box algorithms. This is done in section 3. In section 4, the inversion-rejection principle is presented, which combines the sequential inversion method for discrete random variates with the rejection method. It is demonstrated there that this method can be used for the generation of random variables with a unimodal or monotone density.

2. LOG-CONCAVE DENSITIES.

2.1. Definition.

A density f on R^d is called **log-concave** when $\log f$ is concave on its support. In this section we will obtain universal methods for this class of densities when $d=1$. The class of densities is very important in statistics. A partial list of member densities is given in the table below.

Name of density	Density	Parameter(s)
Normal	$\dfrac{1}{\sqrt{2\pi}}e^{-\frac{x^2}{2}}$	
Gamma (a)	$\dfrac{x^{a-1}e^{-x}}{\Gamma(a)}$ $(x>0)$	$a>1$
Weibull (a)	$ax^{a-1}e^{-x^a}$ $(x>0)$	$a\geq 1$
Beta (a,b)	$\dfrac{x^{a-1}(1-x)^{b-1}}{B(a,b)}$ $(0\leq x\leq 1)$	$a,b\geq 1$
Exponential power (a)	$\dfrac{e^{-|x|^a}}{2\Gamma(1+\frac{1}{a})}$	$a\geq 1$
Perks (a)	$\dfrac{c}{e^x+e^{-x}+a}$	$a>-2$
Logistic	same as above, $a=2$	
Hyperbolic secant	same as above, $a=0$	
Extreme value (k)	$\dfrac{k^k}{(k-1)!}e^{-kx-ke^{-x}}$	$k\geq 1$,integer
Generalized inverse gaussian	$cx^{a-1}e^{-bx-\frac{b*}{x}}$ $(x\geq 0)$	$a\geq 1,b,b*>0$

Important individual members of this family also include the uniform density (as a special case of the beta family), and the exponential density (as a special case of the gamma family). For studies on the less known members, see for example Perks (1932) (for the Perks densities), Talacko (1956) (for the hyperbolic secant density), Gumbel (1958) (for the extreme value distributions) and Jorgensen (1982) (for the generalized inverse gaussian densities).

The family of log-concave densities on R is also important to the mathematical statistician because of a few key properties involving closedness under certain operations: for example, the class is closed under convolutions (Ibragimov (1956), Lekkerkerker (1953)).

The algorithms of this section are based upon rejection. They are of the black box type for all log-concave densities with mode at 0 (note that all log-concave densities are bounded and have a mode, that is, a point x such that f is nonincreasing on $[x,\infty)$ and nondecreasing on $(-\infty,x]$). Thus, the mode must be given to us beforehand. Because of this, we will mainly concentrate on the class $LC_{0,1}$, the class of all log-concave densities with a mode at 0 and $f(0)=1$. The restriction $f(0)=1$ is not crucial: since $f(0)$ can be computed at run-time, we can always rescale the axis after having computed $f(0)$ so that the value of $f(0)$ after rescaling is 1. We define LC_0 as the class of all log-concave densities with a mode at 0.

The bottom line of this section is that there is a rejection-based black box method for LC_0 which takes expected time uniformly bounded over this class if the computation of f at any point and for any f takes one unit of time. The algorithm can be implemented in about ten lines of FORTRAN or PASCAL code. The fundamental inequality needed to achieve this is developed in the next sub-section. All of the results in this section were first published in Devroye (1984).

2.2. Inequalities for log-concave densities.

Theorem 2.1.

Assume that f is a log-concave density on $[0,\infty)$ with a mode at 0, and that $f(0)=1$. Then $f(x)\leq g(x)$ where

$$g(x) = \begin{cases} 1 & (0\leq x \leq 1) \\ \text{the unique solution } t<1 \text{ of } t=e^{-x(1-t)} & (x>1) \end{cases}.$$

The inequality cannot be improved because g is the supremum of all densities in the family.

Furthermore, for any log-concave density f on $[0,\infty)$ with mode at 0,

$$\int_x^\infty f \leq e^{-xf(0)} \quad (x\geq 0).$$

Proof of Theorem 2.1.

We need only consider the case $x > 1$. The density f in the given class which yields the maximal value of $f(x)$ when $x > 1$ is fixed is given by

$$\log f(u) = \begin{cases} -au & (0 \le u \le x) \\ -\infty & (x < u) \end{cases}$$

for some $a > 0$. Thus, $f(u) = e^{-au}$, $0 \le u \le x$. Here a is chosen for the sake of normalization. We must have

$$1 = \frac{1 - e^{-ax}}{a}.$$

Replace $1 - a$ by t.

The second part of the theorem follows by a similar geometrical argument. First fix $x > 0$. Then notice that the tail probability beyond x is maximal for the exponential density, which because of normalization must be of the form $f(0)e^{-yf(0)}$, $y \ge 0$. The tail probability is $e^{-xf(0)}$. ∎

Theorem 2.2.

The function g of Theorem 2.1 can be bounded by two sequences of functions $y_n(x), z_n(x)$ for $x > 1$, where

(i) $0 = z_0(x) \le z_1(x) \le \cdots \le g(x)$;

(ii) $g(x) \le \cdots \le y_1(x) \le y_0(x) = \dfrac{1}{x}$;

(iii) $\lim_{n \to \infty} y_n(x) = g(x)$;

(iv) $\lim_{n \to \infty} z_n(x) = g(x)$;

(v) $y_{n+1}(x) = e^{-x(1 - y_n(x))}$;

(vi) $z_{n+1}(x) = e^{-x(1 - z_n(x))}$.

Proof of Theorem 2.2.

Fix $x > 1$. Consider the functions $f_1(u) = u$ and $f_2(u) = e^{-x(1-u)}$ for $0 \le u \le 1$. We have $f_1(1) = f_2(1) = 1$, $f'_2(1) = x > 1 = f'_1(1)$, $f'_2(0) = xe^{-x} < 1 = f'_1(0)$. Also, f_2 is convex and increases from e^{-x} at $u = 0$ to 1 at $u = 1$. Thus, there exists precisely one solution in $(0,1)$ for the equation $f_1(u) = f_2(u)$. This solution can be obtained by ordinary functional iteration: if one starts with $z_0(x) = 0$, and uses $z_{n+1}(x) = f_2(z_n(x))$, then the unique solution is approached from below in a monotone manner. If we start with $y_0(x)$ at least equal to the value of the solution, then the functional iteration $y_{n+1}(x) = f_2(y_n(x))$ can be used to approach the solution from above in a

monotone way. Since $f(x) \leq \dfrac{1}{x}$ for all monotone densities f on $[0,\infty)$, we have
$g(x) \leq \dfrac{1}{x}$, and thus, we can take $y_0(x) = \dfrac{1}{x}$. ∎

When f is a log-concave density on $[m,\infty)$ with mode at m, then

$$\frac{f\left(m + \dfrac{x}{f(m)}\right)}{f(m)} \leq \min(1, e^{1-x}) \qquad (x \geq 0) .$$

The area under the bounding curve is exactly 2. The inequality applies to all log-concave densities with mode at m (in which case the condition $x > 0$ must be dropped and $1-x$ is replaced by $1 - |x|$). But unfortunately, the area under the dominating curve becomes 4. The two features that make the inequality useful for us are

(i)   The fact that the area under the curve does not depend upon f. (This gives us a uniform guarantee about its performance.)

(ii) The fact that the top curve itself does not depend upon f. (This is a necessary condition for a true black box method.)

2.3. A black box algorithm.

Let us start with the rejection algorithm based upon the inequality

$$\frac{f\left(m + \dfrac{x}{f(m)}\right)}{f(m)} \leq \min(1, e^{1-x}) \qquad (x \geq 0)$$

valid for log-concave densities on $[m,\infty)$ with mode at m:

Rejection algorithm for log-concave densities

[SET-UP](can be omitted)
$c \leftarrow f(m)$
[GENERATOR]
REPEAT
      Generate U uniformly on [0,2] and V uniformly on [0,1].
      IF $U \leq 1$
          THEN $(X,Z) \leftarrow (U,V)$
          ELSE $(X,Z) \leftarrow (1-\log(U-1), V(U-1))$
      $X \leftarrow m + \dfrac{X}{c}$
UNTIL $Z \leq \dfrac{f(X)}{c}$
RETURN X

The validity of this algorithm is quickly verified: just note that the random vector (X,Z) generated in the middle section of the algorithm is uniformly distributed under the curve $\min(1, e^{1-x})$ $(x \geq 0)$. Because of the excellent properties of the algorithm, it is worth pointing out how we can proceed when f is log-concave with support on both sides of the mode m. It suffices to add a random sign to X just after (X,Z) is generated. We should note here that we pay rather heavily for the presence of two tails because the rejection constant becomes 4. A quick fix-up is not possible because of the fact that the sum of two log-concave functions is not necessarily log-concave. Thus, we cannot "add" the left portion of f to the right portion suitably mirrored and apply the given algorithm to the sum. However, when f is symmetric about the mode m, it is possible to keep the rejection constant at 2 by replacing the statement $X \leftarrow m + \dfrac{X}{c}$ by $X \leftarrow m + \dfrac{SX}{2c}$ where S is a random sign.

Let us conclude this section of algorithms with an exponential version of the previous method which should be fast when exponential random variates can be generated cheaply and if the computation of $\log(f)$ can be done efficiently (in most cases, $\log(f)$ can be computed faster than f).

Rejection method for log-concave densities. Exponential version

[SET-UP](can be omitted)

$c \leftarrow f(m)$, $r \leftarrow \log c$

[GENERATOR]

REPEAT

      Generate U uniformly on $[0,2]$. Generate an exponential random variate E.

      IF $U \leq 1$

            THEN $(X,Z) \leftarrow (U,-E)$

            ELSE $(X,Z) \leftarrow (1+E*,-E-E*)$ ($E*$ is a new exponential random variate)

      CASE

          f log-concave on $[m,\infty)$: $X \leftarrow m + \dfrac{X}{c}$

          f log-concave on $(-\infty,\infty)$:

              Generate a random sign S.

              CASE

                  f symmetric: $X \leftarrow m + \dfrac{SX}{2c}$

                  f not known to be symmetric: $X \leftarrow m + \dfrac{SX}{c}$

UNTIL $Z \leq \log f(X) - r$

RETURN X

One of the practical stumbling blocks is that often most of the time spent in the computation of $f(X)$ is spent computing a complicated normalization factor. When f is given analytically, it can be sidestepped by setting up a subprogram for the computation of the ratio $f(x)/f(m)$ since this is all that is needed in the algorithms. For example, for the generalized inverse gaussian distribution, the normalization constant has several factors including the value of the Bessel function of the third kind. The factors cancel out in $f(x)/f(m)$. Note however that we cannot entirely ignore the issue since $f(m)$ is needed in the computation of X. Because m is fixed, we call this a set-up step.

2.4. The optimal rejection algorithm.

In this section, we assume that f is in $LC_{0,1}$. The optimal rejection algorithm uses the best possible uniform bounding curve, that is, the function g of Theorem 2.1. The problem is that g is only defined implicitly. Nevertheless, it is possible to generate random variates with density $g/\int g$ without great difficulty:

Theorem 2.3.

Let E_1, E_2, U, D be independent random variables with the following distributions: E_1, E_2 are exponentially distributed, U is uniformly distributed on $[0,1]$ and D is integer-valued with $P(D=n)=6/(\pi^2 n^2)$, $n \geq 1$. Then

$$(X,Y) = (U\frac{(E_1+E_2)/D}{1-e^{-(E_1+E_2)/D}}, e^{-(E_1+E_2)/D})$$

is uniformly distributed in $\{(x,y) : x \geq 0, 0 \leq y \leq g(x)\}$ where g is defined in Theorem 2.1. In particular, X has density $g/\int g$ and Y is distributed as $Vg(X)$ where V is a uniform $[0,1]$ random variable independent of X.

Proof of Theorem 2.3.

Flip the axes around, and observe that the desired Y should have density proportional to $-\log(y)/(1-y)$, $0 \leq y \leq 1$, and that X should be distributed as $U(-\log(Y)/(1-Y))$ where U is independent of Y. By the transformation $y = e^{-z}$, $Y = e^{-Z}$, we see that Z has density proportional to

$$\frac{ze^{-z}}{1-e^{-z}} = \sum_{n=0}^{\infty} ze^{-(n+1)z}$$

$$= \frac{\pi^2}{6}(\sum_{n=1}^{\infty} (n^2 ze^{-nz})(\frac{6}{\pi^2 n^2})) \quad (z \geq 0),$$

i.e., Z is distributed as $(E_1+E_2)/D$ (since E_1+E_2 has density ze^{-z}, $z \geq 0$). Thus, the couple $(UZ/(1-e^{-Z})), e^{-Z})$ has the correct uniform distribution. ∎

In the proof of Theorem 2.3, we have also shown that

$$\int g = \frac{\pi^2}{6} \approx 1.6433.$$

This is about 18% better than for the algorithms of the previous section. The algorithm based upon Theorem 2.3 is as follows:

Optimal rejection algorithm for log-concave densities

[NOTE: $f \in LC_{0,1}$]

REPEAT

    Generate a uniform [0,1] random variate U.

    Generate iid exponential random variates E_1, E_2. Set $E \leftarrow E_1 + E_2$.

    Generate a discrete random variate D with $P(D = n) = 6/(\pi^2 n^2)$, $n \geq 1$.

    $Z \leftarrow \dfrac{E}{D}$

    $Y \leftarrow e^{-Z}$, $X \leftarrow \dfrac{UZ}{1-Y}$

UNTIL $Y \leq f(X)$

RETURN X

For the generation of D , we could use yet another rejection method such as:

REPEAT

    Generate iid uniform [0,1] random variates U, V.

    IF $U \leq \dfrac{1}{2}$

        THEN $D \leftarrow 1$

        ELSE $D \leftarrow \lceil 1/(2(1-U)) \rceil$

UNTIL $DV \geq 1$

RETURN D

If D is generated as suggested, we have a rejection constant of $\dfrac{12}{\pi^2}$. When used in the former algorithm, this will offset the 18% gain so painstakingly obtained. Since the D generator does not vary with f , it should preferably be implemented based upon a combination of the alias method and a rejection method for the tail of the distribution.

2.5. The mirror principle.

Consider now a normalized log-concave f with two tails, $m = 0$, and $f(0)=1$. In this case, the original algorithms have a rejection constant equal to 4. However, there are two observations of Richard Brent which will considerably improve the performance. The first observation is that if $p = F(m)$ is known (F is the distribution function), then the rejection constant can be reduced to 2 again. This is based upon the following inequality:

Theorem 2.4.

If f is a log-concave density with mode $m = 0$ and $f(0)=1$, then, writing p for $F(0)$, we have

$$f(x) \leq \begin{cases} \min(1, e^{1-\frac{|x|}{1-p}}) & (x \geq 0) \\ \min(1, e^{1-\frac{|x|}{p}}) & (x < 0) \end{cases}.$$

The area under the bounding curve is 2.

Proof of Theorem 2.4.

Note that $\dfrac{f(x)}{1-p}$ is a log-concave density on $(0,\infty)$, and that $\dfrac{f(x)}{p}$ is a log-concave density on $(-\infty,0)$. Since $f(x(1-p))$ is log-concave on $(0,\infty)$, we have

$$f(x(1-p)) \leq \min(1, e^{1-x}) \quad (x \geq 0).$$

The inequality and the statement about the area follow without further work. ∎

The details of the rejection algorithm based upon Theorem 2.4 are left as an exercise. Brent's second observation applies to the case that $F(m)$ is not available. The expected number of iterations in the rejection algorithm can be reduced to between 2.64 and 2.75 at the expense of an increased number of computations of f.

Theorem 2.5.

Let f be a log-concave density on R with mode at 0 and $f(0)=1$. Then, for $x>0$,

$$f(x)+f(-x)\le g(x)=\sup_{p\in(0,1)}(\min(1,e^{1-\frac{x}{1-p}})+\min(1,e^{1-\frac{x}{p}}))$$

$$=\begin{cases} 2 & (0\le x\le\frac{1}{2})\\ 1+e^{2-\frac{1}{1-x}} & (\frac{1}{2}\le x\le 1)\\ e^{1-x} & (x\ge 1)\end{cases}.$$

Furthermore,

$$\int g=\frac{5}{2}+\frac{1}{4}\int_0^\infty\frac{e^{-u}}{(1+\frac{u}{2})^2}\,du<\frac{5}{2}+\frac{1}{4}\int_0^\infty\frac{e^{-u}}{1+u}\,du\approx 2.6491.$$

Define another function $g*$ where $g*=g$ except on $(\frac{1}{2},1)$, where $g*$ is linear with values $g*(\frac{1}{2})=2,g*(1)=1$. Then $g*\ge g$ and $\int g*=\frac{11}{4}$.

Proof of Theorem 2.5.

Let us write once again $p=F(0)$. The first inequality follows directly from Theorem 2.4. We will first rewrite g as $\sup_{0\le p\le\frac{1}{2}}h_p(x)$ where $h_p(x)$ is defined by

$$\begin{cases} 2 & (x\le p)\\ 1+e^{1-\frac{x}{p}} & (p\le x\le 1-p)\\ e^{1-\frac{x}{p}}+e^{1-\frac{x}{1-p}} & (1-p\le x<\infty)\end{cases}.$$

To prove the main statement of Theorem 2.5, we first show that g is at least equal to the right-hand-side of the main equation. For $x\le\frac{1}{2}$, we have $h_{1/2}(x)=2$. For $\frac{1}{2}\le x\le 1$, observe that $h_{1-x}(x)=1+e^{2-1/(1-x)}$. Finally, for $x\ge 1$, we have $h_0(x)=e^{1-x}$. We now show that g is at most equal to the right-hand-side of the main equation. To do this, decompose h_p as $h_{p1}+h_{p2}+h_{p3}$ where $h_{p1}=h_p I_{[0,p]}$, $h_{p2}=h_p I_{(p,1-p)}$, $h_{p3}=h_p I_{[1-p,\infty)}$. Clearly, $h_{p1}\le g$ for all $p\le\frac{1}{2},x\ge 0$. Since $(p,1-p)\subseteq[0,1]$, we have $h_{p2}\le g$ for all $p\le\frac{1}{2},x\ge 0$. It suffices to show that $h_{p\,3} \leq e^{1-x}$ for all $x \geq 1, p \leq \frac{1}{2}$. This follows if for all such p,

$$e^{-\frac{1}{p}} + e^{-\frac{1}{1-p}} \leq \frac{1}{e}$$

because this would imply, for $x \geq 1$,

$$e\left((e^{-\frac{1}{p}})^x + (e^{-\frac{1}{1-p}})^x\right)$$

$$\leq e\,(e^{-\frac{1}{p}} + e^{-\frac{1}{1-p}})^x$$

$$\leq e^{1-x} \ .$$

Putting $u = \frac{1-p}{p}$, we have

$$e\,(e^{-\frac{1}{p}} + e^{-\frac{1}{1-p}}) = e^{-u} + e^{-\frac{1}{u}} \ .$$

The last function has equal maxima at $u = 0$ and $u \uparrow \infty$, and a minimum at $u = 1$. The maximal value is 1 and the minimal value is $\frac{2}{e}$. This concludes the proof of the main equation in the theorem.

Next, $\int g$ is

$$\frac{5}{2} + e^2 \int_{\frac{1}{2}}^1 e^{-\frac{1}{1-x}} \, dx \;=\; \frac{5}{2} + \frac{1}{4} \int_0^\infty (1 + \frac{u}{2})^{-2} e^{-u} \, du$$

where we used the transformation $u = \frac{1}{1-x} - 2$. The rest follows easily. For example, a formula for the exponential integral is used at one point (Abramowitz and Stegun, 1970, p. 231). The last statement of the theorem is a direct consequence of the fact that $h_{p\,2}$ is convex on $[\frac{1}{2}, 1]$. ∎

We conclude this section by mentioning the algorithm derived from Theorem 2.5. It requires on the average 2.75 iterations and 5.5 evaluations of f per random variate. It should be used only when the number of uniform random variates per generated random variate must be kept reasonable.

Rejection method for log-concave densities on the real line

[NOTE]

We assume that f has a mode at 0 and that $f(0)=1$. Otherwise, use a linear transformation to enforce this condition.

[GENERATOR]

REPEAT

    Generate iid uniform [0,1] random variates U,V,W.

    IF $U \leq \dfrac{4}{11}$

        THEN $(X,Y) \leftarrow (\dfrac{W}{2}, 2V)$

        ELSE IF $U \leq \dfrac{7}{11}$

            THEN

                Generate a uniform [0,1] random variate $W*$.

$$(X,Y) \leftarrow (\dfrac{1}{2} + \dfrac{1}{2}\min(W, 2W*), V(1+2(1-X)))$$

            ELSE $(X,Y) \leftarrow (1-\log(W), VW)$

UNTIL $Y \leq f(X) + f(-X)$

Generate a uniform [0,1] random variate Z (this can be done by reuse of the unused portion of U).

IF $Z \leq \dfrac{f(X)}{f(X)+f(-X)}$

    THEN RETURN X

    ELSE RETURN $-X$

2.6. Non-universal rejection methods.

The universal rejection algorithm developed in the previous sections is suboptimal for individual log-concave densities in the following sense: one can find dominating curves which consist of a constant function around the mode and two exponential tails and have at the same time a smaller integral than that of the dominating curves for the universal method. The improvements are individual, because for each density we require additional information about the density not normally available in the black box model. The resulting algorithms are comparable with the ratio-of-uniforms method, where the exponential tails are replaced with quadratic tails. Since log-concave densities have sub-exponential tails, the fit will often be much better than with the ratio-of-uniforms method. More importantly, we can give a very elegant recipe for finding the optimal dominating curve which is valid for all log-concave densities.

By log-concavity, we know that $h = \log(f)$ can be majorized by the derivative of h at any point (the derivative being considered as a line). This corresponds to fitting an exponential curve over f. The problem we have is that of finding points $m + a \geq m$ and $m - b \leq m$ (where m is the mode of f) such that the area under

$$g(x) = \min(f(m), f(m+a)e^{(x-(m+a))h'(m+a)},$$
$$f(m-b)e^{(x-(m-b))h'(m-b)})$$

is minimal. We will formally allow $h'(m+a) = -\infty$ and $h'(m-b) = +\infty$. In those cases, the corresponding terms in the definition of g are either ∞ or 0. This distinction is important for compact support densities where a or b point at the extremal point in the support of f. We can offer the following general principle for finding a and b.

Theorem 2.6.

Let f be decomposed as $f_r + f_l$ where f_r, f_l refer to the parts of of f to the right and left of the mode respectively. The inverses of f_r and f_l are well-defined when evaluated at a point strictly between 0 and $f(m)$. (In case of a continuous f_r, there is no problem. If f_r has a discontinuity at y, then we know that $f_r(x) > 0$ for $x < y$ and $f_r(x) = 0$ for $x > y$. In that case, the inverse, if necessary, is forced to be y.)

The area under g is minimal when

$$m + a = f_r^{-1}(\frac{f(m)}{e}),$$
$$m - b = f_l^{-1}(\frac{f(m)}{e}).$$

The minimal area is given by

$$f(m)(a+b).$$

Furthermore, the minimal area does not exceed $\dfrac{2e}{e-1}$, and can be as small as 1. When in g we use values of $m+a$ and $m-b$ further away from the mode than those given above, the area under g is bounded from above by $f(m)(a+b)$.

Proof of Theorem 2.6.

We will prove the theorem for a monotone density f on $[m, \infty)$ only. The full theorem then follows by a simple combination of antisymmetric results. We begin thus with the inequality

$$g(x) = \min(f(m), f(m+a)e^{(x-(m+a))h'(m+a)}).$$

The cross-over point between the top curves is at a point z between m and $m+a$:

$$z = m+a+\frac{1}{h'(m+a)}\log(\frac{f(m)}{f(m+a)}) .$$

The area under the curve g to the right of m is given by

$$f(m)(z-m)+\int_{z}^{\infty} f(m+a)e^{(x-a)h'(m+a)} dx$$

$$= f(m)(z-m)+\frac{f(m+a)}{-h'(m+a)}e^{(z-(m+a))h'(m+a)}$$

$$= f(m)(z-m-\frac{1}{h'(m+a)})$$

$$= f(m)(a+\frac{1}{h'(m+a)}(h(m)-h(m+a)-1)) .$$

The derivative of this expression with respect to a is

$$\frac{f(m)h''(m+a)(1+h(m+a)-h(m))}{h'^2(m+a)}$$

which is zero for $h(m+a)=h(m)-1$, i.e. $f(m+a)=\dfrac{f(m)}{e}$. Note also that $h''(m+a)\le 0$, and thus that the derivative is nonpositive for values of $m+a$ smaller than this threshold value, and that it is nonnegative for larger values of $m+a$, so that we do indeed have a global minimum for the area under g. At the suggested value of $m+a$, the area is given by $af(m)$. For $m+a$ larger than the suggested value, the area is bounded from above by $af(m)$, since $h'(m+a)\le 0$, $h(m)-h(m+a)-1\ge 0$.

To obtain a distribution-free upper bound for the area $af(m)$ when a is optimally chosen, we use the inequality of Theorem 2.1. If we use the upper bound on f given there, and set it equal to $\dfrac{1}{e}$, then the solution is a number greater than $af(m)$. But that solution is $\dfrac{e}{e-1}$. Thus, for the optimal a,

$$af(m)\le \frac{e}{e-1}. \blacksquare$$

Theorem 2.6 is important. If a lot is known about the density in question, good rejection algorithms can be obtained. Several examples will be given below. If we want to bound f from above by a combination of pieces of exponential functions, then the area can be reduced even further although, as we will see from the examples given below, the reduction is often hardly worth the extra effort since the rejection constant is already good to begin with.

The formal algorithm is as follows:

Rejection with two exponential tails touching at m-b and m+a

[SET-UP]

m is the mode; $a, b \geq 0$ are assumed given.

$\lambda_r \leftarrow -1/h'(m+a), \lambda_l \leftarrow 1/h'(m-b)$ (where $h = \log(f)$).

$f_m \leftarrow f(m)$

$a* \leftarrow a + \lambda_r \log(\dfrac{f(m+a)}{f_m}), \quad b* \leftarrow b + \lambda_l \log(\dfrac{f(m-b)}{f_m})$. ($m+a*$ and $m-b*$ are the thresholds.)

Compute the mixture probabilities: $s \leftarrow \lambda_l + \lambda_r + a* + b*, \quad p_l \leftarrow \lambda_l/s, \quad p_r \leftarrow \lambda_r/s, \quad p_m \leftarrow (a* + b*)/s$.

[GENERATOR]

REPEAT

    Generate iid uniform [0,1] random variates U, V.

    IF $U \leq p_m$ THEN

        Generate a uniform [0,1] random variate Y (which can be done as $Y \leftarrow U/p_m$).

        $X \leftarrow m - b* + Y(a* + b*)$

        Accept $\leftarrow [Vf_m \leq f(X)]$

    ELSE IF $p_m < U \leq p_m + p_r$ THEN

        Generate an exponential random variate E (which can be done as $E \leftarrow -\log(\dfrac{U - p_m}{p_r})$).

        $X \leftarrow m + a* + \lambda_r E$

        Accept $\leftarrow [Vf_m e^{-(X-(m+a*))/\lambda_r} \leq f(X)]$ (which is equivalent to Accept $\leftarrow [Vf_m e^{-E} \leq f(X)]$, or to Accept $\leftarrow [Vf_m \dfrac{U - p_m}{p_r} \leq f(X)]$)

    ELSE

        Generate an exponential random variate E (which can be done as $E \leftarrow -\log(\dfrac{U - (p_m + p_r)}{1 - p_m - p_r})$).

        $X \leftarrow m - b* - \lambda_l E$

        Accept $\leftarrow [Vf_m e^{(X-(m-b*))/\lambda_l} \leq f(X)]$ (which is equivalent to Accept $\leftarrow [Vf_m e^{-E} \leq f(X)]$, or to Accept $\leftarrow [Vf_m \dfrac{U - (p_m + p_r)}{1 - p_m - p_r} \leq f(X)]$)

UNTIL Accept

RETURN X

In most implementations, this algorithm can be considerably simplified. For one thing, the set-up step can be integrated in the algorithm. When the density is monotone or symmetric unimodal, other obvious simplifications are possible.

Example 2.1. The exponential power distribution (EPD).

The EPD density with parameter $\tau > 0$ is

$$f(x) = (2\Gamma(1+\frac{1}{\tau}))^{-1} e^{-|x|^{\tau}}.$$

Generation for this density has been dealt with in Example IV.6.1, by transformations of gamma random variables. For $\tau \geq 1$, the density is log-concave. The values of a, b in the optimal rejection algorithm are easily found in this case: $a = b = 1$. Before giving the details of the algorithm, observe that the rejection constant, the area under the dominating curve, is $f(0)(a+b)$, which is equal to $1/\Gamma(1+\frac{1}{\tau})$. As a function of τ, the rejection constant is a unimodal function with value 1 at the extremes $\tau = 1$ (the Laplace density) and $\tau \uparrow \infty$ (the uniform $[-1,1]$ density), and peak at $\tau = \dfrac{1}{0.4616321449...}$. At the peak, the value is $\dfrac{1}{0.8856031944...}$ (see e.g. Abramowitz and Stegun (1970, p. 259)). Thus, uniformly over all $\tau \geq 1$, the rejection rate is extremely good. For the important case of the normal density ($\tau = 2$) we obtain a value of $1/\Gamma(\frac{3}{2}) = \sqrt{\dfrac{4}{\pi}}$. The algorithm can be summarized as follows:

REPEAT

    Generate a uniform $[0,1]$ random variate U and an exponential random variate $E*$.

    IF $U \leq 1-\dfrac{1}{\tau}$

        THEN

            $X \leftarrow U$ (note that X is uniform on $[0,1-\dfrac{1}{\tau}]$)

            Accept $\leftarrow [\,|\,X\,|^{\tau} \leq E*\,]$

        ELSE

            Generate an exponential random variate E (which can be done as $E \leftarrow -\log(\tau(1-U\,)))$.

            $X \leftarrow 1-\dfrac{1}{\tau}+\dfrac{1}{\tau}E$

            Accept $\leftarrow [\,|\,X\,|^{\tau} \leq E+E*\,]$

UNTIL Accept

RETURN SX where S is a random sign.

The reader will have little difficulty verifying the validity of the algorithm. Consider the monotone density on $[0,\infty)$ given by $\left(\Gamma(1+\dfrac{1}{\tau})\right)^{-1} e^{-x^{\tau}}$. Thus, with $m=0, a=1, h'(1)=-\tau$, we obtain $a*=1-\dfrac{1}{\tau}$. Since we know that $|\,X\,|^{\tau}$ is distributed as a gamma $(\dfrac{1}{\tau})$ random variable, it is easily seen that we have at the same time a good generator for gamma random variates with parameter less than one. For the sake of easy reference, we give the algorithm in full:

Gamma generator with parameter a less than one

REPEAT
    Generate a uniform [0,1] random variate U and an exponential random variate $E*$.
    IF $U \leq 1-a$
        THEN
$$X \leftarrow U^{\frac{1}{a}} \text{ (note that } U \text{ is uniform on } [0,1-a\,])$$
            Accept $\leftarrow [\,|\,X\,| \leq E*\,]$
        ELSE
            Generate an exponential random variate E (which can be done as
$$E \leftarrow -\log(\frac{1-U}{a})).$$
$$X \leftarrow (1-a+aE)^{\frac{1}{a}}$$
            Accept $\leftarrow [\,|\,X\,| \leq E+E*\,]$
UNTIL Accept
RETURN X ∎

Example 2.2. Complicated densities.

For more complicated densities, the equation $f(x)=f(m)/e$ can be difficult to solve explicitly. It is always possible to take the pessimistic, or minimax, approach, by setting a and b both equal to $\dfrac{e}{(e-1)f(m)}$. In some cases, b can be set equal to 0. In the set-up of the algorithm, it is still necessary to evaluate the derivative of $\log(f)$ at the points $m+a$, $m-b$, but this can be done explicitly when f is given in analytic form. This approach can be automated for the beta and generalized inverse gaussian distributions, for example. When $m+a$ or $m-b$ fall outside the support of f, one should consider one-tailed dominating curves with the constant section truncated at the relevant extremal point of the support. For the beta density for example, this leads to an algorithm which resembles in many respects algorithm B2PE of Schmeiser and Babu (1980). ∎

Example 2.3. Algorithm B2PE (Schmeiser and Babu, 1980) for beta random variates.

In 1980, Schmeiser and Babu proposed a highly efficient algorithm for generating beta random variates with parameters a and b when both parameters are at least one. Recall that for these values of the parameters, the beta density is log-concave. Schmeiser and Babu partition the interval $[0,1]$ into three intervals: in the center interval, around the mode $m = \dfrac{a-1}{a+b-2}$, they use as dominating function a constant function $f(m)$. In the tall intervals, they use exponential dominating curves that touch the graph of f at the breakpoints. At the breakpoints, Schmeiser and Babu have a discontinuity. Nevertheless, analysis similar to that carried out in Theorem 2.6 can be used to obtain the optimal placement of the breakpoints. Schmeiser and Babu suggest placing the breakpoints at the inflection points of the density, if they exist. The inflection points are at

$$\max(m-\sigma,0)$$

and

$$\min(m+\sigma,1)$$

where $\sigma = \sqrt{\dfrac{m(1-m)}{a+b-3}}$ if $a+b>3$ and $\sigma = \infty$ otherwise. Two inflection points exist on $[0,1]$ when $m-\sigma$ and $m+\sigma$ both take values in $[0,1]$. In that case, the area under the dominating curve is easily seen to be equal to

$$2\sigma f(m)+f(m)\left(\frac{1}{|h'(m-\sigma)|}+\frac{1}{|h'(m+\sigma)|}\right)$$

$$= f(m)\left(2\sigma+\frac{1}{\sigma(a+b-2)}((m+\sigma)(1-m-\sigma)+(m-\sigma)(1-m+\sigma))\right)$$

$$= f(m)\left(2\sigma+\frac{1}{\sigma(a+b-2)}2m(1-m)(1-\frac{1}{a+b-3})\right)$$

$$= f(m)\left(2\sigma+2\sqrt{\frac{m(1-m)}{a+b-3}}\right)$$

$$= 4f(m)\sigma .$$

Thus, we have the interesting result that the probability mass under the exponential tails equals that under the constant center piece. One or both of the tails could be missing. In those cases, one or both of the contributions $f(m)\sigma$ needs to be replaced by $f(m)m$ or $f(m)(1-m)$. Thus, $4f(m)\sigma$ is a conservative upper bound which can be used in all cases. It can be shown (see exercises) that as $a,b\to\infty$, $4f(m)\sigma\to\sqrt{\dfrac{8}{\pi}}$. Furthermore, a little additional analysis shows that the expected area under the dominating curve is uniformly bounded over all values of $a,b\geq 1$. Even though the fit is far from perfect, the algorithm can be made very fast by the judicious use of the squeeze principle. Another acceleration trick proposed by Schmeiser and Babu (algorithm B4PE) consists of partitioning $[0,1]$ into 5 intervals instead of 3, with a linear dominating curve added ln the new lntervals.

Algorithm B2PE for beta (a,b) random variates

[SET-UP]

$$m \leftarrow \frac{a-1}{a+b-2}$$

IF $a+b>3$ THEN $\sigma \leftarrow \sqrt{\dfrac{m(1-m)}{a+b-3}}$

IF $a<2$

      THEN $x \leftarrow 0, p \leftarrow 0$

      ELSE

          $x \leftarrow m - \sigma$

          $\lambda \leftarrow \dfrac{a-1}{x} - \dfrac{b-1}{1-x}$

          $v \leftarrow e^{(a-1)\log(\frac{x}{a-1})+(b-1)\log(\frac{1-x}{b-1})+(a+b-2)\log(a+b-2)}$

          $p \leftarrow \dfrac{v}{\lambda}$

Now, x is the left breakpoint, p the probability under the left exponential tail, λ the exponential parameter, and v the value of the normalized density f at x.

IF $b<2$

      THEN $y \leftarrow 1, q \leftarrow 0$

      ELSE

          $y \leftarrow m + \sigma$

          $\mu \leftarrow -\dfrac{a-1}{y} + \dfrac{b-1}{1-y}$

          $w \leftarrow e^{(a-1)\log(\frac{y}{a-1})+(b-1)\log(\frac{1-y}{b-1})+(a+b-2)\log(a+b-2)}$

          $q \leftarrow \dfrac{w}{\mu}$

Now, y is the left breakpoint, q the probability under the left exponential tail, μ the exponential parameter, and w the value of the normalized density f at y.

[GENERATOR]

REPEAT

    Generate iid uniform [0,1] random variates U, V. Set $U \leftarrow U(p + q + y - x)$.

    CASE

        $U \leq y - x$:

            $X \leftarrow x + U$ (X is uniformly distributed on $[x, y]$)

            IF $X < m$

                THEN Accept $\leftarrow [V \leq v + \dfrac{(X - x)(1 - v)}{m - x}]$

                ELSE Accept $\leftarrow [V \leq w + \dfrac{(y - X)(1 - w)}{y - m}]$

        $y - x < U \leq y - x + p$:

            $U \leftarrow \dfrac{U - (y - x)}{p}$ (create a new uniform random variate)

            $X \leftarrow x + \dfrac{1}{\lambda}\log(U)$ (X is exponentially distributed)

            Accept $\leftarrow [V \leq \dfrac{\lambda(X - x) + 1}{U}]$

            $V \leftarrow VUv$ (create a new uniform random variate)

        $y - x + p \leq U$:

            $U \leftarrow \dfrac{U - (y - x + p)}{q}$ (create a new uniform random variate)

            $X \leftarrow y - \dfrac{1}{\mu}\log(U)$ (X is exponentially distributed)

            Accept $\leftarrow [V \leq \dfrac{\mu(y - X) + 1}{U}]$

            $V \leftarrow VUw$ (create a new uniform random variate)

    IF NOT Accept THEN

        $T \leftarrow \log(V)$

        IF $T > -2(a + b - 2)(X - m)^2$

            THEN

                Accept $\leftarrow [T \leq (a - 1)\log(\dfrac{X}{a - 1}) + (b - 1)\log(\dfrac{1 - X}{b - 1}) + (a + b - 2)\log(a + b - 2)]$

UNTIL Accept

RETURN X

The algorithm can be improved in many ways. For example, many constants can be computed in the set-up step, and quick rejection steps can be added when X falls outside [0,1]. Note also the presence of another quick rejection step, based upon the following inequality:

$$\log(\frac{f(x)}{f(m)}) \leq -2(a + b - 2)(x - m)^2 .$$

The quick rejection step is useful in situations just like this, i.e. when the fit is not very good. ■

Example 2.4. Tails of log-concave densities.

When f is log-concave, and a random variate from the right tail of f, truncated at $t > m$ where m is the mode of f, is needed, one can always use the exponential majorizing function:

$$f(x) \leq f(t) e^{\frac{f'(t)}{f(t)}(x-t)} \qquad (x \geq t).$$

The first systematic use of these exponential tails can be found in Schmeiser (1980). The expected number of iterations in the rejection algorithm is

$$\frac{f^2(t)}{|f'(t)| \int\limits_t^\infty f} \cdot \blacksquare$$

2.7. Exercises.

1. **The Pearson IV density.** The Pearson IV density on R has two parameters, $m > \dfrac{1}{2}$ and $s \in R$, and is given by

$$f(x) = \frac{c}{(1+x^2)^m} e^{-s \arctan x}.$$

Here c is a normalization constant. For $s = 0$ we obtain the t density. Show the following:

A. If X is Pearson IV (m, s), and $m \geq 1$, then arc $\tan(X)$ has a log-concave density

$$g(x) = c \cos^{2(m-1)}(x) e^{-sx} \qquad (\,|\,x\,| \leq \frac{\pi}{2}\,).$$

B. The mode of g occurs at $\arctan(-\dfrac{s}{2(m-1)})$.

C. Give the complete rejection algorithm (exponential version) for the distribution. For the symmetric case of the t density, give the details of the rejection algorithm with rejection constant 2.

D.   Find a formula for the computation of c .

2.   Prove that a mixture of two log-concave densities is not necessarily log-concave.

3.   Give the details of the rejection algorithm that is based upon the inequality of Theorem 2.4.

4.   Log-concave densities can also occur in R^d . For example, the multivariate normal density is log-concave. The closure under convolutions also holds in R^d (Davidovic et al., 1969), and marginals of log-concave densities are again log-concave (Prekopa, 1973). Unfortunately, it is useless to try to look for a generalization of the inequalities of this section to R^d with $d \geq 2$ because of the following fact which you are asked to show: the supremum over all log-concave densities with mode at 0 and f (0)=1 is the constant function 1.

5.   To speed up the algorithms of this section at the expense of preprocessing, we can compute the normalized log-concave density at $n > 1$ carefully selected points, and use rejection (perhaps combined with squeezing) with a dominating curve consisting of several pieces. Can you give a universal recipe for locating the points of measurement so that the rejection constant is guaranteed to be smaller than a function of n only, and this function of n tends to 1 as $n \to \infty$? Make sure that random variate generation from the dominating density is not difficult, and provide the details of your algorithm.

6.   This is about the area under the dominating curve in algorithm B2PE (Schmeiser and Babu, 1980) for beta random variate generation (Example 2.3). Assume throughout that $a, b \geq 1$.

(i) $\sigma \leq m$ if and only if $a \geq 2$, $\sigma \leq 1-m$ if and only if $b \geq 2$. (Thus, for $a, b \geq 2$, the area under the dominating curve is precisely $4f(m)\sigma$.)

(ii) $\lim\limits_{a,b \to \infty} 4f(m)\sigma = \sqrt{\dfrac{8}{\pi}}$. Use Stirling's approximation.

(iii)   The area under the dominating curve is uniformly bounded over all $a, b \geq 1$. Use sharp inequalities for the gamma function to bound $f(m)$. Consider 3 cases: both $a, b \geq 2$, one of a, b is ≥ 2, and one is < 2, and both a, b are < 2. Try to obtain as good a uniform bound as possible.

(iv)   Prove the quick rejection inequality used in the algorithm:

$$\log\left(\frac{f(x)}{f(m)}\right) \leq -2(a+b-2)(x-m)^2 .$$

3. INEQUALITIES FOR DENSITIES.

3.1. Motivation.

The previous section has shown us the utility of upper bounds in the development of universal methods or black box methods. The strategy is to obtain upper bounds for densities in a large class which

(i)   have a small integral;

(ii)   are defined in terms of quantities that are either computable or present in the definition of the class.

For the log-concave densities with mode at 0 we have for example obtained an upper bound in section VII.2 with integral 4, which requires knowledge of the position of the mode (this is in the definition of the class), and of the value of f (0) (this can be computed). In general, quantities that are known could include:

A. A uniform upper bound for f (called M);
B. The r-th moment μ_r;
C. The value of a functional $\int f^{\alpha}$;
D. A Lipschitz constant;
E. A uniform bound for the s-th derivative;
F. The entire moment generating function $M(t)$, $t \in R$;
G. The entire distribution function $F(x)$, $x \in R$;
H. The support of f.

When this information is combined in various ways, a multitude of useful dominating curves can be obtained. The goodness of a dominating curve is measured in terms of its integral and the ease with which random variates with a density proportional to the dominating curve can be generated. We show by example how some inequalities can be obtained.

3.2. Bounds for unimodal densities.

Let us start with the class of monotone densities on [0,1] which are bounded by M. Note that if M is unknown, it can easily be computed as f (0). Thus, the only true restriction is that we must know that f vanishes off [0,1]. The trivial inequality

$$f(x) \leq MI_{[0,1]}(x)$$

is not very useful, since the integral under the dominating curve is M. There are several ways to increase the efficiency:

1.   Use a table method by evaluating in a set-up step the value of f at many points. Basically, the dominating curve is piecewise constant and hugs the curve of f much better. These methods are very fast but the need for extra storage (usually growing with M) and an additional preprocessing step makes this approach somehow different. It should not be compared with

methods not requiring these extra costs. It will be developed systematically in chapter VIII.

2. Use as much information as possible to improve the bound. For example, in the inequality $f(x) \leq M$, the monotonicity is not used.

3. Ask the user if he has additional knowledge in the form of moments, quantiles, functionals and the like. Then construct good dominating curves.

We will illustrate approaches 2 and 3. For all monotone densities, the following is true:

Theorem 3.1.

    For all monotone densities f on $[0,\infty)$,

$$f(x) \leq \frac{1}{x} .$$

If f is also convex, then

$$f(x) \leq \frac{1}{2x} .$$

Proof of Theorem 3.1.

    Fix $x > 0$. Then, by monotonicity,

$$xf(x) \leq \int_0^x f(y)\, dy \leq 1 .$$

When f is also convex, we can in fact use a geometrical argument: if we wish to find the convex f for which $f(x)$ is maximal, it suffices to consider only triangles. This class is the class of all densities $2a(1-ax)_+$, $0 \leq x \leq \frac{1}{a}$. Thus, we find a for which $f(x)$ is maximal. Setting the derivative with respect to a equal to 0 gives the equation $1 - ax - ax = 0$, i.e. $a = \frac{1}{2x}$. Resubstitution gives the bound. ∎

    The bounds of Theorem 3.1 cannot be improved in the sense that for every x, there exists a monotone (or monotone and convex) f for which the upper bound is attained. If we return now to the class of monotone densities on $[0,1]$ bounded by M, we see that the following inequality can be used:

$$f(x) \leq \min(M, \frac{1}{x})I_{[0,1]}(x) .$$

The area under the dominating curve is $1 + \log(M)$. Clearly, this is always less than M. In most applications the improvement in computer time obtainable by using the last inequality is noticeable if not spectacular. Let us therefore take a moment to give the details of the corresponding rejection algorithm. The dominating density for rejection is

$$g(x) = \frac{1}{1+\log(M)}\min(M,\frac{1}{x})I_{[0,1]}(x) .$$

It has distribution function

$$\begin{cases} \dfrac{Mx}{1+\log(M)} & ,0\le x \le \dfrac{1}{M} \\ \dfrac{1+\log(Mx)}{1+\log(M)} & ,\dfrac{1}{M}\le x \le 1 . \end{cases}$$

Using inversion for generation from g, we obtain

Rejection algorithm for monotone densities on [0,1] bounded by M

REPEAT
    Generate iid uniform [0,1] random variates U,V.
    IF $U \le \dfrac{1}{1+\log(M)}$
        THEN
            $X \leftarrow \dfrac{U}{M}(1+\log(M))$
            IF $VM \le f(X)$ THEN RETURN X
        ELSE
            $X \leftarrow \dfrac{1}{M}e^{U(1+\log(M))-1}$
            IF $V \le Xf(X)$ THEN RETURN X
UNTIL False

When f is also convex, we can use the inequality

$$f(x) \le cg(x)$$

where

$$g(x) = \frac{2}{1+\log(2M)}\min(M,\frac{1}{2x})I_{[0,1]}(x) .$$

It has distribution function

$$\begin{cases} \dfrac{2Mx}{1+\log(2M)} & ,0\le x \le \dfrac{1}{2M} \\ \dfrac{1+\log(2Mx)}{1+\log(2M)} & ,\dfrac{1}{2M}\le x \le 1 . \end{cases}$$

Using inversion for generation from g, we obtain

Rejection algorithm for monotone convex densities on [0,1] bounded by M

REPEAT

Generate iid uniform [0,1] random variates U, V.

IF $U \leq \dfrac{1}{1+\log(2M)}$

    THEN

$$X \leftarrow \frac{U}{2M}(1+\log(2M))$$

        IF $VM \leq f(X)$ THEN RETURN X

    ELSE

$$X \leftarrow \frac{1}{2M} e^{U(1+\log(2M))-1}$$

        IF $V \leq 2Xf(X)$ THEN RETURN X

UNTIL False

The expected number of iterations now is $\dfrac{1+\log(2M)}{2}$, which is for large M roughly speaking half of the expected number of iterations for the nonconvex cases.

The function $\dfrac{1}{x}$ is not integrable on $[1,\infty)$, so that Theorem 3.1 is useless for handling infinite tails of monotone densities. We have to tuck the tails under some integrable function, yet uniformly over all monotone densities we cannot get anything better than $\dfrac{1}{x}$. Thus, additional information is required.

Theorem 3.2.

    Let f be a monotone density on $[0,\infty)$.

A.   If $\int x^r f(x)\, dx \leq \mu_r < \infty$ where $r > 0$, then

$$f(x) \leq \frac{(r+1)\mu_r}{x^{r+1}} \qquad (x > 0).$$

B.   In any case, for all $0 < \alpha \leq 1$,

$$f(x) \leq \frac{(\int f^\alpha)^{\frac{1}{\alpha}}}{x^{\frac{1}{\alpha}}} \qquad (x > 0).$$

Proof of Theorem 3.2.

For part A we proceed as follows:

$$\mu_r \geq \int_0^x y^r f(y)\, dy \geq \frac{f(x)x^{r+1}}{r+1}\ .$$

For part B, we use the trivial observation

$$xf^\alpha(x) \leq \int f^\alpha\ .\ \blacksquare$$

For monotone densities on $[0,\infty)$, bounded by $M=f(0)$, Theorem 3.2 provides us with bounds of the form

$$f(x) \leq \min(M, \frac{A}{x^a})\quad (x>0)$$

where we can take (A,a) as follows:

Information	A	a
$\int x^r f(x)\, dx \leq \mu_r < \infty$	$(r+1)\mu_r$	$r+1$
$(\int f^\alpha)^{\frac{1}{\alpha}} \leq \nu_\alpha < \infty$	ν_α	$\frac{1}{\alpha}$

In all cases, the area under the dominating curve is

$$\frac{a}{a-1} A^{\frac{1}{a}} M^{\frac{a-1}{a}}\ .$$

Furthermore, random variate generation for the dominating density can be done quite easily via the inversion method or the inverse-of-f method (section IV.6.3):

Theorem 3.3.

Let g be the density on $[0,\infty)$ proportional to $\min(M,\dfrac{A}{x^a})$ where $M>0, A>0, a>1$ are parameters. Then the following random variables X have density g:

A. $X = (\dfrac{A}{M})^{\frac{1}{a}}\dfrac{U}{V^{a-1}}$ where U,V are iid uniform $[0,1]$ random variates.

B. Let $x*$ be $(\dfrac{A}{M})^{\frac{1}{a}}$ and let U be uniform on $[0,1]$. Then $X \leftarrow \dfrac{a}{a-1}Ux*$ if $U \le \dfrac{a-1}{a}$, and $X \leftarrow \dfrac{x*}{(aU-(a-1))^{\frac{1}{a-1}}}$ else.

Proof of Theorem 3.3.

By the inverse-of-f method (section IV.6.3), it suffices to note that a random variate with monotone density f can be obtained as $Uf^{-1}(Y)$ where Y has density f^{-1}. It is easy to see that for monotone g not necessarily integrating to one, $Ug^{-1}(Y)$ has density proportional to g if Y has density proportional to g^{-1}. In our case, $g^{-1}(y) = (\dfrac{A}{y})^{\frac{1}{a}}$,$0 \le y \le M$. To generate Y with density proportional to this, we apply the inversion method. Verify that $MV^{\frac{a}{a-1}}$ has distribution function $(\dfrac{y}{M})^{1-\frac{1}{a}}$ on $[0,M]$, which yields a density proportional to g^{-1}. Plugging this Y back into $Ug^{-1}(Y)$ proves part A.

Part B is obtainable by straightforward inversion. Note that $x*$ is the breakpoint where $M = \dfrac{A}{x^a}$, that $\int_0^{x*}g = Mx*$, and that $\int_{x*}^{\infty}g = \dfrac{A}{a-1}x*^{-(a-1)}$. The sum of the two areas is

$$A^{\frac{1}{a}}M^{1-\frac{1}{a}}(1+\dfrac{1}{a-1}) .$$

Thus, with probability $\dfrac{a-1}{a}$, X is distributed uniformly on $[0,x*]$, and with the complementary probability, X is distributed as $\dfrac{x*}{V^{\frac{1}{a-1}}}$ where V is uniformly distributed on $[0,1]$ (the latter random variable has density decreasing as x^{-a} on $[x*,\infty)$). The uniform random variates needed here can be recovered from the uniform random variate U used in the comparison with $\dfrac{a-1}{a}$: given that $U \le \dfrac{a-1}{a}$, $U\dfrac{a}{a-1}$ is again uniform. Given that $U > \dfrac{a-1}{a}$, $aU-(a-1)$ is in turn uniformly distributed on [0,1]. ■

For the sake of completeness, we will now give the rejection algorithm for generating random variates with density f based upon the inequality

$$f(x) \leq \min(M, \frac{A}{x^a}) \quad (x \geq 0) .$$

Rejection method based upon part A of Theorem 3.3

REPEAT

　　Generate iid uniform [0,1] random variates U, V.

　　$Y \leftarrow MV^{\frac{a}{a-1}}$

　　$X \leftarrow U(\frac{A}{Y})^{\frac{1}{a}}$

UNTIL $Y \leq f(X)$

RETURN X

The validity of this algorithm is based upon the fact that $(Y, Ug^{-1}(Y)) = (Y, X)$ is uniformly distributed under the curve of g^{-1}. By swapping coordinate axes, we see that (X, Y) is uniformly distributed under g, and can thus be used in the rejection method. Note that the power operation is unavoidable. Based upon part B, we can use rejection with fewer powers.

Rejection method based upon part B of Theorem 3.3.

REPEAT
> Generate iid uniform [0,1] random variates U, V.
>
> IF $U \leq \frac{a-1}{a}$
> > THEN
> > > $$X \leftarrow \frac{a}{a-1} U x*$$
> > > IF $VM \leq f(X)$ THEN RETURN X
> > ELSE
> > > $$X \leftarrow x* (aU - (a-1))^{-\frac{1}{a-1}}$$
> > > IF $VA \leq X^a f(X)$ THEN RETURN X

UNTIL False

For both implementations, the expected number of computations of f is equal to the expected number of iterations,

$$E(N) = \frac{a}{a-1} A^{\frac{1}{a}} M^{\frac{a-1}{a}} .$$

It is instructive to analyze this measure of the performance in more detail. Consider the moment version for example, where $A = (r+1)\mu_r$, $a = r+1$ and μ_r is the r-th moment of the monotone density. We have

---

Theorem 3.4.

Let $E(N), M, r, A, a, \mu_r$ be as defined above. Then for all monotone densities on $[0,\infty)$,

$$E(N) \geq 1 + \frac{1}{r} .$$

For all monotone densities that are concave on their support,

$$E(N) \leq 2(1+\frac{1}{r})(r+2)^{-\frac{1}{r+1}} \leq 2(1+\frac{1}{r}) .$$

Finally, for all monotone log-concave densities,

$$E(N) \leq (1+\frac{1}{r})(\Gamma(r+2))^{\frac{1}{r+1}} \sim \frac{r+1}{e} \text{(as } r \rightarrow \infty) .$$

Proof of Theorem 3.4.

We start from the expression

$$E(N) = (1+\frac{1}{r})\,((r+1)M^r\mu_r)^{\frac{1}{r+1}}\,.$$

The product $M\mu_r$ is scale invariant, so that we can take $M=1$ without loss of generality. For all such bounded densities, we have $1-F(x)\geq(1-x)_+$. Thus,

$$\mu_r = \int\limits_0^\infty x^r f(x)\, dx = \int\limits_0^\infty rx^{r-1}(1-F(x))\, dx$$

$$\geq \int\limits_0^1 rx^{r-1}(1-x)\, dx$$

$$= 1-\frac{r}{r+1} = \frac{1}{r+1}\,.$$

This proves the first part of the theorem. Note that we have implicitly used the fact that every random variable with a density bounded by 1 on $[0,\infty)$ is stochastically larger than a uniform $[0,1]$ random variate.

For the second part, we use the fact that all random variables with a monotone concave density satisfying $f(0)=M=1$ are stochastically smaller than a random variable with density $(1-\frac{x}{2})_+$ (exercise 3.1). Thus, for this density,

$$\mu_r = \int\limits_0^2 x^r(1-\frac{x}{2})\, dx = 2^{r=1}(\frac{1}{r+1}-\frac{1}{r+2}) = \frac{2^{r+1}}{(r+1)(r+2)}\,.$$

Resubstitution gives us part B for concave densities. Finally, for log-concave densities we need the fact that $f(0)X$ is stochastically smaller than an exponential random variate. Thus, in particular,

$$M^r\mu_r \leq \int\limits_0^\infty y^r e^{-y}\, dy = \Gamma(r+1)\,.$$

This proves the last part of the theorem. ∎

A brief discussion of Theorem 3.4 is in order here. First of all, the inequalities are quite inefficient when r is near 0 in view of the lower bound $E(N)\geq1+\frac{1}{r}$. What is important here is that for important subclasses of monotone densities, the performance is uniformly bounded provided that we know the r-th moment of the density in case. For example, for the log-concave densities, we have the following values for the upper bound for $E(N)$:

r	$E(N) \leq$	Approximate value
1	$\dfrac{4}{\sqrt{3}}$	2.3094...
2	$\dfrac{3}{4^{\frac{1}{3}}}$	1.88988...
3	$\dfrac{8}{3 \, 5^{\frac{1}{4}}}$	1.7833...
4	$\dfrac{5}{2 \, 6^{\frac{1}{5}}}$	1.7470...
5	$\dfrac{12}{5 \, 7^{\frac{1}{6}}}$	1.7352...
6	$\dfrac{7}{3 \, 8^{\frac{1}{7}}}$	1.73366...
7	$\dfrac{16}{7 \, 9^{\frac{1}{8}}}$	1.7367...
$\uparrow\infty$	$\uparrow 2$	

The upper bound is minimal for r near 6. The algorithm is guaranteed to perform at its best when the sixth moment is known. In the exercises, we will develop a slightly better inequality for concave monotone densities. One of the features of the present method is that we do not need any information about the support of f - such information would be required if ordinary rejection from a uniform density is used. Unfortunately, very few important densities are concave on their support, and often we do not know whether a density is concave or not.

The family of log-concave densities is more important. The upper bound for $E(N)$ in Theorem 3.4 has acceptable values for the usual values of r:

r	$E(N) \leq$	Approximate value
1	$\sqrt{8}$	2.82...
2	$\dfrac{3}{2} 6^{\frac{1}{3}}$	2.7256...
3	$\dfrac{4}{3} 24^{\frac{1}{4}}$	2.9511...
$\uparrow\infty$	$\uparrow\infty$	

In this case, the optimal integer value of r is 2. Note that if μ_r is not known, but is replaced in the algorithm and the analysis by its upper bound $\dfrac{\Gamma(r+1)}{M^r}$, then both the algorithm and the performance analysis of Theorem 3.4 remain valid. In that case, we obtain a black box method for all log-concave densities on $[0,\infty)$ with mode at 0, as in the previous section. For $r=2$, the expected number of iterations (about 2.72) is about 36% larger than the algorithm of the previous section which was specially developed for log-concave densities only.

3.3. Densities satisfying a Lipschitz condition.

We say that a function f is **Lipschitz** (C) when

$$\sup_{x \neq y} \frac{|f(x)-f(y)|}{|x-y|} \leq C .$$

When f is absolutely continuous with a.e. derivative f', then we can take $C = \sup |f'|$. Unfortunately, some important functions are not Lipschitz, such as $\sqrt{x}$. However, many of these functions are Lipschitz of order α: formally, we say that f is Lipschitz of order α with constant C (and we write $f \in Lip_\alpha(C)$) when

$$\sup_{x \neq y} \frac{|f(x)-f(y)|}{|x-y|^\alpha} \leq C .$$

Here $\alpha \in (0,1]$ is a constant. It can be shown (exercise 3.6) that the classes $Lip_\alpha(C)$ for $\alpha > 1$ contain no densities. The fundamental inequality for the Lipschitz classes is given below:

Theorem 3.5.

     When f is a density in $Lip_\alpha(C)$ for some $C > 0$, $\alpha \in (0,1]$, then

$$f(x) \leq (\min(F(x),1-F(x))\frac{\alpha+1}{\alpha} C^{\frac{1}{\alpha}})^{\frac{\alpha}{\alpha+1}} .$$

Here F is the distribution function for f. In particular, for $\alpha = 1$, we have

$$f(x) \leq \sqrt{2C \min(F(x),1-F(x))} .$$

Proof of Theorem 3.5.

     Fix x, and define $y = f(x)$. Then fix $z > x$. We clearly have

$$f(z) \geq f(x)-C(z-x)^\alpha .$$

The density f which yields the maximal value for $f(x)$ is equal to the lower bound for $f(z)$ given above. It vanishes beyond

$$z^* = x + (\frac{f(x)}{C})^{\frac{1}{\alpha}} .$$

By integration of the previous inequality we have

$$1 - F(x) \geq \int_x^{z^*} (f(x)-C(z-x)^\alpha)\, dz$$

$$= f(x)(z^*-x) - \frac{C(z^*-x)^{\alpha+1}}{\alpha+1}$$

$$= f(x)(\frac{f(x)}{C})^{\frac{1}{\alpha}} - \frac{C}{\alpha+1}(\frac{f(x)}{C})^{\frac{\alpha+1}{\alpha}}$$

$$= f(x)^{\frac{\alpha+1}{\alpha}} \frac{\alpha}{\alpha+1} C^{-\frac{1}{\alpha}} .$$

By symmetry, the same lower bound is valid for $F(x)$. Rearranging the terms gives us our result. ∎

Theorem 3.5 provides us with an important bridging device. For many distributions, tail inequalities are readily available: standard textbooks usually give Markov's and Chebyshev's inequalities, and these are sometimes supplemented by various exponential inequalities. If f is in $Lip_\alpha(C)$ on $(0,\infty)$ (thus, a discontinuity could occur at 0), then we still have

$$f(x) \le \left[(1-F(x)) \frac{\alpha+1}{\alpha} C^{\frac{1}{\alpha}} \right]^{\frac{\alpha}{\alpha+1}} .$$

Before we proceed with some examples of the use of Theorem 3.5, we collect some of the best known tail inequalities in a lemma:

Lemma 3.1.

Let F be a distribution function of a random variable X. Then the following inequalities are valid:

A. $P(|X| \ge x) \le \dfrac{E(|X|^r)}{|x|^r}$, $r > 0$ (Chebyshev's inequality) .

B. $1-F(x) \le M(t)e^{-tx}$, $t > 0$ where $M(t)=E(e^{tX})$ is the moment generating function (Markov's inequality); note that by symmetry, $F(x) \le M(-t)e^{tx}$, $t > 0$.

C. For log-concave f with mode at 0 and support on $[0,\infty)$, $1-F(x) \le e^{-f(0)x}$.

D. For monotone f on $[0,\infty)$, $1-F(x) \le (\dfrac{r}{r+1})^r \dfrac{E(|X|^r)}{|x|^r}$, $x,r > 0$ (Narumi's inequality).

Proof of Lemma 3.1.

Parts A and B are but special cases of a more general inequality: assume that ψ is a nonnegative function at least equal to one on a set A. Then

$$P(X \in A) = \int_A dF(x) \le \int_A \psi(x) \, dF(x) \le E(\psi(X)) .$$

For part A, take $A =[x,\infty)\cup(-\infty,x]$ and $\psi(y)=\dfrac{|y|^r}{|x|^r}$. For part B, take

$A = [x, \infty)$ and $\psi(y) = e^{t(y-x)}$ for some $t > 0$. Part C follows simply from the fact that for log-concave densities on $[0, \infty)$ with mode at 0, $f(0)X$ is stochastically smaller than an exponential random variable. Thus, only part D seems non-trivial; see exercise 3.7. ■

If inequalities other than those given here are needed, the reader may want to consult the survey article of Savage (1961) or the specialized text by Godwin (1964).

Example 3.1. Convex densities.

When a convex density f on $[0, \infty)$ is in $Lip_1(C)$, we can take $C = f'(0)$. By Narumi's inequality for monotone densities,

$$ f(x) \le \min\left(f(0), \frac{\sqrt{2f'(0)(\frac{r}{r+1})^r \mu_r}}{x^{\frac{r}{2}}}\right), $$

where $\mu_r = E(|X|^r)$. This is of the general form dealt with in Theorem 3.3. It should be noted that for this inequality to be useful, we need $r > 2$. ■

Example 3.2. Densities with known moment generating function.

Patel, Kapadia and Owen (1976) give several examples of the use of moment generating functions $M(t)$ in statistics. Using the exponential version of Markov's inequality, we can bound any $Lip_1(C)$ density as follows:

$$ f(x) \le \begin{cases} \sqrt{2Ce^{-t|x|}M(t)} & , x \ge 0 \\ \sqrt{2Ce^{-t|x|}M(-t)} & , x < 0. \end{cases} $$

Here $t > 0$ is a constant. There is nothing that keeps us from making t depend upon x except perhaps the simplicity of the bound. If we do not wish to upset this simplicity, we have to take one t for all x. When f is also symmetric about the origin, then the bound can be written as follows:

$$ f(x) \le cg(x) $$

where $g(x) = \frac{t}{4}e^{-\frac{t}{2}|x|}$ is the Laplace density with parameter $\frac{t}{2}$, and $c = \sqrt{32\ C\ M(t)/t^2}$ is a constant which depends upon t only. If this bound is used in a rejection algorithm, the expected number of iterations is c. Thus, the best value for t is the value that minimizes $M(t)/t^2$. Note that c increases with C (decreasing smoothness) and with $M(t)$ (increasing size of the tail). Having picked t, the following rejection algorithm can be used:

Rejection method for symmetric Lipschitz densities with known moment generating function

[SET-UP]

$b \leftarrow \sqrt{2CM(t)}$

[GENERATOR]

REPEAT

      Generate E, U, independent exponential and uniform [0,1] random variates.

      $X \leftarrow \dfrac{2}{t} E$

UNTIL $U b e^{-E} \leq f(X)$

RETURN SX where S is a random sign. ■

Example 3.3. The generalized gaussian family.

The generalized gaussian family of distributions contains all distributions for which for some constant $s \geq 0$, $M(t) \leq e^{s^2 t^2/2}$ for all t (Chow, 1966). The mean of these distributions exists and is 0. Also, as shown by Chow (1966), both $1-F(x)$ and $F(-x)$ do not exceed $e^{-x^2/(2s^2)}$ for all $x > 0$. Thus, by Theorem 3.5, when $f \in Lip_1(C)$,

$$f(x) \leq s \sqrt{8C\pi}\left(\frac{1}{s\sqrt{4\pi}} e^{-\frac{x^2}{4s^2}}\right).$$

The function in parentheses is a normal $(0, s\sqrt{2})$ density. The rejection constant is $s\sqrt{8C\pi}$. In its crudest form the algorithm can be summarized as follows:

Rejection algorithm for generalized gaussian distributions with a Lipschitz density

REPEAT

    Generate N, E, independent normal and exponential random variates.

    $X \leftarrow N \delta \sqrt{2}$

UNTIL $-\dfrac{N^2}{2} - E \leq \log(\dfrac{f(X)}{\sqrt{2C}})$

RETURN X ■

Example 3.4. Densities with known moments.

The previous three examples apply to rather small families of distributions. If only the r-th absolute moment μ_r is known, the we have by Chebyshev's inequality,

$$1 - F(x) \leq \frac{\mu_r}{|x|^r}$$

for all $x, r > 0$. This leads to the inequality

$$f(x) \leq \sqrt{2C} \min\left(1, \frac{\sqrt{\mu_r}}{|x|^{\frac{r}{2}}}\right),$$

which is only useful to us for $r > 2$ (otherwise, the dominating function is not integrable). The integral of the dominating curve is $\sqrt{8C} \dfrac{r}{r-2} \mu_r^{\frac{1}{r}}$. Just which r is best depends upon the distribution: $\dfrac{r}{r-2}$ decreases monotonically with r whereas $\mu_r^{\frac{1}{r}}$ is nondecreasing in r (this is known as Lyapunov's inequality, which can be obtained in one line from Jensen's inequality). ■

Example 3.5. Log-concave densities.

Assume that f is log-concave with mode at 0 and support contained in $[0,\infty)$. Using $1-F(x) \le e^{-xf(0)}$, we observe that

$$f(x) \le \frac{\sqrt{8C}}{f(0)} \left(\frac{f(0)}{2} e^{-\frac{xf(0)}{2}} \right) \quad (x>0) .$$

The top bound is $\dfrac{\sqrt{8C}}{f(0)}$ times a Laplace density. It is thus not difficult to see that the following algorithm is useful:

Rejection method for log-concave Lipschitz densities

REPEAT

      Generate iid exponential random variates E_1, E_2

      $X \leftarrow \dfrac{2}{f(0)} E_1$

UNTIL $-E_2 - E_1 \le \log\left(\dfrac{f(X)}{\sqrt{2C}} \right)$

RETURN X ∎

3.4. Normal scale mixtures.

Many distributions in statistics can be written as mixtures of normal densities in which the variance is the mixture parameter. These normal scale mixtures have far-reaching applications ranging from modeling to mathematical statistics. The corresponding random variables X are thus distributed as NY, where N is normal, and Y is a positive-valued random variable. The class of normal scale mixtures is selected here to be contrasted against the class of log-concave densities. It should be clear that we could have picked other classes of mixture distributions.

There are two situations that should be clearly distinguished: in the first case, the distribution of Y is known. In the second case, the distribution of Y is not explicitly given, but it is known nevertheless that X is a normal scale mixture. The first case is trivial: one just generates N and Y and exits with NY. In the table below, some examples are given:

DENSITY OF X	DENSITY OF Y
Cauchy	Density of $1/N$ where N is normal
Laplace	Density of $1/\sqrt{2E}$ where E is exponential
Logistic	Density of $2K$ where K has the Kolmogorov-Smirnov distribution
t_a	Density of $\sqrt{\dfrac{2a}{G}}$ where G is gamma $(\dfrac{a}{2})$
Symmetric stable (α)	Density of $\sqrt{S}$ where S is positive stable $(\dfrac{\alpha}{2})$

This table is far from complete, and all the representations have been known for quite some time. For the inclusion of the symmetric stable, see e.g. Feller (1971), and for the inclusion of the logistic, see e.g. Andrews and Mallows (1974). In fact, it is known that an even density f is a normal scale mixture if and only if the derivatives of $f(\sqrt{x})$ are of alternating sign for all $x > 0$ (Kelker, 1971). Unfortunately, for all the densities given in the table, efficient direct methods of generation are known, so there is no reason why one should use the decomposition.

The more interesting case is the one in which we just know that the distribution is a normal scale mixture. To develop universal rejection methods for this class of distributions, general inequalities are needed. The following inequalities are useful for this purpose:

Theorem 3.6.

   Let f be the density of a normal scale mixture, and let X be a random variable with density f. Then f is symmetric and unimodal, $f(x) \le f(0)$, and for all $a \ge -1$,

$$f(x) \le C_a \frac{\mu_a}{|x|^{1+a}}$$

where

$$\mu_a = E(|X|^a)$$

is the a-th absolute moment of X, and

$$C_a = \left(\frac{1+a}{e}\right)^{\frac{1+a}{2}} \frac{1}{2^{\frac{1+a}{2}} \Gamma(\frac{1+a}{2})}.$$

For $a = 1$ and $a = 2$, we have respectively,

$$f(x) \le \min(f(0), \frac{E(|X|)}{e|x|^2}),$$

$$f(x) \le \min(f(0), (\frac{3}{e})^{\frac{3}{2}} \frac{E(X^2)}{\sqrt{2\pi}|x|^3}).$$

The areas under the dominating curves are respectively, $\frac{4}{\sqrt{e}}\sqrt{f(0)\mu_1}$, and $C(\mu_2 f(0)^2)^{1/3}$ where $C = 3(3/e)^{1/2}(2\pi)^{-1/6}$.

Proof of Theorem 3.6.

   The unimodality is obvious. The upper bounds for f follow directly from similar upper bounds for the normal density. Note that we have, for all $x, \sigma > 0$,

$$e^{-\frac{x^2}{2\sigma^2}} \le (\frac{\sigma}{|x|})^{1+a} (\frac{1+a}{e})^{\frac{1+a}{2}}.$$

Observe that

$$f(x) = E(\frac{1}{\sqrt{2\pi}Y} e^{-\frac{x^2}{2Y^2}})$$

where Y is a random variable used in the mixture (recall that $X = NY$). Using the normal-polynomial bound mentioned above, this leads to the inequality

$$f(x) \le E(Y^a) \frac{1}{\sqrt{2\pi}|x|^{1+a}} (\frac{1+a}{e})^{\frac{1+a}{2}}.$$

But   in   view   of   the   relationship $X = NY$,   we   have
$E(Y^a) = E(|X|^a)/E(|N|^a)$.   Now,   use   the   fact   that
$E(|N|^a)\sqrt{2\pi} = 2^{\frac{1+a}{2}}\Gamma(\frac{1+a}{2})$ (which follows by definition of the gamma
integral). This gives the main inequality. The special cases are easily obtained
from the main inequality, as are the areas under the dominating curves. ∎

The algorithms of section 3.2 are once again applicable. However, we are in
much better shape now. If we had just used the unimodality, we would have
obtained the inequality

$$f(x) \le \min(f(0), \frac{a+1}{2}\frac{\mu_a}{|x|^{a+1}}),$$

which is useful for $a > 0$. See the proof of Theorem 3.2. The area under this dominating curve is larger than the corresponding area for Theorem 3.6, which should
come as no surprise because we are using more information in Theorem 3.6.
Notice that, just as in section 3.2, the areas under the dominating curves are
scale invariant. The choice of a depends of course upon f. Because the class of
normal mixtures contains densities with arbitrarily large tails, we may be forced
to choose a very close to 0 in order to make μ_a finite. Such a strategy is
appropriate for the symmetric stable density.

3.5. Exercises.

1.   Prove the following fact needed in Theorem 3.4: all monotone densities on
     $[0,\infty)$ with value 1 at 0 and concave on their support are stochastically
     smaller than the triangular density $f(x) = (1 - \frac{x}{2})_+$, i.e. their distribution
     functions all dominate the distribution function of the triangular density.

2.   In the rejection algorithm immediately preceding Theorem 3.4, we exit some
     of the time with $X \leftarrow \frac{x*}{\sqrt{aU-(a-1)}}$. The square root is costly. The special
     case $a = 3$ is very important. Show that $\sqrt{3U-2}$ is distributed as
     $\max(3U-2, W)$ where W is another uniform $[0,1]$ random variate.

3.   **Concave monotone densities.** In this exercise, we consider densities f
     which are concave on their support and monotone on $[0,\infty)$. Let us use
     $M = f(0)$, $\mu_r = \int x^r f(x) \, dx$.

     A.   Show that $f(x) \le \min(M, (\frac{2\mu_r(r+1)}{x^{r+1}} - M)_+)$.

     B.   Show that the area under the dominating curve is $2 - 2^{\frac{1}{r+1}}$ times the area under the dominating curve shown in Theorem 3.4. That is, the area is

$$(2-2^{\frac{1}{r+1}})(1+\frac{1}{r})M^{\frac{r}{r+1}}((r+1)\mu_r)^{\frac{1}{r+1}} .$$

C.   Noting that the improvement is most outspoken for $r=1$ $(2-\sqrt{2}\approx0.59)$ and $r=2$ and that it is negligible when r is very large, give the details of the rejection algorithm for these two cases.

4.   Give the strongest counterparts of Theorems 3.1-3.4 you can find for unimodal densities on the real line with a mode at 0. Because this class contains the class dealt with in the section, all the bounds given in the section remain valid for $f(|x|)$, and this leads to performances that are precisely double those of the various theorems. Mimicking the development of section VII.2 for log-concave densities, this can be improved if we know $F(0)$, the value of the distribution function at 0, or are willing to apply Brent's mirror principle (generate a random variate X with density $f(x)+f(-x)$,$x>0$, and exit with X or $-X$ with probabilities $\dfrac{f(x)}{f(x)+f(-x)}$ and $\dfrac{f(-x)}{f(x)+f(-x)}$ respectively). Work out the details.

5.   Compare the rejection constant of Example 3.5 (log-concave densities on $[0,\infty)$) with 2, the rejection constant obtained for the algorithm of section VII.2. Show that it is always at least 2, that is, show that for all log-concave densities on $[0,\infty)$ belonging to $Lip_1(C)$,

$$\frac{\sqrt{8C}}{f(0)} \geq 2 .$$

Hint: fix C, and try to find the density in the class under consideration for which $f(0)$ is maximal. Conclude that one should never use the algorithm of Example 3.5.

6.   Show that the class $Lip_\alpha(C)$ has no densities whenever $\alpha>1$.

7.   Prove Narumi's inequalities (Lemma 3.1, part D).

8.   When f is a normal scale mixture, show that for all $a>0$, the bound of Theorem 3.6 is at least as good as the corresponding bound of Theorem 3.2.

9.   Show that f is an exponential scale mixture if and only if for all $x>0$, the derivatives of f are of alternating sign (see e.g. Feller (1971), Keilson and Steutel (1974)). These mixtures consist of convex densities densities on $[0,\infty)$. Derive useful bounds similar to those of Theorem 3.6.

10. **The z-distribution.** Barndorff-Nielsen, Kent and Sorensen (1982) introduced the class of z-distributions with two shape parameters. The symmetric members of this family have density

$$f(x) = \frac{1}{4^a B_{a,a} \cosh^{2a}(\frac{x}{2})} \qquad (x \in R) ,$$

where $a > 0$ is a parameter. The translation and scale parameters are omitted. For $a = 1/2$, this gives the hyperbolic cosine distribution. For $a = 1$ we have the logistic distribution. For integer a it is also called the generalized logistic distribution (Gumbel, 1944). Show the following:

A.   The symmetric z-distributions are normal scale mixtures (Barndorff-Nielsen, Kent and Sorensen, 1982).

B.   A random variate can be generated as $\log(\dfrac{Y}{1-Y})$ where Y is symmetric beta distributed with parameter a.

C.   If a random variate is generated by rejection based upon the inequalities of Theorem 3.6, the expected time stays uniformly bounded over all values of a.

Additional note: the general z distribution with parameters $a, b > 0$ is defined as the distribution of $\log(\dfrac{Y}{1-Y})$ where Y is beta (a, b).

11.   **The residual life density.** In renewal theory and the study of Poisson processes, one can associate with every distribution function F on $[0, \infty)$ the residual life density

$$f(x) = \frac{1-F(x)}{\mu},$$

where $\mu = \int (1-F)$ is the mean for F. Assume that besides the mean we also know the second moment μ_2. This is the second moment of F, not f. Show the following:

A. $f(x) \leq \mu_2/(\mu(x^2 + \mu_2))$

B.   The black box algorithm shown below is valid and has rejection constant $\pi\sqrt{\mu_2}/\mu$. The rejection constant is at least equal to π, and can be arbitrarily large.

> REPEAT
> > Generate a Cauchy random variate Y, and a uniform $[0,1]$ random variate U.
> > $X \leftarrow \sqrt{\mu_2}\,Y$
> UNTIL $U \leq (1+Y^2)(1-F(X))$
> RETURN X

12.   Assume that f is a monotone density on $[0, \infty)$ with distribution function F. Show that for all $0 \leq t < x$,

$$f(x) \leq \frac{1-F(t)}{x-t}.$$

Derive from this the inequality

$$f(x) \leq f(0)(1 - F(x - \frac{1}{f(0)})) .$$

Note that these inequalities can be used to derive rejection algorithms from tail inequalities for the distribution function.

4. THE INVERSION-REJECTION METHOD.

4.1. The principle.

Assume that f is a density on R, and that we know a few things about f, but not too much. For example, we may know that f is bounded by M, or that $f \in Lip_1(C)$, or that f is unimodal with mode at 0. We have in addition two black boxes, one for computing f, and one for computing the distribution function F. The rejection method is not applicable because we cannot a priori find an integrable dominating curve as for example in the case of log-concave densities. In many cases, this problem can be overcome by the inversion-rejection method (Devroye, 1984). In its most elementary form, it can be put as follows: consider a countable partition of R into intervals $[x_i, x_{i+1})$ where i can take positive and negative values. This partition is fixed but need not be stored: often we can compute the next point x_i from i and/or the previous point. Generate a uniform [0,1] random variate U, and find the index i for which

$$F(x_i) \leq U < F(x_{i+1}) .$$

Thus, interval $[x_i, x_{i+1})$ is chosen with probability $F(x_{i+1}) - F(x_i)$ by inversion. If the x_i's are not stored, then some version of sequential search can be used. After i is selected, return a random variate X with density f restricted to the given interval. What we have gained is the fact that the interval is compact, and that in most cases we can easily find a uniform dominating density and use rejection. For example, if f is known to be bounded by M, then we can use a uniform curve with value M. When $f \in Lip_1(C)$, we can use a triangular dominating curve with value $\min(f(x_i) + C(x - x_i), f(x_{i+1}) + C(x_{i+1} - x))$. When f is unimodal, then a dominating curve with value $\max(f(x_i), f(x_{i+1}))$ can always be used.

There are two contributors to the expected time taken by the inversion-rejection algorithm:

(i) $E(N_s)$: the expected number of computations of F in the sequential search.

(ii) $E(N_r)$: the expected number of iterations in the rejection method. It is not difficult to see that this is the area under the dominating curve.

In the example of a density bounded by M but otherwise arbitrary, the area under the dominating curve is ∞. Thus, $E(N_r) = \infty$. Nevertheless $N_r < \infty$ with probability one. This fact does not come as a surprise considering the magnitude of the class of densities involved. For unimodal f, even with an infinite peak at the mode and two big tails, it is always possible to construct a partition such that the area under the dominating piecewise constant function is finite. Thus, in the analysis of the different cases, it will be important to distinguish between the families of densities.

The inversion-rejection method is of the black-box type. Its main disadvantage is that programs for calculating both f and F are needed. On the positive side, the families that can be dealt with can be gigantic. The method is not recommended when speed is the most important issue.

We look at the three families introduced above in separate sub-sections. A little extra time is spent on the important class of unimodal densities. The analysis is in all cases based upon the distributional properties of N_s and N_r.

4.2. Bounded densities.

As our first example, we take the family of densities f on $[0,\infty)$ bounded by M. There is nothing sacred about the positive half of R, the choice is made for convenience only. Assume that $[0,\infty)$ is partitioned by a sequence

$$0=x_0<x_1<x_2<\cdots .$$

Let us write $p_i=F(x_{i+1})-F(x_i)$, $i\geq 0$. In a black box method, the inversion step should preferably be carried out by sequential search, starting from 0. In that case, we have

$$P(N_s\geq j)=\sum_{i=j-1}^{\infty}p_i=\int_{x_{j-1}}^{\infty}f=1-F(x_{j-1})\quad(j\geq 1) .$$

Also,

$$E(N_s)=1+\sum_{i=0}^{\infty}i\ p_i=\sum_{i=0}^{\infty}(1-F(x_i)) .$$

Given that we have chosen the i-th interval, the number of iterations in the rejection step is geometrically distributed with parameter $p_i/(M(x_{i+1}-x_i))$, $i\geq 0$. Thus,

$$P(N_r\geq j)=\sum_{i=0}^{\infty}p_i(1-\frac{p_i}{M(x_{i+1}-x_i)})^j .$$

Also,

$$E(N_r)=\sum_{i=0}^{\infty}p_i\frac{M(x_{i+1}-x_i)}{p_i}=\infty .$$

Example 4.1. Equi-spaced intervals.

When $x_{i+1}-x_i=\delta>0$, we obtain perhaps the simplest algorithm of the inversion-rejection type. We can summarize its performance as follows:

$$E(N_s) = 1+\sum_{i=0}^{\infty} i \int_{\delta i}^{\delta(i+1)} f \le 1+\frac{1}{\delta}\sum_{i=0}^{\infty} \int_{\delta i}^{\delta(i+1)} xf = 1+\frac{E(X)}{\delta} ;$$

$$E(N_s) \ge \frac{E(X)}{\delta} ;$$

$$P(N_r \ge j) = \sum_{i=0}^{\infty} p_i (1-\frac{1}{M\delta}p_i)^j .$$

The sequential search is intimately linked with the size of the tail of the density (as measured by $E(X)$). It seems reasonable to take $\delta=cE(X)$ for some universal constant c. When we take c too large, the probabilities $P(N_r \ge j)$ could be unacceptably high. When c is too small, $E(N_s)$ is too large. What is needed here is a compromise. We cannot choose c so as to minimize $E(N_s+N_r)$ for example, since this is ∞. Another method of design can be followed: fix j, and minimize $P(N_r \ge j)+P(N_s \ge j)$. This is

$$\sum_{i=0}^{\infty} p_i (1-\frac{p_i}{M\delta})^j + \sum_{i=j-1}^{\infty} p_i$$

$$\le \sum_{i=J}^{\infty} p_i + \frac{JM\delta}{j+1}(\frac{j}{j+1})^j + \sum_{i=j-1}^{\infty} p_i$$

where J is a positive integer to be picked later. We have used the following simple inequality:

$$u(1-\frac{u}{a})^j \le \frac{a}{j+1}\left[1-\frac{\dfrac{a}{j+1}}{a}\right]^j .$$

Since we have difficulty minimizing the original expression and the last upper bound, it seems logical to attempt to minimize yet another bound. This strategy is deliberately suboptimal. What we hope to buy is simplicity and insight. Assume that $\mu=E(X)$ is known. Then the tail sums of p_i's can be bounded from above by Markov's inequality. In particular, using also $(1+\frac{1}{j})^j \ge 2$, $j \ge 1$, the last expression is bounded by

$$\frac{\mu}{\delta J}+\frac{JM\delta}{2(j+1)}+\frac{\mu+2}{\delta(j+1)} .$$

The optimal non-integer J is

$$\sqrt{\frac{2(j+1)\mu}{M\delta^2}}$$

and we will take the ceiling of this. Our upper bound now reads

$$2\sqrt{\frac{M\mu}{2(j+1)}}+\frac{\dfrac{\mu+2}{\delta}+\dfrac{M\delta}{2}}{j+1} .$$

The last thing left to do is to minimize this with respect to δ, the interval width. Notice however that this will affect only the second order term in the upper bound (coefficient of $\dfrac{1}{j+1}$), and not the main asymptotic term. For the choice $\delta = \sqrt{\dfrac{2\mu+4}{M}}$, the second term is

$$\frac{\sqrt{2M(\mu+2)}}{j+1} .$$

The important observation is that for any choice of δ that is independent of j,

$$P(N_s \geq j) + P(N_r \geq j) \leq 2\sqrt{\frac{M\mu}{2(j+1)}} + O\left(\frac{1}{j}\right) .$$

The factor $M\mu$ is scale invariant, and is both a measure of how spread out f is and how difficult f is for the present black box method. For this bound to hold, it is not necessary to know μ. The main term in the upper bound is the contribution from N_r. If we assume the existence of higher moments of the distribution, or the moment-generating function, we can obtain upper bounds which decrease faster than $1/\sqrt{j}$ as $j \to \infty$ (exercise 4.1). ■

There are other obvious choices for interval sizes. For example, we could start with an interval of width δ, and then double the width of consecutive intervals. Because this will be dealt with in greater detail for monotone densities, it will be skipped here. Also, because of the better complexity for monotone densities, it is worthwhile to spend more time there.

4.3. Unimodal and monotone densities.

This entire subsection is an adaptation of Devroye (1984). Let us first reduce the problem to one that is manageable. If we know the position of the mode of a unimodal density, and if we can compute $F(x)$ at all x, which is our standing assumption, then it is obvious that we need only consider monotone densities. These can be conveniently flipped around and/or translated to 0, so that all monotone densities to be considered can be assumed to have a mode at 0 and support on $[0,\infty)$. Unfortunately, compact support cannot be assumed because nonlinear transformations to $[0,1]$ could destroy the monotonicity. One thing we can assume however is that we either have an infinite peak at 0 or an infinite tail but not both. Just use the following splitting device:

Splitting algorithm for monotone densities

[SET-UP]

Choose a number $z > 0$. (If f is known to be bounded, set $z \leftarrow 0$, and if f is known to have compact support contained in $[0, c]$, set $z \leftarrow c$.)

$t \leftarrow F(z)$

[GENERATOR]

Generate a uniform $[0,1]$ random variate U.

IF $U > t$

    THEN generate a random variate X with (bounded monotone) density $f(x)/(1-t)$ on $[z, \infty)$.

    ELSE generate a random variate X with (compact support) density $f(x)/t$ on $[0, z]$.

RETURN X

Thus, it suffices to treat compact support and bounded monotone densities separately. We will provide the reader with three general strategies, two for bounded monotone densities, and one for compact support monotone densities. Undoubtedly, there are other strategies that could be preferable for certain densities, so no claims of optimality are made. The emphasis is on the manner in which the problem is attacked, and on the interaction between design and analysis. As we pointed out in the introduction, the whole story is told by the quantities $E(N_s)$ and $E(N_r)$ when they are finite.

4.4. Monotone densities on $[0,1]$.

In this section, we will analyze the following inversion-rejection algorithm:

Inversion-rejection algorithm with intervals shrinking at a geometrical rate

Generate a uniform $[0,1]$ random variate U.

$X \leftarrow 1$

REPEAT

$$X \leftarrow \frac{X}{r}$$

UNTIL $U \geq F(X)$

REPEAT

      Generate two independent uniform $[0,1]$ random variates, V, W.

      $Y \leftarrow X(1+(r-1)V)$ (Y is uniform on $[X, rX)$)

UNTIL $W \leq \dfrac{f(Y)}{f(X)}$

RETURN Y

The constant $r > 1$ is a design constant. For a first quick understanding, one can take $r = 2$. In the first REPEAT loop, the inversion loop, the following intervals are considered: $[\frac{1}{r}, 1), [\frac{1}{r^2}, \frac{1}{r}), \dots$. For the case $r = 2$, we have interval halving as we go along. For this algorithm,

$$E(N_s) = \sum_{i=1}^{\infty} i \int_{r^{-i}}^{r^{-(i-1)}} f(x)\, dx \ ,$$

$$E(N_r) = \sum_{i=1}^{\infty} \frac{r-1}{r^i} f(r^{-i}) \ .$$

The performance of this algorithm is summarized in Theorem 4.1:

Theorem 4.1.

Let f be a monotone density on [0,1], and define

$$H(f) = \int_0^1 \log(\frac{1}{x})f(x)\,dx \ .$$

Then, for the algorithm described above,

$$\frac{H(f)}{\log(r)} \le E(N_s) \le 1 + \frac{H(f)}{\log(r)}$$

and

$$1 \le E(N_r) \le r \ .$$

The functional $H(f)$ satisfies the following inequalities:

A. $1 \le H(f)$.

B. $\log\left|\dfrac{1}{\displaystyle\int_0^\infty xf(x)\,dx}\right| \le H(f)$ (valid even if f has unbounded support).

C. $H(f) \le 1 + \log(f(0))$.

D. $H(f) \le \dfrac{4}{e} + 2\displaystyle\int_0^1 \log_+ f(x)\,f(x)\,dx$ (valid even if f is not monotone).

Proof of Theorem 4.1.

For the first part, note that on $[r^{-i}, r^{-(i-1)}]$,

$$\frac{\log(x)}{\log(r)} \le i \le 1 + \frac{\log(x)}{\log(r)} \ .$$

Thus, resubstitution in the expression of $E(N_s)$ yields the first inequality. We also see that $E(N_r) \ge 1$. To obtain the upper bound for $E(N_r)$, we use a short geometrical argument:

$$E(N_r) = \sum_{i=1}^\infty \frac{r-1}{r^i} f(r^{-i})$$

$$= \sum_{i=1}^\infty \int_{r^{-i}}^{r^{-(i-1)}} f(r^{-i})\,dx$$

$$\le \sum_{i=1}^\infty \int_{r^{-(i+1)}}^{r^{-i}} f(x)\,dx \times r$$

$$= r\int_0^{\frac{1}{r}} f(x)\,dx$$

$$\leq r .$$

Inequality A uses the fact that $-\log(x)$ and $f(x)$ are both nonincreasing on $[0,1]$, and therefore, by Steffensen's inequality (1925),

$$\int_0^1 -\log(x)f(x)\ dx \ \geq\ \int_0^1 -\log(x)\ dx \int_0^1 f(x)\ dx \ =\ 1 .$$

Inequality B uses the convexity of $-\log(x)$ and Jensen's inequality. If X is a random variable with density f, then

$$H(f) = E(-\log(X)) \geq -\log(E(X)) .$$

Inequality C can be obtained as a special case of another inequality of Steffensen's (1918): in its original form, it states that if $0 \leq h \leq 1$, and if g is nonincreasing and integrable on $[0,1]$, then

$$\int_0^1 g(x)h(x)\ dx \ \leq\ \int_0^a g(x)\ dx$$

where $a = \int_0^1 h(x)\ dx$. Apply this inequality with $g(x) = -\log(x)$, $h(x) = \dfrac{f(x)}{f(0)}$.
Thus, $a = \dfrac{1}{f(0)}$. Therefore,

$$\frac{H(f)}{f(0)} \leq \int_0^{\frac{1}{f(0)}} -\log(x)\ dx$$

$$= \int_{\log(f(0))}^{\infty} ye^{-y}\ dy \ =\ \frac{1}{f(0)}(1+\log(f(0))) .$$

Inequality D is a Young-type inequality which can be found in Hardy, Littlewood and Polya (1952, Theorem 239). ∎

In Theorem 4.1, we have shown that $E(N_s) < \infty$ if and only if $H(f) < \infty$. On the other hand, $E(N_r)$ is uniformly bounded over all monotone f on $[0,1]$. Our main concern is thus with the sequential search. We do at least as well as in the black box method of section 3.2 (Theorem 3.2), where the expected number of iterations in the rejection method was $1+\log(f(0))$. We are guaranteed to have $E(N_s) \leq 1+(1+\log(f(0)))/\log(r)$, and even if $f(0) = \infty$, the inversion-rejection method can have $E(N_s) < \infty$.

Example 4.2. The beta density.

Consider the beta $(1, a+1)$ density $f(x) = (a+1)(1-x)^a$ on $[0,1]$ where $a > 0$ is a parameter. We have $f(0) = a+1$, $E(X) = \dfrac{1}{a+2}$. Thus, by inequalities B and C of Theorem 4.1,

$$\log(a+2) \le H(f) \le 1 + \log(a+1).$$

We have $H(f) \sim \log(a)$ as $a \to \infty$: the average time of the given inversion-rejection algorithm grows as $\log(a)$ as $a \to \infty$. ■

In the absence of extra information about the density, it is recommended that r be set equal to 2. This choice also gives small computational advantages. It is important nevertheless to realize that this choice is not optimal in general. For example, assume that we wish to minimize $E(N_s + N_r)$, a criterion in which both contributions are given equal weight because both N_s and N_r count in effect numbers of computations of f and/or F. The minimization problem is rather difficult. But if we work on a good upper bound for $E(N_s + N_r)$, then it is nevertheless possible to obtain:

Theorem 4.2.

For the inversion-rejection algorithm of this section with design constant $r > 1$, we have

$$\inf_{r > 1} E(N_s + N_r)$$

$$\le 1 + H(f) \left(\frac{1}{\log^2(H(f))} + \frac{1}{\log(H(f)) - 2\log(\log(H(f)))} \right)$$

$$\sim \frac{H(f)}{\log(H(f))}$$

as $H(f) \to \infty$. The bound is attained for

$$r = \frac{H(f)}{\log^2(H(f))}.$$

Proof of Theorem 4.2.

We start from

$$E(N_s + N_r) \leq 1 + r + \frac{H(f)}{\log(r)} .$$

Resubstitution of the value of r given in the theorem gives us the inequality. This value was obtained by functional iteration applied to

$$r = \frac{H(f)}{\log^2(r)} ,$$

an equation which must be satisfied for the minimum of the upper bound (set the derivative of the upper bound with respect to r equal to 0). The functional iteration was started at $r = H(f)$. That the value is not bad follows from the fact that for $H(f) \geq e$,

$$1 + r + \frac{H(f)}{\log(r)} \geq 1 + \frac{H(f)}{\log(H(f))} ,$$

so that at least from an asymptotic point of view no improvement is possible over the given bound. ∎

As a curious application of Theorem 4.2, consider the case again of a monotone density on [0,1] with finite $f(0)$. Recalling that $H(f) \leq 1 + \log(f(0))$, we see that if we take

$$r = \frac{1 + f(0)}{\log^2(1 + f(0))} ,$$

a choice which is indeed implementable, then

$$E(N_s + N_r)$$
$$\leq 1 + 1 + \frac{\log(f(0))}{\log^2(1 + \log(f(0)))} + \frac{\log(f(0))}{\log(1 + \log(f(0))) - 2\log(\log(1 + \log(f(0))))}$$
$$\sim \log \frac{(f(0))}{\log(1 + \log(f(0)))}$$

as $f(0) \to \infty$. This should be compared with the value of $E(N_r) = 1 + \log(f(0))$ for the black box rejection algorithm following Theorem 3.1.

For densities that are also known to be convex, a slight improvement in $E(N_r)$ is possible. See exercise 4.5.

4.5. Bounded monotone densities: inversion-rejection based on Newton-Raphson iterations.

In this section, we assume that f is monotone on $[0,\infty)$ and that $f(0)<\infty$. It is possible that f has a large tail. In an attempt to automatically balance $E(N_s)$ against $E(N_r)$, and thus to avoid the eternal problem of having to find a good design constant, we could determine intervals for sequential search based upon Newton-Raphson iterations started at $x_0=0$. Recall the definition of the hazard rate

$$h(x) = \frac{f(x)}{1-F(x)} \; .$$

If we try to solve $F(x)=1$ for x by Newton-Raphson iterations started at $x_0=0$, we obtain a sequence $x_0 \le x_1 \le x_2 \le \cdots$ where

$$x_{n+1} = x_n + \frac{1-F(x_n)}{f(x_n)} = x_n + \frac{1}{h(x_n)} \; .$$

The x_n's need not be stored. Obviously, storing them could considerably speed up the algorithm.

Inversion-rejection algorithm for bounded densities based upon Newton-Raphson iterations

Generate a uniform $[0,1]$ random variate U.
$X \leftarrow 0$, $R \leftarrow F(X)$, $Z \leftarrow f(X)$
REPEAT
    $X* \leftarrow X + \dfrac{1-R}{Z}$, $R* \leftarrow F(X*)$, $Z* \leftarrow f(X*)$
    IF $U \le R*$
        THEN Accept $\leftarrow$ True
        ELSE $R \leftarrow R*$, $Z \leftarrow Z*$, $X \leftarrow X*$
UNTIL Accept
REPEAT
    Generate two independent uniform $[0,1]$ random variates V,W.
    $Y \leftarrow X+(X*-X)V$, $T \leftarrow WZ$ (Y is uniformly distributed on $[X,X*)$)
    Accept $\leftarrow [T \le Z*]$ (optional squeeze step)
    IF NOT Accept THEN Accept $\leftarrow [T \le f(Y)]$
UNTIL Accept
RETURN Y

One of the differences with the algorithm of the previous section is that in every iteration of the inversion step, one evaluation of both F and f is required as compared to one evaluation of F. The performance of the algorithm is dealt with in Theorem 4.3.

Theorem 4.3.

Let f be a bounded monotone density on $[0,\infty)$ with mode at 0. For the inversion-rejection algorithm given above,

$$E(N_s) = E(N_r) = \sum_{i=0}^{\infty} (1-F(x_i))$$

where $0 = x_0 \le x_1 \le x_2 \le \cdots$ is the sequence of numbers defined by

$$x_{n+1} = x_n + \frac{1-F(x_n)}{f(x_n)} \qquad (n \ge 0).$$

If f is also DHR (has nonincreasing hazard rate), then

$$1 \le E(N_r) = E(N_s) \le 1+E(Xf(0)).$$

If f is also IHR (has nondecreasing hazard rate), then

$$1 \le E(N_r) = E(N_s) \le \frac{e}{e-1}.$$

Proof of Theorem 4.3.

$$E(N_s) = \sum_{i=1}^{\infty} i((1-F(x_{i-1}))-(1-F(x_i))) = \sum_{i=0}^{\infty} (1-F(x_i)),$$

$$E(N_r) = \sum_{i=0}^{\infty} f(x_i)(x_{i+1}-x_i) = \sum_{i=0}^{\infty} (1-F(x_i)).$$

When f is DHR, then

$$E(Xf(0)) = f(0)\int_0^{\infty}(1-F(x))\,dx = \int_0^{\infty}\frac{f(0)}{h(x)}f(x)\,dx \ge 1.$$

For IHR densities, the inequality should be reversed. Thus, for DHR densities,

$$\sum_{i=0}^{\infty}(1-F(x_i)) \le 1+\sum_{i=1}^{\infty}\frac{\int_{x_{i-1}}^{x_i}(1-F(x))\,dx}{x_i-x_{i-1}}$$

$$= 1+\sum_{i=1}^{\infty}\int_{x_{i-1}}^{x_i}(1-F(x))\,dx\,h(x_{i-1})$$

$$\le 1+\int_0^{\infty}f(0)(1-F(x))\,dx = 1+E(Xf(0)).$$

When f is IHR, then

$$1-F(x_{i+1}) = (1-F(x_i))e^{-\int_{x_i}^{x_{i+1}}h(x)\,dx}$$

$$\leq (1-F(x_i))e^{-h(x_i)(x_{i+1}-x_i)}$$

$$= \frac{1-F(x_i)}{e}.$$

Thus,

$$\sum_{i=0}^{\infty} (1-F(x_i)) \leq \sum_{i=0}^{\infty} e^{-i} = \frac{e}{e-1}. \blacksquare$$

We have thus found an algorithm with a perfect balance between the two parts, since $E(N_s)=E(N_r)$. This does not mean that the algorithm is optimal. However, in many cases, the performance is very good. For example, its expected time is uniformly bounded over all IHR densities. Examples of IHR densities on $[0,\infty)$ are given in the table below.

Name	Density f	Hazard rate h	$E(N_s)=E(N_r)$
Halfnormal	$\sqrt{\dfrac{2}{\pi}}e^{-\frac{x^2}{2}}$		$\leq \dfrac{e}{e-1}$
Gamma (a), $a \geq 1$	$\dfrac{x^{a-1}e^{-x}}{\Gamma(a)}$		$\leq \dfrac{e}{e-1}$
Exponential	e^{-x}	1	$\dfrac{e}{e-1}$
Weibull (a), $a \geq 1$	$ax^{a-1}e^{-x^a}$	ax^{a-1}	$\leq \dfrac{e}{e-1}$
Beta $(a,1)$, $a \geq 1$	ax^{a-1} $(0 \leq x \leq 1)$	$\dfrac{x^{a-1}}{1-x^a}$	$\leq \dfrac{e}{e-1}$
Beta $(1,a+1)$, $a \geq 0$	$(a+1)(1-x)^a$ $(0 \leq x \leq 1)$	$\dfrac{a+1}{1-x}$	$\left[1-(1-\dfrac{1}{a+1})^{a+1}\right]^{-1}$
Truncated extreme value, $a > 0$	$\dfrac{1}{a}e^{x-\frac{e^x-1}{a}}$	$\dfrac{e^x}{a}$	$\leq \dfrac{e}{e-1}$

This is not the place to enter into a detailed study of IHR densities. It suffices to state that they are an important family in daily statistics (see e.g. Barlow and Proschan (1965, 1975), and Barlow, Marshall and Proschan (1963)). Some of its salient properties are covered in exercise 4.6. Some entries for $E(N_s)$ in the table given above are explicitly known. They show that the upper bound of Theorem 4.3 is sharp in a strong sense. For example, for the exponential density, we have $x_n=n$, and thus

$$E(N_s) = E(N_r) = \sum_{i=0}^{\infty} (1-F(i)) = \sum_{i=0}^{\infty} e^{-i} = \frac{e}{e-1}.$$

For the beta $(1,a+1)$ density mentioned in the table, we can verify that

$$x_{n+1} = \frac{a}{a+1}x_n + \frac{1}{a+1},$$

and thus,

$$x_n = 1-(\frac{a}{a+1})^n \quad (n \geq 0).$$

Thus,

$$E(N_s) = \sum_{i=0}^{\infty} (1-F(x_i)) = \sum_{i=0}^{\infty} (1-x_i)^{a+1}$$

$$= \sum_{i=0}^{\infty} (\frac{a}{a+1})^{i(a+1)} = \left[1-(1-\frac{1}{a+1})^{a+1}\right]^{-1}.$$

This varies from 1 ($a=0$) to $\frac{e}{e-1}$ ($a \uparrow \infty$) without exceeding $\frac{e}{e-1}$. Thus, once again, the inequality of Theorem 4.3 is tight.

For DHR densities, the upper bound is often very loose, and not as good as the performance bounds obtained for the dynamic thinning method (section VI.2). For example, for the Pareto density $\frac{a}{(1+x)^{a+1}}$ (where $a > 0$ is a parameter), we have a hazard rate $h(x) = \frac{a}{1+x}$, and $E(N_s) = \left[1-(1+\frac{1}{a})^{-a}\right]^{-1}$. This can be seen as follows:

$$(x_{n+1}+1) = (x_n+1)(1+\frac{1}{a}) ;$$

$$(x_n+1) = (1+\frac{1}{a})^n \quad (n \geq 0) ;$$

$$E(N_s) = \sum_{i=0}^{\infty} (1+\frac{1}{a})^{-ia} = \left[1-(1+\frac{1}{a})^{-a}\right]^{-1}.$$

The last expression varies from $\frac{e}{e-1}$ ($a \uparrow \infty$) to 2 ($a=1$) and up to ∞ as $a \downarrow 0$.

4.6. Bounded monotone densities: geometrically increasing interval sizes.

For bounded densities, we can use a sequential search from left to right, symmetric to the method used for unbounded but compact support densities. There are two design parameters: $t > 0$ and $r > 1$, and the consecutive intervals are

$$[0,t),[t,tr),[tr,tr^2),... .$$

A typical choice is $t=1$, $r=2$. General guidelines follow after the performance analysis. Let us begin with the algorithm:

Inversion-rejection method for bounded monotone densities based upon geometrically exploding intervals

Generate a uniform [0,1] random variate U.

$X \leftarrow 0$, $X* \leftarrow t$

WHILE $U > F(X*)$ DO

      $X \leftarrow X*$, $X* \leftarrow rX*$

REPEAT

      Generate two iid uniform [0,1] random variates, V, W.

      $Y \leftarrow X + (X* - X)V$ (Y is uniformly distributed on $[X, X*)$)

UNTIL $W \leq \dfrac{f(Y)}{f(X)}$

RETURN Y

Theorem 4.4.

    Let f be a bounded monotone density, and let $t > 0$ and $r > 1$ be constants. Define

$$H_t(f) = \int_0^\infty \log_+(\frac{x}{t}) f(x) \, dx \ .$$

Then, for the algorithm given above,

$$1 + \frac{H_t(f)}{\log(r)} \leq E(N_s) \leq 2 + \frac{H_t(f)}{\log(r)} \ ,$$

and

$$1 \leq tf(0) + \int_t^\infty f(x) \, dx \leq E(N_r) \leq tf(0) + r \ .$$

Proof of Theorem 4.4.

    We repeatedly use the fact that $tr^{i-1} \leq x < tr^i$ if and only if $i - 1 \leq \log(\frac{x}{t})/\log(r) < i$, $i > 1$. Now,

$$E(N_s) = \int_0^t f(x) \, dx + \sum_{i=1}^\infty (i+1) \int_{tr^{i-1}}^{tr^i} f(x) \, dx = 1 + \sum_{i=1}^\infty i \int_{tr^{i-1}}^{tr^i} f(x) \, dx$$

$$\leq 2 + \int_t^\infty \frac{\log(\frac{x}{t})}{\log(r)} f(x) \, dx = 2 + \frac{H_t(f)}{\log(r)} \ ,$$

and

$$E\,(N_s\,) \geq 1+\int\limits_t^\infty \frac{\log(\frac{x}{t})}{\log(r\,)} f\,(x\,)\,dx \;\; = 1+\frac{H_t\,(f\,)}{\log(r\,)}\;.$$

Also,

$$E\,(N_r\,) = tf\,(0)+\sum_{i\,=1}^\infty (tr\,^i-tr\,^{i-1})f\,(tr\,^{i-1})$$

$$\leq tf\,(0)+\sum_{i\,=1}^\infty \frac{tr\,^i-tr\,^{i-1}}{tr\,^{i-1}-tr\,^{i-2}} \int\limits_{tr\,^{i-2}}^{tr\,^{i-1}} f\,(x\,)\,dx$$

$$\leq tf\,(0)+r\;,$$

and

$$E\,(N_r\,) \geq tf\,(0)+\sum_{i\,=1}^\infty \frac{tr\,^i-tr\,^{i-1}}{tr\,^i-tr\,^{i-1}} \int\limits_{tr\,^{i-1}}^{tr\,^i} f\,(x\,)\,dx$$

$$= tf\,(0)+\int\limits_t^\infty f\,(x\,)\,dx \;\geq 1\;.\;\blacksquare$$

We would like the algorithm to perform at a scale-invariant speed. This can be achieved for $t = \dfrac{1}{f\,(0)}$. In that case, the upper bounds of Theorem 4.4 read:

$$E\,(N_s\,) \leq 2+\frac{H*(f\,)}{\log(r\,)}\;;$$

$$E\,(N_r\,) \leq 1+r\;,$$

where

$$H*(f\,) = \int\limits_0^\infty \log_+(xf\,(0))f\,(x\,)\,dx$$

is the scale invariant counterpart of the quantity $H(f\,)$ defined in Theorem 4.1. $H*(f\,)$ can be considered as the normalized logarithmic moment for the density f. For the vast majority of distributions, $H*(f\,)<\infty$. In fact, one must search hard to find a monotone density for which $H*(f\,)=\infty$. The tail of the density must at least of the order of $1/(x\log^2(x\,))$ as $x\to\infty$, such as is the case for

$$f\,(x\,) = \frac{1}{(x+e\,)\log^2(x+e\,)}\qquad (x>0)\;.$$

With little a priori information, we suggest the choice

$$r = 2$$
$$t = \frac{1}{f(0)}$$

It is interesting to derive a good guiding formula for r. We start from the inequality

$$E(N_s) + E(N_r) \leq 3 + r + \frac{H*(f)}{\log(r)} ,$$

which is minimal for the unique solution $r > 1$ for which $r \log^2(r) = H*(f)$. By functional iteration started at $r = H*(f)$, we obtain the crude estimate

$$r = \frac{H*(f)}{\log^2(H*(f))} .$$

For this choice, we have as $H*(f) \to \infty$,

$$E(N_s) + E(N_r) \leq (1 + o(1)) \frac{H*(f)}{\log(H*(f))} .$$

Example 4.3. Moment known.

A loose upper bound for $H*(f)$ is afforded by Jensen's inequality:

$$H*(f) \leq \int_0^\infty \log(1 + xf(0)) f(x) \, dx \leq \log(1 + E(Xf(0)))$$

where X is a random variable with density f. Thus, the expected time of the algorithm grows at worst as the logarithm of the first moment of the distribution. For example, for the beta $(1, a+1)$ density of Example 4.1, this upper bound is $\log(1 + \frac{a+1}{a+2}) \leq \log(2)$ for all $a > 0$. This is an example of a family for which the first moment, hence $H*(f)$, is uniformly bounded. From this,

$$E(N_s) \leq 2 + \frac{\log(2)}{\log(r)} ;$$

$$E(N_r) \leq 1 + r .$$

The ad hoc choice $r = 2$ makes both upper bounds equal to 3. ∎

4.7. Lipschitz densities on $[0,\infty)$.

The inversion-rejection method can also be used for Lipschitz densities f on $[0,\infty)$. This class is smaller than the class of bounded densities, but very large compared to the class of monotone densities. The black box method of section 3 for this class required knowledge of a moment of the distribution. In contrast, the method presented here works for all densities $f \in Lip_1(C)$ where only C must be given beforehand. The moments of the distribution need not even exist. If the positive half of the real line is partitioned by

$$0 = x_0 < x_1 < x_2 < \cdots ,$$

then, it is easily seen that on $[x_n, x_{n+1}]$,

$$f(x) \leq \min(f(x_n) + C(x - x_n), f(x_{n+1}) + C(x_{n+1} - x)) ,$$

and

$$f(x) \leq \sqrt{2C(1 - F(x_n))}$$

where the last inequality is based upon Theorem 3.5. The areas under the respective dominating curves are

$$E(N_r) = \sum_{n=0}^{\infty} \frac{1}{2C} \left[c \Delta_n (f(x_n) + f(x_{n+1})) - \frac{1}{2}(f(x_n) + f(x_{n+1}))^2 + \frac{C^2 \Delta_n^2}{2} \right]$$

and

$$E(N_r) = \sum_{n=0}^{\infty} \Delta_n \sqrt{2C(1 - F(x_n))} ,$$

where $\Delta_n = x_{n+1} - x_n$. The value of $E(N_s)$ depends only upon the partition, and not upon the inequalities used in the rejection step, and plays no role when the inequalities are compared. Generally speaking, the second inequality is better because it uses more information (the value of F is used). Consider the first inequality. To guarantee that $E(N_r)$ be finite, for the vast majority of Lip_1 densities we need to ask that

$$\sum_{n=0}^{\infty} \Delta_n^2 < \infty .$$

But, since we require a valid partition of R, we must also have

$$\sum_{n=0}^{\infty} \Delta_n = \infty .$$

In particular, we cannot afford to take $\Delta_n = \delta > 0$ for all n. Consider now Δ_n satisfying the conditions stated above. When $\Delta_n \sim n^{-a}$, then it is necessary that $a \in (\frac{1}{2}, 1]$. Thus, the intervals shrink rapidly to 0. Consider for example

$$\Delta_n = \frac{c}{n+1} \quad (n \geq 0) .$$

For this choice, the intervals shrink so rapidly that we spend too much time searching unless f has a very small tail. In particular,

$$E(N_s) = \sum_{n=0}^{\infty} P(X \ge \sum_{i=0}^{n} \Delta_i)$$

$$\le \sum_{n=0}^{\infty} P(X \ge c \log(n+2))$$

$$= \sum_{n=0}^{\infty} P(e^{\frac{X}{c}} \ge n+2)$$

$$\le E(e^{\frac{X}{c}}).$$

A similar lower bound for $E(N_s)$ exists, so that we conclude that $E(N_s) < \infty$ if and only if the moment generating function at $\frac{1}{c}$ is finite, i.e.

$$m(\frac{1}{c}) = E(e^{\frac{X}{c}}) < \infty.$$

In other words, f must have a sub-exponential tail for good expected time. Thus, instead of analyzing the first inequality further, we concentrate on the second inequality.

The algorithm based upon the second inequality can be summarized as follows:

Inversion-rejection algorithm for Lipschitz densities

Generate a uniform [0,1] random variate U.

$X \leftarrow 0$, $R \leftarrow F(X)$

REPEAT

    $X* \leftarrow$ Next (X), $R* \leftarrow F(X*)$ (The function Next computes the next value in the partition.)

    IF $U \le R*$

        THEN Accept $\leftarrow$ True

        ELSE $R \leftarrow R*$, $X \leftarrow X*$

UNTIL Accept

REPEAT

    Generate two independent uniform [0,1] random variates V, W.

    $Y \leftarrow X + V(X*-X)$ (Y is uniformly distributed on $[X, X*]$).

UNTIL $W \sqrt{2C(1-R)} \le f(Y)$

RETURN Y

There are three partitioning schemes that stand out as being either important or practical. These are defined as follows:

A. $x_n = n\delta$ for some $\delta > 0$ (thus, $x_{n+1} - x_n = \delta$).

B. $x_{n+1} = tr^n$ for some $t > 0, r > 1$, $x_1 = t$ (note that $x_{n+1} = rx_n$ for all $n \geq 1$). The intervals grow exponentially fast.

C. $x_{n+1} = x_n + \sqrt{\dfrac{1 - F(x_n)}{2C}}$ (this choice provides a balance between $E(N_s)$ and $E(N_r)$).

Schemes A and B require additional design constants, whereas scheme C is completely automatic. Which scheme is actually preferable depends upon various factors, foremost among these the size of the tail of the distribution. By imposing conditions on the tail, we can derive upper bounds for $E(N_s)$ and $E(N_r)$. These are collected in Theorem 4.5:

Theorem 4.5.

Let $f \in Lip_1(C)$ be a density on $[0,\infty)$. Let $p > 1$ be a constant. When the p-th moment exists, it is denoted by μ_p.

For scheme A,

$$\max(1, \frac{\mu_1}{\delta}) \leq E(N_s) \leq 1 + \frac{\mu_1}{\delta} ;$$

$$\delta\sqrt{2C} \max(1, \frac{1}{\sqrt{\mu_2}}, \frac{\sqrt{\mu_2}}{\delta}) \leq E(N_r) \leq \delta\sqrt{2C} (2 + \frac{p}{p-1} \frac{(\mu_{2p})^{\frac{1}{2p}}}{\delta}).$$

In particular, if $\delta = \sqrt{\dfrac{\mu_1}{\sqrt{8C}}}$, then

$$E(N_s) + E(N_r) \leq 1 + (8C)^{\frac{1}{4}} \sqrt{\mu_1} + \sqrt{8C} (\mu_4)^{\frac{1}{4}} ,$$

and when $\delta = \dfrac{1}{\sqrt{8C}}$,

$$E(N_s) + E(N_r) \leq 2 + \sqrt{8C} ((\mu_4)^{\frac{1}{4}} + \mu_1) \leq 2 + \sqrt{32C} (\mu_4)^{\frac{1}{4}} .$$

For scheme B,

$$E(N_s) \leq 2 + E\left\lceil \frac{\log_+(\frac{X}{t})}{\log(r)} \right\rceil ;$$

$$E(N_r) \leq \sqrt{2C} (t + \frac{\sqrt{\mu_{2p}} r^{p-1}(r-1)}{t^{p-1}(r^{p-1}-1)}).$$

For scheme C,

$$E(N_s) = E(N_r) \leq \sqrt{8C} \int_0^\infty \sqrt{1-F(x)} \, dx \leq \frac{p}{p-1}\sqrt{8C} (\mu_{2p})^{\frac{1}{2p}} .$$

At the same time, even if $\mu_2 = \infty$, the following lower bound is valid:

$$\sqrt{2C\mu_2} \leq \frac{1}{2}\sqrt{8C} \int_0^\infty \sqrt{1-F(x)} \, dx \leq E(N_s) = E(N_r) .$$

Proof of Theorem 4.5.

In this proof, X denotes a random variate with density f. Rewrite $E(N_s)$ as follows:

$$E(N_s) = \sum_{n=0}^{\infty} \int_{\delta n}^{\infty} f(x)\,dx = \int_0^{\infty} \left\lfloor \frac{x}{\delta} + 1 \right\rfloor\,dx \ .$$

This can be obtained by an interchange of the sum and the integral. But then, by Jensen's inequality and trivial bounds,

$$\max(1, \frac{E(X)}{\delta}) \le \int_0^{\infty} \max(1, \frac{x}{\delta}) f(x)\,dx \le E(N_s)$$

$$\le \int_0^{\infty} (\frac{x}{\delta} + 1) f(x)\,dx = 1 + \frac{E(X)}{\delta} \ .$$

Next,

$$E(N_r) = \sum_{n=0}^{\infty} \sqrt{2C(1 - F(\delta n))}\,\delta \ ,$$

so that by Chebyshev's inequality,

$$\frac{E(N_r)}{\delta\sqrt{2C}} \le \sum_{n=0}^{\infty} \min(1, \frac{\sqrt{\mu_{2p}}}{(n\delta)^p})$$

$$\le 1 + \frac{1}{\delta}(\mu_{2p})^{\frac{1}{2p}} + \sum_{n=n_0}^{\infty} \frac{\sqrt{\mu_{2p}}}{(n\delta)^p}$$

where $n_0 = \left\lceil \frac{1}{\delta}(\mu_{2p})^{\frac{1}{2p}} \right\rceil$. By a simple argument, we see that

$$\sum_{n=n_0}^{\infty} n^{-p} \le n_0^{-p} + \int_{n_0}^{\infty} x^{-p}\,dx$$

$$= n_0^{-p} + \frac{1}{p-1} n_0^{-(p-1)} \ .$$

Combining this shows that

$$\frac{E(N_r)}{\delta\sqrt{2C}} \le 1 + \frac{1}{\delta}(\mu_{2p})^{\frac{1}{2p}} + 1 + \frac{1}{(p-1)\delta}(\mu_{2p})^{\frac{1}{2p}}$$

$$= 2 + \frac{p}{p-1}\frac{(\mu_{2p})^{\frac{1}{2p}}}{\delta} \ .$$

This brings us to the lower bounds for scheme A. We have, by the Cauchy-Schwarz inequality,

$$\frac{E(N_r)}{\delta\sqrt{2C}} = \sum_{n=0}^{\infty} \sqrt{\int_{\delta n}^{\infty} f}$$

$$\geq \sum_{n=0}^{\infty} \frac{\int\limits_{\delta n}^{\infty} \sqrt{f}\,(x\sqrt{f}\,)}{\sqrt{\int (x\sqrt{f}\,)^2}}$$

$$= \sum_{n=0}^{\infty} \frac{\int\limits_{\delta n}^{\infty} xf}{\sqrt{\mu_2}}$$

$$\geq \frac{1}{\sqrt{\mu_2}} \int xf\,(x)\max(1,\frac{x}{\delta})\,dx$$

$$\geq \frac{1}{\sqrt{\mu_2}}\max(1,\frac{\mu_2}{\delta})$$

$$= \max(\frac{1}{\sqrt{\mu_2}},\frac{\sqrt{\mu_2}}{\delta})\ .$$

Also,

$$\frac{E(N_r)}{\delta\sqrt{2C}} = \sum_{n=0}^{\infty} \sqrt{\int\limits_{\delta n}^{\infty} f}$$

$$\geq \sum_{n=0\delta n}^{\infty}\int\limits^{\infty} f$$

$$\geq \max(1,\frac{\mu_1}{\delta})\ .$$

For scheme B, we have

$$E(N_s) = 1+ \sum_{n=0}^{\infty}(1-F(tr^n))$$

$$= 1+ \sum_{n=0tr^n}^{\infty}\int\limits^{\infty} f(x)\,dx$$

$$\leq 2+E\left|\frac{\log_+(\frac{X}{t})}{\log(r)}\right|\ .$$

Also,

$$E(N_r) = \sum_{n=0}^{\infty} \sqrt{2C}\ \sqrt{1-F(tr^n)}t\,(r-1)r^n\ +\ \sqrt{2C}\,t$$

$$\leq \sqrt{2C}\,t+\sqrt{2C}\sum_{n=0}^{\infty} t\,(r-1)r^n\ \frac{\sqrt{\mu_{2p}}}{t^p\,r^{np}}$$

$$= \sqrt{2C}\,(t+\frac{\sqrt{\mu_{2p}}\ r^{p-1}(r-1)}{t^{p-1}(r^{p-1}-1)})\ .$$

Finally, we consider scheme C. Consider the graph of $1-\sqrt{1-F(x)}$. Construct for given x_n the triangle with top on the given curve, and base $[x_n, x_{n+1}]$ at height 1. Its area is $\dfrac{1-F(x_n)}{\sqrt{8C}}$. The triangle lies completely above the given curve because the slope of the hypothenusa is $\sqrt{2C}$, which is at least as steep as the derivative of $1-\sqrt{1-F}$ at any point. To see this, note that the latter derivative at x is

$$\frac{f(x)}{2\sqrt{1-F(x)}} \leq \frac{\sqrt{2C(1-F(x))}}{2\sqrt{1-F(x)}} = \sqrt{\frac{C}{2}} .$$

Thus, the sums of the areas of the triangles is not greater than the integral $\int_0^\infty \sqrt{1-F(x)}\, dx$. But this sum is

$$\sum_{n=0}^\infty \frac{1-F(x_n)}{\sqrt{8C}} = \frac{E(N_r)}{\sqrt{8C}} = \frac{E(N_s)}{\sqrt{8C}} .$$

Also, twice the area of the triangles is at least equal to $\int_0^\infty \sqrt{1-F(x)}\, dx$. The bounds in terms of the various moments mentioned are obtained without further trouble. First, by Chebyshev's inequality,

$$\int_0^\infty \sqrt{1-F(x)}\, dx \leq \int_0^\infty \min(1, \frac{\sqrt{\mu_{2p}}}{x^p})\, dx = (\mu_{2p})^{\frac{1}{2p}} + \frac{1}{p-1}(\mu_{2p})^{\frac{1}{2p}} .$$

Also, by the Cauchy-Schwarz inequality,

$$\int_0^\infty \sqrt{\int_x^\infty f}\, dx \geq (\mu_2)^{-\frac{1}{2}} \int_0^\infty \int_x^\infty yf(y)\, dy\, dx$$

$$= (\mu_2)^{-\frac{1}{2}} \int_0^\infty \int_0^y dx\, yf(y)\, dy = \sqrt{\mu_2} . \blacksquare$$

We observe that $\sqrt{C}X$ is a scale-invariant quantity. Thus, one upper bound for scheme A (choice $\delta=\dfrac{1}{\sqrt{8C}}$) and the upper bound for scheme C are scale-invariant: they depend upon the shape of the density only. Scheme C is attractive because no design constants have to be chosen at any time. In scheme A for example, the choice of δ is critical. The geometrically increasing interval sizes of scheme B seem to offer little advantage over the other methods, because $E(N_r)$ is relatively large.

4.8. Exercises.

1. Obtain an upper bound for $P(N_r \geq j)$ in terms of j when equi-spaced intervals are used for bounded densities on $[0,\infty)$ as in Example 4.1. Assume first that the r-th moment μ_r is finite. Assume next that $E(e^{tX})=m(t)<\infty$ for some $t>0$. The interval width δ does not depend upon j. Check that the main term in the upper bound is scale-invariant.

2. Prove inequality D of Theorem 4.1.

3. Give an example of a monotone density on $[0,1]$, unbounded at 0, with $H(f)<\infty$.

4. Inequalities A through C in Theorem 4.1 are best possible: they can be attained for some classes of monotone densities on $[0,1]$. Describe some classes of densities for which we have equality.

5. When f is a monotone convex density on $[0,1]$, then the inversion-rejection algorithm based on shrinking intervals given in the text can be adapted so that rejection is used with a trapezoidal dominating curve joining $[X,f(X)]$ and $[rX,f(rX)]$ where $r>1$ is the shrinkage parameter used in the original algorithm. Such a change would leave N_s the same. It reduces $E(N_r)$ however. Formally, the algorithm can be written as follows:

Inversion-rejection algorithm with intervals shrinking at a geometrical rate

Generate a uniform $[0,1]$ random variate U.

$X \leftarrow 1$

REPEAT

$$X \leftarrow \frac{X}{r}$$

UNTIL $U \geq F(X)$

$Z \leftarrow f(X), Z* \leftarrow f(rX)$

REPEAT

Generate three independent uniform $[0,1]$ random variates, U,V,W.

$$R \leftarrow \min(U, V\frac{Z+Z*}{Z-Z*})$$

$Y \leftarrow X(1+(r-1)R)$ (Y has the given trapezoidal density)

$T \leftarrow W(Z+(Z*-Z)R)$

Accept $\leftarrow [T \leq Z*]$ (optional squeeze step)

IF NOT Accept THEN Accept $\leftarrow [W \leq f(Y)]$

UNTIL Accept

RETURN Y

Prove that $E(N_r) \leq \frac{1}{2}(1+r)$. In other words, for large values of r, this corresponds to an improvement of the order of 50%.

6. **IHR densities.** Prove the following statements:

    A. If X has an IHR density on $[0,\infty)$, then $Xf(0)$ is stochastically smaller than an exponential random variate, i.e. for all $x>0$, $P(Xf(0)>x)\leq e^{-x}$. Conclude that for $r>0$, $E(X^r)\leq\dfrac{\Gamma(r+1)}{f(0)^r}$.

    B. For $r>0$, $E(X^r)\leq\Gamma(r+1)E^r(X)$ (Barlow, Marshall and Proschan, 1963).

    C. The convolution of two IHR densities is again IHR.

    D. Let Y,Z be independent IHR random variables with hazard rates h_Y and h_Z. Then, if h_{Y+Z} is the hazard rate of their sum, $h_{Y+Z}\leq\min(h_Y,h_Z)$.

    E. Construct an IHR density which is continuous, unbounded, and has infinitely many peaks.

7. Show how to choose r and t in the inversion-rejection algorithm with geometrically exploding intervals so as to obtain performance that is sublogarithmic in the first moment of the distribution in the following sense:

    $$E(N_r)+E(N_s)\ \leq\ C\frac{\log(1+\mu f(0))}{\log(\log(e+\mu f(0)))}\ ,$$

    where $\mu=E(X)$, C is some universal constant, and X is a random variable with density f.

8. **Bounded convex monotone densities.** Give an algorithm analogous to that studied in Theorem 4.4 for this class of densities: its sole difference is that the rejection step uses a trapezoidal dominating curve. For this algorithm, in the notation of Theorem 4.4, prove the inequality

    $$E(N_r)\ \leq\ \frac{1}{2}(tf(0)+r+1)\ .$$

9. Prove that if $\Delta_n=\dfrac{c}{n+1}$ in the algorithm for Lipschitz densities, then $E(N_s)<\infty$ if and only if $E(e^{\frac{X}{c}})<\infty$.

10. Suggest good choices for t and r in scheme B of Theorem 4.5. These choices should preferably minimize $E(N_s)+E(N_r)$, or the upper bound for this sum given in the theorem. The resulting upper bound should be scale-invariant.

11. Consider a density f on $[0,\infty)$ which is in $Lip_\alpha(C)$ for some $\alpha\in(0,1]$. Using the inequality of Theorem 3.5 for such densities, give an algorithm generalizing scheme C of Theorem 4.5 for Lip_1 densities. Make sure that $E(N_s)=E(N_r)$ and give an upper bound for $E(N_s)$ which generalizes the upper bound of Theorem 4.5.

12. The lower bound for scheme C in Theorem 4.5 shows that when $\mu_2=\infty$, then $E(N_s)=\infty$. This is a nearly optimal result, in that for most densities with finite second moment, $E(N_s)<\infty$. For example, if $\mu_{2+\epsilon}<\infty$ for some

$\epsilon > 0$, then $E(N_s) < \infty$. Find densities for which $\mu_2 < \infty$, yet $E(N_s) = \infty$.

Chapter Eight
TABLE METHODS FOR
CONTINUOUS RANDOM VARIATES

1. COMPOSITION VERSUS REJECTION.

We have illustrated how algorithms can be sped up if we are willing to compute certain constants beforehand. For example, when a discrete random variate is generated by the inversion method, it pays to compute and store the individual probabilities p_n beforehand. This information can speed up sequential search, or could be used in the method of guide tables. For continuous random variates, the same remains true. Because we know many ultra fast discrete random variate generation methods, but very few fast continuous random variate generation techniques, there is a more pressing need for acceleration in the continuous case. Globally speaking, discretizing the problem speeds generation.

We can for example cut up the graph of f into pieces, and use the composition method. Choosing a piece is a discrete random variate generation problem. Generating a continuous random variate for an individual piece is usually simple because of the shape of the piece which is selected by us. There are only a few drawbacks: first of all, we need to know the areas of the pieces. Typically, this is equivalent to knowing the distribution function. Very often, as with the normal density for example, the distribution function must be computed as the integral of the density, which in our model is an infinite time operation. In particular, the composition method can hardly be made automatic because of this. Secondly, we observe that there usually are several nonrectangular pieces, which are commonly handled via the rejection method. Rectangular pieces are of course most convenient since we can just return a properly translated and scaled uniform random variate. For this reason, the total area of the nonrectangular pieces should be kept as small as possible.

There is another approach which does not require integration of f. If we find a function $g \geq f$, and use rejection, then similar accelerations can be obtained if we cut the graph of g up into convenient pieces. But because g is picked by us, we do of course know the areas (weights) of the pieces, and we can choose g piecewise constant so that each component piece is for example rectangular. One could object that for this method, we need to compute the ratio f / g rather often as part of the rejection algorithm. But this too can be avoided whenever a given piece lies completely under the graph of f . Thus, in the design of pieces, we should try to maximize the area of all the pieces entirely covered by the graph of f .

From this general description, it is seen that all boils down to decompositions of densities into small manageable pieces. Basically, such decompositions account for nearly all very fast methods available today: Marsaglia's rectangle-wedge-tail method for normal and exponential densities (Marsaglia, Maclaren and Bray, 1964; Marsaglia, Ananthanarayanan and Paul, 1976), the method of Ahrens and Kohrt (1981), the alias-rejection-mixture method (Kronmal and Peterson, 1980), and the ziggurat method (Marsaglia and Tsang, 1984). The acceleration can only work well if we have a finite decomposition. Thus, infinite tails must be cut off and dealt with separately. Also, from a didactical point of view, rectangular decompositions are by far the most important ones. We could add triangles, but this would detract from the main points. Since we do care about the generality of the results, it seems pointless to describe a particular normal generator for example. Instead, we will present algorithms which are applicable to large classes of densities. Our treatment differs from that found in the references cited above. But at the same time, all the ideas are borrowed from those same references.

In section 2, we will discuss strip methods, i.e. methods that are based upon the partition of f into parallel strips. Because the strips have unequal probabilities, the strip selection part of the algorithm is usually based upon the alias or alias-urn methods. Partitions into equal parts are convenient because then fast table methods can be used directly. This is further explored in section 3.

2. STRIP METHODS.

2.1. Definition.

The following will be our standing assumptions in this section: f is a bounded density on $[0,1]$; the interval $[0,1]$ is divided into n equal parts (n is chosen by the user); g is a function constant on the n intervals, 0 outside $[0,1]$, and at least equal to f everywhere. We set

$$g(x) = g_i \quad (\frac{i-1}{n} \leq x < \frac{i}{n}) \ (1 \leq i \leq n) \ .$$

Define the strip probabilities

$$p_i = \frac{g_i}{\sum\limits_{j=1}^{n} g_j} \quad (1 \leq i \leq n) \ .$$

Then, the following rejection algorithm is valid for generating a random variate with density f :

REPEAT

    Generate a discrete random variate Z whose distribution is determined by $P(Z=i)=p_i$ $(1 \leq i \leq n)$.

    Generate two iid uniform $[0,1]$ random variate U,V.

$$X \leftarrow \frac{Z-1+V}{n}$$

UNTIL $U g_Z \leq f(X)$

RETURN X

As n increases, the rejection rate should diminish since it is possible to find better and better dominating functions g. But regardless of how large n is picked, there is no avoiding the two uniform random variates and the computation of $f(X)$. Suppose now that each strip is cut into two parts by a horizontal line, and that the bottom part is completely tucked under the graph of f. For part i, the horizontal line has height h_i. We can set up a table of $2n$ probabilities: $p_1, \ldots, p_n$ correspond to the bottom portions, and $p_{n+1}, \ldots, p_{2n}$ to the top portions. Then, random variate generation can proceed as follows:

REPEAT

    Generate a discrete random variate Z whose distribution is determined by $P(Z=i)=p_i$ $(1 \leq i \leq 2n)$.

    Generate a uniform $[0,1]$ random variate V.

$$X \leftarrow \frac{Z-1+V}{n}$$

    IF $Z \leq n$

        THEN RETURN X

        ELSE

            Generate a uniform $[0,1]$ random variate U.

            IF $h_{Z-n} + U(g_{Z-n} - h_{Z-n}) \leq f(X-1)$ THEN RETURN $X-1$

UNTIL False

When the bottom probabilities are dominant, we can get away with generating just one discrete random variate Z and one uniform $[0,1]$ random variate V most of the time. The performance of the algorithm is summarized in Theorem 2.1:

Theorem 2.1.

For the rejection method based upon n split strips of equal width, we have:

1. The expected number of iterations is $\dfrac{1}{n} \sum\limits_{i=1}^{n} g_i$. This is also equal to the expected number of discrete random variates Z per returned random variate X.

2. The expected number of computations of f is $\dfrac{1}{n} \sum\limits_{i=1}^{n} (g_i - h_i)$.

3. The expected number of uniform [0,1] random variates is $\dfrac{1}{n} \sum\limits_{i=1}^{n} g_i + \dfrac{1}{n} \sum\limits_{i=1}^{n} (g_i - h_i)$.

Proof of Theorem 2.1.

The proof uses standard properties of rejection algorithms, together with Wald's equation. ■

The algorithm requires tables for g_i, h_i, $1 \le i \le n$, and p_i, $1 \le i \le 2n$. Some of the $4n$ numbers stored away contain redundant information. Indeed, the p_i's can be computed from the g_i's and h_i's. We store redundant information to increase the speed of the algorithm. There may be additional storage requirements depending upon the discrete random variate generation method: see for example what is needed for the method of guide tables, and the alias and alias-urn methods which are recommended for this application. Recall that the expected time of these generators does not depend upon n.

Thus, we are left only with the computation of the g_i's and h_i's. Consider first the best possible constants:

$$g_i = \sup_{\frac{i-1}{n} \le x < \frac{i}{n}} f(x) ;$$

$$h_i = \inf_{\frac{i-1}{n} \le x < \frac{i}{n}} f(x) .$$

Normally, we cannot hope to compute these values in a finite amount of time. For specially restricted densities f, it is possible however to do so quite easily. Regardless of whether we can actually compute them or not, we have the following important observation:

Theorem 2.2.

Assume that f is a Riemann integrable density on $[0,1]$. Then, if g_i , h_i are defined by:

$$g_i = \sup_{\frac{i-1}{n} \leq x < \frac{i}{n}} f(x) ;$$

$$h_i = \inf_{\frac{i-1}{n} \leq x < \frac{i}{n}} f(x) ,$$

we have:

1. $$\lim_{n \to \infty} \frac{1}{n} \sum_{i=1}^{n} g_i = 1;$$

2. $$\lim_{n \to \infty} \frac{1}{n} \sum_{i=1}^{n} (g_i - h_i) = 0 .$$

Proof of Theorem 2.2.

It suffices to prove the second statement, in view of the fact that

$$\frac{1}{n} \sum_{i=1}^{n} g_i \leq 1 + \frac{1}{n} \sum_{i=1}^{n} (g_i - h_i) .$$

But the second statement is a direct consequence of the definition of Riemann integrability. ∎

Thus, for sufficiently well-behaved densities, if we have optimal bounds g_i, h_i at our disposal, the algorithm becomes very efficient when n grows large.

2.2. Example 1: monotone densities on $[0,1]$.

When f is monotone on $[0,1]$, we can set

$$g_i = f\left(\frac{i-1}{n}\right) ; h_i = f\left(\frac{i}{n}\right) .$$

We also have

$$\frac{1}{n} \sum_{i=1}^{n} (g_i - h_i) = \frac{1}{n} \sum_{i=1}^{n} \left(f\left(\frac{i-1}{n}\right) - f\left(\frac{i}{n}\right) \right)$$

$$= \frac{1}{n}(f(0) - f(1)) \leq \frac{f(0)}{n} .$$

The performance of the algorithm can be summarized quite simply:

1. The expected number of iterations is $\leq 1 + \dfrac{f(0)}{n}$. This is also equal to the expected number of discrete random variates Z per returned random variate X.

2. The expected number of computations of f is $\leq \dfrac{f(0)}{n}$.

3. The expected number of uniform $[0,1]$ random variates is $\leq 1 + \dfrac{2f(0)}{n}$.

We also note that to set up the tables g_i, h_i, it suffices to evaluate f at the $n+1$ mesh points. Furthermore, the extremes of f are reached at the endpoints of the intervals, so that the constants are in this case best possible. The only way to improve the performance of the algorithm would be by considering unequal interval sizes. It should be clear that the interval sizes should become smaller as we approach the origin. The unequal intervals need to be picked with care if real savings are needed. For a fair comparison, we will use n intervals with break-points

$$0 = x_0 < x_1 < x_2 < \ \cdots \ < x_n = 1 \ ,$$

where

$$x_{i+1} - x_i \ = \ \delta b^i \qquad (0 \leq i \leq n-1) \ ,$$
$$\delta \ = \ \frac{b-1}{b^n - 1} \ ,$$

and $b > 1$ is a design constant. The algorithm is only slightly different now because an additional array of x_i's is stored.

[SET-UP]

Choose $b > 1$, and integer $n > 1$. Set $\delta \leftarrow \dfrac{b-1}{b^n-1}$. Set $x_0 \leftarrow 0$.

FOR $i := 1$ TO n DO

$\qquad x_i \leftarrow \delta \dfrac{b^i - 1}{b - 1}$

$\qquad g_i \leftarrow f(x_{i-1}); \ h_i \leftarrow f(x_i)$

$\qquad p_i \leftarrow h_i (x_i - x_{i-1})$

$\qquad p_{n+i} \leftarrow (g_i - h_i)(x_i - x_{i-1})$

Normalize the vector of p_i's.

[GENERATOR]

REPEAT

$\qquad$ Generate a discrete random variate Z whose distribution is determined by $P(Z = i) = p_i \quad (1 \leq i \leq 2n)$.

$\qquad$ Generate a uniform [0,1] random variate V.

$\qquad W \leftarrow (Z-1) \bmod n$

$\qquad X \leftarrow x_W + V(x_{W+1} - x_W))$

$\qquad$ IF $Z \leq n$

$\qquad\qquad$ THEN RETURN X

$\qquad\qquad$ ELSE

$\qquad\qquad\qquad$ Generate a uniform [0,1] random variate U.

$\qquad\qquad\qquad$ IF $h_{Z-n} + U(g_{Z-n} - h_{Z-n}) \leq f(X)$ THEN RETURN X

UNTIL False

Theorem 2.3.

Assume that f is a monotone density on $[0,1]$. Then for the rejection-based strip method shown above,

A. The expected number of iterations does not exceed

$$b + f(0)\frac{b-1}{b^n-1} .$$

B. If $b = 1 + \frac{1}{n}\log(1 + f(0) + f(0)\log(f(0)))$, then the upper bound is of the form

$$1 + \frac{1}{n}(\log(1 + f(0) + f(0)\log(f(0))))(1 + \frac{1}{1 + \log(f(0))} + o(1))$$

as $n \to \infty$. (Note: when $f(0)$ is large, we have approximately $1 + \frac{\log(f(0))}{n}$.)

Proof of Theorem 2.3.

The expected number of iterations is

$$\sum_{i=0}^{n-1} f(x_i)(x_{i+1} - x_i)$$

$$= \sum_{i=0}^{n-1} \delta b^i f(x_i)$$

$$\leq \delta f(0) + \sum_{i=1}^{n-1} b \int_{x_{i-1}}^{x_i} f(y)\, dy$$

$$\leq b + f(0)\frac{b-1}{b^n-1} .$$

When $b = 1 + \frac{c}{n}$ for some constant $c > 0$, then it is easy to see that the upper bound is

$$1 + \frac{1}{n}(c + f(0)\frac{c}{e^c - 1 + o(1)}) .$$

Replace c by $\log(1 + f(0) + f(0)\log(f(0)))$. ∎

What we retain from Theorem 2.3 is that with some careful design, we can do much better than in the equi-spaced interval case. Roughly speaking, we have reduced the expected number of iterations for monotone densities on $[0,1]$ from $1 + \frac{f(0)}{n}$ to $1 + \frac{\log(f(0))}{n}$. Several details of the last algorithm are dealt with in the exercises.

2.3. Other examples.

In the absence of information about monotonicity or unimodality, it is virtually impossible to compute the best possible constants g_i and h_i for the rejection-based table method. Other pieces of information can aid in the derivation of slightly sub-optimal constants. For example, when $f \in Lip_1(C)$, then

$$g_i = \frac{C}{2n} + \frac{f(\frac{i-1}{n}) + f(\frac{i}{n})}{2} ,$$

$$h_i = -\frac{C}{2n} + \frac{f(\frac{i-1}{n}) + f(\frac{i}{n})}{2} ,$$

will do. These numbers can again be computed from the values of f at the $n+1$ mesh points. We can work out the details of Theorem 2.1:

Theorem 2.4.

For the rejection method based upon n split strips of equal width, used on a $Lip_1(C)$ density f on $[0,1]$, we have:

1.   The expected number of iterations is

$$\frac{C}{2n} + \frac{f(0) + 2f(\frac{1}{n}) + \cdots + 2f(\frac{n-1}{n}) + f(1)}{2} \leq 1 + \frac{C}{n} .$$

This is also equal to the expected number of discrete random variates Z per returned random variate X.

2.   The expected number of computations of f is $\leq \frac{C}{n}$.

3.   The expected number of uniform $[0,1]$ random variates is $\leq 1 + \frac{2C}{n}$.

Proof of Theorem 2.4.

The first expression follows directly after resubstitution of the values of g_i and h_i into Theorem 2.1. The upper bound of parts 1 and 2 are obtained by noting that $g_i - h_i = \frac{C}{n}$ for all i. Finally, part 3 is obtained by summing the bounds obtained in parts 1 and 2. ■

Once again, we can control the performance characteristics of the algorithm by our choice of n. The characteristics can be improved slightly if we make use of the fact that for Lipschitz densities known at mesh points, the obvious piecewise linear dominating curve has slightly smaller integral than the piecewise constant dominating curve suggested here. It should be noted that the switch to piecewise linear dominating curves is costly in terms of the number of uniform random variates needed, and in terms of the length of the program. It is much simpler to improve the performance by increasing n.

2.4. Exercises.

1. For the algorithm for monotone densities analyzed in Theorem 2.3, give a good upper bound for the expected number of computations of f, both in terms of general constants $b > 1$ and for the constant actually suggested in Theorem 2.3.

2. When f is monotone and convex on $[0,1]$, then the piecewise linear curve which touches the curve of f at the mesh points can be used as a dominating curve. If n equal intervals are used, show that the expected number of evaluations of f can be reduced by 50% over the corresponding piecewise constant case. Give the details of the algorithm. Compare the expected number of uniform $[0,1]$ random variates for both cases.

3. Develop the details of the rejection-based strip method for Lipschitz densities which uses a piecewise linear dominating curve and n equi-spaced intervals. Compute good bounds for the expected number of iterations, the expected number of computations of f, and the expected number of uniform $[0,1]$ random variates actually required.

4. **Adaptive methods.** Consider a bounded monotone density f on $[0,1]$. When $f(0)$ is known, we can generate a random variate by rejection from a uniform density on $[0,1]$. This corresponds to the strip method with one interval. As random variates are generated, the dominating curve for the strip method can be adjusted by considering a staircase function with breakpoints at the X_i's. This calls for a dynamic data structure for adjusting the probabilities and sampling from a varying discrete distribution. Design such a structure, and prove that the expected time needed per adjustment is $O(1)$ as $n \to \infty$, and that the expected number of f evaluations is $o(1)$ as $n \to \infty$.

5. Let F be a continuous distribution function. For fixed but large n, compute $x_i = F^{-1}(\frac{i}{n})$, $0 \le i \le n$. Select one of the x_i's ($0 \le i < n$) with equal probability $1/n$, and define $X = x_i + U(x_{i+1} - x_i)$ where U is a uniform $[0,1]$ random variate. The random variable X has distribution function G_n which is close to F. It has been suggested as a fast universal table method in a variety of papers; for similar approaches, see Barnard and Cawdery (1974) and Mitchell (1977). When $x_0 = -\infty$ or $x_n = \infty$, define X in a sensible way on the interval in question.

A.   Prove that in all cases, sup $|F - G_n| \to 0$ as $n \to \infty$.

B.   Prove that when F has a density f, then $\int |f - g_n| \to 0$ as $n \to \infty$, where g_n is the density of G_n. This property holds true without exception.

C.   Determine an upper bound on the L_1 error of part B in terms of f' and n whenever f is absolutely continuous with almost everywhere derivative f'.

3. GRID METHODS.

3.1. Introduction.

Some acceleration can be obtained over strip methods if we make sure that all the components boxes (usually rectangles) are of equal area. In that case, the standard (very fast) table methods can be used for generation. The cost can be prohibitive: the boxes must be fine so that they can capture the detail in the outline of the density f, and this forces us to store very many small boxes.

The versatility of the principle is illustrated here on a variety of problems, ranging from the problem of the generation of a uniformly distributed random vector in a compact set of R^d, to avoidance problems, and fast random variate generation.

3.2. Generating a point uniformly in a compact set.

Let us enclose the compact set A of R^d with a hyperrectangle H with sides $h_1, h_2, \ldots, h_d$. Divide each side up into N_i intervals of length $\dfrac{h_i}{N_i}, 1 \le i \le d$.

There are three types of grid rectangles, the good rectangles (entirely contained in A), the bad rectangles (those partially overlapping with A), and the useless rectangles (those entirely outside A). Before we start generating, we need to set up an array of addresses of rectangles, which we shall call a directory. For the time being, we can think of an address of a rectangle as the coordinates of its leftmost vertex (in all directions). The directory (called D) is such that in positions 1 through k we have good rectangles, and in positions $k+1$ through $k+l$, we have bad rectangles. Useless rectangles are not represented in the array. The informal algorithm for generating a uniformly distributed point in A is as follows:

REPEAT

    Generate an integer Z uniformly distributed in $1,2, \ldots, k+l$.

    Generate X uniformly in rectangle $D[Z]$ ($D[Z]$ contains the address of rectangle Z).

    Accept $\leftarrow [Z \leq k]$ (Accept is a boolean variable.)

    IF NOT Accept THEN Accept $\leftarrow [X \in A]$.

UNTIL Accept

RETURN X

The expected number of iterations is equal to

$$\frac{\text{area}(C)}{\text{area}(A)}$$

where C is the union of the good and bad rectangles (if the useless rectangles are not discarded, then $C = H$). If the area of one rectangle is a, then area(C)$= a (k+l)$. For most bounded sets A, this can be made to go to 1 as the grid becomes finer. That this is not always the case follows from this simple example: let A be $[0,1]^d$ union all the rational vectors in $[1,2]^d$. Since the rationals are dense in the real line, any grid cover of A necessarily covers $[0,1]^d$ and $[1,2]^d$, so that the ratio of the areas is always at least 2. Fortunately, for all compact (i.e., closed and bounded) sets A, the given ratio of areas tends to one as the grid becomes finer (see Theorem 3.1).

The speed of the algorithm follows from the fact that when a good rectangle is chosen, no boundary checking needs to be done. Also, there are many more good rectangles than bad rectangles, so that the contribution to the expected time from boundary checking is small. Of course, we must in any case look up an entry in a directory. This is reminiscent of the urn or table look-up method and its modifications (such as the alias method (Walker, 1977) and the alias-urn method (Peterson and Kronmal, 1982)). Finer grids yield faster generators but require more space.

One of the measures of the efficiency of the algorithm is the expected number of iterations. We have to make sure that as the grid becomes finer, this expected number tends to one.

Theorem 3.1.

Let A be a compact set of nonzero area (Lebesgue measure), and let us consider a sequence of grids $G_1, G_2, \ldots$ which is such that as $n \to \infty$, the diameter of the prototype grid rectangle tends to 0. If C_n is the grid cover of A defined by G_n, then the ratio $\dfrac{\text{area}(C_n)}{\text{area}(A)}$ tends to 1 as $n \to \infty$.

Proof of Theorem 3.1.

Let H be an open rectangle covering A, and let B be the intersection of H with the complement of A. Then, B is open. Thus, for every $x \in B$, we know that the grid rectangle in G_n to which it belongs is entirely contained in B for all n large enough. Thus, by the Lebesgue dominated convergence theorem, the Lebesgue measure of the "useless" rectangles tends to the Lebesgue measure of B. But then, the Lebesgue measure of C_n must tend to the Lebesgue measure of A. ■

The directory itself can be constructed as follows: define a large enough array (of size $n = N_1 N_2 \cdots N_d$), initially unused, and keep two stack pointers, one for a top stack growing from position 1 down, and one for a bottom stack growing from the last position up. The two stacks are tied down at the ends of the array and grow towards each other. Travel from grid rectangle to grid rectangle, identify the type of rectangle, and push the address onto the top stack when it corresponds to a good rectangle, and onto the bottom stack when we have a bad rectangle. Useless rectangles are ignored. After this, the array is partially full, and we can move the bottom stack up to fill positions $k+1$ through $k+l$. If the number of useless rectangles is expected to be unreasonably large, then the stacks should first be implemented as linked lists and at the end copied to the directory of size $k+l$. In any case, the preprocessing step takes time equal to n, the cardinality of the grid.

It is important to obtain a good estimate of the size of the directory. We have

$$ k+l \geq \frac{\text{area}(A)}{a} = \frac{\text{area}(A)}{\text{area}(H)}\, n \ . $$

We know from Theorem 3.1 and the fact that $\text{area}(C_n) = (k+l)a$, that

$$ \lim_{n \to \infty} \frac{k+l}{n} = \frac{\text{area}(A)}{\text{area}(H)} \ , $$

provided that as $n \to \infty$, we make sure that $\inf_i N_i \to \infty$ (this will insure that the diameter of the prototype rectangle tends to 0). Upper bounds on the size of the directory are harder to come by in general. Let us consider a few special cases in the plane, to illustrate some points. If A is a convex set for example, then we can look at all N_1 columns and N_2 rows in the grid, and mark the extremal bad rectangles on either side, together with their immediate neighbors on the inside. Thus, in each row and column, we are putting at most 4 marks. Our claim is that unmarked rectangles are either useless or good. For if a bad rectangle is not marked, then it has at least two neighbors due north, south, east and west that are marked. By the convexity of A, it is physically impossible that this rectangle is not completely contained in A. Thus, the number of bad rectangles is at most $4(N_1+N_2)$. Therefore,

$$k+l \leq n\frac{\text{area}(A)}{\text{area}(H)} + 4(N_1+N_2) .$$

If A consists of a union of K convex sets, then a very crude bound for $k+l$ could be obtained by replacing 4 by $4K$ (just repeat the marking procedure for each convex set). We summarize:

Theorem 3.2.

The size of the directory is $k+l$, where

$$\frac{\text{area}(A)}{\text{area}(H)} \leq \frac{k+l}{n} = (1+o(1))\frac{\text{area}(A)}{\text{area}(H)} .$$

The asymptotic result is valid whenever the diameter of the grid rectangle tends to 0. For convex sets A on R^2, we also have the upper bound

$$\frac{k+l}{n} \leq \frac{\text{area}(A)}{\text{area}(H)} + 4\frac{N_1+N_2}{N_1 N_2} .$$

We are left now with the choice of the N_i's. In the example of a convex set in the plane, the expected number of iterations is

$$\frac{(k+l)a}{\text{area}(A)} \leq 1 + \frac{\text{area}(H)}{\text{area}(A)} \frac{4}{n}(N_1+N_2) .$$

The upper bound is minimal for $N_1 = N_2 = \sqrt{n}$ (assume for the sake of convenience that n is a perfect square). Thus, the expected number of iterations does not exceed

$$1 + \frac{\text{area}(H)}{\text{area}(A)} \frac{8}{\sqrt{n}} .$$

This is of the form $1 + \dfrac{\text{constant}}{\sqrt{n}}$ where n is the cardinality of the enclosing grid. By controlling n, we can now control the expected time taken by the algorithm. The algorithm is fast if we avoid the bad rectangles very often. It is easy to see that the expected number of inspections of bad rectangles before halting is the expected number of iterations times $\dfrac{l}{k+l}$, which equals to $\dfrac{l}{n}\dfrac{\text{area}(H)}{\text{area}(A)} = o(1)$

since $\dfrac{l}{n}\rightarrow 0$ (as a consequence of Theorem 3.1). Thus, asymptotically, we spend a negligible fraction of time inspecting bad rectangles. In fact, using the special example of a convex set in the plane with $N_1 = N_2 = \sqrt{n}$, we see that the expected number of bad rectangle inspections is at most

$$\frac{\text{area}(H)}{\text{area}(A)}\,\frac{8}{\sqrt{n}}\ .$$

3.3. Avoidance problems.

In some simulations, usually with geometric implications, one is asked to generate points uniformly in a set A but not in $\cup A_i$ where the A_i's are given sets of R^d. For example, when one simulates the random parking process (cars of length one park at random in a street of length L but should avoid each other), it is important to generate points uniformly in $[0,L]$ minus the union of some intervals of the same length. Towards the end of one simulation run, when the street fills up, it is not feasible to keep generating new points until one falls in a good spot. Here a grid structure will be useful. In two dimensions, similar problems occur: for example, the circle avoidance problem is concerned with the generation of uniform points in a circle given that the point cannot belong to any of a given number of circles (usually, but not necessarily, having the same radius). For applications involving nonoverlapping circles, see Alder and Wainwright (1962), Diggle, Besag and Gleaves (1976), Talbot and Willis (1980), Kelly and Ripley (1976) and Ripley (1977, 1979). Ripley (1979) employs the rejection method for sampling, and Lotwick (1982) triangulates the space in such a way that each triangle has one of the data points as a vertex. The triangulation is designed to make sampling easy, and to improve the rejection constant. Lotwick also investigates the performance of the ordinary rejection method when checking for inclusion in a circle is done based upon an algorithm of Green and Sibson (1978).

We could use the grid method in all the examples given above. Note that unlike the problems dealt with in the previous subsection, avoidance problems are dynamic. We cannot afford to recompute the entire directory each time. Thus, we also need a fast method for updating the directory. For this, we will employ a dual data structure (see e.g. Aho, Hopcroft and Ullman, 1983). The operations that we are interested in are "Select a random rectangle among the good and bad rectangles", and "Update the directory" (which involves changing the status of good or bad rectangles to bad or useless rectangles, because the avoidance region grows continuously). Also, for reasons explained above, we would like to keep the good rectangles together. Assume that we have a d-dimensional table for the rectangles containing three pieces of information:

(1)   The coordinates of the rectangle (usually of vector of integers, one per coordinate).

(ii) The status of the rectangle (good, bad or useless).

(iii) The position of the rectangle in the directory (this is called a pointer to the directory).

The directory is as before, except that it will shrink in size as more and more rectangles are declared useless. The update operation involves changing the status of a number of rectangles (for example, if a new circle to be avoided is added, then all the rectangles entirely within that circle are declared useless, and those that straddle the boundary are declared bad). Since we would like to keep the time of the update proportional to the number of cells involved times a constant, it is obvious that we will have to reorganize the directory. Let us use two lists again, a list of good rectangles tied down at 1 and with top at k, and a list of bad rectangles tied down at n and with top at $n-l+1$ (it has l elements). There are three situations:

(A) A good rectangle becomes bad: transfer from one list to the other. Fill the hole in the good list by filling it with the top element. Update k and l.

(B) A good or bad rectangle becomes useless: remove the element from the appropriate list, and fill the hole as in case (A). Update k or l.

(C) A bad rectangle remains bad: ignore this case.

For generation, there is only a problem when $Z > k$: when this happens, replace Z by $Z+n-l-k$, and proceed as before. This replacement makes us jump to the end of the directory.

Let us turn now to the car parking problem, to see why the grid structure is to be used with care, if at all, in avoidance problems. At first, one might be tempted to think that for fine enough grids, the performance is excellent. Also, the number of cars (N) that are eventually parked on the street cannot exceed L, the length of the street. In fact, $E(N) \sim \lambda L$ as $L \to \infty$ where

$$\lambda = \int\limits_0^\infty e^{-2\int_0^t (1-e^{-u})/u \; du} \; dt = 0.748...$$

(see e.g. Renyi (1958), Dvoretzky and Robbins (1964) or Mannion (1964)). What determines the time of the simulation run is of course the number of uniform [0,1] random variates needed in the process. Let $\mathbf{E}$ be the event

[Car 1 does not intersect [0,1]].

Let T be the time (number of uniforms) needed before we can park a car to the left of the first car. This is infinite on the complement of $\mathbf{E}$, so we will only consider $\mathbf{E}$. The expected time of the entire simulation is at least equal to $P(\mathbf{E})E(T \mid \mathbf{E})$. Clearly, $P(\mathbf{E})=(L-1)/L$ is positive for all $L > 1$. We will show that $E(T \mid \mathbf{E})=\infty$, which leads us to the conclusion that for all $L > 1$, and for all grid sizes n, the expected number of uniform random variates needed is ∞. Recall however that the actual simulation time is finite with probability one.

Let W be the position of the leftmost end of the first car. Then

$$E(T \mid \mathbf{E}) = \frac{L}{L-1}\int\limits_1^L E(T \mid W=t) \; \frac{dt}{L}$$

$$\geq \frac{L}{L-1} \int_1^{1+\frac{1}{n}} E(T \mid W=t) \frac{dt}{L}$$

$$\geq \frac{1}{L-1} \int_1^{1+\frac{1}{n}} \frac{1}{t-1} dt = \infty .$$

Similar distressing results are true for d-dimensional generalizations of the car parking problem, such as the hyperrectangle parking problem, or the problem of parking circles in the plane (Lotwick, 1984)(the circle avoidance problem of figure 3 is that of parking circles with centers in uncovered areas until the unit square is covered, and is closely related to the circle parking problem). Thus, the rejection method of Ripley (1979) for the circle parking problem, which is nothing but the grid method with one giant grid rectangle, suffers from the same drawbacks as the grid method in the car parking problem. There are several possible cures. Green and Sibson (1978) and Lotwick (1984) for example zoom in on the good areas in parking problems by using Dirichlet tessellations. Another possibility is to use a search tree. In the car parking problem, the search tree can be defined very simply as follows: the tree is binary; every internal node corresponds to a parked car, and every terminal node corresponds to a free interval, i.e. an interval in which we are allowed to park. Some parked cars may not be represented at all. The information in one internal node consists of:

> p_l : the total amount of free space in the left subtree
>   of that node;
> p_r : the total amount of free space in the right subtree.

For a terminal node, we store the endpoints of the interval for that node. To park a car, no rejection is used at all. Just travel down the tree taking left turns with probability equal to $p_l/(p_l+p_r)$, and right turns otherwise, until a terminal node is reached. This can be done by using one uniform random variate for each internal node, or by reusing (milking) one uniform random variate time and again. When a terminal node is reached, a car is parked, i.e. the midpoint of the car is put uniformly on the interval in question. This car causes one of three situations to occur:

> 1. The interval of length 2 centered at the midpoint of the car covers the entire original interval.

> 2. The interval of length 2 centered at the midpoint of the car forces the original interval to shrink.

> 3. The interval of length 2 centered at the midpoint of the car splits the original interval in two intervals, separated by the parked car.

In case 1, the terminal node is deleted, and the sibling terminal node is deleted too by moving it up to its parent node. In case 2, the structure of the tree is unaltered. In case 3, the terminal node becomes an internal node, and two new terminal nodes are added. In all cases, the internal nodes on the path from the root to the terminal node in question need to be updated. It can be shown that the expected time needed in the simulation is $O(L \log(L))$ as $L \to \infty$. Intuitively, this can be seen as follows: the tree has initially one node, the root. At the end, it has no nodes. In between, the tree grows and shrinks, but can never have more than L internal nodes. It is known that the random binary search tree has expected depth $O(\log(L))$ when there are L nodes, so that, even though our tree is not distributed as a random binary search tree, it comes as no surprise that the expected time per car parked is bounded from above by a constant time $\log(L)$.

3.4. Fast random variate generators.

It is known that when (X, U) is uniformly distributed under the curve of a density f, then X has density f. This could be a density in R^d, but we will only consider $d = 1$ here. All of our presentation can easily be extended to R^d. Assume that f is a density on $[0,1]$, bounded by M. The interval $[0,1]$ is divided into N_1 equal intervals, and the interval $[0,M]$ for the y-direction is divided into N_2 equal intervals. Then, a directory is set up with k good rectangles (those completely under the curve of f), and l bad rectangles. For all rectangles, we store an integer i which indicates that the rectangle has x-coordinates $[\frac{i}{N_1}, \frac{i+1}{N_1})$. Thus, i ranges from 0 to N_1-1. In addition, for the bad rectangles, we need to store a second integer j indicating that the y coordinates are $[M\frac{j}{N_2}, M\frac{j+1}{N_2})$. Thus, $0 \le j < N_2$. It is worth repeating the algorithm now, because we can re-use some uniform random variates.

Generator for density f on [0,1] bounded by M

(NOTE: $D[1], \ldots, D[k+l]$ is a directory of integer-valued x-coordinates, and $Y[k+1], \ldots, Y[k+l]$ is a directory of integer-valued y-coordinates for the bad rectangles.)

REPEAT

    Generate a uniform [0,1] random variate U.

    $Z \leftarrow \lfloor (k+l)U \rfloor$ (Z chooses a random element in D)

    $\Delta \leftarrow (k+l)U - Z$ (Δ is again uniform [0,1])

    $X \leftarrow \dfrac{D[Z]+\Delta}{N_1}$

    Accept $\leftarrow [Z \leq k]$

    IF NOT Accept THEN

        Generate a uniform [0,1] random variate V.

        Accept $\leftarrow [M(Y[Z]+V) \leq f(X)N_2]$

UNTIL Accept

RETURN X

This algorithm uses only one table-look-up and one uniform random variate most of the time. It should be obvious that more can be gained if we replace the $D[i]$ entries by $\dfrac{D[i]}{N_1}$, and that in most high level languages we should just return from inside the loop. The awkward structured exit was added for readability. Note further that in the algorithm, it is irrelevant whether f is used or cf where c is a convenient constant. Usually, one might want to choose c in such a way that an annoying normalization constant cancels out.

When f is nonincreasing (an important special case), the set-up is facilitated. It becomes trivial to decide quickly whether a rectangle is good, bad or useless. Notice that when f is in a black box, we will not be able to declare a particular rectangle good or useless in our lifetime, and thus all rectangles must be classified as bad. This will of course slow down the expected time quite a bit. Still for nonincreasing f, the number of bad rectangles cannot exceed N_1+N_2. Thus, noting that the area of a grid rectangle is $\dfrac{M}{n}$, we observe that the expected number of iterations does not exceed

$$1+M\frac{N_1+N_2}{n} \ .$$

Taking $N_1 = N_2 = \sqrt{n}$, we note that the bound is $1+O\left(\dfrac{1}{\sqrt{n}}\right)$. We can adjust n to off-set large values of M, the bound on f. But in comparison with strip methods, the performance is slightly worse in terms of n: in strip methods with n equal-size intervals, the expected number of iterations for monotone densities does not exceed $1+\dfrac{M}{n}$. For grid methods, the n is replaced by $\sqrt{n}$. The expected number of computations of f for monotone densities does not exceed

$$\frac{l}{k+l}\left(\frac{(k+l)M}{n}\right) = \frac{lM}{n} \leq \frac{M(N_1+N_2)}{n}.$$

For unimodal densities, a similar discussion can be given. Note that in the case of a monotone or unimodal density, the set-up of the directory can be automated.

It is also important to prove that as the grid becomes finer, the expected number of iterations tends to 1. This is done below.

Theorem 3.3.

For all Riemann integrable densities f on [0,1] bounded by M, we have, as inf $(N_1,N_2)\to\infty$, the expected number of iterations,

$$(k+l)\frac{M}{n}$$

tends to 1. The expected number of evaluations of f is $o(1)$.

Proof of Theorem 3.3.

Given an n-grid, we can construct two estimates of $\int f$,

$$\sum_{i=0}^{N_1-1} \frac{1}{N_1} \sup_{\frac{i}{N_1}\leq x \leq \frac{i+1}{N_1}} f(x),$$

and

$$\sum_{i=0}^{N_1-1} \frac{1}{N_1} \inf_{\frac{i}{N_1}\leq x \leq \frac{i+1}{N_1}} f(x).$$

By the definition of Riemann integrability (Whittaker and Watson, 1927, p.63), these tend to $\int f$ as $N_1\to\infty$. Thus, the difference between the estimates tends to 0. By a simple geometrical argument, it is seen that the area taken by the bad rectangles is at most this difference plus $2N_1$ times the area of one grid rectangle, that is, $o(1)+\dfrac{2M}{N_2}=o(1)$. ∎

Densities that are bounded and not Riemann integrable are somehow peculiar, and less interesting in practice. Let us close this section by noting that extra savings in space can be obtained by grouping rectangles in groups of size m, and putting the groups in an auxiliary directory. If we can do this in such a way that many groups are homogeneous (all rectangles in it have the same value for $D[i]$

and are all good), then the corresponding rectangles in the directory can be discarded. This, of course, is the sort of savings advocated in the multiple table
look-up method of Marsaglia (1963) (see section III.3.2). The price paid for this is
an extra comparison needed to examine the auxiliary directory.

A final remark is in order about the space-time trade-off. Storage is needed
for at most $N_1 + N_2$ bad rectangles and $\dfrac{n}{M}$ good rectangles when f is monotone.
The bound on the expected number of iterations on the other hand is
$1 + \dfrac{M}{n}(N_1 + N_2)$. If $N_1 = N_2 = \sqrt{n}$, then keeping the storage fixed shows that the
expected time increases in proportion to M. The same rate of increase, albeit
with a different constant, can be observed for the ordinary rejection method with
a rectangular dominating curve. If we keep the expected time fixed, then the
storage increases in proportion to M. The product of storage $(1 + 2M/\sqrt{n}$) and
expected time $(2\sqrt{n} + n/M)$ is $4\sqrt{n} + n/M + 4M$. This product is minimal for
$n = 1, M = \sqrt{n}/2$, and the minimal value is 8. Also, the fact that storage times
expected time is at least $4M$ shows that there is no hope of obtaining a cheap
generator when M is large. This is not unexpected since no conditions on f
besides the monotonicity are imposed. It is well-known for example that for
specific classes of monotone or unimodal densities (such as all beta or gamma
densities), algorithms exist which have uniformly bounded (in M) expected time
and storage. On the other hand, table look-up is so fast that grid methods may
well outperform standard rejection methods for many well known densities.

Chapter Nine
CONTINUOUS UNIVARIATE DENSITIES

Chapters IX and X are included for the convenience of a large sub-population of users, the statisticians. The main principles in random variate generation were developed in the first eight chapters. Most particular distributions found here are members of special classes of densities for which universal methods are available. For example, a short algorithm for log-concave densities was developed in section VII.2. When speed is at a premium, then one of the table methods of the previous chapter could be used. This chapter is purely complementary. We are not in the least interested in a historical review of the different methods proposed over the years for the popular densities. Some interesting developments which give us new insight or illustrate certain general principles will be reported. The list of distributions corresponds roughly speaking to the list of distributions in the three volumes of Johnson and Kotz.

1. THE NORMAL DENSITY.

1.1. Definition.

A random variable X is **normally distributed** if it has density

$$f(x) = \frac{1}{\sqrt{2\pi}} e^{-\frac{x^2}{2}} .$$

When X is normally distributed, then $\mu + \sigma X$ is said to be normal (μ, σ^2). The mean μ and the variance σ^2 are uninteresting from a random variate generation point of view.

Comparative studies of normal generators were published by Muller (1959), Ahrens and Dieter (1972), Atkinson and Pearce (1976), Kinderman and Ramage (1976), Payne (1979) and Best (1979). In the table below, we give a general outline of how the available algorithms are related.

Method	References	Speed	Size of code	Section
Inversion	Muller (1959)	Slow	Moderate	II.2.3
Polar method	Box and Muller (1958)	Moderate	Small	V.4.4
	Bell (1968)			
Rejection	Von Neumann (1951)	Moderate	Small	II.3.2
	Sibuya (1962)			
Ratio-of-uniforms	Kinderman and Monahan (1977)	Fast	Small to moderate	IV.7.2
Composition/rejection	Marsaglia and Bray (1964)	Fast	Small to moderate	
	Ahrens and Dieter (1972)			
	Kinderman and Ramage (1976)			
	Sakasegawa (1978)			
Series method		Fast	Small to moderate	IV.5.3
Almost-exact inversion	Wallace (1976)	Moderate	Small	IV.3.3
Table methods	Marsaglia, Maclaren and Bray (1964)	Very fast	Large	
Forsythe's method	Forsythe (1972)	Fast	Moderate	IV.2.1
	Ahrens and Dieter (1973)			
	Brent (1974)			

The list given here is not exhaustive. Many references are missing. What matters are the general trends. We know that table methods are fast, and the rectangle-wedge-tail method of Marsaglia, Maclaren and Bray (1964) is no exception. At the other end of the scale are the small programs of moderate speed, such as the programs for the polar method and some rejection methods. In between are moderate-sized programs that have good speed, such as the ratio-of-uniforms method, the series method, Forsythe's method and the composition/rejection method. Only the inversion method is inadmissible because it is slower and less space efficient than all of the other methods, the table methods excepted. Below, we will mainly focus on the composition/rejection methods which have not been described in earlier chapters. Because we will cut off the tail of the normal density, it seems important to show how random variates with a density proportional to the tail can be generated.

1.2. The tail of the normal density.

In this section, we consider generators for the family of tail densities

$$f(x) = \frac{e^{-\frac{x^2}{2}}}{\Phi(a)} \quad (x > a),$$

where $\Phi(a) = \int_a^\infty e^{-\frac{x^2}{2}}$ is a normalization constant and $a > 0$ is a parameter. Two algorithms will be described:

Marsaglia's method for the tail-of-the-normal density (Marsaglia, 1964)

REPEAT
    Generate iid uniform [0,1] random variates U,V.
    $X \leftarrow \sqrt{a^2 - 2\log(U)}$
UNTIL $VX \leq a$
RETURN X

Marsaglia's method is based upon the trivial inequality

$$ e^{-\frac{x^2}{2}} \leq \frac{x}{a} e^{-\frac{x^2}{2}} \qquad (x \geq a) \ . $$

But $xe^{\frac{a^2-x^2}{2}}$ $(x \geq a)$ is a density having distribution function

$$ F(x) = 1 - e^{\frac{a^2-x^2}{2}} \qquad (x \geq a) \ , $$

which is the tail part of the Rayleigh distribution function. Thus, by inversion, $\sqrt{a^2 - 2\log(U)}$ has distribution function F, which explains the algorithm. The probability of acceptance in the rejection algorithm is

$$ P(VX \leq a) = E(\frac{a}{X}) = \int_a^\infty ae^{\frac{a^2-x^2}{2}} \ dx = ae^{\frac{a^2}{2}} \Phi(a) \rightarrow 1 $$

as $a \rightarrow \infty$. Thus, the rejection algorithm is asymptotically optimal. Even for small values of a, the probability of acceptance is quite high: it is about 66% for $a = 1$ and about 88% for $a = 3$. Note that Marsaglia's method can be sped up somewhat by postponing the square root until after the acceptance:

REPEAT
    Generate iid uniform [0,1] random variates U,V.
    $X \leftarrow c - \log(U)$ (where $c = a^2/2$)
UNTIL $V^2 X \leq c$
RETURN $\sqrt{2X}$

An algorithm which does not require any square roots can be obtained by rejection from an exponential density. We begin with the inequality

$$e^{-\frac{x^2}{2}} \le e^{\frac{a^2}{2}-ax} \qquad (x \ge a),$$

which follows from the observation that $(x-a)^2 \ge 0$. The upper bound is proportional to the density of $a + \dfrac{E}{a}$ where E is exponentially distributed. This yields without further work the following algorithm:

REPEAT
    Generate iid exponential random variates $E, E*$.
UNTIL $E^2 \le 2a^2 E*$
RETURN $X \leftarrow a + \dfrac{E}{a}$

The probability of acceptance is precisely as for Marsaglia's method:

$$P(E* \ge E^2/(2a^2)) = \int_0^\infty e^{-\frac{x^2}{2a^2}} \, dx = ae^{\frac{a^2}{2}} \Phi(a) \to 1 \qquad (a \to \infty).$$

If a fast exponential random variate generator is available, the second rejection algorithm is probably faster than Marsaglia's.

1.3. Composition/rejection methods.

The principle underlying all good composition/rejection methods is the following: decompose the density of f into two parts, $f(x)=pg(x)+(1-p)h(x)$ where $p \in (0,1)$ is a mixture parameter, g is an easy density, and h is a residual density not very often needed when p is close to 1. We rarely stumble upon a good choice for g by accident. But we can always find the optimal g_θ in a family of suitable candidates parametrized by θ. The weight of g_θ in the mixture is denoted by $p(\theta)$:

$$p(\theta) = \inf_x \frac{f(x)}{g_\theta(x)}.$$

The candidates g_θ should preferably be densities of simple transformations of independent uniform [0,1] random variables. Among the simple transformations one might consider, we cite:

(1) $\theta(V_1 + \cdots + V_n)$;
(2) θ median($V_1, \ldots, V_n$);
(3) $\theta_1 V_1 + \theta_2 V_2$;
(4) $\theta_1 V_1 + \theta_2(V_1)^3$.

Here $V_1, V_2, \ldots$ are iid uniform $[-1,1]$ random variates, and $\theta, \theta_1, \theta_2$ are parameters to be selected. Marsaglia and Bray (1964) used the first choice with $n=3$ and with the deliberately suboptimal value $\theta=1$ (because a time-consuming multiplication is avoided for this value). Kinderman and Ramage (1976) optimized θ for choice (1) when $n=2$. And Ahrens and Dieter (1972) proposed to use choice (3). Because the shape of g_θ is trapezoidal, this method is known as the **trapezoidal method**. All three approaches lead to algorithms of about equal length and speed. We will look at choices (1) and (2) in more detail below, and provide enough detail for the reader to be able to reconstruct the algorithms of Marsaglia and Bray (1964) and Kinderman and Ramage (1976).

Theorem 1.1.

The density of θ median($V_1, \ldots, V_{2n+1}$) for n positive and $\theta>0$ is

$$c\left(1-\frac{x^2}{\theta^2}\right)^n \qquad (\mid x \mid \le \theta)$$

where $c = \dfrac{(2n+1)!}{2^{2n+1} n!^2 \theta}$. The maximal value of $p(\theta)$ is reached for $\theta=\sqrt{2n+1}$, and takes the value

$$p = \frac{2^{2n+1} n!^2 \sqrt{n}}{\sqrt{\pi e}\,(2n+1)!}\left(1+\frac{1}{2n}\right)^{n+\frac{1}{2}}.$$

We have

(i) $p = \sqrt{\dfrac{6}{\pi e}} \approx 0.8382112 \qquad (n=1)$;

(ii) $p = \sqrt{\dfrac{125}{18\pi e}} \approx 0.9017717 \qquad (n=2)$;

(iii) $\lim_{n \to \infty} p = 1$.

Proof of Theorem 1.1.

The density can be derived very easily after recalling that the median of $2n+1$ iid uniform $[0,1]$ random variables has a symmetric beta density given by

$$\frac{(2n+1)!}{n!^2}(x(1-x))^n \qquad (0 \le x \le 1) .$$

Define $g_\theta(x) = c(1-(x^2/\theta^2))^n$ ($\mid x \mid \le \theta$), and note that $\log(f/g_\theta)$ attains an extremum at some point x for which the derivative of the logarithm is 0. This yields the equation

$$-x + \frac{2x}{\theta^2} \frac{n}{1 - \frac{x^2}{\theta^2}} = 0 \ ,$$

or,

$$x = 0 \ ; \ x^2 = \theta^2 - 2n \ .$$

When $\theta^2 < 2n$, f/g_θ attains only one minimum, at $x = 0$. When $\theta^2 > 2n$, the function f/g_θ is symmetric around 0: it has a local peak at 0, dips to a minimum, and increases monotonically again to ∞ as $x \uparrow \theta$. Thus, we have

$$p(\theta) = \inf_x \frac{f(x)}{g_\theta(x)} = \begin{cases} \dfrac{1}{\sqrt{2\pi c}} = \dfrac{2^{2n+1} n!^2 \theta}{(2n+1)! \sqrt{2\pi}} & (\theta^2 < 2n) \\[4mm] \dfrac{1}{\sqrt{2\pi c}} \left(\dfrac{e}{2n}\right)^n \theta^{2n} e^{-\frac{\theta^2}{2}} & (\theta^2 > 2n) \ . \end{cases}$$

We still have to maximize this function with respect to θ. The function $p(\theta)$ increases linearly from 0 up to $\theta = \sqrt{2n}$. Then, it increases some more , peaks, and decreases in a bell-shaped fashion. The maximum is attained for some value $\theta > \sqrt{2n}$. Since in that region, $p(\theta)$ is a constant times $\theta^{2n+1} e^{-\theta^2/2}$, the maximum is attained for $\theta = \sqrt{2n+1}$. This gives the desired result. ■

Had we considered the Taylor series expansion of f about 0, given by

$$f(x) = \frac{1}{\sqrt{2\pi}} \left(1 - \frac{x^2}{2} + \frac{x^4}{8} - \frac{x^6}{48} + \cdots \right),$$

which is known to give partial sums that alternately overestimate and underestimate f, then we would have been tempted to choose $g(x) = \frac{3}{4\sqrt{2}}(1 - \frac{x^2}{2})$, because of

$$f(x) \geq \frac{1}{\sqrt{2\pi}}(1 - \frac{x^2}{2}) = pg(x) \quad (\mid x \mid \leq \sqrt{2})$$

where $p = \frac{4}{3\sqrt{\pi}} \approx 0.7522528$ is the weight of g in the mixture. This illustrates the usefulness and the shortcomings of Taylor's series. Simple polynomial bounds are very easy to obtain, but the choice could be suboptimal. From Theorem 1.1 for example, we recall that the optimal g of the inverted parabolic form is a constant times $(1 - \frac{x^2}{3})$ ($\mid x \mid \leq \sqrt{3}$). Sometimes a suboptimal choice of θ is preferable because the residual density h is easier to handle. This is the case for $n = 1$ in Theorem 1.1. The suboptimal choice $\theta = \sqrt{2n}$, which is the choice implicit in

Taylor's series expansion, yields a much cleaner residual density. For $n=2$, we need 5 random variates instead of 3, an increase of 66%, while the gain in efficiency (in value of p) is only of the order of 10%. For this reason, the case $n>1$ is less important in practice. Let us briefly describe the entire algorithm for the case $n=1$, $\theta=\sqrt{2}$. We can decompose f as follows:

$$f(x) = pg(x)+qh(x)+rt(x)$$

where

(1) $\quad g(x) = \dfrac{3}{4\sqrt{2}}(1-\dfrac{x^2}{2});$

$\qquad p = \dfrac{4}{3\sqrt{\pi}} \approx 0.7522528;$

(11) $\quad t(x) = \dfrac{1}{r}\dfrac{1}{\sqrt{2\pi}}e^{-x^2/2} \quad (\mid x \mid >\sqrt{2});$

$\qquad r = \displaystyle\int\limits_{\mid x \mid >\sqrt{2}} \dfrac{1}{\sqrt{2\pi}}e^{-x^2/2}\,dx \approx 0.15729921;$

(111) $\quad h(x) = \dfrac{1}{q}\dfrac{1}{\sqrt{2\pi}}(e^{-x^2/2}-(1-\dfrac{x^2}{2})) \quad (\mid x \mid \leq \sqrt{2});$

$\qquad q = \displaystyle\int\limits_{\mid x \mid \leq \sqrt{2}} \dfrac{1}{\sqrt{2\pi}}(e^{-x^2/2}-(1-\dfrac{x^2}{2})\,dx \approx 0.09044801 \ .$

Sampling from the tail density t has been discussed in the previous sub-section. Sampling from g is simple: just generate three iid uniform $[-1,1]$ random variates, and take $\sqrt{2}$ times the median. Sampling from the residual density h can be done as follows:

```
REPEAT
        Generate V uniformly on [-1,1], and U uniformly on [0,6].
        X←√2V / | V |⁴ᐟ⁵
        Accept ←[U >X²]
        IF NOT Accept THEN
                IF U ≥X²(1-X²/8) THEN
                        Accept ←[(1-U/6)X⁴≤8(e^(-X²/2) -(1-X²/2)))]
UNTIL Accept
RETURN X
```

This is a simple rejection algorithm with squeezing based upon the inequalities

$$e^{-\frac{x^2}{2}}-(1-\dfrac{x^2}{2}) \leq \dfrac{x^4}{8} \quad (\mid x \mid \leq \sqrt{2}) ;$$

$$\frac{x^4}{8}-\frac{x^6}{48}\le e^{-\frac{x^2}{2}}-(1-\frac{x^2}{2})\le \frac{x^4}{8}-\frac{x^6}{48}+\frac{x^8}{384}\ .$$

The reader can easily work out the details. The probability of immediate acceptance in the first iteration is

$$P(X^2<U)=\int_0^1 P(X^2<6x)\ dx\ =\int_0^1 P(\,|\,X\,|\,<\sqrt{6x}\,)\ dx$$

$$=\int_0^{\frac{1}{3}}\frac{5}{4\sqrt{2}}\frac{(\sqrt{6x}\,)^5}{5}\ dx\ +\int_{\frac{1}{3}}^1\ dx$$

$$=\frac{2}{3}+\frac{2}{7}\frac{6^{\frac{5}{2}}}{4\sqrt{2}3^{\frac{7}{2}}}=\frac{6}{7}\ .$$

The same smooth performance for a residual density could not have been
obtained had we not based our decomposition upon the Taylor series expansion.

Let us next look at the density g_θ of $\theta(V_1+V_2+V_3)$ where the V_i's are iid
uniform $[-1,1]$ random variables. For the density of $\theta(V_1+V_2)$, the triangular
density, we refer to the exercises where among other things it is shown that the
optimal θ is 1.1080179... , and that the corresponding value $p(\theta)$ is 0.8840704... .

Theorem 1.2.

The optimal value for θ in the decomposition of the normal density into
$p(\theta)g_\theta(x)$ plus a residual density (where g_θ is the density of $\theta(V_1+V_2+V_3)$ and
the V_i's are iid uniform $[-1,1]$ random variables), is

$$\theta = 0.956668451229...\ .$$

The corresponding optimal value for $p(\theta)$ is 0.962365327... .

Proof of Theorem 1.2.

The density g_θ of $\theta(V_1+V_2+V_3)$ is

$$g_\theta(x) = \begin{cases} \frac{1}{8\theta}(3-(\frac{x}{\theta})^2) & (\,|\,x\,|\,\le\theta) \\ \frac{1}{16\theta}(3-|\,\frac{x}{\theta}\,|)^2 & (\theta\le\,|\,x\,|\,\le3\theta) \\ 0 & (\,|\,x\,|\,>3\theta)\ . \end{cases}$$

The function $h_\theta = f / g_\theta$ can be written as

$$
h_\theta(x) = \begin{cases}
\dfrac{8\theta}{\sqrt{2\pi}} \dfrac{e^{-\frac{x^2}{2}}}{3 - \dfrac{x^2}{\theta^2}} & (0 < x \le \theta) \\[4ex]
\dfrac{16\theta}{\sqrt{2\pi}} \dfrac{e^{-\frac{x^2}{2}}}{(3 - \dfrac{x}{\theta})^2} & (\theta \le x \le 3\theta) ,
\end{cases}
$$

when $x > 0$. We need to find the value of θ for which $\min\limits_{0 < x \le 3\theta} h_\theta(x)$ is maximal.
By setting the derivative of $\log(h_\theta)$ with respect to x equal to 0, and by analyzing the shape of h_θ, we see that the minimum of h_θ belongs to the following set of values: 0, θ, b, c, where

$$
b = \theta \sqrt{3 - \frac{2}{\theta^2}} ;
$$

$$
c = \frac{3\theta}{2} + \frac{\theta}{2} \sqrt{9 - \frac{8}{\theta^2}} .
$$

The following table gives all the local minima together with the values for h_θ.

Local minimum	Value of h_θ at minimum	Local minimum exists when:
0	$\eta = \dfrac{8\theta}{3\sqrt{2\pi}}$	$\theta^2 \le \dfrac{2}{3}$
b	$\psi = \dfrac{4\theta^3}{\sqrt{2\pi}} e^{-\frac{b^2}{2}}$	$1 \ge \theta^2 \ge \dfrac{2}{3}$
θ	$\xi = \dfrac{4\theta}{\sqrt{2\pi}} e^{-\frac{a^2}{2}}$	$\theta = 1$
c	$\phi = \dfrac{16\theta}{\sqrt{2\pi}} \dfrac{e^{-\frac{c^2}{2}}}{(3 - \dfrac{c}{\theta})^2}$	$\theta^2 \ge \dfrac{8}{9}$

The general shape of h_θ is as follows: when $\theta^2 \ge 1$, there is no local minimum on $(0, \theta)$, and h_θ decreases monotonically to reach a global minimum at $x = c$ equal to ϕ, after which it increases again. When $\theta^2 = 1$, the same shape is observed, but a zero derivative occurs at $x = \theta$, although this does not correspond to a local minimum. When $\frac{8}{9} < \theta^2 < 1$, there are two local minima, one on $(0, \theta)$ (at b, of value ψ), and one on $(\theta, 3\theta)$ (at c, of value ϕ). For $\frac{2}{3} < \theta^2 < \frac{8}{9}$, the local minimum at c ceases to exist. We have again a function with one minimum, this time at $b < \theta$, of value ψ. Finally, for $\theta^2 \le \frac{2}{3}$, the function increases monotonically, and its global minimum occurs at $x = 0$ and has value η.

Consider now the behavior of η and ψ as a function of θ. Clearly, η increases linearly with θ. Furthermore, ψ is gamma shaped with global peak at $\theta=1$, and $\eta=\psi$ for $\theta^2=\dfrac{2}{3}$. The value of ϕ on the other hand decreases monotonically on the set $\theta^2\geq\dfrac{8}{9}$. We verify easily that ϕ and ψ cross each other on the segment $\dfrac{8}{9}<\theta^2<1$. It is at this point that $\min\limits_{0<x<3\theta} h_\theta(x)$ is maximal. This cross-over point is precisely the value given in the statement of the theorem. ∎

Theorem 1.2 can be used in the design of a fast composition/rejection algorithm. In particular, the tail beyond the optimal 3θ is very small, having probability 0.004104648... . The residual density on $[-3\theta,3\theta]$ has probability 0.033530022... , but has unfortunately enough five peaks, the largest of which occurs at the origin. It is clear once again that the maximization criterion does not take the complexity of the residual density into account. A suboptimal value for θ sometimes leads to better residual densities. For example, when $\theta=1$, we save one multiplication and end up with a more manageable residual density. This choice was first suggested by Marsaglia and Bray (1964). We conclude this section by giving their algorithm in its entirety.

From the proof of Theorem 1.2, we see that (in the notation of that proof),

$$p\,(\theta) = \phi = \frac{16}{\sqrt{2\pi e}} = 0.86385546...\;.$$

The normal density f can be decomposed as follows:

$$f\,(x) = \sum_{i=1}^{4} p_i\,f_i\,(x)\,,$$

where (p_1,p_2,p_3,p_4) is a probability vector, and the f_i's are densities defined as follows:

(i) $p_1=0.86385546...$, f_1 is the density of $V_1+V_2+V_3$, where the V_i's are iid uniform $[-1,1]$ random variables.

(ii) $p_4=0.002699796063...=\displaystyle\int_{|x|\geq3} f$; f_4 is the tail-of-the-normal density restricted to $|x|\geq3$.

(iii) $f_2(x)=\dfrac{1}{9}(6-4\,|\,x\,|)$ ($|\,x\,|\leq\dfrac{3}{2}$); $p_2=0.1108179673...$.

(iv) $p_3=1-p_1-p_4-p_2=0.02262677245...$; $f_3=\dfrac{1}{p_3}(f-p_1f_1-p_2f_2-p_4f_4)$.

In the design, Marsaglia and Bray decided upon the triangular form of f_2 first, because random variates with this density can be generated simply as $\dfrac{3}{4}(V_4+V_5)$ where the V_i's are again iid uniform $[-1,1]$ random variates. After having picked this simple f_2, it is necessary to choose the best (largest) weight p_2, given by

$$p_2 = \inf_x \frac{f(x) - p_1 f_1(x)}{f_2(x)}.$$

This infimum is found as follows. The derivative of the ratio is 0 at $|x| = 2$ and at $|x| = 0.87386312884\ldots$. Only the latter $|x|$ corresponds to a minimum, and the corresponding value for p_2 is $p_2 = 0.1108179673\ldots$. Having determined random variate generation methods for all parts except f_3, it remains to establish just this for f_3. First, note that f_3 has supremum $0.3181471173\ldots$. If we use rejection from a rectangular density with support on $[-3,3]$, then the expected number of iterations is

$$\frac{6 \times 0.3181471173\ldots}{p_3} = 1.9088827038\ldots .$$

Combining all of this into one algorithm, we have:

Normal generator of Marsaglia and Bray (1964)

[NOTE: This algorithm follows the implementation suggested by Kinderman and Ramage (1977).]

Generate a uniform [0,1] random variate U.

CASE

$0 \leq U \leq 0.8638$:

Generate two iid uniform [–1,1] random variates V, W.

RETURN $X \leftarrow 2.3153508...U - 1 + V + W$

$0.8638 < U \leq 0.9745$:

Generate a uniform [0,1] random variate V.

RETURN $X \leftarrow \frac{3}{2}(V - 1 + 9.0334237...(U - 0.8638))$

$0.9973002... < U \leq 1$:

REPEAT

Generate iid uniform [0,1] random variates V, W.

$X \leftarrow \frac{9}{2} - \log(W)$

UNTIL $XV^2 \leq \frac{9}{2}$

RETURN $X \leftarrow \sqrt{2X}$ sign$(U - 0.9986501...)$

$0.9745 < U \leq 0.9973002...$:

REPEAT

Generate a uniform [–3,3] random variate X and a uniform [0,1] random variate U.

$V \leftarrow | X |$

$W \leftarrow 6.6313339...(3 - V)^2$

Sum $\leftarrow 0$

IF $V < \frac{3}{2}$ THEN Sum $\leftarrow 6.0432809...(\frac{3}{2} - V)$

IF $V < 1$ THEN Sum $\leftarrow$ Sum $+13.2626678...(3 - V^2) - W$

UNTIL $U \leq 49.0024445...e^{-\frac{V^2}{2}} - \text{Sum} - W$

RETURN X

1.4. Exercises.

1. In the trapezoidal method of Ahrens and Dieter (1972), the largest sym-
    metric trapezoid under the normal density is used as the main component in
    the mixture. Show that this trapezoid is defined by the vertices
    $(-\xi,0),(\xi,0),(\eta,\rho),(-\eta,\rho)$ where $\xi=2.1140280833...$ $,\eta=0.2897295736...$,
    $\rho=0.3825445560...$. (Note: the area under the trapezoid is 0.9195444057... .)
    A random variate with such a trapezoidal density can be generated as
    aV_1+bV_2 for some constants $a,b>0$ where V_1,V_2 are iid uniform $[-1,1]$
    random variates. Determine a,b in this case.

2. Show that as $a\uparrow\infty$,

    $$\int\limits_a^\infty e^{-\frac{x^2}{2}} \sim \frac{1}{a}e^{-\frac{a^2}{2}} .$$

3. The optimal probability p in Theorem 1.1 depends upon n. Use Stirling's
    formula to determine a constant c such that $p\geq 1-\dfrac{c}{n}$, valid for all $n\geq 3$.

4. If we want to generate a normal random variate by rejection from the
    exponential density $\dfrac{\lambda}{2}e^{-\lambda|x|}$, the smallest rejection constant is obtained
    when $\lambda=1$. The constant is $\sqrt{\dfrac{2e}{\pi}}$. Show this. Note that the corresponding
    rejection algorithm is:

    > REPEAT
    >> Generate two iid exponential random variates, X,E.
    >
    > UNTIL $2E\leq(X-1)^2$
    > RETURN SX where S is a random sign.

    This algorithm is mentioned in Abramowitz and Stegun (1970), where von
    Neumann is credited. Butcher (1961) attributes it to Kahn. Others have
    rediscovered it later.

5. **Teichroew's distribution.** Teichroew (1957) has shown that the functions
    $\phi(t) = \dfrac{1}{(1+t^2)^a}$ are valid characteristic functions for all values $a>0$ of the
    parameter. Show that random variates from this family can be generated as
    (i) G_1-G_2, where the G_i's are iid gamma (a) random variables;
    (ii) $N\sqrt{2G}$ where N,G are independent random variables with a normal
          and gamma (a) distribution respectively.

6. This question is related to the algorithm of Kinderman and Ramage (1976)
    (programs given in Kinderman and Ramage (1977)). Consider the isosceles

triangular density g_θ of the random variable $\theta(V_1 + V_2)$ where V_1, V_2 are iid uniform $[-1,1]$ random variates. Show that the largest triangle to fit under the normal density f touches f at the origin. Show next that the sides of the largest triangle touch f somewhere else. Conclude that the optimal θ is given by $\theta = 1.1080179...$, and that the corresponding optimal weight of the triangle is $p = 0.88407040...$.

7. **The lognormal density.** When N is normally distributed, then $\theta + e^{\varsigma + \sigma N}$ is lognormal with parameters $\theta, \varsigma, \sigma$, all real numbers. The lognormal distribution has a density with support on (θ, ∞) given by

$$ f(x) = \frac{1}{(x - \theta)\sigma\sqrt{2\pi}} e^{-\frac{(\log(x - \theta) - \varsigma)^2}{2\sigma^2}} \qquad (x > \theta) . $$

Random variate generation requires the exponentiation of a normal random variate, and can be beaten speedwise by the judicious use of a composition/rejection algorithm, or a rejection algorithm with a good squeeze step. Develop just such an algorithm. To help you find a solution, it is instructive to draw several lognormal densities. Consider only the case $\theta = 0$ since θ is a translation parameter. Show also that in that case, the mode is at $e^{\varsigma - \sigma^2}$, the median is at e^ς, and that the r-th moment is $e^{r\varsigma + r^2\sigma^2/2}$ when $r > 0$.

8. In the composition/rejection algorithm of Marsaglia and Bray (1964), we return the sum of three independent uniform $[-1,1]$ random variates about 86% of the time. Schuster (1983) has shown that by considering sums of the form $a_1 V_1 + a_2 V_2 + a_3 V_3$, where the V_i's are iid uniform $[-1,1]$ random variates, it is possible to find coefficients a_1, a_2, a_3 such that we can return the said sum about 97% of the time (note however that the multiplications could actually cause a slowdown). Find these coefficients, and give the entire algorithm.

2. THE EXPONENTIAL DENSITY.

2.1. Overview.

We hardly have to convince the reader of the crucial role played by the exponential distribution in probability and statistics and in random variate generation. We have discussed various generators in the early chapters of this book. No method is shorter than the inversion method, which returns $-\log(U)$ where U is a uniform $[0,1]$ random variate. For most users, this method is satisfactory for their needs. In a high level language, the inversion method is difficult to beat. A variety of algorithms should be considered when the computer does not have a log operation in hardware and one wants to obtain a faster method. These include:

1. The uniform spacings method (section V.3.5).

2. von Neumann's method (section IV.2.2).

3. Marsaglia's exponential generator, or its modifications (discussed below).

4. The ratio-of-uniforms method (section IV.7.2).

5. The series method (section IV.5.3).

6. Table methods.

The methods listed under points 4 and 5 will not be discussed again in this chapter. Methods 2, 3 and 6 are all based upon the memoryless property of the exponential distribution, which states that given that an exponential random variable E exceeds $x > 0$, $E - x$ is again exponentially distributed. This is at the basis of Lemma IV.2.1, repeated here for the sake of readability:

Lemma IV.2.1.

An exponential random variable E is distributed as $(Z-1)\mu + Y$ where Z, Y are independent random variables and $\mu > 0$ is an arbitrary positive number: Z is geometrically distributed with

$$P(Z = i) = \int_{(i-1)\mu}^{i\mu} e^{-x} \, dx = e^{-(i-1)\mu} - e^{-i\mu} \quad (i \geq 1),$$

and Y is a truncated exponential random variable with density

$$f(x) = \frac{e^{-x}}{1 - e^{-\mu}} \quad (0 \leq x \leq \mu).$$

Since Z, Y are independent, exponential random variate generation can truly be considered as the problem of the generation of a discrete random variate plus a continuous random variate with compact support. And because the continuous random variate has compact support, any fast table method can be used .

The uniform spacings method is based upon the fact that $GS_1, \ldots, GS_n$ are iid exponential random variables when G is gamma (n), and $S_1, \ldots, S_n$ are spacings defined by a uniform sample of size $n-1$. For $n = 2$ this is sometimes faster than straightforward inversion:

Generate iid uniform [0,1] random variates U, V, W.
$Y \leftarrow -\log(UV)$
RETURN $WY, (1-W)Y$

Notice that three uniform random variates and one logarithm are needed per couple of exponential random variates. The overhead for the case $n = 3$ is sometimes a drawback. We summarize nevertheless:

Generate iid uniform [0,1] random variates U_1, U_2, U_3, U_4, U_5.
$Y \leftarrow -\log(U_1 U_2 U_3)$
$V \leftarrow \min(U_4, U_5), W \leftarrow \max(U_4, U_5)$
RETURN $VY, (W-V)Y, (1-W)Y$

2.2. Marsaglia's exponential generator.

Marsaglia (1961) proved the following theorem:

Theorem 2.1. (Marsaglia, 1961)

Let $U_1, U_2,...$ be iid uniform $[0,1]$ random variables. Let Z be a truncated Poisson random variate with probability vector

$$P(Z=i) = \frac{1}{e^{\mu}-1} \frac{\mu^i}{i!} \quad (i \geq 1),$$

where $\mu > 0$ is a constant. Let M be a geometric random vector with probability vector

$$P(M=i) = (1-e^{-\mu})e^{-\mu i} \quad (i \geq 0).$$

Then $X \leftarrow \mu(M+\min(U_1, \ldots, U_Z))$ is exponentially distributed. Also,

$$E(M) = \frac{1}{e^{\mu}-1},$$

$$E(Z) = \frac{\mu e^{\mu}}{e^{\mu}-1}.$$

Proof of Theorem 2.1.

We note that for $\mu \geq x > 0$,

$$P(\mu \min(U_1, \ldots, U_Z) \leq x) = \sum_{i=1}^{\infty} P(Z=i)P(\mu \min(U_1, \ldots, U_i) \leq x)$$

$$= \sum_{i=1}^{\infty} \frac{1}{e^{\mu}-1} \frac{\mu^i}{i!}(1-(1-\frac{x}{\mu})^i)$$

$$= 1 - \sum_{i=1}^{\infty} \frac{1}{e^{\mu}-1} \frac{(\mu(1-\frac{x}{\mu}))^i}{i!}$$

$$= 1 - \frac{e^{\mu-x}-1}{e^{\mu}-1}$$

$$= \frac{1-e^{-x}}{1-e^{-\mu}}.$$

Thus, $\mu \min(U_1, \ldots, U_Z)$ has the exponential distribution truncated to $[0,\mu]$. The first part of the theorem now follows directly from Lemma IV.2.1. For the second part, use the fact that $M+1$ is geometrically distributed, so that $E(M+1) = \frac{1}{1-e^{-\mu}}$. Furthermore,

$$E(Z) = \frac{1}{e^{\mu}-1}(\frac{\mu^1}{0!}+\frac{\mu^2}{1!}+\frac{\mu^3}{2!}+ \cdots)$$

$$= \frac{\mu e^{\mu}}{e^{\mu}-1}. \blacksquare$$

We can now suggest an algorithm based upon Theorem 2.1:

Marsaglia's exponential generator

Generate a geometric random variate M defined by $P(M=i)=(1-e^{-\mu})e^{-\mu i}$ $(i \geq 0)$.
$Z \leftarrow 1$
Generate iid uniform $[0,1]$ random variates U,V.
$Y \leftarrow V$
WHILE True Do
    IF $U \leq F(Z)$ (Note: $F(i)=\dfrac{1}{e^{\mu}-1} \sum\limits_{j=1}^{i} \dfrac{\mu^j}{j!}$.)
        THEN RETURN $X \leftarrow \mu(M+Y)$
        ELSE
            $Z \leftarrow Z+1$
            Generate a uniform $[0,1]$ random variate V.
            $Y \leftarrow \min(Y,V)$

For the geometric random variate, the inversion method based upon sequential search seems the obvious choice. This can be sped up by storing the cumulative probabilities, or by mixing sequential search with the alias method. Similarly, the cumulative distribution function F of Z can be partially stored to speed up the second part of the algorithm. The design parameter μ must be found by compromise. Note that if sequential search based inversion is used for the geometric random variate M, then $\dfrac{1}{1-e^{-\mu}}$ comparisons are needed on the average: this decreases from ∞ to 1 as μ varies from 0 to ∞. Also, the expected number of accesses of F in the second part of the algorithm is equal to $E(Z)=\dfrac{\mu}{1-e^{-\mu}}$, and this increases from 1 to ∞ as μ varies from 0 to ∞. Furthermore, the algorithm in its entirety requires on the average $2+E(Z)$ uniform $[0,1]$ random variates. The two effects have to be properly balanced. For most implementations, a value μ in the range $0.40...0.80$ seems to be optimal. This point was addressed in more detail by Sibuya (1961). Special advantages are offered by the choices $\mu=1$ and $\mu=\log(2)$.

The special case $\mu=\log(2)$ allows one to generate the desired geometric random variate by analyzing the random bits in a uniform $[0,1]$ random variate, which can be done conveniently in assembly language by the logical shift operation. This algorithm was proposed by Ahrens and Dieter (1972), where the reader can also find an excellent survey of exponential random variate generation. Again, a table of $F(i)$ values is needed.

Exponential generator of Ahrens and Dieter (1972)

[NOTE: a table of values $F(i) = \sum\limits_{j=1}^{i} \dfrac{(\log(2))^j}{j!}$ is required.]

$M \leftarrow 0$

Generate a uniform $[0,1]$ random variate U.

WHILE $U < \dfrac{1}{2}$ DO $U \leftarrow 2U$, $M \leftarrow M + \log(2)$

> (M is now correctly distributed. It is equal to the number of 0's before the first 1 in the binary expansion of U. Note that $U \leftarrow 2U$ is implementable by a shift operation.)

$U \leftarrow 2U - 1$ (U is again uniform $[0,1]$ and independent of M.)

IF $U < \log(2)$

> THEN RETURN $X \leftarrow M + U$
>
> ELSE
>> $Z \leftarrow 2$
>>
>> Generate a uniform $[0,1]$ random variate V.
>>
>> $Y \leftarrow V$
>>
>> WHILE True Do
>>> Generate a uniform $[0,1]$ random variate V.
>>>
>>> $Y \leftarrow \min(Y,V)$
>>>
>>> IF $U \leq F(Z)$
>>>> THEN RETURN $X \leftarrow M + Y \log(2)$
>>>>
>>>> ELSE $Z \leftarrow Z + 1$

Ahrens and Dieter squeeze the first uniform $[0,1]$ random variate U dry. Because of this, the algorithm requires very few uniform random variates on the average: the expected number is $1 + \log(2)$, which is about 1.69315.

2.3. The rectangle-wedge-tail method.

One of the fastest table methods for the exponential distribution was first published by Maclaren, Marsaglia and Bray (1964). It is ideally suited for implementation in machine language, but even in a high level language it is faster than most other methods described in this section. The extra speed is obtained by principles related to the table method. First, the tail of the density is cut off at some point $n\mu$ where n is a design integer and $\mu > 0$ is a small design constant. The remainder of the graph of f is then divided into n equal strips of width μ. And on interval $[(i-1)\mu, i\mu]$, we divide the graph into a rectangular piece of height $e^{-i\mu}$, and a wedge $f(x) - e^{-i\mu}$. Thus, the density is decomposed into

$2n + 1$ pieces of the following weights:

> one tail of weight $e^{-n\mu}$;
> n rectangles with weights $\mu e^{-i\mu}$, $1 \leq i \leq n$;
> n wedges of weights $e^{-i\mu}(e^{\mu}-1-\mu)$, $1 \leq i \leq n$.

These numbers can be used to set up a table for discrete random variate generation. The algorithm then proceeds as follows:

The rectangle-wedge-tail method

[NOTE: we refer to the $2n + 1$ probabilities defined above.]

$X \leftarrow 0$

REPEAT

> Generate a random integer Z with values in $1, \ldots, 2n + 1$ having the given probability vector.
>
> CASE
>
>> Rectangle i chosen: RETURN $X \leftarrow X + (i-1+U)\mu$ where U is a uniform $[0,1]$ random variate.
>>
>> Wedge i chosen: RETURN $X \leftarrow X + (i-1)\mu + Y$ where Y is a random variate having the wedge density $g(x) = \dfrac{e^{\mu-x}-1}{e^{\mu}-1-\mu}$, $0 \leq x \leq \mu$.
>>
>> Tail is chosen: $X \leftarrow X + n\mu$

UNTIL False

Note that when the tail is picked, we do in fact reject the choice, but keep at the same time track of the number of rejections. Equivalently, we could have returned $n\mu - \log(U)$ but this would have been less elegant since we would in effect rely on a logarithm. The recursive approach followed here seems cleaner. Random variates from the wedge density can be obtained in a number of ways. We could proceed by rejection from the triangular density: note that

$$g(x) = \frac{e^{\mu-x}-1}{e^{\mu}-1-\mu} \leq \frac{\mu-x}{\mu} \frac{e^{\mu}-1}{e^{\mu}-1-\mu}$$

and

$$g(x) \geq \frac{e^{\mu}-xe^{\mu}-1}{e^{\mu}-1-\mu} ,$$

so that the following rejection algorithm is valid:

Wedge generator

REPEAT

Generate two iid uniform [0,1] random variates X, U.

IF $X > U$ THEN $(X, U) \leftarrow (U, X)$ ((X, U) is now uniformly distributed under the triangle with unit sides.)

IF $U \leq 1 - X \dfrac{\mu e^\mu}{e^\mu - 1}$

THEN RETURN μX

ELSE IF $U \leq \dfrac{e^{\mu - \mu X} - 1}{e^\mu - 1}$ THEN RETURN μX

UNTIL False

The wedge generator requires on the average

$$\frac{1}{2} \frac{\mu(e^\mu - 1)}{e^\mu - 1 - \mu}$$

iterations. It is easy to see that this tends to 1 as $\mu \downarrow 0$. The expected number of uniform random variates needed is thus twice this number. But note that this can be bounded as follows:

$$\mu \frac{e^\mu - 1}{e^\mu - 1 - \mu} = \mu(1 + \frac{\mu}{e^\mu - 1 - \mu}) \leq \mu(1 + \frac{2}{\mu}) = \mu + 2 \ .$$

Here we used an inequality based upon the truncated Taylor series expansion. In view of the squeeze step, the expected number of evaluations of the exponential function is of course much less than the expected number of iterations. Having established this, we can summarize the performance of the algorithm by repeated use of Wald's equation:

Theorem 2.2.

This theorem is about the analysis of the rectangle-wedge-tail algorithm shown above.

(i) The expected number of global iterations is $A = \dfrac{1}{1 - e^{-n\mu}}$.

(ii) The expected number of uniform [0,1] random variates needed (excluding the discrete random variate generation portion) is $\dfrac{\mu}{1 - e^{-\mu}}$.

Proof of Theorem 2.2.

Theorem 2.2 is established as follows: we have 1 uniform random variate per rectangle (the probability of this is $\sum_{i=1}^{n} \mu e^{-i\mu} = \mu \dfrac{e^{-\mu} - e^{-(n+1)\mu}}{1-e^{-\mu}}$ in the first itera-

tion). We have $\mu \dfrac{e^{\mu}-1}{e^{\mu}-1-\mu}$ per wedge (the probability of this is

$\sum_{i=1}^{n} e^{-i\mu}(e^{\mu}-1-\mu) = \dfrac{e^{-\mu} - e^{-(n+1)\mu}}{1-e^{-\mu}}(e^{\mu}-1-\mu)$ in the first iteration). Thus, by estab-

lishing the correctness of statement (1), and applying Wald's equation, we observe that the expected number of uniform random variates needed is

$$A\left(\mu \frac{e^{-\mu} - e^{-(n+1)\mu}}{1-e^{-\mu}} + \mu \frac{e^{\mu}-1}{e^{\mu}-1-\mu} \frac{e^{-\mu} - e^{-(n+1)\mu}}{1-e^{-\mu}}(e^{\mu}-1-\mu)\right)$$

$$= A\left(\mu e^{\mu} \frac{e^{-\mu} - e^{-(n+1)\mu}}{1-e^{-\mu}}\right)$$

$$= A\left(\mu \frac{1-e^{-n\mu}}{1-e^{-\mu}}\right)$$

$$= \frac{\mu}{1-e^{-\mu}} \cdot \blacksquare$$

The number of intervals n does not affect the expected number of uniform random variates needed in the algorithm. Of course, the expected number of discrete random variates needed depends very much on n, since it is $\dfrac{1}{1-e^{-n\mu}}$. It is clear that μ should be made very small because as $\mu \downarrow 0$, the expected number of uniform random variates is $1+\dfrac{\mu}{2}+o(\mu)$. But when μ is small, we have to choose n large to keep the expected number of iterations down. For example, if we want the expected number of iterations to be $\dfrac{1}{1-e^{-4}}$, which is entirely reasonable, then we should choose $n = \dfrac{4}{\mu}$. When $\mu = \dfrac{1}{20}$, the table size is $2n+1=161$.

The algorithm given here may differ slightly from the algorithms found elsewhere. The idea remains basically the same: by picking certain design constants, we can practically guarantee that one exponential random variate can be obtained at the expense of one discrete random variate and one uniform random variate. The discrete random variate in turn can be obtained extremely quickly by the alias method or the alias-urn method at the cost of one other uniform random variate and either one or two table look-ups.

2.4. Exercises.

1. It is important to have a fast generator for the truncated exponential density $f(x) = e^{-x}/(1-e^{-\mu})$, $0 \leq x \leq \mu$. From Theorem 2.1, we recall that a random variate with this density can be generated as $\mu \min(U_1, \ldots, U_Z)$ where the U_i's are iid uniform $[0,1]$ random variates and Z is a truncated Poisson variate with probability vector

$$P(Z=i) = \frac{1}{e^{\mu}-1} \frac{\mu^i}{i!} \quad (i \geq 1) .$$

The purpose of this exercise is to explore alternative methods. In particular, compare with a strip table method based upon n equi-sized intervals and with a grid table method based upon n equi-sized intervals. Compare also with rejection from a trapezoidal dominating function, combined with clever squeeze steps.

2. **The Laplace density.** The Laplace density is $f(x) = \frac{1}{2} e^{-|x|}$. Show that a random variate X with this density can be generated as SE or as $E_1 - E_2$ where E, E_1, E_2 are iid exponential random variates, and S is a random sign.

3. Find the density of the sum of two iid Laplace random variables, and verify its bell shape. Prove that such a random variate can be generated as $\log(\frac{U_1 U_2}{U_3 U_4})$ where the U_i's are iid uniform $[0,1]$ random variates. Develop a rejection algorithm for normal random variates with quick acceptance and rejection steps based upon the inequalities:

$$1 - \frac{x^3}{3} \leq \frac{e^{-\frac{x^2}{2}}}{(1+x)e^{-x}} \leq \begin{cases} 1 & ,x > 0 \\ 1 - \frac{x^3}{3}(\frac{23}{27})(1-\frac{x^2}{6}) & ,x > 0 \end{cases} .$$

Prove these inequalities by using Taylor's series expansion truncated at the third term.

3. THE GAMMA DENSITY.

3.1. The gamma family.

A random variable X is **gamma** (a, b) distributed when it has density

$$f(x) = \frac{x^{a-1} e^{-\frac{x}{b}}}{\Gamma(a) b^a} \quad (x \geq 0) .$$

Here $a > 0$ is the shape parameter and $b > 0$ is the scale parameter. We say that X is gamma (a) distributed when it is gamma $(a, 1)$. Before reviewing random variate generation techniques for this family, we will look at some key properties that are relevant to us and that could aid in the design of an algorithm.

The density is unimodal with mode at $(a-1)b$ when $a \geq 1$. When $a < 1$, it is monotone with an infinite peak at 0. The moments are easily computed. For example, we have

$$E(X) = \int_0^\infty x f(x)\, dx = \frac{\Gamma(a+1)b^{a+1}}{\Gamma(a)b^a} = ab \; ;$$

$$E(X^2) = \int_0^\infty x^2 f(x)\, dx = \frac{\Gamma(a+2)b^{a+2}}{\Gamma(a)b^a} = a(a+1)b^2 \; .$$

Thus, $Var(X) = ab^2$.

The gamma family is closed under many operations. For example, when X is gamma (a, b), then cX is gamma (a, bc) when $c > 0$. Also, summing gamma random variables yields another gamma random variable. This is perhaps best seen by considering the characteristic function $\phi(t)$ of a gamma (a, b) random variable:

$$\phi(t) = E(e^{itX}) = \int_0^\infty \frac{x^{a-1} e^{-x(\frac{1}{b}-it)}}{\Gamma(a)b^a}\, dx$$

$$= \frac{(\frac{b}{1-itb})^a}{b^a} \int_0^\infty \frac{x^{a-1} e^{-x(\frac{1}{b}-it)}}{\Gamma(a)(\frac{b}{1-itb})^a}\, dx$$

$$= \frac{1}{(1-itb)^a} \; .$$

Thus, if $X_1, \ldots, X_n$ are independent gamma $(a_1), \ldots,$ gamma (a_n) random variables, then $X = \sum_{i=1}^n X_i$ has characteristic function

$$\phi(t) = \prod_{j=1}^n \frac{1}{(1-it)^{a_j}} = \frac{1}{(1-it)^{\sum_{j=1}^n a_j}} \; ,$$

and is therefore gamma $(\sum_{j=1}^n a_j, 1)$ distributed. The family is also closed under more complicated transformations. To illustrate this, we consider Kullback's result (Kullback, 1934) which states that when X_1, X_2 are independent gamma (a) and gamma $(a + \frac{1}{2})$ random variables, then $2\sqrt{X_1 X_2}$ is gamma $(2a)$.

The gamma distribution is related in innumerable ways to other well-known distributions. The exponential density is a gamma density with parameters $(1,1)$. And when X is normally distributed, then X^2 is gamma $(\frac{1}{2}, 2)$ distributed. This is called the **chi-square** distribution with one degree of freedom. In general, a gamma $(\frac{r}{2},2)$ random variable is called a chi-square random variable with r degrees of freedom. We will not use the chi-square terminology in this section. Perhaps the most important property of the gamma density is its relationship with the beta density. This is summarized in the following theorem:

Theorem 3.1.

   If X_1, X_2 are independent gamma (a_1) and gamma (a_2) random variables, then $\dfrac{X_1}{X_1+X_2}$ and X_1+X_2 are independent beta (a_1,a_2) and gamma (a_1+a_2) random variables. Furthermore, if Y is gamma (a) and Z is beta $(b,a-b)$ for some $b>a>0$, then YZ and $Y(1-Z)$ are independent gamma (b) and gamma $(a-b)$ random variables.

Proof of Theorem 3.1.

   We will only prove the first part of the theorem, and leave the second part to the reader (see exercises). Consider first the transformation $y=x_1/(x_1+x_2)$, $z=x_1+x_2$, which has an inverse $x_1=yz, x_2=(1-y)z$. The Jacobian of the transformation is

$$\begin{vmatrix} \dfrac{\partial x_1}{\partial y} & \dfrac{\partial x_1}{\partial z} \\ \dfrac{\partial x_2}{\partial y} & \dfrac{\partial x_2}{\partial z} \end{vmatrix} = \begin{vmatrix} z & y \\ -z & 1-y \end{vmatrix} = |z| \; .$$

Thus, the density $f(y,z)$ of $(Y,Z)=(\dfrac{X_1}{X_1+X_2}, X_1+X_2)$ is

$$\frac{(yz)^{a_1-1}e^{-yz}}{\Gamma(a_1)} \frac{((1-y)z)^{a_2-1}e^{-(1-y)z}}{\Gamma(a_2)} z$$

$$= \frac{\Gamma(a_1+a_2)y^{a_1-1}(1-y)^{a_2-1}}{\Gamma(a_1)\Gamma(a_2)} \frac{z^{a_1+a_2-1}e^{-z}}{\Gamma(a_1+a_2)} \; ,$$

which was to be shown. ∎

   The observation that for large values of a, the gamma density is close to the normal density could aid in the choice of a dominating curve for the rejection method. This fact follows of course from the observation that sums of gamma random variables are again gamma random variables, and from the central limit theorem. However, since the central limit theorem is concerned with the convergence of distribution functions, and since we are interested in a local central limit theorem, convergence of a density to a density, it is perhaps instructive to give a direct proof of this result. We have:

Theorem 3.2.

If X_a is gamma (a) distributed and if f_a is the density of the normalized gamma random variable $(X_a - a)/\sqrt{a}$, then

$$\lim_{a \uparrow \infty} f_a(x) = \frac{1}{\sqrt{2\pi}} e^{-\frac{x^2}{2}} \quad (x \in R).$$

Proof of Theorem 3.2.

The density of $(X_a - a)/\sqrt{a}$ evaluated at x is

$$\sqrt{a} \frac{(x\sqrt{a} + a)^{a-1} e^{-(x\sqrt{a} + a)}}{\Gamma(a)} \sim \frac{\sqrt{a}\, a^{a-1}(1 + \frac{x}{\sqrt{a}})^{a-1} e^{-a}\, e^{-x\sqrt{a}}}{(\frac{a-1}{e})^{a-1}\sqrt{2\pi(a-1)}}$$

$$\sim \frac{1}{\sqrt{2\pi}} \frac{1}{e}(1 + \frac{1}{a-1})^{a-1} e^{x\sqrt{a} + \frac{(a-1)x}{\sqrt{a}} - \frac{(a-1)x^2}{2a} + O(\frac{1}{\sqrt{a}})}$$

$$= \frac{1}{\sqrt{2\pi}}(1 + o(1))e^{-\frac{x^2}{2} + O(\frac{1}{\sqrt{a}})}$$

$$= \frac{1}{\sqrt{2\pi}} e^{-\frac{x^2}{2}}(1 + o(1)).$$

Here we used Stirling's approximation, and the Taylor series expansion for $\log(1+u)$ when $0 < u < 1$. ∎

3.2. Gamma variate generators.

Features we could appreciate in good gamma generators include

(i)   Uniform speed: the expected time is uniformly bounded over all values of a, the shape parameter.

(ii) Simplicity: short easy programs are more likely to become widely used.

(iii) Small or nonexistent set-up times: design parameters which depend upon a need to be recalculated every time a changes. These recalculations take often more time than the generator.

No family has received more attention in the literature than the gamma family. Many experimental comparisons are available in the general literature: see e.g. Atkinson and Pearce (1976), Vaduva (1977), or Tadikamalla and Johnson

(1980,1981).

For special cases, there are some good recipes: for example, when $a = 1$, we return an exponential random variate. When a is a small integer, we can return either

$$\sum_{i=1}^{a} E_i$$

where the E_i's are iid exponential random variates, or

$$-\log(\prod_{i=1}^{a} U_i)$$

where the U_i's are iid uniform [0,1] random variates. When a equals $\frac{1}{2} + k$ for some small integer k, it is possible to return

$$\frac{1}{2}N^2 + \sum_{i=1}^{k} E_i$$

where N is a normal random variate independent of the E_i's. In older texts one will often find the recommendation that a gamma (a) random variate should be generated as the sum of a gamma ($\lfloor a \rfloor$) and a gamma ($a - \lfloor a \rfloor$) random variate. The former random variate is to be obtained as a sum of independent exponential random variates. The parameter of the second gamma variate is less than 1. All these strategies take time linearly increasing with a; none lead to good gamma generators in general.

There are several successful approaches in the design of good gamma generators: first and foremost are the rejection algorithms. The rejection algorithms can be classified according to the family of dominating curves used. The differences in timings are usually minor: they often depend upon the efficiency of some quick acceptance step, and upon the way the rejection constant varies with a as $a \uparrow \infty$. Because of Theorem 3.2, we see that for the rejection constant to converge to 1 as $a \uparrow \infty$ it is necessary for the dominating curve to approach the normal density. Thus, some rejection algorithms are suboptimal from the start. Curiously, this is sometimes not a big drawback provided that the rejection constant remains reasonably close to 1. To discuss algorithms, we will inherit the names available in the literature for otherwise our discussion would be too verbose. Some successful rejection algorithms include:

GB. (Cheng, 1977): rejection from the Burr XII distribution. To be discussed below.

GO. (Ahrens and Dieter, 1974): rejection from a combination of normal and exponential densities.

GC. (Ahrens and Dieter, 1974): rejection from the Cauchy density.

XG. (Best, 1978): rejection from the t distribution with 2 degrees of freedom.

TAD2.

(Tadikamalla, 1978): rejection from the Laplace density.

Of these approaches, algorithm GO has the best asymptotic value for the rejection constant. This by itself does not make it the fastest and certainly not the shortest algorithm. The real reason why there are so many rejection algorithms around is that the normalized gamma density cannot be fitted under the normal density because its tail decreases much slower than the tail of the normal density. We can of course apply the almost exact inversion principle and find a nonlinear transformation which would transform the gamma density into a density which is very nearly normal, and which among other things would enable us to tuck the new density under a normal curve. Such normalizing transformations include a quadratic transformation (Fisher's transformation) and a cubic transformation (the Wilson-Hilferty transformation): the resulting algorithms are extremely fast because of the good fit. A prototype algorithm of this kind was developed and analyzed in detail in section IV.3.4, Marsaglia's algorithm RGAMA (Marsaglia (1977), Greenwood (1974)). In section IV.7.2, we presented some gamma generators based upon the ratio-of-uniforms method, which improve slightly over similar algorithms published by Kinderman and Monahan (1977, 1978, 1979) (algorithm GRUB) and Cheng and Feast (1979, 1979) (algorithm GBH). Despite the fact that no ratio-of-uniforms algorithm can have an asymptotically optimal rejection constant, they are typically comparable to the best rejection algorithms because of the simplicity of the dominating density. Most useful algorithms fall into one of the categories described above. The universal method for log-concave densities (section VII.2.3) (Devroye, 1984) is of course not competitive with specially designed algorithms.

There are no algorithms of the types described above which are uniformly fast for all a because the design is usually geared towards good performance for large values of a. Thus, for most algorithms, we have uniform speed on some interval $[a*,\infty)$ where $a*$ is typically near 1. For small values of a, the algorithms are often not valid - this is due to the fact that the gamma density has an infinite peak at 0 when $a < 1$, while dominating curves are often taken from a family of bounded densities. We will devote a special section to the problem of gamma generators for values $a < 1$.

Sometimes, there is a need for a very fast algorithm which would be applied for a fixed value of a. What one should do in such case is cut off the tail, and use a strip-based table method (section VIII.2) on the body. Since these table methods can be automated, it is not worth spending extra time on this issue. It is nevertheless worth noting that some automated table methods have table sizes that in the case of the gamma density increase unboundedly as $a \rightarrow \infty$ if the expected time per random variate is to remain bounded, unless one applies a specially designed technique similar to what was done for the exponential density in the rectangle-wedge-tail method. In an interesting paper, Schmeiser and Lal (1980) have developed a semi-table method: the graph of the density is partitioned into about 10 pieces, all rectangular, triangular or exponential in shape, and the set-up time, about five times the time needed to generate one random variate, is reasonable. Moreover, the table size (number of pieces) remains fixed for all values of a. When speed per random variate is at a premium, one should certainly use some sort of table method. When speed is important, and a varies with each call, the almost-exact-inversion method seems to be the winner in most experimental comparisons, and certainly when fast exponential and normal random variate generators are available. The best ratio-of-uniforms methods and the best rejection methods (XG,GO,GB) are next in line, well ahead of all table methods.

Finally, we will discuss random variate generation for closely related distributions such as the Weibull distribution and the exponential power distribution.

3.3. Uniformly fast rejection algorithms for $a \geq 1$.

We begin with one of the shortest algorithms for the gamma density, which is based upon rejection from the t density with 2 degrees of freedom:

$$g(x) = \frac{1}{2\sqrt{2}}(1+\frac{x^2}{2})^{-\frac{3}{2}}$$

This density decreases as x^{-3}, and is symmetric bout 0. Thus, it can be used as a dominating curve of a properly rescaled and translated gamma density. Best's algorithm XG (Best, 1978) is based upon the following facts:

Theorem 3.3.

A. The density g has distribution function

$$G(x) = \frac{1}{2}\left[1 + \frac{\dfrac{x}{\sqrt{2}}}{\sqrt{1 + \dfrac{x^2}{2}}}\right].$$

A random variate with this distribution can be generated as

$$\frac{\sqrt{2}(U - \frac{1}{2})}{\sqrt{U(1-U)}}$$

where U is a uniform $[0,1]$ random variate.

B. Let f be the gamma (a) density, and let g_a be the density of $(a-1) + Y\sqrt{\dfrac{3a}{2} - \dfrac{3}{8}}$ where Y has density g. Then

$$f(x) \le c_a\, g_a(x) = \frac{1}{\Gamma(a)\left[1 + \dfrac{1}{2}\left(\dfrac{x-(a-1)}{\sqrt{\dfrac{3a}{2}-\dfrac{3}{8}}}\right)^2\right]^{\frac{3}{2}}},$$

where the rejection constant is given by

$$c_a = \frac{2\sqrt{3a - \dfrac{3}{4}}}{\Gamma(a)}\left(\frac{a-1}{e}\right)^{a-1}.$$

C. We have $\sup\limits_{a \ge 1} c_a \le e\sqrt{\dfrac{6}{\pi}}$, and $\lim\limits_{a \uparrow \infty} c_a = \sqrt{\dfrac{6}{\pi}}$.

Proof of Theorem 3.3.

The claim about the distribution function G is quickly verified. When U is uniformly distributed on $[0,1]$, then the solution X of $G(X) = U$ is precisely

$$X = \frac{\sqrt{2}(U - \frac{1}{2})}{\sqrt{U(1-U)}}.$$ This proves part A.

Let Y have density g. Then $(a-1) + Y\sqrt{\dfrac{3a}{2} - \dfrac{3}{8}}$ has density

$$\frac{1}{2\sqrt{2}\sqrt{\dfrac{3a}{2}-\dfrac{3}{8}}}\left[1 + \dfrac{1}{2}\left(\dfrac{x-(a-1)}{\sqrt{\dfrac{3a}{2}-\dfrac{3}{8}}}\right)^2\right]^{-\frac{3}{2}}$$

To prove statement B, we need only show that for $x > 0$,

$$x^{a-1}e^{-x} \leq \left(\frac{a-1}{e}\right)^{a-1} \frac{1}{\left[1+\frac{1}{2}\left(\frac{x-(a-1)}{\sqrt{\frac{3a}{2}-\frac{3}{8}}}\right)^2\right]^{\frac{3}{2}}} ,$$

or, after resubstitution $y = x-(a-1)$, that for $y \geq -(a-1)$,

$$e^{-y}\left(1+\frac{y}{a-1}\right)^{a-1} \leq \left(1+\frac{y^2}{3a-\frac{3}{4}}\right)^{-\frac{3}{2}} .$$

Taking logarithms, we see that we must show that

$$h(y) = -y + (a-1)\log(1+\frac{y}{a-1}) + \frac{3}{2}\log\left(1+\frac{y^2}{3a-\frac{3}{4}}\right) \leq 0 .$$

Clearly, $h(0)=0$. It suffices to show that $h'(y) \geq 0$ for $y \leq 0$ and that $h'(y) \leq 0$ for $y \geq 0$. But

$$h'(y) = -1 + \frac{a-1}{(a-1)(1+\frac{y}{a-1})} + \frac{3}{2}\frac{\frac{2y}{3a-\frac{3}{4}}}{1+\frac{y^2}{3a-\frac{3}{4}}}$$

$$= -\frac{y}{a-1+y} + \frac{y}{a-\frac{1}{4}+\frac{y^2}{3}}$$

$$= \frac{y(y-\frac{3}{4}-\frac{y^2}{3})}{(a-1+y)(a-\frac{1}{4}+\frac{y^2}{3})} .$$

The denominator is ≥ 0 for $a \geq \frac{1}{4}$. The numerator is ≥ 0 for $y \leq 0$, and is ≤ 0 for $y \geq 0$ (this can be seen by rewriting it as $-\frac{y}{3}(y-\frac{3}{2})^2$. This concludes the proof of part B.

For part C, we apply Stirling's approximation, and observe that

$$c_a \sim \frac{2\sqrt{3a}}{(\frac{a}{e})^a\sqrt{\frac{2\pi}{a}}}(\frac{a-1}{e})^{a-1}$$

$$= \frac{2e\sqrt{3a}}{\sqrt{2\pi a}}(1-\frac{1}{a})^{a-1}$$

$$\sim \sqrt{\frac{6}{\pi}} .$$

The first $\sim$ is also an upper bound, so that

$$c_a \le \sqrt{\frac{6}{\pi}} e^{\frac{1}{a}}$$

when $a \ge 1$. This proves part C. ■

Based upon Theorem 3.3, we can now state Best's rejection algorithm:

Best's rejection algorithm XG for gamma random variates (Best, 1978)

[SET-UP]

$b \leftarrow a-1, c \leftarrow 3a - \dfrac{3}{4}$

[GENERATOR]

REPEAT

    Generate iid uniform [0,1] random variates U, V.

    $W \leftarrow U(1-U), Y \leftarrow \sqrt{\dfrac{c}{W}}(U - \dfrac{1}{2}), X \leftarrow b + Y$

    IF $X \ge 0$

        THEN

            $Z \leftarrow 64 W^3 V^2$

            Accept $\leftarrow [Z \le 1 - \dfrac{2Y^2}{X}]$

            IF NOT Accept

                THEN Accept $\leftarrow [\log(Z) \le 2(b \log(\dfrac{X}{b}) - Y)]$

UNTIL Accept

RETURN X

The random variate X generated at the outset of the REPEAT loop has density g_a. The acceptance condition is

$$e^{-Y}(1 + \frac{Y}{a-1})^{a-1} \ge V \left(1 + \frac{Y^2}{3a - \frac{3}{4}}\right)^{-\frac{3}{2}}.$$

This can be rewritten in a number of ways: for example, in the notation of the algorithm,

$$e^{-Y}(\frac{X}{b})^b \ge V(4W)^{\frac{3}{2}};$$

$$-Y+b\,\log(\frac{X}{b}) \geq \frac{1}{2}\log(4^3 V^2 W^3) \ ;$$

$$2(-Y+b\,\log(\frac{X}{b})) \geq \log(Z) \ .$$

This explains the acceptance condition used in the algorithm. The squeeze step is derived from the acceptance condition, by noting that

(i) $\log(Z) \leq Z-1;$

(ii) $2(b\,\log(1+\dfrac{Y}{b})-Y) \geq 2Y(-\dfrac{Y}{b+Y}) = -\dfrac{2Y^2}{X}.$

The last inequality is obtained by noting that the left hand side as a function of Y is 0 at $Y=0$, and has derivative $-\dfrac{Y}{b+Y}$. Therefore, by the Taylor series expansion truncated at the first term, we see that for $Y \geq 0$, the left hand side is at least equal to $2(0+Y(-\dfrac{Y}{b+Y}))$. For $Y \leq 0$, the same bound is valid. Thus, when $Z-1 \leq -2Y^2/X$, we are able to conclude that the acceptance condition is satisfied. It should be noted that in view of the rather large rejection constant, the squeeze step is probably not very effective, and could be omitted without a big time penalty.

We will now move on to Cheng's algorithm GB which is based upon rejection from the Burr XII density

$$g(x) = \lambda\mu\frac{x^{\lambda-1}}{(\mu+x^\lambda)^2}$$

for parameters $\mu,\lambda>0$ to be determined as a function of a. Random variates with this density can be obtained as

$$(\frac{\mu U}{1-U})^{\frac{1}{\lambda}}$$

where U is uniformly distributed on [0,1]. This follows from the fact that the distribution function corresponding to g is $x^\lambda/(\mu+x^\lambda), x \geq 0$. We have to choose λ and μ. Unfortunately, minimization of the area under the dominating curve does not give explicitly solvable equations. It is useful to match the curves of f and g, which are both unimodal. Since f peaks at $a-1$, it makes sense to match this peak. The peak of g occurs at

$$x = (\frac{(\lambda-1)\mu}{\lambda+1})^{\frac{1}{\lambda}} \ .$$

If we choose λ large, i.e. increasing with a, then this peak will approximately match the other peak when $\mu=a^\lambda$. Consider now $\log(\dfrac{f}{g})$. The derivative of this function is

$$\frac{a-\lambda-x}{x}+\frac{2\lambda x^{\lambda-1}}{a^\lambda+x^\lambda} \ .$$

This derivative attains the value 0 when $(a+\lambda-x)x^\lambda+(a-\lambda-x)a^\lambda=0$. By analyzing the derivative, we can see that it has a unique solution at $x=0$ when $\lambda=\sqrt{2a-1}$. Thus, we have

$$f(x) \leq cg(x)$$

where

$$c = \frac{a^{a-1}e^{-a}(2a^\lambda)^2}{\Gamma(a)\lambda a^\lambda a^{\lambda-1}}$$

$$= \frac{a^a e^{-a} 4}{\Gamma(a)\lambda}$$

$$\sim \frac{4\sqrt{a}}{\sqrt{2\pi}\lambda} \quad (a\uparrow\infty) .$$

Resubstitution of the value of λ yields the asymptotic value of $\sqrt{\dfrac{4}{\pi}}\approx 1.13$. In fact, we have

$$c \leq \frac{4\sqrt{a}}{\sqrt{2\pi}\lambda} = \sqrt{\frac{4}{\pi}}\sqrt{a/(a-\frac{1}{2})} \leq \sqrt{\frac{8}{\pi}} ,$$

uniformly over $a\geq 1$. Thus, the rejection algorithm suggested by Cheng has a good rejection constant. In the design, we notice that if X is a random variate with density g, and U is a uniform $[0,1]$ random variate, then the acceptance condition is

$$4(\frac{a}{e})^a (\frac{a^\lambda}{X^{\lambda+1}})\frac{X^{2\lambda}}{(a^\lambda+X^\lambda)^2}U \leq X^{a-1}e^{-X} .$$

Equivalently, since $V=X^\lambda/(a^\lambda+X^\lambda)$ is uniformly distributed on $[0,1]$, the acceptance condition can be rewritten as

$$4(\frac{a}{e})^a a^\lambda V^2 U \leq X^{\lambda+a} e^{-X} ,$$

or

$$\log(4)+(\lambda+a)\log(a)-a+\log(UV^2) \leq (\lambda+a)\log(X)-X ,$$

or

$$\log(UV^2) \leq a-\log(4)+(\lambda+a)\log(\frac{X}{a})-X .$$

A quick acceptance step can be introduced which uses the inequality

$$\log(UV^2) \leq d(UV^2)-\log(d)-1$$

which is valid for all d. The value $d=\dfrac{9}{2}$ was suggested by Cheng. Combining all of this, we obtain:

Cheng's rejection algorithm GB for gamma random variates (Cheng, 1977)

[SET-UP]

$b \leftarrow a - \log(4)$, $c \leftarrow a + \sqrt{2a - 1}$

[GENERATOR]

REPEAT

    Generate iid uniform $[0,1]$ random variates U, V.

    $Y \leftarrow a \log(\dfrac{V}{1-V})$, $X \leftarrow ae^{V}$

    $Z \leftarrow UV^2$

    $R \leftarrow b + cY - X$

    Accept $\leftarrow [R \geq \dfrac{9}{2}Z - (1 + \log(\dfrac{9}{2}))]$ (note that $(1 + \log(\dfrac{9}{2})) = 2.5040774...$)

    IF NOT Accept THEN Accept $\leftarrow [R \geq \log(Z)]$

UNTIL Accept

RETURN X

We will close this section with a word about the historically important algorithm GO of Ahrens and Dieter (1974), which was the first uniformly fast gamma generator. It also has a very good asymptotic rejection constant, slightly larger than 1. The authors got around the problem of the tail of the gamma density by noting that most of the gamma density can be tucked under a normal curve, and that the right tail can be tucked under an exponential curve. The breakpoint must of course be to the right of the peak $a - 1$. Ahrens and Dieter suggest the value $(a-1) + \sqrt{6(a + \sqrt{\dfrac{8a}{3}})}$. We recall that if X is gamma (a) distributed, then $\dfrac{(X-a)}{\sqrt{a}}$ tends in distribution to a normal density. Thus, with the breakpoint of Ahrens and Dieter, we cannot hope to construct a dominating curve with integral tending to 1 as $a \uparrow \infty$ (for this, the breakpoint must be at $a-1$ plus a term increasing faster than $\sqrt{a}$). It is true however that we are in practice very close. The almost-exact inversion method for normal random variates yields asymptotically optimal rejection constants without great difficulty. For this reason, we will delegate the treatment of algorithm GO to the exercises.

3.4. The Weibull density.

A random variable has the standard **Weibull density** with parameter $a > 0$ when it has density

$$f(x) = ax^{a-1}e^{-x^a} \quad (x \geq 0).$$

In this, we recognize the density of $E^{\frac{1}{a}}$ where E is an exponential random variable. This fact can also be deduced from the form of its distribution function,

$$F(x) = 1 - e^{-x^a} \quad (x \geq 0).$$

Because of this, it seems hardly worthwhile to design rejection algorithms for this density. But, turning the tables around for the moment, the Weibull density is very useful as an auxiliary density in generators for other densities.

Example 3.1. Gumbel's extreme value distribution.

When X is Weibull (a), then $Y = -a \log(X)$ has the extreme value density

$$f(x) = e^{-x} e^{-e^{-x}} \quad (x \in R).$$

By the fact that X is distributed as $E^{\frac{1}{a}}$, we see of course that the parameter a plays no special role: thus, $-\log(E)$ and $-\log(\log(\frac{1}{U}))$ are both extreme value random variables when E is exponentially distributed, and E is exponentially distributed. ∎

Example 3.2. A compound Weibull distribution.

Dubey (1968) has pointed out that the ratio $W_a / G_b^{\frac{1}{a}}$ has the Pareto-like density

$$f(x) = \frac{abx^{a-1}}{(1+x^a)^{b+1}} \quad (x \geq 0).$$

Here W_a is a Weibull (a) random variable, and G_b is a gamma (b) random variable. As a special case, we note that the ratio of two independent exponential random variables has density $\dfrac{1}{(1+x)^2}$ on $[0,\infty)$. ∎

Example 3.3. Gamma variates by rejection from the Weibull density.

Consider the gamma (a) density f with parameter $0 < a \leq 1$. For this density, random variates can be generated by rejection from the Weibull (a) density (which will be called g). This is based upon the inequality

$$\frac{f(x)}{g(x)} = \frac{e^{x^a - x}}{a\,\Gamma(a)} \leq \frac{e^{b - b^{\frac{1}{a}}}}{\Gamma(a+1)}$$

where

$$b = a^{\frac{a}{1-a}}.$$

A rejection algorithm based upon this inequality has rejection constant

$$\frac{e^{(1-a)a^{\frac{a}{1-a}}}}{\Gamma(1+a)}.$$

The rejection constant has the following properties:

1. It tends to 1 as $a \downarrow 0$, or $a \uparrow 1$.

2. It is not greater than $\dfrac{e}{0.88560}$ for any value of $a \in (0,1]$. This can be seen by noting that $(1-a)b \leq 1-a \leq 1$ and that $\Gamma(1+a) \geq 0.8856031944...$ (the gamma function at $1+a$ is absolutely bounded from below by its value at $1+a = 1.4616321449...$; see e.g. Abramowitz and Stegun (1970, pp. 259)).

This leads to a modified version of an algorithm of Vaduva's (1977):

Gamma generator for parameter smaller than 1

[SET-UP]

$c \leftarrow \dfrac{1}{a}$, $d \leftarrow a^{\frac{a}{1-a}}(1-a)$

[GENERATOR]

REPEAT

       Generate iid exponential random variates Z, E. Set $X \leftarrow Z^c$ (X is Weibull (a)).

UNTIL $Z + E \leq d + X$

RETURN X ∎

3.5. Johnk's theorem and its implications.

Random variate generation for the case $a < 1$ can be based upon a special property of the beta and gamma distributions. This property is usually attributed to Johnk (1964), and has later been rediscovered by others (Newman and Odell, 1971; Whittaker, 1974). We have:

Theorem 3.4. (Johnk, 1964)

Let $a, b > 0$ be given constants, and let U, V be iid uniform [0,1] random variables. Then, conditioned on $U^{\frac{1}{a}} + V^{\frac{1}{b}} \leq 1$, the random variable

$$\frac{U^{\frac{1}{a}}}{U^{\frac{1}{a}} + V^{\frac{1}{b}}}$$

is beta (a, b) distributed.

Theorem 3.5. (Berman, 1971)

Let $a, b > 0$ be given constants, and let U, V be iid uniform [0,1] random variables. Then, conditioned on $U^{\frac{1}{a}} + V^{\frac{1}{b}} \leq 1$, the random variable

$$U^{\frac{1}{a}}$$

is beta $(a, b+1)$ distributed.

Proof of Theorems 3.4 and 3.5.

Note that $X = U^{\frac{1}{a}}$ has distribution function x^a on [0,1]. The density is ax^{a-1}. Thus, the joint density of X and $Y = V^{\frac{1}{b}}$ is

$$f(x, y) = bx^{a-1}y^{b-1} \quad (0 \leq x, y \leq 1) .$$

Consider the transformation $z = x + y, t = \dfrac{x}{x+y}$ with inverse $x = tz, y = (1-t)z$. This transformation has Jacobian

$$\begin{vmatrix} \dfrac{\partial x}{\partial t} & \dfrac{\partial x}{\partial z} \\ \dfrac{\partial y}{\partial t} & \dfrac{\partial y}{\partial z} \end{vmatrix} = \begin{vmatrix} z & t \\ -z & 1-t \end{vmatrix} = |z| .$$

The joint density of $(Z,T)=(X+Y,\dfrac{X}{X+Y})$ is

$$|z| \ f\ (tz,(1-t)z) = zab\,(tz)^{a-1}((1-t)z)^{b-1} \quad (0 \le tz,(1-t)z \le 1)$$
$$= abt^{a-1}(1-t)^{b-1}z^{a+b-1} \quad (0 \le tz,(1-t)z \le 1)\ .$$

The region in the (z,t) plane on which this density is nonzero is $A=\{(z,t):t>0,0<z<\min(\dfrac{1}{t},\dfrac{1}{1-t})\}$. Let A_t be the collection of values z for which $0<z<\min(\dfrac{1}{t},\dfrac{1}{1-t})$. Then, writing $g(z,t)$ for the joint density of (Z,T) at (z,t), we see that the density of T conditional on $Z \le 1$ is given by

$$\frac{\displaystyle\int_{z \le 1, z \in A_t} g(z,t)dz}{\displaystyle\int_A g(z,t)dz\ dt}$$

$$= \frac{1}{c}\frac{ab}{a+b}t^{a-1}(1-t)^{b-1}$$

where $c=\int_A g(z,t)dz\ dt$ is a normalization constant. Clearly,

$$c = P(X+Y \le 1) = \frac{ab}{a+b}\frac{\Gamma(a)\Gamma(b)}{\Gamma(a+b)} = \frac{\Gamma(a+1)\Gamma(b+1)}{\Gamma(a+b+1)}\ .$$

This concludes the proof of Theorem 3.4.

For Berman's theorem, consider the transformation $x=x,z=x+y$ with inverse $x=x,y=z-x$. The joint density of (X,Z) is $f(x,z-x)=abx^{a-1}(z-x)^{b-1}I_B(x,z)$ where B is the set of (z,x) satisfying $0<x<1,0<x<z<x+1$. This is a parallelepid in the (z,x) plane. The density of X conditional on $Z<1$ is equal to a constant times

$$\int_{x<z<1} abx^{a-1}(z-x)^{b-1}\ dz = ax^{a-1}(1-x)^b\ .$$

This concludes the proof of Theorem 3.5. ■

These theorems provide us with recipes for generating gamma and beta variates. For gamma random variates , we observe that YZ is gamma (a) distributed when Y is beta $(a,1-a)$ and Z is gamma (1) (i.e. exponential), or when Y is beta $(a,2-a)$ and Z is gamma (2). Summarizing all of this, we have:

Johnk's beta generator

REPEAT

   Generate iid uniform $[0,1]$ random variates U,V.

   $X \leftarrow U^{\frac{1}{a}}, Y \leftarrow V^{\frac{1}{b}}$

UNTIL $X+Y \leq 1$

RETURN $\dfrac{X}{X+Y}$ (X is beta (a,b) distributed)

Berman's beta generator

REPEAT

   Generate iid uniform $[0,1]$ random variates U,V.

   $X \leftarrow U^{\frac{1}{a}}, Y \leftarrow V^{\frac{1}{b}}$

UNTIL $X+Y \leq 1$

RETURN X (X is beta $(a,b+1)$ distributed)

Johnk's gamma generator

REPEAT

   Generate iid uniform $[0,1]$ random variates U,V.

   $X \leftarrow U^{\frac{1}{a}}, Y \leftarrow V^{\frac{1}{1-a}}$

UNTIL $X+Y \leq 1$

Generate an exponential random variate E.

RETURN $\dfrac{EX}{X+Y}$ (X is gamma (a) distributed)

Berman's gamma generator

REPEAT

    Generate iid uniform $[0,1]$ random variates U, V.

    $X \leftarrow U^{\frac{1}{a}}, Y \leftarrow V^{\frac{1}{1-a}}$

UNTIL $X + Y \leq 1$

Generate a gamma (2) random variate Z (either as the sum of two iid exponential random variates or as $-\log(U*V*)$ where $U*, V*$ are iid uniform $[0,1]$ random variates).

RETURN ZX (X is gamma (a) distributed)

Both beta generators require on the average

$$\frac{1}{P(X+Y \leq 1)} = \frac{\Gamma(a+b+1)}{\Gamma(a+1)\Gamma(b+1)}$$

iterations, and this increases rapidly with a and b. It is however uniformly bounded over all a, b with $0 < a, b \leq 1$. The two gamma generators should only be used for $a \leq 1$. The expected number of iterations is in both cases

$$\frac{1}{\Gamma(1+a)\Gamma(2-a)} .$$

It is known that $\Gamma(a)\Gamma(1-a) = \pi/\sin(\pi a)$. Thus, the expected number of iterations is

$$\frac{\sin \pi a}{\pi a (1-a)} ,$$

which is a symmetric function of a around $\frac{1}{2}$ taking the value 1 near both endpoints ($a \downarrow 0$, $a = 1$), and peaking at the point $a = \frac{1}{2}$: thus, the rejection constant does not exceed $\frac{4}{\pi}$ for any $a \in (0,1]$.

3.6. Gamma variate generators when $a \leq 1$.

We can now summarize the avalaible algorithms for gamma (a) random variate generation when the parameter is less than one. The fact that there is an infinite peak eliminates other time-honored approaches (such as the ratio-of-uniforms method) from contention. We have:

1.   Rejection from the Weibull density (Vaduva, 1977): see section IX.3.7.

2. The Johnk and Berman algorithms (Johnk, 1971; Berman, 1971): see section IX.3.8.

3. The generator based upon Stuart's theorem (see section IV.6.4): $G_{a+1}U^{\frac{1}{a}}$ is gamma (a) distributed when G_{a+1} is gamma ($a+1$) distributed, and U is uniformly distributed on [0,1]. For G_{a+1} use an efficient gamma generator with parameter greater than unity.

4. The Forsythe-von Neumann method (see section IV.2.4).

5. The composition/rejection method, with rejection from an exponential density on $[1,\infty)$, and from a polynomial density on [0,1]. See sections IV.2.5 and II.3.3 for various pieces of the algorithm mainly due to Vaduva (1977). See also algorithm GS of Ahrens and Dieter (1974) and its modification by Best (1983) developed in the exercise section.

6. The transformation of an EPD variate obtained by the rejection method of section VII.2.6.

All of these algorithms are uniformly fast over the parameter range. Comparative timings vary from experiment to experiment. Tadikamalla and Johnson (1981) report good results with algorithm GS but fail to include some of the other algorithms in their comparison. The algorithms of Johnk and Berman are probably better suited for beta random variate generation because two expensive powers of uniform random variates are needed in every iteration. The Forsythe-von Neumann method seems also less efficient time-wise. This leaves us with approaches 1,3,5 and 6. If a very efficient gamma generator is available for $a>1$, then method 3 could be as fast as algorithm GS, or Vaduva's Weibull-based rejection method. Methods 1 and 6 are probably comparable in all respects, although the rejection constant of method 6 certainly is superior.

3.7. The tail of the gamma density.

As for the normal density, it is worthwhile to have a good generator for the tail gamma (a) density truncated at t. It is only natural to look at dominating densities of the form $be^{b(t-x)}$ ($x\geq t$). The parameter b has to be picked as a function of a and t. Note that a random variate with this density can be generated as $t+\dfrac{E}{b}$ where E is an exponential random variate. We consider the cases $a<1$ and $a\geq1$ separately. We can take $b=1$ because the gamma density decreases faster than e^{-x}. Therefore, rejection can be based upon the inequality

$$x^{a-1}e^{-x} \leq t^{a-1}e^{-x} \quad (x\geq t).$$

It is easily seen that the corresponding algorithm is

REPEAT
>   Generate a uniform random variate U and an exponential random variate E. Set
> $X \leftarrow t + E$
UNTIL $XU^{\frac{1}{1-a}} \le a$
RETURN X (X has the gamma density restricted to $[t, \infty)$)

The efficiency of the algorithm is given by the ratio of the integrals of the two functions. This gives

$$\frac{t^{a-1}e^{-t}}{\int_{t}^{\infty} x^{a-1}e^{-x}\ dx}$$

$$= \frac{1}{\int_{t}^{\infty} (\frac{x}{t})^{a-1} e^{t-x}\ dx}$$

$$= \frac{1}{\int_{0}^{\infty} (1+\frac{x}{t})^{a-1} e^{-x}\ dx}$$

$$\le \frac{1}{\int_{0}^{\infty} e^{x(\frac{a-1}{t}-1)}\ dx}$$

$$= 1 + \frac{1-a}{t}$$

$$\rightarrow 1 \quad \text{as } t \rightarrow \infty \ .$$

When $a \ge 1$, the exponential with parameter 1 does not suffice because of the polynomial portion in the gamma density. It is necessary to take a slightly slower decreasing exponential density. The inequality that we will use is

$$(\frac{x}{t})^{a-1} \le e^{(a-1)(\frac{x}{t}-1)}$$

which is easily established by standard optimization methods. This suggests the choice $b = 1 - \frac{a-1}{t}$ in the exponential curve. Thus, we have

$$x^{a-1}e^{-x} \le t^{a-1}e^{(a-1)(\frac{x}{t}-1)-x} \quad .$$

Based on this, the rejection algorithm becomes

REPEAT

    Generate two iid exponential random variates $E, E*$.

$$X \leftarrow t + \frac{E}{1 - \dfrac{a-1}{t}}$$

UNTIL $\dfrac{X}{t} - 1 + \log(\dfrac{t}{X}) \leq \dfrac{E*}{a-1}$

RETURN X (X has the gamma (a) density restricted to $[t, \infty)$.)

The algorithm is valid for all $a > 1$ and all $t > a - 1$ (the latter condition states that the tail should not include the mode of the gamma density). A squeeze step can be included by noting that $\log(\dfrac{X}{t}) = \log(1 + \dfrac{X-t}{t}) \geq 2\dfrac{X-t}{X+t} = \dfrac{2E}{(1-\dfrac{a-1}{t})(X+t)}$. Here we used the inequality $\log(1+u) \geq 2u/(u+2)$. Thus, the quick acceptance step to be inserted in the algorithm is

IF $\dfrac{E^2}{(1-\dfrac{a-1}{t})^2 \, t(X+t)} \leq \dfrac{E*}{a-1}$ THEN RETURN X

We conclude this section by showing that the rejection constant is asymptotically optimal as $t \uparrow \infty$: the ratio of the integrals of the two functions involved is

$$\frac{t^{a-1}e^{-t}}{(1-\dfrac{a-1}{t})\int\limits_{t}^{\infty} x^{a-1}e^{-x} \, dx}$$

$$= \frac{1}{(1-\dfrac{a-1}{t})\int\limits_{0}^{\infty}(1+\dfrac{x}{t})^{a-1}e^{-x} \, dx}$$

which once again tends to 1 as $t \to \infty$. We note here that the algorithms given in this section are due to Devroye (1980). The algorithm for the case $a > 1$ can be slightly improved at the expense of more complicated design parameters. This possibility is explored in the exercises.

3.8. Stacy's generalized gamma distribution.

Stacy (1962) introduced the generalized gamma distribution with two shape parameters, $c, a > 0$: the density is

$$f(x) = \frac{c}{\Gamma(a)} x^{ca-1} e^{-x^c} \qquad (x \geq 0).$$

This family of densities includes the gamma densities ($c = 1$), the halfnormal density ($a = \frac{1}{2}, c = 2$) and the Weibull densities ($a = 1$). Because of the flexibility of having two shape parameters, this distribution has been used quite often in modeling stochastic inputs. Random variate generation is no problem because we observe that $G_a^{\frac{1}{c}}$ has the said distribution where G_a is a gamma (a) random variable.

Tadikamalla (1979) has developed a rejection algorithm for the case $a > 1$ which uses as a dominating density the Burr XII density used by Cheng in his algorithm GB. The parameters μ, λ of the Burr XII density are $\lambda = c \sqrt{2a-1}$, $\mu = a^{\sqrt{2a-1}}$. The rejection constant is a function of a only. The algorithm is virtually equivalent to generating G_a by Cheng's algorithm GB and returning $G_a^{\frac{1}{c}}$ (which explains why the rejection constant does not depend upon c).

3.9. Exercises.

1. Show Kullback's result (Kullback, 1934) which states that when X_1, X_2 are independent gamma (a) and gamma ($a + \frac{1}{2}$) random variables, then $2\sqrt{X_1 X_2}$ is gamma ($2a$).

2. Prove Stuart's theorem (the second statement of Theorem 3.1): If Y is gamma (a) and Z is beta ($b, a-b$) for some $b > a > 0$, then YZ and $Y(1-Z)$ are independent gamma (b) and gamma ($a-b$) random variables.

3. **Algorithm GO (Ahrens and Dieter, 1974).** Define the breakpoint $b = a - 1 + \sqrt{6(a + \sqrt{\frac{8a}{3}})}$. Find the smallest exponentially decreasing function dominating the gamma (a) density to the right of b. Find a normal curve centered at $a-1$ dominating the gamma density to the left of b, which has the property that the area under the dominating curve divided by the area under the leftmost piece of the gamma density tends to a constant as $a \uparrow \infty$. Also, find the similarly defined asymptotic ratio for the rightmost

piece, and establish that it is greater than 1. By combining this, obtain an expression for the limit value of the rejection constant. Having established the bounds, give a rejection method for generating a random variate with the gamma density. Find efficient squeeze steps if possible.

4.   **The Weibull density.** Prove the following properties of the Weibull (a) distribution:

A.   For $a \geq 1$, the density is unimodal with mode at $(1-\frac{1}{a})^{\frac{1}{a}}$. The position of the mode tends to 1 as $a \uparrow \infty$.

B.   The value of the distribution function at $x = 1$ is $1-\frac{1}{e}$ for all values of a.

C.   The r-th moment is $\Gamma(1+\frac{r}{a})$.

D.   The minimum of n iid Weibull random variables is distributed as a constant times a Weibull random variable. Determine the constant and the parameter of the latter random variable.

E.   As $a \uparrow \infty$, the first moment of the Weibull distribution varies as $1-\frac{\gamma}{a}+o\,(\frac{1}{a})$ where $\gamma = 0.57722...$ is Euler's constant. Also, the variance $\sim \pi^2/6a^2$.

5.   Obtain a good uniform upper bound for the rejection constant in Vaduva's algorithm for gamma random variates when $a \leq 1$ which is based upon rejection from the Weibull density.

6.   **Algorithm GS (Ahrens and Dieter, 1974).** The following algorithm was proposed by Ahrens and Dieter (1974) for generating gamma (a) random variates when the parameter a is ≤ 1:

Rejection algorithm GS for gamma variates (Ahrens and Dieter, 1974)

[SET-UP]

$$b \leftarrow \frac{e+a}{e} , c \leftarrow \frac{1}{a}$$

[GENERATOR]

REPEAT

 Generate iid uniform [0,1] random variates U, W. Set $V \leftarrow bU$.

 IF $V \leq 1$

   THEN

     $X \leftarrow V^c$

     Accept $\leftarrow [W \leq e^{-X}]$

   ELSE

     $X \leftarrow -\log(c(b-V))$

     Accept $\leftarrow [W \leq X^{a-1}]$

UNTIL Accept

RETURN X

The algorithm is based upon the inequalities: $f(x) \leq \frac{a}{\Gamma(1+a)} x^{a-1}$ $(0 \leq x \leq 1)$ and $f(x) \leq \frac{a}{\Gamma(1+a)} e^{-x}$ $(x \geq 1)$. Show that the rejection constant is $\frac{e+a}{e\,\Gamma(1+a)}$. Show that the rejection constant approaches 1 as $a \downarrow 0$, that it is $1 + \frac{1}{e}$ at $a = 1$, and that it is uniformly bounded over $a \in (0,1]$ by a number not exceeding $\frac{3}{2}$. Show that in sampling from the composite dominating density, we have probability weights $\frac{e}{e+a}$ for ax^{a-1} $(0 < x \leq 1)$, and $\frac{a}{e+a}$ for e^{1-x} $(x \geq 1)$ respectively.

7. Show that the exponential function of the form ce^{-bx} $(x \geq t)$ of smallest integral dominating the gamma (a) density on $[t, \infty)$ (for $a > 1$, $t > 0$) has parameter b given by

$$b = \frac{t-a+\sqrt{(t-a)^2+4t}}{2t} .$$

Hint: show first that the ratio of the gamma density over e^{-bx} reaches a peak at $x = \frac{a-1}{1-b}$ (which is to the right of t when $b \geq 1 - \frac{a-1}{t}$). Then compute the optimal b and verify that $b \geq 1 - \frac{a-1}{t}$. Give the algorithm for the tail of the gamma density that corresponds to this optimal inequality. Show furthermore that as $t \uparrow \infty$, $b = 1 - \frac{a-1}{t} + o(\frac{1}{t})$, which proves that the choice of b in the text is asymptotically optimal (Dagpunar, 1978).

8. **Algorithm RGS (Best, 1983).** Algorithm GS (of exercise 6) can be optimized by two devices: first, the gamma density f with parameter a can be maximized by a function which is $x^{a-1}/\Gamma(a)$ on $[0,t]$ and $t^{a-1}e^{-x}/\Gamma(a)$ on $[t,\infty)$, where t is a breakpoint. In algorithm GS, the breakpoint was chosen as $t=1$. Secondly, a squeeze step can be added.

A. Show that the optimal breakpoint (in terms of minimization of the area under the dominating curve) is given by the solution of the transcendental equation $t=e^{-t}(1-a+t)$. (Best approximates this solution by $0.07+0.75\sqrt{1-a}$.)

B. Prove the inequalities $e^{-x}\geq(2-x)/(2+x)$ $(x\geq0)$ and $(1+x)^{-c}\geq1/(1+cx)$ $(x\geq0, 1\geq c\geq0)$. (These are needed for the squeeze steps.)

C. Show that the algorithm given below is valid:

Algorithm RGS for gamma variates (Best, 1983)

[SET-UP]

$t\leftarrow0.07+0.75\sqrt{1-a}$, $b\leftarrow1+\dfrac{e^{-t}a}{t}$, $c\leftarrow\dfrac{1}{a}$

[GENERATOR]

REPEAT

    Generate iid uniform [0,1] random variates U,W. Set $V\leftarrow bU$.

    IF $V\leq1$

        THEN

            $X\leftarrow tV^{c}$

            Accept $\leftarrow[W\leq\dfrac{2-X}{2+X}]$

            IF NOT Accept THEN Accept $\leftarrow[W\leq e^{-X}]$

        ELSE

            $X\leftarrow-\log(ct(b-V))$, $Y\leftarrow\dfrac{X}{t}$

            Accept $\leftarrow[W(a+Y-aY)\leq1]$

            IF NOT Accept THEN Accept $\leftarrow[W\leq Y^{a-1}]$

    UNTIL Accept

    RETURN X

9. **Algorithm G4PE (Schmeiser and Lal, 1980).** The graph of the gamma density can be covered by a collection of rectangles, triangles and exponential curves having the properties that (i) all parameters involved are easy to compute; and (ii) the total area under the dominating curve is uniformly bounded over $a\geq1$. One such proposal is due to Schmeiser and Lal (1980): define five breakpoints,

$$t_3 = a - 1$$
$$t_4 = t_3 + \sqrt{t_3}$$
$$t_5 = t_4(1 + 1/(t_4 - t_3))$$
$$t_2 = \max(0, t_3 - \sqrt{t_3})$$
$$t_1 = t_2(1 - 1/(t_3 - t_2))$$

where t_3 is the mode, and t_2, t_4 are the points of inflection of the gamma density. Furthermore, t_1, t_5 are the points at which the tangents of f at t_2 and t_4 cross the x-axis. The dominating curve has five pieces: an exponential tail on $(-\infty, t_1]$ with parameter $1 - t_3/t_1$ and touching f at t_1. On $[t_5, \infty)$ we have a similar exponential dominating curve with parameter $1 - t_3/t_5$. On $[t_1, t_2]$ and $[t_4, t_5]$, we have a linear dominating curve touching the density at the breakpoints. Finally, we have a constant piece of height $f(t_3)$ on $[t_2, t_4]$. All the strips except the two tall sections are partitioned into a rectangle (the largest rectangle fitted under the curve of f) and a leftover piece. This gives ten pieces, of which four are rectangles totally tucked under the gamma density. For the six remaining pieces, we can construct very simple linear acceptance steps.

A. Develop the algorithm.

B. Compute the area under the dominating curve, and determine its asymptotic value.

C. Determine the asymptotic probability that we need only one uniform random variate (the random variate needed to select one of the four rectangles is recycled). This is equivalent to computing the asymptotic area under the four rectangles.

D. With all the squeeze steps defined above in place, compute the asymptotic value of the expected number of evaluations of f.

Hint: obtain the values for an appropriately transformed normal density and use the convergence of the gamma density to the normal density.

10. **The t-distribution.** Show that when $G_{1/2}, G_{a/2}$ are independent gamma random variables, then $\sqrt{aG_{1/2}/G_{a/2}}$ is distributed as the absolute value of a random variable having the t distribution with a degrees of freedom. (Recall that the t density is

$$f(x) = \frac{\Gamma(\frac{a+1}{2})}{\sqrt{\pi a}\ \Gamma(\frac{a}{2})(1 + \frac{x^2}{2})^{\frac{a+1}{2}}} \quad .)$$

In particular, if $G, G*$ are iid gamma $(\frac{1}{2})$ random variables, then $\sqrt{G/G*}$ is Cauchy distributed.

11. **The Pearson VI distribution.** Show that G_a/G_b has density

$$f(x) = \frac{x^{a-1}}{B_{a,b}(1+x)^{b-1}} \quad (x \geq 0)$$

when G_a, G_b are independent gamma random variables with parameters a and b respectively. Here $B_{a,b} = \Gamma(a)\ \Gamma(b)/\Gamma(a+b)$ is a normalization constant. The density in question is the Pearson VI density. It is also called the beta density of the second kind with parameters a and b. b/a times the random variable in question is also called an F distributed random variable with $2a$ and $2b$ degrees of freedom.

4. THE BETA DENSITY.

4.1. Properties of the beta density.

We say that a random variable X on $[0,1]$ is **beta** (a,b) distributed when it has density

$$f(x) = \frac{x^{a-1}(1-x)^{b-1}}{B_{a,b}} \qquad (0 \le x \le 1)$$

where $a, b > 0$ are shape parameters, and

$$B_{a,b} = \int_0^1 x^{a-1}(1-x)^{b-1}\ dx = \frac{\Gamma(a)\Gamma(b)}{\Gamma(a+b)}$$

is a normalization constant. The density can take a number of interesting shapes:

1. When $0 < a, b < 1$, the density is U-shaped with infinite peaks at 0 and 1.

2. When $0 < a < 1 \le b$, the density is said to be J-shaped: it has an infinite peak at 0 and decreases monotonically to a positive constant (when $b = 1$) or to 0 (when $b > 1$).

3. When $a = 1 < b$, the density is bounded and decreases monotonically to 0.

4. When $a = b = 1$, we have the uniform $[0,1]$ density.

5. When $1 < a, b$, the density is unimodal, and takes the value 0 at the endpoints.

The fact that there are two shape parameters makes the beta density a solid candidate for illustrating the various techniques of nonuniform random variate generation. It is important for the design to understand the basic properties. For example, when $a, b > 1$, the mode is located at $\dfrac{a-1}{a+b-2}$. It is also quite trivial to show that for $r > -a$,

$$E(X^r) = \frac{B_{a+r,b}}{B_{a,b}}\ .$$

In particular, $E(X) = \dfrac{a}{a+b}$ and $Var(X) = \dfrac{ab}{(a+b)^2(a+b+1)}$. There are a number of relationships with other distributions. These are summarized in Theorem 4.1:

Theorem 4.1.

This is about the relationships between the beta (a,b) density and other densities.

A. Relationship with the **gamma** density: If G_a, G_b are independent gamma (a), gamma (b) random variables, then $\dfrac{G_a}{G_a+G_b}$ is beta (a,b) distributed.

B. Relationship with the **Pearson VI** (or β_2) density: If X is beta (a,b), then $Y = \dfrac{X}{1-X}$ is $\beta_2(a,b)$, that is, Y is a beta of the second kind, with density $f(x) = \dfrac{x^{a-1}}{B_{a,b}(1+x)^{a+b}}$ $\quad (x \geq 0)$.

C. Relationship with the **(Student's) t** distribution: If X is beta $(\frac{1}{2},\frac{a}{2})$, and S is a random sign, then $S\sqrt{\dfrac{aX}{1-X}}$ is t-distributed with a degrees of freedom, i.e. it has density

$$ f(x) = \frac{\Gamma(\dfrac{a+1}{2})}{\sqrt{\pi a}\;\Gamma(\dfrac{a}{2})(1+\dfrac{x^2}{a})^{\frac{a+1}{2}}} . $$

By the previous property, note that $\sqrt{aY}$ is t-distributed with parameter a when Y is $\beta_2(a,b)$. Furthermore, if X denotes a beta (a,a) random variable, and T denotes a t random variable with $2a$ degrees of freedom, then we have the following equality in distribution: $X = \dfrac{1}{2}+\dfrac{1}{2}\dfrac{T}{\sqrt{2b+T^2}}$, or $T = \dfrac{\sqrt{2a}\,(2X-1)}{2\sqrt{X-X^2}}$. In particular, when U is uniform on $[0,1]$, then $\dfrac{\sqrt{2}(U-\dfrac{1}{2})}{\sqrt{U-U^2}}$ is t with 2 degrees of freedom.

D. Relationship with the **F (Snedecor)** distribution: when X is beta (a,b), then $\dfrac{bX}{a(1-X)}$ is F-distributed with a and b degrees of freedom, i.e. it has density $\dfrac{a}{b}f(\dfrac{ax}{b})$ $(x>0)$, where f is the $\beta_2(\dfrac{a}{2},\dfrac{b}{2})$ density.

E. Relationship with the **Cauchy** density: when X is beta $(\dfrac{1}{2},\dfrac{1}{2})$ distributed (this is called the arc sine distribution), then $\sqrt{\dfrac{X}{1-X}}$ is distributed as the absolute value of a Cauchy random variable.

Proof of Theorem 4.1.

All the properties can be obtained by applying the methods for computing densities of transformed random variables explained for example in section I.4.1.
∎

We should also mention the important connection between the beta distribution and order statistics. When $0 < U_{(1)} < \cdots < U_{(n)}$ are the order statistics of a uniform [0,1] random sample, then $U_{(k)}$ is beta $(k, n-k+1)$ distributed. See section I.4.3.

4.2. Overview of beta generators.

Beta variates can be generated by exploiting special properties of the distribution. The order statistics method, applicable only when both a and b are integer, proceeds as follows:

Order statistics method for beta variates

Generate $a+b-1$ iid uniform [0,1] random variates.
Find the a-th order statistic X (a-th smallest) among these variates.
RETURN X

This method, mentioned as early as 1963 by Fox, requires time at least proportional to $a+b-1$. If standard sorting routines are used to obtain the a-th smallest element, then the time complexity is even worse, possibly $\Omega((a+b-1)\log(a+b-1))$. There are obvious improvements: it is wasteful to sort a sample just to obtain the a-th smallest number. First of all, via linear selection algorithms we can find the a-th smallest in worst case time $O(a+b-1)$ (see e.g. Blum, Floyd, Pratt, Rivest and Tarjan (1973) or Schonhage, Paterson and Pippenger (1976)). But in fact, there is no need to generate the entire sample. The uniform sample can be generated directly from left to right or right to left, as shown in section V.3. This would reduce the time to $O(\min(a,b))$. Except in special applications, not requiring non-integer or large parameters, this method is not recommended.

When property A of Theorem 4.1 is used, the time needed for one beta variate is about equal to the time required to generate two gamma variates. This method is usually very competitive because there are many fast gamma generators. In any case, if the gamma generator is uniformly fast, so will be the beta generator. Formally we have:

Beta variates via gamma variates

Generate two independent gamma random variates, G_a and G_b .

RETURN $\dfrac{G_a}{G_a + G_b}$

Roughly speaking, we will be able to improve over this generator by at most 50%. There is no need to discuss beta variate generators which are not time efficient. A survey of pre-1972 methods can be found in Arnason (1972). None of the methods given there has uniformly bounded expected time. Among the competitive approaches, we mention:

A. Standard rejection methods. For example, we have:
   Rejection from the Burr XII density (Cheng, 1978).
   Rejection from the normal density (Ahrens and Dieter, 1974).
   Rejection from polynomial densities (Atkinson and Whittaker, 1976, 1979; Atkinson, 1979).
   Rejection and composition with triangles, rectangles, and exponential curves (Schmeiser and Babu, 1980).

The best of these methods will be developed below. In particular, we will highlight Cheng's uniformly fast algorithms. The algorithm of Schmeiser and Babu (1980), which is uniformly fast over $a,b \geq 1$, is discussed in section VII.2.6.

B. Forsythe's method, as applied for example by Atkinson and Pearce (1976). This method requires a lot of code and the set-up time is considerable. In comparison with this investment, the speed obtainable via this approach is disappointing.

C. Johnk's method (Johnk, 1964) and its modifications. This too should be considered as a method based upon special properties of the beta density. The expected time is not uniformly bounded in the parameters. It should be used only when both parameters are less than one. See section IX.3.5.

D. Universal algorithms. The beta density is unimodal when both parameters are at least one, and it is monotone when one parameter is less than one and one is at least equal to one. Thus, the universal methods of section VII.3.2 are applicable. At the very least, the inequalities derived in that section can be used to design good (albeit not superb) bounds for the beta density. In any case, the expected time is provably uniform over all parameters a,b with $\max(a,b) \geq 1$.

E. Strip table methods, as developed in section VIII.2.2. We will study below how many strips should be selected as a function of a and b in order to have uniformly bounded expected generation times.

The bottom line is that the choice of a method depends upon the user: if he is not willing to invest a lot of time, he should use the ratio of gamma variates. If he does not mind coding short programs, and a and/or b vary frequently, one of the rejection methods based upon analysis of the beta density or upon universal inequalities can be used. The method of Cheng is very robust. For special cases, such as symmetric beta densities, rejection from the normal density is very competitive. If the user does not foresee frequent changes in a and b, a strip table method or the algorithm of Schmeiser and Babu (1980) are recommended. Finally, when both parameters are smaller than one, it is possible to use rejection from polynomial densities or to apply Johnk's method.

4.3. The symmetric beta density.

In this section, we will take a close look at one of the simplest special cases, the symmetric beta density with parameter a:

$$f(x) = \frac{\Gamma(2a)}{\Gamma^2(a)}(x(1-x))^{a-1} = C(x(1-x))^{a-1} \quad (0 \le x \le 1) .$$

For large values of a, this density is quite close to the normal density. To see this, consider $y = x - \frac{1}{2}$, and

$$\log(f(x)) = \log(C) + (a-1)\log(1+2y) + (a-1)\log(1-2y) - (a-1)\log 4$$
$$= \log(C) - (a-1)\log 4 + (a-1)\log(1-4y^2) .$$

The last term on the right hand side is not greater than $-4(a-1)y^2$, and it is at least equal to $-4(a-1)y^2 - 16(a-1)y^4/(1-4y^2)$. Thus, $\log(f(\frac{1}{2} + \frac{x}{\sqrt{8(a-1)}}))$ tends to $-\log(\sqrt{2\pi}) - \frac{x^2}{2}$ as $a \to \infty$ for all $x \in R$. Here we used Stirling's formula to prove that $\log(C) - (a-1)\log 4$ tends to $-\log(\sqrt{2\pi})$. Thus, if X is beta (a,a), then the density of $\sqrt{8(a-1)}(X - \frac{1}{2})$ tends to the standard normal density as $a \to \infty$. The only hope for an asymptotically optimal rejection constant in a rejection algorithm is to use a dominating density which is either normal or tends pointwise to the normal density as $a \to \infty$. The question is whether we should use the normalization suggested by the limit theorem stated above. It turns out that the best rejection constant is obtained not by taking $8(a-1)$ in the formula for the normal density, but $8(a - \frac{1}{2})$. We state the algorithm first, then announce its properties in a theorem:

Symmetric beta generator via rejection from the normal density

[NOTE: $b = (a-1)\log(1+\dfrac{1}{2a-2}) - \dfrac{1}{2}$.]

[GENERATOR]

REPEAT

    REPEAT

        Generate a normal random variate N and an exponential random variate E.

$$X \leftarrow \frac{1}{2} + \frac{N}{\sqrt{8a-4}}, Z \leftarrow N^2$$

        UNTIL $Z < 2a-1$ (now, $X \in [0,1]$)

        Accept $\leftarrow [E + \dfrac{Z}{2} - \dfrac{(a-1)Z}{2a-1-Z} + b \geq 0]$

        IF NOT Accept THEN Accept $\leftarrow [E + \dfrac{Z}{2} + (a-1)\log(1 - \dfrac{Z}{2a-1}) + b \geq 0]$

    UNTIL Accept

    RETURN X

Theorem 4.2.

Let f be the beta (a) density with parameter $a \geq 1$. Then let $\sigma > 0$ be a constant and let c_σ be the smallest constant such that for all x,

$$f(x) \leq c_\sigma \frac{1}{\sqrt{2\pi}\sigma} e^{-\frac{(x-\frac{1}{2})^2}{2\sigma^2}}.$$

Then c_σ is minimal for $\sigma^2 = \dfrac{1}{8a-4}$, and the minimal value is

$$c_\sigma = \left(\frac{8(a-1)}{4e(8a-4)}\right)^{a-1} \frac{\sqrt{2\pi}}{\sqrt{8a-4}B_{a,a}} e^{\frac{8a-4}{8}}.$$

In the rejection algorithm shown above, the rejection constant is c_σ. The rejection constant is uniformly bounded for $a \in [1,\infty)$: selected values are $\sqrt{\dfrac{\pi e}{6}}$ at $a = 2$, $\sqrt{36\pi e}$ at $a = 3$. We have

$$\lim_{a \to \infty} c_\sigma = 1.$$

and in fact, $c_\sigma \leq e^{\frac{1}{24a} + \frac{1}{2a-1}}$.

Proof of Theorem 4.2.

Let us write $g(x)$ for the normal density with mean $\frac{1}{2}$ and variance σ^2. We first determine the supremum of f/g by setting the derivative of $\log(\frac{f}{g})$ equal to zero. This yields the equation

$$(x-\frac{1}{2})(\sigma^{-2}-\frac{2(a-1)}{x(1-x)}) = 0 \ .$$

One can easily see from this that f/g has a local minimum at $x=\frac{1}{2}$ and two local maxima symmetrically located on either side of $\frac{1}{2}$ at $\frac{1}{2}\pm\frac{1}{2}\sqrt{1-8(a-1)\sigma^2}$. The value of f/g at the maxima is

$$c_\sigma = \left\{\frac{8(a-1)\sigma^2}{4e}\right\}^{a-1} \frac{\sqrt{2\pi}\sigma}{B_{a,a}} e^{\frac{1}{8\sigma^2}} \ .$$

This depends upon σ as follows: $\sigma^{2a-1}e^{\frac{1}{8\sigma^2}}$. This has a unique minimum at $\sigma=1/\sqrt{8a-4}$. Resubstitution of this value gives

$$c_\sigma = \left\{\frac{a-1}{4a-2}\right\}^{a-1} \frac{\sqrt{2\pi}}{\sqrt{8a-4}B_{a,a}} e^{\frac{1}{2}} \ .$$

By well-known bounds on the gamma function (Whittaker ans Watson, 1927, p. 253), we have

$$\frac{1}{B_{a,a}} \le 4^{a-\frac{1}{2}}\sqrt{\frac{a}{\pi}}e^{\frac{1}{24a}} \ ,$$

$$\frac{1}{B_{a,a}} \sim 4^{a-\frac{1}{2}}\sqrt{\frac{a}{\pi}}$$

as $a\to\infty$. Thus,

$$c_\sigma \le \left\{\frac{a-1}{4a-2}\right\}^{a-1}\frac{\sqrt{2\pi}}{\sqrt{8a-4}}4^{a-\frac{1}{2}}\sqrt{\frac{a}{\pi}}e^{\frac{1}{24a}}e^{\frac{1}{2}}$$

$$= \sqrt{ae/(a-\frac{1}{2})}e^{\frac{1}{24a}}(1-\frac{1}{2a-1})^{a-1}$$

$$\le \sqrt{ae/(a-\frac{1}{2})}e^{\frac{1}{24a}}e^{-\frac{a-1}{2a-1}}$$

$$= \sqrt{1+\frac{1}{2a-1}}e^{\frac{1}{24a}+\frac{1}{4a-2}}$$

$$\le e^{\frac{1}{24a}+\frac{1}{2a-1}} \ . \ \blacksquare$$

The algorithm shown above is applicable for all $a \geq 1$. For large values of a, we need about one normal random variate per beta random variate, and the probability that the long acceptance condition has to be verified at all tends to 0 as $a \to \infty$ (exercise 4.1). There is another school of thought, in which normal random variates are avoided altogether, and the algorithms are phrased in terms of uniform random variates. After all, normal random variates are also built from uniform random variates. In the search for a good dominating curve, help can be obtained from other symmetric unimodal long-tailed distributions. There are two examples that have been explicitly mentioned in the literature, one by Best (1978), and one by Ulrich (1984):

Theorem 4.3.

When Y is a t distributed random variable with parameter $2a$, then $X \leftarrow \dfrac{1}{2} + \dfrac{1}{2} \dfrac{Y}{\sqrt{2a + Y^2}}$ is beta (a, a) distributed (Best, 1978).

When U, V are independent uniform $[0,1]$ random variables, then

$$X \leftarrow \frac{1}{2} + \frac{1}{2} \sin(2\pi V) \sqrt{1 - U^{\frac{2}{2a-1}}}$$

is beta (a, a) distributed (Ulrich, 1984).

Proof of Theorem 4.3.

The proof is left as an exercise on transformations of random variables. ∎

If we follow Best, then we need a fast t generator, and we refer to section IX.5 for such algorithms. Ulrich's suggestion is intriguing because it is reminiscent of the polar method. Recall that when X, Y is uniformly distributed in the unit circle with $S = X^2 + Y^2$, then $\left(\dfrac{X}{\sqrt{S}}, \dfrac{Y}{\sqrt{S}} \right)$ and S are independent, and S is uniformly distributed on $[0,1]$. Also, switching to polar coordinates (R, Θ), we see that $XY/S = \cos(\Theta)\sin(\Theta) = 2\sin(2\Theta)$. Thus, since 2Θ is uniformly distributed on $[0, 4\pi]$, we see that the random variable

$$\frac{1}{2} + \frac{XY}{S} \sqrt{1 - S^{\frac{2}{2a-1}}}$$

has a beta (a, a) distribution. We summarize:

Ulrich's polar method for symmetric beta random variates

REPEAT

      Generate U uniformly on $[0,1]$ and V uniformly on $[-1,1]$.

      $S \leftarrow U^2 + V^2$

UNTIL $S \leq 1$

RETURN $X \leftarrow \dfrac{1}{2} + \dfrac{UV}{S}\sqrt{1-S^{\frac{2}{2a-1}}}$

It should be stressed that Ulrich's method is valid for all $a > 0$, provided that for the case $a = 1/2$, we obtain X as $1/2 + UV/S$, that is, X is distributed as a linearly transformed arc sin random variable. Despite the power and the square root needed in the algorithm for general a, its elegance and generality make it a formidable candidate for inclusion in computer libraries.

4.4. Uniformly fast rejection algorithms.

The beta (a,b) density has two shape parameters. If we are to construct a uniformly fast rejection algorithm, it seems unlikely that we can just consider rejection from a density with no shape parameter such as the normal density. This is generally speaking only feasible when there is one shape parameter as in the case of the gamma or symmetric beta families. The trick will then be to find a flexible family of easy dominating densities. In his work, Cheng has repeatedly used the Burr XII density with one scale parameter and one shape parameter with a great deal of success. This density is constructed as follows. If U is uniformly distributed on $[0,1]$, then $\dfrac{U}{1-U}$ has density $(1+x)^{-2}$ on $[0,\infty)$. For $\mu, \lambda > 0$, the density of

$$(\mu \frac{U}{1-U})^{\frac{1}{\lambda}}$$

is

$$g(x) = \frac{\lambda \mu x^{\lambda - 1}}{(\mu + x^\lambda)^2} \quad (x > 0) .$$

This is an infinite-tailed density, of little direct use for the beta density. Fortunately, beta and β_2 random variables are closely related (see Theorem 4.1), so that we need only consider the infinite-tailed β_2 density with parameters (a,b):

$$f(x) = \frac{x^{a-1}}{B_{a,b}(1+x)^{a+b}} \quad (x \geq 0) .$$

The values of μ and λ suggested by Cheng (1978) for good rejection constants are

$$\mu = (\frac{a}{b})^{\lambda} ;$$

$$\lambda = \begin{cases} \min(a,b) & (\min(a,b) \le 1) \\ \sqrt{\dfrac{2ab - (a+b)}{a+b-2}} & (\min(a,b) > 1) \end{cases} .$$

With these choices, it is not difficult to verify that f/g is maximal at $x = a/b$, and that $f \le cg$ where

$$c = \frac{4a^a b^b}{\lambda B_{a,b} (a+b)^{a+b}} .$$

Note that $cg(x)/f(x)$ can be simplified quite a bit. The unadorned algorithm is:

Cheng's rejection algorithm BA for beta random variates (Cheng, 1978)

[SET-UP]

$s \leftarrow a + b$

IF $\min(a,b) \le 1$

      THEN $\lambda \leftarrow \min(a,b)$

      ELSE $\lambda \leftarrow \sqrt{\dfrac{2ab-s}{s-2}}$

$u \leftarrow a + \lambda$

[GENERATOR]

REPEAT

      Generate two iid uniform $[0,1]$ random variates U_1, U_2.

$$V \leftarrow \frac{1}{\lambda} \frac{U_1}{1-U_1}, \quad Y \leftarrow ae^V$$

      UNTIL $s \log(\dfrac{s}{b+Y}) + uV - \log(4) \ge \log(U_1^2 U_2)$

      RETURN $X \leftarrow \dfrac{Y}{b+Y}$

The fundamental property of Cheng's algorithm is that

$$\sup_{a,b>0} c = 4 ; \quad \sup_{a,b \ge 1} c = \frac{4}{e} \approx 1.47 .$$

For fixed a, c is minimal when $b = a$ and increases when $b \downarrow 0$ or $b \uparrow \infty$. The details of the proofs of the various statements about this algorithm are left as an exercise. There exists an improved version of the algorithm for the case that both parameters are greater than 1 which is based upon the squeeze method (Cheng's algorithm BB). Cheng's algorithm is slowest when $\min(a,b) < 1$. In that region of

the parameter space, it is worthwhile to design special algorithms that may or
may not be uniformly fast over the entire parameter space.

4.5. Generators when min(a,b)≤1.

Cheng's algorithm BA is robust and can be used for all values of a, b. However, when both a, b are smaller than one, and $a + b \le 1.5$, Johnk's method is
typically more efficient. When $\min(a, b)$ is very small, and $\max(a, b)$ is rather
large, neither Johnk's method nor algorithm BA are particularly fast. To fill this
gap, several algorithms were proposed by Atkinson and Whittaker (1976, 1979)
and Atkinson (1979). In addition, Cheng (1977) developed an algorithm of his
own, called algorithm BC.

Atkinson and Whittaker (1976,1979) split $[0,1]$ into $[0,t]$ and $[t,1]$, and construct a dominating curve for use in the rejection method based upon the inequalities:

$$x^{a-1}(1-x)^{b-1} \le \begin{cases} x^{a-1}(1-t)^{b-1} & (x \le t) \\ t^{a-1}(1-x)^{b-1} & (x > t) \end{cases}.$$

The areas under the two pieces of the dominating curve are, respectively,
$(1-t)^{b-1}\dfrac{t^a}{a}$ and $t^{a-1}\dfrac{(1-t)^b}{b}$. Thus, the following rejection algorithm can be
used:

First algorithm of Atkinson and Whittaker (1976, 1979)

[SET-UP]

Choose $t \in [0,1]$.

$$p \leftarrow \frac{bt}{bt + a(1-t)}$$

[GENERATOR]

REPEAT

      Generate a uniform $[0,1]$ random variate U and an exponential random variate E.

      IF $U \leq p$

          THEN

$$X \leftarrow t \, (\frac{U}{p})^{\frac{1}{a}}$$

$$\text{Accept} \leftarrow [(1-b)\log(\frac{1-X}{1-t}) \leq E]$$

          ELSE

$$X \leftarrow 1-(1-t)(\frac{1-U}{1-p})^{\frac{1}{b}}$$

$$\text{Accept} \leftarrow [(1-a)\log(\frac{X}{t}) \leq E]$$

UNTIL Accept

RETURN X

Despite its simplicity, this algorithm performs remarkably well when both parameters are less than one, although for $a+b < 1$, Johnk's algorithm is still to be preferred. The explanation for this is given in the next theorem. At the same time, the best choice for t is derived in the theorem.

Theorem 4.4.

Assume that $a \leq 1, b \leq 1$. The expected number of iterations in Johnk's algorithm is

$$c = \frac{\Gamma(a+b+1)}{\Gamma(a+1)\Gamma(b+1)} .$$

The expected number of iterations ($E(N)$) in the first algorithm of Atkinson and Whittaker is

$$c\frac{bt+a(1-t)}{(a+b)t^{1-a}(1-t)^{1-b}} .$$

When $a+b \leq 1$, then for all values of t, $E(N) \geq c$. In any case, $E(N)$ is minimized for the value

$$t_{opt} = \frac{\sqrt{a(1-a)}}{\sqrt{a(1-a)}+\sqrt{b(1-b)}} .$$

With $t = t_{opt}$, we have $E(N) < c$ whenever $a+b > 1$. For $a+b > 1$, $t = \frac{1}{2}$, it is also true that $E(N) < c$.

Finally, $E(N)$ is uniformly bounded over $a, b \leq 1$ when $t = \frac{1}{2}$ (and it is therefore uniformly bounded when $t = t_{opt}$).

Proof of Theorem 4.4.

We begin with the fundamental inequality:

$$x^{a-1}(1-x)^{b-1} \leq \begin{cases} x^{a-1}(1-t)^{b-1} & (x \leq t) \\ t^{a-1}(1-x)^{b-1} & (x > t) \end{cases} .$$

The area under the top curve is $(1-t)^{b-1}\frac{t^a}{a} + t^{a-1}\frac{(1-t)^b}{b}$. The area under the bottom curve is of course $\Gamma(a)\Gamma(b)/\Gamma(a+b)$. The ratio gives us the expression for $E(N)$. $E(N)$ is minimal for the solution t of

$$(1-t)^2 a(a-1)-t^2 b(b-1) = 0 ,$$

which gives us $t = t_{opt}$. For the performance of Johnk's algorithm, we refer to Theorem 3.4. To compare performances for $a+b \leq 1$, we have to show that for all t,

$$(\frac{1}{t})^a (\frac{1}{1-t})^b \leq \frac{1}{a+b}(\frac{b}{1-t}+\frac{a}{t}) .$$

By the arithmetic-geometric mean inequality, the left hand side is in fact not greater than

$$\left[\frac{1}{a+b}(\frac{b}{1-t}+\frac{a}{t})\right]^{a+b}$$

$$\le \frac{1}{a+b}(\frac{b}{1-t}+\frac{a}{t})$$

because $a+b \le 1$, and the argument of the power is a number at least equal to 1. When $a+b>1$, it is easy to check that $E(N)<c$ for $t=\frac{1}{2}$. The statement about the uniform boundedness of $E(N)$ when $t=\frac{1}{2}$ follows simply from

$$E(N) = 2^{1-a-b} c$$

and the fact that c is uniformly bounded over $a,b \le 1$. ∎

Generally speaking, the first algorithm of Atkinson and Whittaker should be used instead of Johnk's when $a,b \le 1$ and $a+b \ge 1$. The computation of t_{opt}, which involves one square root, is only justified when many random variates are needed for the same values of a and b. Otherwise, one should choose $t=\frac{1}{2}$.

When $a \le 1$ and $b \ge 1$, the performance of the first algorithm of Atkinson and Whittaker deteriorates with increasing values of b: for fixed $a<1$, $\lim_{b \to \infty} E(N)=\infty$. The inequalities used to develop the algorithm are altered slightly:

$$x^{a-1}(1-x)^{b-1} \le \begin{cases} x^{a-1} & (x \le t) \\ t^{a-1}(1-x)^{b-1} & (x>t) \end{cases}.$$

The areas under the two pieces of the dominating curve are, respectively, $\frac{t^a}{a}$ and $t^{a-1}\frac{(1-t)^b}{b}$. The following rejection algorithm can be used:

Second algorithm of Atkinson and Whittaker (1976, 1979)

[SET-UP]

Choose $t \in [0,1]$.

$$p \leftarrow \frac{bt}{bt + a(1-t)^b}$$

[GENERATOR]

REPEAT

     Generate a uniform [0,1] random variate U and an exponential random variate E.

     IF $U \leq p$

         THEN

$$X \leftarrow t\,(\frac{U}{p})^{\frac{1}{a}}$$

             Accept $\leftarrow [(1-b)\log(1-X) \leq E\,]$

         ELSE

$$X \leftarrow 1 - (1-t)\,(\frac{1-U}{1-p})^{\frac{1}{b}}$$

             Accept $\leftarrow [(1-a)\log(\frac{X}{t}) \leq E\,]$

UNTIL Accept

RETURN X

Simple calculations show that

$$E(N) = c\,\frac{bt^a + a(1-t)^b\,t^{a-1}}{a+b}$$

where c is the expected number of iterations in Johnk's algorithm (see Theorems 3.4 and 4.4). The optimum value of t is the solution of

$$bt + (a-1)(1-t)^b - bt(1-t)^{b-1} = 0\,.$$

Although this can be solved numerically, most of the time we can not afford a numerical solution just to generate one random variate. We have, however, the following reassuring performance analysis for a choice for t suggested by Atkinson and Whittaker (1976):

Theorem 4.5.

For the second algorithm of Atkinson and Whittaker with $t = \dfrac{1-a}{b+1-a}$,

$$\sup_{a \leq 1, b \geq 1} E(N) < \infty ,$$

$$\lim_{b \to \infty} E(N) = \infty \quad \text{(all } a > 1).$$

4.6. Exercises.

1. For the symmetric beta algorithm studied in Theorem 4.2, show that the quick acceptance step is valid, and that with the quick acceptance step in place, the expected number of evaluations of the full acceptance step tends to 0 as $a \to \infty$.

2. Prove Ulrich's part of Theorem 4.3.

3. Let X be a $\beta_2(a,b)$ random variable. Show that $\dfrac{1}{Y}$ is $\beta_2(b,a)$, and that
$$E(Y) = \frac{a}{b-1} \ (b > 1), \text{ and } Var(Y) = \frac{a(a+b-1)}{(b-1)^2(b-2)} \ (b > 2).$$

4. In the table below, some densities are listed with one parameter $a > 0$ or two parameters $a, b > 0$. Let c be the shorthand notation for $1/B(a,b)$. Show for each density how a random variate can be generated by a suitable transformation of a beta random variate.

$2cx^{2a-1}(1-x^2)^{b-1} \quad (0 \leq x \leq 1)$
$2c\sin^{2a-1}(x)\cos^{2b-1}(x) \quad (0 \leq x \leq \frac{\pi}{2})$
$\dfrac{cx^{a-1}}{(1+x)^{a+b}} \quad (x \geq 0)$
$\dfrac{2cx^{2a-1}}{(1+x^2)^{a+b}} \quad (x \geq 0)$
$c\dfrac{x^{a-1}+x^{b-1}}{(1+x)^{a+b}} \quad (0 \leq x \leq 1)$
$\dfrac{(1-x)^{a-1}}{2^{2a-1}B(a,a)\sqrt{x}} \quad (0 \leq x \leq 1)$
$\dfrac{(1-x^2)^{a-1}}{2^{2a-2}B(a,a)} \quad (0 \leq x \leq 1)$

5. Prove Theorem 4.5.

6. **Grassia's distribution.** Grassia (1977) introduced a distribution which is close to the beta distribution, and can be considered to be as flexible, if not more flexible, than the beta distribution. When X is gamma (a,b), then e^{-X} is Grassia I, and $1-e^{-X}$ is Grassia II. Prove that for every possible combination of skewness and kurtosis achievable by the beta density, there

exists a Grassia distribution with the same skewness and kurtosis (Tadi-kamalla, 1981).

7. A continuation of exercise 6. Use the Grassia distribution to obtain an efficient algorithm for the generation of random variates with density

$$f(x) = \frac{8a^2 x^{a-1} \log(\frac{1}{x})}{\pi^2 (1-x^{2a})} \qquad (0 < x < 1) ,$$

where $a > 0$ is a parameter.

5. THE t DISTRIBUTION.

5.1. Overview.

The **t distribution** plays a key role in statistics. The distribution has a symmetric density with one shape parameter $a > 0$:

$$f(x) = \frac{\Gamma(\frac{a+1}{2})}{\sqrt{\pi a}\ \Gamma(\frac{a}{2})(1+\frac{x^2}{a})^{\frac{a+1}{2}}}$$

This is a bell-shaped density which can be dealt with in a number of ways. As special members, we note the **Cauchy density** ($a = 1$), and the t_3 density ($a = 3$). When a is integer-valued, it is sometimes referred to as the number of degrees of freedom of the distribution. Random variate generation methods for this distribution include:

1. The inversion method. Explicit forms of the distribution function are only available in special cases: for the Cauchy density ($a = 1$), see section II.2.1. For the t_2 density ($a = 2$), see Theorem IX.3.3 in section IX.3.3. For the t_3 density ($a = 3$), see exercise II.2.4. In general, the inversion method is not competitive because the distribution function is only available as an integral, and not as a simple explicit function of its argument.

2. Transformation of gamma variates. When N is a normal random variate, and $G_{a/2}$ is a gamma $(\frac{a}{2})$ random variate independent of N,

$$\frac{\sqrt{2a}\ N}{\sqrt{G_{a/2}}}$$

is t_a distributed. Equivalently, if $G_{1/2}, G_{a/2}$ are independent gamma random variables, then

$$S \sqrt{a}\ \sqrt{\frac{G_{1/2}}{G_{a/2}}}$$

is t_a distributed where S is a random sign. See example I.4.6 for the derivation of this property. Somewhat less useful, but still noteworthy, is the property that if $G_{a/2}, G*_{a/2}$ are iid gamma random variates, then

$$\frac{\sqrt{a}}{2} \frac{G_{a/2} - G*_{a/2}}{\sqrt{G_{a/2} G*_{a/2}}}$$

is t_a distributed (Cacoullos, 1965).

3. Transformation of a symmetric beta random variate. It is known that if X is symmetric beta $(\frac{a}{2}, \frac{a}{2})$, then

$$\sqrt{a} \frac{X - \frac{1}{2}}{\sqrt{X(1-X)}}$$

is t_a distributed. Symmetric beta random variate generation was studied in section IX.4.3. The combination of a normal rejection method for symmetric random variates, and the present transformation was proposed by Marsaglia (1980).

4. Transformation of an F random variate. When S is a random sign and X is $F(1,a)$ distributed, then $S\sqrt{X}$ is t_a distributed (see exercise I.4.6). Also, when X is symmetric F with parameters a and a, then

$$\frac{\sqrt{a}}{2} \frac{1-X}{\sqrt{X}}$$

is t_a distributed.

5. The ratio-of-uniforms method. See section IV.7.2.

6. The ordinary rejection method. Since the t density cannot be dominated by densities with exponentially decreasing tails, one needs to find a polynomially decreasing dominating function. Typical candidates for the dominating curve include the Cauchy density and the t_3 density. The corresponding algorithms are quite short, and do not rely on fast normal or exponential generators. See below for more details.

7. The composition/rejection method, similar to the method used for normal random variate generation. The algorithms are generally speaking longer, more design constants need to be computed for each choice of a, and the speed is usually a bit better than for the ordinary rejection method. See for example Kinderman, Monahan and Ramage (1977) for such methods.

8. The acceptance-complement method (Stadlober, 1981).

9. Table methods.

One of the transformations of gamma or beta random variates is recommended if one wants to save time writing programs. It is rare that additional speed is required beyond these transformation methods. For direct methods, good speed can be obtained with the ratio-of-uniforms method and with the ordinary rejection methods. Typically, the expected time per random variate is uniformly bounded over a subset of the parameter range, such as $[1,\infty)$ or $[3,\infty)$. Not unexpectedly, the small values of a are the troublemakers, because these densities decrease as $x^{-(a+1)}$, so that no fixed exponent polynomial dominating density exists. The large values of a give least problems because it is easy to see that for every x,

$$\lim_{a \to \infty} f(x) = \frac{1}{\sqrt{2\pi}} e^{-\frac{x^2}{2}}.$$

The problem of small a is not important enough to warrant a special section. See however the exercises.

5.2. Ordinary rejection methods.

Let us first start with the development of simple upper bounds for f. For example, when $a \geq 1$, the following inequality is trivially true:

$$\frac{1}{(1+\frac{x^2}{a})^{\frac{a+1}{2}}} \leq \frac{1}{1+\frac{a+1}{2a}x^2}.$$

The top bound is proportional to the density of $\sqrt{\dfrac{2a}{a+1}}C$ where C is a Cauchy random variate. If we want to verify just how good this inequality is, we note that the area under the dominating curve is $\pi\sqrt{\dfrac{2a}{a+1}}$. The area under the curve on the left hand side of the inequality is $\dfrac{\sqrt{\pi a}\ \Gamma(\frac{a}{2})}{\Gamma(\frac{a+1}{2})}$. By the convergence to the normal density, we deduce without computations that this quantity tends to $\sqrt{2\pi}$. Thus, the ratio of the areas, our rejection constant, tends to $\sqrt{\pi}$ as $a \to \infty$. The fit is not very good, except perhaps for a close to 1: for $a=1$, the rejection constant is obviously 1. The details of the rejection algorithm are left to the reader.

Consider next rejection from the t_3 density

$$g(x) = \frac{1}{\sqrt{3}B(\frac{1}{2},\frac{3}{2})(1+\frac{x^2}{3})^2}.$$

Best (1978) has shown the following:

Theorem 5.1.

Let f be the t_a density with $a \geq 3$, and let g be the t_3 density. Then:

$$f(x) \leq cg(x)$$

where

$$c = \frac{8\pi\sqrt{3}}{9\sqrt{a}\, B\left(\frac{1}{2},\frac{a}{2}\right)\left(1+\frac{1}{a}\right)^{\frac{a+1}{2}}}.$$

Also, if

$$T(x) = \frac{f(x)}{cg(x)} = \frac{9}{16}\frac{\left(1+\frac{x^2}{3}\right)^2}{\left(\dfrac{1+\dfrac{x^2}{a}}{1+\dfrac{1}{a}}\right)^{\frac{a+1}{2}}},$$

then

$$T(x) \geq \frac{9}{16}e^{\frac{1}{2}-\frac{x^2}{2}}\left(1+\frac{x^2}{3}\right)^2.$$

Finally,

$$c \leq \sqrt{\frac{32\pi}{27e}}\sqrt{\frac{a}{a+1}}\,e^{\frac{1}{6a+1}}$$

and

$$\lim_{a \to \infty} c = \sqrt{\frac{32\pi}{27e}}.$$

Proof of Theorem 5.1.

Verify that f/g is maximal for $x = \pm 1$. The lower bound for $T(x)$ follows from the inequality

$$\left|\frac{1+\dfrac{x^2}{a}}{1+\dfrac{1}{a}}\right|^{a+1} = \left(1+\frac{x^2-1}{a+1}\right)^{a+1} \leq e^{1-x^2}.$$

Finally, the statement about c follows from Stirling's formula and bounds related to Stirling's formula. For example, the upper bound is obtained as follows:

$$c = \frac{8\pi\sqrt{3}}{9\sqrt{a}\,B\left(\frac{1}{2},\frac{a}{2}\right)\left(1+\frac{1}{a}\right)^{\frac{a+1}{2}}}$$

$$\leq \frac{8\sqrt{3\pi}}{9\sqrt{a}}\left(\frac{a+1}{2e}\right)^{\frac{a+1}{2}}\left(\frac{2e}{a}\right)^{\frac{a}{2}}\sqrt{\frac{a}{a+1}}\,e^{\frac{1}{6(a+1)}}\left(\frac{a}{a+1}\right)^{\frac{a+1}{2}}$$

$$= \frac{8\sqrt{3\pi}}{9\sqrt{2e}}\sqrt{\frac{a}{a+1}}\,e^{\frac{1}{6(a+1)}}$$

$$\to \sqrt{\frac{32\pi}{27e}}\ .$$

A similar lower bound is valid, which establishes the asymptotic result. ∎

The fit with the t_3 dominating density is much better than with the Cauchy density. Also, recalling the ratio-of-uniforms method for generating t_3 random variates in a form convenient to us (see section IV.7.2),

t3 generator based upon the ratio-of-uniforms method

REPEAT

      Generate iid uniform [0,1] random variates U,V. Set $V \leftarrow V - \frac{1}{2}$.

UNTIL $U^2 + V^2 \leq U$

RETURN $X \leftarrow \sqrt{3}\dfrac{V}{U}$

We can summarize Best's algorithm as follows:

t generator based upon rejection from a t3 density (Best, 1978)

REPEAT

    Generate a t_3 random variate X by the ratio-of-uniforms method (see above).

    Generate a uniform $[0,1]$ random variate U.

$$Z \leftarrow X^2 \ , \ W \leftarrow 1 + \frac{Z}{3}$$

$$Y \leftarrow 2 \log \left(\frac{\frac{9}{16} W^2}{U} \right)$$

    Accept $\leftarrow [Y \geq 1 - Z]$

    IF NOT Accept THEN Accept $\leftarrow [Y \geq (a+1) \log(\frac{a+1}{a+Z})]$

UNTIL Accept

RETURN X

The algorithm given above differs slightly from that given in Best (1978). Best adds another squeeze step before the first logarithm.

5.3. The Cauchy density.

The **Cauchy density**

$$f(x) = \frac{1}{\pi(1+x^2)}$$

plays another key role in statistics. It has no shape parameters, and the mean does not exist. Just as for the exponential distribution, it is easily seen that this density causes no problems whatsoever. To start with, the inversion method is applicable because the distribution function is

$$F(x) = \frac{1}{2} + \frac{1}{\pi} \text{arc tan } x \ \ .$$

This leads to the generator $\tan(\pi U)$ where U is a uniform random variate. The tangent being a relatively slow operation, there is hope for improvement. The main property of the Cauchy density is that whenever (X,Y) is a radially distributed random vector in R^2 without an atom at the origin, then $\frac{X}{Y}$ is Cauchy distributed. The proof uses the fact that if (R,Θ) are the polar coordinates for (X,Y), then $\frac{Y}{X} = \tan(\Theta)$, and Θ is distributed as $2\pi U$ where U is a uniform $[0,1]$ random variate. This leads to two straightforward algorithms for generating

Cauchy random variates:

Polar method I for Cauchy random variates

Generate iid normal random variates N_1, N_2.

RETURN $X \leftarrow \dfrac{N_1}{N_2}$

Polar method II for Cauchy random variates

REPEAT
    Generate iid uniform $[-1,1]$ random variates V_1, V_2.
UNTIL $V_1^2 + V_2^2 \leq 1$
RETURN X

Even though the expected number of uniform random variates needed in the second algorithm is $\dfrac{8}{\pi}$, it seems unlikely that the expected time of the second algorithm will be smaller than the expected time of the algorithm based upon the ratio of two normal random variates. Other algorithms have been proposed in the literature, see for example the acceptance-complement method (section II.5.4 and exercise II.5.1) and the article by Kronmal and Peterson (1981).

5.4. Exercises.

1. **Laha's density (Laha, 1958).** The ratio of two independent normal random variates is Cauchy distributed. This property is shared by other densities as well, in the sense that the term "normal" can be replaced by the name of some other distributions. Show first that the ratio of two independent random variables with Laha's density

$$f(x) = \frac{\sqrt{2}}{\pi(1+x^4)}$$

is Cauchy distributed. Give a good algorithm for generating random variates with Laha's density.

2. Let (X, Y) be uniformly distributed on the circle with center (a, b). Describe the density of $\dfrac{X}{Y}$. Note that when $(a, b) = (0,0)$, you should obtain the

Cauchy density.

3. Consider the class of generalized Cauchy densities

$$f(x) = \frac{a \sin(\frac{\pi}{a})}{2\pi(1+|x|^a)},$$

where $a > 1$ is a parameter. The densities in this class are dominated by the Cauchy density times a constant when $a \geq 2$. Use this fact to develop a generator which is uniformly fast on $[2,\infty)$. Can you also suggest an algorithm which is uniformly fast on $(1,\infty)$?

4. The density

$$f(x) = \frac{1}{\pi(1+x)\sqrt{x}} \qquad (x > 0)$$

possesses both a heavy tail and a sharp peak at 0. Suggest a good and short algorithm for the generation of random variates with this density.

5. **Cacoullos's theorem (Cacoullos, 1965).** Prove that when $G, G*$ are iid gamma $(\frac{a}{2})$ random variates, then

$$X \leftarrow \frac{\sqrt{a}}{2} \frac{G - G*}{\sqrt{GG*}}$$

is t_a distributed. In particular, note that when N_1, N_2 are iid normal random variates, then $(N_1 - N_2)/(2\sqrt{N_1 N_2})$ is Cauchy distributed.

6. The following family of densities has heavier tails than any member of the t family:

$$f(x) = \frac{a-1}{x(\log(x))^a} \qquad (x > e).$$

Here $a > 1$ is a parameter. Propose a simple algorithm for generating random variates from this family, and verify that it is uniformly fast over all values $a > 1$.

7. In this exercise, let C_1, C_2, C_3 be iid Cauchy random variables, and let U be a uniform $[0,1]$ random variable. Prove the following distributional properties:

A. $C_1 C_2$ has density $(\log(x^2))/(\pi^2(x^2-1))$ (Feller, 1971, p. 64).

B. $C_1 C_2 C_3$ has density $(\pi^2 + (\log(x^2))^2)/(2\pi^3(1+x^2))$.

C. UC_1 has density $\log(\frac{1+x^2}{x^2})/(2\pi)$.

8. Show that when X, Y are iid random variables with density $\dfrac{2}{\pi(e^x + e^{-x})}$, then $X + Y$ has density

$$g(x) = \frac{4x}{\pi^2(e^x - e^{-x})} = \frac{2}{\pi^2(1 + \frac{x^2}{3!} + \frac{x^4}{5!} + \cdots)}.$$

Hint: find the density of log(| C |) first, where C is a Cauchy random variate, and use the previous exercise. Show how you can generate random variates with density g directly and efficiently by the rejection method (Feller, 1971, p. 64).

9. Develop a composition-rejection algorithm for the t distribution which is based on the inequality

$$\frac{1}{(1+\frac{x^2}{a})^{\frac{a+1}{2}}} \geq e^{-\frac{(a+1)x^2}{2a}}$$

which for large a is close to $e^{-\frac{x^2}{2}}$. Make sure that if the remainder term is majorized for use in the rejection algorithm, that the area under the remainder term is $o(1)$ as $a \to \infty$. Note: the remainder term must have tails which increase at least as $|x|^{-(a+1)}$. Note also that the ratio of the areas under the normal lower bound and the area under the t density tends to 1 as $a \to \infty$.

10. **The tail of the Cauchy density.** We consider the family of tail densities of the Cauchy, with the tail being defined as the interval $[t, \infty)$, where $t > 0$ is a parameter. Show first that

$$X \leftarrow \tan\left(\arctan(t)(1-U) + \frac{\pi U}{2} \right)$$

has such a tail density. (This is the inversion method.) By using the polar properties of the Cauchy density, show that the following rejection method is also valid, and that the rejection constant tends to 1 as $t \to \infty$:

REPEAT
    Generate iid uniform [0,1] random variates U, V.
    $X \leftarrow \frac{t}{U}$
UNTIL $V(1+\frac{1}{X^2}) \leq 1$
RETURN X

11. This exercise is about inequalities for the function

$$f_a(x) = (1+\frac{x^2}{a})^{-\frac{a+1}{2}}$$

which is proportional to the t density with parameter $a \geq 1$. The inequalities have been used by Kinderman, Monahan and Ramage (1977) in the development of several rejection algorithms with squeeze steps:

A. $f_a(x) \leq \min(1, \frac{1}{x^2})$. Using this inequality in the rejection method corresponds to using the ratio-of-uniforms method.

B. $f_a(x) \geq 1 - \frac{|x|}{2}$. The triangular lower bound is the largest such lower bound not depending upon a that is valid for all $a \geq 1$.

C. $f_a(x) \leq \frac{c}{1+x^2}$ where $c = 2(1+\frac{1}{a})^{-\frac{a+1}{2}} \leq \frac{2}{\sqrt{e}}$. If this inequality is used in the rejection method, then the rejection constant tends to $\sqrt{\frac{2\pi}{e}}$ as $a \to \infty$. The bound can also be used as a quick rejection step.

12. A uniformly fast rejection method for the t family can be obtained by using a combination of a constant bound ($f(0)$) and a polynomial tail bound: for the function $(1+\frac{x^2}{a})^{-\frac{a+1}{2}}$, find an upper bound of the form $\frac{c}{x^b}$ where c, b are chosen to keep the area under the combined upper bound uniformly bounded over $a > 0$.

6. THE STABLE DISTRIBUTION.

6.1. Definition and properties.

It is well known that the sum of iid random variables with finite variance tends in distribution to the normal law. When the variance is not finite, the sum tends in distribution to one of the stable laws, see e.g. Feller (1971). Stable laws have thicker tails than the normal distribution, and are well suited for modeling economic data, see e.g. Mandelbrot (1963), Press (1975). Unfortunately, stable laws are not easy to work with because with a few exceptions no simple expressions are known for the density or distribution function of the stable distributions. The stable distributions are most easily defined in terms of their characteristic functions. Without translation and scale parameters, the characteristic function ϕ is usually defined by

$$\log(\phi(t)) = \begin{cases} -|t|^{\alpha}(1 - i\beta \, \text{sgn}(t)\tan(\frac{\alpha\pi}{2})) & (\alpha \neq 1) \\ -|t|(1 + i\beta\frac{2}{\pi}\text{sgn}(t)\log(|t|)) & (\alpha = 1) \end{cases},$$

where $-1 \leq \beta \leq 1$ and $0 < \alpha \leq 2$ are the parameters of the distribution, and $\text{sgn}(t)$ is the sign of t. This will be called Levy's representation. There is another parametrization and representation, which we will call the polar form (Zolotarev, 1959; Feller, 1971):

$$\log(\phi(t)) = -|t|^{\alpha} e^{-i\gamma \operatorname{sgn}(t)}.$$

Here, $0 < \alpha \le 2$ and $|\gamma| \le \frac{\pi}{2}\min(\alpha, 2-\alpha)$ are the parameters. Note however that one should not equate the two forms to deduce the relationship between the parameters because the representations have different scale factors. After throwing in a scale factor, one quickly notices that the α's are identical, and that β and γ are related via the equation $\beta = \tan(\gamma)/\tan(\alpha\pi/2)$. Because γ has a range which depends upon α, it is more convenient to replace γ by $\frac{\pi}{2}\min(\alpha, 2-\alpha)\delta$, where δ is now allowed to vary in $[-1,1]$. Thus, we rewrite the polar form as follows:

$$\log(\phi(t)) = -|t|^{\alpha} e^{-i\frac{\pi}{2}\min(\alpha, 2-\alpha)\delta \operatorname{sgn}(t)}.$$

When we say that a random variable is stable $(1.3, 0.4)$, we are referring to the last polar form with $\alpha = 1.3$ and $\delta = 0.4$. The parameters β, γ and δ are called the skewness parameters. For $\beta = 0$ ($\gamma = 0$, $\delta = 0$), we obtain the symmetric stable distribution, which is by far the most important sub-class of stable distributions. For all forms, the symmetric stable characteristic function is

$$\phi(t) = e^{-|t|^{\alpha}}$$

By using the product of characteristic functions, it is easy to see that if $X_1, \ldots, X_n$ are iid symmetric stable (α), then

$$n^{-\frac{1}{\alpha}} \sum_{i=1}^{n} X_i$$

is again symmetric stable (α). The following particular cases are important: the symmetric stable (1) law coincides with the Cauchy law, and the symmetric stable (2) distribution is normal with zero mean and variance 2. These two representatives are typical: all symmetric stable densities are unimodal (Ibragimov and Chernin, 1959; Kanter, 1975) and in fact bell-shaped with two infinite tails. All moments exist when $\alpha = 2$. For $\alpha < 2$, all moments of order $< \alpha$ exist, and the α-th moment is ∞.

The asymmetric stable laws have a nonzero skewness parameter, but in all cases, α is indicative of the size of the tail(s) of the density. Roughly speaking, the tail or tails drop off as $|x|^{-(1+\alpha)}$ as $|x| \to \infty$. All densities are unimodal, and the existence or nonexistence of moments is as for the symmetric stable densities with the same value of α. There are two infinite tails when $|\delta| \ne 1$ or when $\alpha \ge 1$, and there is one infinite tail otherwise. When $0 < \alpha < 1$, the mode has the same sign as δ. Thus, for $\alpha < 1$, a stable $(\alpha, 1)$ random variable is positive, and a stable $(\alpha, -1)$ random variable is negative. Both are shaped as the gamma density.

There are a few relationships between stable random variates that will be useful in the sequel. It is not necessary to treat negative-valued skewness parameters since minus a stable (α,δ) random variable is stable $(\alpha,-\delta)$ distributed. Next, we have the following basic relationship:

Lemma 6.1.

Let Y be a stable $(\alpha',1)$ random variable with $\alpha'<1$, and let X be an independent stable (α,δ) random variable with $\alpha\neq 1$. Then $XY^{1/\alpha}$ is stable $(\alpha\alpha',\delta\dfrac{\alpha'\min(\alpha,2-\alpha)}{\min(\alpha\alpha',2-\alpha\alpha')})$. Furthermore, the following is true:

A. If N is a normal random variable, and Y is an independent stable $(\alpha',1)$ random variable with $\alpha'<1$, then $N\sqrt{2Y}$ is stable $(2\alpha',0)$.

B. A stable $(\dfrac{1}{2},1)$ random variable is distributed as $1/(2N^2)$ where N is a normal random variable. In other words, it is Pearson V distributed.

C. If $N_1,N_2,...$ are iid normal random variables, then for integer $k\geq 1$,

$$\prod_{j=0}^{k-1}\frac{1}{(2N_j{}^2)^{2^j}}$$

$$=2^{-(2^k-1)}\prod_{j=1}^{k}\frac{1}{N_j{}^{2^j}}$$

is stable $(2^{-k},1)$.

D. For $N_1,N_2,...$, iid normal random variables, and integer $k\geq 1$,

$$N_{k+1}2^{-(2^{k-1}-1)}\prod_{j=1}^{k}\frac{1}{N_j{}^{2^j-1}}$$

is stable $(2^{1-k},0)$.

E. For $N_1,N_2,...$, iid normal random variables , and integer $k\geq 0$,

$$\frac{N_{k+1}}{N_{k+2}}\prod_{j=0}^{k}\left(\frac{1}{2N_j{}^2}\right)^{2^j}$$

$$=\frac{N_{k+1}}{N_{k+2}}2^{-(2^{k+1}-1)}\prod_{j=0}^{k}\left(\frac{1}{N_j{}^2}\right)^{2^j}.$$

is stable $(\dfrac{1}{2^{k+1}},0)$.

Proof of Lemma 6.1.

The first statement is left as an exercise. If in it, we take $\alpha=2$, $\delta=0$, we obtain part A. It is also seen that a symmetric stable (1) is distributed as a symmetric stable (2) random variable times $\sqrt{X}$ where X is stable $(\dfrac{1}{2},1)$. But by the property that stable (1) random variables are nothing but Cauchy random variables, i.e. ratios of two independent normal random variables, we conclude that

X must be distributed as $1/(2N^2)$ where N is normally distributed. This proves part B. Next, again by the main property, if X is as above, and Y is stable $(\alpha',1)$, then XY^2 is stable $(\dfrac{\alpha'}{2},1)$, at least when $\alpha'<1$. If this is applied successively for $\alpha'=\dfrac{1}{2},\dfrac{1}{4},\dfrac{1}{8},...$, we obtain statement C. Statement D follows from statements A and C. Finally, using the fact that a symmetric stable $(1/2^{k+1})$ is distributed as a symmetric stable $(1/2^k)$ times X^{2^k}, where X is stable $(\dfrac{1}{2},1)$, we see that a stable $(1/2^{k+1},0)$ is distributed as a Cauchy random variable times

$$\prod_{j=0}^{k}\left(\frac{1}{2N_j{}^2}\right)^{2^j}.$$

This concludes the proof of part E. ∎

Properties A-E in Lemma 6.1 are all corollaries of the main property given there. The main property is due to Feller (1971). Property A tells us that all symmetric stable random variables can be obtained if we can obtain all positive $(\delta=1)$ stable random variables with parameter $\alpha<1$. Property B is due to Levy (1940). Property C goes back to Brown and Tukey (1946). Property D is but a simple corollary of property C, and finally, property E is a representation of Mitra's (1981). For other similar representations, see Mitra (1982).

There is another property worthy of mention. It states that all stable (α,δ) random variables can be written as weighted sums of two iid stable $(\alpha,1)$ random variables. It was mentioned in chapter IV (Lemma 6.1), but we reproduce it here for the sake of completeness.

Lemma 6.2.

If X and Y are iid stable$(\alpha,1)$, then $Z \leftarrow pX-qY$ is stable(α,δ) where

$$p^{\alpha}=\frac{\sin(\dfrac{\pi \min(\alpha,2-\alpha)(1+\delta)}{2})}{\sin(\pi \min(\alpha,2-\alpha))},$$

$$q^{\alpha}=\frac{\sin(\dfrac{\pi \min(\alpha,2-\alpha)(1-\delta)}{2})}{\sin(\pi \min(\alpha,2-\alpha))}.$$

Proof of Lemma 6.2.

    The characteristic function of Z is

$$\phi(t) = E(e^{itpX})E(e^{-itqY})$$
$$= \psi(pt)\psi(-qt)$$

where ψ is the characteristic function of the stable $(\alpha,1)$ law:

$$\psi(t) = e^{-|t|^{\alpha}e^{-i\frac{\pi}{2}\min(\alpha,2-\alpha)\,\text{sgn}(t)}}.$$

Note next that for $u > 0$, $p^{\alpha}e^{-iu} + q^{\alpha}e^{iu}$ is equal to

$$\cos(u)(p^{\alpha}+q^{\alpha}) - i\sin(u)(p^{\alpha}-q^{\alpha})$$

$$= \frac{1}{\sin(\pi\min(\alpha,2-\alpha))}2(\cos(u)\sin(\frac{\pi}{2}\min(\alpha,2-\alpha))\cos((\frac{\pi}{2}\delta\min(\alpha,2-\alpha))) -$$

$$i\sin(u)\cos(\frac{\pi}{2}\min(\alpha,2-\alpha))\sin((\frac{\pi}{2}\delta\min(\alpha,2-\alpha)))).$$

After replacing u by its value, $\frac{\pi}{2}\min(\alpha,2-\alpha)$, we see that we have

$$\frac{2\cos(u)\sin u}{\sin(2u)}(\cos(\delta u) - i\sin(\delta u)) = e^{-i\delta u}.$$

Resubstitution gives us our result. ∎

6.2. Overview of generators.

    The difficulty with most stable densities and distribution functions is that no simple analytical expression for its computation is available. The exceptions are spelled out in the previous section. Basically, stable random variates with parameter α equal to 2^{-k} for $k \geq 0$, and with arbitrary value for δ, can be generated quite easily by the methods outlined in Lemmas 6.1 and 6.2. One just needs to combine an appropriate number of iid normal random variates. For general α,δ, methods requiring accurate values of the density or distribution function are thus doomed, because these cannot be obtained in finite time. Approximate inversions of the distribution function are reported in Fama and Roll (1968), Dumouchel (1971) and Paulson, Holcomb and Leitch (1975). Paulauskas (1982) suggests another approximate method in which enough iid random variables are summed. Candidates for summing include the Pareto densities. For symmetric stable densities, Bartels (1978) also presents approximate methods. Bondesson (1982) proposes yet another approximate method in which a stable random variable is written as an infinite sum of powers of the event times in a homogeneous Poisson process on $[0,\infty)$. The sum is truncated, and the tail sum is replaced by an appropriately picked normal random variate.

    Fortunately, exact methods do exist. First of all, the stable density can be written as an integral which in turn leads to a simple formula for generating stable random variates as a combination of one uniform and one exponential random variate. These generators were developed in section IV.6.6, and are based upon integral representations of Ibragimov and Chernin (1959) and Zolotarev (1966). The generators themselves were proposed by Kanter (1975) and Chambers, Mallows and Stuck (1976), and are all of the form $g(U)E^{-\frac{1-\alpha}{\alpha}}$ where E is exponentially distributed, and $g(U)$ is a function of a uniform [0,1] random variate U. The sheer simplicity of the representation makes this method very attractive, even though g is a rather complicated function of its argument involving several trigonometric and exponential/logarithmic operations. Unless speed is absolutely at a premium, this method is highly recommended.

For symmetric stable random variates with $\alpha \le 1$, there is another representation: such random variates are distributed as

$$\frac{Y}{(E_1+E_2 I_{[U<\alpha]})^{\frac{1}{\alpha}}}$$

where Y has the Fejer-de la Vallee Poussin density, and E_1, E_2 are iid exponential random variates. This representation is based upon properties of Polya characteristic functions, see section IV.6.7, Theorems IV.6.8, IV.6.9, and Example IV.6.7. Since the Fejer-de la Vallee Poussin density does not vary with α, random variates with this density can be generated quite quickly (remark IV.6.1). This can lead to speeds which are superior to the speed of the method of Kanter and Chambers, Mallows and Stuck.

In the rest of this section we outline how the series method (section IV.5) can be used to generate stable random variates. Recall that the series method is based upon rejection, and that it is designed for densities that are given as a convergent series. For stable densities, such convergent series were obtained by Bergstrom (1952) and Feller (1971). In addition, we will need good dominating curves for the stable densities, and sharp estimates for the tail sums of the convergent series. In the next section, the Bergstrom-Feller series will be presented, together with estimates of the tail sums due to Bartels (1981). Inequalities for the stable distribution which lead to practical implementations of the series method are obtained in the last section. At the same time, we will obtain estimates of the expected time performance as a function of the parameters of the distribution.

6.3. The Bergstrom-Feller series.

The purpose of this section is to get ready for the next section, where the series method for stable random variates is developed. The form of the characteristic function most convenient to us is the first polar form, with parameters α and γ. To obtain series expansions for the stable density function, we consider the Fourier inverse of ϕ, which takes a simple form since $|\phi|$ is absolutely integrable:

$$f(x) = \frac{1}{2\pi} \int_{-\infty}^{\infty} e^{-itx} e^{-|t|^{\alpha} e^{-i\gamma \operatorname{sgn}(t)}} dt$$

$$= \operatorname{Re}\left\{ \frac{1}{\pi} \int_{0}^{\infty} e^{-itx} e^{-t^{\alpha} e^{-i\gamma}} dt \right\}$$

$$= \operatorname{Re}\left\{ \frac{1}{\pi} \int_{0}^{\infty} e^{-txe^{i(\frac{\pi}{2}+\psi)}} e^{-t^{\alpha} e^{i\psi} e^{i(\alpha\psi-\gamma)}} dt \right\}$$

provided that $|\alpha\psi-\gamma| \leq \frac{\pi}{2}$ and that $|\frac{\pi}{2}+\psi| \leq \frac{\pi}{2}$ with at least one of these being a strict inequality. We have used the fact that changing the sign of γ is equivalent to mirroring the density about the origin, and we have considered a contour in the complex plane. The last expression for f will be our starting point. Recall that we need not only a convergent series, but also good bounds for f and for the tail sums. Bergstrom (1952) replaces each of the exponents in the last expression in turn by its Maclaurin series, and integrates (see also Feller (1971)). Bartels (1981) uses Darboux's formula (1876) for the remainder term in the series expansion to obtain good truncation bounds. In Theorem 6.1 below, we present the two Bergstrom-Feller series together with Bartels's bounds. The proof follows Bartels (1981).

Theorem 6.1.

The stable (α,γ) density f can be expanded for values $x \geq 0$ as follows:

$$f(x) = \sum_{j=1}^{n} a_n(x) + A^*_{n+1}(x)$$

where

$$a_j(x) = \frac{1}{\alpha\pi}(-1)^{j-1} \frac{\Gamma(\frac{j}{\alpha})x^{j-1}\sin(j(\frac{\pi}{2}+\frac{\gamma}{\alpha}))}{j-1!},$$

$$|A^*_{n+1}(x)| \leq A_{n+1}(x) = \frac{1}{\alpha\pi} \frac{\Gamma(\frac{n+1}{\alpha})x^n}{n!(\cos(\theta))^{\frac{n+1}{\alpha}}},$$

where $\theta=0$ if $\gamma\leq0$ and $\theta=\gamma$ if $\gamma>0$. For $x<0$, note that the value of the density is equal to $f(-x)$ provided that γ is replaced by $-\gamma$. The expansion converges for $1<\alpha\leq2$. For $0<\alpha<1$, we have a divergent asymptotic series for small $|x|$, i.e., for fixed n, $A_n(x)\to0$ as $|x|\to0$. Note also that

$$f(x) \leq \frac{\Gamma(\frac{1}{\alpha})}{\alpha\pi(\cos(\theta))^{\frac{1}{\alpha}}}.$$

A second expansion for $f(x)$ when $x>0$ is given by

$$f(x) = \sum_{j=1}^{n} b_n(x) + B^*_{n+1}(x),$$

where

$$b_j(x) = \frac{(-1)^{j-1}\Gamma(\alpha j+1)\sin(j(\frac{\alpha\pi}{2}+\gamma))}{\pi j! x^{\alpha j+1}},$$

$$|B^*_{n+1}(x)| \leq B_{n+1}(x) = \frac{\Gamma(\alpha(n+1)+1)}{\pi(n+1)!(x\cos(\theta))^{\alpha(n+1)+1}},$$

with $\theta=\max(0,\frac{\pi}{2}+\frac{1}{\alpha}(\gamma-\frac{\pi}{2}))$. The expansion is convergent for $0<\alpha<1$, and is a divergent asymptotic expansion at $|x|\to\infty$ when $\alpha>1$, i.e. for fixed n, $B_n(x)\to0$ as $|x|\to\infty$. Furthermore, for all α,

$$f(x) \leq \frac{\Gamma(\alpha+1)}{\pi(x\cos(\theta))^{\alpha+1}}.$$

Proof of Theorem 6.1.

The proof is based upon a formula of Darboux (1876), which when applied to e^z with complex z leads to

$$e^z = \sum_{j=0}^{n-1} \frac{z^j}{j!} + \frac{z^n}{n!} M_n \;,$$

where $M_n = \lambda e^{\theta z}$, λ being a complex constant with $|\lambda| \leq 1$, and θ being a real constant in the range $0 \leq \theta < 1$. In particular, for $\mathrm{Re}(z) > 0$, $|M_n| \leq |e^z|$. For $\mathrm{Re}(z) \leq 0$, $|M_n| \leq 1$. Apply this result with $z = -txe^{j(\frac{\pi}{2}+\psi)}$ in the inversion formula for f, and note that $\mathrm{Re}(z) \leq 0$. Take the integrals, and observe that the remainder term can be bounded as follows:

$$|A^*_{n+1}(x)| \leq \frac{x^n}{\pi n!} \int_0^\infty t^n \left| e^{-t^\alpha e^{j(\alpha\psi-\gamma)}} \right| dt$$

$$= \frac{x^n}{\pi n!} \int_0^\infty t^n e^{-t^\alpha \cos(\alpha\psi-\gamma)} dt$$

$$= \frac{1}{\alpha\pi} \frac{\Gamma(\frac{n+1}{\alpha}) x^n}{n!(\cos(\alpha\psi-\gamma))^{\frac{n+1}{\alpha}}} \;.$$

The angle ψ can be chosen within the restrictions put on it, to make the upper bound as small as possible. This leads to the choice $\frac{\gamma}{\alpha}$ when $\gamma \leq 0$, and 0 when $\gamma > 0$. It is easy to verify that for $1 < \alpha \leq 2$, the expansion is convergent. Finally, the upper bound is obtained by noting that $f(x) \leq A_1(x)$.

The second expansion is obtained by applying Darboux's formula to $e^{-t^\alpha e^{j(\alpha\psi-\gamma)}}$ and integrating. Repeating the arguments used for the first expansion, we obtain the second expansion. Using Stirling's formula, it is easy to verify that for $0 < \alpha < 1$, the expansion is convergent. Furthermore, for fixed n, $B_n(x) \to 0$ as $|x| \to \infty$, and $f(x) \leq B_1(x)$. ∎

The convergent series expansion for $\alpha > 1$ requires an increasing number of terms to reach a given truncation error as $|x|$ increases. The asymptotic series increases in accuracy and needs fewer terms as $|x|$ increases. As pointed out by Bartels (1981), the convergent series generally tends to increase first, before converging, and the intermediate values may become so large that the final answer no longer has sufficient significant digits. This drawback occurs mainly for values of α near 1, and large values of $|\gamma|$.

6.4. The series method for stable random variates.

From Theorem 6.1, we deduce the following useful bound for the stable (α,γ) density when $\gamma \geq 0$:

$$f(x) \leq \begin{cases} \dfrac{\Gamma(\frac{1}{\alpha})}{\alpha\pi(\cos(\gamma))^{\frac{1}{\alpha}}} & (x \geq 0) \\[4mm] \dfrac{\Gamma(\alpha+1)}{\pi(x\cos(\eta))^{\alpha+1}} & (x \geq 0) \\[4mm] \dfrac{\Gamma(\frac{1}{\alpha})}{\alpha\pi} & (x < 0) \\[4mm] \dfrac{\Gamma(\alpha+1)}{\pi(-x\cos(\theta))^{\alpha+1}} & (x < 0) \end{cases}$$

where $\theta = \max(0,\frac{\pi}{2}+\frac{1}{\alpha}(-\gamma-\frac{\pi}{2}))$ and $\eta = \max(0,\frac{\pi}{2}+\frac{1}{\alpha}(\gamma-\frac{\pi}{2}))$. The bounds are valid for all values of α. The dominating curve will be used in the rejection algorithm to be presented below. Taking the minimum of the bounds gives basically two constant pieces near the center and two polynomially decreasing tails. There is no problem whatsoever with the generation of random variates with density proportional to the dominating curve. Unfortunately, the bounds provided by Theorem 6.1 are not very useful for asymmetric stable random variates because the mode is located away from the origin. For example, for the positive stable density, we even have $f(0) = 0$. Thus, a constant/polynomial dominating curve does not cap the density very well in the region between the origin and the mode. For a good fit, we would have needed an expansion around the mode instead of two expansions, one around the origin, and one around ∞. The inefficiency of the bound is easily born out in the integral under the dominating curve. We will consider four cases:

$\gamma = 0, \alpha > 1$ (symmetric stable).
$\gamma = 0, \alpha \leq 1$ (symmetric stable).
$\gamma = (2-\alpha)\frac{\pi}{2}, \alpha > 1$ (positive stable).
$\gamma = \alpha\frac{\pi}{2}, \alpha \leq 1$ (positive stable).

The upper bound given to us is of the form $\min(A, Bx^{-(1+\alpha)})$ for $x > 0$. For the symmetric stable density, the dominating curve can be mirrored around the origin, while for the asymmetric cases, we need to replace A, B by values $A*, B*$, and x by $-x$. Recalling that

$$\int_0^\infty \min(A, Bx^{-(1+\alpha)})\, dx = \frac{1+\alpha}{\alpha}A^{\frac{\alpha}{1+\alpha}}B^{\frac{1}{1+\alpha}},$$

It is easy to compute the areas under the various dominating curves. We offer the following table for A,B:

CASE	A	B
1	$\dfrac{\Gamma(\frac{1}{\alpha})}{\pi\alpha}$	$\dfrac{\Gamma(1+\alpha)}{\pi(\sin(\frac{\pi}{2\alpha}))^{\alpha+1}}$
2	$\dfrac{\Gamma(\frac{1}{\alpha})}{\pi\alpha}$	$\dfrac{\Gamma(1+\alpha)}{\pi}$
3	$\dfrac{\Gamma(\frac{1}{\alpha})}{\alpha\pi(\sin((\alpha-1)\frac{\pi}{2}))^{\frac{1}{\alpha}}}$	$\dfrac{\Gamma(\alpha+1)}{\pi(\cos(\frac{\pi}{2\alpha}))^{\alpha+1}}$
4	$\dfrac{\Gamma(\frac{1}{\alpha})}{\alpha\pi(\cos(\frac{\alpha\pi}{2}))^{\frac{1}{\alpha}}}$	$\dfrac{\Gamma(\alpha+1)}{\pi(-\cos(\frac{\pi}{2\alpha}))^{\alpha+1}}$

For example, in case 1, we see that the area under the dominating curve is

$$2\frac{\alpha+1}{\alpha}\left[\frac{\Gamma(\frac{1}{\alpha})}{\pi\alpha}\right]^{\frac{\alpha}{1+\alpha}}\left(\frac{\Gamma(1+\alpha)}{\pi(\sin(\frac{\pi}{2\alpha}))^{\alpha+1}}\right)^{\frac{1}{\alpha+1}}$$

$$\leq\frac{4}{\pi}\left[\Gamma(\frac{1}{\alpha})\right]^{\frac{\alpha}{1+\alpha}}\left(\Gamma(1+\alpha)\right)^{\frac{1}{\alpha+1}}$$

$$\leq\frac{4}{\pi}\pi^{\frac{1}{3}}\sqrt{2}$$

where we used the following inequalities: (i) $(\alpha+1)/\alpha^{\alpha/(1+\alpha)}\leq 2$ ($\alpha\geq 1$); (ii) $\sin(\pi/(2\alpha))\geq 1/\alpha$; (iii) $\Gamma(u)\leq 2$ ($2\leq u\leq 3$); (iv) $\Gamma(u)\leq\Gamma(\frac{1}{2})=\sqrt{\pi}$ ($\frac{1}{2}\leq u\leq 1$). Some of the inequalities are rather loose, so that the actual fit is probably much better than what is predicted by the upper bound. For $\alpha=2$, the normal density, we obtain $32^{1/6}\pi^{-2/3}$. The importance of the good fit is clear: we can now use the dominating curve quite confidently in any rejection type algorithm for symmetric stable random variate generation when $\alpha\geq 1$. The story is not so rosy for the three other cases, because the integral of the dominating curve is not uniformly bounded over the specified parameter ranges. The actual verification of this statement is left as an exercise, but we conclude that it is not worth to use the Bergstrom-Feller series for asymmetric stable random variates. For this reason, we will just concentrate on the symmetric case. The notation a_n,b_n, A_n, B_n is taken from Theorem 6.1. Furthermore, we define a density g and a normalization constant c by

$$cg(x) = \min \begin{cases} \dfrac{\Gamma(\frac{1}{\alpha})}{\alpha\pi} \\ \dfrac{\Gamma(\alpha+1)}{\pi(|x|\sin(\varsigma))^{\alpha+1}} \end{cases}$$

where $\varsigma=0$ for $\alpha<1$, and $\varsigma=\pi/(2\alpha)$ otherwise. The algorithm is of the following form:

Series method for symmetric stable density; case of parameter > 1

REPEAT

    Generate X with density g.

    Generate a uniform $[0,1]$ random variate U.

    $T \leftarrow Ucg(X)$

    $S \leftarrow 0$, $n \leftarrow 0$ (Get ready for series method.)

    REPEAT

        $n \leftarrow n+1, S \leftarrow S + a_n(X)$

    UNTIL $|S-T| \geq A_{n+1}(X)$

UNTIL $T \leq S$

RETURN X

Because of the convergent nature of the series $\sum a_n$, this algorithm stops with probability one. Note that the divergent asymptotic expansion is only used in the definition of cg. It could of course also be used for introducing quick acceptance and rejection steps. But because of the divergent nature of the expansion it is useless in the definition of a stopping rule. One possible use is as indicated in the modified algorithm shown below.

Series method for symmetric stable density; case of parameter > 1

REPEAT

    Generate X with density g .

    Generate a uniform [0,1] random variate U .

    $T \leftarrow U c g(X)$

    $S \leftarrow 0,\ n \leftarrow 0$ (Get ready for series method.)

    $V \leftarrow B_2(X),\ W \leftarrow b_1(X)$

    IF $T \leq W - V$

        THEN RETURN X

        ELSE IF $W - V < T \leq W + V$

            THEN

                REPEAT

                    $n \leftarrow n + 1, S \leftarrow S + a_n(X)$

                  UNTIL $\mid S - T \mid\ \geq A_{n+1}(X)$

UNTIL $T \leq S$ AND $T \leq W + V$

RETURN X

Good speed is obtainable if we can set up some constants for a fixed value of α. In particular, an array of the first m coefficients of x^{j-1} in the series expansion can be computed beforehand. Note that for $\alpha < 1$, both algorithms shown above can be used again, provided that the roles of a_n and b_n are interchanged. For the modified version, we have:

Series method for symmetric stable density; case of parameter less than or equal to one

REPEAT

       Generate X with density g .

       Generate a uniform [0,1] random variate U .

       $T \leftarrow Ucg(X)$

       $S \leftarrow 0$, $n \leftarrow 0$ (Get ready for series method.)

       $V \leftarrow A_2(X)$, $W \leftarrow a_1(X)$

       IF $T \leq W - V$

            THEN RETURN X

            ELSE IF $W - V < T \leq W + V$

                THEN

                    REPEAT

                        $n \leftarrow n + 1, S \leftarrow S + b_n(X)$

                      UNTIL $\mid S - T \mid \geq B_{n+1}(X)$

       UNTIL $T \leq S$ AND $T \leq W + V$

       RETURN X

6.5. Exercises.

1. Prove that a symmetric stable random variate with parameter $\dfrac{1}{2}$ can be obtained as $c(N_1^{-2} - N_2^{-2})$ where N_1, N_2 are iid normal random variates, and $c > 0$ is a constant. Determine c too.

2. The expected number of iterations in the series method for symmetric stable random variates with parameter α ,based upon the inequalities given in the text (based upon the Bergstrom-Feller series), is asymptotic to

$$\frac{2}{\pi e \, \alpha^2}$$

as $\alpha \downarrow 0$.

3. Consider the series method for stable random variates given in the text, without quick acceptance and rejection steps. For all values of α, determine $E(N)$, where N is the number of computations of some term a_n or b_n (note that since a_n or b_n are computed in the inner loop of two nested loops, it is an appropriate measure of the time needed to generate a random variate). For which values, if any, is $E(N)$ finite ?

4. Some approximate methods for stable random variate generation are based upon the following limit law, which you are asked to prove. Assume that $X_1,\ldots$ are iid random variables with common distribution function F satisfying

$$1-F(x) \sim (\frac{b}{x})^\alpha \quad (x \to \infty) ,$$

$$F(-x) \sim (\frac{cb*}{|x|})^\alpha \quad (x \to -\infty) ,$$

for some constants $0 < \alpha < 2$, $b, b* \geq 0$, $b + b* > 0$. Show that there exist normalizing constants c_n such that

$$\frac{1}{n^{\frac{1}{\alpha}}} \sum_{j=1}^{n} X_j - c_n$$

tends in distribution to the stable (α, β) distribution with parameter

$$\beta = \frac{b^\alpha - b*^\alpha}{b^\alpha + b*^\alpha} .$$

(Feller, 1971).

5. This is a continuation of the previous exercise. Give an example of a distribution with a density satisfying the tail conditions mentioned in the exercise, and show how you can generate a random variate. Furthermore, suggest for your example how c_n can be chosen.

6. Prove the first statement of Lemma 6.1.

7. Find a simple dominating curve with uniformly bounded integral for all positive stable densities with parameter $\alpha \geq 1$. Mention how you would proceed with the generation of a random variate with density proportional to this curve.

8. In the spirit of the previous exercise, find a simple dominating curve with uniformly bounded integral for all symmetric stable densities; α can take all values in $(0,2]$.

7. NONSTANDARD DISTRIBUTIONS.

7.1. Bessel function distributions.

The **Polya-Aeppli** distribution is a three-parameter distribution with density

$$f(x) = Cx^{\frac{\lambda-1}{2}} e^{-\theta x} I_{\lambda-1}(\beta\sqrt{x}) \quad (x \geq 0)$$

where $\theta > 0$, $\lambda > 0$, $\beta \geq 0$ are the parameters and $I_a(x)$ is the modified Bessel function of the first kind, formally defined by

$$I_a(x) = \sum_{j=0}^{\infty} \frac{1}{j\,!\Gamma(j+a+1)} \left(\frac{x}{2}\right)^{2j+a} .$$

The normalization constant C is given by

$$C = \left(\frac{2}{\beta}\right)^{\lambda-1} \theta^\lambda e^{-\frac{\beta^2}{4\theta}} .$$

The name Polya-Aeppli is used in many texts such as Ord (1972, p. 125-126). Others prefer the name "type I Bessel function distribution" (Feller, 1971, p. 57). By using the expansion of the Bessel function, it is not difficult to see that if Z is Poisson $\left(\frac{\beta^2}{4\theta}\right)$ distributed, and G is gamma $(\lambda + Z)$ distributed, then $\frac{G}{\theta}$ has the Polya-Aeppli distribution. We summarize:

Polya-Aeppli random variate generator

Generate a Poisson $\left(\frac{\beta^2}{4\theta}\right)$ random variate Z.

Generate a gamma $(\lambda + Z)$ random variate G.

RETURN $X \leftarrow \dfrac{G}{\theta}$

The Polya-Aeppli family contains as a special case the gamma family (set $\beta = 0$, $\theta = 1$). Other distributions can be derived from it without much trouble: for example, if X is Polya-Aeppli $(\beta, \lambda, \frac{\theta}{2})$, then X^2 is a type II Bessel function distribution with parameters (β, λ, θ), i.e. X^2 has density

$$f(x) = Dx^\lambda e^{-\theta \frac{x^2}{2}} I_{\lambda-1}(\beta x) \quad (x \geq 0) ,$$

where $D = \theta^\lambda \beta^{1-\lambda} e^{-\beta^2/(2\theta)}$. Special cases here include the folded normal distribution and the Rayleigh distribution. For more about the properties of type I and II Bessel function distributions, see for example Kotz and Srinivasan (1969), Lukacs and Laha (1964) and Laha (1954).

Bessel functions of the second kind appear in other contexts. For example, the product of two iid normal random variables has density

$$\frac{1}{\pi} K_0(x)$$

where K_0 is the Bessel function of the second kind with purely imaginary argument of order 0 (Springer, 1979, p. 160).

In the study of random walks, the following density appears naturally:

$$f(x) = \frac{r}{x} e^{-x} I_r(x) \quad (x > 0) ,$$

where $r > 0$ is a parameter (see Feller (1971, pp. 59-60,476)). For integer r, this is the density of the time before level r is crossed for the first time in a symmetric random walk, when the time between epochs is exponentially distributed:

$X \leftarrow 0, L \leftarrow 0$
REPEAT
      Generate a uniform $[-1,1]$ random variate U.
      $L \leftarrow L + \text{sign}(U)$
      $X \leftarrow X - \log(\mid U \mid)$
UNTIL $L = r$
RETURN X

Unfortunately, the expected number of iterations is ∞, and the number of iterations is bounded from below by r, so this algorithm is not uniformly fast in any sense. We have however:

Theorem 7.1.

    Let $r > 0$ be a real number. If G, B are independent gamma (r) and beta $(\frac{1}{2}, r + \frac{1}{2})$ random variables, then

$$\frac{G}{2B}$$

has density

$$f(x) = \frac{r}{x} e^{-x} I_r(x) \quad (x > 0) .$$

Proof of Theorem 7.1.

    We use an integral representation of the Bessel function I_r which can be found for example in Magnus et al. (1966, p. 84):

$$f(x) = \frac{r}{x} e^{-x} I_r(x)$$

$$= \frac{1}{\Gamma(r+\frac{1}{2})} \frac{r}{x} e^{-x} \frac{1}{\sqrt{\pi}} (\frac{x}{2})^r \int_{-1}^{1} e^{-zx} (1-z^2)^{r-\frac{1}{2}} dz$$

$$= \frac{1}{\Gamma(r+\frac{1}{2})} \frac{r}{x} \frac{1}{\sqrt{\pi}} (\frac{x}{2})^r 2^{2r} \int_{0}^{1} e^{-2yx} (y(1-y))^{r-\frac{1}{2}} dy \ .$$

The result follows directly from this. ∎

The algorithm suggested by Theorem 7.1 is uniformly fast over all $r > 0$ if uniformly fast gamma and beta generators are used. Of course, we can also use direct rejection. Bounds for f can for example be obtained starting from the integral representation for f given in the proof of Theorem 7.1. The acceptance or rejection has to be decided based upon the series method in that case.

7.2. The logistic and hyperbolic secant distributions.

A random variable has the **logistic distribution** when it has distribution function

$$F(x) = \frac{1}{1+e^{-x}}$$

on the real line. The corresponding density is

$$f(x) = \frac{1}{2+e^x+e^{-x}} \ .$$

For random variate generation, we can obviously proceed by inversion: when U is uniformly distributed on $[0,1]$, then $X \leftarrow \log(\frac{U}{1-U})$ is logistic. To beat this method, one needs either an extremely efficient rejection or acceptance-complement algorithm, or a table method. Rejection could be based upon one of the following inequalities:

A. $f(x) \le e^{-|x|}$: this is rejection from the Laplace density. The rejection constant is 2.

B. $f(x) \le \frac{1}{4+x^2}$: this is rejection from the density of $2C$ where C is a Cauchy random variate. The rejection constant is $\frac{\pi}{2} \approx 1.57$.

A distribution related to the logistic distribution is the **hyperbolic secant distribution** (Talacko, 1956). The density is given by

$$f(x) = \frac{2}{\pi(e^x+e^{-x})} \ .$$

Both the logistic and hyperbolic secant distributions are members of the family of Perks distributions (Talacko, 1956), with densities of the form $c/(a + e^x + e^{-x})$, where $a \geq 0$ is a parameter and c is a normalization constant. For this family, rejection from the Cauchy density can always be used since the density is bounded from above by $c/(a + 2 + x^2)$, and the resulting rejection algorithm has uniformly bounded rejection constant for $a \geq 0$. For the hyperbolic secant distribution in particular, there are other possibilities. One can easily see that it has distribution function

$$F(x) = \frac{2}{\pi}\text{arc} \tan(e^x) .$$

Thus, $X \leftarrow \log(\tan(\frac{\pi}{2} U))$ is a hyperbolic secant random variate whenever U is a uniform [0,1] random variate. We can also use rejection from the Laplace density, based upon the inequality $f(x) \leq \frac{2}{\pi}e^{-|x|}$. This yields a quite acceptable rejection constant of $\frac{4}{\pi}$. The rejection condition can be considerably simplified:

Rejection algorithm for the hyperbolic secant distribution

REPEAT
>   Generate U uniformly on [0,1] and V uniformly on $[-1,1]$.
> $X \leftarrow \text{sign}(V)\log(|V|)$
UNTIL $U(|V| + 1) \leq 1$
RETURN X

Both the logistic and hyperbolic secant distributions are intimately related to a host of other distributions. Most of the relations can be deduced from the inversion method. For example, by the properties of uniform spacings, we observe that $\frac{U}{1-U}$ is distributed as E_1/E_2, the ratio of two independent exponential random variates. Thus, $\log(E_1) - \log(E_2)$ is logistic. This in turn implies that the difference between two iid extreme-value random variables (i.e., random variables with distribution function $e^{-e^{-x}}$) is logistic. Also, $\tan(\frac{\pi}{2} U)$ is distributed as the absolute value of a Cauchy random variable. Thus, if C is a Cauchy random variable, and N_1, N_2 are iid normal random variables, then $\log(|C|)$ and $\log(|N_1|) - \log(|N_2|)$ are both hyperbolic secant.

Many properties of the logistic distribution are reviewed in Olusegun George and Mudholkar (1981).

7.3. The von Mises distribution.

The **von Mises distribution** for points on a circle has become important in the statistical theory of directional data. For its properties, see for example the survey paper by Mardia (1975). The distribution is completely determined by the distribution of the random angle Θ on $[-\pi,\pi]$. There is one shape parameter, $\kappa > 0$, and the density is given by

$$f(\theta) = \frac{e^{\kappa\cos(\theta)}}{2\pi I_0(\kappa)} \quad (\mid \theta \mid \leq \pi) .$$

Here I_0 is the modified Bessel function of the first kind of order 0:

$$I_0(x) = \sum_{j=0}^{\infty} \frac{1}{j!^2} (\frac{x}{2})^{2j} .$$

Unfortunately, the distribution function does not have a simple closed form, and there is no simple relationship between von Mises (κ) random variables and von Mises (1) random variables which would have allowed us to eliminate in effect the shape parameter. Also, no useful characterizations are as yet available. It seems that the only viable method is the rejection method. Several rejection methods have been suggested in the literature, e.g. the method of Seigerstetter (1974) (see also Ripley (1983)), based upon the obvious inequality

$$f(\theta) \leq f(0)$$

which leads to a rejection constant $2\pi f(0)$ which tends quickly to ∞ as $\kappa \to \infty$. We could use the universal bounding methods of chapter 7 for bounded monotone densities since f is bounded, U-shaped (with modes at π and $-\pi$) and symmetric about 0. Fortunately, there are much better alternatives. The leading work on this subject is by Best and Fisher (1979), who, after considering a variety of dominating curves, suggest using the wrapped Cauchy density as a dominating curve. We will just content ourselves with a reproduction of the Best-Fisher algorithm.

We begin with the wrapped Cauchy distribution function with parameter ρ:

$$G(x) = \frac{1}{2\pi} \arccos \left(\frac{(1+\rho^2)\cos(x)-2\rho}{1+\rho^2-2\rho\cos(x)} \right) \quad (\mid x \mid \leq \pi) .$$

For later reference, the density g for G is:

$$g(x) = \frac{1}{2\pi} \frac{1-\rho^2}{1+\rho^2-2\rho\cos(x)} \quad (\mid x \mid \leq \pi) .$$

A random variate with this distribution can easily be generated via the inversion method:

Wrapped Cauchy generator; inversion method

[SET-UP]

$$s \leftarrow \frac{1+\rho^2}{2\rho}$$

[GENERATOR]

Generate a uniform $[-1,1]$ random variate U.

$Z \leftarrow \cos(\pi U)$

$$\text{RETURN } \Theta \leftarrow \frac{\text{sign}(U)}{\cos(\frac{1+sZ}{s+Z})}$$

If the wrapped Cauchy distribution is to be used for rejection, we need to fine tune the distribution, i.e. choose ρ as a function of κ.

Theorem 7.2. (Best and Fisher, 1979)

Let f be the von Mises density with parameter $\kappa > 0$, and let g be the wrapped Cauchy density with parameter $\rho > 0$. Then

$$f(x) \le cg(x) \quad (|x| \le \pi)$$

where c is a constant depending upon κ and ρ only. The constant is minimized with respect to ρ for the value

$$\rho = \frac{r - \sqrt{2r}}{2\kappa}$$

where

$$r = 1 + \sqrt{1 + 4\kappa^2} .$$

The expected number of iterations in the rejection algorithm is

$$c = \frac{\frac{2\rho}{\kappa} e^{\kappa \frac{1+\rho^2}{2\rho} - 1}}{(1-\rho^2) I_0(\kappa)} .$$

Furthermore, $\lim_{\kappa \downarrow 0} c = \infty$ and $\lim_{\kappa \to \infty} c = \sqrt{\frac{2\pi}{e}}$.

Proof of Theorem 7.2.

Consider the ratio

$$h(x) = \frac{f(x)}{g(x)} = \frac{(1+\rho^2-2\rho\cos(x))e^{\kappa\cos(x)}}{I_0(\kappa)(1-\rho^2)}.$$

The derivative of h is zero for $\sin(x)=0$ and for $\cos(x)=(1+\rho^2-\frac{2\rho}{\kappa})/(2\rho)$. By verifying the second derivative of h, we find a local maximum value

$$M_1 = (1-\rho)^2 e^\kappa$$

at $\sin(x)=0$ when

$$\frac{2\rho}{(1-\rho)^2} < \kappa ,$$

and a local maximum value

$$M_2 = \frac{2\rho}{\kappa} e^{\kappa\frac{1+\rho^2}{2\rho}-1}$$

at $\cos(x)=(1+\rho^2-\frac{2\rho}{\kappa})/(2\rho)$ when

$$\frac{2\rho}{(1+\rho)^2} < \kappa < \frac{2\rho}{(1-\rho)^2} .$$

Let ρ_0 and ρ_1 be the roots in $(0,1)$ of $\frac{2\rho}{(1-\rho)^2}=\kappa$ and $\frac{2\rho}{(1+\rho)^2}=\kappa$ respectively. The two intervals for ρ defined by the the two sets of inequalities are nonoverlapping. The two intervals are $(0,\rho_0)$ and $(\rho_0,\min(1,\rho_1))$ respectively. The maximum M is defined as M_1 on $(0,\rho_0)$ and as M_2 on $(\rho_0,\min(1,\rho_1))$.

To find the best value of ρ, it suffices to find ρ for which M as a function of ρ is minimal. First, M_1 considered as a function of ρ is minimal for $\rho=\rho_0$. Next, M_2 considered as a function of ρ is minimal at the solution of

$$-\kappa\rho^4+2\rho^3+2\kappa\rho^2+2\rho-\kappa=0 ,$$

i.e. at $\rho=\rho* =(r-\sqrt{2r})/(2r)$ where $r=1+\sqrt{1+4\kappa^2}$. It can be verified that $\rho* \in(\rho_0,\min(1,\rho_1))$. But because $M_1(\rho_0)=M_2(\rho_0)\geq M_2(\rho*)$, it is clear that the overall minimum is attained at $\rho*$. The remainder of the statements of Theorem 7.2 are left as an exercise. ∎

The rejection algorithm based upon the inequality of Theorem 7.2 is given below:

von Mises generator (Best and Fisher, 1979)

[SET-UP]

$$s \leftarrow \frac{1+\rho^2}{2\rho}$$

[GENERATOR]

REPEAT

    Generate iid uniform $[-1,1]$ random variates U, V.

    $Z \leftarrow \cos(\pi U)$

    $W \leftarrow \dfrac{1+sZ}{s+Z}$

    $Y \leftarrow \kappa(s-W)$

    Accept $\leftarrow [W(2-W)-V \geq 0]$ (Quick acceptance step)

    IF NOT Accept THEN Accept $\leftarrow [\log(\frac{W}{V})+1-W \geq 0]$

UNTIL Accept

RETURN $\Theta \leftarrow \dfrac{\text{sign}(U)}{\cos(W)}$

Two final computational remarks. The cosine in the definition of Z can be avoided by using an appropriate polar method. The cosine in the last statement of the algorithm cannot be avoided.

7.4. The Burr distribution.

In a series of papers, Burr (1942, 1968, 1973) has proposed a versatile family of densities. For the sake of completeness, his original list is reproduced here. The parameters r, k, c are positive real numbers. The fact that k could take non-integer values is bound to be confusing, but at this point it is undoubtedly better to stick to the standard notation. Note that a list of distribution functions, not densities, is provided in the table.

NAME	$F(x)$	RANGE FOR x
Burr I	x	$[0,1]$
Burr II	$(1+e^{-x})^{-r}$	$(-\infty,\infty)$
Burr III	$(1+x^{-k})^{-r}$	$[0,\infty)$
Burr IV	$(1+(\frac{c-x}{x})^{\frac{1}{c}})^{-r}$	$[0,c]$
Burr V	$(1+ke^{-\tan(x)})^{-r}$	$[-\frac{\pi}{2},\frac{\pi}{2}]$
Burr VI	$(1+ke^{-\sinh(x)})^{-r}$	$(-\infty,\infty)$
Burr VII	$2^{-r}(1+\tanh(x))^{r}$	$(-\infty,\infty)$
Burr VIII	$(\frac{2}{\pi}\arctan(e^{x}))^{r}$	$(-\infty,\infty)$
Burr IX	$1-\dfrac{2}{2+k((1+e^{x})^{r}-1)}$	$(-\infty,\infty)$
Burr X	$(1+e^{-x^{2}})^{r}$	$[0,\infty)$
Burr XI	$(x-\frac{1}{2\pi}\sin(2\pi x))^{r}$	$[0,1]$
Burr XII	$1-(1+x^{c})^{-k}$	$[0,\infty)$

Most of the densities in the Burr family are unimodal. In all cases, we can generate random variates directly via the inversion method. By far the most important of these distributions is the Burr XII distribution. The corresponding density,

$$f(x) = \frac{kcx^{c-1}}{(1+x^{c})^{k}} \quad (x \geq 0)$$

with parameters $c,k>0$ can take a variety of shapes. Thus, f is particularly useful as a flexible dominating curve in random variate generation (see e.g. Cheng (1977)). As pointed out by Tadikamalla (1980), the Burr III density is even more flexible. It is called the reciprocal Burr distribution because the reciprocal of a Burr XII with parameters c,k has the Burr III distribution function

$$F(x) = \frac{1}{(1+x^{c})^{k}} \; .$$

The density is

$$f(x) = \frac{kcx^{ck-1}}{(1+x^{c})^{k+1}} \; .$$

It should be noted that a myriad of relationships exist between all the Burr distributions, because of the fact that all are directly related to the uniform distribution via the probability integral transform.

7.5. The generalized inverse gaussian distribution.

The **generalized inverse gaussian**, or GIG, distribution is a three-parameter distribution with density

$$f(x) = \frac{(\frac{\psi}{\chi})^{\frac{\lambda}{2}}}{2K_\lambda(\sqrt{\psi\chi})} x^{\lambda-1} e^{-\frac{1}{2}(\frac{\chi}{x}+\psi x)} \qquad (x > 0).$$

Here $\lambda \in R$, $\chi > 0$, and $\psi > 0$ are the parameters of the distribution, and K_λ is the modified Bessel function of the third kind, defined by

$$K_\lambda(u) = \frac{1}{2} \int_{-\infty}^{\infty} \cosh(\lambda u) e^{-z\cosh(u)} du.$$

A random variable with the density given above will be called a GIG (λ, ψ, χ) random variable. The GIG family was introduced by Barndorff-Nielsen and Halgreen (1977), and its properties are reviewed by Blaesild (1978) and Jorgensen (1982). The individual densities are gamma-shaped, and the family has had quite a bit of success recently because of its applicability in modeling. Furthermore, many well-known distributions are but special cases of GIG distributions. To cite a few:

A. $\chi = 0$: the gamma density.

B. $\psi = 0$: the density of the inverse of a gamma random variable.

C. $\lambda = -\frac{1}{2}$: the inverse gaussian distribution (see section IV.4.3).

Furthermore, the GIG distribution is closely related to the generalized hyperbolic distribution (Barndorff-Nielsen (1977, 1978), Blaesild (1978), Barndorff-Nielsen and Blaesild (1980)), which is of interest in itself. For the relationship, we refer to the exercises.

We begin with a partial list of properties, which show that there are really only two shape parameters, and that for random variate generation purposes, we need only consider the cases of $\chi = \psi$ and $\lambda > 0$.

Lemma 7.4.

   Let GIG (.,.,.) and Gamma (.) denote GIG and gamma distributed random variables with the given parameters, and let all random variables be independent. Then, we have the following distributional equivalences:

A. $\text{GIG}(\lambda,\psi,\chi) = \dfrac{1}{c}\text{GIG}(\lambda,\dfrac{\psi}{c},\chi c)$ for all $c > 0$. In particular,

$$\text{GIG}(\lambda,\psi,\chi) = \sqrt{\frac{\chi}{\psi}}\,\text{GIG}(\lambda,\sqrt{\psi\chi},\sqrt{\psi\chi})\ .$$

B.

$$\text{GIG}(\lambda,\psi,\psi) = \text{GIG}(-\lambda,\psi,\psi) + \frac{2}{\psi}\text{Gamma}(\lambda)\ .$$

C.

$$\text{GIG}(\lambda,\psi,\chi) = \frac{1}{\text{GIG}(-\lambda,\chi,\psi)}\ .$$

For random variate generation purposes, we will thus assume that $\chi=\psi$ and that $\lambda>0$. All the other cases can be taken care of via the equivalences shown in Lemma 7.4. By considering $\log(f\ )$, it is not hard to verify that the distribution is unimodal with mode m at

$$m = \frac{1}{\sqrt{(\frac{\lambda-1}{\psi})^2+1}} - \frac{\lambda-1}{\psi}$$

In addition, the density is log concave for $\lambda \geq 1$. In view of the analysis of section VII.2, we know that this is good news. Log concave densities can be dealt with quite efficiently in a number of ways. First of all, one could employ the universal algorithm for log concave densities given in section VII.2. This has two disadvantages: first, the value of $f\ (m\ )$ has to be computed at least once for every choice of the parameters (recall that this involves computing the modified Bessel function of the third kind); second, the expected number of iterations in the rejection algorithm is large (but not more than 4). The advantages are that the user does not have to do any error-prone computations, and that he has the guarantee that the expected time is uniformly bounded over all $\psi>0$, $\lambda\geq1$. The expected number of iterations can further be reduced by using the non-universal rejection method of section VII.2.6, which uses rejection from a density with a flat part around m, and two exponential tails. In Theorem 2.6, a simple formula is given for the location of the points where the exponential tails should touch f : place these points such that the value of f at the points is $\dfrac{1}{e}f\ (m\ )$. Note that to solve this equation, the normalization constant in f cancels out conveniently.

Because $f(0)=0$, the equation has two well-defined solutions, one on each side of the mode. In some cases, the numerical solution of the equation is well worth the trouble. If one just cannot afford the time to solve the equation numerically, there is always the possibility of placing the points symmetrically at distance $e/((e-1)f(m)$ from m (see section VII.2.6), but this would again involve computing $f(m)$. Atkinson (1979,1982) also uses two exponential tails, both with and without flat center parts, and to optimize the dominating curve, he suggests a crude step search. In any case, the generation process for f can be automated for the case $\lambda \geq 1$.

When $0<\lambda<1$, f is log concave for $x \leq \psi/(1-\lambda)$, and is log convex otherwise. Note that this cut-off point is always greater than the mode m, so that for the part of the density to the left of m, we can use the standard exponential/constant dominating curve as described above for the case $\lambda \geq 1$. The right tail of the GIG density can be bounded by the gamma density (by omitting the $1/x$ term in the exponent). For most choices of $\lambda < 1$ and $\psi > 0$, this is satisfactory.

7.6. Exercises.

1. **The generalized logistic distribution.** When X is beta (a,b), then $\log(\dfrac{X}{1-X})$ is generalized logistic with parameters (a,b) (Johnson and Kotz, 1970; Olusegun George and Ojo, 1980). Give a uniformly fast rejection algorithm for the generation of such random variates when $a=b \geq 1$. Do not use the transformation of a beta method given above.

2. Show that if $L_1, L_2, \ldots$ are iid Laplace random variates, then $\sum\limits_{j=1}^{\infty} \dfrac{L_j}{j^2}$ is logistic. Hint: show first that the logistic distribution has characteristic function $\dfrac{\pi i t}{\sin(\pi i t)} = \Gamma(1-it)\Gamma(1+it)$. Then use a key property of the gamma function.

3. Complete the proof of Theorem 7.2 by proving that for the von Mises generator of Best and Fisher, $\lim\limits_{\kappa \to \infty} c = \sqrt{\dfrac{2\pi}{e}}$.

4. **The Pearson system.** In the beginning of this century, Karl Pearson developed his well-known family of distributions. The Pearson system was, and still is, very popular because the family encompasses nearly all well-known distributions, and because every allowable combination of skewness and kurtosis is covered by at least one member of the family. The family has 12 member distributions, and is described in great detail in Johnson and Kotz (1970). In 1973, McGrath and Irving pointed out that random variates for 11 member distributions can be generated by simple transformations of one or two beta or gamma random variates. The exception is the Pearson IV distribution. Fortunately, the Pearson IV density is log-concave, and can be dealt with quite efficiently using the methods of section VII.2 (see exercise

VII.2.1). The Pearson densities are listed in the table below. In the table, a,b,c,d are shape parameters, and C is a normalization constant. Verify the correctness of the generators, and in doing so, determine the normalization constants C.

PEARSON DENSITIES				
Pearson	$f(x)$	PARAMETERS	SUPPORT	GENERATOR
I	$C(1+\frac{x}{a})^b(1-\frac{x}{c})^d$	$b,d>-1; a,c>0$	$[-a,c]$	$\frac{(a+c)X}{X+Y}-a$ X gamma(b) Y gamma(d)
II	$C(1-(\frac{x}{a})^2)^b$	$b>-1; a>0$	$[-a,a]$	$\frac{a(X-Y)}{X+Y}$ X gamma($b+1$) Y gamma($b+1$)
III	$C(1+\frac{x}{a})^{ba}e^{-bx}$	$ba>-1; b>0$	$[-a,\infty]$	$\frac{X}{b}-a$ X gamma($ba+1$)
IV	$C(1+(\frac{x}{a})^2)^{-b}e^{-c\arctan(\frac{x}{a})}$	$a>0; b>\frac{1}{2}$		
V	$Cx^{-b}e^{-\frac{c}{x}}$	$b>1; c>0$	$[0,\infty)$	$\frac{1}{cX}$ X gamma($b-1$)
VI	$C(x-a)^b x^{-c}$	$c>b+1>0; a>0$	$[a,\infty)$	$a\frac{X+Y}{X}$ X gamma($c-b-1$) Y gamma($b+1$)
VII	$C(1+(\frac{x}{a})^2)^{-b}$	$b>\frac{1}{2}; a>0$		$\frac{aN}{\sqrt{X/2}}$ N normal X gamma($b-\frac{1}{2}$)
VIII	$C(1+\frac{x}{a})^{-b}$	$0\le b\le 1; a>0$	$[-a,0]$	$a(U^{-\frac{1}{b-1}}-1)$ U uniform[0,1]
IX	$C(1+\frac{x}{a})^b$	$b>0; a>0$	$[-a,0]$	$a(U^{\frac{1}{b+1}}-1)$ U uniform[0,1]
X	$\frac{1}{a}e^{-\frac{x}{a}}$	$a>0$	$[0,\infty)$	aE E exponential
XI	$C(\frac{a}{x})^b$	$a>0; b>1$	$[a,\infty)$	$aU^{-\frac{1}{b-1}}$ U uniform[0,1]
XII	$C(\frac{a+x}{b-x})^c$	$0<b<a; 0\le c<1$	$[-a,b]$	$(a+b)X-a$ X beta($c+1,1-c$)

5. **The arcsine distribution.** A random variable X on $[-1,1]$ is said to have an arcsine distribution if its density is of the form $f(x)=(\pi\sqrt{1-x^2})^{-1}$. Show first that when U,V are iid uniform $[0,1]$ random variables, then $\sin(\pi U),\sin(2\pi U)$, $-\cos(2\pi U)$, $\sin(\pi(U+V))$, and $\sin(\pi(U-V))$ are all have

the arcsine distribution. This immediately suggests several polar methods for generating such random variates: prove, for example, that if (X,Y) is uniformly distributed in C_2, then $(X^2-Y^2)/(X^2+Y^2)$ has the arcsine distribution. Using the polar method, show further the following properties for iid arcsine random variables X,Y:

(i) XY is distributed as $\frac{1}{2}(X+Y)$ (Norton, 1978).

(ii) $\frac{1+X}{2}$ is distributed as X^2 (Arnold and Groeneveld, 1980).

(iii) X is distributed as $2X\sqrt{1-X^2}$ (Arnold and Groeneveld, 1980).

(iv) X^2-Y^2 is distributed as XY (Arnold and Groeneveld, 1980).

6.   **Ferreri's system.** Ferreri (1964) suggests the following family of densities:

$$f(x) = \frac{\sqrt{b}}{C\left(c+e^{a+b(x-\mu)^2}\right)} \ ,$$

where a,b,c,μ are parameters, and

$$C = \Gamma(\frac{1}{2}) \sum_{j=1}^{\infty} (-c)^{j-1} e^{-ja} \, j^{-\frac{1}{2}}$$

is a normalization constant. The parameter c takes only the values ± 1. As $a \to \infty$, the density approaches the normal density. Develop an efficient uniformly fast generator for this family.

7.   The family of distributions of the form $aX+bY$ where $a,b \in R$ are parameters, and X,Y are iid gamma random variables was proposed by McKay (1932) and studied by Bhattacharyya (1942). This family has basically two shape parameters. Derive its density, and note that its form is a product of a gamma density multiplied with a modified Bessel function of the second kind when $a,b > 0$.

8.   **Toranzos's system.** Show how you can generate random variates from Toranzos's class (Toranzos, 1952) of bell-shaped densities of the form $Cx^c \, e^{-(a+bx)^2}$ $(x>0)$ (C is a normalization constant) in expected time uniformly bounded over all allowable values of the parameters. Do not use C in the generator, and do not compute C for the proof of the uniform boundedness of the expected time.

9.   **Tukey's lambda distribution.** In 1960, Tukey proposed a versatile family of symmetric densities in terms of the inverse distribution function:

$$F^{-1}(U) = \frac{1}{\lambda}(U^{\lambda}-(1-U)^{\lambda}) \ ,$$

where $\lambda \in R$ is a shape parameter. Clearly, if U is a uniform [0,1] random variate, then $F^{-1}(U)$ has the given distribution. Note that the density is not known in closed form. Tukey's distribution was later generalized in several directions, first by Ramberg and Schmeiser (1972) who added a location and a scale parameter. The most significant generalization was by Ramberg and Schmeiser (1974), who defined

$$F^{-1}(U) = \lambda_1 + \frac{1}{\lambda_2}(U^{\lambda_3} - (1-U)^{\lambda_4}) \; .$$

For yet another generalization, see Ramberg (1975). In the Ramberg-Schmeiser form, λ_1 is a location parameter, and λ_2 is a scale parameter. The merit of this family of distributions is its versatility with respect to its use in modeling data. Furthermore, random variate generation is trivial. It is therefore important to understand which shapes the density can take. Prove all the statements given below.

A.   As $\lambda_3 = \lambda_4 \to 0$, the density tends to the logistic density.

B.   The density is J-shaped when $\lambda_3 = 0$.

C.   When $\lambda_1 = \lambda_3 = 0$, and $\lambda_2 = \lambda_4 \to 0$, the density tends to the exponential density.

D.   The density is U-shaped when $1 \le \lambda_3, \lambda_4 \le 2$.

E.   Give necessary and sufficient conditions for the distribution to be truncated on the left (right).

F.   No positive moments exist when $\lambda_3 < -1$ and $\lambda_4 > 1$, or vice versa.

G.   The density $f(x)$ can be found by computing $1/F^{-1\prime}(u)$, where u is related to x via the equality $x = F^{-1}(u)$. Thus, by letting u vary between 0 and 1, we can compute pairs $(x, f(x))$, and thus plot the density.

H.   Show that for $\lambda_1 = 0$, $\lambda_2 = 0.1975$, $\lambda_3 = \lambda_4 = 0.1349$, the distribution function thus obtained differs from the normal distribution function by at most 0.002.

For a general description of the family, and a more complete bibliography, see Ramberg, Tadikamalla, Dudewicz and Mykytka (1979).

10.   **The hyperbolic distribution.** The hyperbolic distribution, introduced by Barndorff-Nielsen (1977, 1978) has density

$$f(x) = \frac{\varsigma}{2\alpha K_1(\varsigma)} e^{-\alpha\sqrt{1+x^2} + \beta x} \; .$$

Here $\alpha > |\beta|$ are the parameters, $\varsigma = \sqrt{\alpha^2 - \beta^2}$, and K_1 is the modified Bessel function of the third kind. For $\beta = 0$, the density is symmetric. Show the following:

A.   The distribution is log-concave.

B.   If N is normally distributed, and X is GIG $(1, \alpha^2 - \beta^2, 1)$, then $\beta X + N \sqrt{X}$ has the given density.

C.   The parameters for the optimal non-universal rejection algorithm for log-concave densities are explicitly computable. (Compute them, and obtain an expression for the expected number of iterations. Hint: apply Theorem VII.2.6.)

11.   **The hyperbola distribution.** The hyperbola distribution, introduced by Barndorff-Nielsen (1978) has density

$$f(x) = \frac{1}{2K_0(\varsigma)\sqrt{1+x^2}} e^{-\alpha\sqrt{1+x^2}+\beta x} .$$

Here $\alpha > |\beta|$ are the parameters, $\varsigma = \sqrt{\alpha^2-\beta^2}$, and K_0 is the modified Bessel function of the third kind. For $\beta=0$, the density is symmetric. Show the following:

A.   The distribution is not log-concave.

B.   If N is normally distributed, and X is GIG $(0,\alpha^2-\beta^2,1)$, then $\beta X + N\sqrt{X}$ has the given density.

12. **Johnson's system.** Every possible combination of skewness and kurtosis corresponds to one and only one distribution in the Pearson system. Other systems have been designed to have the same property too. For example, Johnson (1949) introduced a system defined by the densities of suitably transformed normal (μ,σ) random variables N: his system consists of the S_L, or lognormal, densities (of e^N), of the S_B densities (of $e^N/(1+e^N)$), and the S_U densities (of $\sinh(N) = \frac{1}{2}(e^N_* - e^{-N})$). This system has the advantage that fitting of parameters by the method of percentiles is simple. Also, random variate generation is simple. In Johnson (1954), a similar system in which N is replaced by a Laplace random variate with center at μ and variance σ^2 is described. Give an algorithm for the generation of a Johnson system random variable when the skewness and kurtosis are given (recall that after normalization to zero mean and unit variance, the skewness is the third moment, and kurtosis is the fourth moment). Note that this forces you in effect to determine the different regions in the skewness-kurtosis plane. You should be able to test very quickly which region you are in. However, your main problem is that the equations linking μ and σ to the skewness and kurtosis are not easily solved. Provide fast-convergent algorithms for their numerical solution.

Chapter Ten
DISCRETE UNIVARIATE DISTRIBUTIONS

1. INTRODUCTION.

1.1. Goals of this chapter.

We will provide the reader with some generators for the most popular families of discrete distributions, such as the geometric, binomial and Poisson distributions. These distributions are the fundamental building blocks in discrete probability. It is impossible to cover most distributions commonly used in practice. Indeed, there is a strong tendency to work more and more with so-called generalized distributions. These distributions are either defined constructively by combining more elementary distributions, or analytically by providing a multi-parameter expression for the probability vector. In the latter case, random variate generation can be problematic since we cannot fall back on known distributions. Users are sometimes reluctant to design their own algorithms by mimicking the designs for similar distributions. We therefore include a short section with universal algorithms. These are in the spirit of chapter VII: the algorithms are very simple albeit not extremely fast, and very importantly, their expected time performance is known. Armed with the universal algorithms, the worked out examples of this chapter and the table methods of chapter VIII, the users should be able to handle most distributions to their satisfaction.

We assume throughout this chapter that the discrete random variables are all integer-valued.

1.2. Generating functions.

Let X be an integer-valued random variable with probability vector

$$p_i = P(X=i) \quad (i \text{ integer}) .$$

An important tool in the study of discrete distributions is the **moment generating function**

$$m(s) = E(e^{sX}) = \sum_i p_i e^{si} .$$

It is possible that $m(s)$ is not finite for some or all values $s > 0$. That of course is the main difference with the characteristic function of X. If $m(s)$ is finite in some open interval containing the origin, then the coefficient of $s^n/n!$ in the Taylor series expansion of $m(s)$ is the n-th moment of X.

A related tool is the **factorial moment generating function**, or simply generating function,

$$k(s) = E(s^X) = \sum_i p_i s^i ,$$

which is usually only employed for nonnegative random variables. Note that the series in the definition of $k(s)$ is convergent for $|s| \leq 1$ and that $m(s) = k(e^s)$. Note also that provided that the n-th factorial moment (i.e., $E(X(X-1) \cdots (X-n+1))$) of X is finite, we have

$$k^{(n)}(1) = E(X(X-1) \cdots (X-n+1)) .$$

In particular $E(X)=k'(1)$ and $Var(X)=k''(1)+k'(1)-k'^2(1)$. The generating function provides us often with the simplest method for computing moments.

It is clear that if $X_1, \ldots, X_n$ are independent random variables with moment generating functions $m_1, \ldots, m_n$, then $\sum X_i$ has moment generating function $\prod m_i$. The same property remains valid for the generating function.

Example 1.1. The binomial distribution.

A **Bernoulli** (p) **random variable** is a $\{0,1\}$-valued random variable taking the value 1 with probability p. Thus, it has generating function $1-p+ps$. A **binomial** (n,p) **random variable** is defined as the sum of n iid Bernoulli (p) random variables. Thus, it has generating function $(1-p+ps)^n$. ∎

Example 1.2. The Poisson distribution.

Often it is easy to compute generating functions by explicitly computing the convergent infinite series $\sum s^i p_i$. This will be illustrated for the Poisson and geometric distributions. X is **Poisson** (λ) when $P(X=i)=\dfrac{\lambda^i}{i!}e^{-\lambda}$ $(i \geq 0)$. By summing $s^i p_i$, we see that the generating function is $e^{-\lambda+\lambda s}$. X is **geometric** (p) when $P(X=i)=(1-p)^i p$ $(i \geq 0)$. The corresponding generating function is $p/(1-(1-p)s)$. ■

If one is shown a generating function, then a careful analysis of its form can provide valuable clues as to how a random variable with such generating function can be obtained. For example, if the generating function is of the form

$$g(k(s))$$

where g, k are other generating functions, then it suffices to take $X_1 + \cdots + X_N$ where the X_i's are iid random variables with generating function k, and N is an independent random variable with generating function g. This follows from

$$g(k(s)) = \sum_{n=0}^{\infty} P(N=n)k^n(s) \quad \text{(definition of } g)$$

$$= \sum_{n=0}^{\infty} P(N=n) \sum_{i=0}^{\infty} P(X_1 + \cdots + X_n = i)s^i$$

$$= \sum_{i=0}^{\infty} \left(s^i \sum_{n=0}^{\infty} P(N=n)P(X_1 + \cdots + X_n = i) \right)$$

$$= \sum_{i=0}^{\infty} s^i P(X_1 + \cdots + X_N = i).$$

Example 1.3.

If $X_1, \ldots$ are Bernoulli (p) random variables and N is Poisson (λ), then $X_1 + \cdots + X_N$ has generating function

$$e^{-\lambda+\lambda(1-p+ps)} = e^{-\lambda p + \lambda p s},$$

i.e. the random sum is Poisson (λp) distributed (we already knew this - see chapter VI). ■

A **compound Poisson distribution** is a distribution with generating function of the form $e^{-\lambda+\lambda k(s)}$, where k is another generating function. By taking

$k(s)=s$, we see that the Poisson distribution itself is a compound Poisson distribution. Another example is given below.

Example 1.4. The negative binomial distribution.

We define the **negative binomial distribution** with parameters (n,p) ($n \geq 1$ is integer, $p \in (0,1)$) as the distribution of the sum of n iid geometric random variables. Thus, it has generating function

$$(\frac{p}{1-(1-p)s})^{n} = e^{-\lambda+\lambda k(s)}$$

where $\lambda = n \log(\frac{1}{p})$ and

$$k(s) = \frac{\log(1-(1-p)s)}{\log(p)}$$

$$= -\frac{1}{\log(p)} \sum_{i=1}^{\infty} \frac{(1-p)^i}{i} s^i \ .$$

The function $k(s)$ is the generating function of the **logarithmic series distribution** with parameter $1-p$. Thus, we have just shown that the negative binomial distribution is a compound Poisson distribution, and that a negative binomial random variable can be generated by summing a Poisson (λ) number of iid logarithmic series random variables (Quenouille, 1949). ■

Another common operation is the mixture operation. Assume that given Y, X has generating function $k_Y(s)$ where Y is a parameter, and that Y itself has some (not necessarily discrete) distribution. Then the unconditional generating function of X is $E(k_Y(s))$. Let us illustrate this once more on the negative binomial distribution.

Example 1.5. The negative binomial distribution.

Let Y be gamma $(n, \frac{1-p}{p})$, and let k_Y be the Poisson (Y) generating function. Then

$$E(k_Y(s)) = \int_0^{\infty} \frac{y^n e^{-\frac{py}{1-p}}}{\Gamma(n)(\frac{1-p}{p})^n} e^{-y+ys} \ dy$$

$$= (\frac{p}{1-(1-p)s})^n \ .$$

We have discovered yet another property of the negative binomial distribution with parameters (n, p), i.e. it can be generated as a Poisson (Y) random variable where Y in turn is a gamma $(n, \frac{1-p}{p})$ random variable. This property will be of great use to us for large values of n, because uniformly fast gamma and Poisson generators are in abundant supply. ∎

1.3. Factorials.

The evaluation of the probabilities p_i frequently involves the computation of one or more factorials. Because our main worry is with the complexity of an algorithm, it is important to know just how we evaluate factorials. Should we evaluate them explicitly, i.e. should $n!$ be computed as $\prod_{i=1}^{n} i$, or should we use a good approximation for $n!$ or $\log(n!)$? In the former case, we are faced with time complexity proportional to n, and with accumulated round-off errors. In the latter case, the time complexity is $O(1)$, but the price can be steep. Stirling's series for example is a divergent asymptotic expansion. This means that for fixed n, taking more terms in the series is bad, because the partial sums in the series actually diverge. The only good news is that it is an asymptotic expansion: for a fixed number of terms in the series, the partial sum thus obtained is $\log(n!)+o(1)$ as $n \to \infty$. An algorithm based upon Stirling's series can only be used for n larger than some threshold n_0, which in turn depends upon the desired error margin.

Since our model does not allow inaccurate computations, we should either evaluate factorials as products, or use squeeze steps based upon Stirling's series to avoid the product most of the time, or avoid the product altogether by using a convergent series. We refer to sections X.3 and X.4 for worked out examples. At issue here is the tightness of the squeeze steps: the bounds should be so tight that the contribution of the evaluation of products in factorials to the total expected complexity is $O(1)$ or $o(1)$. It is therefore helpful to recall a few facts about approximations of factorials (Whittaker and Watson, 1927, chapter 12). We will state everything in terms of the gamma function since $n!=\Gamma(n+1)$.

Lemma 1.1. (Stirling's series, Whittaker and Watson, 1927.)

For $x > 0$, the value of $\log(\Gamma(x)) - (x - \frac{1}{2})\log(x) + x - \frac{1}{2}\log(2\pi)$ always lies between the n-th and $n + 1$-st partial sums of the series

$$\sum_{i=1}^{\infty} \frac{(-1)^{i-1}B_i}{2i(2i-1)x^{2i-1}}$$

where B_i is the i-th Bernoulli number defined by

$$B_n = 4n\int_0^{\infty} \frac{t^{2n-1}}{e^{2\pi t} - 1} \, dt \ .$$

In particular, $B_1 = \frac{1}{6}, B_2 = \frac{1}{30}, B_3 = \frac{1}{42}, B_4 = \frac{1}{30}, B_5 = \frac{5}{66}, B_6 = \frac{691}{2730}, B_7 = \frac{7}{6}.$
We have as special cases the inequalities

$$(x + \frac{1}{2})\log(x+1) - (x+1) + \frac{1}{2}\log(2\pi) \leq \log(\Gamma(x+1))$$

$$\leq (x + \frac{1}{2})\log(x+1) - (x+1) + \frac{1}{2}\log(2\pi) + \frac{1}{12(x+1)} \ .$$

Stirling's series with the Whittaker-Watson lower and upper bounds of Lemma 1.1 is often sufficient in practice. As we have pointed out earlier, we will still have to evaluate the factorial explicitly no matter how many terms are considered in the series, and in fact, things could even get worse if more terms are considered. Luckily, there is a convergent series, attributed by Whittaker and Watson to Binet.

Lemma 1.2. (Binet's series for the log-gamma function.)

For $x > 0$,

$$\log(\Gamma(x)) = (x - \frac{1}{2})\log(x) - x + \frac{1}{2}\log(2\pi) + R(x) ,$$

where

$$R(x) = \frac{1}{2}\left[\frac{c_1}{(x+1)} + \frac{c_2}{2(x+1)(x+2)} + \frac{c_3}{3(x+1)(x+2)(x+3)} + \cdots \right] ,$$

in which

$$c_n = \int_0^1 (u+1)(u+2) \cdots (u+n-1)(2u-1)u \; du .$$

In particular, $c_1 = \frac{1}{6}$, $c_2 = \frac{1}{3}$, $c_3 = \frac{59}{60}$, and $c_4 = \frac{227}{60}$. All terms in $R(x)$ are positive: thus, the value of $\log(\Gamma(x))$ is approached monotonically from below as we consider more terms in $R(x)$. If we consider the first n terms of $R(x)$, then the error is at most

$$C \frac{x+1}{x} \left(\frac{x+1}{x+n+1} \right)^x ,$$

where $C = \frac{5}{48}\sqrt{4\pi} e^{1/6}$. Another upper bound on the truncation error is provided by

$$C(1 + a + \frac{1}{x+1})(\frac{a}{1+a} + \frac{1}{x+1})^{n+1} + C\frac{x+1}{x}(\frac{1}{1+a})^x .$$

where $a \in (0,1]$ is arbitrary (when x is large compared to n, then the value $\frac{n+1}{x}\log(\frac{x}{n+1})$ is suggested).

Proof of Lemma 1.2.

Binet's convergent series is given for example in Whittaker and Watson (1927, p. 253). We need only establish upper bounds for the tail sum in $R(x)$ beginning with the $n+1$-st term. The integrand in c_i is positive for $u > \frac{1}{2}$. Thus, the i-th term is at most

$$\frac{i! \int_{1/2}^1 (2u-1)u \; du}{2i(x+1) \cdots (x+i)} = \frac{5(i-1)!}{48(1+x) \cdots (i+x)}$$

$$= \frac{5\Gamma(i)\Gamma(x+1)}{48\Gamma(i+x+1)}$$

$$\leq \frac{5}{48}\sqrt{\frac{2\pi(x+i+1)}{i(x+1)}} e^{\frac{1}{12i} + \frac{1}{12(x+1)}} (\frac{i}{x+i+1})^i (\frac{x+1}{x+i+1})^{x+1}$$

(by Lemma 1.1)

$$\le \; C(\frac{i}{x+i+1})^i(\frac{x+1}{x+i+1})^{x+1}$$

where $C = \dfrac{5}{48}\sqrt{4\pi}e^{1/6}$ (use the facts that $x>0, i \ge 1$). We obtain a first bound for the sum of all tail terms starting with $i = n+1$ as follows:

$$\sum_{i=n+1}^{\infty} C(\frac{i}{x+i+1})^i(\frac{x+1}{x+i+1})^{x+1} \le \sum_{i=n+1}^{\infty} C(\frac{x+1}{x+i+1})^{x+1}$$

$$\le \int_{n}^{\infty} C(\frac{x+1}{x+t+1})^{x+1}\, dt$$

$$= C\frac{x+1}{x}(\frac{x+1}{x+n+1})^x \; .$$

Another bound is obtained by choosing a constant $a \in (0,1)$, and splitting the tail sum into a sum from $i = n+1$ to $i = m = \lceil a(x+1) \rceil$, and a right-infinite sum starting at $i = m+1$. The first sum does not exceed

$$\sum_{i=n+1}^{m} C(\frac{i}{x+i+1})^i \le \sum_{i=n+1}^{\infty} C(\frac{m}{x+m+1})^i = C\frac{x+m+1}{x+1}(\frac{m}{x+m+1})^{n+1}$$

$$\le \; C(1+a+\frac{1}{x+1})(\frac{a}{1+a}+\frac{1}{x+1})^{n+1} \; .$$

Adding the two sums gives us the following upper bound for the remainder of the series starting with the $n+1$-st term:

$$C(1+a+\frac{1}{x+1})(\frac{a}{1+a}+\frac{1}{x+1})^{n+1} + C\frac{x+1}{x}(\frac{1}{1+a})^x \; . \; \blacksquare$$

The error term given in Lemma 1.2 can be made to tend to 0 merely by keeping n fixed and letting x tend to ∞. Thus, Binet's series is also an asymptotic expansion, just as Stirling's series. It can be used to bypass the gamma function (or factorials) altogether if one needs to decide whether $\log(\Gamma(x)) \le t$ for some real number t. By taking n terms in Binet's series, we have an interval $[a_n, b_n]$ to which we know $\log(\Gamma(x))$ must belong. Since $b_n - a_n \to 0$ as $n \to \infty$, we know that when $t \ne \log(\Gamma(x))$, from a given n onwards, t will fall outside the interval, and the appropriate decision can be made. The convergence of the series is thus essential to insure that this method halts. In our applications, t is usually a uniform or exponential random variable, so that equality $t = \log(\Gamma(x))$ occurs with probability 0. The complexity analysis typically boils down to computing the expected number of terms needed in Binet's series for fixed x. A quantity useful in this respect is

$$\sum_{n=0}^{\infty} n(b_n - a_n) \; .$$

Based upon the error bounds of Lemma 1.2, it can be shown that this sum is $o(1)$ as $x \to \infty$, and that the sum is uniformly bounded over all $x \geq 1$ (see exercise 1.2). As we will see later, this implies that for many rejection algorithms, the expected time spent on the decision is uniformly bounded in x. Thus, it is almost as if we can compute the gamma function in constant time, just as the exponential and logarithmic functions. In fact, there is nothing that keeps us from adding the gamma function to our list of constant time functions, but unless explicitly mentioned, we will not do so. Another collection of inequalities useful in dealing with factorials via Stirling's series is given in Lemma 1.3:

Lemma 1.3. (Knopp, 1964, pp. 543,548)

For integer n, we have

$$\log(n!) = (n + \frac{1}{2})\log(n) - n + \log(\sqrt{2\pi}) + \sum_{j=1}^{k} \frac{(-1)^{j-1} B_j \, n^{-(2j-1)}}{(2j-1)(2j)} + R_{k,n}$$

where $B_1, B_2, \cdots$ are the Bernoulli numbers and

$$| R_{k,n} | \leq \frac{4(2k-1)!}{2\pi(2\pi n)^{2k}}$$

is a residual factor.

1.4. A universal rejection method.

Even when the probabilities p_i are explicitly given, it is often hard to come up with an efficient generator. Quantities such as the mode, the mean and the variance are known, but a useful dominating curve for use in a rejection algorithm is generally not known. The purpose of this section is to go through the mechanics of deriving one acceptable rejection algorithm, which will be useful for a huge class of distributions, the class of all unimodal distributions on the integers for which three quantities are known:

1. m, the location of the mode. If the mode is not unique, i.e. several adjacent integers are all modes, m is allowed to be any real number between the leftmost and rightmost modes.

2. M, an upper bound for the value of p_i at a mode i. If possible, M should be set equal to this value.

3. s^2, an upper bound for the second moment about m. Note that if the variance σ^2 and mean μ are known, then we can take $s^2 = \sigma^2 + (m-\mu)^2$.

The universal algorithm derived below is based upon the following inequalities:

Theorem 1.1.

For all unimodal distributions on the integers,

$$p_i \leq \min(M, \frac{3s^2}{|i-m|^3}) \quad (i \text{ integer}) .$$

In addition, for all integer i and all $x \in [i - \frac{1}{2}, i + \frac{1}{2}]$,

$$p_i \leq g(x) = \min(M, \frac{3s^2}{(|x-m| - \frac{1}{2})_+^3}) .$$

Furthermore,

$$\int g = M + 3(3s^2)^{\frac{1}{3}} M^{\frac{2}{3}} .$$

Proof of Theorem 1.1.

Note that for $i > m$,

$$s^2 = \sum_{j=-\infty}^{\infty} (j-m)^2 p_j \geq \sum_{i \geq j \geq m} (j-m)^2 p_i$$

$$\geq p_i \int_m^i (u-m)^2 \, du = p_i \frac{(i-m)^3}{3} .$$

This establishes the first inequality. The bounding argument for g uses a standard tool for making the transition from discrete probabilities to densities: we consider a histogram-shaped density on the real line with height p_i on $[i - \frac{1}{2}, i + \frac{1}{2})$. This density is bounded by $g(x)$ on the interval in question. Note the adjustment by a translation term of $\frac{1}{2}$ when compared with the first discrete bound. This adjustment is needed to insure that g dominates p_i over the entire interval.

Finally, the area under g is easy to compute. Define $\rho = (3s^2)^{1/3} M^{2/3}$, and observe that the M term in g is the minimum term on $[m - \frac{1}{2} - \frac{\rho}{M}, m + \frac{1}{2} + \frac{\rho}{M}]$. The area under this part is thus $M + 2\rho$. Integrating the two tails of g gives the value ρ. ∎

To understand our algorithm, it helps to go back to the proof of Theorem 1.1. We have turned the problem into a continuous one by replacing the probability vector p_i with a histogram-shaped density of height p_i on $[i - \frac{1}{2}, i + \frac{1}{2})$. Since this histogram is dominated by the function g given in the algorithm, it is clear how to proceed. Note that if Y is a random variable with the said histogram-shaped density, then round(Y) is discrete with probability vector p_i.

Universal rejection algorithm for unimodal distributions

[SET-UP]

Compute $\rho \leftarrow (3s^2)^{\frac{1}{3}} M^{\frac{2}{3}}$.

[GENERATOR]

REPEAT

    Generate U, W uniformly on $[0,1]$ and V uniformly on $[-1,1]$.

    IF $U < \dfrac{\rho}{3\rho + M}$

        THEN

$$Y \leftarrow m + (\frac{1}{2} + \frac{\rho}{M\sqrt{|V|}})\text{sign}(V)$$
$$X \leftarrow \text{round}(Y)$$
$$T \leftarrow WM |V|^{\frac{3}{2}}$$

        ELSE

$$Y \leftarrow m + (\frac{1}{2} + \frac{\rho}{M})V$$
$$X \leftarrow \text{round}(Y)$$
$$T \leftarrow WM$$

UNTIL $T \leq p_X$

RETURN X

In the universal algorithm, no care was taken to reuse unused portions of uniform random variates. This is done mainly to show where independent uniform random variates are precisely needed. The expected number of iterations in the algorithm is precisely $M + 3\rho$. Thus, the algorithm is uniformly fast over a class Q of unimodal distributions with uniformly bounded $(1+s)M$ if p_i can be evaluated in time independent of i and the distribution.

Example 1.6.

For the binomial distribution with parameters n, p, it is known (see section X.4) that the mean μ is np, and that the variance σ^2 is $np(1-p)$. Also, for fixed p, $M \sim 1/(\sqrt{2\pi}\sigma)$, and for all n, p, $M \leq 2/(\sqrt{2\pi}\sigma)$. A mode is at $m = \lfloor (n+1)p \rfloor$. Since $|\mu - m| \leq \min(1, np)$ (exercise 1.4), we can take $s^2 = \sigma^2 + \min(1, np)$. We can verify that

$$\rho^3 \leq \frac{6}{\pi}(1 + \frac{\min(1, np)}{np(1-p)}),$$

and this is uniformly bounded over $n \geq 1, 0 \leq p \leq \frac{1}{2}$. This implies that we can generate binomial random variates uniformly fast provided that the binomial probabilities can be evaluated in constant time. In section X.4, we will see that even this is not necessary, as long as the factorials are taken care of appropriately. We should note that when p remains fixed and $n \to \infty$, $\rho \sim (3/(2\pi))^{1/3}$. The expected number of iterations $\sim 3\rho$, which is about 2.4. Even though this is far from optimal, we should recall that besides the unimodality, virtually no properties of the binomial distribution were used in deriving the bounds. ■

There are important sub-families of distributions for which the algorithm given here is uniformly fast. Consider for example all distributions that are sums of iid integer-valued random variables with maximal probability p and finite variance σ^2. Then the sum of n such random variables has variance $n \sigma^2$. Also, $M \leq \dfrac{1}{\sqrt{n(1-p)}}$ (Rogozin (1961); see Petrov (1975, p. 56)). Thus, if the n-sum is unimodal, Theorem 1.1 is applicable. The rejection constant is

$$3\rho + M \leq 3\left(\frac{3\sigma^2}{1-p}\right)^{1/3} + 1$$

uniformly over all n. Thus, we can handle unimodal sums of iid random variables in expected time bounded by a constant not depending upon n. This assumes that the probabilities can all be evaluated in constant time, an assumption which except in the simplest cases is difficult to support. Examples of such families are the binomial family for fixed p, and the Poisson family.

Let us close this section by noting that the rejection constant can be reduced in special cases, such as for monotone distributions, or symmetric unimodal distributions.

1.5. Exercises.

1.   The discrete distributions considered in the text are all lattice distributions. In these distributions, the intervals between the atoms of the distribution are all integral multiples of one quantity, typically 1. Non-lattice distributions can be considerably more difficult to handle. For example, there are discrete distributions whose atoms form a dense set on the positive real line. One such distribution is defined by

$$P\left(X = \frac{i}{j}\right) = \frac{(e-1)^2}{(e^{i+j}-1)^2} ,$$

where i and j are relatively prime positive integers (Johnson and Kotz, 1969, p. 31). The atoms in this case are the rationals. Discuss how you could efficiently generate a random variate with this distribution.

2. Using Lemma 1.2, show that if ϵ_n is a bound on the error committed when using Binet's series for $\log(\Gamma(x))$ with $n \geq 0$ terms, then

$$\sup_{x \geq 1} \sum_{n=0}^{\infty} n \, \epsilon_n < \infty$$

and

$$\lim_{x \to \infty} \sum_{n=0}^{\infty} n \, \epsilon_n = 0 .$$

3. Assume that all p_i's are at most equal to M, and that the variance is at most equal to σ^2. Derive useful bounds for a universal rejection algorithm which are similar to those given in Theorem 1.1. Show that there exists no dominating curve for this class which has area smaller than a constant times $\sigma \sqrt{M}$, and show that your dominating curve is therefore close to optimal. Give the details of the rejection algorithm. When applied to the binomial distribution with parameters n, p varying in such a way that $np \to \infty$, show that the expected number of iterations grows as a constant times $(np)^{\frac{1}{4}}$ and conclude that for this class the universal algorithm is not uniformly fast.

4. Prove that for the binomial distribution with parameters n, p, the mean μ and the mode $m = \lfloor (n+1)p \rfloor$ differ by at most $\min(1, np)$.

5. Replace the inequalities of Theorem 1.1 by new ones when instead of s^2, we are given the r-th absolute moment about the mean $(r \geq 1)$, and value of the mean. The unimodality is still understood, and values for m, M are as in the Theorem.

6. How can the rejection constant ($\int g$) in Theorem 1.1 be reduced for monotone distributions and symmetric unimodal distributions ?

7. **The discrete Student's t distribution.** Ord (1968) introduced a discrete distribution with parameters $m \geq 0$ (m is integer) and $a \in [0,1], b \neq 0$:

$$p_i = K \prod_{j=0}^{m} \frac{1}{(j+a+i)^2 + b^2} \qquad (-\infty < i < \infty) .$$

Here K is a normalization constant. This distribution on the integers has the remarkable property that all the odd moments are zero, yet it is only symmetric for $a = 0, a = \frac{1}{2}$ and $a = 1$. Develop a uniformly fast generator for the case $m = 0$.

8. **Arfwedson's distribution.** Arfwedson (1951) introduced the distribution defined by

$$p_i = \begin{Bmatrix} k \\ i \end{Bmatrix} \sum_{j=0}^{i} (-1)^j \begin{pmatrix} i \\ j \end{pmatrix} (\frac{i-j}{k})^n \qquad (i \geq 0) ,$$

where k, n are positive integers. See also Johnson and Kotz (1969, p. 251). Compute the mean and variance, and derive an inequality consisting of a flat center piece and two decreasing polynomial or exponential tails having the property that the sum of the upper bound expressions over all i is uniformly bounded over k, n.

9.   Knopp (1964, p. 553) has shown that

$$\sum_{n=1}^{\infty} \frac{1}{c\,(4n^2\pi^2 + t^2)} = 1 ,$$

where $c = \dfrac{1}{2t}\left(\dfrac{1}{e^t - 1} - \dfrac{1}{t} + \dfrac{1}{2}\right)$ and $t > 0$ is a parameter. Give a uniformly fast generator for the family of discrete probability vectors defined by this sum.

2. THE GEOMETRIC DISTRIBUTION.

2.1. Definition and genesis.

X is **geometrically distributed** with parameter $p \in (0,1)$ when

$$P(X = i) = p\,(1-p)^{i-1} \quad (i \geq 1) .$$

The geometric distribution is important in statistics and probability because it is the distribution of the waiting time until success in a sequence of Bernoulli trials. In other words, if $U_1, U_2, \ldots$ are iid uniform [0,1] random variables, and X is the index of the first U_i for which $U_i \leq p$, then X is geometric with parameter p. This property can of course be used to generate X, but to do so has some serious drawbacks because the algorithm is not uniformly fast over all values of p: just consider that the number of uniform random variates needed is itself geometric (p), and the expected number of uniform random variates required is

$$E(X) = \frac{1}{p} .$$

For $p \geq \dfrac{1}{3}$, the method is probably difficult to beat in any programming environment.

2.2. Generators.

The experimental method described in the previous section is summarized below:

Experimental method for geometric random variates

$X \leftarrow 0$
REPEAT
      Generate a uniform [0,1] random variate U.
    $X \leftarrow X + 1$
UNTIL $U \leq p$
RETURN X

This method requires on the average $\dfrac{1}{p}$ uniform random variates and $\dfrac{1}{p}$ comparisons and additions. The number of uniform random variates can be reduced to 1 if we use the inversion method (sequential version):

Inversion by sequential search for geometric random variates

Generate a uniform [0,1] random variate U.
$X \leftarrow 1$
Sum$\leftarrow p$
Prod$\leftarrow p$
WHILE $U >$ Sum DO
     Prod$\leftarrow$Prod$(1-p)$
     Sum$\leftarrow$Sum+Prod
    $X \leftarrow X + 1$
RETURN X

Unfortunately, the expected number of additions is now $\dfrac{2}{p} - 2$, the expected number of comparisons remains $\dfrac{1}{p}$, and the expected number of products is $\dfrac{1}{p} - 1$. Inversion in constant time is possible by truncation of an exponential random variate. What we use here is the property that

$$F(i) = P(X \leq i) = 1 - \sum_{j > i} p (1-p)^{j-1} = 1 - (1-p)^{i} \ .$$

Thus, if U is uniform $[0,1]$ and E is exponential, it is clear that

$$\left\lceil \frac{\log(U)}{\log(1-p)} \right\rceil$$

and

$$\left\lceil \frac{-E}{\log(1-p)} \right\rceil$$

are both geometric (p).

If many geometric random variates are needed for one fixed value of p, extra speed can be found by eliminating the need for an exponential random variate and for truncation. This can be done by splitting the distribution into two parts, a tail carrying small probability, and a main body. For the main body, a fast table method is used. For the tail, we can use the memoryless property of the geometric distribution: given that $X>i$, $X-i$ is again geometric (p) distributed. This property follows directly from the genesis of the distribution.

2.3. Exercises.

1. The quantity $\log(1-p)$ is needed in the bounded time inversion method. For small values of p, there is an accuracy problem because $1-p$ is computed before the logarithm. One can create one's own new function by basing an approximation on the series

    $$-(p+\frac{1}{2}p^2+\frac{1}{3}p^3+\cdots).$$

    Show that the following more quickly convergent series can also be used:

    $$\frac{2}{r}(1+\frac{1}{3}r^{-2}+\frac{1}{5}r^{-4}+\cdots),$$

    where $r=1-\frac{2}{p}$.

2. Compute the variance of a geometric (p) random variable.

3. THE POISSON DISTRIBUTION.

3.1. Basic properties.

X is said to be **Poisson** (λ) distributed when

$$P(X=i) = \frac{\lambda^i}{i!} e^{-\lambda} \quad (i \geq 0) .$$

$\lambda > 0$ is the parameter of the distribution. We do not have to convince the readers that the Poisson distribution plays a key role in probability and statistics. It is thus rather important that a simple uniformly fast Poisson generator be available in any nontrivial statistical software package. Before we tackle the development of such generators, we will briefly review some properties of the Poisson distribution. The Poisson probabilities are unimodal with one mode or two adjacent modes. There is always a mode at $\lfloor \lambda \rfloor$. The tail probabilities drop off faster than the tail of the exponential density, but not as fast as the tail of the normal density. In the design of algorithms, it is also useful to know that as $\lambda \to \infty$, the random variable $(X - \lambda)/\sqrt{\lambda}$ tends to a normal random variable.

Lemma 3.1.

When X is Poisson (λ), then X has characteristic function

$$\phi(t) = E(e^{itX}) = e^{\lambda(e^{it}-1)} .$$

It has moment generating function $E(e^{tX}) = \exp(\lambda(e^t - 1))$, and factorial moment generating function $E(t^X) = e^{\lambda(t-1)}$. Thus,

$$E(X) = Var(X) = \lambda .$$

Also, if X, Y are independent Poisson (λ) and Poisson (μ) random variables, then $X + Y$ is Poisson $(\lambda + \mu)$.

Proof of Lemma 3.1.

Note that

$$E(e^{itX}) = \sum_{j=0}^{\infty} e^{-\lambda} \frac{(\lambda e^{it})^j}{j!} = e^{-\lambda + \lambda e^{it}} .$$

The statements about the moment generating function and factorial moment generating function follow directly from this. Also, if the factorial moment generating function is called k, then $k'(1) = E(X) = \lambda$ and $k''(1) = E(X(X-1)) = \lambda^2$. From this we deduce that $Var(X) = \lambda$. The statement about the sum of two independent Poisson random variables follows directly from the form of the characteristic function. ■

3.2. Overview of generators.

The generators proposed over the years can be classified into several groups:

1. Generators based upon the connection with homogeneous Poisson processes (Knuth, 1969). These generators are very simple, but run in expected time proportional to λ.

2. Inversion methods. Inversion by sequential search started at 0 runs in expected time proportional to λ (see below). If the sequential search is started at the mode, then the expected time is $O(\sqrt{\lambda})$ (Fishman, 1976). Inversion can always be sped up by storing tables of constants (Atkinson, 1979).

3. Generators based upon recursive properties of the distribution (Ahrens and Dieter, 1974). One such generator is known to take expected time proportional to $\log(\lambda)$.

4. Rejection methods. Rejection methods seem to lead to the simplest uniformly fast algorithms (Atkinson, 1979; Ahrens and Dieter, 1980; Devroye, 1981; Schmeiser and Kachitvichyanukul, 1981).

5. The acceptance-complement method with the normal distribution as starting distribution. See Ahrens and Dieter (1982). This approach leads to efficient uniformly fast algorithms, but the computer programs are rather long.

We are undoubtedly omitting a large fraction of the literature on Poisson random variate generation. The early papers on the subject often proposed some approximate method for generating Poisson random variates which was typically based upon the closeness of the Poisson distribution to the normal distribution for large values of λ. It is pointless to give an exhaustive historical survey. The algorithms that really matter are those that are either simple or fast or both. The definition of "fast" may or may not include the set-up time. Also, since our comparisons cannot be based upon actual implementations, it is important to distinguish between computational models. In particular, the availability of the factorial in constant time is a crucial factor.

3.3. Simple generators.

The connection between the Poisson distribution and exponential interarrival times in a homogeneous point process is the following.

Lemma 3.2.

If $E_1, E_2, \ldots$ are iid exponential random variables, and X is the smallest integer such that

$$\sum_{i=1}^{X+1} E_i > \lambda ,$$

then X is Poisson (λ).

Proof of Lemma 3.2.

Let f_k be the gamma (k) density. Then,

$$P(X \leq k) = P(\sum_{i=1}^{k+1} E_i > \lambda) = \int_\lambda^\infty f_{k+1}(y) \, dy .$$

Thus, by partial integration,

$$P(X=k) = P(X \leq k) - P(X \leq k-1)$$

$$= \int_\lambda^\infty (f_{k+1}(y) - f_k(y)) \, dy$$

$$= \int_\lambda^\infty (y-k) \frac{y^{k-1}}{k!} e^{-y} \, dy$$

$$= \frac{1}{k!} \int_\lambda^\infty d(-y^k e^{-y})$$

$$= e^{-\lambda} \frac{\lambda^k}{k!} . \blacksquare$$

The algorithm based upon this property is:

Poisson generator based upon exponential inter-arrival times

$X \leftarrow 0$
Sum$\leftarrow 0$
WHILE True DO
  Generate an exponential random variate E.
  Sum$\leftarrow$Sum$+E$
  IF Sum$<\lambda$
    THEN $X \leftarrow X+1$
    ELSE RETURN X

Using the fact that a uniform random variable is distributed as e^{-E}, it is easy to see that Lemma 3.2 is equivalent to Lemma 3.3, and that the algorithm shown above is equivalent to the algorithm following Lemma 3.3:

Lemma 3.3.

  Let $U_1, U_2, ...$ be iid uniform $[0,1]$ random variables, and let X be the smallest integer such that

$$\prod_{i=1}^{X+1} U_i < e^{-\lambda}.$$

Then X is Poisson (λ).

Poisson generator based upon the multiplication of uniform random variates

$X \leftarrow 0$
Prod$\leftarrow 1$
WHILE True DO
  Generate a uniform $[0,1]$ random variate U.
  Prod$\leftarrow$Prod U
  IF Prod$>e^{-\lambda}$ (the constant should be computed only once)
    THEN $X \leftarrow X+1$
    ELSE RETURN X

The expected number of iterations is the same for both algorithms. However, an addition and an exponential random variate are replaced by a multiplication and a uniform random variate. This replacement usually works in favor of the multiplicative method. The expected complexity of both algorithms grows linearly with λ.

Another simple algorithm requiring only one uniform random variate is the inversion algorithm with sequential search. In view of the recurrence relation

$$\frac{P(X=i+1)}{P(X=i)} = \frac{\lambda}{i+1} \quad (i \geq 0),$$

this gives

Poisson generator based upon the inversion by sequential search

$X \leftarrow 0$
Sum $\leftarrow e^{-\lambda}$, Prod $\leftarrow e^{-\lambda}$
Generate a uniform $[0,1]$ random variate U.
WHILE $U >$ Sum DO
$\qquad X \leftarrow X+1$
$\qquad$ Prod $\leftarrow \dfrac{\lambda}{X}$Prod
$\qquad$ Sum $\leftarrow$ Sum+Prod
RETURN X

This algorithm too requires expected time proportional to λ as $\lambda \to \infty$. For large λ, round-off errors proliferate, which provides us with another reason for avoiding large values of λ. Speed-ups of the inversion algorithm are possible if sequential search is started near the mode. For example, we could compare U first with $b = P(X \leq \lfloor \lambda \rfloor)$, and then search sequentially upwards or downwards. If b is available in time $O(1)$, then the algorithm takes expected time $O(\sqrt{\lambda})$ because $E(|X - \lfloor \lambda \rfloor |) = O(\sqrt{\lambda})$. See Fishman (1976). If b has to be computed first, this method is hardly competitive. Atkinson (1979) describes various ways in which the inversion can be helped by the judicious use of tables. For small values of λ, there is no problem. He then custom builds fast table-based generators for all λ's that are powers of 2, starting with 2 and ending with 128. For a given value of λ, a sum of independent Poisson random variates is needed with parameters that are either powers of 2 or very small. The speed-up comes at a tremendous cost in terms of space and programming effort.

3.4. Rejection methods.

To see how easy it is to improve over the algorithms of the previous section, it helps to get an idea of how the probabilities vary with λ. First of all, the peak at $\lfloor \lambda \rfloor$ varies as $1/\sqrt{\lambda}$:

Lemma 3.4.

The value of $P(X = \lfloor \lambda \rfloor)$ does not exceed

$$\frac{1}{\sqrt{2\pi \lfloor \lambda \rfloor}} \ ,$$

and $\sim 1/\sqrt{2\pi\lambda}$ as $\lambda \to \infty$.

Proof of Lemma 3.4.

We apply the inequality $i! \geq i^i e^{-i} \sqrt{2\pi i}$, valid for all integer $i \geq 1$. Thus,

$$e^{-\lambda} \frac{\lambda^{\lfloor \lambda \rfloor}}{\lambda!} \leq e^{-(\lambda - \lfloor \lambda \rfloor)} (\frac{\lambda}{\lfloor \lambda \rfloor})^{\lfloor \lambda \rfloor} \frac{1}{\sqrt{2\pi \lfloor \lambda \rfloor}}$$

$$\leq \frac{1}{\sqrt{2\pi \lfloor \lambda \rfloor}} \ .$$

Furthermore, by Stirling's approximation, it is easy to establish the asymptotic result as well. ∎

We also have the following inequality by monotonicity:

Lemma 3.5.

$$P(X = \lfloor \lambda \rfloor \pm i) \leq \frac{2(\sqrt{\lambda}+1)}{i(i+1)} \quad (i > 0).$$

Proof of Lemma 3.5.

We will argue for the positive side only. Writing p_i for $P(X = i)$, we have by unimodality,

$$\sqrt{\lambda} + 1 \geq E(|X - \lambda|) + 1$$
$$\geq E(|X - \lfloor \lambda \rfloor|) \geq \sum_{j \geq \lfloor \lambda \rfloor} |j - \lfloor \lambda \rfloor| \, p_j$$
$$\geq p_{i + \lfloor \lambda \rfloor} \sum_{j=0}^{i} j$$

$$= \frac{i(i+1)}{2} p_{i+\lfloor \lambda \rfloor} \cdot \blacksquare$$

If we take the minimum of the constant upper bound of Lemma 3.4 and the quadratically decreasing upper bound of Lemma 3.5, it is not difficult to see that the cross-over point is near $\lambda \pm c \sqrt{\lambda}$ where $c = (8\pi)^{1/4}$. The area under the bounding sequence of numbers is $O(1)$ as $\lambda \to \infty$. It is uniformly bounded over all values $\lambda \geq 1$. We do not imply that one should design a generator based upon this dominating curve. The point is that it is very easy to construct good bounding sequences. In fact, we already knew from Theorem 1.1 that the universal rejection algorithm of section 1.4 is uniformly fast. The dominating curves of Theorem 1.1 and Lemmas 3.4 and 3.5 are similar, both having a flat center part. Atkinson (1979) proposes a logistic majorizing curve, and Ahrens and Dieter (1980) propose a double exponential majorizing curve. Schmeiser and Kachitvichyanukul (1981) have a rejection method with a triangular hat and two exponential tails. We do not describe these methods here. Rather, we will describe an algorithm of Devroye (1981) which is based upon a normal-exponential dominating curve. This has the advantage that the rejection constant tends to 1 as $\lambda \to \infty$. In addition, we will illustrate how the factorial can be avoided most of the time by the judicious use of squeeze steps. Even if factorials are computed in linear time, the overall expected time per random variate remains uniformly bounded over λ. For large values of λ, we will return a truncated normal random variate with large probability.

Some inequalities are needed for the development of tight inequalities for the Poisson probabilities. These are collected in the next Lemma:

Lemma 3.6.

Assume that $u \geq 0$ and all the arguments of the logarithms are positive in the list of inequalities shown below. We have:

(i) $\log(1+u) \leq u$

(ii) $\log(1+u) \leq u - \dfrac{1}{2}u^2 + \dfrac{1}{3}u^3$

(iii) $\log(1+u) \geq u - \dfrac{1}{2}u^2$

(iv) $\log(1+u) \geq \dfrac{2u}{2+u}$

(v) $\log(1-u) \leq - \displaystyle\sum_{i=1}^{k} \dfrac{1}{i}u^i \qquad (k \geq 1)$

(vi) $\log(1-u) \geq - \displaystyle\sum_{i=1}^{k-1} \dfrac{1}{i}u^i - \dfrac{u^k}{k(1-u)} \qquad (k \geq 2)$

Most of these inequalities are well-known. The other ones can be obtained without difficulty from Taylor's theorem (Whittaker and Watson, 1927, is a good source of information). We assume that $\lambda \geq 1$. Since we will use rejection algorithms, it can't harm to normalize the Poisson probabilities. Instead of the probabilities p_i, we will use the normalized log probabilities

$$q_j = \log(p_{\mu+j}) + \log(\mu!) - \mu \log(\lambda) + \lambda$$

where $\mu = \lfloor \lambda \rfloor$. This can conveniently be rewritten as follows:

$$q_j = j \log(\frac{\lambda}{\mu}) + j \log(\mu) - \log(\frac{(\mu+j)!}{\mu!})$$

$$= j \log(\frac{\lambda}{\mu}) + \begin{cases} -\log(\displaystyle\prod_{i=1}^{j}(1+\frac{i}{\mu})) & (j>0) \\[2ex] 0 & (j=0) \\[2ex] -\log(\displaystyle\prod_{i=0}^{-j-1}(1-\frac{i}{\mu})) & (j<0) \end{cases} \quad .$$

Lemma 3.7.

Let us use the notation j_+ for $\max(j, 0)$. Then, for all integer $j \geq -\mu$,

$$q_j \leq \frac{j_+}{\mu} - \frac{j(j+1)}{2\mu + j_+} \, .$$

Proof of Lemma 3.7.

Use (iv) and (v) of Lemma 3.6, together with the identity

$$\sum_{i=1}^{j} i = \frac{j(j+1)}{2} \, . \blacksquare$$

The inequality of Lemma 3.7 can be used as the starting point for the development of tight dominating curves. The last term on the right hand side in the upper bound is not in a familiar form. On the one hand, it suggests a normal bounding curve when j is small compared to μ. On the other hand, for large values of $|j|$, an exponential bounding curve seems more appropriate. Recall that the Poisson probabilities cannot be tucked under a normal curve because they drop off as $e^{-cj\log(j)}$ for some c as $j \to \infty$. In Lemma 3.8 we tuck the Poisson probabilities under a normal main body and an exponential right tail.

Lemma 3.8.

Assume that $\mu \geq 6$ and that δ is an integer satisfying

$$6 \leq \delta \leq \mu \, .$$

Then

$$q_j \leq -\frac{j(j+1)}{2\mu} \leq -\frac{j^2}{2\mu} \quad (j \leq 0)$$

$$q_0 \leq 0$$

$$q_1 \leq \frac{1}{\mu(2\mu+1)} \leq \frac{1}{78}$$

$$q_j \leq -\frac{(j-1)^2}{2\mu+\delta} + \frac{1}{2\mu+\delta} \quad (0 \leq j \leq \delta)$$

$$q_j \leq -\frac{\delta}{2\mu+\delta}\left(\frac{j}{2}+1\right) \quad (j \geq \delta) \, .$$

Proof of Lemma 3.8.

The first three inequalities follow without work from Lemma 3.7. For the fourth inequality, we observe that for $2 \leq j \leq \delta$,

$$q_j \leq \frac{j + \frac{j}{2}}{\mu + \frac{j}{2}} - \frac{j(j+1)}{2(\mu + \frac{j}{2})} \quad \text{(since } j \leq \delta \leq \mu)$$

$$= \frac{2j - j^2}{2\mu + j}$$

$$\leq \frac{2j - j^2}{2\mu + \delta} \quad \text{(since } 2 \leq j \leq \delta) .$$

The fourth inequality is also valid for $j = 0$. For $j = 1$, a quick check shows that $1/\mu(2\mu+1) \leq 1/(2\mu+\delta)$ because $\delta \leq \mu$. This leaves us with the fifth and last inequality. We note that $\delta \geq 6 \geq \frac{4\mu}{\mu-2}$. Thus,

$$q_j \leq \frac{j}{\mu} - \frac{\delta}{2\mu+\delta}(j+1)$$

$$= -\frac{\delta}{2\mu+\delta} + j\left(\frac{1}{\mu} - \frac{\delta}{2\mu+\delta}\right)$$

$$\leq -\frac{\delta}{2\mu+\delta}\left(1 + \frac{j}{2}\right) . \quad \blacksquare$$

Based on these inequalities, we can now give a first Poisson algorithm:

Rejection method for Poisson random variates

[SET-UP]

$\mu \leftarrow \lfloor \lambda \rfloor$

Choose δ integer such that $6 \leq \delta \leq \mu$.

$c_1 \leftarrow \sqrt{\pi \mu / 2}$

$c_2 \leftarrow c_1 + \sqrt{\pi(\mu + \delta/2)/2}\, e^{\frac{1}{2\mu + \delta}}$

$c_3 \leftarrow c_2 + 1$

$c_4 \leftarrow c_3 + e^{\frac{1}{78}}$

$c \leftarrow c_4 + \frac{2}{\delta}(2\mu + \delta) e^{-\frac{\delta}{2\mu + \delta}(1 + \frac{\delta}{2})}$

[NOTE]

The function q^*_j is defined as $q_j - j\log(\frac{\lambda}{\mu}) = j\log(\mu) - \log((\mu + j)!/\mu!)$.

[GENERATOR]

REPEAT

    Generate a uniform $[0,c]$ random variate U and an exponential random variate E.
Accept $\leftarrow$ False.

    CASE

        $U \leq c_1$:

            Generate a normal random variate N.

            $Y \leftarrow -\mid N \mid \sqrt{\mu}$

            $X \leftarrow \lfloor Y \rfloor$

            $W \leftarrow -\frac{N^2}{2} - E - X\log(\frac{\lambda}{\mu})$

            IF $X \geq -\mu$ THEN $W \leftarrow \infty$

        $c_1 < U \leq c_2$:

            Generate a normal random variate N.

            $Y \leftarrow 1 + \mid N \mid \sqrt{\mu + \frac{\delta}{2}}$

            $X \leftarrow \lceil Y \rceil$

            $W \leftarrow \frac{-Y^2 + 2Y}{2\mu + \delta} - E - X\log(\frac{\lambda}{\mu})$

            IF $X \leq \delta$ THEN $W \leftarrow \infty$

        $c_2 < U \leq c_3$:

            $X \leftarrow 0$

            $W \leftarrow -E$

        $c_3 < U \leq c_4$:

            $X \leftarrow 1$

            $W \leftarrow -E - \log(\frac{\lambda}{\mu})$

        $c_4 < U$:

            Generate an exponential random variate V.

$$Y \leftarrow \delta + V \frac{2}{\delta}(2\mu + \delta)$$

$$X \leftarrow \lceil Y \rceil$$

$$W \leftarrow -\frac{\delta}{2\mu + \delta}(1 + \frac{Y}{2}) - E - X \log(\frac{\lambda}{\mu})$$

Accept $\leftarrow [W \leq q_X^*]$

UNTIL Accept

RETURN $X + \mu$

Observe the careful use of the floor and ceiling functions in the algorithm to insure that the continuous dominating curves exceed the Poisson staircase function at every point of the real line, not just the integers ! The monotonicity of the dominating curves is exploited of course. The function

$$q_x = x \log(\lambda) - \log(\frac{(\mu + x)!}{\mu!})$$

is evaluated in every iteration at some point x. If the logarithm of the factorial is available at unit cost, then the algorithm can run in uniformly bounded time provided that δ is carefully picked. Thus, the first issue to be dealt with is that of the relationship between the expected number of iterations and δ.

Lemma 3.9.

If δ depends upon λ in such a way that

$$\delta = o(\mu) \ , \quad \frac{\delta}{\sqrt{\mu}} \to \infty \ ,$$

then the expected number of iterations $E(N)$ tends to one as $\lambda \to \infty$. In particular, the expected number of iterations remains uniformly bounded over $\lambda \geq 6$.

Furthermore,

$$\inf_{\delta} E(N) = 1 + (1 + o(1)) \sqrt{\frac{\log(\mu)}{32\mu}} \quad \text{as } \lambda \to \infty$$

where the infimum is reached if we choose

$$\delta \sim \sqrt{2\mu \log(\frac{128\mu}{\pi})} \ .$$

Proof of Lemma 3.9.

In a preliminary computation, we have to evaluate

$$\sum_{j \geq -\mu} e^{q_j}$$

since this is the total weight of the normalized Poisson probabilities. It is easy to see that this gives

$$\sum_{j=0}^{\infty} p_j \, e^{\lambda} \mu! \lambda^{-\mu}$$

$$\sim e^{\lambda} (\frac{\mu}{e \lambda})^{\mu} \sqrt{2\pi\mu}$$

$$\sim \sqrt{2\pi\mu}$$

where we used the fact that $\log(\lambda/\mu) = \log(1+(\lambda-\mu)/\mu) = (\lambda-\mu)/\mu + O(\mu^{-2})$. Thus, the expected number of iterations is the total area under the dominating curve (with the atoms at 0 and 1 having areas one and $e^{\frac{1}{78}}$ respectively) divided by $(1+o(1))\sqrt{2\pi\mu}$. The area under the dominating curve is, taking the five contributors from left to right,

$$\sqrt{\pi\mu/2} + 1 + e^{\frac{1}{78}} + \sqrt{\pi(\mu+\frac{\delta}{2})/2} \, e^{\frac{1}{2\mu+\delta}} + \frac{2(2\mu+\delta)}{\delta} e^{-\frac{\delta}{2\mu+\delta}(\frac{\delta}{2}+1)} .$$

If δ is not $o(\mu)$, this can not $\sim \sqrt{2\pi\mu}$. If $\delta \leq c\sqrt{\mu}$ for some constant c, then the last term is at least $\sim \frac{4}{c} e^{-c^2/4} \sqrt{\mu}$, while it should really be $o(\sqrt{\mu})$. Thus, the conditions imposed on δ are necessary for $E(N) \to 1$. That they are also sufficient can be seen as follows. The fifth term in the area under the dominating curves is $o(\sqrt{\mu})$, and so are the constant second and third terms. The fourth term $\sim \sqrt{\pi\mu/2}$, which establishes the result.

To minimize $E(N)-1$ in an asymptotically optimal fashion, we have to consider some sort of expansion of the area in terms of decreasing asymptotic importance. Using the Taylor series expansion for $\sqrt{1+u}$ for u near 0, we can write the first four terms as

$$\sqrt{\pi\mu/2} \left[1 + O(\mu^{-\frac{1}{2}}) + 1 + \frac{\delta}{4\mu} + O((\frac{\delta}{\mu})^2) \right] .$$

The main term in excess of $\sqrt{2\pi\mu}$ is

$$\sqrt{\pi\mu/2} \frac{\delta}{4\mu} .$$

We can also verify easily that the contribution from the exponential tail is

$$\frac{4\mu}{\delta} (1+o(1)) e^{-\frac{\delta^2}{2(2\mu+\delta)}} .$$

To obtain a first (but as we will see, good) guess for δ, we will minimize

$$\sqrt{\pi\mu/2}\,\frac{\delta}{4\mu}+\frac{4\mu}{\delta}e^{-\frac{\delta^2}{2(2\mu+\delta)}}\ .$$

This is equivalent to solving

$$(2+\frac{4\mu}{\delta^2})e^{-\frac{\delta^2}{4\mu}}=\sqrt{\frac{\pi}{32\mu}}\ .$$

If we ignore the $o\,(1)$ term $\dfrac{4\mu}{\delta^2}$, we can solve this explicitly and obtain

$$\delta=\sqrt{2\mu\,\log(\frac{128\mu}{\pi})}\ .$$

A plugback of this value in the original expression for the area under the dominating curve shows that it increases as

$$\sqrt{2\pi\mu}+(1+o\,(1))\frac{\sqrt{\pi}}{4}\sqrt{\log(\mu)}\ .$$

The constant terms are absorbed in $o\,(1)$; the exponential tail contribution is $O\,(1/\sqrt{\log(\mu)})$. If we replace δ by $\delta(1+\epsilon)$ where ϵ is allowed to vary with μ but is bounded from below by $c>0$, then the area is asymptotically larger because the $\sqrt{\log(\mu)}$ term should be multiplied by at least $1+c$. If we replace δ by $\delta(1-\epsilon)$, then the contribution from the exponential tail is at least $\Omega(\mu^{c/2}/\sqrt{\log(\mu)})$. This concludes the proof of the Lemma. ■

We have to insure that δ falls within the limits imposed on it when the dominating curves were derived. Thus, the following choice should prove failsafe in practice:

$$\delta=\max(6,\min(\mu,\sqrt{2\mu\,\log(\frac{128\mu}{\pi})}))\ .$$

We have now in detail dealt with the optimal design for our Poisson generator. If the log-factorial is available at unit cost, the rejection algorithm is uniformly fast, and asymptotically, the rejection constant tends to one. δ was picked to insure that the convergence to one takes place at the best possible rate. For the optimal δ, the algorithm basically returns a truncated normal random variate most of the time. The exponential tail becomes asymptotically negligible.

We may ask what would happen to our algorithm if we were to compute all products of successive integers explicitly ? Disregarding the horrible accuracy problems inherent in all repeated multiplications, we would also face a breakdown in our complexity. The computation of

$$q_X=X\log(\frac{\lambda}{\mu})+X\log(\mu)-\log(\frac{(X+\mu)!}{\mu}!)$$

can be done in time proportional to $1+|X|$. Now, X is with high probability normal with mean 0 and variance approximately equal to $\sqrt{\mu}$. Since q is computed only once with probability tending to one, it is clear that the expected time complexity now grows as $\sqrt{\mu}$. If we had perfect squeeze curves, i.e. squeeze curves in which the top and bottom bounds are equal, then we would get our uniform speed back. The same is true for very tight but imperfect squeeze curves. A class of such squeeze curves is presented below. Note that we are no longer concerned with the dominating curves. The squeeze curves given below are also not derived from the inequalities for Stirling's series or Binet's series for the log gamma function (see section 1). We could have used those, but it is instructive to show yet another method of deriving good bounds. See however exercise 3.9 for the application of Stirling's series in squeeze curves for Poisson probabilities.

Lemma 3.10.

Define

$$t_j = q_j - j \log(\frac{\lambda}{\mu}) + \frac{j(j+1)}{2\mu} \ .$$

Then for integer $j \geq 0$,

$$t_j \begin{cases} \geq \max\left(0, \dfrac{j(j+1)(2j+1)}{12\mu^2} - \dfrac{j^2(j+1)^2}{12\mu^3}\right) \\[2ex] \leq \dfrac{j(j+1)(2j+1)}{12\mu^2} \end{cases} .$$

Furthermore, for integer $-\mu \leq j \leq 0$, the converse is almost true:

$$t_j \begin{cases} \geq \dfrac{j(j+1)(2j+1)}{12\mu^2} - \dfrac{j^2(j+1)^2}{12\mu^2(\mu+j+1)} \\[2ex] \leq \min\left(0, \dfrac{j(j+1)(2j+1)}{12\mu^2}\right) \end{cases} .$$

Proof of Lemma 3.10.

The proof is based upon Lemma 3.6, the identities

$$\sum_{i=1}^{k} i = \frac{k(k+1)}{2} \ , \quad \sum_{i=1}^{k} i^2 = \frac{k(k+1)(2k+1)}{6} \ , \quad \sum_{i=1}^{k} i^3 = \frac{k^2(k+1)^2}{4} \ ,$$

and the fact that q_j can be rewritten as follows:

$$q_j - j \log(\frac{\lambda}{\mu}) = \begin{cases} -\log(\prod\limits_{i=1}^{j}(1+\dfrac{i}{\mu})) & (j>0) \\[2ex] 0 & (j=0) \\[2ex] \log(\prod\limits_{i=0}^{-j-1}(1-\dfrac{i}{\mu})) & (j<0) \end{cases} \quad . \blacksquare$$

The algorithm requires of course little modification. Only the line

$$\text{Accept} \leftarrow [W \leq q *_X]$$

needs replacing. The replacement looks like this:

$$T \leftarrow \frac{X(X+1)}{2\mu}$$
$$\text{Accept} \leftarrow [W \leq -T] \cap [X \geq 0]$$
IF NOT Accept THEN
$$Q \leftarrow T(\frac{2X+1}{6\mu}-1)$$
$$P \leftarrow Q - \frac{T^2}{3(\mu+(X+1)_-)}$$
$$\text{Accept} \leftarrow [W \leq Q]$$
IF NOT Accept AND $[W \leq P]$ THEN Accept $\leftarrow [W \leq q *_X]$

It is interesting to go through the expected complexity proof in this one example because we are no longer counting iterations but multiplications.

Lemma 3.11.

The expected time taken by the modified Poisson generator is uniformly bounded over $\lambda \geq 6$ when δ is chosen as in Lemma 3.10, even when factorials are explicitly evaluated as products.

Proof of Lemma 3.11.

It suffices to establish the uniform boundedness of

$$E(|X| I_{[Q < W < P]})$$

where we use the notation of the algorithm. Note that this statement implicitly uses Wald's equation, and the fact that the expected number of iterations is uniformly bounded. The expression involving $|X|$ is arrived at by looking at the time needed to evaluate $q *_X$. The expected value will be split into five parts according to the five components in the distribution of X. The atomic parts

$X=0, X=1$ are easy to take care of. The contribution from the normal portions can be bounded from above by a constant times

$$E(|X|(P-Q)) \le E(|X| \frac{X^2(X+1)^2}{12\mu^2(\mu+(X+1)_-)}) .$$

Here we have used the fact that W consists of a sum of some random variable and an exponential random variable. When $X \ge 0$, the last upper bound is in turn not greater than a constant times $E(|X|^5)/\mu^3 = O(\mu^{-1/2})$. The case $X < 0$ is taken care of similarly, provided that we first split off the case $X < -\frac{\mu}{2}$. The split-off part is bounded from above by

$$O(\mu^3)P(X < -\frac{\mu}{2}) \le O(\mu^3)\frac{E(X^2)}{\mu^2} = O(1) .$$

For the exponential tail part, we need a uniform bound for

$$E(|X|^5\mu^{-3})(\log(\mu))^{-\frac{1}{2}}$$

where we have used a fact shown in the proof of Lemma 3.10, i.e. the probability that X is exponential decreases as a constant times $\log^{-1/2}(\mu)$. Verify next that given that X is from our exponential tail, $E(|X|^5)=O(\delta^5)$. Combining all of this shows that our expression in question is

$$O(\frac{\log^2(\mu)}{\sqrt{\mu}}) .$$

This concludes the proof of Lemma 3.11. ∎

The computations of the previous Lemma reveal other interesting facets of the algorithm. For example, the expected time contribution of the evaluations of factorials is $O(\frac{\log^2(\mu)}{\sqrt{\mu}})$. In other words, it is asymptotically negligible. Even so, the main contribution to this $o(1)$ expected time comes from the exponential tail. This suggests that it is possible to obtain a new value for δ which would minimize the expected time spent on the evaluation of factorials, and that this value will differ from that obtained by minimizing the expected number of iterations.

3.5. Exercises.

1. Atkinson (1979) has developed a Poisson (λ) generator based upon rejection from the logistic density

    $$f(x) = \frac{1}{b} e^{-\frac{x-a}{b}} \left(1 + e^{-\frac{x-a}{b}}\right)^{-2},$$

    where $a = \lambda$ and $b = \sqrt{3\lambda}/\pi$. A random variate with this density can be generated as $X \leftarrow a + b \log(\frac{1-U}{U})$ where U is uniform [0,1].

    A. Find the distribution of $\left\lfloor X + \frac{1}{2} \right\rfloor$.

    B. Prove that X has the same mean and variance as the Poisson distribution.

    C. Determine a rejection constant c for use with the distribution of part A.

    D. Prove that c is uniformly bounded over all values of λ.

2. **A recursive generator.** Let n be an integer somewhat smaller than λ, and let G be a gamma (n) random variable. Show that the random variable X defined below is Poisson (λ): If $G > \lambda$, X is binomial ($n-1, \lambda/G$); if $G \le \lambda$, then X is n plus a Poisson ($\lambda-G$) random variable. Then, taking $n = \lfloor 0.875\lambda \rfloor$, use this recursive property to develop a recursive Poisson generator. Note that one can leave the recursive loop either when at one point $G > \lambda$ or when λ falls below a fixed threshold (such as 10 or 15). By taking n a fixed fraction of λ, the value of λ falls at a geometric rate. Show that in view of this, the expected time complexity is $O(1 + \log(\lambda))$ if a constant expected time gamma generator is used (Ahrens and Dieter, 1974).

3. Prove all the inequalities of Lemma 3.6.

4. Prove that for any λ and any $c > 0$, $\lim_{j \to \infty} p_j / e^{-cj^2} = \infty$. Thus, the Poisson curve cannot be tucked under any normal curve.

5. **Poisson variates in batches.** Let $X_1, \ldots, X_n$ be a multinomial $(Y, p_1, \ldots, p_n)$ random vector (i.e., the probability of attaining the value $i_1, \ldots, i_n$ is 0 when $\sum i_j$ is not Y and is

    $$\frac{Y!}{i_1! \cdots i_n!} p_1^{i_1} \cdots p_n^{i_n}$$

    otherwise. Show that if Y is Poisson (λ), then $X_1, \ldots, X_n$ are independent Poisson random variables with parameters $\lambda p_1, \ldots, \lambda p_n$ respectively. (Moran, 1951; Patil and Seshadri, 1964; Bolshev, 1965; Tadikamalla, 1979).

6. Prove that as $\lambda \to \infty$, the distribution of $(X-\lambda)/\sqrt{\lambda}$ tends to the normal distribution by proving that the characteristic function tends to the characteristic function $e^{-t^2/2}$ of the normal distribution.

7. Show that for the rejection method developed in the text, the expected time complexity is $O(\sqrt{\lambda})$ and $\Omega(\sqrt{\lambda})$ as $\lambda \to \infty$ when no squeeze steps are used and the factorial has to be evaluated explicitly.

8. Give a detailed rejection algorithm based upon the constant upper bound of Lemma 3.4 and the quadratically decreasing tails of Lemma 3.5.

9. Assume that factorials are avoided by using the zero-term and one-term Stirling approximations (Lemma 1.1) as lower and upper bounds in squeeze steps (the difference between the zero-term and one-term approximations of $\log(\Gamma(n))$ is the term $1/(12n)$). Show that this suffices for the following rejection algorithms to be uniformly fast:

   A. The universal algorithm of section 1.

   B. The algorithm based upon Lemmas 3.4 and 3.5 (and developed in Exercise 8).

   C. The normal-exponential rejection algorithm developed in the text.

10. Repeat exercise 9, but assume now that factorials are avoided altogether by evaluating an increasing number of terms in Binet's convergent series for the log gamma function (Lemma 1.2) until an acceptance or rejection decision can be made. Read first the text following Lemma 1.2.

11. **The matching distribution.** Suppose that n cars are parked in front of Hanna's rubber skin suit shop, and that each of Hanna's satisfied customers leaves in a randomly picked car. The number N of persons who leave in their own car has the matching distribution with parameter n:

$$P(N=i) = \frac{1}{i!} \sum_{j=0}^{n-i} \frac{(-1)^j}{j!} \quad (0 \le i \le n) .$$

   A. Show this by invoking the inclusion exclusion principle.

   B. Show that $\lim_{n \to \infty} P(N=i) = \dfrac{1}{e \, i!}$, i.e. that the Poisson (1) distribution is the limit (Barton, 1958).

   C. Show that $P(N=i) \le \dfrac{1}{i!}$, i.e. rejection from the Poisson (1) distribution can be used with rejection constant e not depending upon n.

   D. Show that the algorithm given below is valid, and that its expected complexity is uniformly bounded in n.

WHILE True DO

Generate a Poisson (1) random variate X, and a uniform [0,1] random variate U.

IF $X \leq n$ THEN

$k \leftarrow 1, j \leftarrow 0, s \leftarrow 1$

WHILE $j \leq n - X$ AND $U \leq s$ DO

$j \leftarrow j + 1, k \leftarrow -jk, s \leftarrow s + \dfrac{1}{k}$

IF $j \leq n - X$ AND $U < s$

THEN RETURN X

ELSE $j \leftarrow j + 1, k \leftarrow -jk, s \leftarrow s + \dfrac{1}{k}$

12. **The Borel-Tanner distribution.** A distribution important in queuing theory, with parameters $n \geq 1$ (n integer) and $\alpha \in (0,1)$ was discovered by Borel and Tanner (Tanner, 1951). The probabilities p_i are defined by

$$p_i = \frac{n}{(i-n)!} i^{i-n-1} \alpha^{i-n} e^{-\alpha i} \qquad (i \geq n).$$

Show that the mean is $\dfrac{n}{1-\alpha}$ and that the variance is $\dfrac{n\alpha}{(1-\alpha)^3}$. The distribution has a very long positive tail. Develop a uniformly fast generator.

4. THE BINOMIAL DISTRIBUTION.

4.1. Properties.

X is **binomially distributed** with parameters $n \geq 1$ and $p \in [0,1]$ if

$$P(X=i) = \binom{n}{i} p^i (1-p)^{n-i} \qquad (0 \leq i \leq n).$$

We will say that X is binomial (n, p).

Lemma 4.1. (Genesis.)

Let X be the number of successes in a sequence of n Bernoulli trials with success probability p, i.e.

$$X = \sum_{i=1}^{n} I_{[U_i < p]},$$

where $U_1, \ldots, U_n$ are iid uniform [0,1] random variables. Then X is binomial (n, p).

Lemma 4.2.

The binomial distribution with parameters n, p has generating function $(1-p+ps)^n$. The mean is np, and the variance is $np(1-p)$.

Proof of Lemma 4.2.

The factorial moment generating function of X (or simply generating function) is

$$k(s) = E(s^X) = \prod_{i=1}^{n} E(s^{I_{[U_i < p]}}),$$

where we used the Lemma 4.1 and its notation. Each factor in the product is obviously equal to $1-p+ps$. This concludes the proof of the first statement. Next, $E(X) = k'(1) = np$, and $E(X(X-1)) = k''(1) = n(n-1)p^2$. Hence, $Var(X) = E(X^2)-E^2(X) = E(X(X-1))+E(X)-E^2(X) = np(1-p)$. ∎

From Lemma 4.1, we can conclude without further work:

Lemma 4.3.

If $X_1, \ldots, X_k$ are independent binomial $(n_1, p), \ldots, (n_k, p)$ random variables, then $\sum_{i=1}^{k} X_i$ is binomial $(\sum_{i=1}^{k} n_i, p)$.

Lemma 4.4.(First waiting time property.)

Let $G_1, G_2,...$ be iid geometric (p) random variables, and let X be the smallest integer such that

$$\sum_{i=1}^{X+1} G_i > n \ .$$

Then X is binomial (n,p).

Proof of Lemma 4.4.

G_1 is the number of Bernoulli trials up to and including the first success. Thus, by the independence of the G_i's, $G_1 + \cdots + G_{X+1}$ is the number of Bernoulli trials up to and including the $X+1$-st success. This number is greater than n if and only if among the first n Bernoulli trials there are at most X successes. Thus,

$$P(X \leq k) = P(\sum_{i=1}^{k+1} G_i > n) = \sum_{j=0}^{k} \binom{n}{j} p^j (1-p)^{n-j} \quad \text{(integer } k\text{)}. \blacksquare$$

Lemma 4.5. (Second waiting time property.)

Let $E_1, E_2,...$ be iid exponential random variables, and let X be the smallest integer such that

$$\sum_{i=1}^{X+1} \frac{E_i}{n-i+1} > -\log(1-p) \ .$$

Then X is binomial (n,p).

Proof of Lemma 4.5.

Let $E_{(1)} < E_{(2)} < \cdots < E_{(n)}$ be the order statistics of an exponential distribution. Clearly, the number of $E_{(i)}$'s smaller than $-\log(1-p)$ is binomially distributed with parameters n and $P(E_1 < -\log(1-p)) = 1 - e^{\log(1-p)} = p$. Thus, if X is the smallest integer such that $E_{(X+1)} \geq -\log(1-p)$, then X is binomial (n,p). Lemma 4.5 now follows from the fact (section V.2) that $(E_{(1)}, \ldots, E_{(n)})$ is distributed as

$$(\frac{E_1}{n}, \frac{E_1}{n} + \frac{E_2}{n-1}, \ldots, \frac{E_1}{n} + \frac{E_2}{n-1} + \cdots + \frac{E_n}{1}). \blacksquare$$

4.2. Overview of generators.

The binomial generators can be partitioned into a number of classes:

A. The simple generators. These generators are based upon the direct application of one of the lemmas of the previous section. Typically, the expected complexity grows as n or as np, the computer programs are very short, and no additional workspace is required.

B. Uniformly fast generators based upon the rejection method (Fishman (1979), Ahrens and Dieter (1980), Kachitvichyanukul (1982), Devroye and Naderisamani (1980)). We will not bother with older algorithms which are not uniformly fast. Fishman's method is based upon rejection from the Poisson distribution, and is explored in exercise 4.1. The universal rejection algorithm derived from Theorem 1.1 is also uniformly fast, but since it was not specifically designed for the binomial distribution, it is not competitive with tailor-made rejection algorithms. To save space, only the algorithm of Devroye and Naderisamani (1980) will be developed in detail. Although this algorithm may not be the fastest on all computers, it has two desirable properties: the dominating curve is asymptotically tight because it exploits convergence to the normal distribution, and it does not require a subprogram for computing the log factorial in constant time.

C. Table methods. The finite number of values make the binomial distribution a good candidate for the table methods. To obtain uniformly fast speed, the table size has to grow in proportion to n, and a set-up time proportional to n is needed. It is generally accepted that the marginal execution times of the alias or alias-urn methods are difficult to beat. See sections III.3 and III.4 for details.

D. Generators based upon recursion (Relles (1972), Ahrens and Dieter (1974)). The problem of generating a binomial (n,p) random variate is usually reduced in constant time to that of generating another binomial random variate with much smaller value for n. This leads to $O(\log(n))$ or $O(\log\log(n))$ expected time algorithms. In view of the superior performance of the generators in classes B and C, the principle of recursion will be described very briefly, and most details can be found in the exercises.

4.3. Simple generators.

Lemma 4.1 leads to the

Coin flip method

$X \leftarrow 0$

FOR $i := 1$ TO n DO

    Generate a random bit B (B is 1 with probability p, and can be obtained by generating a uniform $[0,1]$ random variate U and setting $B = I_{[U \leq p]}$).

    $X \leftarrow X + B$

RETURN X

This simple method requires time proportional to n. One can use n uniform random variates, but it is often preferable to generate just one uniform random variate and recycle the unused portion. This can be done by noting that a random bit and an independent uniform random variate can be obtained as $(I_{[U < p]}, \min(\dfrac{U}{p}, \dfrac{1-U}{1-p}))$. The coin flip method with recycling of uniform random variates can be rewritten as follows:

[NOTE: We assume that $p \leq 1/2$.]

$X \leftarrow 0$

Generate a uniform $[0,1]$ random variate U.

FOR $i := 1$ TO n DO

    $B \leftarrow I_{[U > 1-p]}$

    $U \leftarrow \dfrac{U - (1-p)B}{pB + (1-p)(1-B)}$ (reuse the uniform random variate)

    $X \leftarrow X + B$

RETURN X

For the important case $p = \dfrac{1}{2}$, it suffices to generate a random uniformly distributed computer word of n bits, and to count the number of ones in the word. In machine language, this can be implemented very efficiently by the standard bit operations.

Inversion by sequential search takes as we know expected time proportional to $E(X) + 1 = np + 1$. We can avoid tables of probabilities because of the recurrence relation

$$p_{i+1} = p_i \frac{(n-i)p}{(i+1)(1-p)} \quad (0 \leq i < n),$$

where $p_i = P(X=i)$. The algorithm will not be given here. It suffices to mention that for large n, the repeated use of the recurrence relation could also lead to accuracy problems. These problems can be avoided if one of the two waiting time algorithms (based upon Lemmas 4.4 and 4.5) is used:

First waiting time algorithm

$X \leftarrow -1$
Sum $\leftarrow 0$
REPEAT
      Generate a geometric (p) random variate G.
      Sum $\leftarrow$ Sum $+G$
      $X \leftarrow X+1$
UNTIL Sum $>n$
RETURN X

Second waiting time method

[SET-UP]
$q \leftarrow -\log(1-p)$
[GENERATOR]
$X \leftarrow 0$
Sum $\leftarrow 0$
REPEAT
      Generate an exponential random variate E.
      Sum $\leftarrow$ Sum $+\dfrac{E}{n-X}$ (Note: Sum is allowed to be ∞.)
      $X \leftarrow X+1$
UNTIL Sum $>q$
RETURN $X \leftarrow X-1$

Both waiting time methods have expected time complexities that grow as $np+1$.

4.4. The rejection method.

To develop good dominating curves, it helps to recall that by the central limit theorem, the binomial distribution tends to the normal distribution as $n \to \infty$ and p remains fixed. When p varies with n in such a way that $np \to c$, a positive constant, then the binomial distribution tends to the Poisson (c) distribution, which in turn is very close to the normal distribution for large values of c. It seems thus reasonable to consider the normal density as our dominating curve. Unfortunately, the binomial probabilities do not decrease quickly enough for one single normal density to be useful as a dominating curve. We cover the binomial tails with exponential curves and make use of Lemma 3.6. To keep things simple, we assume:

1. $\lambda = np$ is a nonzero integer.

2. $p \leq \dfrac{1}{2}$.

So as not to confuse p with $p_i = P(X=i)$, we use the notation

$$b_i = \binom{n}{i} p^i (1-p)^{n-i} \quad (0 \leq i \leq n).$$

The second assumption is not restrictive because a binomial (n,p) random variable is distributed as n minus a binomial ($n,1-p$) random variable. The first assumption is not limiting in any sense because of the following property.

Lemma 4.6.

If Y is a binomial (n,p') random variable with $p' \leq p$, and if conditional on Y, Z is a binomial ($n-Y, \dfrac{p-p'}{1-p'}$) random variable, then $X \leftarrow Y+Z$ is binomial (n,p).

Proof of Lemma 4.6.

The lemma is based upon the decomposition

$$X = \sum_{i=1}^{n} I_{[U_i \leq p]} = \sum_{i=1}^{n} I_{[U_i \leq p']} + \sum_{i=1}^{n} I_{[p' < U_i \leq p]} = Y+Z,$$

where $U_1, \ldots, U_n$ are iid uniform [0,1] random variables. ∎

To recapitulate, we offer the following generator for general values of n,p, but $0 < p \leq \dfrac{1}{2}$:

Splitting algorithm for binomial random variates

[NOTE: t is a fixed threshold, typically about 7. For $np \leq t$, one of the waiting time algorithms is recommended. Assume thus that $np > t$.]

$$p' \leftarrow \frac{1}{n} \lfloor np \rfloor$$

Generate a binomial (n, p') random variate Y by the rejection method in uniformly bounded expected time.

Generate a binomial $(n - Y, \frac{p - p'}{1 - p'})$ random variate Z by one of the waiting time methods.

RETURN $X \leftarrow Y + Z$

The expected time taken by this generator when $np > t$ is bounded from above by $c_1 + c_2 n \frac{p - p'}{1 - p'} \leq c_1 + 2c_2$ for some universal constants c_1, c_2. Thus, it can't harm to impose assumption 1.

Lemma 4.7.

For integer $0 \leq i \leq n(1-p)$ and integer $\lambda = np \geq 1$, we have

$$\log(\frac{b_{\lambda+i}}{b_\lambda}) \leq -\frac{i(i-1)}{2n(1-p)} - \frac{i(i+1)}{2np + i}$$

and

$$\log(\frac{b_{\lambda+i}}{b_\lambda}) + \frac{i^2 + ((1-p) - p)i}{2np(1-p)} \begin{cases} \leq s \\ \geq s - t \end{cases}$$

where

$$s = \frac{i(i+1)(2i+1)}{12n^2 p^2} - \frac{(i-1)i(2i-1)}{12n^2(1-p)^2}$$

and

$$t = \frac{i^2(i-1)^2}{12n^2(1-p)^2(n(1-p)-i+1)} + \frac{i^2(i+1)^2}{12n^3 p^3} .$$

For all integer $0 \leq i \leq np$, $\log(\frac{b_{\lambda-i}}{b_\lambda})$ satisfies the same inequalities provided that p is replaced throughout by $1-p$ in the various expressions.

Proof of Lemma 4.7.

For $i=0$, the statements are obviously true because equality is reached. Assume thus that $0 < i \le n(1-p)$. We have

$$\frac{b_{\lambda+i}}{b_\lambda} = \frac{\binom{n}{\lambda+i} p^{\lambda+i}(1-p)^{n-\lambda-i}}{\binom{n}{\lambda} p^\lambda (1-p)^{n-\lambda}} = \left(\frac{p}{1-p}\right)^i \frac{\binom{n}{\lambda+i}}{\binom{n}{\lambda}}$$

$$= \left(\frac{p}{1-p}\right)^i \frac{(n-\lambda)!\lambda!}{(n-\lambda-i)!(\lambda+i)!}$$

$$= \frac{\prod_{j=0}^{i-1}\left(1-\frac{j}{n(1-p)}\right)}{\prod_{j=0}^{i}\left(1+\frac{j}{np}\right)}.$$

Thus,

$$\log\left(\frac{b_{\lambda+i}}{b_\lambda}\right) = \sum_{j=0}^{i-1}\log\left(1-\frac{j}{n(1-p)}\right) - \sum_{j=0}^{i}\log\left(1+\frac{j}{np}\right)$$

$$\le -\sum_{j=0}^{i-1}\frac{j}{n(1-p)} - \sum_{j=0}^{i}\frac{2j}{2np+j}$$

$$\le -\frac{i(i-1)}{2n(1-p)} - \frac{i(i+1)}{2np+i}.$$

Here we used Lemma 3.6. This proves the first statement of the lemma. Again by Lemma 3.6, we see that

$$\log\left(\frac{b_{\lambda+i}}{b_\lambda}\right) \le \sum_{j=0}^{i-1}\left(-\frac{j}{n(1-p)} - \frac{j^2}{2n^2(1-p)^2}\right) + \sum_{j=0}^{i}\left(-\frac{j}{np} + \frac{j^2}{2n^2p^2}\right)$$

$$= -\frac{i^2+((1-p)-p)i}{2np(1-p)} + s.$$

Furthermore,

$$\log\left(\frac{b_{\lambda+i}}{b_\lambda}\right) \ge \sum_{j=0}^{i-1}\left(-\frac{j}{n(1-p)} - \frac{j^2}{2n^2(1-p)^2} - \frac{j^3}{3n^3(1-p)^3\left(1-\frac{i-1}{n(1-p)}\right)}\right)$$

$$+ \sum_{j=0}^{i}\left(-\frac{j}{np} + \frac{j^2}{2n^2p^2} - \frac{j^3}{3n^3p^3}\right)$$

$$= -\frac{i^2+((1-p)-p)i}{2np(1-p)} + s - t.$$

This concludes the proof of the first part of Lemma 4.7. For integer $0 < i \le np$, we have

$$\frac{b_{\lambda-i}}{b_{\lambda}} = (\frac{p}{1-p})^{-i} \frac{\binom{n}{\lambda-i}}{\binom{n}{\lambda}}$$

$$= \frac{\prod_{j=0}^{i-1}(1-\frac{j}{np})}{\prod_{j=0}^{i}(1+\frac{j}{n(1-p)})} .$$

This is formally the same as an expression used as starting point above, provided that p is replaced throughout by $1-p$. ■

Lemma 4.7 is used in the construction of a useful function $g(x)$ with the property that for all $x \in [i, i+1)$, and all allowable i ($-np \leq i \leq n(1-p)$),

$$g(x) \geq \log(\frac{b_{\lambda+i}}{b_{\lambda}}) .$$

The algorithm is of the form:

REPEAT
    Generate a random variate Y with density proportional to e^g .
    Generate an exponential random variate E .
    $X \leftarrow \lfloor Y \rfloor$ (this is truncation to the left, even for negative values of Y)
UNTIL $[-np \leq X \leq n(1-p)]$ AND $[g(Y) \leq \log(\frac{b_{\lambda+X}}{b_{\lambda}})+E]$
RETURN $X \leftarrow \lambda + X$

The normal-exponential dominating curve e^g suggested earlier is defined in Lemma 4.8:

Lemma 4.8.

Let $\delta_1 \geq 1$, $\delta_2 \geq 1$ be given integers. Define furthermore

$$\sigma_1 = \sqrt{np(1-p)}(1+\frac{\delta_1}{4np}) \, ,$$

$$\sigma_2 = \sqrt{np(1-p)}(1+\frac{\delta_2}{4n(1-p)}) \, ,$$

$$c = \frac{2\delta_1}{np} \, .$$

Then the function g can be chosen as follows:

$$g(x) = \begin{cases} c - \dfrac{x^2}{2\sigma_1^2} & (0 \leq x < \delta_1) \\[2ex] \dfrac{\delta_1}{n(1-p)} - \dfrac{\delta_1 x}{2\sigma_1^2} & (\delta_1 < x) \\[2ex] -\dfrac{x^2}{2\sigma_2^2} & (-\delta_2 < x < 0) \\[2ex] -\dfrac{\delta_2 x}{2\sigma_2^2} & (x \leq -\delta_2) \end{cases}$$

Proof of Lemma 4.8.

For $i = 0$ we need to show that $c \geq 1/(2\sigma_1^2)$. This follows from

$$2c\,\sigma_1^2 = \frac{4\delta_1}{np}np(1-p)(1+\frac{\delta_1}{4np})^2 \geq 4\delta_1(1-p) \geq 2\delta_1 \geq 2 \, .$$

When $0 < i < \delta_1$, we have

$$-\frac{i(i-1)}{2n(1-p)} \leq -\frac{(x-1)(x-2)}{2n(1-p)} \, ,$$

$$-\frac{i(i+1)}{2np+i} \leq -\frac{x(x-1)}{2np+\delta_1} \, .$$

By Lemma 4.7,

$$\log(\frac{b_{np+i}}{b_{np}}) \leq -\frac{(x-1)(x-2)}{2n(1-p)} - \frac{x(x-1)}{2np+\delta_1}$$

$$= -(\frac{1}{2n(1-p)}+\frac{1}{2np+\delta_1})x^2+(\frac{3}{2n(1-p)}+\frac{1}{2np+\delta_1})x-\frac{1}{n(1-p)}$$

$$\leq -(\frac{1}{2n(1-p)}+\frac{1}{2np+\delta_1})x^2+\frac{2\delta_1}{np}$$

$$\leq -\frac{x^2}{2\sigma_1^2}+\frac{2\delta_1}{np} \, .$$

The last step follows by application of the inequality $\sqrt{1+u} < 1 + \frac{u}{2}$, valid for $u > 0$, in the following chain of inequalities:

$$\frac{1}{2n(1-p)} + \frac{1}{2np+\delta_1} = \frac{1+\dfrac{\delta_1}{2n}}{2np(1-p)(1+\dfrac{\delta_1}{2np})}$$

$$\geq \frac{1}{2np(1-p)(1+\dfrac{\delta_1}{2np})}$$

$$\geq \frac{1}{(\sqrt{2np(1-p)}(1+\dfrac{\delta_1}{4np}))^2} = \frac{1}{2\sigma_1^2} .$$

When $i \geq \delta_1$, we have

$$-\frac{i(i-1)}{2n(1-p)} \leq -\frac{\delta_1(x-2)}{2n(1-p)} \; ; \; -\frac{i(i+1)}{2np+i} \leq -\frac{\delta_1 x}{2np+\delta_1} .$$

By Lemma 4.7,

$$\log(\frac{b_{np+i}}{b_{np}}) \leq -\frac{\delta_1(x-2)}{2n(1-p)} - \frac{\delta_1 x}{2np+\delta_1}$$

$$= -(\frac{1}{2np} + \frac{1}{2np+\delta_1})\delta_1 x + \frac{\delta_1}{n(1-p)}$$

$$\leq -\frac{\delta_1 x}{2\sigma_1^2} + \frac{\delta_1}{n(1-p)} .$$

When $0 > i \geq -\delta_2$, we have

$$\log(\frac{b_{np+i}}{b_{np}}) \leq -\frac{i(i+1)}{2np} - \frac{i(i-1)}{2n(1-p)-i}$$

$$= -(\frac{1}{2np} + \frac{1}{2n(1-p)+\delta_2})i^2 - \frac{i}{2np} + \frac{i}{2n(1-p)+\delta_2}$$

$$\leq -(\frac{1}{2np} + \frac{1}{2n(1-p)+\delta_2})x^2$$

$$\leq -\frac{x^2}{2\sigma_2^2} .$$

Finally, when $i < -\delta_2$, we see that

$$-\frac{i(i+1)}{2np} \leq \frac{\delta_2 x}{2np} \; ;$$

$$-\frac{i(i-1)}{2n(1-p)-i} \leq \frac{\delta_2(i-1)}{2n(1-p)+\delta_2} \leq \frac{\delta_2(x-1)}{2n(1-p)+\delta_2} .$$

Therefore,

$$\log\left(\frac{b_{np+i}}{b_{np}}\right) \le -\frac{\delta_2 x}{2np} + \frac{\delta_2(x-1)}{2n(1-p)+\delta_2}$$

$$= \left(\frac{1}{2np} + \frac{1}{2n(1-p)+\delta_2}\right)\delta_2 x - \frac{\delta_2}{2n(1-p)+\delta_2}$$

$$\le \frac{\delta_2 x}{2\sigma_2^2} \cdot \blacksquare$$

The dominating curve e^g suggested by Lemma 4.8 consists of four pieces, one piece per interval. The integrals of e^g over these intervals are needed by the generator. These are easy to compute for the exponential tails, but not for the normal center intervals. Not much will be lost if we replace the two normal pieces by halfnormals on the positive and negative real line respectively, and reject when the normal random variates fall outside $[-\delta_2, \delta_1]$. This at least allows us to work with the integrals of halfnormal curves. We will call the areas under the different components of e^g a_i ($1 \le i \le 4$). Thus,

$$a_1 = \int_0^\infty e^{c - \frac{x^2}{2\sigma_1^2}} \, dx = \frac{1}{2} e^c \sigma_1 \sqrt{2\pi} \, ,$$

$$a_2 = \frac{1}{2}\sigma_2\sqrt{2\pi} \, ,$$

$$a_3 = \int_{\delta_1}^\infty e^{\frac{\delta_1}{n(1-p)} - \frac{\delta_1 x}{2\sigma_1^2}} \, dx = e^{\frac{\delta_1}{n(1-p)}} \frac{2\sigma_1^2}{\delta_1} e^{-\frac{\delta_1^2}{2\sigma_1^2}} \, ,$$

$$a_4 = \frac{2\sigma_2^2}{\delta_2} e^{-\frac{\delta_2^2}{2\sigma_2^2}} \, .$$

We can now summarize the algorithm:

A rejection algorithm for binomial random variates

[SET-UP]

$\sigma_1 \leftarrow \sqrt{np\,(1-p)}(1+\delta_1/(4np\,)), \sigma_2 \leftarrow \sqrt{np\,(1-p)}(1+\delta_2/(4n\,(1-p\,))),\ c \leftarrow 2\delta_1/(np\,)$

$a_1 \leftarrow \dfrac{1}{2} e^c \sigma_1 \sqrt{2\pi}\ , a_2 \leftarrow \dfrac{1}{2} \sigma_2 \sqrt{2\pi}$

$a_3 \leftarrow e^{\frac{\delta_1}{n\,(1-p)}}\dfrac{2\sigma_1{}^2}{\delta_1}e^{-\frac{\delta_1{}^2}{2\sigma_1{}^2}}$

$a_4 \leftarrow \dfrac{2\sigma_2{}^2}{\delta_2}e^{-\frac{\delta_2{}^2}{2\sigma_2{}^2}}$

$s \leftarrow a_1 + a_2 + a_3 + a_4$

[GENERATOR]

REPEAT

    Generate a uniform $[0,s\,]$ random variate U.

    CASE

        $U \leq a_1$:

            Generate a normal random variate N; $\ Y \leftarrow \sigma_1 \mid N\mid$

            Reject $\leftarrow [Y \geq \delta_1]$

            IF NOT Reject THEN $X \leftarrow \lfloor Y\rfloor, V \leftarrow -E - \dfrac{N^2}{2} + c$ where E is an exponential random variate.

        $a_1 < U \leq a_1 + a_2$:

            Generate a normal random variate N; $\ Y \leftarrow \sigma_2 \mid N\mid$

            Reject $\leftarrow [Y \geq \delta_2]$

            IF NOT Reject THEN $X \leftarrow \lfloor -Y\rfloor, V \leftarrow -E - \dfrac{N^2}{2}$ where E is an exponential random variate.

        $a_1 + a_2 < U \leq a_1 + a_2 + a_3$:

            Generate two iid exponential random variates E_1, E_2.

            $Y \leftarrow \delta_1 + 2\sigma_1{}^2 E_1/\delta_1$

            $X \leftarrow \lfloor Y\rfloor, V \leftarrow -E_2 - \delta_1 Y/(2\sigma_1{}^2) + \delta_1/(n\,(1-p\,))$

            Reject $\leftarrow$ False

        $a_1 + a_2 + a_3 < U$:

            Generate two iid exponential random variates E_1, E_2.

            $Y \leftarrow \delta_2 + 2\sigma_2{}^2 E_1/\delta_2$

            $X \leftarrow \lfloor -Y\rfloor, V \leftarrow -E_2 - \delta_2 Y/(2\sigma_2{}^2)$

            Reject $\leftarrow$ False

      Reject $\leftarrow$ Reject OR $[X < -np\,]$ OR $[X > n\,(1-p\,)]$

      Reject $\leftarrow$ Reject OR $[V > \log(b_{np+X}/b_{np}\,)]$

UNTIL NOT Reject

RETURN X

We need only choose δ_1, δ_2 so that the expected number of iterations is approximately minimal. This is done in Lemma 4.9.

Lemma 4.9.

Assume that $p \le \dfrac{1}{2}$ and that as $\lambda = np \to \infty$, we have uniformly in p, $\delta_1 = o(\lambda), \delta_2 = o(n)$, $\delta_1/\sqrt{\lambda} \to \infty$, $\delta_2/\sqrt{np} \to \infty$. Then the expected number of iterations is uniformly bounded over $n \ge 1, 0 \le p \le \dfrac{1}{2}$, and tends to 1 uniformly in p as $\lambda \to \infty$.

The conditions on δ_1, δ_2 are satisfied for the following (nearly optimal) choices:

$$\delta_1 = \left\lfloor \max\left(1, \sqrt{np(1-p)\log(\frac{128np}{81\pi(1-p)})}\right) \right\rfloor ,$$

$$\delta_2 = \left\lfloor \max\left(1, \sqrt{np(1-p)\log(\frac{128n(1-p)}{\pi p})}\right) \right\rfloor .$$

Proof of Lemma 4.9.

We first observe that under the stated conditions on δ_1, δ_2, we have

$$\sigma_1 = \sqrt{np(1-p)}(1+o(1)) , \quad \sigma_2 = \sqrt{np(1-p)}(1+o(1)) ,$$

$$c = o(1) ,$$

$$a_1 = \sqrt{\frac{\pi np(1-p)}{2}}(1+o(1)) , \quad a_2 = \sqrt{\frac{\pi np(1-p)}{2}}(1+o(1)) ,$$

$$a_3 = \frac{2np(1-p)}{\delta_1}(1+o(1))e^{-\frac{\delta_1^2(1+o(1))}{2np(1-p)}} ,$$

$$a_4 = \frac{2np(1-p)}{\delta_2}(1+o(1))e^{-\frac{\delta_2^2(1+o(1))}{2np(1-p)}} ,$$

$$a_1+a_3 \sim a_1 , \quad a_2+a_4 \sim a_2 ,$$

$$a_1+a_2+a_3+a_4 \sim \sqrt{2\pi np(1-p)} .$$

The expected number of iterations in the algorithm is $(a_1+a_2+a_3+a_4)b_{np} \sim \sqrt{2\pi np(1-p)}/\sqrt{2\pi np(1-p)} = 1$. All $o(.)$ and $\sim$ symbols inherit the uniformity with respect to p, as long as $\lambda \to \infty$. The uniform boundedness of the expected number of iterations follows from this.

The particular choices for δ_1, δ_2 are easily seen to satisfy the convergence conditions. That they are nearly optimal (with respect to the minimization of the expected number of iterations) is now shown. The minimization of a_1+a_3 would provide us with a good value for δ_1. In the asymptotic expansions for a_1, a_3, it is now necessary to consider the first two terms, not just the main term. In particular, we have

$$a_1 = \sqrt{\frac{\pi np(1-p)}{2}} e^c (1+\frac{\delta_1}{4np}) = \sqrt{\frac{\pi np(1-p)}{2}}(1+\frac{(9+o(1))\delta_1}{4np}),$$

$$a_3 = \frac{2np(1-p)}{\delta_1} e^{-\frac{(1+o(1))\delta_1^2}{2np(1-p)}} \approx \frac{2np(1-p)}{\delta_1} e^{-\frac{\delta_1^2}{2np(1-p)}}.$$

Setting the derivative of the sum of the two right-hand-side expressions equal to zero gives the equation

$$\frac{\delta_1^2}{np(1-p)} e^{\frac{\delta_1^2}{2np(1-p)}} = (1+\frac{\delta_1^2}{np(1-p)})\sqrt{np(1-p)}\frac{8\sqrt{2}}{9(1-p)\sqrt{\pi}}.$$

Disregarding the term "1" with respect to $\frac{\delta_1^2}{np(1-p)}$ and solving with respect to δ_1 gives

$$\delta_1 = \sqrt{np(1-p)\log(\frac{128np}{81\pi(1-p)})}.$$

A suitable expression for δ_2 can be obtained by a similar argument. Indeed,

$$a_2+a_4 = \sqrt{\frac{\pi np(1-p)}{2}}(1+\frac{\delta_2}{4n(1-p)})$$
$$+ (1+o(1))\frac{2np(1-p)}{\delta_2} e^{-\frac{(1+o(1))\delta_2^2}{2np(1-p)}}.$$

Disregard the $o(1)$ term, and set the derivative of the resulting expression with respect to δ_2 equal to zero. This gives the equation

$$\frac{e^{\frac{\delta_2^2}{2np(1-p)}}}{4n(1-p)} = 2(np(1-p)+\delta_2^2)\sqrt{\frac{2}{\pi np(1-p)}} \sim \sqrt{\frac{8}{\pi np(1-p)}}\delta_2^2.$$

If $\sim$ is replaced by equality, then the solution with respect to δ_2 is

$$\delta_2 = \sqrt{np(1-p)\log(\frac{128n(1-p)}{\pi p})}. \blacksquare$$

Lemma 4.9 is crucial for us. For large values of np, the rejection constant is nearly 1. Also, since δ_1 and δ_2 are large compared to the standard deviation $\sqrt{np(1-p)}$ of the distribution, the exponential tails float to infinity as $np \to \infty$. In other words, we exit most of the time with a properly scaled normal random variate. At this point we leave the algorithm. The interested readers can find more information in the exercises. For example, the evaluation of b_{np+i}/b_{np}

takes time proportional to $1+|i|$. This implies that the expected complexity grows as $\sqrt{np(1-p)}$ when $np\to\infty$. It can be shown that the expected complexity is uniformly bounded if we do one of the following:

A.   Use squeeze steps suggested in Lemma 4.7, and evaluate b_{np+i}/b_{np} explicitly when the squeeze steps fail.

B.   Use squeeze steps based upon Stirling's series (Lemma 1.1), and evaluate b_{np+i}/b_{np} explicitly when the squeeze steps fail.

C.   Make all decisions involving factorials based upon sequentially evaluating more and more terms in Binet's convergent series for factorials (Lemma 1.2).

D.   Assume that the log gamma function is a unit cost function.

4.5. Recursive methods.

The recursive methods are all based upon the connection between the binomial and beta distributions given in Lemma 4.6. This is best visualized by considering the order statistics $U_{(1)}<\cdots<U_{(n)}$ of iid uniform $[0,1]$ random variables, and noting that the number of $U_{(i)}$'s in $[0,p]$ is binomial (n,p). Let us call this quantity X. Furthermore, $U_{(i)}$ itself is beta $(i,n+1-i)$ distributed. Because $U_{(i)}$ is approximately $\dfrac{i}{n+1}$, we can begin with generating a beta $(i,n+1-i)$ random variate Y with $i=\lfloor(n+1)p\rfloor$. Y should be close to p. In any case, we have gone a long way toward solving our problem. Indeed, if $Y\leq p$, we note that X is equal to i plus the number of $U_{(j)}$'s in the interval $(Y,p]$, which we know is binomial $(n-i,\dfrac{p-Y}{1-Y})$ distributed. By symmetry, if $Y>p$, X is equal to i minus a binomial $(i-1,\dfrac{Y-p}{Y})$ random variate. Thus, the following recursive program can be used:

Recursive binomial generator

[NOTE: n and p will be destroyed by the algorithm.]

$X \leftarrow 0, S \leftarrow +1$ (S is a sign)

REPEAT

    IF $np < t$ (t is a design constant)

        THEN

            Generate a binomial (n, p) random variate B by a simple method such as the waiting time method.

            RETURN $X \leftarrow X + SB$

        ELSE

            Generate a beta $(i, n+1-i)$ random variate Y with $i = \lfloor (n+1)p \rfloor$.

            $X \leftarrow X + Si$

            IF $Y \leq p$

                THEN $n \leftarrow n - i, p \leftarrow \dfrac{p - Y}{1 - Y}$

                ELSE $S \leftarrow -S, n \leftarrow i - 1, p \leftarrow \dfrac{Y - p}{Y}$

    UNTIL False

In this simple algorithm, we use a uniformly fast beta generator. The simple binomial generator alluded to should be such that its expected time is $O(np)$. Note however that it is not crucial: the algorithm works fine even if we set $t = 0$ and thus bypass the simple binomial generator. The algorithm halts when $n = 0$, which happens with probability one.

Let us give an informal outline of the proof of the claim that the expected time taken by the algorithm is bounded by a constant times $\log(\log(n))$. By the properties of the beta distribution, $Y - p$ is of the order of $\sqrt{\dfrac{i(n-i)}{n^3}}$, i.e. it is approximately $\sqrt{p(1-p)/n}$. Since Y itself is close to p, we see that the new values for (n, p) are either about $(n(1-p), \sqrt{p/((1-p)n)})$ or about $(np, \sqrt{(1-p)/(pn)})$. The new product np is thus of the order of magnitude of $\sqrt{np(1-p)}$. We see that np gets replaced at worst by about $\sqrt{np}$ in one iteration. In k iterations, we have about

$$(np)^{2^{-k}}.$$

Since we stop when this reaches t, our constant, the number of iterations should be of the order of magnitude of

$$\log\left(\frac{\log(np)}{\log(t)}\right).$$

This argument can be formalized, and the mathematically inclined reader is urged to do so (exercise 4.7). Since the loglog function increases very slowly, the recursive method can be competitive depending upon the beta generator. It was precisely the latter point, poor speed of the pre-1975 beta generators, which prompted Relles (1972) and Ahrens and Dieter (1974) to propose slightly different recursive generators in which i is not chosen as $\lfloor (n+1)p \rfloor$, but rather as $(n+1)/2$ when n is odd. This implies that all beta random variates needed are symmetric beta random variates, which can be generated quite efficiently. Because n gets halved at every iteration, their algorithm runs in $O(\log(n))$ time.

4.6. Symmetric binomial random variates.

The purpose of this section is to point out that in the case $p = \frac{1}{2}$ a single normal dominating curve suffices in the rejection algorithm, and to present and analyze the following simple rejection algorithm:

Rejection method for symmetric binomial random variates

[NOTE: This generator returns a binomial $(2n, \frac{1}{2})$ random variate.]

[SET-UP]

$s \leftarrow 1/\sqrt{2(n^{-1}-(2n^2)^{-1})}, \sigma \leftarrow s + \frac{1}{4}, c \leftarrow 2/(1+8s)$

[GENERATOR]

REPEAT

    Generate a normal random variate N and an exponential random variate E.

    $Y \leftarrow \sigma N$, $X \leftarrow \text{round}(Y)$

    $T \leftarrow -E + c - \frac{1}{2}N^2 + \frac{1}{n}X^2$

    Reject $\leftarrow [\,|X| > n\,]$

    IF NOT Reject THEN

$$\text{Accept} \leftarrow [T < -\frac{X^4}{6n^3(1-(\frac{|X|-1}{n})^2)}]$$

        IF NOT Accept THEN

$$\text{Reject} \leftarrow [T > \frac{X^2}{2n^2}]$$

            IF NOT Reject THEN

$$\text{Accept} \leftarrow [T > \log(\frac{b_n + X}{b_n}) + \frac{X^2}{n}]$$

UNTIL NOT Reject AND Accept

RETURN $X \leftarrow n + X$

The algorithm has one quick acceptance step and one quick rejection step designed to reduce the probability of having to evaluate the final acceptance step which involves computing the logarithms of two binomial probabilities. The validity of the algorithm follows from the following Lemma.

Lemma 4.10.

Let $b_0, \ldots, b_{2n}$ be the probabilities of a binomial $(2n, p)$ distribution. Then, for any $\sigma > s$,

$$\log\left(\frac{b_{n+i}}{b_n}\right) \leq c - \frac{\left(|i| + \frac{1}{2}\right)^2}{2\sigma^2} \qquad \text{(integer } i \text{ , } |i| \leq n \text{)},$$

where $c = 1/(8(\sigma^2 - s^2))$. Also, for all $n > i > 0$,

$$-\frac{i^4}{6n^3\left(1 - \left(\frac{i-1}{n}\right)^2\right)} \leq \log\left(\frac{b_{n+i}}{b_n}\right) + \frac{i^2}{n} \leq \frac{i^2}{2n^2} \; .$$

Proof of Lemma 4.10.

We will use repeatedly the following fact: for $1 > x > 0$,

$$-2x - \frac{2x^3}{3(1-x^2)} < \log\left(\frac{1-x}{1+x}\right) < -2x - \frac{2x^3}{3} \; ,$$

$$-\frac{1}{2}x^2 < \log(1+x) - x < 0 \; .$$

The first inequality follows from the fact that $\log\left(\frac{1-x}{1+x}\right)$ has series expansion $-2\left(x + \frac{1}{3}x^3 + \frac{1}{5}x^5 + \cdots\right)$. Thus, for $n > i > 0$,

$$\log\left(\frac{b_{n+i}}{b_n}\right) = \log\left(\frac{n!\,n!}{(n+i)!(n-i)!}\right) = \log\left(\prod_{j=1}^{i-1} \frac{1 - \frac{j}{n}}{1 + \frac{j}{n}} \frac{1}{1 + \frac{i}{n}}\right)$$

$$= \sum_{j=1}^{i-1}\left(\log\left(\frac{1-\frac{j}{n}}{1+\frac{j}{n}}\right) + \frac{2j}{n}\right) - \left(\log\left(1+\frac{i}{n}\right) - \frac{i}{n}\right) - \frac{i^2}{n}$$

$$= c_i + d_i - \frac{i^2}{n} \; .$$

We have

$$0 + \frac{1}{2}\left(\frac{i}{n}\right)^2 \geq c_i + d_i$$

$$\geq -\sum_{j=1}^{i-1} \frac{2}{3}\left(\frac{j}{n}\right)^3\left(1 - \left(\frac{j}{n}\right)^2\right)^{-1} + 0$$

$$\geq -\frac{2}{3}\left(1 - \left(\frac{i-1}{n}\right)^2\right)^{-1}\sum_{j=1}^{i-1}\left(\frac{j}{n}\right)^3$$

$$\geq -\frac{2}{3}(1-(\frac{i-1}{n})^2)^{-1}\frac{i^4}{4n^3} .$$

Thus,

$$\log(\frac{b_{n+i}}{b_n}) \leq -\frac{i^2}{n}+\frac{i^2}{2n^2} = -\frac{i^2}{2s^2} \leq c-\frac{(\,|\,i\,|+\frac{1}{2})^2}{2\sigma^2} \quad (\,|\,i\,|\leq n),$$

where

$$c = \sup_{u>0}\frac{(u+\frac{1}{2})^2}{2\sigma^2}-\frac{u^2}{2s^2} .$$

Assuming that $\sigma>s$, this supremum is reached for

$$u = \frac{s^2}{2(\sigma^2-s^2)} , \quad c = \frac{1}{8(\sigma^2-s^2)} . \blacksquare$$

The dominating curve suggested by Lemma 4.11 is a centered normal density with variance σ^2. The best value for σ is that for which the area $\sqrt{2\pi}\sigma e^c$ is minimal. Setting the derivative with respect to σ of the logarithm of this expression equal to 0 gives the equation

$$\sigma^2-\frac{1}{2}\sigma-s^2 = 0.$$

The solution is $\sigma = \frac{1}{4}+s\sqrt{1+1/(16s^2)} = \frac{1}{4}+s+o(1)$. It is for this reason that the value $\sigma=s+\frac{1}{4}$ was taken in the algorithm. The corresponding value for c is $2/(1+8s)$.

The expected number of iterations is $b_n\sqrt{2\pi}\sigma e^c \sim \frac{1}{\sqrt{\pi n}}\sqrt{2\pi}\sqrt{\frac{n}{2}} = 1$ as $n\to\infty$. Assuming that b_{n+i}/b_n takes time $1+|\,i\,|$ when evaluated explicitly, it is clear that without the squeeze steps, we would have obtained an expected time which would grow as $\sqrt{n}$ (because the i is distributed as σ times a normal random variate). The efficiency of the squeeze steps is highlighted in the following Lemma.

Lemma 4.11.

The algorithm shown above is uniformly fast in n when the quick acceptance step is used. If in addition a quick rejection step is used, then the expected time due to the explicit evaluation of b_{n+i}/b_n is $O(1/\sqrt{n})$.

Proof of Lemma 4.11.

Let $p(x)$ be the probability that the inequality in the quick acceptance step is not satisfied for fixed $X=x$. We have $P(|X|\geq 1+n\sqrt{5/6})=O(r^{-n})$ for some $r>1$. For $|x|\leq 1+n\sqrt{5/6}$, we have in view of $|Y^2-x^2|\leq(|x|+\frac{1}{2})/2$,

$$p(x)\leq P(-E+c-\frac{(x^2-\frac{1}{4}-\frac{|x|}{2})}{2\sigma^2}+\frac{x^2}{n}>-\frac{x^4}{n^3})$$

$$\leq P(E<c+\frac{1}{8\sigma^2}+\frac{|x|}{4\sigma^2}+x^2(\frac{1}{n}-\frac{1}{2\sigma^2})+\frac{x^4}{n^3})$$

$$\leq 2c+\frac{|x|}{4\sigma^2}+x^2(\frac{1}{n}-\frac{1}{2\sigma^2})+\frac{x^4}{n^3}$$

$$=O(n^{-\frac{1}{2}})+|x|O(n^{-1})+x^2O(n^{-\frac{3}{2}})+x^4O(n^{-3}).$$

Thus, the probability that a couple (X,E) does not satisfy the quick acceptance condition is $E(p(X))$. Since $E(|X|)=O(\sigma)=O(\sqrt{n})$, $E(X^2)=O(n)$ and $E(X^4)=O(n^2)$, we conclude that $E(p(X))=O(1/\sqrt{n})$. If every time we rejected, we were to start afresh with a new couple (X,E), the expected number of such couples needed before halting would be $1+O(1/\sqrt{n})$. Using this, it is also clear that in the algorithm without quick rejection step, the expected time is bounded by a constant times $1+E(|X|p(X))$. But

$$E(|X|p(X))\leq E(|X|I_{[|X|>1+n\sqrt{5/6}]})+E(|X|)O(n^{-\frac{1}{2}})$$

$$+E(X^2)O(n^{-1})+E(|X|^3)O(n^{-\frac{3}{2}})+E(|X|^5)O(n^{-3})$$

$$=O(1).$$

This concludes the proof of the first statement of the Lemma. If a quick rejection step is added, and $q(x)$ is the probability that for $X=x$, both the quick acceptance and rejection steps are failed, then, arguing as before, we see that for $|x|\leq 1+n\sqrt{5/6}$,

$$q(x)\leq\frac{x^4}{n^3}+\frac{x^2}{n^2}.$$

Thus, the probability that both inequalities are violated is

$$E(q(X))\leq\frac{E(X^4)}{n^3}+\frac{E(X^2)}{n^2}+P(|X|\geq 1+n\sqrt{5/6})=O(\frac{1}{n}).$$

The expected time spent on explicitly evaluating factorials is bounded by a constant times $1+E(|X|q(X))=O(1/\sqrt{n})$. ∎

4.7. The negative binomial distribution.

In section X.1, we introduced the negative binomial distribution with parameters (n, p), where $n \geq 1$ is an integer and $p \in (0,1)$ is a real number as the distribution of the sum of n iid geometric random variables. It has generating function

$$\left(\frac{p}{1-(1-p)s}\right)^n \ .$$

Using the binomial theorem, and equating the coefficients of s^i with the probabilities p_i for all i shows that the probabilities are

$$P(X=i) = p_i = \binom{-n}{i} p^n (-1+p)^i = \binom{n+i-1}{i} p^n (1-p)^i \quad (i \geq 0) \ .$$

When $n=1$, we obtain the geometric (p) distribution. For $n=1$, X is distributed as the number of failures in a sequence of independent experiments, each having success probability p, before the n-th success is encountered. From the properties of the geometric distribution, we see that the negative binomial distribution has mean $\dfrac{n(1-p)}{p}$ and variance $\dfrac{n(1-p)}{p^2}$.

Generation by summing n iid geometric p random variates yields at best an algorithm taking expected time proportional to n. The situation is even worse if we employ Example 1.4, in which we showed that it suffices to sum N iid logarithmic series $(1-p)$ random variates where N itself is Poisson (λ) and $\lambda = n \log(\frac{1}{p})$. Here, at best, the expected time grows as $E(N) = n \log(\frac{1}{p})$.

The property that one can use to construct a uniformly fast generator is obtained in Example 1.5: a negative binomial random variate can be generated as a Poisson (Y) random variate where Y in turn is a gamma $(n, \frac{1-p}{p})$ random variate. The same can be achieved by designing a uniformly fast rejection algorithm from scratch.

4.8. Exercises.

1. **Binomial random variates from Poisson random variates.** This exercise is motivated by an idea first proposed by Fishman (1979), namely to generate binomial random variates by rejection from Poisson random variates. Let b_i be the probability that a binomial (n, p) random variable takes the value i, and let p_i be the probability that a Poisson $((n+1)p)$ random variable takes the value i.

A. Prove the crucial inequality $\sup_i b_i / p_i \leq e^{1/(12(n+1))} / \sqrt{1-p}$, valid for all n and p. Since we can without loss of generality assume that $p \leq \dfrac{1}{2}$, this implies that we have a uniformly fast binomial generator if we have a uniformly fast Poisson generator, and if we can handle the evaluation of b_i/p_i in uniformly bounded time. To prove the inequality, start with inequalities for the factorial given in Lemma 1.1, write i as $(n+1)p+x$, note that $x \leq (n+1)(1-p)$, and use the inequality $1+u \geq e^{u/(1+u)}$, valid for all $u > -1$.

B. Give the details of the rejection algorithm, in which factorials are squeezed by using the zero-term and one-term bounds of Lemma 1.1, and are explicitly evaluated as products when the squeezing fails.

C. Prove that the algorithm given in B is uniformly fast over all $n \geq 1, p \leq 1/2$ if Poisson random variates are generated in uniformly bounded expected time (not worst case time).

2. **Bounds for the mode of the binomial distribution.** Consider a binomial (n,p) distribution in which np is integer. Then the mode m is at np, and

$$\binom{n}{m} p^m (1-p)^{n-m} \leq \frac{e^{\frac{1}{12(n+1)} + \frac{1}{n^2 p(1-p)+n+1}}}{\sqrt{2\pi np(1-p)}} \leq \frac{2}{\sqrt{2\pi np(1-p)}} .$$

Prove this inequality by using the Stirling-Whittaker-Watson inequality of Lemma 1.1, and the inequalities $e^{u/(1+u)} \leq 1+u \leq e^u$, valid for $u \geq 0$ (Devroye and Naderisamani, 1980).

3. Add the squeeze steps suggested in the text to the normal-exponential algorithm, and prove that with this addition the expected complexity of the algorithm is uniformly bounded over all $n \geq 1, 0 < p \leq \frac{1}{2}$, np integer (Devroye and Naderisamani, 1980).

4. A continuation of the previous exercise. Show that for fixed $p \leq \frac{1}{2}$, the expected time spent on the explicit evaluation of b_{np+i}/b_{np} is $O(1/\sqrt{np(1-p)})$ as $n \to \infty$. (This implies that the squeeze steps of Lemma 4.7 are very powerful indeed.)

5. Repeat exercise 3 but use squeeze steps based upon bounds for the log gamma function given in Lemma 1.1.

6. **The hypergeometric distribution.** Suppose an urn contains N balls, of which M are white and $N-M$ are black. If a sample of n balls is drawn at random without replacement from the urn, then the number (X) of white balls drawn is hypergeometrically distributed with parameters n,M,N. We have

$$P(X=i) = \frac{\binom{M}{i}\binom{N-M}{n-i}}{\binom{N}{n}} \qquad (\max(0,n-N+M) \leq i \leq \min(n,M)) .$$

Note that the same distribution is obtained when n and M are interchanged. Note also that if we had sampled with replacement, we would have obtained the binomial $(n, \frac{M}{N})$ distribution.

A. Show that if a hypergeometric random variate is generated by rejection from the binomial $(n, \frac{M}{N})$ distribution, then we can take $(1-\frac{n}{N})^{-n}$ as rejection constant. Note that this tends to 1 as $n^2/N \to 0$.

B. Using the facts that the mean is $n\frac{M}{N}$, that the variance σ^2 is $\frac{N-n}{N-1}n\frac{M}{N}(1-\frac{M}{N})$, and that the distribution is unimodal with a mode at $\left[(n+1)\frac{M+1}{N+2}\right]$, give the details for the universal rejection algorithm of section X.1. Comment on the expected time complexity, i.e. on the maximal value for $(\sigma B)^{2/3}$ where B is an upper bound for the value of the distribution at the mode.

C. Find a function $g(x)$ consisting of a constant center piece and two exponential tails, having the properties that the area under the function is uniformly bounded, and that the function has the property that for every i and all $x \in [i-\frac{1}{2}, i+\frac{1}{2})$, $g(x) \geq P(X=i)$. Give the corresponding rejection algorithm (hint: recall the universal rejection algorithm of section X.1) (Kachitvichyanukul, 1982; Kachitvichyanukul and Schmeiser, 1985).

7. Prove that for all constant $t > 0$, there exists a constant C only depending upon t such that the expected time needed by the recursive binomial algorithm given in the text is not larger than $C \log(\log(n+10))$ for all n and p. The term "10" is added to make sure that the loglog function is always strictly positive. Show also that for a fixed $p \in (0,1)$ and a fixed $t > 0$, the expected time of the algorithm grows as a constant times $c \log(\log(n))$ as $n \to \infty$, where c depends upon p and t only. If time is equated with the number of beta random variates needed before halting, determine c.

5. THE LOGARITHMIC SERIES DISTRIBUTION.

5.1. Introduction.

A random variable X has the **logarithmic series distribution** with parameter $p \in (0,1)$ if

$$P(X=i) = p_i = \frac{a}{i}p^i \quad (i=1,2,...) ,$$

where $a = -1/\log(1-p)$ is a normalization constant. In the tail, the probabilities decrease exponentially. Its generating function is

$$a \sum_{i=1}^{\infty} \frac{1}{i} p^i s^i = \frac{\log(1-ps)}{\log(1-p)} .$$

From this, one can easily find the mean $ap/(1-p)$ and second moment $ap/(1-p)^2$.

5.2. Generators.

The material in this section is based upon the fundamental work of Kemp (1981) on logarithmic series distributions. The problems with the logarithmic series distribution are best highlighted by noting that the obvious inversion and rejection methods are not uniformly fast.

If we were to use sequential search in the inversion method, using the recurrence relation

$$p_i = (1-\frac{1}{i})pp_{i-1} \quad (i \geq 2) ,$$

the inversion method could be implemented as follows:

Inversion by sequential search

[SET-UP]
Sum $\leftarrow -p/\log(1-p)$
[GENERATOR]
Generate a uniform $[0,1]$ random variate U.
$X \leftarrow 1$
WHILE $U > $ Sum DO
$\qquad U \leftarrow U - $ Sum
$\qquad X \leftarrow X + 1$
$\qquad$ Sum $\leftarrow$ Sum $\dfrac{p(X-1)}{X}$
RETURN X

The expected number of comparisons required is equal to the mean of the distribution, $ap/(1-p)$, and this quantity increases monotonically from 1 ($p \downarrow 0$) to ∞ ($p \uparrow \infty$). For $p < 0.95$, it is difficult to beat this simple algorithm in terms of expected time. Interestingly, if rejection from the geometric distribution $(1-p)p^i$ ($i \geq 1$) is used, the expected number of geometric random variates required is again equal to the same mean. But because the geometric random variates themselves are rather costly, the sequential search method is to be preferred at this stage.

We can obtain a one-line generator based upon the following distributional property:

Theorem 5.1. (Kendall (1948), Kemp (1981))

Let U, V be iid uniform $[0,1]$ random variables. Then

$$X \leftarrow \left\lfloor 1 + \frac{\log(V)}{\log(1-(1-p)^U)} \right\rfloor$$

has the logarithmic series distribution with parameter p.

Proof of Theorem 5.1.

The logarithmic series distribution is the distribution of a geometric $(1-Y)$ random variate X (i.e. $P(X=i \mid Y) = Y(1-Y)^{i-1}$ $(i \geq 1)$), provided that Y has distribution function

$$F(y) = \int_0^y \frac{1}{(z-1)\log(1-p)} dz = \frac{\log(1-y)}{\log(1-p)} \quad (0 \leq y \leq p) .$$

This can be seen from the integral

$$\int_0^p \frac{s(1-y)}{(1-ys)(y-1)\log(1-p)} dy = \frac{\log(1-ps)}{\log(1-p)}$$

and from the fact that the generating function of a geometric $(1-Y)$ random variate is $\dfrac{s(1-Y)}{(1-Ys)}$. A random variable Y with distribution function F can be obtained by the inversion method as $Y \leftarrow 1-(1-p)^U$ where U is a uniform $[0,1]$ random variable. ∎

Kemp (1981) has suggested two clever tricks for accelerating the algorithm suggested by Theorem 5.1. First, when $V > p$, the value $X \leftarrow 1$ is delivered because

$$V > p \geq 1-(1-p)^U .$$

For small p, the savings thus obtained are enormous. We summarize:

Kemp's generator with acceleration

[SET-UP]

$r \leftarrow \log(1-p)$

[GENERATOR]

$X \leftarrow 1$

Generate a uniform [0,1] random variate V.

IF $V \geq p$

    THEN RETURN X

    ELSE

        Generate a uniform [0,1] random variate U.

$$\text{RETURN } X \leftarrow \left\lfloor 1+\frac{\log(V)}{\log(1-e^{rU})} \right\rfloor$$

Kemp's second trick involves taking care of the values 1 and 2 separately. He notes that $X=1$ if and only if $V \geq 1-e^{rU}$, and that $X \in \{1,2\}$ if and only if $V \geq (1-e^{rU})^2$ where r is as in the algorithm shown above. The algorithm incorporating this is given below.

Kemp's second accelerated generator

[SET-UP]

$r \leftarrow \log(1-p)$

[GENERATOR]

$X \leftarrow 1$

Generate a uniform [0,1] random variate V.

IF $V \geq p$

    THEN RETURN X

    ELSE

        Generate a uniform [0,1] random variate U.

        $q \leftarrow 1-e^{rU}$

        CASE

$$V \leq q^2 : \text{RETURN } X \leftarrow \left\lfloor 1+\frac{\log(V)}{\log(q)} \right\rfloor$$

$$q^2 < V \leq q : \text{RETURN } X \leftarrow 1$$

$$V > q : \text{RETURN } X \leftarrow 2$$

5.3. Exercises.

1. The following logarithmic series generator is based upon rejection from the geometric distribution:

Logarithmic series generator based upon rejection

REPEAT

Generate a uniform $[0,1]$ random variate U and an exponential random variate E.

$$X \leftarrow \left\lceil -\frac{E}{\log(p)} \right\rceil$$

UNTIL $UX < 1$

RETURN X

Show that the expected number of exponential random variates needed is equal to the mean of the logarithmic series distribution, i.e. $-p/((1-p)\log(1-p))$. Show furthermore that this number increases monotonically to ∞ as $p \uparrow 1$.

2. **The generalized logarithmic series distribution.** Patel (1981) has proposed the following generalization of the logarithmic series distribution with parameter p :

$$p_i = \frac{p^i (1-p)^{bi-i} \Gamma(bi)}{-i \log(1-p) \Gamma(i) \Gamma(bi-i+1)} \quad (i \geq 1) .$$

Here $b \geq 1$ is a new parameter satisfying the inequality

$$0 < pb \left(\frac{b-bp}{b-1}\right)^{b-1} < 1 .$$

Suggest one or more efficient generators for this two-parameter family.

3. Consider the following discrete distribution:

$$p_i = \frac{1}{ci} \quad (1 \leq i \leq k) ,$$

where the integer k can be considered as a parameter, and c is a normalization constant. Show that the following bounded workspace algorithm generates random variates with this distribution:

REPEAT

    Generate iid uniform $[0,1]$ random variates U,V.

    $Y \leftarrow (k+1)^U$

    $X \leftarrow \lfloor Y \rfloor$

UNTIL $2VX < Y$

RETURN X

Analyze the expected number of iterations as a function of k. Suggest at least one effective improvement.

6. THE ZIPF DISTRIBUTION.

6.1. A simple generator.

In linguistics and social sciences, the **Zipf distribution** is frequently used to model certain quantities. This distribution has one parameter $a > 1$, and is defined by the probabilities

$$p_i = \frac{1}{\varsigma(a)i^a} \quad (i \geq 1)$$

where

$$\varsigma(a) = \sum_{i=1}^{\infty} \frac{1}{i^a}$$

is the Riemann zeta function. Simple expressions for the zeta function are known in special cases. For example, when a is integer, then

$$\varsigma(2a) = \frac{2^{2a-1}\pi^{2a}}{(2a)!}B_a$$

where B_a is the a-th Bernoulli number (Titchmarsh, 1951, p. 20). Thus, for $a = 2,4,6$ we obtain the probability vectors $\{6/(\pi i)^2\}, \{90/(\pi i)^4\}$ and $\{945/(\pi i)^6\}$ respectively.

To generate a random Zipf variate in uniformly bounded expected time, we propose the rejection method. Consider for example the distribution of the random variable $Y \leftarrow \left\lfloor U^{-1/(a-1)} \right\rfloor$ where U is uniformly distributed on $[0,1]$:

$$P(Y=i) = \frac{1}{(i+1)^{a-1}}((1+\frac{1}{i})^{a-1}-1) \quad (i \geq 1).$$

This distribution is a good candidate because the probabilities vary as $(a-1)i^{-a}$ as $i \to \infty$. For the sake of simplicity, let us define $q_i = P(Y=i)$. First, we note that the rejection constant c is

$$c = \sup_{i \geq 1} \frac{p_i}{q_i} = \frac{p_1}{q_1} = \frac{2^{a-1}}{\zeta(a)(2^{a-1}-1)} .$$

Hence, the following rejection algorithm can be used:

A Zipf generator based upon rejection

[SET-UP]
$b \leftarrow 2^{a-1}$
[GENERATOR]
REPEAT

       Generate iid uniform [0,1] random variates U, V.

$$X \leftarrow \left\lfloor U^{-\frac{1}{a-1}} \right\rfloor$$

$$T \leftarrow (1 + \frac{1}{X})^{a-1}$$

UNTIL $VX\dfrac{T-1}{b-1} \leq \dfrac{T}{b}$

RETURN X

Lemma 6.1.

    The rejection constant c in the rejection algorithm shown above satisfies the following properties:

A. $\sup\limits_{a \geq 2} c \leq \dfrac{12}{\pi^2}$.

B. $\sup\limits_{1 < a \leq 2} c \leq \dfrac{2}{\log(2)}$.

C. $\lim\limits_{a \to \infty} c = 1$.

D. $\lim\limits_{a \downarrow 1} c = \dfrac{1}{\log(2)}$.

Proof of Lemma 6.1.

Part A follows from

$$c \le \frac{2^{a-1}}{2^{a-1}-1} \frac{6}{\pi^2} \le \frac{12}{\pi^2} .$$

Part B follows from

$$c \le \frac{2^{a-1}}{(2^{a-1}-1) \int\limits_{1}^{\infty} x^{-a} \, dx} = \frac{(a-1)2^{a-1}}{2^{a-1}-1}$$

$$\le \frac{(a-1)2^{a-1}}{(a-1)\log(2)} = \frac{2^{a-1}}{\log(2)} \le \frac{2}{\log(2)} .$$

Part C follows by observing that $\varsigma(a) \to 1$ as $a \uparrow \infty$. Finally, part D uses the fact that $\varsigma(a) \sim \dfrac{1}{a-1}$ as $a \downarrow 1$ (in fact, $\varsigma(a) - \dfrac{1}{a-1} \to \gamma$, Euler's constant (Whittaker and Watson, 1927, p. 271). ∎

6.2. The Planck distribution.

The Planck distribution is a two-parameter distribution with density

$$f(x) = \frac{b^{a+1}}{\Gamma(a+1)\varsigma(a+1)} \frac{x^a}{e^{bx}-1} \quad (x > 0) .$$

Here $a > 0$ is a shape parameter and $b > 0$ is a scale parameter (Johnson and Kotz, 1970). The density f can be written as a mixture:

$$f(x) = \sum_{i=1}^{\infty} \frac{1}{i^{a+1}\varsigma(a+1)} \frac{x^a e^{-ibx} (ib)^{a+1}}{\Gamma(a+1)} .$$

In view of this, the following algorithm can be used to generate a random variate with the Planck distribution.

Planck random variate generator

Generate a gamma $(a+1)$ random variate G.

Generate a Zipf $(a+1)$ random variate Z.

RETURN $X \leftarrow \dfrac{G}{bZ}$.

6.3. The Yule distribution.

Simon (1954,1960) has suggested the Yule distribution as a better approximation of word frequencies than the Zipf distribution. He defined the discrete distribution by the probabilities

$$p_i = c(a) \int_0^1 (1-u)^{i-1} u^{a-1} du \qquad (i \geq 1) ,$$

where $c(a)$ is a normalization constant and $a > 1$ is a parameter. Using the fact that this is a mixture of the geometric distribution with parameter $e^{-Y/(a-1)}$ where Y is exponentially distributed, we conclude that a random variate X with the Yule distribution can be generated as

$$X \leftarrow \left\lceil -\frac{E}{\log(1-e^{-\frac{E*}{a-1}})} \right\rceil ,$$

where $E, E*$ are iid exponential random variates.

6.4. Exercises.

1. **The digamma and trigamma distributions.** Sibuya (1979) introduced two distributions, termed the digamma and trigamma distributions. The digamma distribution has two parameters, a, c satisfying $c > 0, a > -1$, $a + c > 0$. It is defined by

$$p_i = \frac{1}{\psi(a+c)-\psi(c)} \frac{a(a+1)\cdots(a+i-1)}{i(a+c)(a+c+1)\cdots(a+c+i-1)} \qquad (i \geq 1) .$$

Here ψ is the derivative of the log gamma function, i.e. $\psi = \Gamma'/\Gamma$. When we let $a \downarrow 0$, the trigamma distribution with parameter $c > 0$ is obtained:

$$p_i = \frac{1}{\psi'(c)} \frac{(i-1)!}{ic(c+1)\cdots(c+i-1)} \qquad (i \geq 1) .$$

For $c = 1$ this is a zeta distribution. Discuss random variate generation for this family of distributions, and provide a uniformly fast rejection algorithm.

Chapter Eleven
MULTIVARIATE DISTRIBUTIONS

1. GENERAL PRINCIPLES.

1.1. Introduction.

In section V.4, we have discussed in great detail how one can efficiently generate random vectors in R^d with radially symmetric distributions. Included in that section were methods for generating random vectors uniformly distributed in and on the unit sphere C_d of R^d. For example, when $N_1, \ldots, N_d$ are iid normal random variables, then

$$(\frac{N_1}{N}, \ldots, \frac{N_d}{N})$$

where $N = \sqrt{N_1^2 + \cdots + N_d^2}$, is uniformly distributed on the surface of C_d. This uniform distribution is the building block for all radially symmetric distributions because these distributions are all scale mixtures of the uniform distribution on the surface of C_d. This sort of technique is called a special property technique: it exploits certain characteristics of the distribution. What we would like to do here is give several methods of attacking the generation problem for d-dimensional random vectors, including many special property techniques.

The material has little global structure. Most sections can in fact be read independently of the other sections. In this introductory section several general principles are described, including the conditional distribution method. There is no analog to the univariate inversion method. Later sections deal with specific subclasses of distributions, such as uniform distributions on compact sets, elliptically symmetric distributions (including the multivariate normal distribution), bivariate uniform distributions and distributions on lines.

1.2. The conditional distribution method.

The conditional distribution method allows us to reduce the multivariate generation problem to d univariate generation problems, but it can only be used when quite a bit of information is known about the distribution.

Assume that our random vector $\mathbf{X}$ has density

$$f(x_1, \ldots, x_d) = f_1(x_1) f_2(x_2 \mid x_1) \cdots f_d(x_d \mid x_1, \ldots, x_{d-1}),$$

where the f_i's are conditional densities. Generation can proceed as follows:

Conditional distribution method

FOR $i := 1$ TO d DO
       Generate X_i with density $f_i(. \mid X_1, \ldots, X_{i-1})$. (For $i = 1$, use $f_1(.)$.)
    RETURN $\mathbf{X} = (X_1, \ldots, X_d)$

It is necessary to know all the conditional densities. This is equivalent to knowing all marginal distributions, because

$$f_i(x_i \mid x_1, \ldots, x_{i-1}) = \frac{f_i^*(x_1, \ldots, x_i)}{f_{i-1}^*(x_1, \ldots, x_{i-1})}$$

where f_i^* is the marginal density of the first i components, i.e. the density of $(X_1, \ldots, X_i)$.

Example 1.1. The multivariate Cauchy distribution.

The multivariate Cauchy density f is given by

$$f(x) = \frac{c}{(1 + \|x\|^2)^{\frac{d+1}{2}}},$$

where $c = \Gamma(\frac{d+1}{2})/\pi^{(d+1)/2}$. Here $\|.\|$ is the standard L_2 Euclidean norm. It is known that X_1 is univariate Cauchy, and that given $X_1, \ldots, X_{i-1}$, the random variable X_i is distributed as $T(1 + \sum_{j=1}^{i-1} X_j)/\sqrt{i}$ where T has the t distribution with i degrees of freedom (Johnson and Kotz, 1970). ∎

Example 1.2. The normal distribution.

Assume that f is the density of the zero mean normal distribution on R^2, with variance-covariance matrix $\mathbf{A} = \{a_{ij}\}$ where $a_{ij} = E(X_i X_j)$:

$$f(x) = \frac{1}{2\pi\sqrt{|\mathbf{A}|}} e^{-\frac{1}{2}x^t \mathbf{A}^{-1} x}.$$

In this case, the conditional density method yields the following algorithm:

Conditional density method for normal random variates

Generate N_1, N_2, iid normal random variates.

$X_1 \leftarrow N_1 \sqrt{a_{11}}$

$X_2 \leftarrow \dfrac{a_{21}}{a_{11}} X_1 + N_2 \sqrt{\dfrac{a_{22} a_{11} - a_{21}^2}{a_{11}}}$

RETURN (X_1, X_2)

This follows by noting that X_1 is zero mean normal with variance a_{11}, and computing the conditional density of X_2 given X_1 as a ratio of marginal densities. ■

Example 1.3.

Let f be the uniform density in the unit circle C_2 of R^2. The conditional density method is easily obtained:

Generate X_1 with density $f_1(x) = \dfrac{2}{\pi}\sqrt{1-x^2}$ $(|x| \leq 1)$.

Generate X_2 uniformly on $[-\sqrt{1-X_1^2}, \sqrt{1-X_1^2}]$.

RETURN (X_1, X_2) ■

In all three examples, we could have used alternative methods. Examples 1.1 and 1.2 deal with easily treated radially symmetric distributions, and Example 1.3 could have been handled via the ordinary rejection method.

1.3. The rejection method.

It should be clear that the rejection method is not tied to a particular space. It can be used in multivariate random variate generation problems, and is probably the most useful general purpose technique here. A few traps to watch out for are worth mentioning. First of all, rejection from a uniform density on a rectangle of R^d often leads to a rejection constant which deteriorates quickly as d increases. A case in point is the rejection method for generating points uniformly in the unit sphere of R^d (see section V.4.3). Secondly, unlike in R^1, upper bounds for certain densities are not easily obtainable. For example, the information that f is unimodal with a mode at the origin is of little use, whereas in R^1, the same information allows us to conclude that $f(x) \leq 1/|x|$. Similarly, combining unimodality with moment conditions is not enough. Even the fact that f is log-concave is not sufficient to derive universally applicable upper bounds (see section VII.2).

In general, the design of an efficient rejection method is more difficult than in the univariate case.

1.4. The composition method.

The composition method is not tied to a particular space such as R^1. A popular technique for obtaining dependence from independence is the following: define a random vector $\mathbf{X} = (X_1, \ldots, X_d)$ as $(SY_1, \ldots, SY_d)$ where the S_i's are iid random variables, and S is a random scale. In such cases, we say that the distribution of $\mathbf{X}$ is a **scale mixture**. If Y_1 has density f, then $\mathbf{X}$ has a density given by

$$E\left[\prod_{i=1}^{d} \left[\frac{1}{S} f\left(\frac{x_i}{S}\right) \right] \right] .$$

If Y_1 has distribution function $F = 1 - G$, then

$$P(X_1 > x_1, \ldots, X_d > x_d) = E\left(\prod_{i=1}^{d} G\left(\frac{x_i}{S}\right) \right) .$$

Example 1.4. The multivariate Burr distribution.

When Y_1 is Weibull with parameter a (i.e. $G(y) = e^{-y^a}$ $(y > 0)$), and S is gamma (b), then $(SY_1, \ldots, SY_d)$ has distribution function determined by

$$P(X_1 > x_1, \ldots, X_d > x_d) = E(\prod_{i=1}^{d} e^{-(x_i/S)^a})$$

$$= \int_0^\infty \frac{s^{b-1}e^{-s}}{\Gamma(b)} e^{-s^{-a}(\sum_{i=1}^{d} x_i^a)} \, ds$$

$$= \frac{1}{(1 + \sum_{i=1}^{d} x_i^a)^b} \qquad (x_i > 0 , \ i = 1, 2, \ldots, d) .$$

This defines the multivariate Burr distribution of Takahasi (1965). From this relation it is also easily seen that all univariate or multivariate marginals of a multivariate Burr distribution are univariate or multivariate Burr distributions. For more examples of scale mixtures in which S is gamma, see Hutchinson (1981). ∎

Example 1.5. The multinomial distribution.

The conditional distribution method is not limited to continuous distributions. For example, consider the **multinomial distribution** with parameters $n, p_1, \ldots, p_d$ where the p_i's form a probability vector and n is a positive integer. A random vector $(X_1, \ldots, X_d)$ is multinomially distributed with these parameters when

$$P((X_1, \ldots, X_d) = (i_1, \ldots, i_d)) = \frac{n!}{\prod_{j=1}^{d} i_j!} \prod_{j=1}^{d} p_j^{i_j}$$

$$(i_j \geq 0 , \ j = 1, \ldots, d ; \ \sum_{j=1}^{d} i_j = n) .$$

This is the distribution of the cardinalities of d urns into which n balls are thrown at random and independently of each other. Urn number j is selected with probability p_j by every ball. The ball-in-urn experiment can be mimicked, which leads us to an algorithm taking time $O(n+d)$ and $\Omega(n+d)$. Note however that X_1 is binomial (n, p_1), and that given X_1, the vector $(X_2, \ldots, X_d)$ is multinomial $(n - X_1, q_2, \ldots, q_d)$ where $q_j = p_j/(1-p_1)$. This recurrence relation is nothing but another way of describing the conditional distribution method for this case. With a uniformly fast binomial generator we can proceed in expected time $O(d)$ uniformly bounded in n :

Multinomial random vector generator

[NOTE: the parameters $n, p_1, \ldots, p_d$ are destroyed by this algorithm. Sum holds a cumulative sum of probabilities.]

Sum $\leftarrow 0$

FOR $i := 1$ TO d DO

> Generate a binomial $(n, \dfrac{p_i}{S})$ random vector X_i.
>
> $n \leftarrow n - X_i$
>
> Sum $\leftarrow$ Sum - p_i

For small values of n, it is unlikely that this algorithm is very competitive, mainly because the parameters of the binomial distribution change at every call. ∎

1.5. Discrete distributions.

Consider the problem of the generation of a random vector taking only values on d-tuples of nonnegative integers. One of the striking differences with the continuous multivariate distributions is that the d-tuples can be put into one-to-one correspondence with the nonnegative integers on the real line. This one-to-one mapping can be used to apply the inversion method (Kemp, 1981; Kemp and Loukas, 1978) or one of the table methods (Kemp and Loukas, 1981). We say that the function which transforms d-tuples into nonnegative integers is a **coding function**. The inverse function is called the **decoding function**.

Coding functions are easy to construct. Consider $d = 2$. Then we can visit all 2-tuples in the positive quadrant in cross-diagonal fashion. Thus, first we visit (0,0), then (0,1) and (1,0), then (0,2),(1,1) and (2,0), etcetera. Note that we visit all the integers (i, j) with $i + j = k$ before visiting those with $i + j = k + 1$. Since we visit $k(k-1)/2$ 2-tuples with $i + j < k$, we see that we can take as coding function

$$h(i, j) = \frac{(i + j)(i + j - 1)}{2} + i .$$

This can be generalized to d-tuples (exercise 1.4), and a simple decoding function exists which allows us to recover (i, j) from the value of $h(i, j)$ in time $O(1)$ (exercise 1.4). There are other orders of traversal of the 2-tuples. For example, we could visit 2-tuples in order of increasing values of $\max(i, j)$.

In general one cannot visit all 2-tuples in order of increasing values of i, its first component, as there could be an infinite number of 2-tuples with the same value of i. It is like trying to visit all shelves in a library, and getting stuck in the first shelf because it does not end. If the second component is bounded, as it often is, then the library traversal leads to a simple coding function. Let M be the maximal value for j. Then we have

$$h(i,j) = (M+1)i + j.$$

One should be aware of some pitfalls when the univariate connection is exploited. Even if the distribution of probability over the d-tuples is relatively smooth, the corresponding univariate probability vector is often very oscillatory, and thus unfit for use in the rejection method. Rejection should be applied almost exclusively to the original space.

The fast table methods require a finite distribution. Even though on paper they can be applied to all finite distributions, one should realize that the number of possible d-tuples in such distributions usually explodes exponentially with d. For a distribution on the integers in the hypercube $\{1,2, \ldots, n\}^d$, the number of possible values is n^d. For this example, table methods seem useful only for moderate values of d. See also exercise 1.5.

Kemp and Loukas (1978) and Kemp (1981) are concerned with the inversion method and its efficiency for various coding functions. Recall that in the univariate case, inversion by sequential search for a nonnegative integer-valued random variate X takes expected time (as measured by the expected number of comparisons) $E(X)+1$. Thus, with the coding function h for $X_1, \ldots, X_d$, we see without further work that the expected number of comparisons is

$$E(h(X_1, \ldots, X_d)+1).$$

Example 1.6.

Let us apply inversion for the generation of (X_1, X_2), and let us scan the space in cross diagonal fashion (the coding function is $h(i,j) = \dfrac{(i+j)(i+j-1)}{2} + i$). Then the expected number of comparisons before halting is

$$E\left[\frac{(X_1+X_2)(X_1+X_2-1)}{2} + X_1 + 1\right].$$

This is at least proportional to either one of the marginal second moments, and is thus much worse than one would normally have expected. In fact, in d dimensions, a similar coding function leads to a finite expected time if and only if $E(X_i^d) < \infty$ for all $i = 1, \ldots, d$ (see exercise 1.6). ∎

Example 1.7.

Let us apply inversion for the generation of (X_1, X_2), where $0 \leq X_2 \leq M$, and let us perform a library traversal (the coding function is $h(i,j) = (M+1)i + j$). Then the expected number of comparisons before halting is

$$E\left((M+1)X_1 + X_2 + 1\right) \ .$$

This is finite when only the first moments are finite, but has the drawback that M figures explicitly in the complexity. ∎

We have made our point. For large values of d, ordinary generation methods are often not feasible because of time or space inefficiencies. One should nearly always try to convert the problem into several univariate problems. This can be done by applying the conditional distribution method. For the generation of X_1, X_2, we first generate X_1, and then generate X_2 conditional on the given value of X_1. Effectively, this forces us to know the marginal distribution of X_1 and the joint two-dimensional distribution. The marginal distribution of X_2 is not needed. To see how this improves the complexities, consider using the inversion method in both stages of the algorithm. The expected number of comparisons in the generation of X_2 given X_1 is $E(X_2 \mid X_1) + 1$. The number of comparisons in the generation of X_1 is $X_1 + 1$. Summing and taking expected values shows that the expected number of comparisons is

$$E(X_1 + X_2 + 2)$$

(Kemp and Loukas, 1978). Compare with Examples 1.6 and 1.7.

In the conditional distribution method, we can improve the complexity even further by employing table methods in one, some or all of the stages. If $d = 2$ and both components have infinite support, we cannot use tables. If only the second component has infinite support, then a table method can be used for X_1. This is the ideal situation. If both components have finite support, then we are tempted to apply the table method in both stages. This would force us to set up many tables, one for each of the possible values of X_1. In that case, we could as well have set up one giant table for the entire distribution. Finally, if the first component has infinite support, and the second component has finite support, then the incapability of storing an infinite number of finite tables forces us to set up the tables as we need them, but the time spent doing so is prohibitively large.

If a distribution is given in analytic form, there usually is some special property which can be used in the design of an efficient generator. Several examples can be found in section 3.

1.6. Exercises.

1. Consider the density $f(x_1,x_2) = 5x_1 e^{-x_1 x_2}$ defined on the infinite strip $0.2 \leq x_1 \leq 0.4$, $0 \leq x_2$. Show that the first component X_1 is uniformly distributed on $[0.2,0.4]$, and that given X_1, X_2 is distributed as an exponential random variable divided by X_1 (Schmeiser, 1980).

2. Show how you would generate random variates with density

$$\frac{6}{(1+x_1+x_2+x_3)^4} \qquad (x_1,x_2,x_3 \geq 0) .$$

Show also that $X_1+X_2+X_3$ has density $3x^2/(1+x)^4$ $(x \geq 0)$ (Springer, 1979, p.87).

3. Prove that for any distribution function F on R^d, there exists a measurable function $g:[0,1] \to R^d$ such that $g(U)$ has distribution function F, where U is uniformly distributed on $[0,1]$. This can be considered as a generalization of the inversion method. Hint: from U we can construct d iid uniform $[0,1]$ random variables by skipping bits. Then argue via conditioning.

4. Consider the coding function for 2-tuples of nonnegative integers (i,j) given by $h(i,j) = \dfrac{(i+j)(i+j-1)}{2} + i + 1.$

   A. Generalize this coding function to d-tuples. The generalization should be such that all d-tuples with sum of the components equal to some integer k are grouped together, and the groups are ordered according to increasing values for k. Within a group, this rule should be applied recursively to groups of $d-1$-tuples with constant sum.

   B. Give the decoding function for the two-dimensional h shown above, and indicate how it can be evaluated in time $O(1)$ (independent of the size of the argument).

5. Consider the multinomial distribution with parameters $n,p_1, \ldots , p_d$, which assigns probability

$$\frac{n!}{i_1! \cdots i_d!} \prod_{j=1}^{d} p_j^{i_j}$$

to all d-tuples with $i_j \geq 0$, $\sum_{j=1}^{d} i_j = n$. Let the total number of possible values be $N(n,d)$. For fixed n, find a simple function $\psi(d)$ with the property that

$$\lim_{d \to \infty} \frac{N(n,d)}{\psi(d)} = 1 .$$

This gives some idea about how quickly $N(n,d)$ grows with d.

6. Show that when a cross-diagonal traversal is followed in d dimensions for inversion by sequential search of a discrete probability distribution on the nonnegative integers of R^d, then the expected time required by the inversion is finite if and only if $E(X_i^d) < \infty$ for all $i=1, \ldots , d$ where $X_1, \ldots , X_d$ is a d-dimensional random vector with the given distribution.

7. **Relationship between multinomial and Poisson distributions.** Show
    that the algorithm given below in which the sample size parameter is used as
    a mixing parameter delivers a sequence of d iid Poisson (λ) random vari-
    ables.

> Generate a Poisson $(d\lambda)$ random variate N.
>
> RETURN a multinomial $(N, \frac{1}{d}, \ldots, \frac{1}{d})$ random vector $(X_1, \ldots, X_d)$.

Hint: this can be proved by explicitly computing the probabilities, by working with generating functions, or by employing properties of Poisson point
processes.

8. **A bivariate extreme value distribution.** Marshall and Olkin (1983)
    have studied multivariate extreme value distributions in detail. One of the
    distributions considered by them is defined by

$$P(X_1 > x_1, X_2 > x_2) = e^{-(e^{-x_1} + e^{-x_2} - (e^{x_1} + e^{x_2})^{-1})} \qquad (x_1 \geq 0, x_2 \geq 0) .$$

How would you generate a random variate with this distribution?

9. Let f be an arbitrary univariate density on $(0, \infty)$. Show that
    $f(x_1 + x_2)/(x_1 + x_2)$ $(x_1 > 0, x_2 > 0)$ is a bivariate density (Feller, 1971,
    p.100). Exploiting the structure in the problem to the fullest, how would
    you generate a random vector with the given bivariate density?

2. LINEAR TRANSFORMATIONS. THE MULTINORMAL DISTRI-BUTION.

2.1. Linear transformations.

When an R^d-valued random vector $\mathbf{X}$ has density $f(\mathbf{x})$, then the random
vector $\mathbf{Y}$ defined as the solution of $\mathbf{X} = \mathbf{H}\mathbf{Y}$ has density

$$g(\mathbf{y}) = |\mathbf{H}| f(\mathbf{H}\mathbf{y}), \mathbf{y} \in R^d ,$$

for all nonsingular $d \times d$ matrices $\mathbf{H}$. The notation $|\mathbf{H}|$ is used for the absolute
value of the determinant of $\mathbf{H}$. This property is reciprocal, i.e. when Y has density g, then $\mathbf{X} = \mathbf{H}\mathbf{Y}$ has density f.

The linear transformation $\mathbf{H}$ deforms the coordinate system. Particularly
important linear deformations are rotations: these correspond to orthonormal
transformation matrices $\mathbf{H}$. For random variate generation, linear transformations
are important in a few special cases:

A. The generation of points uniformly distributed in d-dimensional simplices or hyperellipsoids.

B. The generation of random vectors with a given dependence structure, as measured by the covariance matrix.

These two application areas are now dealt with separately.

2.2. Generators of random vectors with a given covariance matrix.

The covariance matrix of an R^d-valued random vector $\mathbf{Y}$ with mean 0 is defined as $\Sigma = E(\mathbf{YY'})$ where $\mathbf{Y}$ is considered as a column vector, and $\mathbf{Y'}$ denotes the transpose of $\mathbf{Y}$. Assume first that we wish to generate a random vector $\mathbf{Y}$ with zero mean and covariance matrix Σ and that we do not care for the time being about the form of the distribution. Then, it is always possible to proceed as follows: generate a random vector $\mathbf{X}$ with d iid components $X_1, \ldots, X_d$ each having zero mean and unit variance. Then define $\mathbf{Y}$ by $\mathbf{Y} = \mathbf{HX}$ where $\mathbf{H}$ is a nonsingular $d \times d$ matrix. Note that

$$E(\mathbf{Y}) = \mathbf{H}E(\mathbf{X}) = 0 ,$$

$$E(\mathbf{YY'}) = \mathbf{H}E(\mathbf{XX'})\mathbf{H'} = \mathbf{HH'} = \Sigma .$$

We need a few facts now from the theory of matrices. First of all, we recall the definition of positive definiteness. A matrix $\mathbf{A}$ is positive definite (positive semidefinite) when $\mathbf{x'Ax} > 0 \ (\geq 0)$ for all nonzero R^d-valued vectors $\mathbf{x}$. But we have

$$\mathbf{x'\Sigma x} = E(\mathbf{x'YY'x}) = E(\mid \mid \mathbf{x'Y} \mid \mid) \geq 0$$

for all nonzero $\mathbf{x}$. Here $\mid \mid . \mid \mid$ is the standard L_2 norm in R^d. Equality occurs only if the Y_i's are linearly dependent with probability one, i.e. $\mathbf{x'Y} = 0$ with probability one for some $\mathbf{x} \neq 0$. In that case, $\mathbf{Y}$ is said to have dimension less than d. Otherwise, $\mathbf{Y}$ is said to have dimension d. Thus, all covariance matrices are positive semidefinite. They are positive definite if and only if the random vector in question has dimension d.

For symmetric positive definite matrices Σ, we can always find a nonsingular matrix $\mathbf{H}$ such that

$$\mathbf{HH'} = \Sigma .$$

In fact, such matrices can be characterized by the existence of a nonsingular $\mathbf{H}$. We can do even better. One can always find a lower triangular nonsingular $\mathbf{H}$ such that

$$\mathbf{HH'} = \Sigma .$$

We have now turned our problem into one of decomposing a symmetric positive definite matrix Σ into a product of two lower triangular matrices. The algorithm can be summarized as follows:

Generator of a random vector with given covariance matrix

[SET-UP]

Find a matrix $\mathbf{H}$ such that $\mathbf{HH'}=\Sigma$.

[GENERATOR]

Generate d independent zero mean unit variance random variates $X_1, \ldots, X_d$.

RETURN $\mathbf{Y}=\mathbf{HX}$

The set-up step can be done in time $O(d^3)$ as we will see below. Since $\mathbf{H}$ can have up to $\Omega(d^2)$ nonzero elements, there is no hope of generating $\mathbf{Y}$ in less than $\Omega(d^2)$. Note also that the distributions of the X_i's are to be picked by the users. We could take them iid and biatomic: $P(X_1=1)=P(X_1=-1)=\dfrac{1}{2}$. In that case, Y is atomic with up to 2^d atoms. Such atomic solutions are rarely adequate. Most applications also demand some control over the marginal distributions. But these demands restrict our choices for X_1. Indeed, if our method is to be universal, we should choose $X_1, \ldots, X_d$ in such a way that all linear combinations of these independent random variables have a given distribution. This can be assured in several ways, but the choices are limited. To see this, let us consider iid random variables X_i with common characteristic function ϕ, and assume that we wish all linear combinations to have the same distribution up to a scale factor. The sum $\sum a_j X_j$ has characteristic function

$$\prod_{j=1}^{d} \phi(a_j t) \ .$$

This is equal to $\phi(at)$ for some constant a when ϕ has certain functional forms. Take for example

$$\phi(t) = e^{-|t|^\alpha}$$

for some $\alpha \in (0,2]$ as in the case of a symmetric stable distribution. Unfortunately, the only symmetric stable distribution with a finite variance is the normal distribution ($\alpha=2$). Thus, the property that the normal distribution is closed under the operation "linear combination" is what makes it so attractive to the user. If the user specifies non-normal marginals, the covariance structure is much more difficult to enforce. See however some good solutions for the bivariate case as developed in section XI.3.

A computational remark about $\mathbf{H}$ is in order here. There is a simple algorithm known as the **square root method** for finding a lower triangular $\mathbf{H}$ **with** $\mathbf{HH'} = \Sigma$ (Faddeeva, 1959; Moonan, 1957; Graybill, 1969). We give the relationship between the matrices here. The elements of Σ are called σ_{ij}, and those of the lower triangular solution matrix $\mathbf{H}$ are called h_{ij}.

$$h_{i1} = \sigma_{i1}/\sqrt{\sigma_{11}} \quad (1 \leq i \leq d)$$

$$h_{ii} = \sqrt{\sigma_{ii} - \sum_{j=1}^{i-1} h_{ij}^{2}} \quad (1 < i \leq d)$$

$$h_{ij} = \frac{\sigma_{ij} - \sum_{k=1}^{j-1} h_{ik}h_{jk}}{h_{jj}} \quad (1 < j < i \leq d)$$

$$h_{ij} = 0 \quad (i < j \leq d)$$

2.3. The multinormal distribution.

The **standard multinormal distribution** on R^d has density

$$f(\mathbf{x}) = (2\pi)^{-\frac{d}{2}} e^{-\frac{1}{2}\mathbf{x}'\mathbf{x}}$$

$$= (2\pi)^{-\frac{d}{2}} e^{-\frac{1}{2}||\mathbf{x}||^2} \quad (\mathbf{x} \in R^d) .$$

This is the density of d iid normal random variables. When $\mathbf{X}$ has density f, $\mathbf{Y} = \mathbf{HX}$ has density

$$g(\mathbf{y}) = |\mathbf{H}^{-1}| f(\mathbf{H}^{-1}\mathbf{y}) , \mathbf{y} \in R^d .$$

But we know that $\Sigma = \mathbf{HH}'$, so that $|\mathbf{H}^{-1}| = |\Sigma|^{-1/2}$. Also, $||\mathbf{H}^{-1}\mathbf{y}||^2 = \mathbf{y}'\Sigma^{-1}\mathbf{y}$, which gives us the density

$$g(\mathbf{y}) = (2\pi)^{-\frac{d}{2}} |\Sigma|^{-\frac{1}{2}} e^{-\frac{1}{2}\mathbf{y}'\Sigma^{-1}\mathbf{y}} \quad (\mathbf{y} \in R^d) .$$

This is the density of the multinormal distribution with zero mean and nonsingular covariance matrix Σ. We note without work that the i-th marginal distribution is zero mean normal with variance given by the i-th diagonal element of Σ. In the most general form of the normal distribution, we need only add a translation parameter (mean) to the distribution.

Random variate generation for the normal distribution can be done by the linear transformation of d iid normal random variables described in the previous section. This involves decomposition of Σ into a product of the form $\mathbf{HH}'$. This method has been advocated by Scheuer and Stoller (1962) and Barr and Slezak (1972). Deak (1979) gives other methods for generating multinormal random vectors. For the conditional distribution method in the case $d = 2$, we refer to Example 1.2. In the general case, see for example Scheuer and Stoller (1962).

An important special case is the bivariate multinormal distribution with zero mean, and covariance matrix

$$\begin{vmatrix} 1 & \rho \\ \rho & 1 \end{vmatrix}$$

where $\rho \in [-1,1]$ is the correlation between the two marginal random variables. It is easy to see that if (N_1, N_2) are iid normally distributed random variables, then

$$(N_1, \rho N_1 + \sqrt{1-\rho^2} N_2)$$

has the said distribution. The multinormal distribution can be used as the starting point for creating other multivariate distributions, see section XI.3. We will also exhibit many multivariate distributions with normal marginals which are not multinormal. To keep the terminology consistent throughout this book, we will refer to all distributions having normal marginals as multivariate normal distributions. Multinormal distributions form only a tiny subclass of the multivariate normal distributions.

2.4. Points uniformly distributed in a hyperellipsoid.

A hyperellipsoid in R^d is defined by a symmetric positive definite $d \times d$ matrix $\mathbf{A}$: it is the collection of all points $\mathbf{y} \in R^d$ with the property that

$$\mathbf{y}'\mathbf{A}\mathbf{y} \leq 1 .$$

A random vector uniformly distributed in this hyperellipsoid can be generated by a linear transformation of a random vector $\mathbf{X}$ distributed uniformly in the unit hypersphere C_d of R^d. Such random vectors can be generated quite efficiently (see section V.4). Recall that linear transformations cannot destroy uniformity. They can only alter the shape of the support of uniform distributions. The only problem we face is that of the determination of the linear transformation in function of $\mathbf{A}$.

Let us define $\mathbf{Y} = \mathbf{H}\mathbf{X}$ where $\mathbf{H}$ is our $d \times d$ transformation matrix. The set defined by

$$\mathbf{y}'\mathbf{A}\mathbf{y} \leq 1$$

corresponds to the set

$$\mathbf{x}'\mathbf{H}'\mathbf{A}\mathbf{H}\mathbf{x} \leq 1 .$$

But since this has to coincide with $\mathbf{x}'\mathbf{x} \leq 1$ (the definition of C_d), we note that

$$\mathbf{H}'\mathbf{A}\mathbf{H} = \mathbf{I}$$

where $\mathbf{I}$ is the unit $d \times d$ matrix. Thus, we need to take $\mathbf{H}$ such that $\mathbf{A}^{-1} = \mathbf{H}\mathbf{H}'$. See also Rubinstein (1982).

2.5. Uniform polygonal random vectors.

A **convex polytope** of R^d with vertices $\mathbf{v}_1, \ldots, \mathbf{v}_n$ is the collection of all points in R^d that are obtainable as convex combinations of $\mathbf{v}_1, \ldots, \mathbf{v}_n$. Every point $\mathbf{x}$ in this convex polytope can be written as

$$\mathbf{x} = \sum_{i=1}^{n} a_i \mathbf{v}_i$$

for some $a_1, \ldots, a_n$ with $a_i \geq 0$, $\sum_{i=1}^{n} a_i = 1$. The set $\mathbf{v}_1, \ldots, \mathbf{v}_n$ is minimal for the convex polytope generated by it when all $\mathbf{v}_i$'s are distinct, and no $\mathbf{v}_i$ can be written as a strict convex combination of the $\mathbf{v}_j$'s. (A strict convex combination is one which has at least one a_i not equal to 0 or 1.)

We say that a set of vertices $\mathbf{v}_1, \ldots, \mathbf{v}_n$ is in general position if no three points are on a line, no four points are in a plane, etcetera. Thus, if the set of vertices is minimal for a convex polytope P, then it is in general position.

A **simplex** is a convex polytope with $d+1$ vertices in general position. Note that d points in general position in R^d define a hyperplane of dimension $d-1$. Thus, any convex polytope with fewer than $d+1$ vertices must have zero d-dimensional volume. In this sense, the simplex is the simplest nontrivial object in R^d.

We can define a basic simplex by the origin and d points on the positive coordinate axes at distance one from the origin.

There are two distinct generation problems related to convex polytopes. We could be asked to generate a random vector uniformly distributed in a given polytope (see below), or we could be asked to generate a random collection of vertices defining a convex polytope. The latter problem is not dealt with here. See however Devroye (1982) and May and Smith (1982).

Random vectors distributed uniformly in an arbitrary simplex can be obtained by linear transformations of random vectors distributed uniformly in the basic simplex. Fortunately, we do not have to go through the agony of factorizing a matrix as in the case of a given covariance matrix structure. Rather, there is a surprisingly simple direct solution to the general problem.

Theorem 2.1.

Let $(S_1, \ldots, S_{d+1})$ be the spacings generated by a uniform sample of size d on [0,1]. (Thus, $S_i \geq 0$ for all i, and $\sum S_i = 1$.) Then

$$\mathbf{X} = \sum_{i=1}^{d+1} S_i \mathbf{v}_i$$

is uniformly distributed in the polytope P generated by $\mathbf{v}_1, \ldots, \mathbf{v}_{d+1}$, provided that $\mathbf{v}_1, \ldots, \mathbf{v}_{d+1}$ are in general position.

Proof of Theorem 2.1.

Let $\mathbf{S}$ be the column vector $S_1, \ldots, S_d$. We recall first that $\mathbf{S}$ is uniformly distributed in the basic simplex B where

$$B = \{(x_1, \ldots, x_d) : x_i \geq 0, \sum_i x_i \leq 1\}.$$

If all $\mathbf{v_i}$'s are considered as column vectors, and $\mathbf{A}$ is the matrix

$$\left[\mathbf{v_1} - \mathbf{v_{d+1}} \quad \mathbf{v_2} - \mathbf{v_{d+1}} \quad \cdots \quad \mathbf{v_d} - \mathbf{v_{d+1}} \right],$$

then we can write $\mathbf{X}$ as follows:

$$\mathbf{X} = \mathbf{v_{d+1}} + \sum_{i=1}^{d} (\mathbf{v_i} - \mathbf{v_{d+1}}) S_i = \mathbf{v_{d+1}} + \mathbf{S'A}.$$

It is clear that $\mathbf{X}$ is uniformly distributed, since it can be obtained by a linear transformation of $\mathbf{S}$. The support Supp $(\mathbf{X})$ of the distribution of $\mathbf{X}$ is the collection of all points which can be written as $\mathbf{v_{d+1}} + \mathbf{a'A}$ where $\mathbf{a} \in B$ is a column vector. First, assume that $\mathbf{x} \in P$. Then, $\mathbf{x} = \sum_{i=1}^{d+1} a_i \mathbf{v_i}$ for some probability vector $a_1, \ldots, a_{d+1}$. This can be rewritten as follows:

$$\mathbf{x} = \mathbf{v_{d+1}} + \sum_{i=1}^{d} a_i (\mathbf{v_i} - \mathbf{v_{d+1}}) = \mathbf{v_{d+1}} + \mathbf{a'A},$$

where $\mathbf{a}$ is the vector formed by $a_1, \ldots, a_d$. Thus, $P \subseteq$ Supp $(\mathbf{X})$. Next, assume $\mathbf{x} \in$ Supp $(\mathbf{X})$. Then, for some column vector $\mathbf{a} \in B$,

$$\mathbf{x} = \mathbf{v_{d+1}} + \mathbf{a'A} = \mathbf{v_{d+1}} + \sum_{i=1}^{d} a_i (\mathbf{v_i} - \mathbf{v_{d+1}})$$

$$= \sum_{i=1}^{d+1} a_i \mathbf{v_i},$$

which implies that $\mathbf{x}$ is a convex combination of the $\mathbf{v_i}$'s, and thus $\mathbf{x} \in P$. Hence Supp $(\mathbf{X}) \subset P$, and hence Supp $(\mathbf{X}) = P$, which concludes the proof of Theorem 2.1. ∎

Example 2.1. Triangles.

The following algorithm can be used to generate random vectors uniformly distributed in the triangle defined by $\mathbf{v_1}, \mathbf{v_2}, \mathbf{v_3}$ of R^2:

Generator for uniform distribution in triangle

Generate iid uniform [0,1] random variates U, V.
IF $U > V$ then swap U and V.
RETURN $(U\mathbf{v}_1 + (V - U)\mathbf{v}_2 + (1 - V)\mathbf{v}_3)$

See also exercise 2.1. ∎

Example 2.2. Convex polygons in the plane.

Convex polygons on R^2 with $n > 3$ vertices can be partitioned into $n - 2$ disjoint triangles. This can always be done by connecting all vertices with a designated root vertex. Triangulation of a polygon is of course always possible, even when the polygon is not convex. To generate a point uniformly in a triangulated polygon, it suffices to generate a point uniformly in the i-th triangle (see e.g. Example 2.1), where the i-th triangle is selected with probability proportional to its area. It is worth recalling that the area of a triangle formed by $(v_{11}, v_{12}), (v_{21}, v_{22}), (v_{31}, v_{32})$ is

$$\frac{1}{2} \left| \sum_{i < j} (v_{i1} v_{j2} - v_{j1} v_{i2}) \right| . ∎$$

We can deal with all simplices in all Euclidean spaces via Theorem 2.1. Example 2.2 shows that all polygons in the plane can be dealt with too, because all such polygons can be triangulated. Unfortunately, decomposition of d-dimensional polytopes into d-dimensional simplices is not always possible, so that Example 2.2 cannot be extended to higher dimensions. The decomposition is possible for all convex polytopes however. A decomposition algorithm is given in Rubin (1984), who also provides a good survey of the problem. Theorem 2.1 can also be found in Rubinstein (1982). Example 2.2 describes a method used by Hsuan (1982). The present methods which use decomposition and linear transformations are valid for polytopes. For sets with unusual shapes, the grid methods of section VIII.3.2 should be useful.

We conclude this section with the simple mention of how one can attack the decomposition of a convex polytope with n vertices into simplices for general Euclidean spaces. If we are given an ordered polytope, i.e. a polytope with all its faces clearly identified, and with pointers to neighboring faces, then the partition is trivial: choose one vertex, and construct all simplices consisting of a face (each face has d vertices) and the picked vertex. For selection of a simplex, we also need the area of a simplex with vertices v_i, $i = 1, 2, \ldots, d+1$. This is given by

$$\frac{|A|}{d!}$$

where A is the $d \times d$ matrix with as columns $v_1 - v_{d+1}, \ldots, v_d - v_{d+1}$. The complexity of the preprocessing step (decomposition, computation of areas) depends upon m, the number of faces. It is known that $m = O(n^{\lfloor d/2 \rfloor})$ (McMullen, 1970). Since each area can be computed in constant time (d is kept fixed, n varies), the set-up time is $O(m)$. The expected generation time is $O(1)$ if a constant time selection algorithm is used.

The aforementioned ordered polytopes can be obtained from an unordered collection of n vertices in worst-case time $O(n \log(n) + n^{\lfloor (d+1)/2 \rfloor})$ (Seidel, 1981), and this is worst-case optimal for even dimensions under some computational models.

2.6. Time series.

The generation of random time series with certain specific properties (marginal distributions, autocorrelation matrix, etcetera) is discussed by Schmeiser (1980), Franklin (1965), Price (1976), Hoffman (1979), Li and Hammond (1975), Lakhan (1981), Polge, Holliday and Bhagavan (1973), Mikhailov (1974), Fraker and Rippy (1974), Badel (1979), Lawrance and Lewis (1977, 1980, 1981), and Jacobs and Lewis (1977).

2.7. Singular distributions.

Singular distributions in R^d are commonplace. Distributions that put all their mass on a line or curve in the plane are singular. So are distributions that put all their mass on the surface of a hypersphere of R^d. Computer generation of random vectors on such hyperspheres is discussed by Ulrich (1984), who in particular derives an efficient generator for the Fisher-von Mises distribution in R^d.

A line in R^d can be given in many forms. Perhaps the most popular form is the parametric one, where $x = h(z)$ and $z \in R$ is a parameter. An example is the circle in R^2, determined by

$$x_1 = \cos(2\pi z),$$
$$x_2 = \sin(2\pi z).$$

Now, if Z is a random variable and $\mathbf{h}$ is a Borel measurable function, then $\mathbf{X} = \mathbf{h}(Z)$ is a random vector which puts all its mass on the line defined by $\mathbf{x} = \mathbf{h}(z)$. In other words, $\mathbf{X}$ has a line distribution. For a one-to-one mapping $\mathbf{h} : R \to R^d$, which is also continuous, we can define a line density $f(z)$ at the point $\mathbf{x} = \mathbf{h}(z)$ via the relationships

$$P\left(\mathbf{X} = \mathbf{h}(z) \text{ for some } z \in [a,b]\right) = \int_a^b f(z)\psi(z)\, dz \qquad (\text{all } [a,b]) \,,$$

where $\psi(z) = \sqrt{\sum_{i=1}^{d} h'_i{}^2(z)}$ is the norm of the tangent of $\mathbf{h}$ at z, and h_i is the i-th component of $\mathbf{h}$. But since this must equal $P(a \leq Z \leq b) = \int_a^b g(z)\, dz$ where g is the density of Z, we see that

$$f(z) = \frac{g(z)}{\psi(z)} \ .$$

For a uniform line density, we need to take g proportional to ψ.

As a first example, consider a function in the plane determined by the equation $y = \chi(x)$ $(0 \leq x \leq 1)$. A point with uniform line density can be obtained by considering the x-coordinate as our parameter z. This yields the algorithm

Generate a random variate X with density $c\sqrt{1 + \chi'^2(x)}$.
RETURN $(X, \chi(X))$

This could be called the projection method for obtaining random variates with certain line densities. The converse, projection from a line to the x-axis is much less useful, since we already have many techniques for generating real-line-valued random variates.

2.8. Exercises.

1. Consider a triangle with vertices $\mathbf{v}_1, \mathbf{v}_2, \mathbf{v}_3$, and let U, V be iid uniform $[0,1]$ random variables.

    A. Show that if we set $\mathbf{Y} \leftarrow \mathbf{v}_2 + (\mathbf{v}_3 - \mathbf{v}_2)U$, and $\mathbf{X} \leftarrow \mathbf{v}_1 + (\mathbf{Y} - \mathbf{v}_1)V$, then $\mathbf{X}$ is not uniformly distributed in the given triangle. This method is misleading, as $\mathbf{Y}$ is uniformly distributed on the edge $(\mathbf{v}_2, \mathbf{v}_3)$, and $\mathbf{X}$ is uniformly distributed on the line joining $\mathbf{v}_1$ and $\mathbf{Y}$.

B. Show that **X** in part A is uniformly distributed in the said triangle if we replace V in the algorithm by $\max(V, V*)$ where $V, V*$ are iid uniform $[0,1]$ random variables.

2. Define a simple boolean function which returns the value true if and only if **x** belongs to the a triangle in R^2 with three given vertices.

3. Consider a triangle ABC where AB has length one, BC has length b, and the angle ABC is θ. Let **X** be uniformly distributed in the triangle, and let **Y** be the intersection of the lines A**X** and BC. Let Z be the distance between **Y** and B. Show that Z has density

$$\frac{1}{\sqrt{z^2 - 2z\cos(\theta) + 1}} \qquad (0 < z < b).$$

Compare the geometric algorithm for generating Z given above with the inversion method.

3. DEPENDENCE. BIVARIATE DISTRIBUTIONS.

3.1. Creating and measuring dependence.

In many experiments, a controlled degree of dependence is required. Sometimes, users want distributions with given marginals and a given dependence structure as measured with some criterion. Sometimes, users know precisely what they want by completely specifying a multivariate distribution. In this section, we will mainly look at problems in which certain marginal distributions are needed together with a given degree of dependence. Usually, there are very many multivariate distributions which satisfy the given requirements, and sometimes there are none. In the former case, we should design generators which are efficient and lead to distributions which are not unrealistic.

For a clear treatment of the subject, it is best to emphasize bivariate distributions. A number of different measures of association are commonly used by practicing statisticians. First and foremost is the **correlation coefficient** ρ (also called Pearson product moment correlation coefficient) defined by

$$\rho = \frac{E((X_1 - \mu_1)(X_2 - \mu_2))}{\sigma_1 \sigma_2},$$

where μ_1, μ_2 are the means of X_1, X_2, and σ_1, σ_2 are the corresponding standard deviations. The key properties of ρ are well-known. When X_1, X_2 are independent, $\rho = 0$. Furthermore, by the Cauchy-Schwarz inequality, it is easy to see that $|\rho| \leq 1$. When $X_1 = X_2$, we have $\rho = 1$, and when $X_1 = -X_2$, we have $\rho = -1$. Unfortunately, there are a few enormous drawbacks related to the correlation coefficient. First, it is only defined for distributions having marginals with finite variances. Furthermore, it is not invariant under monotone transformations of the coordinate axes. For example, if we define a bivariate uniform distribution with a given value for ρ and then apply a transformation to get certain specific marginals, then the value of ρ could (and usually does) change. And most importantly, the value of ρ may not be a solid indicator of the dependence. For one thing, $\rho=0$ does not imply independence.

Measures of association which are invariant under monotone transformations are in great abundance. For example, there is **Kendall's tau** defined by

$$\tau = 2P\,((X_1-X_2)(X'_1-X'_2)>0)-1$$

where (X_1,X_2) and (X'_1,X'_2) are iid. The invariance under strictly monotone transformations of the coordinate axes is obvious. Also, for all distributions, τ exists and takes values in $[-1,1]$, and $\tau=0$ when the components are independent and nonatomic. The **grade correlation** (also called **Spearman's rho** or the **rank correlation**) ρ_g is defined as $\rho(F_1(X_1),F_2(X_2))$ where ρ is the standard correlation coefficient, and F_1,F_2 are the marginal distribution functions of X_1,X_2 (see for example Gibbons (1971)). ρ_g always exists, and is invariant under monotone transformations. τ and ρ_g are also called ordinal measures of association since they depend upon rank information only (Kruskal, 1958). Unfortunately, $\tau=0$ or $\rho_g=0$ do not imply independence (exercise 3.4). It would be desirable for a good measure of association or dependence that it be zero only when the components are independent.

The two measures given below satisfy all our requirements (universal existence, invariance under monotone transformations, and the zero value implying independence):

A.   **The sup correlation** (or maximal correlation) $\rho*$ defined by Gebelein (1941) and studied by Sarmanov (1962,1963) and Renyi (1959):

$$\bar{\rho}(X_1,X_2) = \sup\ \rho(g_1(X_1),g_2(X_2))$$

where the supremum is taken over all Borel-measurable functions g_1,g_2 such that $g_1(X_1),g_2(X_2)$ have finite positive variance, and ρ is the ordinary correlation coefficient.

B.   **The monotone correlation** $\rho*$ introduced by Kimeldorf and Sampson (1978), which is defined as $\bar{\rho}$ except that the supremum is taken over monotone functions g_1,g_2 only.

Let us outline why these measures satisfy our requirements. If $\rho*=0$, and X_1,X_2 are nondegenerate, then X_2 is independent of X_1 (Kimeldorf and Sampson, 1978). This is best seen as follows. We first note that for all s,t,

$$\rho(I_{(-\infty,s]}(X_1),I_{(-\infty,t]}(X_2)) = 0$$

because the indicator functions are monotone and $\rho*=0$. But this implies

$$P(X_1\leq s,X_2\leq t) = P(X_1\leq s)P(X_2\leq t)\,,$$

which in turn implies independence. For $\bar{\rho}$, we refer to exercise 3.6 and Renyi (1959). Good general discussions can be found in Renyi (1959), Kruskal (1958), Kimeldorf and Sampson (1978) and Whitt (1976). The measures of dependence are obviously interrelated. We have directly from the definitions,

$$| \rho | \leq \rho* \leq \overline{\rho} \leq 1 .$$

There are examples in which we have equality between all correlation coefficients (multivariate normal distribution, exercise 3.5), and there are other examples in which there is strict inequality. It is perhaps interesting to note when $\rho*$ equals one. This is for example the case when X_2 is monotone dependent upon X_1, i.e. there exists a monotone function g such that $P(X_2 = g(X_1)) = 1$, and X_1, X_2 are nonatomic (Kimeldorf and Sampson (1978)). This follows directly from the fact that $\rho*$ is invariant under monotone transformations, so that we can assume without loss of generality that the distribution is bivariate uniform. But then g must be the identity function, and the statement is proved, i.e. $\rho* = 1$. Unfortunately, $\rho* = 1$ does not imply monotone dependence.

For continuous marginals, there is yet another good measure of dependence, based upon the distance between probability measures. It is defined as follows:

$$L = \sup_A \ | P((X_1, X_2) \in A) - P((X_1, X'_2) \in A) |$$

$$= \frac{1}{2} \int | f(x_1, x_2) - f_1(x_1) f_2(x_2) | \ dx_1 dx_2 ,$$

where A is a Borel set of R^2, X'_2 is distributed as X_2, but is independent of X_1, f is the density of (X_1, X_2), and f_1, f_2 are the marginal densities. The supremum in the definition of L measures the distance between the given bivariate probability measure and the artificial bivariate probability measure constructed by taking the product of the two participating marginal probability measures. The invariance under strictly monotone transformations is clear. The integral form for L is Scheffe's theorem in disguise (see exercise 3.9). It is only valid when all the given densities exist.

Example 3.1.

It is clearly possible to have uniform marginals and a singular bivariate distribution (consider $X_2 = X_1$). It is even possible to find such a singular distribution with $\rho = \rho_g = 0$ (consider a carefully selected distribution on the surface of the unit circle; or consider $X_2 = SX_1$ where S takes the values $+1$ and -1 with equal probability). However, when we take A equal to the support of the singular distribution, then A has zero Lebesgue measure, and therefore zero measure for any absolutely continuous probability measure. Hence, $L = 1$. In particular, when X_2 is monotone dependent on X_1, then the bivariate distribution is singular, and therefore $L = 1$. ∎

Example 3.2.

X_1, X_2 are independent if and only if $L = 0$. The if part follows from the fact that for all A, the product measure of A is equal to the given bivariate probability measure of A. Thus, both probability measures are equal. The only if part is trivially true. ■

In the search for good measures of association, there is no clear winner. Probability theoretical considerations lead us to favor L over ρ_g, $\rho*$ and $\overline{\rho}$. On the other hand, as we have seen, approximating the bivariate distribution by a singular distribution, always gives $L = 1$. Thus, L is extremely sensitive to even small local deviations. The correlation coefficients are much more robust in that respect.

We will assume that what the user wants is a distribution with given absolutely continuous marginal distribution functions, and a given value for one of the transformation-invariant measures of dependence. We can then construct a bivariate uniform distribution with the given measure of dependence, and then transform the coordinate axes as in the univariate inversion method to achieve given marginal distributions (Nataf, 1962; Kimeldorf and Sampson, 1975; Mardia, 1970). If we can choose between a family of bivariate uniform distributions, then it is perhaps possible to pick out the unique distribution, if it exists, with the given measure of dependence. In the next section, we will deal with bivariate uniform distributions in general.

3.2. Bivariate uniform distributions.

We say that a distribution is bivariate uniform (exponential, gamma, normal, Cauchy, etcetera) when the univariate marginal distributions are all uniform (exponential, gamma, normal, Cauchy, etcetera). Distributions of this form are extremely important in mathematical statistics in the context of testing for dependence between components. First of all, if the marginal distributions are continuous, it is always possible by a transformation of both axes to insure that the marginal distributions have any prespecified density such as the uniform [0,1] density. If after the transformation to uniformity the joint density is uniform on $[0,1]^2$, then the two component random variables are independent. In fact, the joint density after transformation provides a tremendous amount of information about the sort of dependence.

There are various ways of obtaining bivariate distributions with specified marginals from bivariate uniform distributions, which make these uniform distributions even more important. Good surveys are provided by Johnson (1976), Johnson and Tenenbein (1979) and Marshall and Olkin (1983). The following theorem comes closest to generalizing the univariate properties which lead to the inversion method.

Theorem 3.1.

Let (X_1, X_2) be bivariate uniform with joint density g. Let f_1, f_2 be fixed univariate densities with corresponding distribution functions F_1, F_2. Then the density of $(Y_1, Y_2) = (F^{-1}_1(X_1), F^{-1}_2(X_2))$ is

$$f(y_1, y_2) = f_1(y_1) f_2(y_2) g(F_1(y_1), F_2(y_2)) .$$

Conversely, if (Y_1, Y_2) has density f given by the formula shown above, then Y_1 has marginal density f_1 and Y_2 has marginal density f_2. Furthermore, $(X_1, X_2) = (F_1(Y_1), F_2(Y_2))$ is bivariate uniform with joint density

$$g(x_1, x_2) = \frac{f(F^{-1}_1(x_1), F^{-1}_2(x_2))}{f_1(F^{-1}_1(x_1)) f_2(F^{-1}_2(x_2))} \qquad (0 \leq x_1, x_2 \leq 1) .$$

Proof of Theorem 3.1.

   Straightforward. ∎

There are many recipes for cooking up bivariate distributions with specified marginal distribution functions F_1, F_2. We will list a few in Theorem 3.2. It should be noted that if we replace $F_1(x_1)$ by x_1 and $F_2(x_2)$ by x_2 in these recipes, then we obtain bivariate uniform distribution functions. Recall also that the bivariate density, if it exists, can be obtained from the bivariate distribution function by taking the partial derivative with respect to $\partial x_1 \partial x_2$.

Theorem 3.2.

Let $F_1 = F_1(x_1), F_2 = F_2(x_2)$ be univariate distribution functions. Then the following is a list of bivariate distribution functions $F = F(x_1, x_2)$ having as marginal distribution functions F_1 and F_2:

A. $F = F_1 F_2(1 + a(1 - F_1)(1 - F_2))$. Here $a \in [-1, 1]$ is a parameter (Farlie (1960), Gumbel (1958), Morgenstern (1956)). This will be called Morgenstern's family.

B. $F = \dfrac{F_1 F_2}{1 - a(1 - F_1)(1 - F_2)}$. Here $a \in [-1, 1]$ is a parameter (Ali, Mikhail and Haq (1978)).

C. F is the solution of $F(1 - F_1 - F_2 + F) = a(F_1 - F)(F_2 - F)$ where $a \geq 0$ is a parameter (Plackett, 1965).

D. $F = a \max(0, F_1 + F_2 - 1) + (1 - a) \min(F_1, F_2)$ where $0 \leq a \leq 1$ is a parameter (Frechet, 1951).

E. $(-\log(F))^m = (-\log(F_1))^m + (-\log(F_2))^m$ where $m \geq 1$ is a parameter (Gumbel, 1960).

Proof of Theorem 3.2.

To verify that F is indeed a distribution function, we must verify that F is nondecreasing in both arguments, and that the limits as $x_1, x_2 \to -\infty$ and $\to \infty$ are 0 and 1 respectively. To verify that the marginal distribution functions are correct, we need to check that

$$\lim_{x_2 \to \infty} F(x_1, x_2) = F_1(x_1)$$

and

$$\lim_{x_1 \to \infty} F(x_1, x_2) = F_2(x_2) .$$

The latter relations are easily verified. ∎

It helps to visualize these recipes. We begin with Frechet's inequalities (Frechet, 1951), which follow by simple geometric arguments in the plane:

Theorem 3.3. Frechet's inequalities.

For any two univariate distribution functions F_1, F_2, and any bivariate distribution function F having these two marginal distribution functions,

$$\max(0, F_1(x) + F_2(y) - 1) \leq F(x, y) \leq \min(F_1(x), F_2(y)) .$$

Proof of Theorem 3.3.

For fixed (x_1, x_2) in the plane, let us denote by $Q_{SE}, Q_{NE}, Q_{SW}, Q_{NW}$ the four quadrants centered at x, y where equality is resolved by including boundaries with the south and west halfplanes. Thus, (x_1, x_2) belongs to Q_{SW} while the vertical line at x_1 belongs to $Q_{SW} \cup Q_{NW}$. It is easy to see that at x_1, x_2,

$$F_1(x_1) = P(Q_{SW} \cup Q_{NW}),$$
$$F_2(x_2) = P(Q_{SW} \cup Q_{SE}),$$
$$F(x_1, x_2) = P(Q_{SW}).$$

Clearly, $F \leq \min(F_1, F_2)$ and $1 - F \leq 1 - F_1 + 1 - F_2$. ∎

These inequalities are valid for all bivariate distribution functions F with marginal distribution functions F_1 and F_2. Interestingly, both extremes are also valid distribution functions. In fact, we have the following property which can be used for the generation of random vectors with these distribution functions.

Theorem 3.4.

Let U be a uniform $[0,1]$ random variable, and let F_1, F_2 be continuous univariate distribution functions. Then

$$(F^{-1}_1(U), F^{-1}_2(U))$$

has distribution function $\min(F_1, F_2)$. Furthermore,

$$(F^{-1}_1(U), F^{-1}_2(1 - U))$$

has distribution function $\max(0, F_1 + F_2 - 1)$.

Proof of Theorem 3.4.

   We have

$$P(F^{-1}_1(U)\le x_1, F^{-1}_2(U)=x_2) = P(U \le \min(F_1(x_1), F_2(x_2))) .$$

Also,

$$P(F^{-1}_1(U)\le x_1, F^{-1}_2(1-U)=x_2) = P(U \le F_1(x_1), 1-U \le F_2(x_2)) . \blacksquare$$

   Frechet's extremal distribution functions are those for which maximal positive and negative dependence are obtained respectively. This is best seen by considering the bivariate uniform case. The upper distribution function $\min(x_1, x_2)$
puts its mass uniformly on the 45 degree diagonal of the first quadrant. The bottom distribution function $\max(0, x_1+x_2-1)$ puts its mass uniformly on the -45
degree diagonal of $[0,1]^2$. Hoeffding (1940) and Whitt (1976) have shown that
maximal positive and negative correlation are obtained for Frechet's extremal distribution functions (see exercise 3.1). Note also that maximally correlated random
variable are very important in variance reduction techniques in Monte Carlo
simulation. Theorem 3.4 shows us how to generate such random vectors. We have
thus identified a large class of applications in which the inversion method seems
essential (Fox, 1980). For Frechet's bivariate family (case D in Theorem 3.2), we
note without work that it suffices to consider a mixture of Frechet's extremal distributions. This is often a poor way of creating intermediate correlation. For
example, in the bivariate uniform case, all the probability mass is concentrated
on the two diagonals of $[0,1]^2$.

   The list of examples in Theorem 3.2 is necessarily incomplete. Other examples can be found in exercises 3.2 and 3.3. Random variate generation is usually
taken care of via the conditional distribution method. The following example
should suffice.

Example 3.3. Morgenstern's family.

   Consider the uniform version of Morgenstern's bivariate family with parame-
ter $|a| \le 1$ given by part A of Theorem 3.2. It is easy to see that for this fam-
ily, there exists a density given by

$$f(x_1, x_2) = 1 + a(2x_1-1)(2x_2-1) .$$

Here we can generate X_1 uniformly on $[0,1]$. Given X_1, X_2 has a trapezoidal density which is zero outside $[0,1]$ and varies from $1-a(2X_1-1)$ at $x_2=0$ to
$1+a(2X_1-1)$ at $x_2=1$. If U,V are iid uniform $[0,1]$ random variables, then X_2

can be generated as

$$\min(U, -\frac{V}{a(2X_1-1)}) \quad X_1 < \frac{1}{2}$$

$$\max(U, 1-\frac{V}{a(2X_1-1)}) \quad X_1 \geq \frac{1}{2} \quad \blacksquare$$

There are other important considerations when shopping around for a good bivariate uniform family. For example, it is useful to have a family which contains as members, or at least as limits of members, Frechet's extremal distributions, plus the product of the marginals (the independent case). We will call such families **comprehensive**. Examples of comprehensive bivariate families are given in the table below. Note that the comprehensiveness of a family is invariant under strictly monotone transformations of the coordinate axes (exercise 3.11), so that the marginals do not really matter.

Distribution function	Reference
F is the solution of $F(1-F_1-F_2+F) = a(F_1-F)(F_2-F)$ where $a \geq 0$ is a parameter	Plackett (1965)
$F = \dfrac{a^2(1-a)}{2}\max(0, F_1+F_2-1)$ $+\dfrac{a^2(1+a)}{2}\min(F_1,F_2)+(1-a^2)F_1F_2$ where $\mid a \mid \leq 1$ is a parameter	Frechet (1958)
$\dfrac{1}{2\pi\sqrt{1-r^2}}e^{-\frac{x_1^2+x_2^2-2rx_1x_2}{2(1-r^2)}}$ where $\mid r \mid \leq 1$ is a measure of association	Bivariate normal (see e.g. Mardia, 1970)

From this table, one can create other comprehensive families either by monotone transformations, or by taking mixtures. Note that most families, including Morgenstern's family, are not comprehensive.

Another issue is that of the range spanned by the family in terms of the values of a given measure of dependence. For example, for Morgenstern's bivariate uniform family of Example 3.3, the correlation coefficient is $-a/3$. Therefore, it can take all the values in $[-\frac{1}{3}, \frac{1}{3}]$, but no values outside this interval. Needless to say, full ranges for certain measures of association are an asset. Typically, this goes hand in hand with comprehensiveness.

Example 3.4. Full correlation range families.

Plackett's bivariate family with parameter $a \geq 0$ and arbitrary continuous marginal distribution functions has correlation coefficient

$$\rho = \frac{-(1-a^2)-2a \log(a)}{(1-a)^2} ,$$

which can be shown to take the values $1,0,-1$ when $a \rightarrow \infty$, $a = 1$ and $a = 0$ respectively (see e.g. Barnett, 1980). Since ρ is a continuous function of a, all values of ρ can be achieved.

The bivariate normal family can also achieve all possible values of correlation. Since for this family, $\rho = \rho* = \overline{\rho}$, we also achieve the full range for the sup correlation and the monotone correlation. ∎

Example 3.5. The Johnson-Tenenbein families.

Johnson and Tenenbein (1981) proposed a general method of constructing bivariate families for which τ and ρ_g can attain all possible values in $(-1,1)$. The method consists simply of taking $(X_1, X_2) = (U, H(cU + (1-c)V))$, where U, V are iid random variables with common distribution function F, $c \in [0,1]$ is a weight parameter, and H is a monotone function chosen in such a way that $H(cU + (1-c)V)$ also has distribution function F. To take a simple example, let U, V be iid normal random variables. Then we should take $H(u) = u / \sqrt{c^2 + (1-c)^2}$. The resulting two-dimensional random vector is easily seen to be bivariate normal, as it is a linear combination of iid normal random variables. Its correlation coefficient is

$$\frac{c}{\sqrt{c^2 + (1-c)^2}} ,$$

which can take all values in $[0,1]$. Moreover,

$$\rho_g = \frac{6}{\pi} \arcsin\left(\frac{c}{2\sqrt{c^2 + (1-c)^2}}\right) ,$$

$$\tau = \frac{2}{\pi} \arcsin\left(\frac{c}{\sqrt{c^2 + (1-c)^2}}\right) .$$

It is easy to see that these measures of association can also take all values in $[0,1]$ when we vary c. Negative correlations can be achieved by considering $(-U, H(cU + (1-c)V))$. Recall next that τ and ρ_g are invariant under strictly monotone transformations of the coordinate axes. Thus, we can now construct bivariate families with specified marginals and given values for ρ_g or τ. ∎

3.3. Bivariate exponential distributions.

We will take the bivariate exponential distribution as our prototype distribution for illustrating just how we can construct such distributions directly. At the same time, we will discuss random variate generators. There are two very different approaches:

A. The analytic method: one defines explicitly a bivariate density or distribution function, and worries about generators later. An example is Gumbel's bivariate exponential family (1960) described below. Another example is the distribution of Nagao and Kadoya (1971) dealt with in exercise 3.10.

B. The empiric method: one constructs a pair of random variables known to have the correct marginals, and worries about the form of the distribution function later. Here, random variate generation is typically a trivial problem. Examples include distributions proposed by Johnson and Tenenbein (1981), Moran (1967), Marshall and Olkin (1967), Arnold (1967) and Lawrance and Lewis (1983).

The distinction between A and B is often not clear-cut. Families can also be partitioned based upon the range for given measures of association, or upon the notion of comprehensiveness. Let us start with Gumbel's family of bivariate exponential distribution functions:

$$1-e^{-x_1}-e^{-x_2}+e^{-x_1-x_2-ax_1x_2} \qquad (x_1,x_2>0) \ .$$

Here $a \in [0,1]$ is the parameter. The joint density is

$$e^{-x_1-x_2-ax_1x_2}\left((1+ax_1)(1+ax_2)-a\right) \ .$$

Notice that the conditional density of X_2 given $X_1=x_1$ is

$$e^{-(1+ax_1)x_2}\left((1+ax_1)(1+ax_2)-a\right)$$

$$= \frac{a}{\theta}\left(\theta^2 x_2 e^{-\theta x_2}\right) + \frac{\theta-a}{\theta}\left(\theta e^{-\theta x_2}\right) \ ,$$

where $\theta=1+ax_1$. In this decomposition, we recognize a mixture of a gamma (2) and a gamma (1) density. Random variates can easily be generated via the conditional distribution method, where the conditional distribution of X_2 given X_1 can be handled by composition (see below). Unfortunately, the family contains only none of Frechet's extremal distributions, which suggests that extreme correlations cannot be obtained.

Gumbel's bivariate exponential distribution with parameter a

Generate iid exponential random variates X_1, X_2.

Generate a uniform [0,1] random variate U.

IF $U \le \dfrac{a}{1+aX_1}$

      THEN

            Generate an exponential random variate E.

        $X_2 \leftarrow X_2 + E$

RETURN $(X_1, \dfrac{X_2}{1+aX_1})$

Generalizations of Gumbel's distribution have been suggested by various authors. In general, one can start from a bivariate uniform distribution function F, and define a bivariate exponential distribution function by

$$F(1-e^{-x_1}, 1-e^{-x_2}) \ .$$

For a generator, we need only consider $(-\log(U), -\log(V))$ where U, V is bivariate uniform with distribution function F. For example, if we do this for Morgenstern's family with parameter $|a| \le 1$, then we obtain the bivariate exponential distribution function

$$(1-e^{-x_1})(1-e^{-x_2})(1+ae^{-x_1-x_2}) \quad (x_1, x_2 \ge 0).$$

This distribution has also been studied by Gumbel (1960). Both Gumbel's exponential distributions and other possible transformations of bivariate uniform distributions are often artificial.

In the empiric (or constructive) method, one argues the other way around, by first defining the random vector. In the table shown below, a sampling of such bivariate random vectors is given. We have taken what we consider are good didactical examples showing a variety of approaches. All of them exploit special properties of the exponential distribution, such as the fact that the sum of squares of iid normal random variables is exponentially distributed, or the fact that the minimum of independent exponential random variables is again exponentially distributed.

(X_1, X_2)	Reference
$\left(\min(\dfrac{E_1}{\lambda_1}, \dfrac{E_3}{\lambda_3}), \min(\dfrac{E_2}{\lambda_2}, \dfrac{E_3}{\lambda_3})\right)$	Marshall and Olkin (1967)
$\left(\beta_1 E_1 + S_1 E_2, \beta_2 E_2 + S_2 E_1\right),$ $P(S_i = 1) = 1 - P(S_i = 0) = 1 - \beta_i \ (i = 1, 2)$	Lawrance and Lewis (1983)
$\left(E_1, -\log((1-c)e^{-\frac{E_2}{1-c}} + ce^{-\frac{E_2}{c}}) + \log(1-2c)\right),$ $c \in [0, 1]$	Johnson and Tenenbein (1981)
$\left(\dfrac{1}{2}(N_1{}^2 + N_2{}^2), \dfrac{1}{2}(N_3{}^2 + N_4{}^2)\right),$ $(N_1, N_3), (N_2, N_4)$ iid multinormal with correlation ρ	Moran (1967)

In this table, E_1, E_2, E_3 are iid exponential random variates, and $\lambda_1, \lambda_2, \lambda_3 \geq 0$ are parameters with $\lambda_1 \lambda_2 + \lambda_3 > 0$. The N_i's are normal random variables, and c, β_1, β_2 are [0,1]-valued constants. A special property of the marginal distribution, closure under the operation **min**, is exploited in the definition. To see this, note that for $x > 0$,

$$P(X_1 > x) = P(E_1 > \lambda_1 x, E_3 > \lambda_3 x)$$
$$= e^{-(\lambda_1 + \lambda_3)x} \quad (x > 0).$$

Thus, X_1 is exponential with parameter $\lambda_1 + \lambda_3$. The joint distribution function is uniquely determined by the function $G(x_1, x_2)$ defined by

$$G(x_1, x_2) = P(X_1 > x_1, X_2 > x_2) = e^{-\lambda_1 x_1 - \lambda_2 x_2 - \lambda_3 \max(x_1, x_2)}.$$

The distribution is a mixture of a singular distribution carrying weight $\lambda_3 / (\lambda_1 + \lambda_2 + \lambda_3)$, and an absolutely continuous part (exercise 3.6). Also, it is unfortunate that when (X_1, X_2) has the given bivariate exponential distribution, then $(a_1 X_1, a_2 X_2)$ is bivariate exponential in the case $a_1 = a_2$ only. On the positive side, we should note that the family includes the independent case ($\lambda_3 = 0$), and one of Frechet's extremal cases ($\lambda_1 = \lambda_2 = 0$). In the latter case, note that

$$G(x_1, x_2) = P(X_1 > x_1, X_2 > x_2) = e^{-\lambda_3 \max(x_1, x_2)}.$$

The Lawrance-Lewis bivariate exponential is just one of a long list of bivariate exponentials constructed by them. The one given in the table is particularly flexible. We can quickly verify that the marginals are exponential via characteristic functions. The characteristic function of X_1 is

$$\phi(t) = E(e^{itX_1}) = E(e^{\beta_1 it E_1})(\beta_1 + (1 - \beta_1)E(e^{it E_2}))$$
$$= \frac{1}{1 - it\beta_1}(\beta_1 + \frac{(1 - \beta_1)}{1 - it}) = \frac{1}{1 - it}.$$

The correlation $\rho=2\beta_1(1-\beta_2)+\beta_2(1-\beta_1)$, valid for $0\leq\beta_1\leq\beta_2\leq 1$, can take all values between 0 and 1. To create negative correlation, one can replace E_1,E_2 in the formulas for X_2 by two other exponential random variables, $h(E_1),h(E_2)$ where $h(x)=-\log(1-e^{-x})$ (Lawrance and Lewis, 1983).

The Johnson and Tenenbein construction is almost as simple as the Lawrance-Lewis construction. Interestingly, by varying the parameter c, all possible nonnegative values for ρ_g, τ and ρ are achievable.

Finally, in Moran's bivariate distribution, good use is made of yet another property of exponential random variables. His distribution has correlation ρ^2 where ρ is the correlation of the underlying bivariate normal distribution. Again, random variate generation is extremely simple, and the correlation spans the full nonnegative range. Difficulties arise only when one needs to compute the exact value of the density at some points, but then again, these same difficulties are shared by most empiric methods.

3.4. A case study: bivariate gamma distributions.

We have seen how bivariate distributions with any given marginals can be constructed from bivariate uniform distributions or bivariate distributions with other continuous marginals, via transformations of the coordinate axes. These transformations leave ρ_g,τ and other ordinal measures of association invariant, but generally speaking not ρ. Furthermore, the inversion of the marginal distribution functions (F_1,F_2) required to apply these transformations is often unfeasible. Such is the case for the gamma distribution. In this section we will look at these new problems, and provide new solutions.

To clarify the problems with inversion, we note that if X_1,X_2 is bivariate gamma (a_1,a_2), where a_i is the parameter for X_i, then maximum and minimum correlation are obtained for the Frechet bounds, i.e.

$$X_2 = F_2^{-1}(F_1(X_1)) \, ,$$
$$X_2 = F_2^{-1}(1-F_1(X_1))$$

respectively (Moran (1967), Whitt (1976)). Direct use of Frechet's bounds is possible but not recommended if generator efficiency is important. In fact, it is not recommended to start from any bivariate uniform distribution. Also, the method of Johnson and Tenenbein (1981) illustrated on the bivariate uniform, normal and exponential distributions in the previous sections requires an inversion of a gamma distribution function if it were to be applied here.

We can also obtain help from the composition method, noting that the random vector (X_1,X_2) defined by

$$(X_1,X_2) = \begin{cases} (Y_1,Y_2) & \text{,with probability } p \\ (Z_1,Z_2) & \text{,with probability } 1-p \end{cases}$$

has the right marginal distributions if both random vectors on the right hand side have the same marginals. Also, (X_1, X_2) has correlation coefficient $p\, \rho_Y + (1-p)\rho_Z$ where ρ_Y, ρ_Z are the correlation coefficients of the two given random vectors. One typically chooses ρ_Y and ρ_Z at the extremes, so that the entire range of ρ values is covered by adjusting p. For example, one could take $\rho_Y = 0$ by considering iid random variables Y_1, Y_2. Then ρ_Z can be taken maximal by using the Frechet maximal dependence as in $(Z_1, Z_2) = (Z_1, F_2^{-1}(1 - F_1(Z_1))$ where Z_1 is gamma (a_1). Doing so leads to a mixture of a continuous distribution (the product measure) and a singular distribution, which is not desirable.

The gamma distribution shares with many distributions the property that it is closed under additions of independent random variables. This has led to inversion-free methods for generating bivariate gamma random vectors, now known as **trivariate reduction methods** (Cherian, 1941; David and Fix, 1961; Mardia, 1970; Johnson and Ramberg, 1977; Schmeiser and Lal, 1982). The name is borrowed from the principle that two dependent random variables are constructed from three independent random variables. The application of the principle is certainly not limited to the gamma distribution, but is perhaps best illustrated here. Consider independent gamma random variables G_1, G_2, G_3 with parameters a_1, a_2, a_3. Then the random vector

$$(X_1, X_2) = (G_1 + G_3, G_2 + G_3)$$

is bivariate gamma. The marginal gamma distributions have parameters $a_1 + a_3$ and $a_2 + a_3$ respectively. Furthermore, the correlation is given by

$$\rho = \frac{a_3}{\sqrt{(a_1 + a_3)(a_2 + a_3)}} \, .$$

If ρ and the marginal gamma parameters are specified beforehand, we have one of two situations: either there is no possible solution for a_1, a_2, a_3, or there is exactly one solution. The limitation of this technique, which goes back to Cherian (1941) (see Schmeiser and Lal (1980) for a survey), is that

$$0 \le \rho \le \frac{\min(\alpha_1, \alpha_2)}{\sqrt{\alpha_1 \alpha_2}}$$

where α_1, α_2 are the marginal gamma parameters. Within this range, trivariate reduction leads to one of the fastest algorithms known to date for bivariate gamma distributions.

Trivariate reduction for bivariate gamma distribution

[NOTE: ρ is a given correlation, α_1, α_2 are given parameters for the marginal gamma distributions. It is assumed that $0 \leq \rho \leq \dfrac{\min(\alpha_1, \alpha_2)}{\sqrt{\alpha_1 \alpha_2}}$.]

[GENERATOR]

Generate a gamma $(\alpha_1 - \rho\sqrt{\alpha_1\alpha_2})$ random variate G_1.

Generate a gamma $(\alpha_2 - \rho\sqrt{\alpha_1\alpha_2})$ random variate G_2.

Generate a gamma $(\rho\sqrt{\alpha_1\alpha_2})$ random variate G_3.

RETURN $(G_1 + G_3, G_2 + G_3)$

Ronning (1977) generalized this principle to higher dimensions, and suggested several possible linear combinations to achieve desired correlations. Schmeiser and Lal (1982) (exercise 3.19) fill the void by extending the trivariate reduction method in two dimensions, so that all theoretically possible correlations can be achieved in bivariate gamma distributions. But we do not get something for nothing: the algorithm requires the inversion of the gamma distribution function, and the numerical solution of a set of nonlinear equations in the set-up stage.

3.5. Exercises.

1. Prove that over all bivariate distribution functions with given marginal univariate distribution functions F_1, F_2, the correlation coefficient ρ is minimized for the distribution function $\max(0, F_1(x) + F_2(y) - 1)$. It is maximized for the distribution function $\min(F_1(x), F_2(y))$ (Whitt, 1976; Hoeffding, 1940).

2. **Plackett's bivariate uniform family** (Plackett (1965). Consider the bivariate uniform family defined by part C of Theorem 3.2, with parameter $a \geq 0$. Show that on $[0,1]^2$, this distribution has a density given by

$$f(x_1, x_2) = \frac{a(a-1)(x_1 + y_1 - 2x_1 x_2) + a}{\left(((a-1)(x_1 + x_2) + 1)^2 - 4a(a-1)x_1 x_2\right)^{3/2}}.$$

For this distribution, Mardia (1970) has proposed the following generator:

Mardia's generator for Plackett's bivariate uniform family

Generate two iid uniform $[0,1]$ random variables U, V.

$X_1 \leftarrow U$

$Z \leftarrow V(1-V)$

$X_2 \leftarrow \dfrac{2Z(a^2 X_1 + 1 - X_1) + a(1 - 2Z) - (1 - 2V)\sqrt{a(a + 4ZX_1(1 - X_1)(1-a)^2)}}{a + Z(1-a)^2}$

RETURN (X_1, X_2)

Show that this algorithm is valid.

3. Suggest generators for the following bivariate uniform families of distributions:

Density	Parameter(s)	Reference
$1 + a((m+1)x_1^m - 1)((n+1)x_2^n - 1)$	$\dfrac{1}{mn} \leq a \leq \max(m,n)$, $m,n \geq 0$	Farlie (1960)
$\dfrac{a(x_1^{1-a} + x_2^{1-a} - 1)^{\frac{2a-1}{1-a}}}{(x_1 x_2)^a}$	$a > 1$	(derived from multivariate Pareto)
$\dfrac{\pi(1+u^2)(1+v^2)}{2 + u^2 + v^2}$ where $u = 1/\tan^2(\pi x_1)$, $v = 1/\tan^2(\pi x_2)$		Mardia (1970) (derived from multivariate Cauchy)
$1 + a(2x_1 - 1)(2x_2 - 1) + b(3x_1^2 - 1)(3x_2^2 - 1)$	$\mid a \mid \leq \dfrac{1}{2}$, $\mid b \mid \leq \dfrac{1}{8}$	Kimeldorf and Sampson (1975)

4. This is about various measures of association. Construct a bivariate uniform distribution for which $\rho = \rho_g = \tau = 0$, and $X_2 = g(X_1)$ for some function g (i.e. X_2 is completely dependent on X_1, see e.g. Lancaster, 1963).

5. Show that for the normal distribution in R^2, $\mid \rho \mid = \rho* = \bar{\rho}$.

6. Prove that $\bar{\rho} = 0$ implies independence of components (Renyi, 1959).

7. Recall the definition of complete dependence of exercise 3.4. Construct a sequence of bivariate uniform distributions in which for every n, the second coordinate is completely dependent on the first coordinate. The sequence should also tend in distribution to the independent bivariate uniform distribution (Kimeldorf and Sampson, 1978). Conclude that the notion of complete dependence is peculiar.

8. The phenomenon described in exercise 7 cannot happen for monotone dependent sequences. If a sequence of random bivariate uniform random vectors in

which the second component is monotone dependent on the first component for all n, tends in distribution to a random vector, then this new random vector is bivariate uniform, and the second component is monotone dependent on the first component (Kimeldorf and Sampson, 1978).

9. One measure of association for bivariate distributions is

$$L = \sup_{A} |P((X_1, X_2) \in A) - P((X_1, X'_2) \in A)|$$

$$= \frac{1}{2} \int |f(x_1, x_2) - f_1(x_1) f_2(x_2)| \, dx_1 dx_2,$$

where A is a Borel set of R^2, X'_2 is distributed as X_2, but is independent of X_1, f is the density of (X_1, X_2) and f_1, f_2 are the marginal densities. The second equality is valid only if the densities involved in the right-hand-side exist. Prove the second equality (Scheffe, 1947).

10. Nagao and Kadoya (1971) studied the following bivariate exponential density:

$$f(x_1, x_2) = \frac{e^{-\frac{1}{1-r}(\frac{x_1}{\sigma_1} + \frac{x_2}{\sigma_2})} I_0\left[\frac{2}{1-r}\sqrt{\frac{r x_1 x_2}{\sigma_1 \sigma_2}}\right]}{\sigma_1 \sigma_2 (1-r)}$$

where $r \in [0,1)$ is a measure of dependence, $\sigma_1, \sigma_2 > 0$ are constants (parameters), and I_0 is a modified Bessel function of the first kind. Obtain the parameters of the marginal exponential distributions. Compute the correlation coefficient ρ. Finally indicate how you would generate random vectors in uniformly bounded expected time.

11. Show that the property of comprehensiveness of a bivariate family is invariant under strictly monotone transformations of the coordinate axes (Kimeldorf and Sampson, 1975).

12. Show that Plackett's bivariate family with parameter $a \geq 0$ is comprehensive. Show in particular that Frechet's extremal distributions are attained for $a = 0$ and $a \to \infty$, and that the product of the marginals is obtained for $a = 1$.

13. Show that the standard bivariate normal family (i.e., the normal distribution in the plane) with variable correlation is comprehensive.

14. Show that Morgenstern's bivariate family is not comprehensive.

15. Consider the Johnson-Tenenbein family of Example 3.4, with parameter $c \in [0,1]$. Let U and V have uniform [0,1] densities.

    A. Find H such that the distribution is bivariate uniform. Hint: H is parabolic on $[0,b]$ and $[1-b,1]$, and linear in between, where $b = \min(c, 1-c)$.

    B. Find ρ, τ and ρ_g as a function of c. In particular, prove that

$$\tau = \begin{cases} \dfrac{4c - 5c^2}{6(1-c)^2} & 0 < c < \dfrac{1}{2} \\[3mm] \dfrac{11c^2 - 6c + 1}{6c^2} & \dfrac{1}{2} < c < 1 \end{cases},$$

$$\rho_g = \begin{cases} \dfrac{10c - 13c^2}{10(1-c)^2} & 0 < c < \dfrac{1}{2} \\[4mm] \dfrac{3c^3 + 16c^2 - 11c + 2}{10c^3} & \dfrac{1}{2} < c < 1 \end{cases}.$$

Conclude that all nonnegative values for ρ, τ and ρ_g are achievable by adjusting c (Johnson and Tenenbein, 1981).

16. Show that for Gumbel's bivariate exponential family with parameter $a \in [0,1]$, the correlation reaches a minimum for $a = 1$, and this minimum is $-0.40365\ldots$ Show that the correlation is a decreasing function of a, taking the maximal value 0 at $a = 0$.

17. Consider the following pair of random variables: $\beta_1 E_1 + S_1 E_2, \beta_2 E_2 + S_2 E_1$ where $P(S_i = 1) = 1 - P(S_i = 0) = 1 - \beta_i$ $(i = 1,2)$ and E_1, E_2 are iid exponential random variables (Lawrance and Lewis (1983)). Does this family contain one of Frechet's extremal distributions?

18. Compute ρ, ρ_g and τ for the bivariate exponential distribution of Johnson and Tenenbein (1981), defined as the distribution of $E_1, -\log((1-c)e^{-\frac{E_2}{1-c}} + ce^{-\frac{E_2}{c}}) + \log(1-2c)$ where $c \in [0,1]$ and E_1, E_2 are iid exponential random variables.

19. Schmeiser and Lal (1982) proposed the following method for generating a bivariate gamma random vector: let G_1, G_2, G_3 be independent gamma random variables with respective parameters a_1, a_2, a_3, let U, V be an independent bivariate uniform random vector with $V = U$ or $V = 1 - U$, let F_b denote the gamma distribution function with parameter b, and let b_1, b_2 be two nonnegative numbers. Define

$$(X_1, X_2) = (F_{b_1}^{-1}(U) + G_1 + G_3, F_{b_2}^{-1}(V) + G_2 + G_3).$$

A. Show that this random vector is bivariate gamma.

B. Show constructively that the five-parameter family is comprehensive, i.e. for every possible combination of specified marginal gamma distributions, give the values of the parameters needed to obtain the Frechet extremal distributions and the product distribution. Indicate also whether $V = U$ or $V = 1 - U$ is needed each time.

C. Show that by varying the five parameters, we can cover all theoretically possible combinations for the correlation coefficient and the marginal gamma parameters.

D. Consider the simplified three parameter model

$$(X_1, X_2) = (F_{b_1}^{-1}(U) + G_1, F_{\alpha_2}^{-1}(V))$$

for generating a bivariate gamma random vector with marginal parameters (α_1, α_2) and correlation ρ. Show that this family is still comprehensive. There are two equations for the two free parameters (b_1 and a_1).

Suggest a good numerical algorithm for finding these parameters.

20. **A bivariate Poisson distribution.** (X_1, X_2) is said to be bivariate Poisson with parameters $\lambda_1, \lambda_2, \lambda_3$, when it has characteristic function

$$\phi(t_1, t_2) = e^{\lambda_1(e^{it_1}-1) + \lambda_2(e^{it_2}-1) + \lambda_3(e^{it_1+it_2}-1)} .$$

   A. Show that this is indeed a bivariate Poisson distribution.

   B. Apply the trivariate reduction principle to generate a random vector with the given distribution.

   C. (Kemp and Loukas, 1978). Show that we can generate the random vector as $(Z+W, X_2)$ where X_2 is Poisson $(\lambda_1+\lambda_3)$, and given X_2, Z, W are independent Poisson (λ_2) and binomial $(X_2, \lambda_3/(\lambda_1+\lambda_3))$ random variables. Hint: prove this via generating functions.

21. **The Johnson-Ramberg bivariate uniform family.** Let U_1, U_2, U_3 be iid uniform $[0,1]$ random variables, and let $b \geq 0$ be a parameter of a family of bivariate uniform random vectors defined by

$$(X_1, X_2) = (\frac{U_1 U_3^b - b U_1^{\frac{1}{b}} U_3}{1-b} , \frac{U_2 U_3^b - b U_2^{\frac{1}{b}} U_3}{1-b}) .$$

   This construction can be considered as trivariate reduction. Show that the full range of nonnegative correlations is possible, by first showing that the correlation is

$$\frac{b^2(2b^2+9b+6)}{(1+b)^2(1+2b)(2+b)} .$$

   Show also that one of the Frechet extremal distributions can be approximated arbitrarily closely from within the family. For $b=1$, the defining formula is invalid. By what should it be replaced? (Johnson and Ramberg, 1977)

22. Consider a family of univariate distribution functions $\{1-(1-F)^a , a>0\}$, where F is a distribution function. Families of this form are closed under the operation $\min(X_1, X_2)$ where X_1, X_2 are independent random variables with parameters a_1, a_2: the parameter of the minimum is $a_1 + a_2$. Use this to construct a bivariate family via trivariate reduction, and compute the correlations obtainable for bivariate exponential, geometric and Weibull distributions obtained in this manner (Arnold, 1967).

23. **The bivariate Hermite distribution.** A univariate Hermite distribution $\{p_i , i \geq 0\}$ with parameters $a, b > 0$ is a distribution on the nonnegative integers which has generating function (defined as $\sum_i p_i s^i$)

$$e^{a(s-1)+b(s^2-1)} .$$

The bivariate Hermite distribution with parameters $a_i > 0$, $i = 1, 2, \ldots, 5$, is defined on all pairs of nonnegative integers and has bivariate generating function (defined as $E(s_1{}^{X_1}s_2{}^{X_2})$ where (X_1,X_2) is a bivariate Hermite random vector)

$$e^{a_1(s_1-1)+a_2(s_1{}^2-1)+a_3(s_2-1)+a_4(s_2{}^2-1)+a_5(s_1s_2-1)}$$

(Kemp and Kemp (1965,1966); Kemp and Papageorgiou (1976)).

A. How can you generate a univariate Hermite (a,b) random variate using only Poisson random variates in uniformly bounded expected time?

B. Give an algorithm for the efficient generation of bivariate Hermite random variates. Hint: derive first the generating function of (X_1+X_3,X_2+X_3) where X_1,X_2,X_3 are independent random variables with generating functions g_1,g_2,g_3.

This exercise is adapted from Kemp and Loukas (1978).

24. Write an algorithm for computing the probabilities of a bivariate discrete distribution on $\{1,2,\ldots,K\}^2$ with specified marginal distributions, and achieving Frechet's inequality. Repeat for both of Frechet's extremal distributions.

4. THE DIRICHLET DISTRIBUTION.

4.1. Definitions and properties.

Let $a_1,\ldots,a_{k+1}$ be positive numbers. Then $(X_1,\ldots,X_k)$ has a **Dirichlet distribution** with parameters $(a_1,\ldots,a_{k+1})$, denoted $(X_1,\ldots,X_k) \sim D(a_1,\ldots,a_{k+1})$, if the joint distribution has density

$$f(x_1,\ldots,x_k) = cx_1{}^{a_1-1} \cdots x_k{}^{a_k-1}(1-x_1-\cdots-x_k)^{a_{k+1}-1}$$

over the k-dimensional simplex S_k defined by the inequalities $x_i > 0$ $(i=1,2,\ldots,k)$, $\sum_{i=1}^{k} x_i < 1$. Here c is a normalization constant. Basically, the X_i's can be thought of as a_i-spacings in a uniform sample of size $\sum a_j$ if the a_i's are all positive integers. The only novelty is that the a_i's are now allowed to take non-integer values. The interested reader may want to refer back to section V.2 for the properties of spacings and to section V.3 for generators. The present section is only a refinement of sorts.

Theorem 4.1.

Let $Y_1, \ldots, Y_{k+1}$ be independent gamma random variables with parameters $a_i > 0$ respectively. Define $Y = \sum Y_i$ and $X_i = Y_i / Y$ $(i = 1, 2, \ldots, k)$. Then $(X_1, \ldots, X_k) \sim D(a_1, \ldots, a_{k+1})$ and $(X_1, \ldots, X_k)$ is independent of Y.

Conversely, if Y is gamma $(\sum a_i)$, and Y is independent of $(X_1, \ldots, X_k) \sim D(a_1, \ldots, a_{k+1})$, then the random variables $YX_1, \ldots, YX_k, Y(1 - \sum_{i=1}^{k} X_i)$ are independent gamma random variables with parameters $a_1, \ldots, a_{k+1}$.

Proof of Theorem 4.1.

The joint density of the Y_i's is

$$f(y_1, \ldots, y_{k+1}) = c \prod_{i=1}^{k+1} y_i^{a_i - 1} e^{-\sum_{i=1}^{k+1} y_i}$$

where c is a normalization constant. Consider the transformation $y = \sum, x_i = y_i / y$ $(i \leq k)$, which has as reverse transformation $y_i = y x_i$ $(i \leq k), y_{k+1} = y(1 - \sum_{i=1}^{k} x_i)$. The Jacobian of the transformation is y^k. Thus, the joint density of $Y, X_1, \ldots, X_k)$ is

$$g(y, x_1, \ldots, x_k) = c \prod_{i=1}^{k} x_i^{a_i - 1} (1 - \sum_{i=1}^{k} x_i)^{a_{k+1} - 1} y^{\sum_{i=1}^{k+1} a_i - 1} e^{-y} .$$

This proves the first part of the Theorem. The proof of the second part is omitted. ■

Theorem 4.1 suggests a generator for the Dirichlet distribution via gamma generators. There are important relationships with the beta distribution as well, which are reviewed by Wilks (1962), Aitchison (1963) and Basu and Tiwari (1982). Here we will just mention the most useful of these relationships.

Theorem 4.2.

Let $Y_1, \ldots, Y_k$ be independent beta random variables where Y_i is beta $(a_i, a_{i+1} + \cdots + a_{k+1})$. Then $(X_1, \ldots, X_k) \sim D(a_1, \ldots, a_{k+1})$ where the X_i's are defined by

$$X_i = Y_i \prod_{j=1}^{i-1} Y_j .$$

Conversely, when $(X_1, \ldots, X_k) \sim D(a_1, \ldots, a_{k+1})$, then the random variables $Y_1, \ldots, Y_k$ defined by

$$Y_i = \frac{X_i}{1 - X_1 - \cdots - X_{i-1}}$$

are independent beta random variables with parameters given in the first statement of the Theorem.

Theorem 4.3.

Let $Y_1, \ldots, Y_k$ be independent random variables, where Y_i is beta $(a_1 + \cdots + a_i, a_{i+1})$ for $i < k$ and Y_k is gamma $(a_1 + \cdots + a_k)$. Then the following random variables are independent gamma random variables with parameters $a_1, \ldots, a_k$:

$$X_i = (1 - Y_{i-1}) \prod_{j=i}^{k} Y_j \quad (i = 1, 2, \ldots, k) .$$

To avoid trivialities, set $Y_0 = 0$.

Conversely, when $X_1, \ldots, X_k$ are independent gamma random variables with parameters $a_1, \ldots, a_k$, then the Y_i's defined by

$$Y_i = \frac{X_1 + \cdots + X_i}{X_1 + \cdots + X_{i+1}} \quad (i = 1, 2, \ldots, k-1)$$

and

$$Y_k = X_1 + \cdots + X_k$$

are independent. Here Y_i is beta $(a_1 + \cdots + a_i, a_{i+1})$ for $i < k$ and Y_k is gamma $(a_1 + \cdots + a_k)$.

The proofs of Theorems 4.2 and 4.3 do not differ substantially from the proof of Theorem 4.1, and are omitted. See however the exercises. Theorem 4.2 tells us how to generate a Dirichlet random vector by transforming a sequence of beta random variables. Typically, this is more expensive than generating a Dirichlet random vector by transforming a sequence of gamma random variables, as is suggested by Theorem 4.1.

Theorem 4.2 also tells us that the marginal distributions of the Dirichlet distribution are all beta. In particular, when $(X_1, \ldots, X_k) \sim D(a_1, \ldots, a_{k+1})$, then X_i is beta $(a_i, \sum_{j \neq i} a_j)$.

Theorem 4.1 tells us how to relate independent gammas to a Dirichlet random vector. Theorem 4.2 tells us how to relate independent betas to a Dirichlet distribution. These two connections are put together in Theorem 4.3, where independent gammas and betas are related to each other. This offers the exciting possibility of using simple transformations to transform long sequences of gamma random variables into equally long sequences of beta random variables. Unfortunately, the beta random variables do not have equal parameters. For example, consider k iid gamma (a) random variables $X_1, \ldots, X_k$. Then the second part of Theorem 4.3 tells us how to obtain independent random variables distributed as beta (a,a), beta $(2a,a)$, $\ldots$, beta $((k-1)a,a)$ and gamma (ka) random variables respectively. When $a=1$, this reduces to a well-known property of spacings given in section V.2.

We also deduce that $BG, (1-B)G$ are independent gamma (a), gamma (b) random variables when G is gamma $(a+b)$ and B is beta (a,b) and independent of G. In particular, we obtain Stuart's theorem (Stuart, 1962), which gives us a very fast method for generating gamma (a) random variates when $a<1$: a gamma (a) random variate can be generated as the product of a gamma $(a+1)$ random variate and an independent beta $(a,1)$ random variate (the latter can be obtained as $e^{-E/a}$ where E is exponentially distributed).

4.2. Liouville distributions.

Sivazlian (1981) introduced the class of Liouville distributions, which generalizes the Dirichlet distributions. These distributions have a density on R^k given by

$$c \, \psi(\sum_{i=1}^{k} x_i) \prod_{i=1}^{k} x_i^{a_i - 1} \quad (x_i \geq 0, \; i = 1,2, \ldots, k),$$

where ψ is a Lebesgue measurable nonnegative function, $a_1, \ldots, a_k$ are positive constants (parameters), and c is a normalization constant. The functional form of ψ is not fixed. Note however that not all nonnegative functions ψ can be substituted in the formula for the density because the integral of the unnormalized density has to be finite. A random vector with the density given above is said to be Liouville $L_k(\psi, a_1, \ldots, a_k)$. Sivazlian (1981) calls this distribution a **Liou-**

ville distribution of the first kind.

Example 4.1. Independent gamma random variables.

When $X_1, \ldots, X_k$ are independent gamma random variables with parameters $a_1, \ldots, a_k$, then $(X_1, \ldots, X_k)$ is $L_k(e^{-x}, a_1, \ldots, a_k)$. ■

Example 4.2.

A random variable X with density $c\,\psi(x)\,x^{a-1}$ on $[0,\infty)$ is $L_1(\psi,a)$. This family of distributions contains all densities on the positive halfline. ■

We are mainly interested in generating random variates from multivariate Liouville distributions. It turns out that two key ingredients are needed here: a Dirichlet generator, and a generator for univariate Liouville distributions of the form given in Example 4.2. The key property is given in Theorem 4.4.

Theorem 4.4. (Sivazlian, 1981)

The normalization constant c for the Liouville $L_k(\psi, a_1, \ldots, a_k)$ density is given by

$$\frac{\Gamma(\sum_{i=1}^{k} a_i)}{\prod_{i=1}^{k} \Gamma(a_i) \int_{0}^{\infty} \psi(x) x^{a-1} \, dx}$$

where $a = \sum_{i=1}^{k} a_i$.

Let $(X_1, \ldots, X_k)$ be $L_k(\psi, a_1, \ldots, a_k)$, and let $(Y_1, \ldots, Y_k)$ be defined by

$$Y_i = \frac{X_i}{X_1 + \cdots + X_k} \quad (1 \leq i < k),$$

$$Y_k = X_1 + \cdots + X_k .$$

Then $(Y_1, \ldots, Y_{k-1})$ is Dirichlet $(a_1, \ldots, a_k)$, and Y_k is independent of this Dirichlet random vector and $L_1(\psi, \sum_{i=1}^{k} a_i)$.

Conversely, if $(Y_1, \ldots, Y_{k-1})$ is Dirichlet $(a_1, \ldots, a_k)$, and Y_k is independent of this Dirichlet random vector and $L_1(\psi, \sum_{i=1}^{k} a_i)$, then the random vector $(X_1, \ldots, X_k)$ defined by

$$X_i = Y_i Y_k \quad (1 \leq i < k),$$

$$X_k = (1 - Y_1 - \cdots - Y_{k-1}) Y_k .$$

is $L_k(\psi, a_1, \ldots, a_k)$.

Proof of Theorem 4.4.

The constant c is given by

$$\frac{1}{c} = \int_{0}^{\infty} \cdots \int_{0}^{\infty} \psi(\sum_{i=1}^{k} x_i) \prod_{i=1}^{k} x_i^{a_i - 1} dx_1 \cdots dx_k$$

$$= \frac{\prod_{i=1}^{k} \Gamma(a_i)}{\Gamma(\sum_{i=1}^{k} a_i)} \int_{0}^{\infty} \psi(x) x^{a-1} \, dx ,$$

where a property of Liouville multiple integrals is used (Sivazlian, 1981). This proves the first part of the Theorem.

Assume next that $(X_1, \ldots, X_k)$ is $L_k(\psi, a_1, \ldots, a_k)$, and that $(Y_1, \ldots, Y_k)$ is obtained via the transformation given in the statement of the Theorem. This transformation has Jacobian $Y_k{}^{k-1}$. The joint density of $(Y_1, \ldots, Y_k)$ is

$$c y_k{}^{k-1} \psi(y_k) \prod_{i=1}^{k-1} (y_i y_k)^{a_i - 1} ((1 - \sum_{i=1}^{k-1} y_i) y_k)^{a_k - 1}$$

$$= c\, \psi(y_k) y_k{}^{a-1} \prod_{i=1}^{k-1} y_i{}^{a_i - 1} (1 - \sum_{i=1}^{k-1} y_i)^{a_k - 1} \qquad (y_i \geq 0\ (i = 1, 2, \ldots, k-1),\ \sum_{i=1}^{k-1} y_i \leq 1).$$

In this we recognize the product of an $L_1(\psi, a)$ density (for Y_k), and a Dirichlet $(a_1, \ldots, a_k)$ density (for $(Y_1, \ldots, Y_{k-1})$). This proves the second part of the Theorem.

For the third part, we argue similarly, starting from the last density shown above. After the transformation to $(X_1, \ldots, X_k)$, which has Jacobian $(\sum_{i=1}^{k} X_i)^{k-1}$, we obtain the $L_k(\psi, a_1, \ldots, a_k)$ density again. ∎

Dirichlet generators are described in section 4.1, while $L_1(\psi, a)$ generators can be handled individually based upon the particular form for ψ. Since this is a univariate generation problem, we won't be concerned with the associated problems here.

4.3. Exercises.

1. Prove Theorems 4.2 and 4.3.

2. Prove the following fact: when $(X_1, \ldots, X_k) \sim D(a_1, \ldots, a_{k+1})$, then $(X_1, \ldots, X_i) \sim D(a_1, \ldots, a_i, \sum_{j=i+1}^{k+1} a_j)$, $i < k$.

3. **The generalized Liouville distribution.** A random vector $(X_1, \ldots, X_k)$ is generalized Liouville (Sivazlian, 1981) when it has a density which can be written as

$$c\, \psi(\sum_{i=1}^{k} (\frac{x_i}{c_i})^{b_i}) \prod_{i=1}^{k} x_i{}^{a_i - 1} \qquad (x_i \geq 0).$$

Here $a_i, b_i, c_i > 0$ are parameters, ψ is a nonnegative Lebesgue measurable function, and c is a normalization constant. Generalize Theorem 4.4 to this distribution. In particular, show how you can generate random vectors with this distribution when you have a Dirichlet generator and an $L_1(\psi, a)$ generator at your disposal.

4.   In the proof of Theorem 4.4, prove the two statements made about the Jaco-
     bian of the transformation.

5. SOME USEFUL MULTIVARIATE FAMILIES.

5.1. The Cook-Johnson family.

Cook and Johnson (1981) consider the multivariate uniform distribution
defined as the distribution of

$$(X_1, \ldots, X_d) = ((1+\frac{E_1}{S})^{-a}, \ldots, (1+\frac{E_d}{S})^{-a}),$$

where $E_1, \ldots, E_d$ are iid exponential random variables, S is an independent
gamma (a) random variable, and $a > 0$ is a parameter. This family is interesting
from a variety of points of view:

A.   Random variate generation is easy.

B.   Many multivariate distributions can be obtained by appropriate monotone
     transformations of the components, such as the multivariate logistic distribu-
     tion (Satterthwaite and Hutchinson, 1978; Johnson and Kotz, 1972, p. 291),
     the multivariate Burr distribution (Takahasi, 1965; Johnson and Kotz, 1972,
     p. 289), and the multivariate Pareto distribution (Johnson and Kotz, 1972,
     p. 286).

C.   For $d = 2$, the full range of nonnegative correlations can be achieved. The
     independent bivariate uniform distribution and one of Frechet's extremal
     distributions (corresponding to the case $X_2 = X_1$) are obtainable as limits.

Theorem 5.1.

The Cook-Johnson distribution has distribution function

$$F(x_1, \ldots, x_d) = \left[\sum_{i=1}^{d} x_i^{-\frac{1}{a}} - (d-1) \right]^{-a} \qquad (0 < x_i \leq 1, \ i = 1, 2, \ldots, d)$$

and density

$$f(x_1, \ldots, x_d) = \frac{\Gamma(a+d)}{\Gamma(a) a^d} \prod_{i=1}^{d} x_i^{-\frac{1}{a}-1} \left[\sum_{i=1}^{d} x_i^{-\frac{1}{a}} - (d-1) \right]^{-(a+d)}$$

$$(0 < x_i \leq 1, \ i = 1, 2, \ldots, d).$$

The distribution is invariant under permutations of the coordinates, and is multivariate uniform. Furthermore, as $a \to \infty$, the distribution function converges to $\prod_{i=1}^{d} x_i$ (the independent case), and as $a \downarrow 0$, it converges to $\min(x_1, \ldots, x_d)$ (the totally dependent case).

Proof of Theorem 5.1.

The distribution function is derived without difficulty. The density is obtained by differentiation. The permutation invariance follows by inspection. The marginal distribution function of the first component is $F(x_1, 1, \ldots, 1) = x_1$ for $0 < x_1 \leq 1$. Thus, the distribution is multivariate uniform. The limit of the distribution function as $a \downarrow 0$ is $\min(x_1, \ldots, x_d)$. Similarly, for $0 < \min(x_1, \ldots, x_d) \leq \max(x_1, \ldots, x_d) < 1$, as $a \to \infty$,

$$F(x_1, \ldots, x_d) = \left[\sum_{i=1}^{d} e^{-\frac{\log(x_i)}{a}} - (d-1) \right]^{-a}$$

$$= \left[\sum_{i=1}^{d} (1 - \frac{\log(x_i)}{a} + O(a^{-2})) - (d-1) \right]^{-a}$$

$$= \left[1 - \frac{\log(\prod_{i=1}^{d} x_i)}{a} + O(a^{-2}) \right]^{-a}$$

$$\sim e^{\log(\prod_{i=1}^{d} x_i)} = \prod_{i=1}^{d} x_i \quad \blacksquare$$

Let us now turn to a collection of other distributions obtainable from the Cook-Johnson family with parameter a by simple transformations of the X_i's. Some transformations to be applied to each X_i are shown in the next table.

Transformation for X_i	Parameters	Resulting distribution	Reference
$-\log(X_i^{-\frac{1}{a}}-1)$		Gumbel's bivariate logistic ($d=2$) and the multivariate logistic ($a=1$)	Satterthwaite and Hutchinson (1978), Johnson and Kotz (1972, p. 291)
$(d_i\,(X_i^{-\frac{1}{a}}-1))^{\frac{1}{c_i}}$	$c_i,d_i>0$	multivariate Burr	Takahasi (1965), Johnson and Kotz (1972, p. 286)
$a_i\,X_i^{-\frac{1}{a}}$	$a_i>0$	multivariate Pareto	Johnson and Kotz (1972, p. 286)
$\Phi^{-1}(X_i)$	None. Φ is the normal distribution function.	multivariate normal without elliptical contours	Cook and Johnson (1981)

Example 5.1. The multivariate logistic distribution.

In 1961, Gumbel proposed the bivariate logistic distribution, a special case of the generalized multivariate logistic distribution with distribution function

$$\left(1+\sum_{i=1}^{d} e^{-x_i}\right)^{-a} \qquad (x_i>0\,,\ i=1,2,\ldots,d)\ .$$

For $a=1$ this reduces to the multivariate logistic distribution given by Johnson and Kotz (1972, p. 293). Note that from the form of the distribution function, we can deduce immediately that all univariate and multivariate marginals are again multivariate logistic. Transformation of a Cook-Johnson random variate leads to the following simple recipe for generating multivariate logistic random variates:

Multivariate logistic generator

Generate iid exponential random variates $E_1,\ldots,E_{d+1}$.

RETURN $(\log(\frac{E_1}{E_{d+1}}),\ \ldots,\ \log(\frac{E_d}{E_{d+1}}))$ ∎

Example 5.2.

The multivariate normal distribution in the table has nonelliptical contours. Kowalski (1973) provides other examples of multivariate normal distributions with nonnormal densities. ■

5.2. Multivariate Khinchine mixtures.

Bryson and Johnson (1982) proposed the family of distributions defined constructively as the distributions of random vectors in R^d which can be written as

$$(Z_1 U_1, \ldots, Z_d U_d)$$

where the $Z_1, \ldots, Z_d$ is independent of the multivariate uniform random vector $U_1, \ldots, U_d$, and has a distribution which is such that certain given marginal distributions are obtained. Recalling Khinchine's theorem (section IV.6.2), we note that all marginal distributions have unimodal densities.

Controlled dependence can be introduced in many ways. We could introduce dependence in $U_1, \ldots, U_d$ by picking a multivariate uniform distribution based upon the multivariate normal density or the Cook-Johnson distribution. Two models for the Z_i's seem natural:

A. The identical model: $Z_1 = \cdots = Z_d$.

B. The independent model: $Z_1, \ldots, Z_d$ are iid.

These models can be mixed by choosing the identical model with probability p and the independent model with probability $1-p$.

Example 5.3.

To achieve exponential marginals, we can take all Z_i's gamma (2). In the identical bivariate model, the joint bivariate density is

$$\int_{\max(x_1, x_2)}^{\infty} \frac{e^{-t}}{t} \, dt \ .$$

In the independent bivariate model, the joint density is

$$\frac{x_1 x_2}{(x_1 + x_2)^3} (2 + 2(x_1 + x_2) + (x_1 + x_2)^2) e^{-(x_1 + x_2)} \ .$$

Unfortunately, the correlation in the first model is $\frac{1}{2}$, and that of the second model is $\frac{1}{3}$. By probability mixing, we can only cover correlations in the small range $[\frac{1}{3}, \frac{1}{2}]$. Therefore, it is useful to replace the independent model by the totally independent model (with density $e^{-(x_1+x_2)}$), thereby enlarging the range to $[0,\frac{1}{2}]$. ■

Example 5.4. Nonnormal bivariate normal distributions.

For symmetric marginals, it is convenient to take the U_i's uniform on $[-1,1]$. It is easy to see that in order to obtain normal marginals, the Z_i's have to be distributed as the square roots of chi-square random variables with 3 degrees of freedom. If (U_1,U_2) has bivariate density h on $[-1,1]^2$, then (Z_1U_1,Z_1U_2) has joint density

$$\int\limits_{\max(|x_1|,|x_2|)}^{\infty} \Gamma^{-1}(\frac{3}{2})2^{-\frac{5}{2}} e^{-\frac{t^2}{2}} h(\frac{1}{2}+\frac{y_1}{2t},\frac{1}{2}+\frac{y_2}{2t}) \, dt \ .$$

This provides us with a rich source of examples of bivariate distributions with normal marginals, zero correlations and non-normal densities. At the same time, random variate generation for these examples is trivial (Bryson and Johnson, 1982). ■

5.3. Exercises.

1. **The multivariate Pareto distribution.** The univariate Pareto density with parameter $a>0$ is defined by a/x^{a+1} $(x\geq1)$. Johnson and Kotz (1972, p. 286) define a multivariate Pareto density on R^d with parameter a by

$$\frac{a(a+1)\cdots(a+d-1)}{(\sum\limits_{i=1}^{d} x_i-(d-1))^{a+d}} \qquad (x_i\geq1 \ , \ i=1,2,\ldots,d) \ .$$

A. Show that the marginals are all univariate Pareto with parameter a.

B. In the bivariate case, show that the correlation is $\frac{1}{a}$. Since the marginal variance is finite if and only if $a>2$, we see that all correlations between 0 and $\frac{1}{2}$ can be achieved.

C. Prove that a random vector can be generated as $(X_1^{-\frac{1}{a}},\ldots,X_d^{-\frac{1}{a}})$ where $(X_1,\ldots,X_d)$ has the Cook-Johnson distribution with parameter a. Equivalently, it can be generated as $(1+\frac{E_1}{S},\ldots,1+\frac{E_d}{S})$, where $E_1, \ldots, E_d$ are iid exponential random variables, and S is an independent gamma (a) random variable.

6. RANDOM MATRICES.

6.1. Random correlation matrices.

To test certain statistical methods, one should be able to create random test problems. In several applications, one needs a random correlation matrix. This problem is equivalent to that of the generation of a random covariance matrix if one asks that all variances be one. Unfortunately, posed as such, there are infinitely many answers. Usually, one adds structural requirements to the correlation matrix in terms of expected value of elements, eigenvalues, and distributions of elements. It would lead us too far to discuss all the possibilities in detail. Instead, we just kick around a few ideas to help us to better understand the problem. For a recent survey, consult Marsaglia and Olkin (1984).

A correlation matrix is a symmetric positive semi-definite matrix with ones on the diagonal. It is well known that if $\mathbf{H}$ is a $d \times n$ matrix with $n \geq d$, then $\mathbf{HH'}$ is a symmetric positive semi-definite matrix. To make it a correlation matrix, it is necessary to make the rows of $\mathbf{H}$ of length one (this forces the diagonal elements to be one). Thus, we have the following property, due to Marsaglia and Olkin (1984):

Theorem 6.1.

$\mathbf{HH'}$ is a random correlation matrix if and only if the rows of $\mathbf{H}$ are random vectors on the unit sphere of R^n.

Theorem 6.1 leads to a variety of algorithms. One still has the freedom to choose the random rows of $\mathbf{H}$ according to any recipe. It seems logical to take the rows as independent uniformly distributed random vectors on the surface of C_n, the unit sphere of R^n, where $n \geq d$ is chosen by the user. For this case, one can actually compute the explicit form of the marginal distributions of $\mathbf{HH'}$. Marsaglia and Olkin suggest starting from any $d \times n$ matrix of iid random variables, and to normalize the rows. They also suggest in the case $n = d$ starting from lower triangular $\mathbf{H}$, thus saving about 50% of the variates.

The problem of the generation of a random correlation matrix with a given set of eigenvalues is more difficult. The diagonal matrix $\mathbf{D}$ defined by

$$\begin{vmatrix} \lambda_1 & 0 & \cdots & 0 \\ 0 & \lambda_2 & \cdots & 0 \\ \cdots & \cdots & \cdots & \cdots \\ 0 & 0 & \cdots & \lambda_d \end{vmatrix}$$

has eigenvalues $\lambda_1, \ldots, \lambda_d$. Also, eigenvalues do not change when $\mathbf{D}$ is pre and post multiplied with an orthogonal matrix. Thus, we need to make sure that there exist many orthogonal matrices $\mathbf{H}$ such that $\mathbf{HDH'}$ is a correlation matrix. Since the trace of our correlation matrix must be d, we have to start with a matrix $\mathbf{D}$ with trace d. For the construction of random orthogonal $\mathbf{H}$ that satisfy the given collection of equations, see Chalmers (1975), Bendel and Mickey (1978) and Marsaglia and Olkin (1984). See also Johnson and Welch (1980), Bendel and Afifi (1977) and Ryan (1980).

In a third approach, designed to obtain random correlation matrices with given mean $\mathbf{A}$, Marsaglia and Olkin (1984) suggest forming $\mathbf{A+H}$ where $\mathbf{H}$ is a perturbation matrix. We have

---

Theorem 6.2.

Let $\mathbf{A}$ be a given $d \times d$ correlation matrix, and let $\mathbf{H}$ be a random symmetric $d \times d$ matrix whose elements are zero on the diagonal, and have zero mean off the diagonal. Then $\mathbf{A+H}$ is a random correlation matrix with expected value $\mathbf{A}$ if and only if the eigenvalues of $\mathbf{A+H}$ are nonnegative.

Proof of Theorem 6.2.

The expected value is obviously correct. Also, $\mathbf{A+H}$ is symmetric. Furthermore, the diagonal elements are all one. Finally, $\mathbf{A+H}$ is positive semi-definite when its eigenvalues are nonnegative. ∎

We should also note that the eigenvalues of $\mathbf{A+H}$ and those of $\mathbf{A}$ differ by at most

$$\Delta = \max\left(\sqrt{\sum_{i,j} h_{ij}^2}, \max_i \sum_j |h_{ij}|\right),$$

where h_{ij} is an element of $\mathbf{H}$. Thus, if Δ is less than the smallest eigenvalue of $\mathbf{A}$, then $\mathbf{A+H}$ is a correlation matrix. Marshall and Olkin (1984) use this fact to suggest two methods for generating $\mathbf{H}$:

A. Generate all h_{ij} for $i < j$ with zero mean and support on $[-b_{ij}, b_{ij}]$ where the b_{ij}'s form a zero diagonal symmetric matrix with Δ smaller than the smallest eigenvalue of $\mathbf{A}$. Then for $i > j$, define $h_{ij} = h_{ji}$. Finally, $h_{ii} = 0$.

B. Generate $h_{12}, h_{13}, \ldots, h_{d-1,d}$ with a radially symmetric distribution in or on the $d(d-1)/2$ sphere of radius $\lambda/\sqrt{2}$ where λ is the smallest eigenvalue of $\mathbf{A}$. Define the other elements of $\mathbf{H}$ by symmetry.

6.2. Random orthogonal matrices.

An orthonormal $d \times d$ matrix can be considered as a rotation of the coordinate axes in R^d. In such a rotation, there are $d(d-1)/2$ degrees of freedom. To see this, we look at where the points $(1,0,0,\ldots,0),\ldots,(0,0,\ldots,0,1)$ are mapped to by the orthonormal transformation. These points are mapped to other points on the unit sphere. In turn, the mapped points define the rotation. We can choose the first point (d coordinates). Given the first point, the second point should be in a hyperplane perpendicular to the line joining the origin and the first point. Here we have only $d-1$ degrees of freedom. Continuing in this fashion, we see that there are $d(d-1)/2$ degrees of freedom in all.

Heiberger (1978) (correction by Tanner and Thisted (1982)) gives an algorithm for generating an orthonormal matrix which is uniformly distributed. This means that the first point is uniformly distributed on the unit sphere of R^d, that the second point is uniformly distributed on the unit sphere of R^d intersected with the hyperplane which is perpendicular to the line from the origin to the first point, and so forth.

His algorithm requires $d(d+1)/2$ independent normal random variables, while the total time is $O(d^3)$. It is perhaps worth noting that no heavy matrix computations are necessary at all if one is willing to spend a bit more time. To illustrate this, consider performing $\binom{d}{2}$ random rotations of two axes, each rotation keeping the $d-2$ other axes fixed. A random rotation of two axes is easy to carry out, as we will see below. The global random rotation boils down to $\binom{d}{2}$ matrix multiplications. Luckily, each matrix is nearly diagonal: there are four random elements on the intersections of two given rows and columns. The remainder of each matrix is purely diagonal with ones on the diagonal. This structure implies that the time needed to compute the global (product) rotation matrix is $O(d^3)$.

A **random uniform rotation** of R^2 can be generated as

$$\begin{vmatrix} X & Y \\ -SY & SX \end{vmatrix}$$

where (X,Y) is a point uniformly distributed on C_2, and S is a random sign. A random rotation in R^3 in which the z-axis remains fixed is

$$\begin{vmatrix} X & Y & 0 \\ -SY & SX & 0 \\ 0 & 0 & 1 \end{vmatrix} .$$

Thus, by the threefold combination (i.e., product) of matrices of this type, we can obtain a random rotation in R^3. If $\mathbf{A}_{12}, \mathbf{A}_{23}, \mathbf{A}_{13}$ are three random rotations of two axes with the third one fixed, then the product

$$\mathbf{A}_{12}\mathbf{A}_{23}\mathbf{A}_{13}$$

is a random rotation of R^3.

6.3. Random $R \times C$ tables.

A two-way contingency table with r rows and c columns is a matrix of non-negative integer-valued numbers. It is also called an $R \times C$ table. Typically, the integers represent the frequencies with which a given pair of integers is observed in a sample of size n. The purpose of this section is to explore the generation of a random $R \times C$ table with given sample size (sum of elements) n. Again, this is an ill-posed problem unless we impose more structure on it. The standard restrictions are:

A.   Generate a random table for sample size n, such that all tables are equally likely.

B.   Generate a random table for sample size n, with given row and column totals. The row totals are called r_i ,$1 \leq i \leq r$. The column totals are c_i ,$1 \leq i \leq c$.

Let us just consider problem B. In a first approach, we take a ball-in-urn strategy. Consider balls numbered $1,2,\ldots,n$. Of these, the first c_1 are class one balls, the next c_2 are class two balls, and so forth. Think of classes as different colors. Generate a random permutation of the balls, and put the first r_1 balls in row 1, the next r_2 balls in row 2, and so forth. Within a given row, class i balls should all be put in column i. This ball-in-urn method, first suggested by Boyett (1979), takes time proportional to n, and is not recommended when n is much larger than rc, the size of the matrix.

Ball-in-urn method

[NOTE: N is an $r \times c$ array to be returned. $B[1], \ldots, B[n]$ is an auxiliary array.]

Sum $\leftarrow 0$

FOR $j := 1$ TO c DO

    FOR $i := \text{Sum}+1$ TO $\text{Sum}+c_j$ DO $B[i] \leftarrow j$

    Sum $\leftarrow$ Sum $+ c_j$

Randomly permute the array B.

Set N to all zeroes.

Sum $\leftarrow 0$

FOR $j := 1$ TO r DO

    FOR $i := \text{Sum}+1$ TO $\text{Sum}+r_j$ DO $N[j,B[i]] \leftarrow N[j,B[i]]+1$

    Sum $\leftarrow$ Sum $+ r_j$

RETURN N

Patefield (1980) uses the conditional distribution method to reduce the dependence of the performance upon n. The conditional distribution of an entry N_{ij} given the entries in the previous rows, and the previous entries in the same row i is given by

$$P(N_{ij} = k) = \frac{\alpha\beta\gamma\delta}{\epsilon\eta\varsigma\theta k!}$$

where

$$\alpha = (r_i - \sum_{l<j} N_{il})! \, ,$$

$$\beta = (n - \sum_{m \leq i} r_m - \sum_{m < j} c_m + \sum_{l<j, m \leq i} N_{ml})! \, ,$$

$$\gamma = (c_j - \sum_{m<i} N_{mj})! \, ,$$

$$\delta = \left(\sum_{l>j} (c_l - \sum_{m<i} N_{ml}) \right)! \, ,$$

$$\epsilon = (r_i - \sum_{l<j} N_{il} - k)! \, ,$$

$$\eta = (n - \sum_{m \leq i} r_m - \sum_{m \leq j} c_m + \sum_{l<j, m \leq i} N_{ml} + \sum_{m<i} N_{mj} + k)! \, ,$$

$$\varsigma = (c_j - \sum_{m<i} N_{mj} - k)! \, ,$$

$$\theta = \left(\sum_{l \geq j} (c_l - \sum_{m<i} N_{ml}) \right)! \, .$$

The range for k is such that all factorial terms are nonnegative. Although the expression for the conditional probabilities appears complicated, we note that quite a bit of regularity is present, which makes it possible to adjust the partial sums "on the fly". As we go along, we can quickly adjust all terms. More precisely, the constants needed for the computation of the probabilities of the next entry in the same row can be computed from the previous one and the value of the current element N_{ij} in constant time. Also, there is a simple recurrence relation for the probability distribution as a function of k, which makes the distribution tractable by the sequential inversion method (as suggested by Patefield, 1980). However, the expected time of this procedure is not bounded uniformly in n for fixed values of r, c.

6.4. Exercises.

1. Let $\mathbf{A}$ be a $d \times d$ correlation matrix, and let $\mathbf{H}$ be a symmetric matrix. Show that the eigenvalues of $\mathbf{A}+\mathbf{H}$ differ by at most Δ from the eigenvalues of $\mathbf{A}$, where

$$\Delta = \max\left(\sqrt{\sum_{i,j} h_{ij}{}^2}, \max_i \sum_j |h_{ij}|\right).$$

2. Generate $h_{12}, h_{13}, \ldots, h_{d-1,d}$ with a radially symmetric distribution in or on the $d(d-1)/2$ sphere of radius $\lambda/\sqrt{2}$ where λ is the smallest eigenvalue of $\mathbf{A}$. Define the other elements of $\mathbf{H}$ by symmetry. Put zeroes on the diagonal of $\mathbf{H}$. Then $\mathbf{A}+\mathbf{H}$ is a correlation matrix when $\mathbf{A}$ is. Show this.

3. Consider Patefield's conditional distribution method for generating a random $R \times C$ table. Show the following:

   A. The conditional distribution as given in the text is correct.

   B. (Difficult.) Design a constant expected time algorithm for generating one element in the $r \times c$ matrix. The expected time should be uniformly bounded over all conditions, but with r and c fixed.

Chapter Twelve
RANDOM SAMPLING

1. INTRODUCTION.

In this chapter we consider the problem of the selection of a random sample of size k from a set of n objects. This is also called sampling without replacement since duplicates are not allowed. There are several issues here which should be clarified in this, the introductory section.

1. Some users may wish to generate an ordered random sample. Not unexpectedly, it is easier to generate unordered random samples. Thus, algorithms that produce ordered random samples should not be compared on an equal basis with other algorithms.

2. Sometimes, n is not known, and we are asked to grab each object in turn and make an instantaneous decision whether to include it in our random sample or not. This can best be visualized by considering the objects as being given in a linked list and not an array.

3. In nearly all cases, we worry about the expected time complexity as a function of k and n. In typical situations, n is much larger than k, and we would like to have expected time complexities which are bounded by a constant times k, uniformly over n.

4. The space required by an algorithm is defined as the space required outside the original array of n records or objects and outside the array of k records to be returned. Some of the algorithms in this chapter are bounded workspace algorithms, i.e. the space requirements are $O(1)$.

The strategies for sampling can be partitioned as follows: (i) classical sampling: generate random objects and include them in the sample if they have not already been picked; (ii) sequential sampling: generate the sample by traversing the collection of objects once and making instantaneous decisions during that one pass; (iii) oversampling: by means of a simple technique, obtain a random sample (usually of incorrect size), and in a second phase, adjust the sample so that it has the right size. Each of these strategies has some very competitive algorithms, so that no strategy should a priori be excluded from contention.

We assume that the set of objects is $\{1,2, \ldots, n\}$. If the objects are different, then these integers should be considered as pointers (indices) to the objects in an array.

2. CLASSICAL SAMPLING.

2.1. The swapping method.

Assume that the objects are given in array form: $A[1], \ldots, A[n]$. Then, if we are allowed to permute the objects, random sampling is extremely simple. We can choose an object uniformly and at random, and swap it with the last object. If we need another object, we choose one uniformly from among the first $n-1$ objects, and swap with the $n-1$st object, and so forth. This algorithm takes time proportional to k, and $O(1)$ extra space is needed. The disadvantage is that the sample is not ordered. Also, record swapping is sometimes not allowed. We are allowed to swap pointers though, but this would then require $\Theta(n)$ extra space for pointers. If there are no records to begin with, then the space requirement is $\Omega(n)$. Formally we have:

Swapping method

FOR $i := n$ DOWNTO $n-k+1$ DO

      Generate a uniform $[0,1]$ random variate U.

      $X \leftarrow \lceil iU \rceil$

      Swap $(A[X], A[i])$

RETURN $A[n-k+1], \ldots, A[n]$

The swapping method is very convenient. If we set $k = n$, then the returned array is a random permutation. Thus, the swapping method is based upon the principle that generating a random subset of size k is equivalent to generating the first k entries in a random permutation.

2.2. Classical sampling with membership checking.

If we are not allowed to swap information, then we are forced to check whether a certain element is not already picked. The checking can be achieved in a number of ways via different data structures. Regardless of the data structure, we can formulate the algorithm:

Classical sampling with membership checking

$S \leftarrow \emptyset$ (S will be the set of random integers to be returned)

FOR $i := 1$ TO k DO

    REPEAT

        Generate a random integer Z in $\{1, \ldots, n\}$.

        UNTIL NOT Member (Z) (Member returns true if an integer is already picked, and false otherwise.)

    $S \leftarrow S \cup \{Z\}$

RETURN S

The data structure used for S should support the following operations: initialize empty set, insert, member. Among the tens of possible data structures, the following are perhaps most representative:

A. The bit-vector implementation. Define an array of n bits, which are initially set to false, and which are switched to true upon insertion of an element.

B. An unordered array of chosen elements. Elements are added at the end of the array.

C. A binary search tree of chosen elements. The expected depth of the k-th element added to the tree is $\sim 2\log(k)$. The worst-case depth can be as large as k.

D. A height-balanced binary tree or 2-3 tree of chosen elements. The worst-case depth of the tree with k elements is $O(\log(k))$.

E. A bucket structure (open hashing with chaining). Partition $\{1, \ldots, n\}$ into k about equal intervals, and keep for each interval (or: bucket) a linked list of all elements chosen until now.

F. Closed hashing into a table of size a bit larger than k.

It is perhaps useful to give a list of expected complexities of the various operations needed on these data structures. We also include the space requirements, with the convention that the array of k integers to be returned is in any case

Included in the space requirements.

DATA STRUCTURE	Initialize	Insert	Member	Space requirements
Bit vector	n	1	1	n
Unordered array	1	1	k	k
Binary search tree	1	$\log(k)$	$\log(k)$	k
Height-balanced tree	1	$\log(k)$	$\log(k)$	k
Buckets	k	1	1	k
Closed hashing	k	1	1	k

Timewise, none of the suggested data structures is better than the bit-vector data structure. The problem with the bit-vector implementation is not so much the extra storage proportional to n, because we can often use the already existing records and use common programming tricks (such as changing signs etcetera) to store the extra bits. The problem is the re-initialization necessary after a sample has been generated. At the very least, this will force us to consider the selected set S, and turn the k bits off again for all elements in S. Of course, at the very beginning, we need to set all n bits to false.

The first important quantity is the expected number of iterations in the sampling algorithm.

Theorem 2.1.

The expected number of iterations in classical sampling with membership checking is

$$\sum_{i=1}^{k} \frac{n}{n-i+1} .$$

For $k=n$, this is $n\sum_{i=1}^{n} \frac{1}{i} \sim n\,\log(n)$. When $k \leq \left\lceil \frac{n}{2} \right\rceil$, this number is $\leq 2k$.

Proof of Theorem 2.1.

Observe that to generate the i-th random integer, we carry out a series of independent experiments, each having probability of success $\frac{n-i+1}{n}$. This yields the given expected value. The asymptotic result when $k=n$ is trivially true. The general upper bound is obtained by a standard integral argument: bound the sum from above by

$$n \sum_{i=n-k+1}^{n} \frac{1}{i}$$

$$\leq n \left(\frac{1}{n-\left\lceil \dfrac{n}{2} \right\rceil+1} + \int_{n-k+1}^{n} \frac{1}{x}\,dx \right)$$

$$= n \left(\frac{1}{n - \left\lceil \dfrac{n}{2} \right\rceil + 1} + \log(\frac{n}{n-k+1}) \right)$$

$$\leq 2 + n \log(1 + \frac{k-1}{n-k+1})$$

$$\leq 2 + \frac{n}{n-k+1}(k-1)$$

$$\leq 2 + 2(k-1) = 2k \quad . \blacksquare$$

What matters here is that the expected time increases no faster than $O(k)$ when k is at most half of the sample. Of course, when k is larger than $\frac{n}{2}$, one should really sample the complement set. In particular, the expected time for the bit-vector implementation is $O(k)$. For the tree methods, we obtain $O(k \log(k))$. If we work with ordered or unordered lists, then the generation procedure takes expected time $O(k^2)$. Finally, with the hash structure we have expected time $O(k)$ provided that we can show that the expected time of an insert or a delete is $O(1)$ (apply Wald's equation). Assume that we have a bucket structure with mk equal-sized intervals, where $m \geq 1$ is a design integer usually equal to 1. The interval number is an integer between 1 and mk, and integer $x \in \{1, \ldots, n\}$ is hashed to interval $\left\lceil \frac{x}{n} mk \right\rceil$. Thus, if the hash table has k elements, then every interval has about $\frac{1}{m}$ elements. The expected number of comparisons needed to check the membership of a random integer in a hash table containing i elements is bounded from above by $E(1 + n_Z)$ where n_Z is equal to the number of elements in the interval Z, and Z is a random interval index, chosen with probability proportional to the cardinality of the interval. The "1" accounts for the comparison spent checking the endmarker in the chain. Thus, the expected number of comparisons is not greater than

$$1 + \sum_{j=1}^{mk} \frac{(\frac{n}{mk} + 1)}{n} n_j$$

$$\leq 1 + i \frac{(\frac{n}{mk} + 1)}{n}$$

$$= 1 + \frac{i}{mk} + \frac{i}{n} .$$

In the worst case $(i = k)$, this upper bound is $1 + \frac{1}{m} + \frac{k}{n} \leq 2 + \frac{1}{m}$. The upper bound is very loose. Nevertheless, we have an upper bound which is clearly $O(1)$. Also, if we can afford the space, it pays to take m as large as possible. One possible hashing algorithm is given below:

Classical sampling with membership checking based on a hash table

This algorithm uses three arrays of integers of size k. An array of headers Head
$[1], \ldots$, Head $[k]$ is initially set to 0. An array pointers to successor elements
Next$[1], \ldots$, Next$[k]$ is also set to 0. The array $A[1], \ldots, A[k]$ will be returned.

FOR $i := 1$ TO k DO
      Accept $\leftarrow$ False
      REPEAT
            Generate a random integer Z uniformly distributed on $\{1, \ldots, n\}$.
            Bucket $\leftarrow 1 + \left\lfloor \dfrac{k(Z-1)}{n} \right\rfloor$
            Top $\leftarrow$ Head [Bucket]
            IF Top$=0$
                THEN
                    Head [Bucket] $\leftarrow i$
                    $A[k] \leftarrow Z$
                    Accept $\leftarrow$ True
                ELSE
                    WHILE $A[\text{Top}] \neq Z$ AND Top$\neq 0$ DO
                        (Top, Top∗) $\leftarrow$ (Next [Top], Top)
                    IF Top$=0$ THEN
                      $A[i] \leftarrow Z$
                      Next [Top∗] $\leftarrow i$
                      Accept $\leftarrow$ True
          UNTIL Accept
        RETURN $A[1], \ldots, A[k]$

The hashing algorithm requires $2k$ extra storage space. The array returned is not
sorted, but sorting can be done in linear expected time. We give a short formal
proof of this fact. It is only necessary to travel from bucket to bucket and sort
the elements within the buckets (because an order-preserving hash function was
used). If this is done by a simple quadratic method such as bubble sort or selection sort, then the overall expected time complexity is $O(k)$ (for the overhead
costs) plus a constant times

$$E\left(\sum_{i=1}^{mk} n_i^2\right).$$

But n_i is hypergeometric with parameters n, l, k where l is the number of
integers in the i-th bucket (this is about $\dfrac{n}{mk}$), i.e. for each j,

$$P(n_i = j) = \frac{\begin{pmatrix} l \\ j \end{pmatrix} \begin{pmatrix} n-l \\ k-j \end{pmatrix}}{\begin{pmatrix} n \\ k \end{pmatrix}} .$$

We know that $E(n_i) = \dfrac{kl}{n}$, and this tends to $\dfrac{1}{m}$ as $k, n \to \infty$, and it does not exceed $\dfrac{1}{m} + \dfrac{k}{n}$ in any case. Simple computations show that

$$Var(n_i) = \frac{n-k}{n-1} \frac{n-l}{n} \frac{kl}{n}$$

which in turn tends to $\dfrac{1}{m}$ as $k, n \to \infty$, without exceeding $\dfrac{1}{m} + \dfrac{k}{n}$ for any value of k, m, n. Combining this, we see that the the expected time complexity is a constant times

$$\sim k(1 + \frac{1}{m}) .$$

It is not greater than a constant times

$$mk\left[(\frac{1}{m} + \frac{k}{n})^2 + (\frac{1}{m} + \frac{k}{n}) \right] = k(1 + \frac{km}{n})(1 + \frac{1}{m} + \frac{k}{n}) .$$

These expressions show that it is important to take m large. One should not fall into the trap of letting m increase with k, n because the set-up time is proportional to mk, the number of buckets. The hashing method with chaining, as given here, was implicitly given by Muller (1958) and studied by Ernvall and Nevalainen (1982). Its space inefficiency is probably its greatest drawback. Closed hashing with a table of size k has been suggested by Nijenhuis and Wilf (1975). Ahrens and Dieter (1985) consider closed hashing tables of size mk where now m is a number, not necessarily integer, greater than 1. See also Teuhola and Nevalainen (1982). It is perhaps instructive to give a brief description of the algorithm of Nijenhuis and Wilf (1975). An unordered sample $A[1], \ldots, A[k]$ will be generated, and an auxiliary vector Next[1], $\ldots$, Next[k] of links is needed in the process. A pointer p points to the largest index i for which $A[i]$ is not yet specified.

Algorithm of Nijenhuis and Wilf

[SET-UP]

$p \leftarrow k + 1$

FOR $i := 1$ TO k DO $A[i] \leftarrow 0$

[GENERATOR]

REPEAT

    Generate a random integer X uniformly distributed on $1, \ldots, n$. Set Bucket$\leftarrow X$ mod $k + 1$.

    IF A [Bucket]$= 0$

        THEN A [Bucket]$\leftarrow X$, Next[Bucket]$\leftarrow 0$

        ELSE

            WHILE A [Bucket]$\neq X$ DO

              IF Next[Bucket]$= 0$

                THEN

                    REPEAT $p \leftarrow p - 1$ UNTIL $p = 0$ OR $A[p] = 0$

                    Next[Bucket]$\leftarrow p$

                    Bucket$\leftarrow p$

                ELSE Bucket$\leftarrow$Next[Bucket]

UNTIL $p = 0$

RETURN $A[1], \ldots, A[k]$

The algorithm of Nijenhuis and Wilf differs slightly from standard closed hashing schemes because of the vector of links. The links actually create small linked lists within the table of size k. When we look at the cost associated with the algorithm, we note first that the expected number of uniform random variates needed is at the same as for all other classical sampling schemes (see Theorem 2.1). The search for an empty space ($p \leftarrow p - 1$) takes time $O(k)$. The search for the end of the linked list (inner WHILE loop) takes on the average fewer than 2.5 link accesses per random variate X, independent of when X is generated and how large k and n are (Knuth, 1969, pp. 513-518). Thus, both expected time and space are $O(k)$.

2.3. Exercises.

1. The number of elements n_1 that end up in a bucket of capacity l in the
    bucket method is hypergeometrically distributed with parameters n,k,l.
    That is,

    $$P(n_1=i) = \frac{\binom{l}{i}\binom{n-l}{k-i}}{\binom{n}{k}}, \quad 0 \le i \le \min(k,l).$$

    In the text, we needed the expected value and variance of n_1. Derive these
    quantities.

2. Prove that the expected time in the algorithm of Nijenhuis and Wilf is
    $O(k)$.

3. **Weighted sampling without replacement.** Assume that we wish to gen-
    erate a random sample of size k from $\{1,\ldots,n\}$, where the integers
    $1,\ldots,n$ have weights w_i. Drawing an integer from a set of integers is to
    be done with probability proportional to the weight of the integer. Using
    classical sampling, this involves dynamically updating a selection probability
    vector. Wong and Easton (1980) suggest setting up a binary tree of height
    $O(\log(n))$ in time $O(n)$ in a preprocessing step, and using this tree in the
    inversion method. Generating a random integer takes time $O(\log(n))$, while
    updating the tree has a similar cost. This leads to a method with worst-case
    time $O(k\log(n)+n)$. The space requirement is proportional to n (space is
    less critical because the vector of weights must be stored anyway). Develop
    a dynamic structure based upon the alias method or the method of guide
    tables, which has a better expected time performance for all vectors of
    weights.

3. SEQUENTIAL SAMPLING.

3.1. Standard sequential sampling.

In sequential sampling, we want an ordered sample of size k drawn from
$1,\ldots,n$. An unordered sample can always be obtained by one of the methods
described in the previous section, and in many cases (e.g. the hashing methods),
sorting can be done extremely efficiently in expected time $O(k)$. What we will do
in this chapter is different. The methods described here are fundamentally one
pass methods in which the random sample is constructed in order. There are two
possible strategies: first, we could grab each integer in $1,\ldots,n$ in turn, and
decide whether to take it or leave it. It turns out, as we will see below, that for
each decision, we need only compare a new uniform random variate with a certain threshold. Unfortunately, this standard sequential sampling algorithm takes

time proportional to n: it becomes particularly inefficient when k is much smaller than n. The second strategy circumvents this problem by generating the spacings between successive integers. Assume for a moment that each spacing can be generated in expected time $O(1)$ uniformly over all parameter values. Then the spacings method takes expected time $O(k)$. The problem here is that the distribution of the spacings is rather complicated; it also depends upon the partially generated sample.

In the standard sequential sampling algorithm of Jones (1962) and Fan, Muller and Rezucha (1962), the probability of selection of an integer depends upon only two quantities: the number of integers remaining to be selected, and the number of integers not yet processed. Initially, these quantities are k and n. To keep the notation simple, we will let k decrease during execution of the algorithm.

Standard sequential sampling

FOR $i := 1$ TO n DO

    Generate a uniform [0,1] random variate U.

    IF $U \leq \dfrac{k}{n-i+1}$ THEN select i, $k \leftarrow k-1$

Integer 1 is selected with probability $\dfrac{k}{n}$ as can easily be seen from the following argument: there are

$$\binom{n}{k}$$

ways of choosing a subset of size k from $1, \ldots, n$. Furthermore, of these,

$$\binom{n-1}{k-1}$$

include integer 1. The probability of inclusion of 1 should therefore be the ratio of these two numbers, or k/n. Note that this argument uses only k, the number of remaining integers to be selected, and n, the number of integers not yet processed. It can be used inductively to prove that the algorithm is correct. Note for example that if at any time in the algorithm $k = n$, then each of the remaining n integers in the file is selected with probability one. If at some point $k = 0$, no more integers are selected. The time taken by the algorithm is proportional to n, but no extra space is needed. For small values of n, the standard sequential algorithm has little competition.

3.2. The spacings method for sequential sampling.

We say that a random variable X has the distribution $D(k,n)$ when X is distributed as the minimal integer in a random subset of size k drawn from $\{1, \ldots, n\}$. The spacings method for sequential sampling is defined as follows:

The spacings method for sequential sampling

$Y \leftarrow 0$ (Y is a running pointer)
REPEAT
      Generate a random integer X with distribution $D(k,n)$.
      $k \leftarrow k-1$, $n \leftarrow n-X$ (update parameters).
      Select $Y+X$, set $Y \leftarrow Y+X$
UNTIL $k=0$

In the algorithm, the original values of k and n are destroyed - this saves us the trouble of having to introduce two new symbols. If we can generate $D(k,n)$ random variates in expected time $O(1)$ uniformly over k and n, then the spacings method takes expected time $O(k)$. The space requirements depend of course on what is needed for the generation of $D(k,n)$. There are many possible algorithms for generating a $D(k,n)$ random variable. We discuss the following approaches:

1.   The inversion method (Devroye and Yuen, 1981; Vitter, 1984).

2.   The ghost sample method (Devroye and Yuen, 1981).

3.   The rejection method (Vitter, 1983, 1984).

The three methodologies will be discussed in different subsections. All techniques require a considerable programming effort when implemented. In cases 1 and 3, most of the energy is spent on numerical problems such as the evaluation of ratios of factorials. Case 2 avoids the numerical problems at the expense of some additional storage (not exceeding $O(k)$). We will first state some properties of $D(k,n)$.

Theorem 3.1.

Let X have distribution $D(k,n)$. Then

$$P(X>i) = \frac{\binom{n-i}{k}}{\binom{n}{k}} \quad , 0 \le i \le n-k \; ,$$

$$P(X=i) = \frac{\binom{n-i}{k-1}}{\binom{n}{k}} \quad , 1 \le i \le n-k+1 \; .$$

Proof of Theorem 3.1.

Argue by counting the number of subsets of k out of n, the number of sub-sets of k out of $n-i$, and the number of subsets of $k-1$ out of $n-i$. ■

Theorem 3.2.

The random variable $X = \min(X_1, \ldots, X_k)$ is $D(k,n)$ distributed when-ever $X_1, \ldots, X_k$ are independent random variables and each X_i is uniformly distributed on $\{1, \ldots, n-k+i\}$.

Proof of Theorem 3.2.

For $0 \le i \le n-k$, notice that

$$P(Y>i) = \prod_{j=1}^{k} \frac{n-k+i-j}{n-k+i} = \prod_{j=0}^{k-1} \frac{n-i-j}{n-j} = \frac{\binom{n-i}{k}}{\binom{n}{k}} \; ,$$

which was to be shown. ■

From Theorem 3.2, we deduce without further work:

Theorem 3.3.

Let X be $D(k,n)$ distributed, and let Y be the minimum of k iid uniform $\{1, \ldots, n-k+1\}$ random variables. Then X is stochastically greater than Y, that is,

$$P(X>i) \geq P(Y>i) \quad ,\text{all } i .$$

Furthermore, related to the closeness of X and Y is the following collection of inequalities.

Theorem 3.4.

Let X and Y be as in Theorem 3.3. Then

$$\frac{n+1}{k+1} = E(X) \geq E(Y) \geq \frac{n-k+1}{k+1} .$$

In particular,

$$0 \leq E(X)-E(Y) \leq 1 .$$

Proof of Theorem 3.4.

In the proof, we let $U_1, \ldots, U_k$ be iid uniform $[0,1]$ random variables. Note that

$$E(X) = \frac{1}{\binom{n}{k}} \sum_{i=1}^{n-k+1} i\binom{n-i}{k-1} = \frac{\binom{n+1}{k+1}}{\binom{n}{k}} = \frac{n+1}{k+1} .$$

Also,

$$E(Y) \geq (n-k+1)E(\min(U_1, \ldots, U_k)) = \frac{n-k+1}{k+1} .$$

Clearly,

$$E(X)-E(Y) \leq \frac{k}{k+1} . \blacksquare$$

3.3. The inversion method for sequential sampling.

The distribution function F for a $D(k,n)$ random variable X is

$$F(i) = P(X \le i) = 1 - \frac{\binom{n-i}{k}}{\binom{n}{k}} \qquad 0 \le i \le n-k .$$

Thus, if U is a uniform $[0,1]$ random variable, the unique integer X with the property that

$$F(X-1) < U \le F(X)$$

has distribution function F, and is thus $D(k,n)$ distributed. The solution can be obtained sequentially by computing $F(1)$, $F(2)$,... until for the first time U is exceeded. The expected number of iterations is $E(X) = \dfrac{n+1}{k+1}$. The expected time complexity depends upon how F is computed. If $F(i)$ is computed from scratch (Fan, Muller and Rezucha, 1962), then time proportional to $k+1$ is needed, and X is generated in expected time proportional to n. This is unacceptable as it would lead to an $O(nk)$ sampling algorithm. Luckily, we can compute F recursively by noting that

$$\frac{1-F(i+1)}{1-F(i)} = \frac{\binom{n-i-1}{k}}{\binom{n-i}{k}} = \frac{n-i-k}{n-i} .$$

Using this, plus the fact that $1-F(0)=1$, we see that X can be generated in expected time proportional to $\dfrac{n+1}{k+1}$, and that a random sample can thus be generated in expected time proportional to n. This is still rather inefficient. Moreover, the recursive computation of F leads to unacceptable round-off errors for even moderate values of k and n. If F is recomputed from scratch, one must be careful in the handling of ratios of factorials so as not to introduce large cancelation errors in the computations. Thus, help can only come if we take care of the two key stumbling blocks:

1.   The efficient computation of F.

2.   The reduction of the number of iterations in the solution of $F(X-1) < U \le F(X)$.

These issues are dealt with in the next section, where an algorithm of Devroye and Yuen (1981) is given.

3.4. Inversion-with-correction.

A reduction in the number of iterations for solving the inversion inequalities is only possible if we can guess the solution pretty accurately. This is possible thanks to the closeness of X to Y as defined in Theorems 3.3 and 3.4. The random variable Y introduced there has distribution function G where

$$G(i) = P(Y \leq i) = 1 - \left(\frac{n-k+1-i}{n-k+1}\right)^k , \quad 0 \leq i \leq n-k .$$

Recall that $F \leq G$ and that $0 \leq E(X-Y) \leq 1$. By inversion of G, Y can be generated quite simply as

$$Y \leftarrow \left\lfloor (1-(1-U)^{\frac{1}{k}})(n-k+1)+1 \right\rfloor$$

where U is the same uniform $[0,1]$ random variate that will be used in the inversion inequalities for X. Because X is at least equal to Y, it suffices to start looking for a solution by trying $Y, Y+1, Y+2, \ldots$. This, of course, is the principle of inversion-with-correction explained in more detail in section III.2.5. The algorithm can be summarized as follows:

Inversion-with-correction (Devroye and Yuen, 1981)

IF $n = k$

      THEN RETURN $X \leftarrow 1$

      ELSE

            Generate a uniform $[0,1]$ random variate U.

            $X \leftarrow \left\lfloor (1-(1-U)^{\frac{1}{k}})(n-k+1)+1 \right\rfloor$

            $T \leftarrow 1-F(X)$

            WHILE $1-U \leq T$ DO

                  $T \leftarrow T\dfrac{n-k-X}{n-X}$

                  $X \leftarrow X+1$

            RETURN X

The point here is that the expected number of iterations in the WHILE loop is $E(X-Y)$, which is less than or equal to 1. Therefore, the expected time taken by the algorithm is a constant plus the expected time needed to compute F at one point. In the worst possible scenario, F is computed as a ratio of products of integers since

$$1-F(i) = \prod_{j=0}^{k-1} \frac{n-i-j}{n-j} .$$

This takes time proportional to k. The random sampling algorithm would therefore take expected time proportional to k^2. Interestingly, if F can be computed in time $O(1)$, then X can be generated in expected time $O(1)$, and the random sampling algorithm takes expected time $O(k)$. Furthermore, the algorithm requires bounded workspace.

If we accept the logarithm of the gamma function as a function that can be computed in constant time, then F can be computed in time $O(1)$ via:

$$\log(1-F(i)) = \log(\Gamma(n-i+1)) + \log(\Gamma(n-k+1))$$
$$-\log(\Gamma(n-i-k+1)) + \log(\Gamma(n+1)) .$$

Of course, here too we are faced with some cancelation error. In practice, if one wants a certain fixed number of significant digits, there is no problem computing $\log(\Gamma)$ in constant time. From Lemma X.1.3, one can easily check that for $n \geq 8$, the series truncated at $k=3$ gives 7 significant digits. For $n < 8$, the logarithm of n can be computed directly. There are other ways for obtaining a certain accuracy. See for example Hart et al. (1968) for the computation of $\log(\Gamma)$ as a ratio of two polynomials. See also section X.1.3 on the computation of factorials in general.

A final point about cancelation errors in the computation of $1-(1-U)^{1/k}$ when k is large. When E is an exponential random variable, the following two random variables are both distributed as $1-(1-U)^{1/k}$:

$$1-e^{-\frac{E}{k}}$$

$$\frac{\tanh(\frac{E}{2k})}{1+\tanh(\frac{E}{2k})} .$$

The second random variable is to be preferred because it is less susceptible to cancelation error.

3.5. The ghost point method.

Random variables with distribution $D(k,n)$ can also be generated by exploiting special properties such as Theorem 3.2. Recall that X is distributed as

$$1 + \left\lfloor \min((n-k+1)U_1, (n-k+2)U_2, \ldots, (n-k+k)U_k) \right\rfloor$$

where $U_1, \ldots, U_k$ are independent uniform [0,1] random variables. Direct use of this property leads of course to an algorithm taking time $\Theta(k)$. Therefore, the random sampling algorithm corresponding to it would take time proportional to k^2. What distinguishes the algorithm from the inversion algorithms is that no heavy computations are involved. In the ghost point (or ghost sample) method, developed in Devroye and Yuen (1981), the fact that X is almost distributed as the minimum of k iid random variables is exploited. The expected time per random variate is bounded from above uniformly over all $k \leq \rho n$ for some constant $\rho \in (0,1)$. Unfortunately, extra storage proportional to k is needed.

We coined the term "ghost point" because of the following embedding argument, in which X is written as the minimum of k independent random variables, which are linked to k iid random variables provided that we treat some of the iid random variables as non-existent. The iid random variables are $X_1, \ldots, X_k$, each uniformly distributed on $\{1, \ldots, n-k+1\}$. If we were to define X as the minimum of the X_i 's, we would obtain an incorrect result. We can correct however by treating some of the X_i 's as ghost points: define independent Bernoulli random variables $Z_1, \ldots, Z_k$ where $P(Z_i = 1) = \dfrac{i-1}{n-k+i}$. The X_i 's for which $Z_i = 1$ are to be deleted. Thus, we can define an updated collection of random variables, $X_1', \ldots, X_k'$, where

$$X_i' = \begin{cases} X_i & \text{if } Z_i = 0 \\ n-k+1 & \text{if } Z_i = 1 \end{cases}.$$

Theorem 3.5.

For the construction given above,

$$X = \min(X_1', \ldots, X_k')$$

is $D(k,n)$ distributed.

Proof of Theorem 3.5.

Fix $0 \leq i \leq n-k$. Then,

$$P(X > i) = \prod_{j=1}^{k} P(X_j' > i)$$

$$= \prod_{j=1}^{k} (P(Z_i = 1) + P(Z_i = 0)P(X_i > k))$$

$$= \prod_{j=1}^{k} (\frac{j-1}{n-k+i} + \frac{n-k+1}{n-k+j} \cdot \frac{n-k+1-i}{n-k+1})$$

$$= \prod_{j=1}^{k} \frac{n-k+j-i}{n-k+j}$$

$$= \frac{\dbinom{n-i}{k}}{\dbinom{n}{k}} . \blacksquare$$

Every X_i has an equal probability of being the smallest. Thus, we can keep generating uniformly random integers from $1, \ldots, k$, without replacement of course, until we find one for which $Z_i = 0$, i.e. until we find an index for which the X_i is not a ghost point. Assume that we have skipped over m ghost points in the process. Then the X_i in question is distributed as the $m + 1$-st smallest of the original sequence $X_1, \ldots, X_k$. The point is that such a random variable can be generated in expected time $O(1)$ because beta random variates can be generated in $O(1)$ expected time. Before proceeding with the expected time analysis, we give the algorithm:

The ghost point method

[SET-UP]

An auxiliary linked list L is needed, which is initially empty. The maximum list size is k. The stack size is Size.

Size $\leftarrow 0$.

[GENERATION]

REPEAT

    REPEAT

        Generate an integer W uniformly distributed on $\{1, \ldots, k\}$.

    UNTIL W is not in L

    Add W to L, Size $\leftarrow$ Size $+1$.

    Generate a uniform [0,1] random variate U.

UNTIL $U \geq \dfrac{W-1}{n-k+W}$

Generate a beta (Size,k–Size+1) random variable B (note that B is distributed as the "Size" smallest of k iid uniform [0,1] random variables.)

RETURN $X \leftarrow \lfloor 1 + B(n-k+1) \rfloor$

We refer to the section on beta random variate generation for uniformly fast generators. If a beta variate generator is not locally available, one can always generate B as $\dfrac{G}{G+G'}$ where G, G' are independent gamma (W) and gamma $(k-W+1)$ random variables respectively.

For the analysis, we assume that $k \leq \rho n$ where $\rho \in (0,1)$ is a constant. Let N denote the number of W random variates generated in the inner REPEAT loop. It will appropriately measure the complexity of the algorithm provided that we can check membership in list L in constant time.

Theorem 3.6.

For the ghost point algorithm, we have

$$E(N) \le c\frac{1+\rho}{(1-\rho)^2}$$

where $c > 0$ is a universal constant and $k \le \rho n$ where $\rho \in (0,1)$. Furthermore, the expected length of the list L, i.e. the expected value of Size, does not exceed $\frac{1}{1-\rho}$.

Proof of Theorem 3.6.

If T is the eventual value of Size, then

$$E(N \mid T) = \sum_{i=1}^{T} \frac{k}{k-i+1} .$$

Therefore, for constant $a \in (0,1)$,

$$E(N) = E(\sum_{i=1}^{T} \frac{k}{k-i+1}) = \sum_{i=1}^{k} \frac{k}{k-i+1} P(T \ge i) \quad \text{(by a change of } \int \text{)}$$

$$\le E(T^2) \sum_{i=1}^{k} \frac{k}{i^2(k-i+1)}$$

$$\le E(T^2)(\frac{k}{k - \lfloor ak \rfloor +1} \sum_{i=1}^{\infty} \frac{1}{i^2} + k \sum_{i > \lfloor ak \rfloor} \frac{1}{i^2})$$

$$\le E(T^2)(\frac{\pi^2}{6(1-a)} + k(\frac{1}{(ak)^2} + \int_{ak}^{\infty} \frac{1}{x^2} \, dx))$$

$$= E(T^2)(\frac{\pi^2}{6(1-a)} + \frac{1}{ka^2} + \frac{1}{a})$$

which is approximately minimal when

$$a = \frac{\sqrt{6}}{\pi + \sqrt{6}} .$$

The upper bound is thus not greater than a constant times $E(T^2)$. But T is stochastically smaller than a geometric random variable with probability of success $\frac{n-k+1}{n} \ge 1-\rho$. Thus, $E(T) \le 1/(1-\rho)$ and

$$E(T^2) \le (\frac{1}{1-\rho})^2 + \frac{\rho}{(1-\rho)^2} = \frac{1+\rho}{(1-\rho)^2} . \blacksquare$$

The value of the constant c can be deduced from the proof. However, no attempt was made to obtain the best possible constant there. The assumption that membership checking in L can be done in constant time requires that a bit vector of k flags be used, indicating for each integer whether it is included in L or not. Setting up the bit vector takes time proportional to k. However, this cost is to be born just once, for after one variate X is generated, the flags can be reset by emptying the list L. The expected time taken by the reset operation is thus equal to a constant plus the expected length of the list, which, as we have shown in Theorem 6, is bounded by $1/(1-\rho)$. For the global random sampling algorithm, the total expected cost of setting and resetting the bit vector does not exceed a constant times k.

Fortunately, we can avoid the bit vector of flags altogether. Membership checking in list L can always be done in time not exceeding the length of the list. Even with this grotesquely inefficient implementation, one can show (see exercises) that the expected time for generating X is bounded uniformly over all $k \leq \rho n$.

The issue of membership checking can be sidestepped if we generate integers without replacement by the swapping method. This would require an additional vector initially set to $1, \ldots, k$. After X is generated, this vector is slightly permuted - its first "Size" members for example constitute our list L. This does not matter, as long as we keep track of where integer k is. To get ready for generating a $D(k-1,n)$ random variate, we need only swap k with the last element of the vector, so that the first $k-1$ components form a permutation of $1, \ldots, k-1$. Thus, fixing the vector between random variates takes a constant time. Note also that to generate X, the expected time is now bounded by a constant times the expected length of the list, which we know is not greater than $1/(1-\rho)$. This is due to the fact that the inner loop of the algorithm is now replaced by one loopless section of code.

When $k > \rho n$, one should use another algorithm, such as the following piece taken from the standard sequential sampling algorithm:

$X \leftarrow 0$

REPEAT

      Generate a uniform random variate U.

    $X \leftarrow X + 1$

UNTIL $U \leq \dfrac{k}{n-X+1}$

RETURN X

The expected number of uniform $[0,1]$ random variates needed by this algorithm is $E(X) = \dfrac{n+1}{k+1} \leq \dfrac{n}{k} \leq \dfrac{1}{\rho}$. The combination of the two algorithms depending upon the relative sizes of k and n yields an $O(1)$ expected time algorithm for generating X. The optimal value of the threshold ρ will vary from implementation to implementation. Note that if a membership swap vector is used, it is best to reset the vector after each X is generated by traversing the list in LIFO order.

3.6. The rejection method.

The generation of $D(k,n)$ random variates by the rejection method creates special problems, because the probabilities p_i contain ratios of factorials. Whenever we evaluate p_i, we can use one of two approaches: p_i is evaluated in constant time (this, in fact, assumes that the logarithm of the Γ function is available in constant time, and that we do give up our infinite accuracy because a Stirling series approximation is used), and p_i is computed in time proportional to $k+1$ (i.e. the factorials are evaluated explicitly). With the latter model, called the explicit factorial model, it does not suffice to find a dominating probability vector q_i which satisfies

$$p_i \leq c q_i$$

for some constant c independent of k,n. We could indeed still end up with an expected time complexity that is not uniformly bounded over k,n. Thus, in the explicit factorial model, we have to find good dominating and squeeze curves which will allow us to effectively avoid computing p_i except perhaps about $O(\frac{1}{k})$ percent of the time. Because $D(k,n)$ is a two-parameter family, the design is quite a challenge. We will not be concerned with all the details here, just with the flavor of the problem. The detailed development can be found in Vitter (1984). Nearly all of this section is an adaptation of Vitter's results. Gehrke (1984) and Kawarasaki and Sibuya (1982) have also developed rejection algorithms, similar to the ones discussed in this section.

At the very heart of the design is once again a collection of inequalities. Recall that for a $D(k,n)$ random variable X,

$$p_i = P(X=i) = \frac{\binom{n-i}{k-1}}{\binom{n}{k}} \qquad (1 \leq i \leq n-k+1) .$$

Theorem 3.7.

We have

$$h_1(i) \leq p_i \leq c_1 g_1(i+1)$$

where

$$h_1(i) = \frac{k}{n}\left(1 - \frac{i-1}{n-k+1}\right)^{k-1} \qquad (1 \leq i \leq n-k+1) ,$$

$$c_1 = \frac{n}{n-k+1} ,$$

$$g_1(x) = \frac{k}{n}\left(1 - \frac{x-1}{n}\right)^{k-1} \qquad (1 \leq x \leq n+1) .$$

Also,

$$h_2(i) \leq p_i \leq c_2 g_2(i+1)$$

where

$$h_2(i) = \frac{k}{n}\left(1 - \frac{k-1}{n-i+1}\right)^{i-1} \qquad (1 \leq i \leq n-k+1) ,$$

$$c_2 = \frac{k}{k-1}\frac{n-1}{n} ,$$

$$g_2(i) = \frac{k-1}{n-1}\left(1 - \frac{k-1}{n-1}\right)^{i-1} \qquad (i \geq 1) .$$

Note that g_1 is a density in x, and that g_2 is a probability vector in i.

Proof of Theorem 3.7.

Note that

$$p_i = \frac{k}{n-k+1}\prod_{j=0}^{k-2}\frac{n-i-j}{n-j}$$

$$\leq \frac{k}{n-k+1}\left(\frac{n-i}{n}\right)^{k-1}$$

$$= \frac{k}{n-k+1}(1 - \frac{i}{n})^{k-1}$$

$$= c_1 g_1(i+1) .$$

Furthermore,

$$h_1(i) = \frac{k}{n}\left(1 - \frac{i-1}{n-k+1}\right)^{k-1}$$

$$\leq \frac{k}{n}\prod_{j=0}^{k-2}\frac{n-k-i+2+j}{n-k+1+j}$$

$$= \frac{k}{n} \prod_{j=0}^{k-2} \frac{n-i-j}{n-1-j}$$

$$= p_i \ .$$

This concludes the first half of the proof. For the second half, we argue similarly. Indeed, for $i \geq 1$,

$$p_i = \frac{k}{n} \prod_{j=0}^{i-2} \frac{n-k-j}{n-1-j}$$

$$\leq \frac{k}{n} \left(\frac{n-k}{n-1} \right)^{i-1}$$

$$= \frac{k}{k-1} \frac{n-1}{n} \frac{k-1}{n-1} \left(1 - \frac{k-1}{n-1} \right)^{i-1}$$

$$= c_2 g_2(i) \ .$$

Furthermore,

$$h_2(i) = \frac{k}{n} \left(\frac{n-k-i+2}{n-i+1} \right)^{i-1} \leq \frac{k}{n} \prod_{j=0}^{i-2} \frac{n-k-j}{n-1-j} = p_i \ . \blacksquare$$

Random variate generators based upon both groups of inequalities are now easy to find, because g_1 is basically a transformed beta density, and g_2 is a geometric probability vector. In the case of g_1, we need to use rejection from a continuous density of course. The expected number of iterations in case 1 is $c_1 = n/(n-k+1)$ (which is uniformly bounded over all k,n with $k \leq \rho n$, where $\rho \in (0,1)$ is a constant). In case 2, we have $c_2 = \frac{k}{k-1} \frac{n-1}{n}$, and this is uniformly bounded over all $k \geq 2$ and all $n \geq 1$.

First rejection algorithm

REPEAT

    Generate two iid uniform [0,1] random variates U, V.

    $Y \leftarrow 1 + n\,(1 - U^{\frac{1}{k}})$ (Y has density g_1)

    $X \leftarrow \lfloor Y \rfloor$

    IF $X \leq n - k + 1$

        THEN

$$\text{Accept} \leftarrow [V \leq \frac{n-k+1}{n} \left(\frac{1 - \dfrac{X-1}{n-k+1}}{1 - \dfrac{Y-1}{n}} \right)^{k-1}]$$

        IF NOT Accept THEN

$$\text{Accept} \leftarrow [V \leq \frac{p_X}{c_1 g_1(Y)}]$$

UNTIL Accept

RETURN X

Second rejection algorithm

REPEAT

    Generate an exponential random variate E and a uniform [0,1] random variate V.

    $X \leftarrow \left\lceil -E / \log(1 - \dfrac{k-1}{n-1}) \right\rceil$ (X has probability vector g_2)

    IF $X \leq n - k + 1$

        THEN

$$\text{Accept} \leftarrow [V \leq \left(\frac{1 - \dfrac{k-1}{n-X+1}}{1 - \dfrac{k-1}{n-1}} \right)^{X-1}]$$

        IF NOT Accept THEN

$$\text{Accept} \leftarrow [V \leq \frac{p_X}{c_2 g_2(X)}]$$

UNTIL Accept

RETURN X

3.7. Exercises.

1. Assume that in the standard sequential sampling algorithm, each element is chosen with equal probability $\frac{k}{n}$. The sample size is a binomial $(n,\frac{k}{n})$ random variable N. Show that as $k\rightarrow\infty, n\rightarrow\infty, n-k\rightarrow\infty$, we have

$$P(N=k)\sim\sqrt{\frac{n}{2\pi k(n-k)}}\ .$$

2. Assume that $k\leq\rho n$ for some fixed $\rho\in(0,1)$. Show that if the ghost point algorithm is used to generate a random sample of size k out of n, the expected time is bounded by a function of ρ only. Assume that a vector of membership flags is used in the algorithm, but do not switch to the standard sequential method when during the generation process, the current value of k temporarily exceeds ρ times the current value of n (as is suggested in the text).

3. Assume that in the ghost point algorithm, membership checking is done by traversing the list L. Show that to generate a random variate X with distribution $D(k,n)$, the algorithm takes expected time bounded by a function of $\frac{k}{n}$ only.

4. If X is $D(k,n)$ distributed, then

$$Var(X)=\frac{(n+1)(n-k)k}{(k+2)(k+1)^2}\ .$$

5. Consider the explicit factorial model in the rejection algorithm. Noting that the value of p_X can be computed in time $\min(k,X+1)$, find good upper bounds for the expected time complexity of the two rejection algorithms given in the text. In particular, prove that for the first algorithm, the expected time complexity is uniformly bounded over $k\leq\rho n$ where $\rho\in(0,1)$ is a constant (Vitter, 1984).

4. OVERSAMPLING.

4.1. Definition.

If we are given a random sequence of k uniform order statistics, and transform it via truncation into a random sequence of ordered integers in $\{1,\ldots,n\}$, then we are almost done. Unfortunately, some integers could appear more than once, and it is necessary to generate a few more observations. If we had started with $k_1>k$ uniform order statistics, then with some luck we could have ended up with at least k different integers. The probability of this increases rapidly with k_1. On the other hand, we do not want to take k_1 too large, because then we will be left with quite a bit of work trying to eliminate some values to obtain a sample of precisely size k. This method is called oversampling. The main issue at stake is the choice of k_1 as a function of k and n so that not only the total expected time is $O(k)$, but the total expected time is approximately minimal. One additional feature that makes oversampling attractive is that we will obtain an ordered random sample. Because the method is basically a two step method (uniform sample generator, followed by excess eliminator), it is not included in the section on sequential methods.

The oversampling algorithm

REPEAT

> Generate $U_{(1)} < \cdots < U_{(k_1)}$, the order statistics of a uniform sample of size k_1 on $[0,1]$.
>
> Determine $X_i \leftarrow \left\lfloor 1 + nU_{(i)} \right\rfloor$ for all i, and construct, after elimination of duplicates, the ordered array $X_{(1)}, \ldots, X_{(K_1)}$.

UNTIL $K_1 \geq k$

Mark a random sample of size $K_1 - k$ of the sequence $X_{(1)}, \ldots, X_{(K_1)}$ by the standard sequential sampling algorithm.

RETURN the sequence of k unmarked X_i's.

The amount of extra storage needed is $K_1 - k$. Note that this is always bounded by $k_1 - k$. For the expected time analysis of the algorithm, we observe that the uniform sample generation takes expected time $c_u k_1$, and that the elimination step takes expected time $c_e K_1$. Here c_u and c_e are positive constants. If the standard sequential sampling algorithm is replaced by classical sampling for elimination (i.e., to mark one integer, generate random integers on $\{1, \ldots, K_1\}$ until a nonmarked integer is found), then the expected time taken by the elimination algorithm is

$$c_e \sum_{i=1}^{K_1-k} \frac{K_1}{K_1 - i + 1}$$

$$\leq c_e (K_1 - k) \frac{K_1}{k+1} \ .$$

What we should also count in the expected time complexity is the probability of accepting a sequence. The results are combined in the following theorem:

Theorem 4.1.

Let c_u, c_e be as defined above. Assume that $n > k$ and that

$$k_1 = k + (k+a)/\log(\frac{n}{k})$$

for some constant $a > 0$. Then the expected time spent on the uniform sample is

$$E(N)c_u k_1$$

where $E(N)$ is the expected number of iterations. We have the following inequality:

$$E(N) = \frac{1}{P(K_1 \geq k)} \leq \frac{1}{1-e^{-a}} \ .$$

The expected time spent marking does not exceed $c_e k_1$, which, when $a = O(k)$, $\frac{k}{n} \to 0$, is asymptotic to $c_e k$. If classical sampling is used for marking, then it is not greater than

$$\frac{k_1}{k+1} \frac{k+a}{\log(\frac{n}{k})} \ .$$

Proof of Theorem 4.1.

The expression for the expected time spent generating order statistics is based upon Wald's equation. Furthermore, $E(N) = 1/P(K_1 \geq k)$. But

$$P(K_1 < k) \leq \binom{n}{k}\left\{\frac{k}{n}\right\}^{k_1} \leq \left\{\frac{en}{k}\right\}^{k_1}$$

$$= \left\{\frac{n}{k}\right\}^{k-k_1} e^k$$

$$= e^{-a} \ .$$

The only other statement in the theorem requiring some explanation is the statement about the marking scheme with classical sampling. The expected time spent doing so does not exceed c_e times

$$E\left((K_1 - k)\frac{K_1}{k+1} \mid K_1 \geq k\right)$$

$$\leq \frac{(k_1 - k)k_1}{k+1} \ . \blacksquare$$

Once again, we see that uniformly over $k \leq \rho n$, the expected time is bounded by a constant times k, for all fixed $\rho \in (0,1)$ and for all choices of a that are either fixed or vary with k in such a manner that $a = O(k)$. We recommend that a be taken large but fixed, say $a = 10$. Note that in the special case that $\frac{n}{k} \to \infty$, $a = O(k)$, $k_1 \sim k$. Thus, the expected time of the marking section based upon classical sampling is $o(k)$, i.e. it is asymptotically negligible. Also, if $a \to \infty$, $E(N) \to 1$ for all choices of n, k. In those cases, the main contributions to the expected time complexity come from the generation of the k_1 uniform order statistics, and the elimination of the marked values (not the marking itself).

4.2. Exercises.

1. Show that for the choice of k_1 given in Theorem 4.1, we have $E(N) \to 1$ as $n, k \to \infty$, $\frac{k}{n} \to \rho \in (0,1)$. Do this by proving the existence of a universal constant A depending upon ρ only such that $E(N) \leq 1 + \frac{A}{\sqrt{n}}$.

5. RESERVOIR SAMPLING

5.1. Definition.

There is one particular sequential sampling problem deserving special attention, namely the problem of sampling records from large (presumably external) files with an unknown total population. While k is known, n is not. Knuth (1969) gives a particularly elegant solution for drawing such a random sample called the reservoir method. See also Vitter (1985). Imagine that we associate with each of the records an independent uniform $[0,1]$ random variable U_i. If the object is simply to draw a random set of size k, it suffices to pick those k records that correspond to the k largest values of the U_i's. This can be done sequentially:

Reservoir sampling

[NOTE: S is a set of pairs (i, U_i).]

FOR $i := 1$ TO k DO

      Generate a uniform [0,1] random variate U_i, and add (i, U_i) to S. Keep track of the pair (m, U_m) with the smallest value for the uniform random variate.

$i \leftarrow k + 1$ (i is a record counter)

WHILE NOT end of file DO

      Generate a uniform [0,1] random variate U_i.

      IF $U_i \geq U_m$

          THEN

              Delete (m, U_m) from S.

              Insert (i, U_i) in S.

              Find a new smallest pair (m, U_m).

      $i \leftarrow i + 1$

RETURN all integers i for which $(i, U_i) \in S$.

The general algorithm of reservoir sampling given above returns integers (indices); it is trivial to modify the algorithm so that actual records are returned. It is clear that n uniform random variates are needed. In addition, there is a cost for updating S. The expected number of deletions in S (which is equal to the number of insertions minus k) is

$$\sum_{i=k+1}^{n} P((i, U_i) \text{ is inserted in } S)$$

$$= \sum_{i=k+1}^{n} \frac{k}{i}$$

$$= k \log(\frac{n}{k}) + o(1)$$

as $k \rightarrow \infty$. Here we used the fact that the first n terms of the harmonic series are $\log(n) + \gamma + o(1/n)$ where γ is Euler's constant. There are several possible implementations for the set S. Because we are mainly interested in ordinary insertions and deletions of the minimum, the obvious choice should be a heap. Both the expected and worst-case times for a delete operation in a heap of size k are proportional to $\log(k)$ as $k \rightarrow \infty$. The overall expected time complexity for deletions is proportional to

$$k \log(\frac{n}{k}) \log(k)$$

as $k \rightarrow \infty$. This may or may not be larger than the $\theta(n)$ contribution from the uniform random variate generator. With ordered or unordered linked lists, the time complexity is worse. In the exercise section, a hash structure exploiting the fact that the inserted elements are uniformly distributed is explored.

5.2. The reservoir method with geometric jumps.

In some applications, such as when records are stored on a sequential access device (e.g., a magnetic tape), there is no way that we can avoid traversing the entire file. When the records are in RAM or on a random access device, it is possible to skip over any number of records in constant time: in those cases, it should be possible to get rid of the $\theta(n)$ term in the time complexity. Given (m, U_m), we know that the waiting time until the occurrence of a uniform value greater than U_m is geometrically distributed with success probability $1-U_m$. It can be generated as $\left\lceil -E/\log(U_m) \right\rceil$ where E is an exponential random variate. The corresponding record-breaking value is uniformly distributed on $[U_m, 1]$. Thus, the reservoir method with geometric jumps can be summarized as follows:

Reservoir sampling with geometric jumps

[NOTE: S is a set of pairs (i, U_i).]

FOR $i := 1$ TO k DO

    Generate a uniform $[0,1]$ random variate U_i, and add (i, U_i) to S. Keep track of the pair (m, U_m) with the smallest value for the uniform random variate.

$i \leftarrow k$ (i is a record counter)

WHILE True DO

    Generate an exponential random variate E.

    $i \leftarrow i + \left\lceil -E/\log(U_m) \right\rceil$.

    IF i not outside file

        THEN

            Generate a uniform $[U_m, 1]$ random variate U_i.

            Delete (m, U_m) from S.

            Insert (i, U_i) in S.

            Find a new smallest pair (m, U_m).

        ELSE RETURN all integers i for which $(i, U_i) \in S$.

The analysis of the previous section about the expected time spent updating S remains valid here. The difference is that the $\theta(n)$ has disappeared from the picture, because we only generate uniform random variates when insertions in S are needed.

5.3. Exercises.

1. Design a bucket-based dynamic data structure for the set S, which yields a total expected time complexity for N insertions and deletions that is $o(N\log(k))$ when $N, k \rightarrow \infty$. Note that inserted elements are uniformly distributed on $[U_m, 1]$ where U_m is the minimal value present in the set. Initially, S contains k iid uniform $[0,1]$ random variates. For the heap implementation of S, the expected time complexity would be $\theta(N\log(k))$.

Chapter Thirteen
RANDOM COMBINATORIAL OBJECTS

1. GENERAL PRINCIPLES.

1.1. Introduction.

Some applications demand that random combinatorial objects be generated: by definition, a combinatorial object is an object that can be put into one-to-one correspondence with a finite set of integers. The main difference with discrete random variate generation is that the one-to-one mapping is usually complicated, so that it may not be very efficient to generate a random integer and then determine the object by using the one-to-one mapping. Another characteristic is the size of the problem: typically, the number of different objects is phenomenally large. A final distinguishing feature is that most users are interested in the uniform distribution over the set of objects.

In this chapter, we will discuss general strategies for generating random combinatorial objects, with the understanding that only uniform distributions are considered. Then, in different subsections, particular combinatorial objects are studied. These include random graphs, random free trees, random binary trees, random search trees, random partitions, random subsets and random permutations. This is a representative sample of the simplest and most frequently used combinatorial objects. It is hoped that for more complicated objects, the readers will be able to extrapolate from our examples. A good reference text is Nijenhuis and Wilf(1978).

1.2. The decoding method.

Since we want to generate only one of a finite number of objects, it is possible to find a function f such that for every pair of objects (ξ,ς) in the collection of objects Ξ, we have

$$f(\xi) \neq f(\varsigma) \in \{1, \ldots, n\} \,,$$

where n is an integer, which is usually equal to $|\Xi|$, the number of elements in Ξ. Such a function will be called a coding function. By $f^{-1}(i)$, we define the object ξ in Ξ for which $f(\xi) = i$ (if this object exists). When $|\Xi| = n$, the following decoding algorithm is valid.

The decoding method

[NOTE: f is a coding function.]
Generate a uniform random integer $X \in \{1, \ldots, n\}$.
RETURN $f^{-1}(X)$

The expected time taken by this algorithm is the average time needed for decoding f :

$$\frac{1}{n} \sum_{i=1}^{n} \text{TIME}(f^{-1}(i)) \,.$$

The advantage of the method is that only one uniform random variate is needed per random combinatorial object. The decoding method is optimal from a storage point of view, since each combinatorial object corresponds uniquely to an integer in $1, \ldots, n$. Thus, about $\log_2 n$ bits are needed to store each combinatorial object, and this cannot be improved upon. Thus, the coding functions can be used to store data in compact form. The disadvantages usually outweigh the advantages:

1. Except in the simplest cases, $|\Xi|$ is too large to be practical. For example, if this method is to be used to generate a random permutation of $1, \ldots, 40$, we have $|\Xi| = 40!$, so that multiple precision arithmetic is necessary. Recall that $12! < 2^{35} < 13!$.

2. The expected time taken by the decoding algorithm is often unacceptable. Note that the time taken by the uniform random variate generator is negligible compared to the time needed for decoding.

3. The method can only be used when for the given value of n, we are able to count the number of objects. This is not always the case. However, if we use rejection (see below), the counting problem can be avoided.

Example 1.1. Random permutations.

Assume that $\Xi = \{$ all permutations of $1, \ldots, n$ $\}$. There are a number of possible coding functions. For example, we could use the factorial representation of Lehmer (1964), where a permutation $\sigma_1, \ldots, \sigma_n$ is uniquely described by a sequence of $n-1$ integers $a_1, \ldots, a_{n-1}$ (where $0 \le a_i \le n-i$) according to the following rule: start with $1, \ldots, n$. Let σ_1 be the a_1+1-st integer from this list, and delete this number. Let σ_2 be the a_2+1-st number of the remaining numbers, and so forth. Then, define

$$f(a_1, \ldots, a_{n-1}) = a_1(n-1)! + a_2(n-2)! + \cdots + a_{n-1}1! + 1$$

It is easy to see that f is a proper coding function giving all values between 1 and n!. Just observe that

$$n! = (n-1)!n = (n-1)!(n-1) + (n-1)!$$
$$= (n-1)!(n-1) + (n-2)!(n-2) + \cdots + 1!1 + 1 .$$

The algorithm consists of generating a random integer X between 1 and n!, determining $a_1, \ldots, a_{n-1}$ from X, and determining the random permutation $\sigma_1, \ldots, \sigma_n$ from the a_i sequence. First, the a_i's are obtained by repeated divisions by $(n-1)!, (n-2)!$, etcetera. The σ_i's can be obtained by an exchange algorithm. Formally, we have:

Random permutation generator

Generate a random integer X uniformly distributed on $\{1, \ldots, n!\}$. $X \leftarrow X - 1$.
FOR $i := 1$ TO $n-1$ DO

$$(a_i, X) \leftarrow (\left\lfloor \frac{X}{(n-i)!} \right\rfloor, X \bmod (n-i)!) \text{ (This determines all the } a_i \text{'s.)}$$

Set $\sigma_1, \ldots, \sigma_n \leftarrow 1, \ldots, n$.
FOR $i := 1$ TO $n-1$ DO
    Exchange (swap) σ_{a_i+1} and σ_{n-i+1}
RETURN $\sigma_1, \ldots, \sigma_n$.

In the exchange step of the algorithm, we exchange a randomly picked element with the last element in every iteration. The time taken by the algorithm is $O(n)$. ∎

Sometimes simple coding functions can be found with the property that $n > |\Xi|$, that is, some of the integers in $1, \ldots, n$ do not correspond to any combinatorial object in Ξ. When n is not much larger than $|\Xi|$, this is not a big problem, because we can apply the rejection principle:

Decoding with rejection

REPEAT

      Generate a random integer X with a uniform distribution on $\{1, \ldots, n\}$.

      Accept $\leftarrow [f(\xi)=X$ for some $\xi \in \Xi]$

UNTIL Accept

RETURN $f^{-1}(X)$

Just how one determines quickly whether $f(\xi)=X$ for some $\xi \in \Xi$ depends upon the circumstances. Usually, because of the size of $|\Xi|$, it is impossible or not practical to store a vector of flags, flagging the bad values of X. If $|\Xi|$ is moderately small, then one could consider doing this in a preprocessing step. Most of the time, it is necessary to start decoding X, until in the process of decoding one discovers that there is no combinatorial object for the given value of X. In any case, the expected number of iterations is $\dfrac{n}{|\Xi|}$. What we have bought here is (i) simplicity (the decoding function can be simpler if we allow gaps in our enumeration) and (ii) convenience (it is not necessary to count $|\Xi|$; in fact, this value need not be known at all !).

1.3. Generation based upon recurrences.

Most combinatorial objects can be counted indirectly via recurrence relations. Direct counting, as in the case of random permutations, addresses itself to the decoding method. Counting via recurrences can be used to obtain alternative generators. The idea has been around for some time. It was first developed thoroughly by Wilf (1977) (see also Nijenhuis and Wilf (1978)).

We need to have two things:

1. A formula for the number of combinatorial objects with a certain parameter (or parameters) k in terms of the number of combinatorial objects with smaller parameter(s). This will be called our recurrence relation.

2. A good understanding of the recurrence relation, so that the relation itself can be linked in a constructive way to a combinatorial object.

For example, consider Ξ_n, the collection of permutations of $1, \ldots, n$. We have

$$|\Xi_n| = n |\Xi_{n-1}| .$$

The meaning of this relation is clear: we can obtain $\xi \in \Xi_n$ by considering all permutations Ξ_{n-1}, padding them with the single element n (in the last position), and then swapping the n-th element with one of the n elements. The swapping operation gives us the factor n in the recurrence relation. We will rewrite the recurrence relation as follows:

$$\Xi_n(1,2,\ldots,n)=\bigcup_{i=1}^{n}\Xi_{n-1}(1,2,\ldots,i-1,i+1,\ldots,n).i$$

where $\Xi_{n-1}(1,2,\ldots,i-1,i+1,\ldots,n)$ is the collection of all permutations of the given $n-1$ elements, and . is the concatenation operator. To generate a random element from Ξ_n, it suffices to choose a random term in the union (with probability proportional to the cardinality of the chosen term), and to construct the part of the combinatorial object that corresponds to this choice. In the case of the random permutations, each of the n terms in the union shown in the recurrence relation has equal cardinality, and should thus be chosen with equal probability. But choosing the i-th term corresponds to putting the i-th element of the n-vector at the end of the permutation, and generating a random permutation for the $n-1$ remaining elements. This leads quite naturally to the swapping method for random permutations:

The swapping method for random permutations

Set $\sigma_1,\ldots,\sigma_n\leftarrow 1,\ldots,n$.
FOR $i:=n$ DOWNTO 2 DO
      Generate X uniformly in $1,\ldots,i$.
      Swap σ_X and σ_i.
RETURN $\sigma_1,\ldots,\sigma_n$.

There are obviously more complicated situations: see for example the subsections on random partitions and random binary trees in the corresponding subsections. For now, we will merely apply the technique to the generation of random subsets of size k out of n elements, and see that it reduces to the sequential method in random sampling.

There are

$$\binom{n}{k}=\binom{n-1}{k}+\binom{n-1}{k-1}$$

sets of size $k\geq 1$ consisting of different integers picked from $\{1,\ldots,n\}$, where $n\geq k$. Clearly, as boundary conditions, we have

$$\binom{n}{n}=1\ ;\ \binom{n}{1}=1\ .$$

The recurrence can be interpreted as follows: k integers can be drawn from $2,\ldots,n$ (thus, ignoring 1), or by choosing 1 and choosing a random subset of size $k-1$ from $2,\ldots,n$ (thus, including 1). The probability of inclusion of 1 is therefore

$$\frac{\binom{n-1}{k-1}}{\binom{n}{k}} = \frac{k}{n} .$$

This leads directly to the following algorithm:

Random subset of size k from 1,...,n

$S \leftarrow \emptyset$ (set to be returned is empty)
FOR $i := 1$ TO n DO
    Generate a uniform [0,1] random variate U.
    IF $U \le \dfrac{k}{n-i+1}$ THEN $S \leftarrow S \cup \{i\}$; $k \leftarrow k-1$
RETURN S

We can also look at the method of recurrences as some sort of composition method. Typically, Ξ_n is split into a number of subsets of objects, each having a special property. Let us write

$$\Xi_n = \bigcup_{i=1}^{k} \Xi_n(i)$$

where the sets $\Xi_n(i)$ are non-overlapping. If an integer i is picked with probability

$$\frac{|\Xi_n(i)|}{|\Xi_n|} \quad (1 \le i \le k) ,$$

and if we generate a uniformly distributed object in $\Xi_n(i)$, then the random object is uniformly distributed over Ξ_n. Of course, we are allowed to apply the same decomposition principle to the individual subsets in turn. The subsets have generally speaking some property which allows us to construct part of the solution, as was illustrated with random permutations and random subsets.

2. RANDOM PERMUTATIONS.

2.1. Simple generators.

The decoding method of section XIII.1.2 requires only one uniform random variate per random permutation of $1, \ldots, n$. It was suggested in a number of papers (see e.g. Robinson (1967), Jansson (1966), de Balbine (1967), and the survey paper by Plackett (1968)). Given an arbitrary array of length n, and one uniformly distributed random integer on $1, \ldots, n!$, the decoding method constructs in time $O(n)$ one random permutation of $1, \ldots, n$. The algorithm of section XIII.1.2 is a two-pass algorithm. Robson (1969) has pointed out that there is a simple one-pass algorithm based upon decoding:

Robson's decoding algorithm

[NOTE: This algorithm assumes that some permutation $\sigma_1, \ldots, \sigma_n$ of $1, \ldots, n$ is given. Usually, this permutation is a previously generated random permutation.]

Generate a random integer X uniformly on $1, \ldots, n!$.

FOR $i := n$ DOWNTO 2 DO

$$(X, Z) \leftarrow (\left\lfloor \frac{X}{i} \right\rfloor, X \bmod i + 1)$$

    Swap σ_i and σ_Z

RETURN $\sigma_1, \ldots, \sigma_n$

Despite the obvious improvement over the algorithm of section XIII.1.2, the decoding method remains of limited value because $n!$ increases too quickly with n.

The exchange method of section XIII.1.3 on the other hand does not have this drawback. It is usually attributed to Moses and Oakford (1963) and to Durstenfeld (1964). The method requires $n-1$ independent uniform random variates per random permutation, but it is extremely simple in conception, requiring only one pass and no multiplications, divisions or truncations.

2.2. Random binary search trees.

Random permutations are useful in a number of applications. As we have pointed out earlier, the swapping method can be stopped after a given number of iterations to yield a method for generating a random subset of $1, \ldots, n$ of size $k < n$. This was dealt with in chapter XII on random sampling. Another application deals with the generation of a random binary search tree.

A random binary search tree with n nodes is defined as a binary search tree constructed from a random permutation, where each permutation is equally likely. It is easy to see that different permutations can yield a tree of the same shape, so all trees are not equally likely (but the permutations are !). It is clear that if we proceed by inserting the elements of a random permutation in turn, starting from an empty tree, then the expected time of the algorithm can be measured by

$$\sum_{i=1}^{n} E(D_i)$$

where D_i is the depth (path length from root to node) of the i-th node when inserted into a binary search tree of size $i-1$ (the depth of the root is 0). The following result is well-known, but is included here because of its short unorthodox proof, based upon the theory of records (see Glick (1978) for a recent survey):

Lemma 2.1. In a random binary search tree,
$$E(D_n) \leq 2(\log(n)+1) .$$

In fact $E(D_n) \sim 2 \log(n)$. Based upon Lemma 2.1, it is not difficult to see that the expected time for the generator is $O(n \log(n))$. Since $E(D_n) \sim 2 \log(n)$, the expected time is also $\Omega(n \log(n))$.

Proof of Lemma 2.1.

D_n is equal to the number of left turns plus the number of right turns on the path from the root to the node corresponding to the n-th element. By symmetry, $E(D_n)$ is twice the expected number of right turns. These right turns can conveniently be counted as follows. Consider the random permutation of $1, \ldots, n$, and extract the subsequence of all elements smaller than the last element. In this subsequence (of length at most $n-1$), flag the records, i.e. the largest values seen thus far. Note that the first element always represents a record. The second element is a record with probability one half, and the i-th element is a record with probability $1/i$. Each record corresponds to a right turn and vice versa. This can be seen by noting that elements following a record which are not records themselves are in a left subtree of a node on the path to the record, whereas the n-th original element is in the right subtree. Thus, these elements cannot have any influence on the level of the n-th element. The subsequence has length between 0 and $n-1$, and to bound the expected number of records from above, it suffices to consider subsequences of length equal to $n-1$. Therefore, the expected depth of the last node is not more than

$$2\sum_{i=1}^{n-1}\frac{1}{i} \leq 2(1+\int_1^n \frac{1}{x} \, dx) = 2(1+\log(n)) . \blacksquare$$

But just as with the problem of the generation of an ordered random sample, there is an important short-cut, which allows us to generate the random binary search tree in linear expected time. The important fact here is that if the root is fixed (say, its integer value is i), then the left subtree has cardinality $i-1$, and the right subtree has cardinality $n-i$. Furthermore, the value of the root itself is uniformly distributed on $1, \ldots, n$. These properties allow us to use recursion in the generation of the random binary search tree. Since there are n nodes, we need no more than n uniform random variates, so that the total expected time is $O(n)$. A rough outline follows:

Linear expected time algorithm for generating a random binary search tree with n nodes

[NOTE: The binary search tree consists of cells, having a data field "Data", and two pointer fields, "Left" and "Right". The algorithm needs a stack S for temporary storage.]

MAKENULL (S) (stack S is initially empty).

Grab an unused cell pointed to by pointer p .

PUSH $[p,1,n]$ onto S .

WHILE NOT EMPTY (S) DO

    POP S , yielding the triple $[p,l,r]$.

    Generate a random integer X uniformly distributed on $l, \ldots, r$.

    $p \uparrow$.Data$\leftarrow X$, $p \uparrow$.Left$\leftarrow$NIL, $p \uparrow$.Right$\leftarrow$NIL

    IF $X < r$ THEN

        Grab an unused cell pointed to by $q*$.

        $p \uparrow$.Right$\leftarrow q*$ (make link with right subtree)

        PUSH $[q*,X+1,r]$ onto stack S (remember for later)

    IF $X > l$ THEN

        Grab an unused cell pointed to by q .

        $p \uparrow$.Left$\leftarrow q$ (make link with left subtree)

        PUSH $[q,l,X-1]$ onto stack S (remember for later)

2.3. Exercises.

1.   Consider the following putative swapping method for generating a random permutation:

Start with an arbitrary permutation $\sigma_1, \ldots, \sigma_n$ of $1, \ldots, n$.

FOR $i := 1$ TO n DO

    Generate a random integer X on $1, \ldots, n$ (note that the range does not depend upon i).

    Swap σ_i and σ_X

RETURN $\sigma_1, \ldots, \sigma_n$

Show that this algorithm does not yield a valid random permutation (all permutations are not equally likely). Hint: there is a three line combinatorial proof(de Balbine, 1967).

2. The distribution of the height H_n of a random binary search tree is very complicated. To simulate H_n, we can always generate a random binary search tree and find H_n. This can be done in expected time $O(n)$ as we have seen. Find an algorithm for the generation of H_n in sublinear expected time. The closer to constant expected time, the better.

3. Show why Robson's decoding algorithm is valid.

4. Show that for a random binary search tree, $E(D_n) \sim 2 \log(n)$ by employing the analogy with records explained in the proof of Lemma 2.1.

5. Give a linear expected time algorithm for constructing a random trie with n elements. Recall that a trie is a binary tree in which left edges correspond to zeroes and right edges correspond to ones. The n elements can be considered as n independent infinite sequences of zeroes and ones, where all zeroes and ones are obtained by perfect coin tosses. This yields an infinite tree in which there are precisely n paths, one for each element. The trie defined by these elements is obtained by truncating all these paths to the point that any further truncation would lead to two identical paths. Thus, all internal nodes which are fathers of leaves have two children.

6. **Random heap.** Give a linear expected time algorithm for generating a random heap with elements $1, \ldots, n$ so that each heap is equally likely. Hint: associate with integer i the i-th order statistic of a uniform sample of size n, and argue in terms of order statistics.

3. RANDOM BINARY TREES.

3.1. Representations of binary trees.

A binary tree consists of a root, or a root and a left and/or a right subtree, and each of the subtrees in turn is a binary tree. Two binary trees are similar if they have the same shape. They are equivalent if they are similar, and if the corresponding nodes contain the same information. The distinction between similarity and equivalence is thus based upon the absence or presence of labels for the nodes. If there are n nodes, then every permutation of the labels of the nodes yields another labeled binary tree, and all such trees are similar.

A random binary tree with n nodes is a random unlabeled binary tree which is uniformly distributed over all nonsimilar binary trees with n nodes. The uniform distribution on the n nodes causes some problems, as we can see from the following simple example: there are 5 different binary trees with 3 nodes. Yet, if we generate such a tree either by generating a random permutation of 1,2,3 and constructing a binary search tree from this permutation, or by growing the tree via uniform replacements of NIL pointers by new nodes, then the resulting trees are not equally likely. For example, the complete binary tree with 3 nodes has probability $\frac{1}{3}$ in both schemes, instead of $\frac{1}{5}$ as is required. The uniformity condition will roughly speaking stretch the binary trees out, make them appear more unbalanced, because less likely shapes (under standard models) become equally likely.

In this section, we look at some handy representations of binary trees which can be useful further on.

Theorem 3.1.

Let $p_1, p_2, \ldots, p_{2n}$ be a balanced sequence of parentheses, i.e. each p_i belongs to $\{(,)\}$, for every partial sequence $p_1, p_2, \ldots, p_i$, the number of opening parentheses is at least equal to the number of closing parentheses, and in the entire sequence, there are an equal number of opening and closing parentheses.

Then there exists a one-to-one correspondence between all such balanced sequences of $2n$ parentheses and all binary trees with n nodes.

Proof of Theorem 3.1.

We will prove this constructively. Consider an inorder traversal of a binary tree, i.e. a traversal whereby each node is visited after its left subtree has been visited, but before its right subtree is visited. In the traversal, a stack S is used. Initially the root is pushed onto the stack. Then, a move to the left down the tree corresponds to another push. If there is no left subtree, we pop the stack and go the the right subtree if there is one (this requires yet another push). If there is no right subtree either, then we pop again, and so forth until we try to pop an empty stack. The algorithm is as follows:

Inorder stack traversal of a binary tree

[NOTE: The binary tree consists of n cells, each having a left and a right pointer field. S is a stack, and $p_1, \ldots, p_{2n}$ is the sequence of pushes (opening parentheses) and pops (closing parentheses) to be returned.]

$p \leftarrow$ root of tree (p is a pointer)

$i \leftarrow 2$ (i is a counter)

MAKENULL (S)

PUSH p onto S ; $p_1 \leftarrow ($

REPEAT

    IF $p \uparrow .\text{Left} \neq \text{NIL}$

        THEN PUSH $p \uparrow .\text{Left}$ onto S ; $p \leftarrow p \uparrow .\text{Left}$; $p_i \leftarrow ($; $i \leftarrow i + 1$

        ELSE

           REPEAT

               POP S , yielding p ; $p_i \leftarrow)$; $i \leftarrow i + 1$

           UNTIL $i > 2n$ OR $p \uparrow .\text{Right} \neq \text{NIL}$

           IF $i \leq 2n$

               THEN PUSH $p \uparrow .\text{Right}$ onto S ; $p \leftarrow p \uparrow .\text{Right}$; $p_i \leftarrow ($; $i \leftarrow i + 1$

UNTIL $i > 2n$

RETURN $p_1, \ldots, p_{2n}$

Different sequences of pushes and pops correspond to different binary trees. Also, every partial sequence of pushes and pops is such that the number of pushes is at least equal to the number of pops. Upon exit from the algorithm, both numbers are of course equal. Thus, if a push is identified with an opening parenthesis, and a pop with a closing parenthesis, then the equivalence claimed in the theorem is obvious. ∎

For example, the sequence ()()()()() $\cdots$ () corresponds to a binary tree in which all nodes have only right subtrees. And the sequence ((((($\cdots$))))) corresponds to a binary tree in which all nodes have only left subtrees. The representation of a binary tree in terms of a balanced sequence of parentheses comes in very handy. There are other representations that can be derived from Theorem 3.1.

Theorem 3.2.
There is a one-to-one correspondence between a balanced sequence of $2n$ parentheses and a random walk of length $2n$ which starts at the origin and returns to the origin without ever crossing the zero axis.

Proof of Theorem 3.2.

Let every opening parenthesis correspond to a step of size "+1" in the random walk, and let every closing parenthesis correspond to a step of size "-1" in the random walk. Obviously, such a random walk returns to the origin if the string of parentheses is balanced. Also, it does not take any negative values. ■

Theorem 3.2 can be used to obtain a short proof for counting the number of different (i.e., nonsimilar) binary trees with n nodes.

Theorem 3.3.
There are $$\frac{1}{n+1}\binom{2n}{n}$$ different binary trees with n nodes.

Proof of Theorem 3.3.

The proof uses the celebrated mirror principle (Feller, 1965). Consider a random walk starting at $(2k,0)$ ($2k \geq 0$ is the initial value; 0 is the initial time): in one time unit, the value of the random walk either increases by 1 or decreases by 1. The number of paths ending up at $(0,2n)$ which take at least one negative value is equal to the number of unrestricted paths from $(2k,0)$ to $(-2,2n)$. This can most easily be seen by the following argument: there is a one-to-one correspondence between the given restricted and unrestricted paths. Note that each restricted path must take the value -1 at some point in time. Let t be the first time that this happens. From the restricted path to $(0,2n)$, construct an unrestricted path to $(-2,2n)$ as follows: keep the initial segment up to time t, and flip the tail segment between time t and time $2n$ around, so that the path ends up at $(-2,2n)$. Each different restricted path yields a different unrestricted path. Vice versa, since the unrestricted paths must all cross the horizontal line at -1, time t is well defined, and each unrestricted path corresponds to a restricted path.

The number of paths from $(2k,0)$ to $(0,2n)$ which do not cross the zero axis equals the total number of unrestricted paths minus the number of paths that do cross the zero axis, i.e.

$$\binom{2n}{k+n} - \binom{2n}{k+n+1},$$

which is easily seen by using a small argument involving numbers of possible sub-sets. In particular, if we set $k=0$, we see that the total number of binary trees (or the total number of nonnegative paths from $(0,0)$ to $(0,2n)$) is

$$\binom{2n}{n} - \binom{2n}{n+1} = \frac{1}{n+1}\binom{2n}{n}. \quad \blacksquare$$

The number of binary trees with n nodes grows very quickly with n (see table below).

n	Number of binary trees with n nodes
1	1
2	2
3	5
4	14
5	42
6	132
7	429
8	3430

One can show (see exercises) that this number $\sim 4^n/(\sqrt{\pi}n^{3/2})$. Because of this, the decoding method seems once again impractical except perhaps for n smaller than 15, because of the wordsize of the integers involved in the computations.

3.2. Generation by rejection.

Random binary trees or random strings of balanced parentheses can be generated by the rejection method. This could be done for example by generating a random permutation of n opening parentheses and n closing parentheses, and accepting only if the resulting string satisfies the property that all partial sub-strings have at least as many opening parentheses as closing parentheses. There are

$$\binom{2n}{n}$$

initial strings, all equally likely. By Theorem 3.3, the probability of acceptance of a string is thus $\frac{1}{n+1}$. Furthermore, to decide whether a string has the said property takes expected time proportional to n. Thus, the expected time taken by the algorithm varies as n^2. For this reason, the rejection method is not recommended.

3.3. Generation by sequential sampling.

It is possible to generate a random binary tree with n nodes in time $O(n)$ by first generating a random string of balanced parentheses of length $2n$ in time $O(n)$ and then reconstructing the tree by mimicking the inorder traversal given in the proof of Theorem 3.1. The string can be generated in one pass, from left to right, similar to the sequential sampling method for generating a random subset. It is perhaps best to consider the analogy with random walks once again. We start at $(0,0)$, and have to end up at $(0,2n)$. At each point, say (k,t), we decide to generate a (with probability equal to the ratio of the number of nonnegative paths from $(k+1,t+1)$ to $(0,2n)$ to the number of nonnegative paths from (k,t) to $(0,2n)$. We generate a) otherwise. It is clear that this method uses a recurrence relation for binary trees, but the explanation given here in terms of random walks is perhaps more insightful. The number of nonnegative paths from (k,t) to $(0,2n)$ is (see the proof of Theorem 3.3):

$$\binom{2n-t}{\frac{k+2n-t}{2}} - \binom{2n-t}{\frac{k+2+2n-t}{2}} = \binom{2n-t}{\frac{k+2n-t}{2}} \frac{2k+2}{2n-t+k+2} .$$

The probability of a (at (k,t) is thus

$$\frac{\binom{2n-t-1}{\frac{k+2n-t}{2}} \frac{2k+4}{2n-t+k+2}}{\binom{2n-t}{\frac{k+2n-t}{2}} \frac{2k+2}{2n-t+k+2}} = \frac{k+2}{k+1} \frac{2n-t-k}{2(2n-t)} .$$

The resulting algorithm for generating a random string of balanced parentheses is due to Arnold and Sleep (1980):

Sequential method for generating a random string of balanced parentheses

[NOTE: The string generated by us is returned in $p_1, \ldots, p_{2n}$.]

$X \leftarrow 0$ (X holds the current "value" of the corresponding random walk.)

FOR $t := 0$ TO $2n - 1$ DO

    Generate a uniform $[0,1]$ random variate U.

    IF $U \leq \dfrac{X+2}{X+1} \dfrac{2n-t-X}{2(2n-t)}$

        THEN $X \leftarrow X+1$, $p_{t+1} \leftarrow ($

        ELSE $X \leftarrow X-1$, $p_{t+1} \leftarrow)$

RETURN $p_1, \ldots, p_{2n}$

It is relatively straightforward to check that the random walk cannot take negative values because when $X = 0$, the probability of generating (in the algorithm is 1. It is also not possible to overshoot the origin at time $2n$ because whenever $X = 2n - t$, the probability that a (is generated is 0.

The reconstruction in linear time of a binary tree from a string of balanced parentheses is left as an exercise to the reader. Basically, one should mimic the algorithm of Theorem 3.1 where such a string is constructed given a binary tree.

3.4. The decoding method.

There are a number of sophisticated coding functions for binary trees, which can be decoded in linear time, but all of them require extra storage space for auxiliary constants. See e.g. Knott (1977), Ruskey (1978), Ruskey and Hu (1977) and Trojanowski (1978). See also Tinhofer and Schreck (1984).

3.5. Exercises.

1. Show that the number of binary trees with n nodes $\sim \dfrac{4^n}{\sqrt{\pi n}^{\frac{3}{2}}}$.

2. Consider an arbitrary (unrestricted) random walk from $(0,0)$ to $(0,2n)$ (this can be generated by generating a random permutation of n 1's and n -1's). Define another random walk by taking the absolute value of the unrestricted random walk. This random walk does not take negative values, and corresponds therefore to a string of balanced parentheses of length $2n$. Show that the random strings obtained in this manner are not uniformly

distributed.

3.   Give a linear time algorithm for reconstructing a binary tree from a string of balanced parentheses of length $2n$ using the correspondence established in Theorem 3.1.

4.   **Random rooted trees.** A rooted tree with n vertices consists of a root and an ordered collection of nonempty rooted subtrees when $n > 1$. When $n = 1$, it consists of just a root. The vertices are unlabeled. Thus, there are 5 different rooted trees when $n = 4$. There are a number of representations of rooted trees, such as:

   A.   A vector of degrees: write down for each node the number of children (nonempty subtrees) when the tree is traversed in preorder or level order.

   B.   A vector of levels: traverse the tree in preorder or postorder and write down the level number of each node when it is visited.

   We can call these vectors of length n codewords. There are other more storage-efficient codewords: find a codeword of length $2n$ consisting of bits only, which uniquely represents a rooted tree. Show that all codewords for representing rooted trees or binary trees must take at least $(2 + o(1))n$ bits of storage. Generating a codeword is equivalent to generating a rooted tree. Pick any codeword you like, and give a linear time algorithm for generating a valid random codeword such that all codewords are equally likely to be generated. Hint: notice the connection between rooted trees and binary trees.

5.   Let us grow a tree by replacing on a sequential basis all NIL pointers by new nodes, where the choice of a NIL pointer is uniform over the set of such pointers (see section 3.1). Note that there are $n + 1$ NIL pointers if the tree has n nodes. Let us generate another tree by generating a random permutation and constructing a binary search tree. Are the two trees similar in distribution, i.e. is it true that for each shape of a tree with n nodes, and for all n, the probability of a tree with that shape is the same under both schemes ? Prove or disprove.

6.   Find a coding function for binary trees which can be decoded in time $O(n)$.

4. RANDOM PARTITIONS.

4.1. Recurrences and codewords.

   Many problems can be related to the generation of random partitions of $\{1, \ldots, n\}$ into k nonempty subsets. We know that there are $\left\{ {n \atop k} \right\}$ such partitions, where $\{.\}$ denotes the Stirling number of the second kind. Rather than give a formula for the Stirling numbers in terms of a series, we will employ the recursive definition:

$$\begin{Bmatrix} n \\ k \end{Bmatrix} = k \begin{Bmatrix} n-1 \\ k \end{Bmatrix} + \begin{Bmatrix} n-1 \\ k-1 \end{Bmatrix} \quad (0 < k < n),$$

$$\begin{Bmatrix} n \\ 1 \end{Bmatrix} = 1 \; ; \; \begin{Bmatrix} n \\ n \end{Bmatrix} = 1 \, .$$

Using this, we can form a table of Stirling numbers, just as we can form a table (Pascal's triangle) from the well-known recursion for binomial numbers. We have:

$n =$	1	2	3	4	5	6
$k =$						
1	1	1	1	1	1	1
2		1	3	7	15	31
3			1	6	25	90
4				1	10	65
5					1	15
6						1

The recursion has a physical meaning: we can form a partition into k nonempty subsets by considering a partition of $\{1, \ldots, n-1\}$ and adding one number, n. That number n can be considered as a new singleton set in the partition (this explains the contribution

$$\begin{Bmatrix} n-1 \\ k-1 \end{Bmatrix}$$

in the recursion). It can also be added to one of the sets in the partition of $\{1, \ldots, n-1\}$. In this case, we can add it to one of the k sets in the latter partition. To have a unique way of addressing these sets, we order the sets according to the value of their smallest elements, and label the sets $1,2,3, \ldots, k$. The addition of n to set i implies that we must include

$$\begin{Bmatrix} n-1 \\ k \end{Bmatrix}$$

in the recursion.

Before we proceed with the generation of a random partition based upon this recursion, it is perhaps useful to describe one kind of codeword for random partitions. Consider the case $n = 5$ and $k = 3$. Then, the partition $(1,2,5),(3),(4)$ can be represented by the n-tuple 11231 where each integer in the n-tuple represents the set to which each element belongs. By convention, the sets are ordered according to the values of their smallest elements. So it is easy to see that different codewords yield different partitions, and vice versa, that all n-tuples of integers from $\{1, \ldots, k\}$ (such that each integer is used at least once) having this ordering property correspond to some partition into k nonempty subsets. Thus, generating random codewords or random partitions is equivalent. Also, one can be constructed from the other in time $O(n)$.

4.2. Generation of random partitions.

The generator described below produces a random codeword, uniformly distributed over the collection of all possible codewords. It is based upon the recursion explained above. To add n to a partition of $\{1, \ldots, n-1\}$, we should define a singleton set $\{n\}$ (in which case it must have set number k) with probability

$$\frac{\left\{ {n-1 \atop k-1} \right\}}{\left\{ {n \atop k} \right\}}$$

and add it to a randomly picked set from among $1, \ldots, k$ with probability

$$\frac{\left\{ {n-1 \atop k} \right\}}{\left\{ {n \atop k} \right\}}$$

each. Obviously, we have to generate the random codeword backwards.

Random partition generator based upon recurrence relation for Stirling numbers

[NOTE: n and k are given and will be destroyed.]
REPEAT
    Generate a uniform [0,1] random variate U.

$$\text{IF } U \le \frac{\left\{ {n-1 \atop k-1} \right\}}{\left\{ {n \atop k} \right\}}$$

        THEN $X_n \leftarrow k$, $k \leftarrow k-1$
        ELSE Generate X_n uniformly on $1, \ldots, k$
    $n \leftarrow n-1$
UNTIL $n=0$
RETURN the codeword $X_1, X_2, \ldots, X_n$

If the Stirling numbers can be computed in time $O(1)$ (for example, if they are stored in a two-dimensional table), then the algorithm takes time $O(n)$ per codeword. The storage requirements are proportional to nk. The preprocessing time needed to set up the table of size n by k is also proportional to nk if we use the fundamental recursion.

We conclude this section by noting that the algorithm given above is a slightly modified version of an algorithm given in Wilf (1977) and Nijenhuis and Wilf (1978).

4.3. Exercises.

1. Define a coding function for random partitions, and find an $O(n)$ decoding algorithm.

2. **Random partitions of integers.** Let $p(n,k)$ be the number of partitions of an integer n such that the largest part is k. The following recurrence holds:

$$p(n,k) = p(n-1,k-1) + p(n-k,k) .$$

The first term on the right-hand side represents those partitions of n whose largest part is k and whose second largest part is less than k (because such partitions can be obtained from one of $n-1$ whose largest part is $k-1$ by adding 1 to the largest part). The partitions of n whose largest two parts are both k come from partitions of $n-k$ of largest part k by replicating the largest part. Arguing as in Wilf (1977), a partition is a series of decisions "add 1 to the largest part" or "adjoin another copy of the largest part".

A. Give an algorithm for the generation of such a random partition (all partitions should be equally likely of course), based upon the given recurrence relation.

B. Find a coding function for these partitions. Hint: base your function on the parts of the partition given in descending order.

C. How would you generate an unrestricted partition of n ? Here, unrestricted means that no bound is given for the largest part in the partition.

D. Find a recurrence relation similar to the one given above for the number of partitions of n with parts less than or equal to k.

E. For the combinatorial objects of part D, find a coding function and a decoding algorithm for generating a random object. See also McKay (1965).

5. RANDOM FREE TREES.

5.1. Prufer's construction.

A free tree is a connected graph with no cycles. If there are n nodes, then there are $n-1$ edges. The distinction between labeled and unlabeled free trees is important. Note however that unlike other trees, free trees do not have a given root. All nodes are treated equally. We will however keep using the term leaf for nodes with degree one.

The generation of a random free tree can be based upon the following theorem:

Theorem 5.1.

Cayley's theorem. There are exactly n^{n-2} labeled free trees with n nodes.

Prufer's construction. There exists a one-to-one correspondence between all $(n-2)$-tuples ("codewords") of integers $a_1, \ldots, a_{n-2}$, each taking values in $\{1, \ldots, n\}$, and all labeled free trees with n nodes. The relationship is given in the proof below.

Proof of Theorem 5.1.

Cayley's theorem follows from Prufer's construction. Let the nodes of the labeled free tree have labels $1, \ldots, n$. From a labeled free tree a codeword can be constructed as follows. Let a_1 be the label of the neighbor of the leaf with the smallest label. Delete the corresponding edge. Since one of the endpoints of the edge is a leaf, removal of the edge will leave us with a labeled free tree of size $n-1$. Repeat this process until $n-2$ components of the codeword have been calculated. At the end, we have a labeled free tree with just 2 nodes, which can be discarded. For example, for the labeled free tree with 6 nodes and edges (1,2), (2,3), (4,3), (5,3), (6,3), the codeword (2,3,3,3) is obtained.

Conversely, from each codeword, we can construct a free tree having the property that if we use the construction given above, the initial codeword is obtained again. This is all that is needed to establish the one-to-one correspondence. For the construction of the tree from a given codeword, we begin with three lists:

A. The codeword: $a_1, \ldots, a_{n-2}$.

B. A list of n flags: $f_1, \ldots, f_n$, where $f_i = 1$ indicates that node i is available. Initially, all flags are 1. Flag i is set to 0 only when i is a leaf, and the edge connected to i is suddenly removed from the tree.

C. A list of n flags indicating whether a node is a leaf or not: $l_1, \ldots, l_n$. $l_i = 1$ indicates that node i is a leaf. Note that this list is redundant, since a node is a leaf if and only if its label can be found in the codeword. The initialization of this list of flags is simple.

The construction proceeds by first recreating the $n-2$ edges that correspond to the $n-2$ components of the codeword. This is done simply as follows: choose node a_1 (this is not a leaf, since it is in the codeword), and choose the smallest leaf v (flag $l_v = 1$ and availability flag $f_v = 1$). Return the edge (a_1, v), and set the flag of v to 0, which effectively eliminates v. If a_1 cannot be found in the remainder of the codeword, then a_1 becomes a leaf in the new free tree, and the flag l_{a_1} must be set to 1. This process can be repeated until $a_1, \ldots, a_{n-2}$ is exhausted. The last $(n-1\text{-st})$ edge at the end is simply found by taking the only two nodes whose availability flags are still 1. This concludes the construction. It is easy to verify that if the tree is used to construct a codeword, the initial codeword is obtained. ∎

The degree of a node is one plus the number of occurrences of the node in the codeword, at least if codewords are translated into free trees via Prufer's construction. To generate a random labeled free tree with n nodes (such that all such trees are equally likely), one can proceed as follows:

Random labeled free tree generator

FOR $i := 1$ TO $n-2$ DO
    Generate a_i uniformly on $\{1, \ldots, n\}$.
Translate the codeword into a labeled free tree via Prufer's construction.

A careless translation of the codeword could be inefficient. For example, the verification of whether an internal node becomes a leaf during construction, when done by traversing the leftover part of the codeword, yields an $\Omega(n^2)$ contribution to the total time. Using linear search to find the smallest available leaf would give a contribution of $\Omega(n^2)$ to the total time. Even if a heap were used for this, we would still be facing a contribution of $\Omega(n \log(n))$ to the total time. In the next section, a linear time translation algorithm due to Klingsberg (1977) is presented.

5.2. Klingsberg's algorithm.

The purpose of this section is to explain Klingsberg's $O(n)$ algorithm for translating a codeword $a_1, \ldots, a_{n-2}$ into a labeled free tree. His solution requires one additional array $T[1], \ldots, T[n]$, which is used to return the edges and to keep information about the availability flags and about the leaf flags (see proof of Theorem 1). The edges returned are

$$(1, T[1]), (2, T[2]), \ldots, (n-1, T[n-1]).$$

The other uses of this array are:

A. $T[i]$=available_not_leaf means that node i is still available and is not a leaf. The constant is set to -1 in Klingsberg's work.

B. $T[i]$=available_leaf means that node i is an available leaf. The constant is set to 0 in Klingsberg's work.

C. $T[i]=j>0$ indicates that node i is no longer available, and in fact that (i,j) is an edge of the labeled free tree.

In the example of codeword (2,3,3,3) given in section 5.1, the array T would initially be set to (available_leaf , available_not_leaf , available_not_leaf , available_leaf , available_leaf , available_leaf) since only nodes 2 and 3 are internal nodes.

To speed up the determination of when an internal node becomes a leaf, we merely flag the last occurrence of every node in the codeword. This can be conveniently be done by changing the signs of these entries. In our example, the codeword would initially be replaced by $(-2,3,3,-3)$.

Finally, to find the smallest available leaf quickly, we note that in the construction, these leaf labels increase except when a new leaf is added, and its label is smaller than the current smallest leaf label. This can be managed with the aid of two moving pointers: there is a masterpointer which moves up monotonically from 1 to n; in addition, there is a temporary pointer, which usually moves with the masterpointer, except in the situation described above, when it is temporarily set to a value smaller than that of the masterpointer. The temporary pointer always points at the smallest available leaf. It is this ingenious device which permitted Klingsberg to obtain an $O(n)$ algorithm. We can now summarize his algorithm.

Klingsberg's algorithm for constructing a labeled free tree from a codeword

[PREPARATION.]

FOR $i := 1$ TO n DO $T[i] \leftarrow$ available_leaf

FOR $i := n-2$ DOWNTO 1 DO

    IF $T[a_i] =$ available_leaf THEN

        $T[a_i] =$ available_not_leaf; $a_i \leftarrow -a_i$

Master $\leftarrow 1$

$a_{n-1} \leftarrow n$ (for convenience in defining last edge)

Master $\leftarrow \min(j : T[j] =$ available_leaf)

Temp $\leftarrow$ Master

[TRANSLATION.]

FOR $i := 1$ TO $n-1$ DO

    Select $\leftarrow |a_i|$ (select internal node)

    $T[\text{Temp}] \leftarrow$ Select (return edge)

    IF $i < n-1$ THEN

        IF $a_i > 0$

            THEN

                Master $\leftarrow \min(j : T[j] =$ available_leaf)

                Temp $\leftarrow$ Master

            ELSE

                $T[\text{Select}] \leftarrow$ available_leaf

                IF Select $\leq$ Master THEN Temp $\leftarrow$ Select   (temporary step up)

RETURN $(1, T[1]), \ldots, (n-1, T[n-1])$

The linearity of the algorithm follows from the fact that the masterpointer can only increase, and that all the operations in every iteration that do not involve the masterpointer are constant time operations.

5.3. Free trees with a given number of leaves.

Assume next that we wish to generate a labeled free tree with n nodes and l leaves where $2 \leq l \leq n-1$. For the solution of this problem, we recall Prufer's codeword. The codeword contains the labels of all internal nodes. Thus, it is necessary to generate only codewords in which precisely $n-l$ labels are present. The actual labels can be put in by selecting $n-l$ labels from n labels by one of the random sampling algorithms. Thus, we have:

Generator of a labeled free tree with n nodes and l leaves

Generate a random subset of $n-l$ labels from $1, \ldots, n$.

Perform a random permutation on these labels (this may not be necessary, depending upon the random subset algorithm.)

Generate a random partition of $n-2$ elements into $n-l$ non-empty subsets, and assign the first label to the first subset, etcetera. This yields a codeword of length $n-2$ with precisely $n-l$ different labels.

Translate the codeword into a labeled free tree (preferably using Klingsberg's algorithm).

In this algorithm, we need algorithms for random subsets, random partitions and random permutations. It goes without saying that some of these algorithms can be combined. Another by-product of the decomposition of the problem into manageable sub-problems is that it is easy to count the number of combinatorial objects. We obtain, in this example:

$$\binom{n}{l}(n-l)!\left\{\begin{matrix} n-2 \\ n-l \end{matrix}\right\} = \frac{n!}{l!}\left\{\begin{matrix} n-2 \\ n-l \end{matrix}\right\} \;.$$

5.4. Exercises.

1. Let $d_1, \ldots, d_n$ be the degrees of the nodes $1, \ldots, n$ in a free tree. (Note that the sum of the degrees is $2n-2$.) How would you generate such a free tree ? Hint: generate a random Prufer codeword with the correct number of occurrences of all labels. The answer is extremely simple. Derive also a simple formula for the number of such labeled free trees.

2. Give an algorithm for computing the Prufer codeword for a labeled free tree with n nodes in time $O(n)$.

3. Prove that the number of free trees that can be built with n labeled edges (but unlabeled nodes) is $(n+1)^{n-2}$. Hint: count the number of free trees with n labeled nodes and $n-1$ labeled edges first.

4. Give an $O(n)$ algorithm for the generation of a random free tree with n labeled edges and $n+1$ unlabeled nodes. Hint: try to use Klingsberg's algorithm by reducing the problem to one of generating a labeled free tree.

5. **Random unlabeled free trees with n vertices.** Find the connection between unlabeled free trees with n vertices and rooted trees with n vertices. Exploit the connection to generate random unlabeled free trees such that all trees are equally likely (Wilf, 1981).

6. RANDOM GRAPHS.

6.1. Random graphs with simple properties.

Graphs are the most general combinatorial objects dealt with in this chapter. They have applications in nearly all fields of science and engineering. It is quite impossible to give a thorough overview of the different subclasses of graphs, and how objects in these subclasses can be generated uniformly and at random. Instead, we will just give a superficial treatment, and refer the reader to general principles or specific articles in the literature whenever necessary.

We will use the notation n for the number of nodes in a graph, and e for the number of edges in a graph. A random graph with a certain property P is such that all graphs with this property are equally likely to occur. Perhaps the simplest property is the property: "Graph G has n nodes". We know that there are

$$2^{\binom{n}{2}}$$

objects with this property. This can easily be seen by considering that each of the $\binom{n}{2}$ possible edges can either be present or absent. Thus, we should include each edge in a random graph with this property with probability $1/2$.

The number of edges chosen is binomially distributed with parameters n and $1/2$. It is often necessary to generate sparser graphs, where roughly speaking e is $O(n)$ (or at least not $\Omega(n^2)$). This can be done in two ways. If we do not require a specific number of edges, then the simplest solution is to select all edges independently and with probability p. Note that the expected number of edges is $p\binom{n}{2}$. This is most easily implemented, especially for small p, by using the fact that the waiting time between two selected edges is geometrically distributed with parameter p, where by "waiting time" we mean the number of edges we must manipulate before we see another selected edge. This requires a linear ordering on the edges, which can be done by the coding function given below.

If the property is "Graph G has n nodes and e edges", then we should first select a random subset of e edges for the set of $\binom{n}{2}$ possible edges. This property is simple to deal with. The only slight problem is that of establishing a simple coding function for the edges, which is easy to decode. This is needed since we have to access the endpoints of the edges some of the time (e.g., when returning edges), and the coded edges most of the time (e.g., when random sampling based upon hashing). One possibility is shown below:

Node u	Node v	Coded version of edge (u,v)
1	2	1
1	3	2
$\cdots$	$\cdots$	$\cdots$
$\cdots$	$\cdots$	$\cdots$
1	n	$n-1$
2	3	$(n-1)+1$
$\cdots$	$\cdots$	$\cdots$
$\cdots$	$\cdots$	$\cdots$
2	n	$(n-1)+(n-2)$
$\cdots$	$\cdots$	$\cdots$
$\cdots$	$\cdots$	$\cdots$
$n-1$	n	$(n-1)+(n-2)+\cdots+2+1$

The coding function for this scheme is

$$f(u,v) = (u-1)n - \frac{u(u-1)}{2} + (v-u) .$$

Interestingly, this function can be decoded in time $O(1)$ (see exercise 6.1). Whether random sampling should be done on coded integers with decoding only at the very end, or on sets of edges (u,v) without any decoding, depends upon the sampling scheme. In classical sampling schemes for example, it is necessary to verify whether a certain edge has already been selected. The verification can be based upon a vector of flags (which can be done here by using a lower triangular n by n matrix of flags). When a heap or a tree structure is used, there is no need ever for coding. When hashing is used, coding seems appropriate. In sequential sampling, no coding is needed, as long as we can easily implement the function NEXT(u,v)   (IF $v=n$ THEN   NEXT(u,v)$\leftarrow(u+1,u+2)$ ELSE NEXT(u,v)$\leftarrow(u,v+1)$). However, if sequential sampling is accelerated by taking giant steps, then coding the edges seems the wise thing to do.

6.2. Connected graphs.

Most random graphs that people want to generate should be of the connected type. From the work of Erdos and Renyi (1959, 1960), we know that if e is much larger than $\frac{1}{2}n\log(n)$ (or if p is much larger than $\frac{\log(n)}{n}$), then the probability that a random graph with e (or binomial (n,p)) edges is connected tends to 1 as $n \to \infty$. In those situations, it is clear that we could use the rejection algorithm:

Rejection method for generating a connected random graph with n nodes and e edges

REPEAT

   Generate a random graph G with e edges and n nodes.

UNTIL G is connected

RETURN G

To verify that a graph is connected is a standard operation: if we use depth first search, this can be done in time $O\left(\max(n,e)\right)$ (Aho, Hopcroft and Ullman, 1983). Thus, the expected time needed by the algorithm is $O\left(\max(n,e)\right)$ when

$$\lim_{n\to\infty}\inf\frac{e}{n\log(n)} > \frac{1}{2}\ .$$

In fact, since in those cases the probability of acceptance tends to 1 as $n\to\infty$, the expected time taken by the algorithm is $(1+o(1))$ times the expected time needed to check for connectedness and to generate a random graph with e edges. Unfortunately, the condition given above is asymptotic, and it is difficult to verify whether for given values of e and n, we have a good rejection constant. Also, there is a gap for precisely the most interesting sorts of graphs, the very sparse graphs when e is of the order of n. This can be done via a general graph generation technique of Tinhofer's (1978,1980), which is explained in the next section. In it, we recognize ingredients of Wilf's recurrence based method.

6.3. Tinhofer's graph generators.

In two publications, Tinhofer (1978,1980) has proposed useful random graph generators, with applications to connected graphs (with or without a specific number of edges), digraphs, bichromatic graphs, and acyclic connected graphs. His algorithms require in all cases that we can count certain subclasses of graphs, and they run fastest if tables of these counts can be set up beforehand. We will merely give the general outline, and refer to Tinhofer's work for the details.

Let us represent a graph by a sequence of adjacency lists, with the property that each edge should appear in only one adjacency list. The adjacency list for node i will be denoted by A_i. Thus, the graph is completely determined by the sequence

$$A_1 A_2 \cdots A_n\ .$$

We will generate the adjacency lists in some (usually random) order, $A_{v_1}, A_{v_2}, \ldots$, where $v_1, v_2, \ldots, v_n$ is a permutation of $1, \ldots, n$. To avoid the duplication of nodes, we require that all nodes in adjacency list A_{v_j} fall outside $\bigcup_{i=1}^{j} \{v_i\}$. The following sets of nodes will be needed:

1. The set U_j of all nodes in $\bigcup_{i=1}^{j} A_{v_i}$ with label not in $v_1, \ldots, v_j$. This set contains all neighbors of the first j nodes outside $v_1, \ldots, v_j$.

2. The set V_j which consists of all nodes with label outside $v_1, \ldots, v_j$ that are not in U_j.

3. The special sets $U_0 = \{1\}$, $V_0 = \{2,3, \ldots, n\}$.

When the adjacency lists are being generated, it is also necessary to do some counting: define the quantity N_j as the total number of graphs with the desired property, having fixed adjacency lists $A_{v_1}, \ldots, A_{v_j}$. Sometimes we will write $N_j(A_{v_1}, \ldots, A_{v_j})$ to make the dependence explicit. Given $A_{v_1}, \ldots, A_{v_{j-1}}$, we should of course generate A_{v_j} according to the following distribution:

$$ P(A_{v_j} = A) = \frac{N_j(A_{v_1}, \ldots, A_{v_{j-1}}, A)}{N_{j-1}(A_{v_1}, \ldots, A_{v_{j-1}})} . $$

It is easy to see that this is indeed a probability vector in A. We are now ready to give Tinhofer's general algorithm.

Tinhofer's random graph generator

$U_0 \leftarrow \{1\}$; $V_0 \leftarrow \{2, \ldots, n\}$
FOR $j := 1$ TO n DO
    IF EMPTY (U_{j-1})
        THEN $v_j \leftarrow \min(i : i \in V_{j-1})$
        ELSE $v_j \leftarrow \min(i : i \in U_{j-1})$
    Generate a random subset A_{v_j} on $U_{j-1} \cup V_{j-1} - \{v_j\}$ according to the probability distribution given above.
    $U_j \leftarrow U_{j-1} \cup A_j - \{v_j\}$
    $V_j \leftarrow V_{j-1} - A_j - \{v_j\}$
RETURN $A_{v_1}, A_{v_2}, \ldots, A_{v_{n-1}}$

The major problem in this algorithm is to compute (on-line) the probability distribution for A_{v_j}. In many examples, the probabilities depend only upon the cardinalities of U_{j-1} and V_{j-1} and possibly some other sets, and not upon the actual structure of these sets. This is the case for the class of all connected graphs with n nodes, or all connected graphs with n nodes and e edges (see Tinhofer, 1980). Nevertheless, we still have to count, and run into numerical problems when n or e are large.

6.4. Bipartite graphs.

A **bipartite graph** is a graph in which we can color all vertices with two colors (baby pink and mustard yellow) such that no two vertices with the same color are adjacent. There exists a useful connection with matrices which makes bipartite graphs a manageable class of graphs. If there are b baby vertices and m mustard vertices, then a bipartite graph is completely defined by a $b \times m$ incidence matrix of 0's and 1's. At this point we may recall the algorithms of section XI.6.3 for generating a random $R \times C$ table with given row and column totals. This leads directly to a rejection algorithm for generating a random bipartite graph with given degrees for all vertices:

Bipartite graph generator

[NOTE: This algorithm returns a $b \times m$ incidence matrix defining a random bipartite graph with b baby vertices and m mustard vertices. The row totals are r_i , $1 \leq i \leq b$, and the column totals are c_j , $1 \leq j \leq m$.]

REPEAT

Generate a random $R \times C$ matrix of dimension $b \times m$ with the given row and column totals.

UNTIL all elements in the matrix are 0 or 1

RETURN the matrix

The reduction to a random $R \times C$ matrix was suggested by Wormald (1984). By Wald's equation, we know that the expected time taken by the algorithm is equal to the product of the expected time needed to generate one random $R \times C$ matrix and the expected number of iterations. For example, if we use the ball-in-urn method of section XI.6.3, then a random $R \times C$ matrix can be obtained in time proportional to e , the total number of edges (which is also equal to $\sum r_i$ and to $\sum c_j$). The analysis of the expected number of iterations is also due to Wormald (1984):

Theorem 6.1.

Assume that all r_i 's and c_j 's are at most equal to k . The expected number of iterations in the rejection algorithm is

$$(1+o(1))\exp\left[\frac{2}{e^2}\sum_{i=1}^{b}\binom{r_i}{2}\sum_{j=1}^{m}\binom{c_j}{2}\right] ,$$

where e is the total number of edges, and $o(1)$ denotes a function tending to 0 as $e \to \infty$ which depends only upon k and not on the r_i 's and b_j 's.

As a corollary of this Theorem, we see that the expected number of iterations is uniformly bounded over all bipartite graphs whose degrees are uniformly
bounded by some number k.

Bipartite graphs play a crucial role in graph theory partly because of the following connection. Consider a $b \times m$ incidence matrix for a bipartite graph in
which all baby vertices have degree 2, i.e. all r_i's are equal to 2. This defines a
graph on m nodes in the following manner: each pair of edges connected to a
baby vertex defines an edge in the graph on m nodes. Thus, the new graph has
b edges. We can now generate a random graph with given collection of degrees
as follows:

Random graph generator

[NOTE: This algorithm returns an array of b edges defined on a graph with vertices
$\{1, \ldots, m\}$. The degree sequence is $c_1, \ldots, c_m$.]

REPEAT

  Generate a random $b \times m$ bipartite graph with degrees all equal to two for the baby
  vertices ($r_i = 2$), and degrees equal to $c_1, \ldots, c_m$ for the mustard vertices.

UNTIL no two baby vertices share the same two neighbors

RETURN $(k_1, l_1), \ldots, (k_m, l_m)$ where k_i and l_i are the columns of the two 1's found in
the i-th row of the incidence matrix of the bipartite graph.

Again we use the rejection principle, in the hope that for many graphs the
rejection constant is not unreasonable. Note that we need to check that there are
no duplicate edges in the graph. This is done by checking that no two rows in the
bipartite graph's incidence matrix are identical. It can be verified that the procedure takes expected time $O(b+m)$ where b is the number of edges in the
graph, provided that all degrees of the vertices in the graph are bounded by a
constant k (Wormald, 1984). In particular, the method seems to be ideally suited
for generating random **r-regular graphs**, i.e. graphs in which all degrees are
equal to r. It can be shown that the expected number of $R \times C$ matrices needed
before halting is roughly speaking $e^{(r^2-1)/4}$. This increases rapidly with r. Wormald also gives a particular algorithm for generating 3-regular, or cubic, graphs.

6.5. Exercises.

1. Find a simple $O(1)$ decoding rule for the coding function for edges in a graph given in the text.

2. Prove Theorem 6.1.

3. Prove that if random graphs with b edges and m vertices are generated by Wormald's method, then, provided that all degrees are bounded by k, the expected time is $O(b+m)$. Give the details of all the data structures involved in the solution.

4. **Event simulators.** We are given n events with the following dependence structure. Each individual event has probability p of occurring, and each pair of events has probability q of occurring. All triples carry probability zero. Determine the allowable values for p, q. Also indicate how you would handle one simulation. Note that in one simulation, we have to report all the indices of events that are supposed to occur. Your procedure should have constant expected time.

5. **Random strings in a context-free language.** Let S be the set of all strings of length n generated by a given context-free grammar. Assume that the grammar is unambiguous. Using at most $O(n^{r+1})$ space where r is the number of nonterminals in the grammar, and using any amount of preprocessing time, find a method for generating a uniformly distributed random string of length n in S in linear expected time. See also Hickey and Cohen (1983).

Chapter Fourteen
PROBABILISTIC SHORTCUTS
AND ADDITIONAL TOPICS

A probabilistic shortcut in random variate generation is a method for reducing the expected time in a simulation by recognizing a certain structure in the problem. This principle can be illustrated in hundreds of ways. Indeed, there is not a single example that could be called "typical". It should be stressed that the efficiency is derived from the problem itself, and is probabilistic in nature. This distinguishes these shortcuts from certain techniques that are based upon clever data structures or fast algorithms for certain sub-tasks. We will draw our examples from three sources: the simulation of maxima and sums of iid random variables, and the simulation of regenerative processes.

Other topics briefly touched upon include the problem of the generation of random variates under incomplete information (e.g. one just wants to generate random variates with a unimodal density having certain given moments) and the generation of random variates when the distribution is indirectly specified (e.g. the characteristic function is given). Finally, we will briefly deal with the problem of the design of efficient algorithms for large simulations.

1. THE MAXIMUM OF IID RANDOM VARIABLES.

1.1. Overview of methods.

In this section, we will look at methods for generating $X = \max(X_1, \ldots, X_n)$, where the X_i's are iid random variables with common density f (the corresponding distribution function will be called F). We will mainly be interested in the expected time as a function of n. For example, the naive method takes time proportional to n, and should be avoided whenever possible. Because X has distribution function F^n, it is easy to see that the following algorithm is valid:

Inversion method

Generate a uniform [0,1] random variate U.

RETURN $X \leftarrow F^{-1}(U^{\frac{1}{n}})$.

The problem with this approach is that for large n, $U^{1/n}$ is close to 1, so that in regular wordsize arithmetic, there could be an accuracy problem (see e.g. Devroye, 1980). This problem can be alleviated if we use $G = 1 - F$ instead of F and proceed as follows:

Inversion method with more accuracy

Generate an exponential random variate E and a gamma (n) random variate G_n.

RETURN $X \leftarrow G^{-1}(\frac{E}{E + G_n})$.

Unless the distribution function is explicitly invertible, both inversion-based algorithms are virtually useless. In the remaining sections, we present two probabilistic shortcuts, one based upon the quick elimination principle, and one on the use of records. The expected times of these methods usually increase as $\log(n)$. This is not as good as the constant time inversion method, but a lot better than the naive method. The advantages over the inversion method are measured in terms of accuracy and flexibility (fewer things are needed in order to be able to apply the shortcuts).

1.2. The quick elimination principle.

In the quick elimination principle, we generate the maximum of a sequence of iid random variables after having eliminated all but a few of the X_i's without ever generating them. We need a threshold point t and the tail probability $p = 1 - F(t)$. These are picked before application of the algorithm. Typically, p is of the order of $(\log(n))/n$. The number of X_i's that exceed t is binomial (n, p). Thus, the following algorithm is guaranteed to work:

The quick elimination algorithm (Devroye, 1980)

Generate a binomial (n,p) random variate Z.

IF $Z=0$

    THEN

        RETURN $X \leftarrow \max(X_1, \ldots, X_n)$ where the X_i's are iid random variates with density $f/(1-p)$ on $(-\infty, t]$.

    ELSE

        RETURN $X \leftarrow \max(X_1, \ldots, X_Z)$ where the X_i's are iid random variates with density f/p on $[t, \infty)$.

To analyze the expected time complexity, observe that the binomial (n,p) random variate can be generated in expected time proportional to np as $np \rightarrow \infty$ by the waiting time method. Obviously, we could use $O(1)$ expected time algorithms too, but there is no need for this here. Assume furthermore that every X_i in the algorithm is generated in one unit of expected time, uniformly over all values of p. It is easy to see that the expected time of the algorithm is $T + o(np)$ where we define $T = aP(Z=0)n + b(1-P(Z=0))np + cnp$ for some constants $a, b, c > 0$.

Lemma 1.1.

$$\inf_{0 < p < 1} T \sim (b+c)\log(n) \quad (n \rightarrow \infty) .$$

If we set

$$p = \frac{\log(n) + \delta_n}{n} ,$$

then $T \sim (b+c)\log(n)$ provided that the sequence of real numbers δ_n is chosen so that

$$\lim_{n \rightarrow \infty} \delta_n + \log(\log(n)) = \infty , \quad \delta_n = o(\log(n)) .$$

Proof of Lemma 1.1.

Note that

$$T = na(1-p)^n + bnp(1-(1-p)^n) + cnp$$
$$\leq (b+c)np + ane^{-np} .$$

The upper bound is convex in p with one minimum. Setting the derivative with respect to p equal to zero and solving for p gives the solution

$$p = \frac{1}{n}\log(\frac{an}{b+c}) .$$

Resubstitution in the upper bound for T shows that

$$T \leq (b+c)\log(\frac{ane}{b+c}) .$$

When $p = (\log(n)+\delta_n)/n$, then the upper bound for T is

$$ae^{-\delta_n} + (b+c)(\log(n)+\delta_n) .$$

This $\sim (b+c)\log(n)$ if $\delta_n = o(\log(n))$ and $e^{-\delta_n} = o(\log(n))$. The latter condition is satisfied when $\delta_n + \log(\log(n)) \to \infty$.

Finally, it suffices to work on a lower bound for T. We have for every $\epsilon > 0$ and all n large enough, since the optimal p tends to zero:

$$T \geq (na-bnp)e^{-\frac{np}{1-p}} + (b+c)np$$
$$\geq na(1-\epsilon)e^{-\frac{np}{1-\epsilon}} + (b+c)np.$$

We have already minimized such an expression with respect to p above. It suffices to formally replace n by $n/(1-\epsilon)$, a by $a(1-\epsilon)^2$, and $(b+c)$ by $(b+c)(1-\epsilon)$. Thus,

$$\inf_{0<p<1} T \geq (1-\epsilon)(b+c)\log(\frac{ane}{b+c})$$

for all n large enough. This concludes the proof of Lemma 1.1. ∎

A good choice for δ_n in Lemma 1.1 is $\delta_n = \log(\frac{a}{b+c})$. When $Z=0$ in the algorithm, iid random variates from the density $f/(1-p)$ restricted to $(-\infty, t]$ can be generated by generating random variates from f until n values less than or equal to t are observed. This would force us to replace the term $aP(Z=0)n$ in the definition of T by $aP(Z=0)n/(1-p)$. However, all the statements of Lemma 1.1 remain valid.

The main problem is that of the computation of a pair (p,t). For if we start with a value for p, such as the value suggested by Lemma 1.1, then the value for t is given by $F^{-1}(1-p)$ (or $G^{-1}(p)$ where $G=1-F$, if numerical accuracy is of concern). This is unfortunately possible only when the inverse of the distribution function is known. But if the inverse of the distribution were known, we would have been able to generate the maximum quite efficiently by the inversion method. There is a subtle difference though: for here, we need one inversion, even if we would need to generate a million iid random variables all distributed as the maximum X. With the inversion method, a million inversions would be required. If on the other hand we were to start with a value for t, then p would have to be set equal to $\int_t^\infty f = G(t) = 1-F(t)$. This requires knowledge of the distribution function but not of its inverse. The value of t we start with should be such that p satisfies the conditions of Lemma 1.1. Typically, t is picked on theoretical grounds as is now illustrated for the normal density.

Example 1.1.

For the normal density it is known that $G(x)\sim f(x)/x$ as $x\to\infty$. A first approximate solution of $f(t)/t = p$ is $t=\sqrt{2\log(1/p)}$, but even if we substitute the value $p=(\log(n))/n$ in this formula, the value of $G(t)$ would be such that the expected time taken by the algorithm far exceeds $\log(n)$. A second approximation is

$$ t = \sqrt{2\log(\frac{1}{p})} - \frac{\log(4\pi)+\log(\log(\frac{1}{p}))}{2\sqrt{2\log(\frac{1}{p})}} , $$

with $p=(\log(n))/n$. It can be verified that with this choice, $T=O(\log(n))$. ∎

For other densities, one can use similar arguments. For the gamma (a) density for example, we have $G(x)\sim f(x)$ as $x\to\infty$, and $f(x)\le G(x)\le f(x)/(1-(a/x))$ for $a>1,x>a-1$. This helps in the construction of a useful value for t.

The computation of $G(t)$ is relatively straightforward for most distributions. For the normal density, see the series of papers published after the book of Kendall and Stuart (1977) (Cooper (1968), Hill (1969), Hitchin (1973)), the paper by Adams (1969), and an improved version of Adams's method, called algorithm AS66 (Hill (1973)). For the gamma density, algorithm AS32 (Bhattacharjee (1970)) is recommended: it is based upon a continued fraction expansion given in Abramowitz and Stegun (1965).

1.3. The record time method.

In some process simulations one needs a sequence $(Z_{n_1}, \ldots, Z_{n_k})$ of maxima that correspond to one realization of the experiment, where $n_1 < n_2 < \cdots < n_k$. In other words, for all i, we have $Z_i = \max(X_1, \ldots, X_i)$ where the X_i's are iid random variables with common density f. The inversion method requires k inversions, and can be implemented as follows:

Inversion method

$n_0 \leftarrow 0, Z \leftarrow -\infty$

FOR $i := 1$ TO k DO

         Generate Z, the maximum of $n_i - n_{i-1}$ iid random variables with common density f.

         $Z_{n_i} \leftarrow \max(Z_{n_{i-1}}, Z)$

The record time method introduced in this section requires on the average about $\log(n_k)$ exponential random variates and evaluations of the distribution function. In addition, we need to report the k values Z_{n_i}. When $\log(n_k)$ is small compared to k, the record time method can be competitive. It exploits the fact that in a sequence of n iid random variables with common density f, there are about $\log(n)$ records, where we call the n-th observation a record if it is the largest observation seen thus far. If the n-th observation is a record, then the index n itself is called a record time. It is noteworthy that given the value V_i of the i-th record, and given the record time T_i of the i-th record, $T_{i+1} - T_i$ and V_{i+1} are independent: $T_{i+1} - T_i$ is geometrically distributed with parameter $G(V_i)$:

$$P(T_{i+1} - T_i = j \mid T_i, V_i) = G(V_i)(1 - G(V_i))^{j-1} \quad (j \geq 1) .$$

Also, V_{i+1} has conditional density $f / G(V_i)$ restricted to $[V_i, \infty)$. An infinite sequence of records and record times $\{(V_i, T_i), i \geq 1\}$ can be generated as follows:

The record time method (Devroye, 1980)

$T_1 \leftarrow 1, i \leftarrow 1$

Generate a random variate V_1 with density f .

$p \leftarrow G(V_1)$

WHILE True DO

    $i \leftarrow i+1$

    Generate an exponential random variate E .

    $T_i \leftarrow T_{i-1} + \lceil -E/\log(1-p)\rceil$

    Generate V_i from the tail density $\dfrac{f(x)}{1-p} I_{[x \geq V_{i-1}]}$.

    $p \leftarrow G(V_i)$

It is a straightforward exercise to report the Z_{n_i} values given the sequence of records and record times. We should exit from the loop when $T_i \geq n_k$. The expected number of loops before halting is thus equal to the expected number of records in a sequence of length n_k , i.e. it is

$$\sum_{i=1}^{n_k} \frac{1}{i} = \log(n_k) + \gamma + o(1)$$

where $\gamma = 0.5772...$ is Euler's constant. We note that the most time consuming operation in every iteration is the evaluation of G . If the inverse of G is available, the lines

    Generate V_i from the tail density $\dfrac{f(x)}{1-p} I_{[x \geq V_{i-1}]}$.

    $p \leftarrow G(V_i)$

can be replaced by

Generate a uniform [0,1] random variate U.

$p \leftarrow pU$

$V_i \leftarrow G^{-1}(p)$

A final remark is in order here. If we assume that G can be computed in one unit of time for all distributions, then the (random) time taken by the algorithm is an invariant, because the distribution of record times is distribution-free.

1.4. Exercises.

1. **Tail of the normal density.** Let f be the normal density, let $t > 0$ and define $p = G(t)$ where $G = 1-F$ and F is the normal distribution function. Prove the following statements:

   A. **Gordon's inequality.** (Gordon (1941), Mitrinovic (1970)).

   $$\frac{t}{t^2+1} f(t) \leq p \leq \frac{1}{t} f(t).$$

   B. As $t \to \infty$, $G(t) \sim f(t)/t$.

   C. If $t = \sqrt{2\log(n/\log(n))}$, then for the quick elimination algorithm, $T = \Omega(n^{1-\epsilon})$ for every $\epsilon > 0$ as $n \to \infty$.

   D. If $t = s - \frac{1}{2s}(\log(4\pi) + \log(\log(\frac{n}{\log}(n))))$, where s is as in point C, then for the quick elimination algorithm, $T = O(\log(n))$. Does $T \sim (b+c)\log(n)$ if b, c are the constants in the definition of T (see Lemma 1.1)?

2. Let $T_1, T_2, \ldots$ be the record times in a sequence of iid uniform [0,1] random variables. Prove that $E(T_2) = \infty$. Show furthermore that $\log(T_n) \sim n$ in probability as $n \to \infty$.

2. RANDOM VARIATES WITH GIVEN MOMENTS

2.1. The moment problem.

The classical moment problem can be formulated as follows. Let $\{\mu_i, 1 \leq i\}$ be a collection of moments. Determine whether there is at least one distribution which gives rise to these moments; if so, construct such a distribution and determine whether it is unique. Solid detailed treatments of this problem can be found in Shohat and Tamarkin (1943) and Widder (1941). The main result is the following.

Theorem 2.1.

If there exists a distribution with moments μ_i, $1 \leq i$, then

$$\begin{vmatrix} 1 & \mu_1 & \cdots & \mu_s \\ \mu_1 & \mu_2 & & \mu_{s+1} \\ \cdot & & & \cdot \\ \cdot & & & \cdot \\ \mu_s & \cdots & & \mu_{2s} \end{vmatrix} \geq 0$$

for all integers s with $s \geq 1$. The inequalities hold strictly if the distribution is nonatomic. Conversely, if the matrix inequality holds strictly for all integers s with $s \geq 1$, then there exists a nonatomic distribution matching the given moments.

Proof of Theorem 2.1.

We will only outline why the matrix inequality is necessary. Considering the fact that

$$E\left((c_0 + c_1 X + \cdots + c_s X^s)^2\right) \geq 0$$

for all values of $c_0, \ldots, c_s$, we have by a standard result from linear algebra (Mirsky (1955, p. 400)) that

$$\begin{vmatrix} 1 & \mu_1 & \cdots & \mu_s \\ \mu_1 & \mu_2 & & \mu_{s+1} \\ \cdot & & & \cdot \\ \cdot & & & \cdot \\ \mu_s & \cdots & & \mu_{2s} \end{vmatrix} \geq 0 . \blacksquare$$

Theorem 2.2.

If there exists a distribution on $[0,\infty)$ with moments μ_i , $1 \leq i$, then

$$
\begin{vmatrix}
1 & \mu_1 & \cdots & \mu_s \\
\mu_1 & \mu_2 & & \mu_{s+1} \\
\cdot & & & \cdot \\
\cdot & & & \cdot \\
\mu_s & \cdots & \cdots & \mu_{2s}
\end{vmatrix} \geq 0 \, ,
$$

$$
\begin{vmatrix}
\mu_1 & \mu_2 & \cdots & \mu_{s+1} \\
\mu_2 & \mu_3 & & \mu_{s+2} \\
\cdot & & & \cdot \\
\cdot & & & \cdot \\
\mu_{s+1} & \cdots & \cdots & \mu_{2s+1}
\end{vmatrix} \geq 0 \, ,
$$

for all integers $s \geq 0$. The inequalities hold strictly if the distribution is nonatomic. Conversely, if the matrix inequality holds strictly for all integers $s \geq 0$, then there exists a nonatomic distribution matching the given moments.

The determinants in Theorems 2.1, 2.2 are called Hankel determinants. What happens when one or more of them are zero is more complicated (see e.g. Widder (1941)). The problem of the uniqueness of a distribution is covered by Theorem 2.3.

Theorem 2.3.

Let $\mu_1, \mu_2, \ldots$ be the moment sequence of at least one distribution. Then this distribution is unique if Carleman's condition holds, i.e.

$$\sum_{i=0}^{\infty} |\mu_{2i}|^{-\frac{1}{2i}} = \infty .$$

If we have a distribution on the positive halfline, then a sufficient condition for uniqueness is

$$\sum_{i=0}^{\infty} (\mu_i)^{-\frac{1}{2i}} = \infty .$$

When the distribution has a density f, then a necessary and sufficient condition for uniqueness is

$$\int_{-\infty}^{\infty} \frac{\log(f(x))}{1+x^2} \, dx = -\infty$$

(Krein's condition).

For example, normal distributions or distributions on compact sets satisfy Carleman's condition and are thus uniquely determined by their moment sequence. In exercises 2.2 and 2.3, examples are developed of distributions having identical infinite moment sequences, but widely varying densities. In exercise 2.2, a unimodal discrete distribution is given which has the same moments as the log-normal distribution.

The problem that we refer to as the moment problem is that of the generation of a random variate with a given collection of moments $\mu_1, \mu_2, \ldots, \mu_n$, where n can be ∞. Note that if we expand the characteristic function ϕ of a random variable in its Taylor series about 0, then

$$\phi(t) = \phi(0) + \frac{t}{1!}\phi^{(1)}(0) + \cdots + \frac{t^{k-1}}{(k-1)!}\phi^{(k-1)}(0) + R_k$$

where the remainder term satisfies

$$|R_k| \leq \mu_k \frac{|t|^k}{k!} .$$

This uses the fact that if $|\mu_k| < \infty$, the k-th derivative of ϕ exists, and is a continuous function given by $E((iX)^k e^{itX})$. In particular, the k-th derivative is in absolute value not greater than $E(|X|^k)$. See for example Feller (1971, pp. 512-514). The remainder term R_k tends to 0 in a neighborhood of the origin when

$$\limsup \frac{|\mu_k|^{1/k}}{k} < \infty .$$

Thus, the Taylor series converges in those cases. It follows that ϕ is analytic in a neighborhood of the origin, and hence completely determined by its power series about the origin. The condition given above is thus sufficient for the moment sequence to uniquely determine the distribution. One can verify that the condition is weaker, but not much weaker, than Carleman's condition. The point of all this is that if we are given an infinite moment sequence which uniquely determines the distribution, we are in fact given the characteristic function in a special form. The problem of the generation of a random variate with a given characteristic function will be dealt with in section 3. Here we will mainly be concerned with the finite moment case. This is by far the most important case in practice, because researchers usually worry about matching the first few moments, and because the majority of distributions have only a finite number of finite moments. Unfortunately, there are typically an infinite number of distributions sharing the same first n moments. These include discrete distributions and distributions with densities. If some additional constraints are satisfied by the moments, it may be possible to pick a distribution from relatively small classes of distributions. These include:

A.   The class of all unimodal densities, i.e. uniform scale mixtures.

B.   The class of normal scale mixtures.

C.   Pearson's system of densities.

D.   Johnson's system of densities.

E.   The class of all histograms.

F.   The class of all distributions of random variables of the form $a + bN + cN^2 + dN^3$ where N is normally distributed.

The list is incomplete, but representative of the attempts made in practice by some statisticians. For example, in cases C,D and F, we can match the first four moments with those of exactly one member in the class except in case F, where some combinations of the first four moments have no match in the class. The fact that a match always occurs in the Pearson system has contributed a lot to the early popularity of the system. For a description and details of the Pearson system, see exercise IX.7.4. Johnson's system (exercise IX.7.12) is better for quantile matching than moment matching. We also refer the reader to the Burr family (section IX.7.4) and other families given in section IX.7.5. These families of distributions are usually designed for matching up to four moments. This of course is their main limitation. What is needed is a general algorithm that can be used for arbitrary $n > 4$. In this respect, it may first be worthwhile to verify whether there exists a uniform or normal scale mixture having the given set of moments. If this is the case, then one could proceed with the construction of one such distribution. If this attempt fails, it may be necessary to construct a matching histogram or discrete distribution (note that discrete distributions are limits of histograms). Good references about the moment problem include Widder (1941), Shohat and Tamarkin (1943), Godwin (1964), von Mises (1964), Hill (1969) and Springer (1979).

2.2. Discrete distributions.

Assume that we want to match the first $2n-1$ moments with those of a discrete distribution having n atoms located at $x_1, \ldots, x_n$, with respective weights $p_1, \ldots, p_n$. We know that we should have

$$\sum_{i=1}^{n} p_i (x_i)^j = \mu_j \quad (0 \le j \le 2n-1) .$$

This is a system of $2n$ equalities with $2n$ unknowns. It has precisely one solution if at least one distribution exists with the given moments (von Mises, 1964). In particular, if the locations x_i are known, then the p_i's can be determined from the first n linear equations. The locations can first be obtained as the n roots of the equation

$$x^n + c_{n-1} x^{n-1} + \cdots + c_1 x + c_0 = 0 ,$$

where the c_i's are the solutions of

$$\begin{vmatrix} \mu_0 & \cdot & \mu_{n-1} \\ \mu_1 & \cdot & \mu_n \\ \cdot & & \cdot \\ \mu_{n-1} & \cdot & \mu_{2n-2} \end{vmatrix} \begin{vmatrix} c_0 \\ c_1 \\ \cdot \\ c_{n-1} \end{vmatrix} = - \begin{vmatrix} \mu_n \\ \mu_{n+1} \\ \cdot \\ \mu_{2n-1} \end{vmatrix} .$$

To do this could take some valuable time, but at least we have a minimal solution, in the sense that the distribution is as concentrated as possible in as few atoms as possible. One could argue that this yields some savings in space, but n is rarely large enough to make this the deciding factor. On the other hand, it is impossible to start with $2n$ locations of atoms and solve the $2n$ equations for the weights p_i, because there is no guarantee that all p_i's are nonnegative.

If an even number of moments is given, say $2n$, then we have $2n+1$ equations. If we consider $n+1$ atom locations with $n+1$ weights, then there is an excess of one variable. We can thus choose one item, such as the location of one atom. Call this location a. Shohat and Tamarkin (1943) (see also Royden, 1953) have shown that if there exists at least one distribution with the given moments, then there exists at least one distribution with at most $n+1$ atoms, one of them located at a, sharing the same moments. The locations $x_0, \ldots, x_n$ of the atoms are the zeros of

$$\begin{vmatrix} 1 & 1 & \mu_0 & \cdot & \mu_{n-1} \\ x & a & \mu_1 & \cdot & \mu_n \\ \cdot & & \cdot & & \cdot \\ \cdot & & & \cdot & \cdot \\ x^{n+1} & a^{n+1} & \mu_{n+1} & \cdot & \mu_{2n} \end{vmatrix} = 0 .$$

The weights $p_0, p_1, \ldots, p_n$ are linear combinations of the moments:

$$p_i = \sum_{j=0}^{n} c_{ji} \mu_j .$$

The coefficients c_{ji} in turn are defined by the identity

$$\sum_{j=0}^{n} c_{ji} x^{j} \equiv \prod_{j \neq i} \frac{x - x_j}{x_i - x_j} \quad (0 \leq i \leq n) .$$

When the distribution puts all its mass on the nonnegative real line, a slight modification is necessary (Royden, 1953). Closely related to discrete distributions are the histograms: these can be considered as special cases of distributions with densities

$$f(x) = \sum_{i=1}^{n} \frac{p_i}{h_i} K(\frac{x - x_i}{h_i}) ,$$

where K is a fixed form density (such as the uniform $[-1,1]$ density in the case of a histogram), x_i is the center of the i-th component, p_i is the weight of the i-th component, and h_i is the "width" of the i-th component. Densities of this form are well-known in the nonparametric density estimation literature: they are the kernel estimates. Archer (1980) proposes to solve the moment equations numerically for the unknown parameters in the histogram. We should point out that the density f shown above is the density of $x_Z + h_Z Y$ where Y has density K, and Z has probability vector $p_1, \ldots, p_n$ on $\{1, \ldots, n\}$. This greatly facilitates the computations and the visualization process.

2.3. Unimodal densities and scale mixtures.

A random variable X has a unimodal distribution if and only if there exists a random variable Y such that X is distributed as YU where U is a uniform $[0,1]$ random variable independent of Y (Khinchine's theorem). If U is not uniform and Y is arbitrary then the distribution of X is called a scale mixture for U. Of particular importance are the normal scale mixtures, which correspond to the case when U is normally distributed. For us it helps to be able to verify whether for a given collection of n moments, there exists a unimodal distribution or a scale mixture which matches these moments. Usually, we have a particular scale mixture in mind. Assume for example that U has moments $\nu_1, \nu_2, \ldots$. Then, because $E(X^i) = E(Y^i) E(U^i)$, we see that Y has i-th moment μ_i / ν_i. Thus, the existence problem is solved if we can find at least one distribution having moments μ_i / ν_i.

Applying Theorem 2.1, then we observe that a sufficient condition for the moment sequence μ_i to correspond to a U scale mixture is that the determinants

$$\begin{vmatrix} 1 & \mu_1/\nu_1 & . & . & . & \mu_s/\nu_s \\ \mu_1/\nu_1 & \mu_2/\nu_2 & & & & \mu_{s+1}/\nu_{s+1} \\ . & & & & & . \\ . & & & & & . \\ \mu_s/\nu_s & & . & . & . & \mu_{2s}/\nu_{2s} \end{vmatrix} \geq 0$$

are all positive for $2s < n$, n odd. This was first observed by Johnson and Rogers (1951). For uniform mixtures, i.e. unimodal distributions, we should replace ν_i by $1/(i+1)$ in the determinants. Having established the existence of a scale mixture with the given moments, it is then up to us to determine at least one Y with moment sequence μ_i/ν_i. This can be done by the methods of the previous section.

By insisting that a particular scale mixture be matched, we are narrowing down the possibilities. By this is meant that fewer moment sequences lead to solutions. The advantage is that if a solution exists, it is typically "nicer" than in the discrete case. For example, if Y is discrete with no atom at 0, and U is uniform, then X has a unimodal staircase-shaped density with mode at the origin and breakpoints at the atoms of Y. If U is normal, then X is a superposition of a few normal densities centered at 0 with different variances. Let us illustrate briefly how restrictive some scale mixtures are. We will take as example the case of four moments, with normalized mean and variance, $\mu_1=0, \mu_2=1$. Then, the conditions of Theorem 2.1 imply that we must always have

$$\begin{vmatrix} 1 & 0 & 1 \\ 0 & 1 & \mu_3 \\ 1 & \mu_3 & \mu_4 \end{vmatrix} \geq 0 \, .$$

Thus, $\mu_4 \geq (\mu_3)^2+1$. It turns out that for all μ_3, μ_4 satisfying the inequality, we can find at least one distribution with these moments. Incidentally, equality occurs for the Bernoulli distribution. When the inequality is strict, a density exists. Consider next the case of a unimodal distribution with zero mean and unit variance. The existence of at least one distribution with the given moments is guaranteed if

$$\begin{vmatrix} 1 & 0 & 3 \\ 0 & 3 & 4\mu_3 \\ 3 & 4\mu_3 & 5\mu_4 \end{vmatrix} \geq 0 \, ,$$

In other words, $\mu_4 \geq \dfrac{9}{5}+\dfrac{16}{15}(\mu_3)^2$. It is easy to check that in the (μ_3, μ_4) plane, a smaller area gets selected by this condition. It is precisely the (μ_3, μ_4) plane which can help us in the fast construction of moment matching distributions. This is done in the next section.

2.4. Convex combinations.

If Y and Z are random variables with moment sequences μ_i and ν_i respectively, then the random variable X which equals Y with probability p and Z with probability $1-p$ has moment sequence $p\,\mu_i+(1-p\,)\nu_i$, in other words, it is the convex combination of the original moment sequences. Assume that we want to match four normalized moments. Recall that the allowable area in the (μ_3,μ_4) plane is the area above the parabola $\mu_4\geq(\mu_3)^2+1$. Every point (μ_3,μ_4) in this area lies on a horizontal line at height μ_4 which intersects the parabola at the points $(-\sqrt{\mu_4-1},\mu_4)$, $(\sqrt{\mu_4-1},\mu_4)$. In other words, we can match the moments by a simple convex combination of two distributions with third and fourth moments $(-\sqrt{\mu_4-1},\mu_4)$ and $(\sqrt{\mu_4-1},\mu_4)$ respectively.

The weight in the convex combination is determined quite easily since we must have, attaching weight p to the distribution with positive third moment,

$$(p-(1-p\,))\sqrt{\mu_4-1} = \mu_3 \ .$$

Thus, it suffices to take

$$p = \frac{1+\dfrac{\mu_3}{\sqrt{\mu_4-1}}}{2} \ .$$

It is also easy to verify that for a Bernoulli $(q\,)$ random variable, we have normalized fourth moment

$$\frac{3q^2-3q+1}{q\,(1-q\,)}$$

and normalized third moment

$$\frac{1-2q}{\sqrt{q\,(1-q\,)}} \ .$$

Notice that this distribution always falls on the limiting parabola. Furthermore, by letting q vary from 0 to 1, all points on the parabola are obtained. Given the fourth moment μ_4, we can determine q via the equation

$$q = \frac{1}{2}(1\pm\sqrt{\frac{\mu_4-1}{\mu_4+3}}) \ ,$$

where the plus sign is chosen if $\mu_3\geq0$, and the minus sign is chosen otherwise. Let us call the solution with the plus sign q. The minus sign solution is $1-q$. If B is a Bernoulli $(q\,)$ random variable, then $(B-q\,)/\sqrt{q\,(1-q\,)}$ and $-(B-q\,)/\sqrt{q\,(1-q\,)}$ are the two random variables corresponding to the two intersection points on the parabola. Thus, the following algorithm can be used to generate a general random variate with four moments $\mu_1,\ \ldots\ ,\mu_4$:

Generator matching first four moments

Normalize the moments: $\sigma \leftarrow \sqrt{\mu_2 - (\mu_1)^2}$,

$$(\mu_3, \mu_4) \leftarrow \left(\frac{\mu_3 - 3\mu_2\mu_1 + 2(\mu_1)^3}{\sigma^3}, \frac{\mu_4 - 4\mu_3\mu_1 + 6\mu_2(\mu_1)^2 - 3(\mu_1)^4}{\sigma^4} \right)$$

$$q \leftarrow \frac{1}{2}\left(1 + \sqrt{\frac{\mu_4 - 1}{\mu_4 + 3}}\right)$$

$$p \leftarrow \frac{1 + \dfrac{\mu_3}{\sqrt{\mu_4 - 1}}}{2}$$

Generate a uniform [0,1] random variate U.

IF $U \leq p$

    THEN

           $X \leftarrow I_{[U \leq pq]}$ (X is Bernoulli (q))

           RETURN $X \leftarrow \mu_1 + \sigma \dfrac{X - q}{\sqrt{q(1-q)}}$

    ELSE

           $X \leftarrow I_{[U \leq p + (1-p)q]}$ (X is Bernoulli (q))

           RETURN $X \leftarrow \mu_1 - \sigma \dfrac{X - q}{\sqrt{q(1-q)}}$

The algorithm shown above can be shortened by a variety of tricks. As it stands, one uniform random variate is needed per returned random variate. The point of this example is that it is very simple to generate random variates that match four moments if one is not picky. Indeed, few users will be pleased with the convex combination of two Bernoulli distributions used in the example. But interestingly, the example can also be used in the construction of the distribution of Y in scale mixtures of the form YU discussed in the previous section. In that respect, the algorithm becomes more useful, because the returned distributions are "nicer". The algorithm for unimodal distributions with mode at 0 is given below.

Simple unimodal distribution generator matching four moments

Readjustment of moments: $\mu_1 \leftarrow 2\mu_1$, $\mu_2 \leftarrow 3\mu_2$, $\mu_3 \leftarrow 4\mu_3$, $\mu_4 \leftarrow 5\mu_5$.

Generate a random variate Y having the readjusted moments (e.g. by the algorithm given above).

Generate a uniform [0,1] random variate U.

RETURN $X \leftarrow YU$.

The algorithms for other scale mixtures are similar.

One final remark about moment matching is in order here. Even with a unimodality constraint, there are many distributions with widely varying densities but identical moments up to the n-th moment. One should therefore always ask the question whether it is a good thing at all to blindly go ahead and generate random variates with a certain collection of moments. Let us make this point with two examples.

Example 2.1.(Godwin, 1964)

The following two densities have identical infinite moment sequences:

$$f(x) = \frac{1}{4}e^{-|x|^{\frac{1}{2}}} \quad (x \in R).$$

$$g(x) = \frac{1}{4}e^{-|x|^{\frac{1}{2}}}(1 + \cos(\sqrt{|x|}) \quad (x \in R)$$

(Kendall and Stuart (1977), see exercise 2.3). Thus, noting that

$$\int_A f = 0.4656... \; ; \int_A g = 0.7328... ,$$

where $A = [-\pi^2/4, \pi^2/4]$, we observe that

$$\int |f - g| \geq 0.5344... .$$

Considering that the L_1 distance between two densities is at most 2, the distance 0.5344... is phenomenally large. ∎

Example 2.2.

The previous example involves a unimodal and an oscillating density. But even if we enforce unimodality on our counterexamples, not much changes. See for example Leipnik's example described in exercise 2.2. Another way of illustrating this is as follows: for any symmetric unimodal density f with moments μ_2, μ_4, it is true that

$$\sup_g \int | f -g | \geq \omega^2(1-\omega)$$

where the supremum is taken over all symmetric unimodal g with the same second and fourth moments, and $\omega=\sqrt{(3\mu_2)^2/(5\mu_4)}$. It should be noted that $0\leq\omega\leq1$ in all cases (this follows from the nonnegativity of the Hankel determinants applied to unimodal distributions). When f is normal, $\omega=\sqrt{3/5}$ and the lower bound is $\frac{3}{5}(1-\sqrt{\frac{3}{5}})$, which is still quite large. For some combinations of moments, the lower bound can be as large as $\frac{4}{27}$. There are two differences with Example 2.1: we are only matching the first four moments, not all moments, and the counterexample applies to any symmetric unimodal f , not just one density picked beforehand for convenience. Example 2.2 thus reinforces the belief that the moments contain surprisingly little information about the distribution. To prove the inequality of this example, we will argue as follows: let f ,g ,h be three densities in the given class of densities. Clearly,

$$\max(\int | f -h | ,\int | f -g |) \geq \frac{1}{2}(\int | f -h | +\int | f -g |)$$

$$\geq \frac{1}{2}\int | h -g | \ .$$

Thus it suffices to prove twice the lower bound for $\int | h -g |$ for two particular densities h ,g . Consider densities of random variables YU where U is uniformly distributed on $[0,1]$ and Y is independent of U and has a symmetric discrete distribution with atoms at $\pm b$,$\pm c$, where $0 < b < c < \infty$. The atom at c has weight p /2, and the atom at b has weight $(1-p$)/2. For h and g we will consider different choices of b ,c ,p . First, any choice must be consistent with the moment restrictions:

$$(1-p)b^2+pc^2 = 3\mu_2 ,$$

$$(1-p)b^4+pc^4 = 5\mu_4 .$$

Solving for p gives

$$1-p = \frac{5\mu_4-3\mu_2c^2}{b^4-b^2c^2} \ .$$

Forcing $p \in[0,1]$ gives us the constraints $0\leq3\mu_2c^2-5\mu_4\leq b^2(c^2-b^2)$. It is to our advantage to take the extreme values for c . In particular, for g we will take $c =\sqrt{(5\mu_4)/(3\mu_2)}$, $b =0$, $p =\omega^2$. It should be noted that this not yield a density g since there will be an atom at the origin. Thus, we use an approximating argument with a sequence g_n approaching g in the sense that the atom at 0 is approached by an atom at $\epsilon_n \to 0$. Next, for h, we take the limit of the sequence h_n where as $n \to \infty$, $b \to \sqrt{3\mu_2}$, $p \to 0$, and $c \to \infty$. This is the case in which the rightmost atom escapes to infinity but has increasingly negligible weight p. Since $p \to 0$, the contribution of the rightmost atom to the L_1 distance is also $o(1)$. Thus, h can be considered as having one atom at $\sqrt{3\mu_2}$ of weight $1/2$. We obtain by simple geometrical considerations,

$$\lim_{n \to \infty} \int |g_n - h_n| = 4(\sqrt{(5\mu_4)/(3\mu_2)} - \sqrt{3\mu_2})(\frac{1}{2}\omega^2 \frac{1}{\sqrt{(5\mu_4)/(3\mu_2)}})$$
$$= 2\omega^2(1-\omega) .$$

Since the sequences h_n, g_n are entirely in our class, we see that the lower bound for $\sup_g \int |f - g|$ is at least $\omega^2(1-\omega)$. ∎

2.5. Exercises.

1. Show that for the normal density, the $2i$-th moment is

$$\mu_{2i} = (2i-1)(2i-3) \cdots (3)(1) \quad (i \geq 2) .$$

Show furthermore that Carleman's condition holds.

2. **The lognormal density.** In this exercise, we consider the lognormal density

$$f(x) = \frac{1}{\sqrt{2\pi}\sigma x} e^{-\frac{(\log(x))^2}{2\sigma^2}} \quad (x > 0) .$$

Show first that this density fails both Carleman's condition and Krein's condition. Hint: show first that the r-th moment is $\mu_r = e^{\sigma^2 r^2/2}$. Thus, there exist other distributions with the same moments. We will construct a family of such distributions, referred to hereafter as Heyde's family (Heyde (1963), Feller (1971, p. 227)): let $-1 \leq a \leq 1$ be a parameter, and define the density

$$f_a(x) = f(x)(1 + a \sin(2\pi \log(x))) \quad (x > 0) .$$

To show that f_a is a density, and that all the moments are equal to the moments of $f_0 = f$, it suffices to show that

$$\int_0^\infty x^k f(x) \sin(2\pi \log(x)) \, dx = 0$$

for all integer $k \geq 0$. Show this. Show also the following result due to Leipnik (1981): there exists a family of discrete unimodal random variables X having the same moments as a lognormal random variable. It suffices to let X take the value $ae^{\sigma i}$ with probability $ca^{-i}e^{-\sigma^2 i^2/2}$ for $i = 0, \pm 1, \pm 2, ...$, where $a > 0$ is a parameter, and c is a normalization constant.

3. **The Kendall-Stuart density.** Kendall and Stuart (1977) introduced the density

$$f(x) = \frac{1}{4}e^{-|x|^{\frac{1}{2}}} \quad (x \in R) .$$

Following Kendall and Stuart, show that for all real a with $|a| \leq 1$,

$$f_a(x) = \frac{1}{4}e^{-|x|^{\frac{1}{2}}}(1+a \, \cos(\sqrt{|x|})) \quad (x \in R)$$

are densities with moments equal to those of f .

4. Yet another family of densities sharing the same moment sequence is given by

$$f_a(x) = e^{-x^{\frac{1}{4}}}\frac{(1-a \, \sin(x^{\frac{1}{4}}))}{24} \quad (x > 0) ,$$

where $a \in [0,1)$ is a parameter. Show that f_0 violates Krein's condition and that all moments are equal to those of f_0. This example is due to Stieltjes (see e.g. Widder (1941, pp. 125-126)).

5. Let $p \in (0, \frac{1}{2})$ be a parameter, and let $c = (p \cos(p \pi))^{1/p} / \Gamma(1/p)$ be a constant. Show that the following two densities on $(0, \infty)$ have the same moments:

$$f(x) = c \, e^{-x^p \cos(p \pi)} ,$$

$$g(x) = f(x) \, (1+\sin(x^p \sin(p \pi)))$$

(Lukacs (1970, p. 20)).

6. **Fleishman's family of distributions.** Consider all random variables of the form $a + bN + cN^2 + dN^3$ where N is a normal random variable, and a, b, c, d are constants. Many distributions are known to be approximately normal, and can probably be modeled by distributions of random variables of the form given above. This family of distributions, studied by Fleishman (1978), has the advantage that random variate generation is easy once the constants are determined. To compute the constants, the first four moments can be matched with fixed values $\mu_1, \mu_2, \mu_3, \mu_4$. For the sake of simplicity, let us normalize as follows: $\mu_1 = 0, \mu_2 = 1$. Show that b, d can be found by solving

$$1 = b^2 + 6bd + 15d^2 + 2c^2 ,$$

$$\mu_4 - 3 = 24(bd + c^2(1 + b^2 + 28bd) + d^2(12 + 48bd + 141c^2 + 255d^2)) ,$$

where

$$c = \frac{\mu_3}{2(b^2 + 24bd + 105d^2 + 2)} .$$

Furthermore, $a = -c$. Show that not all combinations of normalized moments of distributions (i.e. all pairs (μ_3, μ_4) with $\mu_4 \geq (\mu_3)^2 + 1$) lead to a solution. Determine the region in the (μ_3, μ_4) plane of allowable pairs. Finally, prove that there exist combinations of constants for which the density is not unimodal, and determine the form of the distribution in these cases.

7. Assume that we wish to match the first six moments of a symmetric distribution (all odd moments are zero). We normalize by forcing μ_2 to be 1. Show first that the allowable region in the (μ_4, μ_6) plane is defined by the inequalities $\mu_4 \geq 1$, $\mu_6 \geq (\mu_4)^2$. Find simple families of distributions which cover the borders of this region. Rewrite each point in the plane as the convex combination of two of these simple distributions, and give the corresponding generator, i.e. the generator for the distribution that corresponds to this point.

8. Let the a-th and b-th absolute moments of a unimodal symmetric distribution with a density be given. Find a useful lower bound for

$$\inf_f \sup_g \int | f - g | ,$$

where the infimum and supremum is over all symmetric unimodal densities having the given absolute moments. The lower bound should coincide with that of Example 2.2 in the case $a = 2, b = 4$.

3. CHARACTERISTIC FUNCTIONS.

3.1. Problem statement.

In many applications, a distribution is best described by its characteristic function ϕ. Sometimes, it is outright difficult to invert the characteristic function to obtain a value for the density or distribution function. One might ask whether in those cases, it is still possible to generate a random variate X with the given distribution. An example of such a distribution is the stable distribution. In particular, the symmetric stable distribution with parameter $\alpha \in (0,2]$ has the simple characteristic function $e^{-|t|^\alpha}$. Yet, except for $\alpha \in \{\frac{1}{2}, 1, 2\}$, no convenient analytic expression is known for the corresponding density f ; the density is best computed with the help of a convergent series or a divergent asymptotic expansion (section IX.6.3). For random variate generation in this simple case, we refer to section IX.6. For $\alpha \in (0,1]$ the characteristic function can be written as a mixture of triangular characteristic functions. This property is shared by all real (thus, symmetric) convex characteristic functions, also called Polya characteristic functions. The mixture property can be used to obtain generators (Devroye, 1984; see also section IV.6.7). In a black box method one only assumes that ϕ belongs to a certain class of characteristic functions, and that $\phi(t)$ can be computed in finite time for every t. Thus, making use of the mixture property of Polya characteristic functions cannot lead to a black box method because ϕ has to be given explicitly in analytic form.

Under certain regularity conditions, upper bounds for the density can be obtained in terms of quantities (functionals, suprema, and so forth) defined in terms of the characteristic function (Devroye, 1981). These upper bounds can in turn be used in a rejection algorithm. This simple approach is developed in section 3.2. Unfortunately, one now needs to compute f in every iteration of the rejection algorithm. This requires once again an inversion of ϕ, and may not be feasible. One should note however that this can be avoided if we are able to use the series method based upon a convergent series for f. This series could be based upon the inversion formula.

A genuine black box method for a large subclass of Polya characteristic functions was developed in Devroye (1985). Another black box method based upon the series method will be studied in section 3.3.

3.2. The rejection method for characteristic functions.

General rejection algorithms can be based upon the following inequality:

Theorem 3.1.

   Assume that a given distribution has two finite moments, and that the characteristic function ϕ has two absolutely integrable. Then the distribution has a density f bounded as follows:

$$
f(x) \leq
\begin{cases}
\dfrac{1}{2\pi} \int |\phi| \\[2mm]
\dfrac{1}{2\pi x^2} \int |\phi''|
\end{cases}.
$$

The area under the minimum of the two bounding curves is $\dfrac{2}{\pi} \sqrt{\int |\phi| \int |\phi''|}$.

Proof of Theorem 3.1.

Since ϕ is absolutely integrable, f can be computed as follows from ϕ:

$$f(x) = \frac{1}{2\pi} \int \phi(t) e^{-itx} \, dt \ .$$

Furthermore, because the first absolute moment is finite, ϕ' exists and

$$f(x) = \frac{1}{2\pi i x} \int \phi'(t) e^{-itx} \, dt \ .$$

Because the second moment is finite, ϕ'' exists and

$$f(x) = -\frac{1}{2\pi x^2} \int \phi''(t) e^{-itx} \, dt$$

(Loeve, 1963, p. 199). From this, all the inequalities follow trivially. ∎

The integrability condition on ϕ implies that f is bounded and continuous. The integrability condition on ϕ'' translates into a strong tail condition: the tail of f can be tucked under a quickly decreasing curve. This explains why f can globally be tucked under a bounded integrable curve. Based upon Theorem 3.1, we can now formulate a first general rejection algorithm for characteristic functions satisfying the conditions of the Theorem.

General rejection algorithm for characteristic functions

[SET-UP]

$a \leftarrow \dfrac{1}{2\pi} \int |\phi| \ , \ b \leftarrow \dfrac{1}{2\pi} \int |\phi''|$

[GENERATOR]

REPEAT

    Generate two iid uniform $[-1,1]$ random variates U, V.

    IF $U < 0$

        THEN $X \leftarrow \sqrt{\dfrac{b}{a}} V \ , \ T \leftarrow |U| a$

        ELSE $X \leftarrow \sqrt{\dfrac{b}{a}} \dfrac{1}{V} \ , \ T \leftarrow \dfrac{|U| b}{X^2}$

        (Note that this is $|U| aV^2$.)

UNTIL $T \leq f(X)$

RETURN X

Various simplifications are possible in this rudimentary algorithm. What matters is that f is still required in the acceptance step.

Remark 3.1.

The expected number of iterations is $\frac{2}{\pi}\sqrt{\int |\phi| \int |\phi''|}$. This is a scale invariant quantity: indeed, let X have characteristic function ϕ. Then, under the conditions of Theorem 3.1, $\phi(t)=E(e^{itX})$, $\phi''(t)=E(-X^2 e^{itX})$. For the scaled random variable aX, we obtain respectively $\phi(at)$ and $a^2\phi''(at)$. The product of the integrals of the last two functions does not depend upon a. Unfortunately, the product is not translation invariant. Noting that $X+c$ has characteristic function $\phi(t)e^{itc}$, we see that $\int |\phi|$ is translation invariant. However,

$$\int |\phi''| = \int |E(-(X-c)^2 e^{itX})|$$

is not. From the quadratic form of the integrand, one deduces quickly that the integral is approximately minimal when $c=E(X)$, i.e. when the distribution is centered at the mean. This is a common sense observation, reinforced by the symmetric form of the dominating curve. Let us finally note that in Theorem 3.1 we have implicitly proved the inequality

$$\int |\phi| \int |\phi''| \geq \frac{\pi^2}{4},$$

which is of independent interest in mathematical statistics. ∎

If the evaluation of f is to be avoided, then we must find at the very least a converging series for f. Assume first that ϕ is absolutely integrable, symmetric and nonnegative. Then $f(x)$ is sandwiched between consecutive partial sums in the series

$$f(0)-\frac{x^2}{2!}f'(0)+\frac{x^4}{4!}f''(0)-\cdots.$$

This can be seen as follows: since $\cos(tx)$ is sandwiched between consecutive partial sums in its Taylor series expansion, and since

$$f(x) = \frac{1}{2\pi}\int \phi(t)\cos(tx)\, dt,$$

we see that by our assumptions on ϕ, $f(x)$ is sandwiched between consecutive partial sums in

$$\nu_0-\frac{x^2}{2!}\nu_2+\frac{x^4}{4!}\nu_4-\cdots,$$

where

$$\nu_{2n} = \frac{1}{2\pi} \int t^{2n} \phi(t) \, dt \ .$$

If $\int t^{2n} \phi(t) \, dt$ is finite, then $f^{(2n)}$ exists, and its value at 0 is equal to it. This gives the desired collection of inequalities. Note thus that for an inequality involving $f^{(2n)}$ to be valid, we need to ask that

$$\int t^{2n} \phi(t) \, dt \ < \ \infty \ .$$

This moment condition on ϕ is a smoothness condition on f. For extremely smooth f, all moments can be finite. Examples include the normal density, the Cauchy density and all symmetric stable densities with parameter at least equal to one. Also, all characteristic functions with compact support are included, such as the triangular characteristic function. If furthermore the series $x^{2n} \nu_{2n} /(2n)!$ is summable for all $x > 0$, we see that f is determined by all its derivatives at 0. A sufficient condition is

$$\nu_{2n}^{\frac{1}{2n}} = o(n) \ .$$

This class of densities is enormously smooth. In addition, these densities are unimodal with a unique mode at 0 (see exercises). Random variate generation can thus be based upon the alternating series method. As dominating curve, we can use any curve available to us. If Theorem 3.1 is used, note that $\int |\phi| = \int \phi = f(0)$.

Series method for very smooth densities

[NOTE: This algorithm is valid for densities with a symmetric real nonnegative characteristic function for which the value of f is uniquely determined by the Taylor series expansion of f about 0.]

[SET-UP]

$$a \leftarrow \frac{1}{2\pi} \int |\phi| \ (=f(0)) \ , \ b \leftarrow \frac{1}{2\pi} \int |\phi''| \ .$$

[GENERATOR]

REPEAT

Generate a uniform [0,1] random variate U, and a random variate X with density proportional to $g(x) = \min(a, b/x^2)$.

$T \leftarrow Ug(X)$

$S \leftarrow f(0)$, $n \leftarrow 0$, $Q \leftarrow 1$ (prepare for series method)

WHILE $T \leq S$ DO

$\quad n \leftarrow n+1$, $Q \leftarrow -QX^2/(2n(2n-1))$

$\quad S \leftarrow S + Qf^{(n)}(0)$

$\quad$ IF $T \leq S$ THEN RETURN X

$\quad n \leftarrow n+1$, $Q \leftarrow -QX^2/(2n(2n-1))$, $S \leftarrow S + Qf^{(n)}(0)$

UNTIL False

This algorithm could have been presented in the section on the series method, or in the section on universal algorithms. It has a place in this section because it shows how one can avoid inverting the characteristic function in a general rejection method for characteristic functions.

3.3. A black box method.

When ϕ is absolutely integrable, the value of the density f can be computed by the inversion formula

$$f(x) = \frac{1}{2\pi} \int \phi(t)e^{-itx} \ dt \ = \ \int \psi(t) \ dt \ .$$

This integral can be approximated in a number of ways, by using well-known techniques from numerical integration. If such approximations are to be useful, it is essential that we have good explicit estimates of the error. The approximations include the **rectangular rule**

$$r_n(x) = \frac{b-a}{n} \sum_{j=0}^{n-1} \psi(a + (b-a)\frac{j}{n}) \ ,$$

where $[a,b]$ is a finite interval. Other popular rules are the **trapezoidal rule**

$$t_n(x) = \frac{b-a}{n} \sum_{j=1}^{n} (\frac{1}{2}\psi(a + \frac{(j-1)(b-a)}{n}) + \frac{1}{2}\psi(a + \frac{j(b-a)}{n})) ,$$

and **Simpson's rule**

$$s_n(x) = \frac{b-a}{n} \sum_{j=1}^{n} (\frac{1}{6}\psi(a + \frac{(j-1)(b-a)}{n})$$

$$+ \frac{4}{6}\psi(a + \frac{(j-\frac{1}{2})(b-a)}{n}) + \frac{1}{6}\psi(a + \frac{j(b-a)}{n})) .$$

These are the first few rules in an infinite sequence of rules called the **Newton-Cotes integration formulas.** The simple trapezoidal rule integrates linear functions on $[a,b]$ exactly, and Simpson's rule integrates cubics exactly. The next few rules, listed for example in Davis and Rabinowitz (1975, p. 63-64), integrate higher degree polynomials exactly. For example, **Boole's rule** is

$$b_n(x) = \frac{b-a}{n} \sum_{j=1}^{n} (\frac{7}{90}\psi(a + \frac{(j-1)(b-a)}{n}) + \frac{32}{90}\psi(a + \frac{(j-\frac{3}{4})(b-a)}{n})$$

$$+ \frac{12}{90}\psi(a + \frac{(j-\frac{1}{2})(b-a)}{n}) + \frac{32}{90}\psi(a + \frac{(j-\frac{1}{4})(b-a)}{n})$$

$$+ \frac{7}{90}\psi(a + \frac{j(b-a)}{n})) .$$

The error committed by these rules is very important to us. In general ψ is a complex-valued function; and so are the estimates r_n, t_n, etcetera. A little care should be taken when we use only the real parts of these estimates. The main tools are collected in Theorem 3.2:

Theorem 3.2.

Let $[-a, a]$ be a finite interval on the real line, let n be an arbitrary integer, and let the density $f(x)$ be approximated by $f_n(x)$ where $f_n(x)$ is $\text{Re}(r_n(x))$, $\text{Re}(t_n(x))$, $\text{Re}(s_n(x))$, or $\text{Re}(b_n(x))$. Let X be a random variable with density f and j-th absolute moment μ_j. Define the absolute difference $E_n = |f(x) - f_n(x)|$, and the tail integral

$$T_n = \frac{1}{2\pi} \left(\int_{-\infty}^{-a} |\phi| + \int_a^{\infty} |\phi| \right).$$

Then:

A.   If r_n is used and $\mu_1 < \infty$, then

$$E_n \leq T_n + \frac{(2a)^2}{4\pi n} \left(|x| + \mu_1 \right).$$

B.   If t_n is used and $\mu_2 < \infty$, then

$$E_n \leq T_n + \frac{(2a)^3}{24\pi n^2} \left(|x| + \mu_2^{\frac{1}{2}} \right)^2.$$

C.   If s_n is used and $\mu_4 < \infty$, then

$$E_n \leq T_n + \frac{(2a)^5}{360\pi n^4} \left(|x| + \mu_4^{\frac{1}{4}} \right)^4.$$

D.   If b_n is used and $\mu_6 < \infty$, then

$$E_n \leq T_n + \frac{(2a)^7}{3870720\pi n^6} \left(|x| + \mu_6^{\frac{1}{6}} \right)^6.$$

Before proving Theorem 3.2, it is helpful to point out the following inequalities:

Lemma 3.1.

Let ϕ be a characteristic function, and let ψ be defined by

$$\psi(t) = \phi(t)e^{-itx} \ .$$

Assume that the absolute moments for the distribution corresponding to ϕ are denoted by μ_j. Then, if the j-th absolute moment is finite,

$$\sup_t |\psi^{(j)}(t)| \leq \left(|x| + \mu_j^{\frac{1}{j}} \right)^j ,$$

where $j = 0,1,2,\dots$.

Proof of Lemma 3.1.

Note that $\psi^{(j)} = g_j e^{-itx}$ for some function g_j. It can be verified by induction that

$$g_j = \sum_{k=0}^{j} \binom{j}{k} (-ix)^k \phi^{(j-k)} \ .$$

When $\mu_j < \infty$, $\phi^{(j)}$ is a bounded continuous function given by $\int (ix)^j e^{itx} f(x) \, dx$. In particular, $|\phi^{(j)}| \leq \mu_j$. If we also use the inequalities

$$\mu_k \leq \mu_j^{\frac{k}{j}} \quad (k \leq j) ,$$

then we obtain

$$|\psi^{(j)}| \leq |g_j| \leq \sum_{k=0}^{j} \binom{j}{k} |x|^k \mu_{j-k}$$

$$\leq \sum_{k=0}^{j} \binom{j}{k} |x|^k \mu_j^{\frac{j-k}{j}}$$

$$= \left(|x| + \mu_j^{\frac{1}{j}} \right)^j . \blacksquare$$

Proof of Theorem 3.2.

Let us define $\psi(t) = \dfrac{1}{2\pi} \phi(t)e^{-itx}$. Then by Lemma 3.1,

$$2\pi |\psi^{(j)}| \leq \left(|x| + \mu_j^{\frac{1}{j}} \right)^j ,$$

where μ_j is the finite j-th absolute moment of the distribution. Next, we need some estimates from numerical analysis. In particular,

$$| f(x) - f_n(x) | \leq T_n + | \int_{-a}^{a} \text{Re}(\psi(t)) \, dt - f_n(x) | .$$

To the last term, which is an error term in the estimation of the integral of $\text{Re}(\psi)$ over a finite interval, we can apply estimates such as those given in Davis and Rabinowitz (1975, pp. 40-64). To apply these estimates, we recall that, when $\mu_j < \infty$, ψ is a bounded continuous function on the real line. If r_n is used and $\mu_1 < \infty$, then the last term does not exceed

$$\frac{(2a)^2}{2n} \sup | \text{Re}(\psi)' | \leq \frac{(2a)^2}{2n} \sup | \psi^{(1)} |$$

$$\leq \frac{(2a)^2}{4\pi n} \left(| x | + \mu_1 \right) .$$

If t_n is used and $\mu_2 < \infty$, then the last term does not exceed

$$\frac{(2a)^3}{12n^2} \sup | \psi^{(2)} | \leq \frac{(2a)^3}{24\pi n^2} \left(| x | + \mu_2^{\frac{1}{2}} \right)^2 .$$

If s_n is used and $\mu_4 < \infty$, then the last term does not exceed

$$\frac{(2a)^5}{180n^4} \sup | \psi^{(4)} | \leq \frac{(2a)^5}{360\pi n^4} \left(| x | + \mu_4^{\frac{1}{4}} \right)^4 .$$

If b_n is used and $\mu_6 < \infty$, then the last term does not exceed

$$\frac{(2a)^7}{1935360n^6} \sup | \psi^{(6)} | \leq \frac{(2a)^7}{3870720\pi n^6} \left(| x | + \mu_6^{\frac{1}{6}} \right)^6 . \blacksquare$$

The bounds of Theorem 3.2 allow us to apply the series method. There are two key problems left to solve:

A.   The choice of a as a function of n.

B.   The selection of a dominating curve g for rejection.

It is wasteful to compute $t_n, t_{n+1}, t_{n+2}, \dots$ when trying to make an acceptance or rejection decision. Because the error decreases at a polynomial rate with n, it seems better to evaluate t_{c^k} for some $c > 1$ and $k = 1,2,\dots$. Additionally, it is advantageous to use the standard dyadic "trick" of computing only t_2, t_4, t_8, etcetera. When computing t_{2n}, the computations made for t_n can be reused provided that we align the cutpoints. In other words, if a_n is the constant a with the dependence upon n made explicit, it is necessary to demand that

$$\frac{a_{2^t}}{2^k}$$

be equal to

$$\frac{a_{2^{k+1}}}{2^{k+1}}$$

or to

$$\frac{a_{2^{k+1}}}{2^{k}} .$$

Thus, $a_{2^{k+1}}$ is equal to $a_{2^{k}}$ or to twice that value. Note that for the estimates f_n in Theorem 3.2 to tend to $f(x)$, it is necessary that $a_n \to \infty$ (unless the characteristic function has compact support), and that $a_n = o(n^{\overline{j+1}})$ where j is 1,2,4 or 6 depending upon the estimator used. Thus, it does not hurt to choose a_n monotone and of the form

$$a_{2^{k}} = a_0 2^{c_k}$$

where c_k is a positive integer sequence satisfying $c_{k+1} - c_k \in \{0,1\}$, and a_0 is a constant.

The problem of the selection of a dominating curve has a simple solution in many cases. To be able to use Theorem 3.2, we need upper bounds for μ_j and $\int_a^{\infty} |\phi|$. Luckily, this is also sufficient for the design of good upper bounds. To make this point, we consider several examples, after an auxiliary lemma.

Lemma 3.2.

Let ϕ be a characteristic function with continuous absolutely integrable n-th derivative $\phi^{(n)}$ where n is a nonnegative integer. Then ϕ has a density f where

$$f(x) \leq \frac{\int |\phi^{(n)}|}{2\pi |x|^n} .$$

If $\int |t| |\phi(t)| dt < \infty$, then ϕ has a Lipschitz density f with Lipschitz constant not exceeding

$$\frac{\int |t| |\phi(t)| dt}{2\pi} .$$

Proof of Lemma 3.2.

When ϕ has a continuous absolutely integrable n-th derivative $\phi^{(n)}$, then a density f exists, and the following inversion formula is valid:

$$(ix)^n f(x) = \frac{1}{2\pi} \int \phi^{(n)}(t) e^{-tx} \, dt \quad .$$

The first inequality follows directly from this. Next, assume that $\int |t| \, |\phi(t)| \, dt < \infty$. Once again, a density f exists, and because f can be computed by the standard inversion formula, we have

$$|f(x) - f(y)| = \frac{1}{2\pi} \left| \int (e^{-itx} - e^{-ity}) \, \phi(t) \, dt \right|$$

$$\leq \frac{1}{2\pi} \int \left| e^{-it(y-x)} - 1 \right| \, |\phi(t)| \, dt$$

$$\leq \frac{1}{2\pi} |y-x| \int |t| \, |\phi(t)| \, dt \quad . \blacksquare$$

Example 3.1. Characteristic functions with compact support.

Assume that ϕ is known to vanish outside $[-A, A]$ for some finite value A. It should be stressed that this is a very strong condition of smoothness for the density f of this distribution. From Lemma 3.2, we know that f is a bounded density:

$$f(x) \leq \frac{A}{\pi} \quad .$$

Furthermore, f is Lipschitz with Lipschitz constant C not exceeding $A^2/(2\pi)$. The densities in this class can have arbitrarily large tails, and can not be uniformly bounded without imposing some sort of tail condition. For a detailed discussion of this, we refer to section VII.3.3, and in particular to Example VII.3.4, where a dominating curve for a Lipschitz (C) density on the positive real line with absolute moment μ_j $(j > 2)$ is given. The area under that dominating curve is

$$2\sqrt{8C} \, \frac{j}{j-2} \mu_j^{\frac{1}{j}} \quad .$$

Here the factor 2 allows for the extension of the bound to the entire real line. Note that with $C = A^2/(2\pi)$, the rejection constant becomes

$$\frac{4A}{\sqrt{\pi}} \frac{j}{j-2} \mu_j^{\frac{1}{j}} \quad ,$$

which is scale invariant.

We suggest that a be picked constant and equal to A, since $T_n = 0$ in Theorem 3.2 when $a \geq A$. ∎

Example 3.2. Unimodal densities.

For unimodal densities with mode at 0, a variety of good dominating curves were given in section VII.3.2. These required a bound on the value of $f(0)$ and one additional piece of information, such as an upper bound for μ_j. For the bound at the mode, we can use

$$f(x) \leq \frac{\int |\phi|}{2\pi} .$$

It is difficult to verify the unimodality of a density from a characteristic function, so this example is not as strong as Example 3.1. Also, the choice of a causes a few extra problems. See Example 3.3 below. ∎

Example 3.3. Optimization of parameter a.

Using a Chebyshev type inequality applied to characteristic functions,

$$\int_a^\infty |\phi| \leq \frac{\int_0^\infty |t|^r |\phi(t)| \, dt}{a^r} ,$$

we can obtain upper bounds of the form $ca^k + da^{-r}$ for the error E_n in Theorem 3.2, where c, d, k, r are positive constants, and c depends upon n. Considered as a function of a, this has one minimum at

$$a = \left(\frac{dr}{ck} \right)^{\frac{1}{k+r}} .$$

The minimal value is

$$c^{\frac{r}{k+r}} d^{\frac{k}{k+r}} \left(\left(\frac{r}{k}\right)^{\frac{k}{k+r}} + \left(\frac{k}{r}\right)^{\frac{r}{k+r}} \right) .$$

What matters here is that the only factor depending upon n is the first one, and that it tends to 0 at the rate $c^{r/(k+r)}$. Since c varies typically as $n^{-(k-1)}$ for the estimators given in Theorem 3.2, we obtain the rate

$$n^{-\frac{r(k-1)}{k+r}}.$$

This rate is necessarily sublinear when $r=1$, regardless of how large k is. Note that it decreases quickly when $r \geq 2$ for all usual values of k. For example, with $r=2$ and Simpson's rule ($k=5$), our rate is $n^{-8/7}$. With $r=3$ and the trapezoidal rule ($k=3$), our rate is $n^{-3/2}$. ∎

Example 3.4. Sums of iid uniform random variables.

The uniform density on $[-1,1]$ has characteristic function $\phi(t) = \sin(t)/t$. The sum of m iid uniform $[-1,1]$ random variables has characteristic function

$$\phi_m(t) = \left(\frac{\sin(t)}{t}\right)^m.$$

The corresponding density is unimodal, which should be of help in the derivation of bounds for the density. By taking consecutive derivatives of ϕ_m, it is easily established that the second moment μ_2 is $\frac{m}{3}$, and that the fourth moment μ_4 is $\frac{m^2}{3} - \frac{2m}{15}$. Furthermore, the mode, which occurs at zero, has value

$$\frac{1}{2\pi}\int \phi_m(t)\,dt$$

$$\leq \frac{1}{2\pi}\int \min\left((1-\frac{t^2}{6}+\frac{t^4}{120})^m, |t|^{-m}\right) dt$$

$$\leq \frac{1}{2\pi}\int \min\left(e^{-\frac{m}{6}t^2(1-\frac{t^2}{20})}, |t|^{-m}\right) dt$$

$$\leq \frac{2}{2\pi(m-1)}+\int \frac{1}{2\pi}e^{-\frac{m}{6}\frac{19}{20}t^2}\,dt$$

$$= \frac{1}{\pi(m-1)}+\sqrt{\frac{60}{19m}} = M\ ,$$

where we split the integral over the intervals $[-1,1]$ and its complement. We now refer to Theorem VII.3.2 for symmetric unimodal densities bounded by M and having r-th absolute moment μ_r. Such densities are bounded by $\min(M, (r+1)\mu_r/|x|^{r+1})$, and the dominating curve has integral

$$\frac{r+1}{r}((r+1)\mu_r)^{\frac{1}{r+1}} M^{\frac{r}{r+1}}.$$

For example, for $r = 4$, we obtain in our example

$$\frac{5}{4}(5\mu_4)^{\frac{1}{5}} M^{\frac{4}{5}} \sim \frac{5}{4}(\frac{5}{3})^{\frac{1}{5}}(\frac{60}{19})^{\frac{2}{5}}$$

as $m \rightarrow \infty$. In other words, as $m \rightarrow \infty$, the rejection constant tends to a fixed value. One can verify that this same property holds true for all values of $r > 0$. This example is continued in Example 3.6. ■

This leaves us with the black box algorithm and its analysis. We assume that a dominating curve cg is known, where g is a density, that another function h is known having the property that

$$\frac{1}{2\pi}\left(\int_{-\infty}^{-a} |\phi| + \int_{a}^{\infty} |\phi|\right) \le h(a) \quad (a > 0),$$

and that integrals will be evaluated only for the subsequence $a_0 2^k$, $k \ge 0$, where a_0 is a given integer. Let f_n denote a numerical integral estimating ψ such as r_n, s_n, t_n or b_n. This estimate uses as interval of integration $[-l(n,x), l(n,x)]$ for some function l which normally diverges as n tends to ∞.

Series method based upon numerical integration

REPEAT
    Generate a random variate X with density g.
    Generate a uniform [0,1] random variate U.
    Compute $T \leftarrow Ucg(X)$ (recall that $f \le cg$).
    $n \leftarrow a_0 2^k$, $a \leftarrow l(n,X)$ (prepare for integration)
    REPEAT
        $W \leftarrow f_n(X)$ (f_n is an integral estimate of $f = \int \psi$ with parameter n on interval $[-a, a]$; the number of evaluations of ϕ required is proportional to n)
        Compute an upper bound on the error, E. (Use the bounds of Theorem 3.2 plus $h(a)$.)
        $n \leftarrow 2n$
    UNTIL $|T - W| > E$
UNTIL $T < W$
RETURN X

The first issue is that of correctness of the algorithm. This boils down to verifying whether the algorithm halts with probability one. We have:

Theorem 3.3.

The algorithm based upon the series method given above is correct, i.e. halts with probability one, when

$$\lim_{n \to \infty} l(n,x) = \infty \quad \text{(all } x \text{)} ,$$

$$\lim_{a \to \infty} h(a) = \infty$$

(this forces ϕ to be absolutely integrable), and one of the following conditions holds:

A. r_n is used, $\mu_1 < \infty$, and $l(n,x) = o(n^{1/2})$ for all x.

B. t_n is used, $\mu_2 < \infty$, and $l(n,x) = o(n^{2/3})$ for all x.

C. s_n is used, $\mu_4 < \infty$, and $l(n,x) = o(n^{4/5})$ for all x.

D. b_n is used, $\mu_6 < \infty$, and $l(n,x) = o(n^{6/7})$ for all x.

Here μ_j is the j-th absolute moment for f.

Proof of Theorem 3.3.

We need only verify that the error bound used in the algorithm tends to 0 as $n \to \infty$ for all x. Theorem 3.3 is a direct corollary of Theorem 3.2. ∎

Theorem 3.3 is reassuring. Under very mild conditions on the density, a valid algorithm indeed exists. We have to know μ_j for some j and we need also an explicit expression for the tail bound $h(a)$. The theorem just states that given this information, we can choose a function $l(n,x)$ and an estimator f_n which guarantee the validity. Unfortunately, there is a snake in the grass. The function $l(n,x)$ has a profound impact on the time before halting. In many examples, the expected time is ∞. Thus, let us consider the expected number of evaluations of ψ (or ϕ) before halting. This can't possibly be given without discussing how large $h(.)$ is, and which function $l(.,.)$ is picked. Perhaps the best thing to do at this stage is to offer a helpful lemma, and then to illustrate it on a few examples.

Lemma 3.3.

    Consider the series method given above, and assume that for the given functions h and l, we have an inequality of the type

$$| f(x) - f_n(x) | \leq C(x) n^{-\alpha} \qquad (n \geq 1 \text{ , all } x) \text{ ,}$$

where C is a positive function and $\alpha > 1$ is a constant. If $a_0 = 1$ and f_n requires $\beta n + 1$ evaluations of ψ for some constant β (for t_n, $\beta = 1$, and for s_n, $\beta = 2$), then the expected number of evaluations of ψ before halting does not exceed

$$\leq c(\beta + 1) + 2^\gamma c^{1-\gamma} \int C^\gamma g^{1-\gamma} \frac{2\beta + 1}{1 - 2^{1-\gamma\alpha}}$$

$$\leq c(\beta + 1) + 2^\gamma c^{1-\gamma} \frac{2\beta + 1}{1 - 2^{1-\gamma\alpha}} \left(\int Cg \right)^{1-\gamma} \left[\int C^{2 - \frac{1}{\gamma}} \right]^\gamma \text{ ,}$$

where γ is a number satisfying

$$\alpha\gamma > 1 \text{ , } \gamma \leq 1 \text{ .}$$

Proof of Lemma 3.3.

    By Wald's equation, our expected number is equal to c times the expected number of evaluations in the first iteration (regardless of acceptance or rejection). Let us first condition on $X = x$ with density g. For f_1, we use up $\beta + 1$ evaluations in all cases. The probability of having to evaluate f_2 does not exceed $2C(x)1^{-\alpha}/cg(x)$. Continuing in this fashion, it is easily seen that the expected number of evaluations of ψ is not greater than

$$\sum_{k=0}^{\infty} \left[(\beta 2^{k+1} + 1) \min\left(\frac{2C(x)(2^k)^{-\alpha}}{cg(x)}, 1 \right) \right] + \beta + 1 \text{ .}$$

Taking expectations with respect to $g(x) \, dx$ and multiplying with c gives the unconditional upper bound

$$c(\beta + 1) + \sum_{k=0}^{\infty} \left((\beta 2^{k+1} + 1) \int \min(2C(x)(2^k)^{-\alpha}, cg(x)) \, dx \right)$$

$$\leq c(\beta + 1) + \sum_{k=0}^{\infty} \left((\beta 2^{k+1} + 1) \int \min(2C(x)(2^k)^{-\alpha}, cg(x)) \, dx \right)$$

$$\leq c(\beta + 1) + \int (2C(x))^\gamma (cg(x))^{1-\gamma} \, dx \sum_{k=0}^{\infty} 2^{-k\gamma\alpha}(\beta 2^{k+1} + 1)$$

$$= c(\beta + 1) + 2^\gamma c^{1-\gamma} \int C^\gamma g^{1-\gamma} \left(\frac{2\beta}{1 - 2^{1-\gamma\alpha}} + \frac{1}{1 - 2^{-\gamma\alpha}} \right)$$

$$\leq c(\beta + 1) + 2^\gamma c^{1-\gamma} \int C^\gamma g^{1-\gamma} \frac{2\beta + 1}{1 - 2^{1-\gamma\alpha}} \text{ ,}$$

where γ is a number satisfying

$$\alpha\gamma > 1 \, , \gamma \le 1 \, .$$

By Holder's inequality, the integral in the last expression does not exceed

$$\left(\int Cg\right)^{1-\gamma}\left[\int C^{2-\frac{1}{\gamma}}\right]^{\gamma} . \blacksquare$$

Lemma 3.3 reveals the extent to which the efficiency of the algorithm is affected by c , $C(x)$, $g(x)$ and μ_j .

Example 3.5. Characteristic functions with compact support.

Assume that the characteristic function vanishes outside $[-A \, , A \,]$. If we take $l(n,x)=A$, then $h \equiv 0$ in the algorithm. Note that this choice violates the consistency conditions of Theorem 3.3, but leads nevertheless to a consistent procedure. With t_n , we have $\beta=1, \alpha=2$ and an error

$$E_n \le C(x)n^{-\alpha}$$

where

$$C(x) = \frac{(2A)^3}{24\pi}(\,\mid x \mid + \sqrt{\mu_2})^2 \, .$$

With s_n , we have $\beta=2$, $\alpha=4$ and

$$C(x) = \frac{(2A)^5}{360\pi}(\,\mid x \mid + \mu_4^{\frac{1}{4}})^4 \, .$$

With both error bounds, $\int C = \infty$, so we can't take $\gamma=1$ in Lemma 3.3. Also,

$$\int C^{2-\frac{1}{\gamma}} < \infty$$

when $\frac{1}{\gamma} > 2 + \frac{1}{\alpha}$. Thus, for the bound of Lemma 3.3 to be useful, we need to choose

$$\frac{1}{\alpha} < \gamma < \frac{\alpha}{2\alpha+1} \, .$$

This yields the intervals $(\frac{1}{2}, \frac{2}{5})$ and $(\frac{1}{4}, \frac{4}{9})$ respectively. Of course, the former interval is empty. This is due to the fact that the last inequality in Lemma 3.3 (combined with Theorem 3.2) never leads to a finite upper bound for the trapezoidal rule. Let us further concentrate therefore on s_n . Note that

$$\int Cg \ \leq \ \frac{(2A\ )^5}{360\pi} \ \int g\ (x\ )(\mid x\ \mid +\mu_4{}^{\frac{1}{4}})^4 \ dx$$

$$\leq \ \frac{(2A\ )^5}{360\pi} \ 8\int g\ (x\ )(\mid x\ \mid ^4+\mu_4)\ dx$$

$$= \ \frac{32A\ ^5(\mu\ {}^*_4+\mu_4)}{45\pi}\ ,$$

where $\mu{}^*_4$ is the fourth absolute moment for g. Typically, when g is close to f, the fourth moment is close to that of f. We won't proceed here with the explicit computation of the full bound of Lemma 3.3. It suffices to note that the bound is large when either A or μ_4 is large. In other words, it is large when the support of ϕ is large (the density is less smooth) and/or the tail of the density is large. Let us conclude this section by repeating the algorithm:

Series method based upon numerical integration

[NOTE: The characteristic function ϕ vanishes off $[-A\ ,A\ ]$, and the fourth absolute moment does not exceed μ_4.]

REPEAT

    Generate a random variate X with density g.

    Generate a uniform $[0,1]$ random variate U.

    Compute $T \leftarrow Ucg\ (X\ )$ (recall that $f \leq cg$).

    $n \leftarrow a_0$ (prepare for integration)

    REPEAT

        $W \leftarrow \text{Re}(s_n\ (X\ ))$ (s_n is Simpson's integral estimate of $f =\int \psi$ with parameter n on interval $[-A\ ,A\ ]$; the number of evaluations of ϕ required is $2n +1$)

        $E \leftarrow \frac{(2A\ )^5}{360\pi}(\mid X\ \mid +\mu_4{}^{\frac{1}{4}})^4 n^{-4}$

        $n \leftarrow 2n$

    UNTIL $\mid T -W\ \mid >E$

UNTIL $T < W$

RETURN X

For dominating curves cg, there are numerous possibilities. See for example Lemma 3.2. In Example 3.1, a dominating curve based upon an inequality for Lipschitz densities (section VII.3.4) was developed. The rejection constant c for that example is

$$\frac{8}{\sqrt{\pi}}A\ \mu_4{}^{\frac{1}{4}}\ .\ \blacksquare$$

Example 3.6. Sums of iid uniform random variables.

This is a continuation of Example 3.4, where a good dominating density was found for use in the rejection algorithm. What is left here is mainly the choice of h and l for use in the algorithm. Let us start with the decision to estimate f by Simpson's rule s_n. This is based upon a quick preliminary analysis which shows that the trapezoidal rule for example just isn't good enough to obtain finite expected time.

The function $h(a)$ can be chosen as

$$h(a) = \frac{1}{\pi a^{m-1}(m-1)}$$

where m is the number of uniform $[-1,1]$ random variables that are summed. To see this, note that

$$2 \int_a^\infty \frac{1}{2\pi} \left| \frac{\sin(t)}{t} \right|^m dt \le \frac{1}{\pi} \int_a^\infty |t|^{-m} dt = h(a).$$

Given $X=x$ in the algorithm, we see that with s_n, the error E_n is not greater than

$$E_n \le h(a) + \frac{(2a)^5(|x|+\mu_4^{1/4})^4}{360\,\pi n^4},$$

where a determines the integration interval (Theorem 3.2). Optimization of the upper bound with respect to a is simple and leads to the value

$$a = \left(\frac{9n^4}{4(|x|+\mu_4^{1/4})} \right)^{\frac{1}{m+4}}.$$

With this value for a (or $l(n,x)$), we obtain

$$E_n \le C(x)n^{-\alpha}$$

for $\alpha = 4(m-1)/(m+4)$ and

$$C(x) = \frac{m}{m-1} \frac{1}{\pi} \left(\frac{4}{9} \right)^{\frac{m-1}{m+4}} (|x|+\mu_4^{1/4})^\alpha.$$

This is all the users need to implement the algorithm. We can now apply Lemma 3.3 to obtain an idea of the expected complexity of the algorithm. We will show that the expected time is better than $O(m^{(5+\epsilon)/8})$ for all $\epsilon > 0$. A brief outline of the proof should suffice at this point. In Lemma 3.3, we need to pick a constant γ. The conditions $\alpha\gamma > 1$ and $\int C^{2-\frac{1}{\gamma}} < \infty$ force us to impose the conditions

$$\frac{m+4}{4m-4} < \gamma < \frac{4m-4}{9m-4}.$$

Both inequalities can be satisfied simultaneously for all $m \geq 9$. After fixing γ, compute all quantities in the upper bound of Lemma 3.3. Since $C(x) = (C_0 + o(1))(\mid x \mid + \mu_4^{1/4})^\alpha$ with $C_0 = 4/(9\pi)$, it is easy to see that

$$\int Cg = (C_0 + o(1))E((\mid X \mid + \mu_4^{\frac{1}{4}})^\alpha)$$

where X is a random variable with density g, and $\alpha = 4(m-1)/(m+4)$. We can choose g such that $E(\mid X \mid^\alpha)$ is close to $\mu_4^{\alpha/4}$ (e.g., in Example 3.4, take $r = 6$ or larger in the bound for unimodal densities; taking $r = 4$ isn't good enough because for $r = 4$, $E(\mid X \mid^4) = \infty$). Noting next that $\mu_4^{1/4} \sim \sqrt{m}/3^{1/4}$ as $m \to \infty$, we note that $\int Cg$ increases as a constant times $m^{\alpha/2}$. Next, $\int C^{2 - \frac{1}{\gamma}}$ increases as a constant times

$$\mu_4^{\frac{1 + \alpha(2 - \frac{1}{\gamma})}{4}}$$

which in turn increases as $m^{\frac{9}{2} - \frac{2}{\gamma}}$. The upper bound in Lemma 3.3 increases as

$$m^{2 - 2\gamma + \frac{9\gamma}{2} - 2} = m^{\frac{5\gamma}{2}}.$$

The smallest allowable value for γ is $1/\alpha \sim 1/4$. Thus, the upper bound on the expected complexity is of the order of magnitude of $m^{5/8}$. ∎

3.4. Exercises.

1. Show that when a characteristic function ϕ is absolutely integrable, then the distribution has a bounded continuous density f. Is the density also uniformly continuous?

2. Construct a symmetric real characteristic function for a distribution with a density, having the property that ϕ takes negative and positive values.

3. Consider symmetric nonnegative characteristic functions ϕ, and define $\nu_{2n} = \int t^{2n} \phi(t)\, dt$.

    A. Show that $\nu_{2n}^{1/(2n)} = o(n)$ implies that $(x^{2n} \nu_{2n})/(2n)!$ is summable for all $x > 0$.

    B. Show that f is unimodal and has a unique mode at 0 (Feller, 1971, p. 528).

    C. In the alternating series algorithm for this class of densities given in the text, why can we take $b = \mu_1$ or $b = \sigma$ in the formula for the dominating curve where μ_1 is the first absolute moment for f and σ is the standard deviation for f ?

D.   A continuation of part C. If all operations in the algorithm take one unit of time, give a useful sufficient condition on ϕ for the expected time of the algorithm to be finite.

4.   The following is an important symmetric nonnegative characteristic function:

$$\phi(t) = \sqrt{\frac{\sqrt{2t}}{\sinh(\sqrt{2t})}} = \frac{1}{\sqrt{1 + 2\dfrac{|t|}{3!} + 2\dfrac{|t|^2}{5!} + \cdots}}$$

(see e.g. Anderson and Darling, 1952). Near $t=0$, ϕ varies as $1 - |t|/6$. This implies that the first absolute moment is infinite. Find a dominating curve for this particular characteristic function, verify that the density f is determined by its Taylor series about 0, and give all the details of the alternating series method for this distribution.

5.   The following characteristic function appears as the limit of a sequence of characteristic functions in mathematical statistics (Anderson and Darling, 1952):

$$\phi(t) = \left[\frac{-2\pi it}{\cos(\dfrac{\pi}{2}\sqrt{1+8it})} \right]^{\frac{1}{2}}.$$

Give a finite time random variate generator for this distribution. Ignore efficiency issues (e.g., the expected time is allowed to be infinite).

6.   Give the full details of the proof that the expected number of evaluations of ϕ in the series method for generating the sum of m iid uniform $[-1,1]$ random variables (Example 3.6) is $O(m^{(5+\epsilon)/8})$ for all $\epsilon > 0$.

7.   How can you improve on the expected complexity in Example 3.6?

4. THE SIMULATION OF SUMS.

4.1. Problem statement.

Let X be a random variable with density f on the real line. In this section we consider the problem of the simulation of $S_n = X_1 + \cdots + X_n$ where $X_1, \ldots, X_n$ are iid random variables distributed as X. The naive method

Naive method

$S \leftarrow 0$
FOR $i := 1$ TO n DO
      Generate X with density f .
      $S \leftarrow S + X$
RETURN S

takes worst-case or expected time proportional to n depending upon whether X can be generated in constant worst-case or constant expected time. We say that a generator is uniformly fast when the expected time $E(T_n)$ needed to generate S_n satisfies

$$\sup_{n \geq 1} E(T_n) < \infty .$$

This supremum is allowed to depend upon f . Note that the uniformity is with respect to n and not to f . This differs from our standard notion of uniformity over a class of distributions.

In trying to develop uniformly fast generators, we should get a lot of help from the central limit theorem, which states that under some conditions on the distribution of X, the sum S_n, properly normalized, tends in distribution to one of the stable laws. Ideally, a uniformly fast generator should return such a stable random variate most of the time. What complicates matters is that the distribution of S_n is not easy to describe. For example, in a rejection based method, the computation of the value of the density of S_n at one point usually requires time increasing with n. Needless to say, it is this hurdle which makes the problem both challenging and interesting.

In a first approach, we will cheat a bit: recall that if ϕ is the characteristic function of X, then S_n has characteristic function ϕ^n. If we have a uniformly fast generator for the family $\{\phi, \phi^2, \ldots, \phi^n, \ldots\}$, then we are done. In other words, we reduce the problem to that of the generation of random variates with a given characteristic function, discussed in section 3. The reason why we call this cheating is that ϕ is usually not available, only f .

In the second approach, the problem is tackled head on. We will first derive inequalities which relate the density of S_n to the normal density. In proving the inequalities, we have to rederive a so-called local central limit theorem. The inequalities allow us to design uniformly fast rejection algorithms which return a stable random variate with high probability. The tightness of the bounds allows us to obtain this result despite the fact that the density of S_n can't usually be computed in constant time. When the density can be computed in constant time, the algorithm is extremely efficient. This is the case when the density of S_n has a relatively simple analytic form, as in the case of the exponential density when

S_n is gamma (n).

Other solutions are suggested in the exercises and in later sections, but the most promising generally applicable strategies are definitely the two mentioned above.

4.2. A detour via characteristic functions.

S_n has characteristic function ϕ^n when X has characteristic function ϕ. This fact can be used to generate S_n efficiently provided that all the ϕ_n's belong to a family of characteristic functions for which a good efficient generator is available.

One such family is the family of Polya characteristic functions dealt with in section IV.6.7. In particular, if ϕ is Polya, so is ϕ^n. Based upon Theorems IV.6.8 and IV.6.9, we can conclude the following:

Theorem 4.1.

If ϕ is a Polya characteristic function, then $X \leftarrow \dfrac{Y}{Z}$ has characteristic function ϕ^n when Y, Z are independent random variables, Y has the FVP density (defined in Theorem IV.6.9), and Z has distribution function

$$F(s) = 1 - \phi^n + sn\,\phi'(s)\phi^{n-1}(s) \quad (s > 0).$$

Here ϕ' is the right-hand derivative of ϕ. When F is absolutely continuous, then it has density

$$s^2 n(n-1)\phi'^2(s)\phi^{n-2}(s) + s^2 n\,\phi''(s)\phi^{n-1}(s) \quad (s > 0).$$

When ϕ is explicitly given, and it often is, this method should prove to be a formidable competitor. For one thing, we have reduced the problem to one of generating a random variate with an explicitly given distribution function or density, i.e. we have taken the problem out of the domain of characteristic functions.

The principle outlined here can be extended to a few other classes of characteristic functions, but we are still far away from a generally applicable technique, let alone a universal black box method. The approach outlined in the next section is better suited for this purpose.

4.3. Rejection based upon a local central limit theorem.

We assume that f is a zero mean density with finite variance σ^2. Summing n iid random variables with this density is known to give a random variable with approximately normal $(0, n\sigma^2)$ distribution. The study of the closeness of this approximation is the subject of the classical central limit theory. The only things that can be of use to us are precise (i.e., not asymptotic) inequalities which clarify just how close the density of S_n is to the normal $(0, n\sigma^2)$ density. For a smooth treatment, we put two further restrictions on f :

A.   The density f has an absolutely integrable characteristic function ϕ. Recall that this implies among other things that f is bounded and continuous.

B.   The random variable X has finite third absolute moment not exceeding β: $E(|X|^3) \le \beta < \infty$.

Condition A allows us to use the simple inversion formula for characteristic functions, while condition B guarantees us that the error term is $O(1/\sqrt{n})$. Densities f satisfying all the conditions outlined above are called **regular**. Clearly, most zero mean densities occurring in practice are regular. There is only one large class of exceptions, the distributions in the domain of attraction of stable laws. By forcing the variance to be finite, we can only have convergence to the normal distribution. In exercise 4.1, which is more a research project than an exercise, the reader is challenged to repeat this section for distributions whose sums converge to symmetric stable laws with parameter $\alpha < 2$. For once we will do things backwards, by giving the results and their implications before the proofs, which are deferred to next section.

The fundamental result upon which this entire section rests is the following form of a local central limit theorem:

Theorem 4.2.

Let f be a regular density, and let f_n be the density of $S_n/(\sigma\sqrt{n})$. Let g be the standard normal density. There exist sequences a_n and b_n only depending upon f such that

$$|f_n(x) - g(x)| \le h_n(x) = \min(a_n, \frac{b_n}{x^2}),$$

and

$$\max(a_n, b_n) = O(\frac{1}{\sqrt{n}}).$$

For a proof and references, see section 4.4. Explicit values for a_n and b_n follow. It is important to note that

$$g - h_n \le f_n \le g + h_n,$$

where $\int h_n = O\left(1/\sqrt{n}\right)$. In other words, the inequality is eminently suited for use in a rejection algorithm with squeezing. Both g and h_n can be considered as very easy densities from a random variate generation point of view. Furthermore, the obvious rejection algorithm, described in Example II.3.6, has rejection constant $1+\int h_n$ tending to 1 as $n \rightarrow \infty$. There is even more good news: if the lower bound is used for squeezing, then the expected number of evaluations of f is at most $2\int h_n = O\left(1/\sqrt{n}\right) = o\left(1\right)$. The cumbersome part is the evaluation of f_n.

There are essentially two possibilities when it comes to evaluating f_n: first, f_n is explicitly known. This is for example the case when f is an exponential density centered around its mean, and f_n is the density of a linearly transformed gamma (n) density. In the case of the gamma density, we can easily compute the different constants in the bound of Theorem 4.2. as is done in exercise 4.2. Another example for the sums of uniform random variables follows in a separate section.

To compute f_n via convolutions is all but impossible. The only other alternative is to write f_n as a series based upon the inversion formula for ϕ^n, and to apply the series method. Here too the hurdles are formidable.

4.4. A local limit theorem.

It is the purpose of this section to prove Theorem 4.2. The proof is quite long, and is given in full because we require explicit knowledge of the bounding sequence, and a careful derivation of the bounds to keep the constants as small as possible. Local limit theorems of the type needed by us have been derived in a number of papers, see e.g. Inzevitov (1977), Survila (1964) and Maejima (1980). An excellent general reference is Petrov (1975). For example, Survila (1964) has obtained the existence of a constant C depending upon f only such that for regular f ,

$$\left| f_n\left(x\right) - g\left(x\right)\right| \leq \frac{C}{1+x^2} \ .$$

Ibragimov and Linnik (1971) have obtained an upper bound of the type $\dfrac{C}{\sqrt{n}}$. Note that Survila's bound does not tend to zero with n. The Ibragimov-Linnik upper bound is called a uniform estimate in the local central limit theorem. Such uniform estimates are useless to us because the upper bound when integrated with respect to x is not finite. The bound which we derive here uses well-known tricks of the trade, documented for example in Petrov (1975) and Maejima (1980).

Let us start slowly with a few key lemmas.

Lemma 4.1.

For any real t,

$$\left| e^{it} - \sum_{j=0}^{n-1} \frac{(it)^j}{j!} \right| \leq \frac{t^n}{n!} \quad (n \geq 0).$$

Lemma 4.2.

Let ϕ be the characteristic function for a regular density f. Then the following inequalities are valid:

$$\left| \phi(t)-1+\frac{\sigma^2 t^2}{2} \right| \leq \frac{\beta |t|^3}{6},$$

$$\left| \phi'(t)+t\sigma^2 \right| \leq \frac{\beta}{2}t^2,$$

$$\left| \phi''(t)+\sigma^2 \right| \leq \beta |t|.$$

Proof of Lemma 4.2.

Since three absolute moments exist, we notice that the first three derivatives of ϕ exist and are continuous functions given by the formulas (Feller, 1971, p. 512)

$$\phi^{(j)}(t) = \int e^{itx} (ix)^j f(x)\, dx \quad (j=0,1,2,3).$$

Observe that

$$\left| \phi(t)-1+\frac{\sigma^2 t^2}{2} \right| \leq \int \left| e^{itu}-1-itu+\frac{t^2 u^2}{2} \right| f(u)\, du$$

$$\leq \int \left| \frac{|t|^3 |u|^3}{6} \right| f(u)\, du = \frac{\beta}{6} |t|^3.$$

Next,

$$\phi'(t)+\frac{\sigma^2 t}{2} = \int (e^{itu}-1-itu) iu f(u)\, du.$$

Thus,

$$\left| \phi'(t)+\frac{\sigma^2 t}{2} \right| \leq \int \left| \frac{t^2 u^2}{2} \right| |u| f(u)\, du \leq \frac{\beta}{2}t^2.$$

Finally,

$$\phi''(t)+\sigma^2 = -\int (e^{itu}-1)u^2 f(u)\, du.$$

Thus,

$$| \phi''(t) + \sigma^2 | \leq \int | tu | u^2 f(u) \, du \leq \beta | t | \quad \blacksquare$$

Lemma 4.3.

Consider the absolute differences

$$A_m(t) = | (1 - \frac{t^2}{2n})^m - e^{-\frac{t^2}{2}} | \qquad (m = n-2, n-1, n) \, .$$

For $t^2 \leq n$, we have

$$A_n(t) \leq \frac{t^4}{4n} e^{-\frac{t^2}{2}} \, ,$$

$$A_{n-1}(t) \leq \frac{1}{2(n-1)} e^{\frac{1}{2(n-1)}} e^{-\frac{t^2}{2}} \, ,$$

$$A_{n-2}(t) \leq \frac{2}{n-2} e^{\frac{2}{n-2}} e^{-\frac{t^2}{2}} \, .$$

If all integrals shown below are over $\{ | t | \leq \sqrt{n} \}$, then we have

$$\int A_n(t) \, dt \leq \frac{3}{4n} \sqrt{2\pi} \, , \quad \int t^2 A_n(t) \, dt \leq \frac{15}{4n} \sqrt{2\pi} \, ,$$

$$\int A_{n-1}(t) \, dt \leq \frac{3}{4n} \sqrt{2\pi} + \frac{\sqrt{2\pi}}{2(n-1)} e^{\frac{1}{2(n-1)}} \, ,$$

$$\int t^2 A_{n-1}(t) \, dt \leq \frac{15}{4n} \sqrt{2\pi} + \frac{\sqrt{2\pi}}{2(n-1)} e^{\frac{1}{2(n-1)}} \, ,$$

$$\int A_{n-2}(t) \, dt \leq \frac{3}{4n} \sqrt{2\pi} + \frac{2\sqrt{2\pi}}{n-2} e^{\frac{2}{n-2}} \, ,$$

$$\int t^2 A_{n-2}(t) \, dt \leq \frac{15}{4n} \sqrt{2\pi} + \frac{2\sqrt{2\pi}}{n-2} e^{\frac{2}{n-2}} \, .$$

Proof of Lemma 4.3.

First,

$$e^{-\frac{t^2}{2}} - (1 - \frac{t^2}{2n})^{n-2} \leq e^{-\frac{t^2}{2}} - (1 - \frac{t^2}{2n})^{n-1}$$

$$\leq e^{-\frac{t^2}{2}} - (1 - \frac{t^2}{2n})^n$$

$$\leq e^{-\frac{t^2}{2}}\left(1-e^{-\frac{nt^4}{8n^2(1-\frac{t^2}{2n})}}\right)$$

$$\leq e^{-\frac{t^2}{2}}\left(1-e^{-\frac{nt^4}{4n^2}}\right)$$

$$\leq e^{-\frac{t^2}{2}}\frac{t^4}{4n}.$$

Here we used the inequality $\log(1-u)\geq-u-u^2/(2(1-u))\geq-u-u^2$ valid for $0\leq u\leq 1/2$. Since

$$0\leq e^{-\frac{t^2}{2}}-(1-\frac{t^2}{2n})^n,$$

the bound for A_n is proved. For the other bounds, consider A_m in general. Clearly,

$$(1-\frac{t^2}{2n})^m-e^{-\frac{t^2}{2}}\leq e^{-\frac{t^2}{2}}\left(e^{t^2(\frac{1}{2}-\frac{m}{2n})-t^4\frac{m}{8n^2}}-1\right).$$

For $m=n-i$, the exponent is at most $t^2i/(2n)-t^4(n-i)/(8n^2)$. This function is at most $i^2/(2(n-i))$. By the inequality $e^u-1\leq ue^u$ valid for $u\geq 0$, we finally conclude that the expression on the right hand side of the last inequality is at most

$$e^{-\frac{t^2}{2}}\frac{i^2}{2(n-i)}e^{\frac{i^2}{2(n-i)}}.$$

This proves all the pointwise inequalities for A_m. The integral inequalities are obtained by integrating the pointwise inequalities over the whole real line (this can only make the upper bounds larger). One needs the facts that for a normal random variable N, $E(N^2)=1, E(N^4)=3$, and $E(N^6)=15$. ∎

Lemma 4.4.

For regular f, and $|t| \leq \dfrac{3\sigma^3 \sqrt{n}}{4\beta}$, we have

$$| \phi^n (\frac{t}{\sigma \sqrt{n}}) - e^{-\frac{t^2}{2}} | \leq \frac{\beta |t|^3}{3\sigma^3 \sqrt{n}} e^{-\frac{t^2}{4}} + | A_n (t) | .$$

Integrated over the given interval for t, we have

$$\int | \phi^n (\frac{t}{\sigma \sqrt{n}}) - e^{-\frac{t^2}{2}} | \ dt \leq \frac{16\beta}{3\sigma^3 \sqrt{n}} + \frac{3}{4n} \sqrt{2\pi} .$$

Proof of Lemma 4.4.

Note that

$$| \phi^n (\frac{t}{\sigma \sqrt{n}}) - e^{-\frac{t^2}{2}} | \leq | \phi^n (\frac{t}{\sigma \sqrt{n}}) - (1 - \frac{t^2}{2n})^n | + | A_n (t) | .$$

The last term is taken care of by applying Lemma 4.3. Here we need the fact that the given interval for t is always included in $[-\sqrt{n}, \sqrt{n}]$, so that the bounds of Lemma 4.3 are indeed applicable. By Lemma 4.2, the first term can be written as

$$(1 - \frac{t^2}{2n})^n \ | (1 + \frac{\theta \beta |t|^3}{6\sigma^3 n^{\frac{3}{2}} (1 - \frac{t^2}{2n})})^n - 1 |$$

where $|\theta| \leq 1$. Using the fact that $(1+u)^n - 1 \leq n |u| e^{n|u|}$ for all $n > 0$, and all $u \in R$, this can be bounded from above by

$$e^{-\frac{t^2}{2}} \frac{\beta |t|^3}{3\sigma^3 \sqrt{n}} e^{\frac{\beta |t|^3}{3\sigma^3 \sqrt{n}}} \leq e^{-\frac{t^2}{4}} \frac{\beta |t|^3}{3\sigma^3 \sqrt{n}} .$$

To obtain the integral inequality, use Lemma 4.3 again, and note that $\int |t|^3 e^{-t^2/4} \ dt = 16$. ∎

Lemma 4.5.

For regular f,

$$\sup_x \; |\, f_n(x) - g(x)\,| \; \leq \; a_n$$

where

$$a_n \;=\; \frac{1}{2\pi}\,\frac{16\beta}{3\sigma^3\sqrt{n}}\,(1 + \frac{1}{2}\,e^{-\frac{9\sigma^6 n}{32\beta^2}})$$

$$+ \; \frac{1}{2\pi}\,\frac{3}{4n}\,\sqrt{2\pi} + \frac{1}{2\pi}\,\sup_{|\,t\,| \geq \frac{4\sigma^2}{3\beta}} \; |\,\phi(t)\,|^{\,n-1}\sigma\sqrt{n}\int |\,\phi\,| \; ,$$

f_n is the density of $S_n/(\sigma\sqrt{n}\,)$ and g is the normal density. Also,

$$a_n \; \sim \; \frac{8\beta}{3\pi\sigma^3\sqrt{n}}$$

as $n \to \infty$.

Proof of Lemma 4.5.

By the inversion formula for absolutely integrable characteristic functions, we see that

$$2\pi\,|\,f_n(x) - g(x)\,| \; \leq \; \int |\,\phi^n(\frac{t}{\sigma\sqrt{n}}) - e^{-\frac{t^2}{2}}\,|$$

$$\leq \int_D |\,\phi^n(\frac{t}{\sigma\sqrt{n}}) - e^{-\frac{t^2}{2}}\,| \; dt \; + \int_{D^c}\left(\,|\,\phi^n(\frac{t}{\sigma\sqrt{n}})\,| + e^{-\frac{t^2}{2}}\,\right)\,dt$$

where D is the interval defined by the condition $|\,t\,| \leq \dfrac{3\sigma^3\sqrt{n}}{4\beta}$, and D^c is the complement of D. The integral over D is bounded in Lemma 4.4 by

$$\frac{16\beta}{3\sigma^3\sqrt{n}} + \frac{3}{4n}\,\sqrt{2\pi} \; .$$

The integral over D^c does not exceed

$$\sup_{|\,t\,| \geq \frac{4\sigma^2}{3\beta}} \; |\,\phi(t)\,|^{\,n-1}\sigma\sqrt{n}\int |\,\phi\,| \; + \; \frac{8\beta}{3\sigma^3\sqrt{n}}\,e^{-\frac{9\sigma^6 n}{32\beta^2}} \; ,$$

where we used a well-known inequality for the tail of the normal distribution, i.e.
$\int_u^\infty g \leq g(u)/u$. This concludes the proof of Lemma 4.5. ∎

Lemma 4.6.

For regular f , and

$$| t | \le \frac{3\sigma^3 \sqrt{n}}{4\beta} ,$$

we have

$$| \phi^{n-1}(\frac{t}{\sigma\sqrt{n}}) - e^{-\frac{t^2}{2}} | \le \frac{\beta | t |^3}{3\sigma^3 \sqrt{n}} e^{-\frac{t^2}{4}} .$$

Integrated over the given interval for t , we have

$$\int | \phi^{n-1}(\frac{t}{\sigma\sqrt{n}}) - e^{-\frac{t^2}{2}} | \; dt \le \frac{16\beta}{3\sigma^3 \sqrt{n}} + \frac{3}{4n} \sqrt{2\pi} .$$

Proof of Lemma 4.6.

Note that

$$| \phi^{n-1}(\frac{t}{\sigma\sqrt{n}}) - e^{-\frac{t^2}{2}} | \le | \phi^{n-1}(\frac{t}{\sigma\sqrt{n}}) - (1-\frac{t^2}{2n})^{n-1} | + | A_{n-1}(t) | .$$

The last term is taken care of by applying Lemma 4.3. Here we need the fact that the given interval for t is always included in $[-\sqrt{n} , \sqrt{n}]$, so that the bounds of Lemma 4.3 are indeed applicable. By Lemma 4.2, the first term can be written as

$$(1-\frac{t^2}{2n})^{n-1} | \left[1+ \frac{\theta\beta | t |^3}{6\sigma^3 n^{\frac{3}{2}} (1-\frac{t^2}{2n})} \right]^{n-1} -1 |$$

where $| \theta | \le 1$. Using the fact that $(1+u)^{n-1}-1 \le n | u | e^{n | u |}$ for all $n > 0$, and all $u \in R$, this can be bounded from above by

$$\frac{e^{-(n-1)t^2}}{2n} \frac{\beta | t |^3}{3\sigma^3 \sqrt{n}} e^{\frac{\beta | t |^3}{3\sigma^3 \sqrt{n}}} \le e^{-(1-\frac{1}{2n})\frac{t^2}{4}} \frac{\beta | t |^3}{3\sigma^3 \sqrt{n}} .$$

To obtain the integral inequality, use Lemma 4.3 again, and note that $\int | t |^3 e^{-t^2/4} \; dt = 16$. ∎

Lemma 4.7.

Let g be the normal density and let f_n be the density of the normalized sum $S_n/(\sigma\sqrt{n})$ for iid random variables with a regular density f. Let ϕ be the characteristic function for f. Then

$$|f_n(x)-g(x)| \le \frac{b_n}{x^2},$$

where

$$b_n = \frac{4\beta}{3\pi\sigma^3\sqrt{n}}e^{-\frac{9\sigma^6 n}{32\beta^2}} + \frac{\sqrt{8}}{\pi}e^{-\frac{9\sigma^6 n}{64\beta^2}}$$

$$+ \frac{1}{2\pi}\rho^{n-2}\sigma\sqrt{n}\int|\phi| + \frac{1}{2\pi}\rho^{n-3}\sigma^3 n^{\frac{3}{2}}\int t^2|\phi|$$

$$+ \frac{1}{2\pi}\left(\frac{208\beta}{3\sigma^3\sqrt{n}} + \frac{18\sqrt{2\pi}}{4n}\right)$$

$$+ \frac{1}{\sqrt{4\pi n(n-1)}} + \frac{3}{(n-2)\sqrt{2\pi}}$$

$$+ \frac{1}{n\sigma^2\sqrt{2\pi}} + \frac{\beta}{\sigma^3\sqrt{n}}\left(\frac{1}{\sigma^2\sqrt{8\pi}}+2\right).$$

Here $\rho = \sup\limits_{|t| \ge \frac{3\sigma^2}{4\beta}}|\phi(t)|$. Note that as $n\to\infty$, $b_n \sim \dfrac{b}{\sqrt{n}}$ where

$$b = \frac{\beta}{\sigma^3}\left(\frac{1}{\sigma^2\sqrt{8\pi}}+2+\frac{208}{6\pi}\right).$$

Proof of Lemma 4.7.

As in Lemma 4.5, we define the interval D by the condition $|t| \le \dfrac{3\sigma^3\sqrt{n}}{4\beta}$, and let D^c be the complement of D. Let I be the interval defined by $|t| \le \dfrac{3\sigma^2}{4\beta}$, and let I^c be the complement of I. By Lemma 4.2, it is easy to see that for $t\in I$, $|\phi(t)| \le 1-\sigma^2 t^2/4$. Thus,

$$\frac{1}{2\pi}\int_D \left|\phi^n\left(\frac{t}{\sigma\sqrt{n}}\right)-\phi^{n-1}\left(\frac{t}{\sigma\sqrt{n}}\right)\right|\,dt$$

$$\le \frac{1}{2\pi}\int_D \left|1-\phi\left(\frac{t}{\sigma\sqrt{n}}\right)\right|\,\left|\phi\left(\frac{t}{\sigma\sqrt{n}}\right)\right|^{n-1}\,dt$$

$$\le \frac{1}{2\pi}\int \frac{t^2}{2n}e^{-\frac{(n-1)t^2}{4n}}\,dt$$

$$= \frac{1}{4\pi n}\sqrt{2\pi}\sqrt{\frac{2n}{n-1}}$$

$$= \frac{1}{\sqrt{4\pi n\,(n-1)}} \; .$$

Similarly,

$$\frac{1}{2\pi}\int_D t^2 \mid \phi^n\left(\frac{t}{\sigma\sqrt{n}}\right) - \phi^{n-2}\left(\frac{t}{\sigma\sqrt{n}}\right) \mid \; dt$$

$$\leq \frac{1}{2\pi}\int_D t^2 \mid 1-\phi^2\left(\frac{t}{\sigma\sqrt{n}}\right)\mid \; \mid \phi\left(\frac{t}{\sigma\sqrt{n}}\right)\mid^{n-2} \; dt$$

$$\leq \frac{1}{2\pi}\int \frac{t^4}{n} e^{-\frac{(n-2)t^2}{4n}} \; dt$$

$$= \frac{1}{2\pi n}\sqrt{2\pi}\,3\frac{2n}{n-2}$$

$$= \frac{3}{(n-2)\sqrt{2\pi}} \; .$$

So far for the preliminary computations. We begin with the observation that

$$x^2(f_n(x)-g(x)) = \frac{1}{2\pi}\int ((t^2-1)e^{-\frac{t^2}{2}} - \phi_n''(t))e^{-itx} \; dt$$

where ϕ_n is the characteristic function corresponding to f_n. Obviously,

$$x^2 \mid f_n(x)-g(x)\mid \; \leq \frac{1}{2\pi}\int \mid (t^2-1)e^{-\frac{t^2}{2}} - \phi_n''(t)\mid \; dt \; .$$

The second derivative of the n-th power of $\phi(t/(\sigma\sqrt{n}))$ is

$$\frac{n-1}{\sigma^2}\phi'^2\phi^{n-2} + \frac{1}{\sigma^2}\phi''\phi^{n-1} \; ,$$

where all the omitted arguments are $t/(\sigma\sqrt{n})$. By the triangle inequality, we obtain

$$x^2 \mid f_n(x)-g(x)\mid \; \leq \frac{1}{2\pi}\int \mid (t^2-1)e^{-\frac{t^2}{2}} - \phi_n''(t)\mid \; dt$$

$$\leq \frac{1}{2\pi}\left(\int \mid e^{-\frac{t^2}{2}} - \phi^{n-1}(t/(\sigma\sqrt{n}))\mid \; dt + \int t^2\mid e^{-\frac{t^2}{2}} - \phi^{n-2}(t/(\sigma\sqrt{n}))\mid \; dt\right.$$

$$+ \int e^{-\frac{t^2}{2}}\mid \frac{n-1}{\sigma^2}\phi'^2(t/(\sigma\sqrt{n})) - t^2\mid \; dt$$

$$\left.+ \int e^{-\frac{t^2}{2}}\mid \sigma^{-2}\phi''(t/(\sigma\sqrt{n}))+1\mid \; dt\right)$$

$$= J_1+J_2+J_3+J_4 \; .$$

From Lemma 4.2, we recall

$$\left| \frac{\sqrt{n}\ \phi'}{\sigma} + t \right| \le \frac{\beta t^2}{2\sigma^3\sqrt{n}}\ ,$$

$$\left| \frac{\phi''}{\sigma^2} + 1 \right| \le \frac{\beta\ |\ t\ |}{\sigma^3\sqrt{n}}\ .$$

Using the fact that $|\ \phi'(t/(\sigma\sqrt{n}\ ))\ | \le E(\ |\ X\ |)/(\sigma\sqrt{n}\ ) \le 1/\sqrt{n}$, we have

$$\left| \frac{n\ \phi'^2}{\sigma^2} - t^2 \right| \le \left| \frac{\sqrt{n}\ \phi'}{\sigma} - t \right| \left| \frac{\sqrt{n}\ \phi'}{\sigma} + t \right|$$

$$\le (\frac{1}{\sigma^2} + |\ t\ |)\frac{\beta t^2}{2\sigma^3\sqrt{n}}\ .$$

Using the fact that $\int |\ t\ |^i\ e^{-t^2/2}\ dt$ takes the values $\sqrt{2\pi}, 2, \sqrt{2\pi}$ and 4 for $i = 0,1,2,3$ respectively, we see that

$$J_3 + J_4 \le \int |\ \sigma^{-2}\phi'^2 e^{-\frac{t^2}{2}}\ |\ dt + \frac{1}{\sqrt{2\pi}}\frac{\beta}{2\sigma^5\sqrt{n}} + \frac{2\beta}{\pi\sigma^3\sqrt{n}}$$

$$\le \frac{1}{n\ \sigma^2\sqrt{2\pi}} + \frac{\beta}{\sigma^3\sqrt{n}}(\frac{1}{\sigma^2\sqrt{8\pi}} + 2)\ .$$

This leaves us with J_1 and J_2. Here we will split the integrals over D and D^c. First of all,

$$\frac{1}{2\pi}\left(\int_D |\ e^{-\frac{t^2}{2}} - \phi^{n-1}(t/(\sigma\sqrt{n}\ ))\ |\ dt + \int_D t^2\ |\ e^{-\frac{t^2}{2}} - \phi^{n-2}(t/(\sigma\sqrt{n}\ ))\ |\ dt \right)$$

$$\le \frac{1}{2\pi}\left(\int_D |\ e^{-\frac{t^2}{2}} - \phi^n\ (t/(\sigma\sqrt{n}\ ))\ |\ dt + \int_D t^2\ |\ e^{-\frac{t^2}{2}} - \phi^n\ (t/(\sigma\sqrt{n}\ ))\ |\ dt \right)$$

$$+ \frac{1}{2\pi}\left(\int_D |\ \phi^{n-1}(t/(\sigma\sqrt{n}\ )) - \phi^n\ (t/(\sigma\sqrt{n}\ ))\ |\ dt \right.$$

$$+ \left. \int_D t^2\ |\ \phi^{n-2}(t/(\sigma\sqrt{n}\ )) - \phi^n\ (t/(\sigma\sqrt{n}\ ))\ |\ dt \right)\ .$$

The last two terms were bounded from above earlier on in the proof by

$$\frac{1}{\sqrt{4\pi n\ (n-1)}} + \frac{3}{(n-2)\sqrt{2\pi}}\ .$$

By Lemma 4.4, we have for $t \in D$,

$$|\ \phi^n\ (\frac{t}{\sigma\sqrt{n}}) - e^{-\frac{t^2}{2}}\ | \le \frac{\beta\ |\ t\ |^3}{3\sigma^3\sqrt{n}}e^{-\frac{t^2}{4}} + |\ A_n\ (t)\ |\ .$$

Thus, by Lemma 4.3, and the following integrals:

$$\int |\ t\ |^3 e^{-\frac{t^2}{4}}\ dt\ =\ 16\ ,$$

$$\int |t|^5 e^{-\frac{t^2}{4}} \, dt = 192 \, ,$$

$$\int |t|^4 e^{-\frac{t^2}{2}} \, dt = 3\sqrt{2\pi} \, ,$$

$$\int |t|^6 e^{-\frac{t^2}{4}} \, dt = 15\sqrt{2\pi} \, ,$$

we have

$$\frac{1}{2\pi} \int_D (1+t^2) \, | \, \phi^n \, (\frac{t}{\sigma\sqrt{n}}) - e^{-\frac{t^2}{2}} \, | \, dt$$

$$\leq \frac{1}{2\pi} \int (1+t^2) \left[\frac{\beta \, |t|^3}{3\sigma^3\sqrt{n}} e^{-\frac{t^2}{4}} + \frac{t^4}{4n} e^{-\frac{t^2}{2}} \right] dt$$

$$= \frac{1}{2\pi} \left(\frac{208\beta}{3\sigma^3\sqrt{n}} + \frac{18\sqrt{2\pi}}{4n} \right) \, .$$

Finally, we have to evaluate the integrals in J_1+J_2 taken over D^c. These are estimated from above by

$$\frac{1}{2\pi} \int_{D^c} (1+t^2) e^{-\frac{t^2}{2}} \, dt + \frac{1}{2\pi} \rho^{n-2} \sigma\sqrt{n} \int |\phi| + \frac{1}{2\pi} \rho^{n-3} \sigma^3 n^{\frac{3}{2}} \int t^2 \, |\phi|$$

where $\rho = \sup_{I^c} |\phi|$. The region D^c is defined by the condition $|t| > c$ for some constant c. The first term in the last expression can thus be rewritten as

$$\frac{1}{\pi} \int_{u > c^2/2} ((2u)^{\frac{1}{2}} + \sqrt{2u} \,) e^{-u} \, du$$

$$\leq \frac{1}{c\pi} e^{-\frac{c^2}{2}} + \frac{\sqrt{8}}{\pi} e^{-\frac{c^2}{4}}$$

$$= \frac{4\beta}{3\pi\sigma^3\sqrt{n}} e^{-\frac{9\sigma^6 n}{32\beta^2}} + \frac{\sqrt{8}}{\pi} e^{-\frac{9\sigma^6 n}{64\beta^2}} \, .$$

Collecting bounds gives the desired result. ∎

For the bound of Lemma 4.7 to be useful, it is necessary that f not only be regular, but also that its characteristic function satisfy

$$\int t^2 \, |\phi(t)| \, dt < \infty \, .$$

This implies that f has two bounded continuous derivatives tending to 0 as $|x| \to \infty$, and in fact

$$f''(x) = -\frac{1}{2\pi} \int e^{-itx} t^2 \phi(t) \, dt \ .$$

(see e.g. Kawata, 1972, pp. 438-439). This smoothness condition is rather restrictive and can be considerably weakened. The asymptotic bound $b/(x^2\sqrt{n}\ )$ remains valid if $\int t^2 \mid \phi(t) \mid^k < \infty$ for some positive integer k (exercise 4.4). Lemmas 4.5 and 4.7 together are but special cases of more general local limit theorems, such as those found in Maejima (1980) and Inzevitov (1977), except that here we explicitly compute the universal constants in the bounds.

4.5. The mixture method for simulating sums.

When a density f can be written as a mixture

$$f(x) = \sum_{i=1}^{\infty} p_i f_i(x) ,$$

where the f_i's are simple densities, then simulation of the sum S_n of n iid random variables with density f can be carried out as follows.

The mixture method for simulating sums

Generate a multinomial $(n, p_1, p_2,...)$ random sequence $N_1, N_2,...$ (note that the N_i's sum to n). Let K be the index of the largest nonzero N_i.
$X \leftarrow 0$
FOR $i := 1$ TO K DO
    Generate S, the sum of N_i iid random variables with common density f_i.
    $X \leftarrow X + S$
RETURN X

The validity of the algorithm is obvious. The algorithm is put in its most general form, allowing for infinite mixtures. A multinomial random sequence is of course defined in the standard way: imagine that we have an infinite number of urns, and that n balls are independently thrown in the urns. Each ball lands with probability p_i in the i-th urn. The sequence of cardinalities of the urns is a multinomial $(n, p_1, p_2,...)$ random sequence. To simulate such a sequence, note that N_1 is binomial (n, p_1), and that given N_1, N_2 is binomial $(n-N_1, p_2/(1-p_1))$, etcetera. If K is the index of the last occupied urn, then it is easy to see that the multinomial sequence can be generated in expected time $O(E(K))$.

The mixture method is efficient if sums of iid random variables with densities f_i are easy to generate. This would for example be the case if f were a finite mixture of stable, gamma, exponential or normal random variables. Perhaps the most intriguing decomposition is that of a unimodal density: every unimodal density can be written as a countable mixture of uniform densities. This statement is intuitively clear, because subtracting a function of the form $cI_{[a,b]}(x)$ from f leaves a unimodal piece on $[a,b]$ and two unimodal tails. This can be repeated for all pieces individually, and at the same time the integral of the leftover function can be made to tend to zero by the judicious choice of rectangular functions (see exercise 4.5). If we can generate sums of iid uniform random variables uniformly fast (with respect to n), then the expected time taken by the mixture method is $O(E(K))$. A few remarks about generating uniform sums are given in the next section.

4.6. Sums of independent uniform random variables.

In this section we consider the distribution of

$$S_n = \sum_{i=1}^n U_i ,$$

where $U_1, \ldots, U_n$ are iid uniform $[-1,1]$ random variables. The distribution can be described in a variety of ways:

Theorem 4.3.

The characteristic function of S_n is

$$\left(\frac{\sin(t)}{t} \right)^n .$$

For all $n \geq 2$, the density f_n can be obtained by the inversion formula

$$f_n(x) = \frac{1}{2\pi} \int \left(\frac{\sin(t)}{t} \right)^n \cos(tx) \, dt .$$

This yields

$$f_n(x) = \frac{1}{(i-1)!} \frac{1}{2} \sum_{k=0}^{i-1} (-1)^k \binom{i}{k} (x - (2k-n))^{i-1}$$

where $2i-2-n < x < 2i-n$; $i = 1, 2, \ldots, n$.

Proof of Theorem 4.3.

The characteristic function is obtained by using the definition. Since the characteristic function of S_n for all $n \geq 2$ is absolutely integrable, f_n can be obtained by the given inversion integral. There is also a direct way of computing the distribution function F_n and density of S_n; its derivation goes back to the nineteenth century (see e.g. Cramer (1951, p. 245)). Different proofs include the geometric approach followed by us in Theorem I.4.4 (see also Hall (1927) and Roach (1963)), an induction argument (Olds, 1952), and an application of the residue theorem (Lusk and Wright, 1982). Taking the derivative of F_n given in Theorem I.4.4 gives the formula

$$\frac{1}{(n-1)!}\left(x_+^{n-1} - n(x-1)_+^{n-1} + \binom{n}{2}(x-2)_+^{n-1} - \cdots + (-1)^n \binom{n}{n}(x-n)_+^{n-1} \right)$$

for the density of the sum of n iid uniform [0,1] random variables. The the density of sums of symmetric uniform random variables is easily obtained by the transformation formula for densities. ∎

It is easy to see that the local limit theorems developed in Lemmas 4.5 and 4.7 are applicable to this case. There is one small technical hurdle since the characteristic function of a uniform random variable is not absolutely integrable. This is easily overcome by noting that the square of the characteristic function is absolutely integrable. If we recall the rejection algorithm of section 4.3, we note that the expected number of iterations is $O(1/\sqrt{n})$ and that the expected number of evaluations of f_n is $O(1/\sqrt{n})$. Unfortunately, this is not good enough, since the evaluation of $f_n(x)$ by the last formula of Theorem 4.3 takes time roughly proportional to n for nearly all x of interest. This would yield a global expected time roughly increasing as $\sqrt{n}$. The formula for f_n is thus of limited value. There are two solutions: either one uses the series method based upon a series expansion for f_n which is tailored around the normal density, or one uses a local limit theorem with $O(1/n)$ error by using as main component the normal density plus the first term in the asymptotic expansion which is a normal density multiplied with a Hermite polynomial (see e.g. Petrov, 1975). The latter approach seems the most promising at this point (see exercise 4.6).

4.7. Exercises.

1. Let f be a density, whose normalized sums tend in distribution to the symmetric stable (α) density. Assume that the stable density can be evaluated exactly in one unit of time at every point. Derive first some inequalities for the difference between the density of the normalized sum and the stable density. These non-uniform inequalities should be such that the integral of the error bound with respect to x tends to 0 as $n \to \infty$. Hint: look for error terms of the form $\min(a_n, b_n \mid x \mid^{-c})$ where c is a positive constant, and a_n, b_n are positive number sequences tending to 0 with n. Mimic the derivation of the local limit theorem in the case of attraction to the normal law.

2. **The gamma density.** The zero mean exponential density has characteristic function $\phi = e^{-it}/(1-it)$. In the notation of this chapter, derive for this distribution the following quantities:

    A. $\sigma = 1$, $\beta = \dfrac{12}{e} - 2$.

    B. $\int \mid \phi \mid = \infty$, $\int \mid \phi \mid^2 = \pi$.

    C. $\sup\limits_{\mid t \mid \geq c} \mid \phi(t) \mid = 1/\sqrt{1+c^2}$ $(c > 0)$.

    Note that the bounds in the local limit theorems are not directly applicable since $\int \mid \phi \mid = \infty$. However, this can be overcome by bounding $\int \mid \phi \mid^n$ by $s \int \mid \phi \mid^2$ where s is the supremum of $\mid \phi \mid$ over the domain of integration, to the power $n - 2$. Using this device, derive the rejection constant from the thus modified local limit theorem as a function of n.

3. A continuation of exercise 2. Let f_a be the normalized (zero mean, unit variance) gamma (a) density, and let g be the normal density. By direct means, find sequences a_n, b_n such that for all $a \geq 1$,

    $$\mid f_a(x) - g(x) \mid \leq \min(a_n, \frac{b_n}{x^2}),$$

    and compare your constants with those obtained in exercise 2. (They should be dramatically smaller.)

4. Prove the claim that in Lemma 4.7, $b_n \sim b/(x^2\sqrt{n})$ when the condition $\int t^2 \mid \phi(t) \mid dt < \infty$ is relaxed to

    $$\int t^2 \mid \phi(t) \mid^k dt < \infty$$

    where $k > 0$ is a fixed integer.

5. Consider a monotone density f on $[0,\infty)$. Give a constructive completely automatic rule for decomposing this density as a countable mixture of uniform densities, i.e. the decomposition should be obtainable even if f is only given in black box format, and the countable mixture should give us the monotone density again in the sense that the L_1 distance between the two densities is zero (this allows the functions to be different on possibly uncountable sets of zero measure). Can you make a statement about the rate of decrease of p_i for the following subclasses of monotone densities: the log-

concave densities, the concave densities, the convex densities? Prove that when $p_i \leq ce^{-bi}$ for some $b,c > 0$ and all i, then $E(K) = O(\log(n))$, where K is the largest integer in a sample of size n drawn from probability vector $p_1, p_2, \ldots$. Conclude that for important classes of densities, we can generate sums of n iid random variates in expected time $O(\log(n))$.

6. **Gram-Charlier series.** The standard approximation for the density f_n of $S_n/(\sigma\sqrt{n})$ where S_n is the sum of n iid zero mean random variables with second moment σ^2 is g where g is the normal density. The closeness is covered by local central limit theorems, and the errors are of the order of $1/\sqrt{n}$. To obtain errors of the order of $1/n$ it is necessary to user a finer approximation. For example, one could use an extra term in the Gram-Charlier series (see e.g. Ord (1972, p. 26)). This leads to the approximation by

$$\frac{1}{\sqrt{2\pi}} e^{-\frac{x^2}{2}} \left(1 + \frac{\mu_3}{6\sigma^3\sqrt{n}}(x^3 - x)\right),$$

where μ_3 is the third moment for f. For symmetric distributions, the extra correction term is zero. This suggests that the local limit theorems of section 4.3 can be improved. For the symmetric uniform density, find constants a,b such that $|f_n - g| \leq \frac{1}{n}\min(a, bx^{-2})$. Use this to design a uniformly fast generator for sums of symmetric uniform random variables.

7. A continuation of the previous exercise. Let $a \in R$ be a constant. Give a random variate generator for the following class of densities related to the Gram-Charlier series approximation of the previous exercise:

$$g(x) = c\left[\frac{1}{\sqrt{2\pi}} e^{-\frac{x^2}{2}}(1 + a(x^3 - x))\right]_+,$$

where c is a normalization constant.

5. DISCRETE EVENT SIMULATION.

5.1. Future event set algorithms.

Several complex systems evolving in time fall into the following category: they can be characterized by a state, and the state changes only at discrete times. Systems falling into this category include most queueing systems such as those appearing in banks, elevators, computer networks, computer operating systems and telephone networks. Systems not included in this category are those which change state continuously, such as systems driven by differential equations (physical or chemical processes, traffic control systems). In discrete event simulation of such systems, one keeps a subset of all the future events in a future event set, where an event is defined as a change of state, e.g. the arrival or departure of a person in a bank. By taking the next event from the future event set, we can make time advance with big jumps. After having grabbed this event, it is necessary to update the state and if necessary schedule new future events. In other words, the future event set can shrink and grow in its lifetime. What matters is that no event is missed. All future event set algorithms can be summarized as follows:

Future event set algorithm

Time ←0.
Initialize State (the state of the system).
Initialize FES (future event set) by scheduling at least one event.
WHILE NOT EMPTY (FES) DO
       Select the minimal time event in FES, and remove it from FES.
       Time ← time of the selected event, i.e. make time progress.
       Analyze the selected event, and update State and FES accordingly.

For worked out examples, we refer the readers to more specialized texts such as Bratley, Fox and Schrage (1983), Banks and Carson (1984) or Law and Kelton (1982). Our main concern is with the complexity aspect of future event set algorithms. It is difficult to get a good general handle on the time complexity due to the state updates. On the other hand, the contribution to the time complexity of all operations involving FES, the future event set, is amenable to analysis. These operations include

A.   INSERT a new event in FES.

B.   DELETE the minimal time event from FES.

C.   CANCEL a particular event (remove it from FES).

There are two kinds of INSERT: INSERT based upon the time of the event, and INSERT based upon other information related to the event. The latter INSERT is required when a simulation demands information retrieval from the FES other than selection of the minimal time event. This is the case when cancelations can occur, i.e. deletions of events other than the minimal time event. It can always be avoided by leaving the event to be canceled in FES but marking it "canceled", so that when it is selected at some point as the minimal time event, it can immediately be discarded. In most cases we have to use a dual data structure which allows us to implement the operations INSERT, DELETE and either CANCEL or MARK efficiently. Typically, one part of the data structure consists of a dictionary (ordered according to keys used for canceling or marking), and another part is a priority queue (see Aho, Hopcroft and Ullman (1983) for our terminolgy). Since the number of elements in FES grows and shrinks with time, it is difficult to uniformize the analysis. For this reason, sometimes the following assumptions are made:

A. The future event set has n events at all times. This implies that when the minimum time event is deleted, the empty slot is immediately filled by a new event, i.e. the DELETE and INSERT operations always go together.

B. Initially, the future event set has n events, with random times, all iid with common distribution function F on $[0,\infty)$.

C. When an event with event time t is deleted from FES, the new event replacing it in FES has time $t+T$, where T also has distribution function F.

These three assumptions taken together form the basis of the so-called hold model, coined after the SIMULA HOLD operation, which combines our DELETE and INSERT operations. Assumptions B and C are of a stochastic nature to facilitate the expected time analysis. They are motivated by the fact that in homogeneous Poisson processes, the inter-event times are independent exponentially distributed. Therefore, the distribution function F is typically the exponential distribution. The quantity of interest to us is the expected time needed to execute a HOLD operation. Unfortunately, this quantity depends not only upon n, but also on F and the time instant at which the expected time analysis is needed. This is due to the fact that the times of the events in the FES have distributions that vary. It is true that relative to the minimum time in the FES, the distribution of the $n-1$ non-minimal times approaches a limit distribution, which depends upon F and n. Analysis based upon this limit distribution is at times risky because it is difficult to pinpoint in complex systems when the steady state is almost reached. What complicates matters even more is the dependence of the limit distribution upon n. The limit of the limit distribution with respect to n, a double limit of sorts, has density $(1-F(x))/\mu$ $(x>0)$ where μ is the mean for F (Vaucher, 1977). The analyses are greatly facilitated if this limit distribution is used as the distribution of the relative event times in FES. The results of these analyses should be handled with great care. Two extensive reports based upon this model were carried out by Kingston (1985) and McCormack and Sargent (1981). An alternative model was proposed by Reeves (1984). He also works with this limiting distribution, but departs from the HOLD model, in that events are inserted, or scheduled, in the FES according to a homogeneous Poisson process. This implies that the size of the FES is no longer fixed at a given level n, but hovers around a mean value n. It seems thus safer to perform a worst-case time analysis, and to include an expected time analysis only where exact calculations can be carried out. Luckily, for the important exponential distribution, this can be done.

Theorem 5.1.

If assumptions A-C hold, and F is the exponential (λ) distribution, if k HOLD operations have been carried out for any integer k, if $X*$ is the minimal event time in the FES, and $X_1, X_2, \ldots, X_{n-1}$ are the $n-1$ non-minimal event times in the FES (unordered, but in order of their insertion in the FES), then $X_1-X*, \ldots, X_{n-1}-X*$ are iid exponential (λ) random variables.

Proof of Theorem 5.1.

This is best proved inductively. Initially, we have n exponentially distributed times. The assertion is certainly true, by the memoryless property of the exponential distribution. Now, take the minimum time, say M, remove it, and insert the time $M+E$ in the FES, where E is exponential (λ). Clearly, all n times in the FES are now iid with an exponential (λ) distribution on $[M,\infty)$. We are thus back where we started from, and can apply the memoryless property again. ∎

Reeves's model allows for a simple direct analysis for all distribution functions F. Because of its importance, we will briefly study his model in a separate section, before moving on to the description of a few possible data structures for the FES.

5.2. Reeves's model.

In Reeves's model, the FES is initially empty. Insertions occur at random times, which correspond to a homogeneous Poisson process with rate λ. The time of an inserted event is the insertion time plus a delay time which has distribution function F. A few properties will be needed further on, and these are collected in Theorem 5.2:

Theorem 5.2.

Let $0 < T_1 < T_2 < \cdots$ be a homogeneous Poisson process with rate $\lambda > 0$ (the T_i's are the insertion times), and let $X_1, X_2,...$ be iid random variables with common distribution function F on $[0,\infty)$. Then

A. The random variables $T_i + X_i$, $1 \leq i$, form a nonhomogeneous Poisson process with rate function $\lambda F(t)$.

B. If N_t is the number of events in FES at time t, then N_t is Poisson $(\lambda \int_0^t (1-F))$. N_t is thus stochastically smaller than a Poisson $(\lambda \mu)$ random variable where $\mu = \int_0^\infty (1-F)$ is the mean for F.

C. Let $V_i, i \leq N_t$, be the event times for the events in FES at time t. Then the random variables $V_i - t$ form a nonhomogeneous Poisson process with rate function $\lambda(F(t+u) - F(u))$, $u \geq 0$.

Proof of Theorem 5.2.

Most of the theorem is left as an exercise on Poisson processes. The main task is to verify the Poisson nature of the defined processes by checking the independence property for nonoverlapping intervals. We will mainly point out how the various rate functions are obtained.

For part A, let L be the number of insertions up to time t, a Poisson (λt) random variable, and let M be the number of $T_i + X_i$'s not exceeding t. Clearly, by the uniform distribution property of homogeneous Poisson processes, M is distributed as

$$\sum_{i=0}^{L} I_{[tU_i + X_i \leq t]} \, ,$$

where the U_i's are iid uniform $[0,1]$ random variables. Note that this is a Poisson sum of iid Bernoulli random variables. As we have seen elsewhere, such sums are again Poisson distributed. The parameter is $\lambda t p$ where $p = P(tU_1 + X_1 \leq t)$. The parameter can be rewritten as

$$\lambda t P(X_1 \leq tU_1) = \lambda t \int_0^1 F(tu) \, du$$

$$= \lambda \int_0^t F(u) \, du \, .$$

For part B, the rate function can be obtained similarly by writing N_t as a Poisson (λt) sum of iid Bernoulli random variables with success probability $p = P(tU_1 + X_1 > t)$. This is easily seen to be Poisson ($\lambda \int_0^t (1-F)$). For the second statement of part B, recall that the mean for distribution function F is $\int_0^\infty (1-F)$.

Finally, consider part C. Here again, we argue analogously. Let M be the number of events in FES at time t with event times not exceeding $t + u$. Then M is a Poisson (λt) sum of iid Bernoulli random variables with success parameter p given by

$$P(t \leq tU_1 + X_1 < t + u) = \int_0^1 (F(tz + u) - F(tz)) \, dz$$

$$= \frac{1}{t} \int_0^t (F(z + u) - F(z)) \, dz \, .$$

The statement about the rate function follows directly from this. ∎

The asymptotics in Reeves's model should not be with respect to N_t, the size of the FES, because this oscillates randomly. Rather, it should be with respect to t, the time. The first important observation is that the expected size of the FES at time t is $\lambda \int_0^t (1-F) \rightarrow \lambda \mu$ as $t \rightarrow \infty$, where μ is the mean for F. If μ is small, the FES is small because events spend only a short time in FES. On the other hand, if $\mu = \infty$, then the expected size of the FES tends to ∞ as $t \rightarrow \infty$, i.e. we would need infinite space in order to be able to carry out an unlimited time simulation. The situation is also bad when $\mu < \infty$, although not as bad as in the case $\mu = \infty$: it can be shown (see exercises) that $\lim\sup_{t \rightarrow \infty} N_t = \infty$ almost surely.

Thus, in all cases, an unlimited memory would be required. This should be viewed as a serious drawback of Reeves's model. But the insight we gain from his model is invaluable, as we will find out in the next section on linear lists.

5.3. Linear lists.

The oldest and simplest structure for an FES is a linear list in which the elements are kept according to increasing event times. For what follows, it is all but irrelevant whether a linked list implementation or an array implementation is chosen. Deletion is obviously a constant time operation. Insertion of an element in the i-th position takes time proportional to i if we start searching from the front (small event times) of the list, and to $n-i+1$ if we start from the back and n is the cardinality of the FES. We can't say that the time is $\min(i, n-i+1)$ because the value of i is unknown beforehand. Thus, one of the questions to be studied is whether we should start the search from the front or the back.

By Theorem 5.2, part C, we observe that at time t_0, the expected value of the number of events exceeding the currently inserted element (called M_{t_0}) is

$$E(M_{t_0}) = \lambda \int_0^\infty \int_t^\infty (F(t_0+u)-F(u))\ du\ \ dF(t)$$

$$= \lambda \int_0^\infty (F(t_0+u)-F(u)) \int_0^u dF(t)\ du$$

$$= \lambda \int_0^\infty F(u)(F(t_0+u)-F(u))\ du\ .$$

Here we used a standard interchange of integrals. Since the expected number of elements in the FES is $\lambda \int_0^\infty (F(t_0+u)-F(u))\ du$, the expected value of the number of event times at most equal to the event time of the currently inserted element (called L_{t_0}) is

$$E(L_{t_0}) = \lambda \int_0^\infty (1-F(u))(F(t_0+u)-F(u))\ du\ .$$

We should search from the back when $E(M_{t_0}) < E(L_{t_0})$, and from the front otherwise. In an array implementation, the search can always be done by binary search in logarithmic time, but the updating of the array calls for the shift by one position of the entire lower or upper portion of the array. If one imagines a circular array implementation with free wrap-around, of the sort used to implement queues (Standish, 1980), then it is always possible to move only the smaller portion. The same is true for a linked list implementation if we keep pointers to the front, rear and middle elements in the linked list and use double linking to allow for the two types of search. The middle element is first compared with the inserted element. The outcome determines in which half we should insert, where the search should start from, and how the middle element should be updated. The last operation would also require us to keep a count of the number of elements in the linked list. We can thus conclude that for a linear list insertion, we can find an implementation taking time bounded by $\min(M_{t_0}, L_{t_0})$. By Jensen's inequality, the expected time for insertion does not exceed

$$\min(E(M_{t_0}), E(L_{t_0})) \ .$$

The fact that all the formulas for expected values encountered so far depend upon the current time t_0 could deprive us from some badly needed insight. Luckily, as $t_0 \to \infty$, a steady state is reached. In fact, this is the only case studied by Reeves (1984). We summarize:

Theorem 5.3.

In Reeves's model, we have

$$E(M_{t_0}) \uparrow \lambda \int_0^\infty F(1-F) \text{ as } t_0 \to \infty \ ,$$

$$E(L_{t_0}) \uparrow \lambda \int_0^\infty (1-F)^2 \text{ as } t_0 \to \infty \ .$$

Proof of Theorem 5.3.

We will only consider the first statement. Note that $E(M_{t_0})$ is monotone $\uparrow$ in t_0, and that for every t_0, the value does not exceed $\lambda \int_0^\infty F(1-F)$. Also, by Fatou's lemma,

$$\liminf_{t_0 \to \infty} E(M_{t_0}) \geq \lambda \int_0^\infty \liminf_{t_0 \to \infty} F(u)(F(t_0+u)-F(u)) \, du = \lambda \int_0^\infty F(1-F) \ . \blacksquare$$

Remark 5.1. Front or back search.

From Theorem 5.3, we deduce that a front search is indicated when $\int (1-F)^2 < \int F(1-F)$. It is perhaps interesting to note that equality is reached for the exponential distribution. Barlow and Proschan (1975) define the NBUE (NWUE) distributions as those distributions for which for all $t > 0$,

$$\int_t^\infty (1-F) \leq (\geq) \mu(1-F(t)),$$

where μ is the mean for F. Examples of NBUE (new better than used in expectation) distributions include the uniform, normal and gamma distributions for parameter at least one. NWUE distributions include mixtures of exponentials and gamma distributions with parameter at most one. By our original change of integral we note that for NBUE distributions,

$$\lambda \int_0^\infty F(1-F) = \lambda \int_0^\infty \left[\int_t^\infty (1-F) \right] dF(t)$$

$$\leq \lambda \mu \int_0^\infty (1-F(t)) \, dF(t) = \frac{\lambda \mu}{2}.$$

Since the asymptotic expected size of the FES is $\lambda \mu$, we observe that for NBUE distributions, a back search is to be preferred. For NWUE distributions, a front search is better. In all cases, the trick with the median pointer (for linked lists) or the median comparison (for circular arrays) automatically selects the best search mode. ∎

Remark 5.2. The HOLD model.

In the HOLD model, the worst-case insertion time can be as poor as n. For the expected insertion time, the computations are simple for the exponential distribution function. In view of Theorem 5.1, it is easy to see that the expected number of comparisons in a forward scan is $\dfrac{n+2}{2} - \dfrac{1}{n+1} = \dfrac{n}{2} + \dfrac{n}{n+1}$. The expected number of backward scans is equal to this, by symmetry. For all distributions F having a density, the expected insertion time grows linearly with n (see exercises). ∎

A brief historical remark is in order. Linear lists have been used extensively in the past. They are simple to implement, easy to analyze and use minimal storage. Among the possible physical implementations, the doubly linked list is perhaps the most popular (Knuth, 1969). The asymptotic expected insertion time for front and back search under the HOLD model was obtained by Vaucher (1977) and Englebrecht-Wiggans and Maxwell (1978). Reeves (1984) discusses the same thing for his model. Interestingly, if the size n in the HOLD model is replaced by the asymptotic value of the expected size of the FES, $\lambda\mu$, the two results coincide. In particular, Remark 5.1 applies to both models. The point about NBUE distributions in that remark is due to McCormack and Sargent (1981). The idea of using a median pointer or a median comparison goes back to Pritsker (1976) and Davey and Vaucher (1980). For more analysis involving linear lists, see e.g. Jonassen and Dahl (1975).

The simple linear list has been generalized and improved upon in many ways. For example, a number of algorithms have been proposed which keep an additional set of pointers to selected events in the FES. These are known as multiple pointer methods, and the implementations are sometimes called indexed linear list implementations. The pointers partition the FES into smaller sets containing a few events each. This greatly facilitates insertion. For example, Vaucher and Duval (1975) space pointer events (events pointed to by these pointers) equal amounts of time (Δ) apart. In view of this, we can locate a particular subset of the FES very quickly by making use of the truncation operation. The subset is then searched in the standard sequential manner. Ideally, one would like to have a constant number of events per interval, but this is difficult to enforce. In Reeves's model, the analysis of the Vaucher-Duval bucket structure is easy. We need only concern ourselves with insertions. Furthermore, the time needed to locate the subset (or bucket) in which we should insert is constant. The buckets should be thought of as small linked lists. They actually need not be globally concatenated, but within each list, the events are ordered. The global time interval is divided into intervals $[0,\Delta),[\Delta,2\Delta),....$. Let A_j be the j-th interval, and let $F(A_j)$ denote the probability of the j-th interval. For the sake of simplicity, let us assume that the time spent on an insertion is equal to the number of events already present in the interval into which we need to insert. In any case, ignoring a constant access time, this will be an upper bound on the actual insertion time. The expected number of events in bucket $A_j = [(j-1)\Delta, j\Delta)$ under Reeves model at time t is given by

$$\int_{A_j-t} \lambda\ (F(t+u)-F(u))\ du$$

where $A_j - t$ means the obvious thing. Let J be the collection of all indices for which A_j overlaps with $[t,\infty)$, and let B_j be $A_j \cup [t,\infty)$. Then the expected time is

$$\sum_{j \in J} \int_{B_j-t} \lambda\ (F(t+u)-F(u))\ du\ F(B_j-t)\ .$$

In Theorem 5.4, we derive useful upper bounds for the expected time.

Theorem 5.4.

Consider the bucket based linear list structure of Vaucher and Duval with bucket width Δ. Then the expected time for inserting (scheduling) an event at time t in the FES under Reeves's model is bounded from above by

A. $\lambda\mu$.

B. $\lambda\Delta$.

C. $\lambda C \mu\Delta$, where C is an upper bound for the density f for F (this point is only applicable when a density exists).

In particular, for any t and F, taking $\Delta \leq \dfrac{c}{\lambda}$ for some constant c guarantees that the expected time spent on insertions is bounded by c.

Proof of Theorem 5.4.

Bound A is obtained by noting that each $F(B_j - t)$ in the sum is at most equal to 1, and that $F(t+u) \leq 1$. Bound B is obtained by bounding

$$\int_{B_j - t} \lambda \ (F(t+u) - F(u)) \ du$$

by $\lambda\Delta$, and observing that the terms $F(B_j - t)$ summed over $j \in J$ yield the value 1. Finally inequality C uses the fact that $F(B_j - t) \leq C \Delta$ for all j. ■

Theorem 5.4 is extremely important. We see that it is possible to have constant expected time deletions and insertions, uniformly over all F, t and λ, provided that Δ is taken small enough. The bound on Δ depends upon λ. If λ is known, there is no problem. Unfortunately, λ has to be estimated most of the time. Recall also that we are in Reeves's idealized model. The present analysis does not extend beyond this model. As a rule of thumb, one can take Δ equal to $1/\lambda$ where λ is the expected number of points inserted per unit of time. This should insure that every bucket has at most one point on the average. Taking Δ too small is harmful from a space point of view because the number of intervals into which the FES is cut up is

$$\left\lceil (\max(Y_i) - t)/\Delta \right\rceil$$

where the Y_i's are the scheduled event times at time t. Taking the expected value, we see that this is bounded from above by

$$1 + \frac{E(\max(Y_1, \ldots, Y_N))}{\Delta} ,$$

where N is Poisson $(\lambda\mu)$. Recall that for an upper bound the Y_i's can be considered as iid random variables with density $(1-F)/\mu$ on $[0,\infty)$. This allows us to get a good idea of the expected number of buckets needed as a function of the expected FES size, or λ. We offer two quantitative results.

Theorem 5.5.

The expected number of buckets needed in Reeves's model does not exceed

$$1+\frac{\sqrt{\dfrac{\lambda}{3}E(X^3)}}{\Delta},$$

where X has distribution function F. If $\Delta\sim\dfrac{c}{\lambda}$ as $\lambda\to\infty$ for some constant c, then this upper bound $\sim$

$$\frac{1}{c\sqrt{3}}\sqrt{E((\lambda X)^3)}.$$

Furthermore, if $E(e^{uX})<\infty$ for some $u>0$, and Δ is as shown above, then the expected number of buckets is $O(\lambda\log(\lambda))$.

Proof of Theorem 5.5.

For the first part of the Theorem, we can assume without loss of generality that X has finite third moment. We argue as follows:

$$E(\max(Y_1,\ldots,Y_N)) \le E\left(\sqrt{\sum_{i\le N}Y_i^2}\right)$$

$$\le \sqrt{E(N)E(Y_1^2)} \quad \text{(Jensen's inequality)}$$

$$= \sqrt{\lambda\mu E(X^3)/(3\mu)} = \sqrt{\lambda E(X^3)/3}.$$

The last step follows from the simple observation that

$$\int_0^\infty x^2\frac{1-F(x)}{\mu}\,dx = \int_0^\infty x^2\int_x^\infty\frac{1}{\mu}\,dF(t)\,dx$$

$$= \int_0^\infty\frac{1}{\mu}\int_0^t x^2\,dx\,dF(t)$$

$$= \frac{1}{3\mu}E(X^3).$$

The second statement of the Theorem follows in three lines. Let u be a fixed constant for which $E(e^{uX})=a<\infty$. Then, using $X_1,\ldots,X_n$ to denote an iid sample with distribution function F,

$$E(\max(Y_1,\ldots,Y_n)) \le E(\max(X_1,\ldots,X_n))$$

$$\le E\left(\frac{1}{t}\log\left(\sum_{i\le N}e^{uX_i}\right)\right)$$

$$\le \frac{1}{t}\log(E(N)E(e^{uX_i})) = \frac{1}{t}\log(\lambda\mu a).$$

This concludes the proof of Theorem 5.5. ■

Except when F has compact support, the expected number of buckets needed grows superlinearly with λ, when Δ is picked as a constant over λ. The situation is worse when Δ is picked even smaller. This is a good example of the time-space trade-off, because taking Δ larger than $1/\lambda$ effectively decreases the space requirements but slows down the algorithm. However, large Δ's are uninteresting since we will see that there are nonlinear data structures which will run in expected or even worst-case time $O(\log(\lambda))$. Thus, there is no need to study cases in which the Vaucher-Duval structure performs worse than this. Vaucher and Duval (1975) and Davey and Vaucher (1980) circumvent the superlinear (in λ) storage need by collapsing many buckets in one big bucket, called an overflow bucket, or overflow list. Denardo and Fox (1979) consider a hierarchy of bucket structures where bucket width decreases with the level.

Various other multiple pointer structures have been proposed, such as the structures of Franta and Maly (1977, 1978) and Wyman (1976). They are largely similar to the Vaucher-Duval bucket structure. One nice new idea surfacing in these methods is the following. Assume that one wants to keep the cardinality of all sublists about equal and close to a number m, and assume that the FES has about n elements. Therefore, about n/m pointers are needed, which in turn can be kept in a linear list, to be scanned sequentially from left to right or right to left. The time needed for an insertion cannot exceed a constant times $\dfrac{n}{m}+m$ where the last term accounts for the sequential search into the selected sublist. The optimal choice for m is thus about $\sqrt{n}$, and the resulting complexity of an insertion grows also as $\sqrt{n}$. The difficulty with theses structures is the dynamic balancing of the sublist cardinalities so that all sublists have about m elements. Henriksen (1977) proposes to keep about m events per sublist, but the pointer records are now kept in a balanced binary search tree, which is dynamically adjusted. The complexity of an insertion is not immediately clear since the updating of the pointer tree requires some complicated work. Without the updating, we would need time about equal to $\log(\dfrac{n}{m})+m$ just to locate the point of insertion of one event. This expression is minimal for constant m ($m=4$ is the usual recommendation for Henriksen's algorithm (Kingston, 1984)). The complexity of insertion without updating is $\Theta(\log(n))$. For a more detailed expected time analysis, see Kingston (1984). In the next section, we discuss $O(\log(n))$ worst-case structures which are much simpler to implement than Henriksen's structure, and perform about equally well in practice.

5.4. Tree structures.

If the event times are kept in a binary search tree, then one would suspect that after a while the tree would be skewed to the right, because elements are deleted from the left and added mostly to the right. Interestingly, this is not always the case, and the explanation parallels that for the forward and backward scanning methods in linear lists. Consider for example an exponential F in the HOLD model. As we have seen in Theorem 5.1, all the relative event times in the FES are iid exponentially distributed. Thus, the binary search tree at every point in time is distributed as for any binary search tree constructed from a random permutation of $1, \ldots, n$. The properties of these trees are well-known. For example, the expected number of comparisons needed for an insertion of a new element, distributed as the n other elements, and independent of it, is $\sim 2\log(n)$ (see e.g. Knuth (1973) or Standish (1980)). The expected time needed to delete the smallest element is $O(\log(n))$. First, we need to locate the element at the bottom left, and then we need to restore the binary tree in case the deleted element had right descendants, by putting the bottom left descendant of these right descendants in its place. Unfortunately, one cannot count on F being exponential, and some distributions could lead to dangerous unbalancing, either to the left or the right. This was for example pointed out by Kingston (1985).

For robust performance, it is necessary to look at worst-case insertion and deletion times. They are $O(\log(n))$ for such structures as the 2–3 tree, the AVL tree and the heap. Of these, the heap is the easiest to implement and understand. The overhead with the other trees is excessive. Suggested for the FES by Floyd in a letter to Fox in the late sixties, and formalized by Gonnet (1976), the heap compares favorably in the extensive timing studies of McCormack and Sargent (1981), Ulrich (1978) and Reeves (1984). However, in isolated applications, it is clearly inferior to the bucket structures (Franta and Maly, 1978). This should come as no surprise since properly designed bucket structures have constant expected time insertions and deletions. If robustness is needed such as in a general purpose software package, the heap structure is warmly recommended (see also Ulrich (1978) and Kingston (1985)).

It is possible to streamline the heap for use in discrete event simulation. The first modification (Franta and Maly, 1978) consists of combining the DELETE and INSERT operations into one operation, the HOLD operation. Since a deletion calls for a replacement of the root of the heap, it would be a waste of effort to replace it by the last element in the heap, fix the heap, then insert a new element in the last position, and finally fix the heap again. In the HOLD operation, the root position can be filled by the new element directly. After this, the heap needs only be fixed once. This improvement is most marked when the number of HOLD operations is relatively large compared to the number of bare DELETE or INSERT operations. A second improvement, suggested by Kingston (1985), consists of using an m-ary heap instead of a binary heap. Good experimental results were obtained by him for the ternary heap. This improvement is based on the fact that insertions are more efficient for large values of m, while deletions become only slightly more time-consuming.

5.5. Exercises.

1. Prove Theorem 5.2.

2. Consider Reeves's model. Show that when $\mu < \infty$, $\lim\sup\limits_{t \to \infty} N_t = \infty$ almost surely.

3. Show that the gamma (a) ($a \geq 1$) and uniform $[0,1]$ distributions are NBUE. Show that the gamma (a) ($a \leq 1$) distribution is NWUE.

4. Generalize Theorem 5.5 as follows. For $r \geq 1$, the expected number of buckets needed in Reeves's model does not exceed

$$1 + \frac{\left(\dfrac{\lambda}{r+1} E\left(X^{r+1}\right) \right)^{\frac{1}{r}}}{\Delta} \; ,$$

where X has distribution function F. If $\Delta \sim \dfrac{c}{\lambda}$ as $\lambda \to \infty$ for some constant c, then this upper bound $\sim$

$$\frac{1}{c} \left[\frac{E\left((\lambda X)^{r+1}\right)}{r+1}) \right]^{\frac{1}{r}} .$$

5. Assume that F is the absolute normal distribution function. Prove that if Δ is $1/\lambda$ in the Vaucher-Duval bucket structure, then the expected number of buckets needed is $O\left(\lambda\sqrt{\log(\lambda)}\right)$ and $\Omega(\lambda\sqrt{\log(\lambda)})$ as $\lambda \to \infty$.

6. In the HOLD model, show that whenever F has a density, the expected time needed for insertion of a new element in an ordered doubly linked list is $\Omega(n)$ and $O(n)$.

7. Consider the binary heap under the HOLD model with an exponential distribution F. Show that the expected time needed for inserting an element at time t in the FES is $O(1)$.

8. Give a heap-based data structure for implementing the operations DELETE, INSERT and CANCEL in $O(\log(n))$ worst-case time.

9. Consider the HOLD model with an ordinary binary search tree implementation. Find a distribution F for which the expected insertion time of a new element at time $t > 0$ is $\Omega(\psi(n))$ for some function ψ increasing faster than a logarithm: $\lim\limits_{n \to \infty} \psi(n)/\log(n) = \infty$.

6. REGENERATIVE PHENOMENA.

6.1. The principle.

Many processes in simulation are repetitive, i.e. one can identify a null state, or origin, to which a system evolving in time returns, and given that the system is in the null state at a certain time, the future evolution does not depend upon what has happened up to that point. Consider for example a simple random walk in which at each time unit, one step to the right or left is taken with equal probability 1/2. When the random walk returns to the origin, we start from scratch. The future of the random walk is independent of the history up to the point of return to the origin. In some simulations of such processes, we can efficiently skip ahead in time by generating the waiting time until a return occurs, at least when this waiting time is a proper random variable. Systems in which the probability of a return is less than one should be treated differently.

The gain in efficiency is due to the fact that the waiting time until the first return to the origin is sometimes easy to generate. We will work through the example of the simple random walk in the next section. Regenerative phenomena are ubiquitous: they occur in queueing systems (see section 6.3), in Markov chains, and renewal processes in general. Heyman and Sobel (1982) provide a solid study of many stochastic processes of practical importance and pay particular attention to regenerative phenomena.

6.2. Random walks.

The one-dimensional random walk is defined as follows. Let $U_1, U_2, \ldots$ be iid $\{-1,1\}$-valued random variables where $P(U_1=1)=P(U_1=-1)=\dfrac{1}{2}$. Form the partial sums

$$S_n = \sum_{i=1}^{n} U_i .$$

Here S_n can be considered as a gambler's gain of coin tossing after n tosses provided that the stake is one dollar; n is the time. Let T be the time until a first return to the origin. If we need to generate S_n, then it is not necessary to generate the individual U_i's. Rather, it suffices to proceed as follows:

```
X←0
WHILE X ≤ n DO
        Generate a random variate T (distributed as the waiting time for the first return to
        the origin).
        X←X+T
V←X-T , Y←0
WHILE V < n DO
        Generate a random {1,-1}-valued step U.
        Y←Y+U , V←V+1
        IF Y=0 THEN V←X-T (reset V by rejecting partial random walk)
RETURN Y
```

The principle is clear: we generate all returns to the origin up to time n, and simulate the random walk explicitly from the last return onwards, keeping in mind that from the last return onwards, the random walk is conditional: no further returns to the origin are allowed. If another return occurs, the partial random walk is rejected. The example of the simple random walk is rather unfortunate in two respects: first, we know that S_n is binomial $(n, \frac{1}{2})$. Thus, there is no need for an algorithm such as the one described above, which cannot possibly run in uniformly bounded time. But more importantly, the method described above is intrinsically inefficient because random walks spend most of their time on one of the two sides of the origin. Thus, the last return to the origin is likely to be $\Omega(n)$ away from n, so that the probability of acceptance of the explicitly generated random walk, which is equal to the probability of not returning to the origin, is $O(\frac{1}{n})$. Even if we could generate T in zero time, we would be looking at an overall expected time complexity of $\theta(n^2)$. Nevertheless, the example has great didactical value.

The distribution of the time of the first return to the origin is given in the following Theorem.

Theorem 6.1.

In a symmetric random walk, the time T of the first return to the origin satisfies

$$P(T=2n) = p_{2n} = \frac{1}{n\,2^{2n-1}}\binom{2n-2}{n-1} \quad (n \geq 1),$$

$$P(T=2n+1) = 0 \quad (n \geq 0).$$

If q_{2n} is the probability that the random walk returns to the origin at time $2n$, then we have

A. $p_{2n} = q_{2n}/(2n-1)$;

B. $p_{2n} \sim 1/(2\sqrt{\pi}n^{3/2})$;

C. $E(T) = \infty$;

D. $p_{2n} = q_{2n-2} - q_{2n}$;

E. $p_2 = \dfrac{1}{2}$, $p_{2n+2} = p_{2n}(1-\dfrac{1}{2n})(1+\dfrac{1}{n})$.

Proof of Theorem 6.1.

This proof will be given in full, because it is a beautiful illustration of how one can compute certain renewal time distributions via generating functions. We begin with the generating function $G(s)$ for the probabilities $q_{2i}=P(S_{2i}=0)$ where S_{2i} is the value of the random walk at time $2i$. We have

$$G(s) = \sum_i q_{2i}\,s^i = \sum_{i=1}^{\infty} 2^{-2i}\binom{2i}{i}s^i$$

$$= \sum_{i=1}^{\infty}\binom{-\frac{1}{2}}{i}(-s)^i = \frac{1}{\sqrt{1-s}}-1.$$

Let us now relate p_{2n} to q_{2i}. It is clear that

$$q_{2n} = p_{2n} + \sum_{i=1}^{n-1} p_{2n-2i}\,q_{2i}.$$

If $H(s)$ is the generating function for p_{2n}, then we have

$$H(s) = \sum_{n=1}^{\infty} q_{2n}\,s^n$$

$$= \sum_{n=1}^{\infty}\left[p_{2n}\,s^n + \sum_{i=1}^{n-1} p_{2n-2i}\,s^{n-i}\,q_{2i}\,s^i \right]$$

$$= H(s) + \sum_{i=1}^{\infty}\sum_{n=i+1}^{\infty} p_{2n-2i}\,s^{n-i}\,q_{2i}\,s^i$$

$$= H(s) + \sum_{i=1}^{\infty} q_{2i} s^{i} \sum_{n=1}^{\infty} p_{2n} s^{n} = H(s) + G(s)H(s) .$$

Therefore,

$$H(s) = \frac{G(s)}{1+G(s)} = 1 - \sqrt{1-s} = \sum_{i=1}^{\infty} \binom{\frac{1}{2}}{i} (-1)^{i-1} s^{i} .$$

Equating the coefficient of s^{i} with p_{2i} gives the distribution of T. Statement A is easily verified. Statement B follows by using Stirling's formula. Statement C follows directly from B. Finally, D and E are obtained by simple computations. ∎

Even though T has a unimodal distribution on the even integers with peak at 2, generation by sequential inversion started at 2 is not recommended because $E(T) = \infty$. We can proceed by rejection based upon the following inequalities:

Lemma 6.1.

The probabilities p_{2n} satisfy for $n \geq 1$,

$$1 - \frac{1}{2n} \leq \frac{p_{2n}}{\dfrac{1}{2\sqrt{\pi}(n-\frac{1}{2})^{\frac{3}{2}}}} \leq e^{\frac{1}{12(2n-1)}} \leq e^{\frac{1}{12}} .$$

Proof of Lemma 6.1.

We rewrite p_{2n} as follows:

$$p_{2n} = \frac{\Gamma(2n-1)}{2n \, 2^{2n-2} \Gamma^{2}(n)}$$

$$= \frac{e^{-(2n-1)}(2n-1)^{2n-1} \sqrt{2\pi/2n-1} \, e^{\frac{\theta}{12(2n-1)}}}{2n \, 2^{2n-2} e^{-2n} n^{2n} \dfrac{2\pi}{n}}$$

$$= \frac{e \, (1-\frac{1}{2n})^{2n-1} e^{\frac{\theta}{12(2n-1)}}}{n \sqrt{2\pi(2n-1)}}$$

for some $0 < \theta < 1$. An upper bound is provided by

$$= \frac{e^{\frac{1}{12(2n-1)}}}{(n-\frac{1}{2})^{\frac{3}{2}} \sqrt{4\pi}} .$$

A lower bound is provided by

$$= \frac{e\,(1-\frac{1}{2n})^{2n}}{(n-\frac{1}{2})^{\frac{3}{2}}\sqrt{4\pi}}$$

$$\geq \frac{(1+\frac{1}{2n})^{2n}\,(1-\frac{1}{2n})^{2n}}{(n-\frac{1}{2})^{\frac{3}{2}}\sqrt{4\pi}}$$

$$\geq \frac{(1-\frac{1}{4n^2})^{2n}}{(n-\frac{1}{2})^{\frac{3}{2}}\sqrt{4\pi}}$$

$$\geq \frac{(1-\frac{1}{2n})}{(n-\frac{1}{2})^{\frac{3}{2}}\sqrt{4\pi}} \cdot \blacksquare$$

Generation can now be dealt with by truncation of a continuous random variate. Note that $p_{2n} \leq cg\,(x)$ where

$$cg\,(x) = \begin{cases} \dfrac{1}{2} & (n=1,\ n-1<x<n) \\[3mm] \dfrac{e^{\frac{1}{12}}}{\sqrt{4\pi}(x-\frac{1}{2})^{\frac{3}{2}}} & (n>1,\ n-1<x<n) \end{cases}$$

where

$$c = \frac{1}{2} + \frac{2e^{\frac{1}{12}}}{\sqrt{\pi}}\ .$$

Random variates with density g can quite easily be generated by inversion. The algorithm can be summarized as follows:

Generator for first return to origin in simple random walk

[SET-UP]

$$c \leftarrow \frac{1}{2} + \frac{2e^{\frac{1}{12}}}{\sqrt{\pi}}$$

[GENERATOR]

REPEAT

    Generate a uniform $[0, c]$ random variate U.

    IF $U \leq \dfrac{1}{2}$

        THEN RETURN $X \leftarrow 2$

        ELSE

            Generate a uniform $[0,1]$ random variate V.

$$Y \leftarrow \frac{1}{2} + \frac{1}{2 - (U - \frac{1}{2})\sqrt{\pi} e^{-\frac{1}{12}}} \quad (Y \text{ has density } g \text{ restricted to } [1, \infty)).$$

$$T \leftarrow V e^{\frac{1}{12}} / (\sqrt{\pi}(Y - \frac{1}{2})^{3/2}) \text{ (prepare for rejection)}$$

$$X \leftarrow \lceil Y \rceil$$

$$W \leftarrow 1 / (\sqrt{\pi}(X - \frac{1}{2})^{3/2}) \text{ (prepare for squeeze steps)}$$

            IF $T / W \leq 1 - \dfrac{1}{2X}$ (quick acceptance)

                THEN RETURN $2X$

                ELSE IF $T / W \leq e^{\frac{1}{12(2X-1)}}$ (quick rejection)

                    THEN IF $T \leq p_{2X}$ THEN RETURN $2X$

UNTIL False

The rejection constant c is a good indicator of the expected time spent before halting provided that p_{2X} can be evaluated in constant time uniformly over all X. However, if p_{2X} is computed directly from its definition, i.e. as a ratio of factorials, then the computation takes time roughly proportional to X. Assume that it is exactly X. Without squeeze steps, the expected time spent computing p_{2X} would be c times $E(X)$ where X has density g. This is ∞ (exercise 6.1). However, with the squeeze steps, the probability of evaluating p_{2X} explicitly for fixed value of X decreases as $\dfrac{1}{X}$ as $X \to \infty$. This implies that the overall expected time of the algorithm is finite (exercise 6.2).

6.3. Birth and death processes.

A birth and death process is a process with states 0,1,2,3,..., in which the time spent in state i is distributed as an exponential random variate divided by $\lambda_i + \mu_i$, at which time the system jumps to state $i+1$ (a birth) with probability $\lambda_i/(\lambda_i + \mu_i)$, and to state $i-1$ (a death) otherwise. Simple examples include

A.   The Poisson process: $\lambda_i \equiv \lambda > 0$, $\mu_i \equiv 0$. Births correspond essentially to events such as arrivals in a bank.

B.   The Yule process: $\lambda_i \equiv \lambda i > 0$, $\mu_i \equiv 0$. Here we also require that at time 0, the state be 1. This is a particular case of a pure birth process. The state can be identified with the size of a given population in which no deaths can occur.

C.   The M/M/1 queue: $\lambda_i \equiv \lambda > 0$, $\mu_i \equiv \mu > 0$, $\mu_0 = 0$. Here the state can be identified with the size of a queue, a birth with an arrival, and a death with a departure. The condition $\mu_0 = 0$ is natural since nobody can leave the queue when the queue is empty.

In all these examples, simulation can often be accelerated by making use of first-passage-time random variables. Formally, we define the first passage time from i to j $(j > i)$, T_{ij}, by

$$T_{ij} = \inf \{t : X_t = j \mid X_0 = i\} \; .$$

Here X_t is the state of the system (an integer) at time t, and X_0 is the initial state. Let us consider the M/M/1 queue. The busy period of such a queue is T_{10}. If the system starts in state 0 (empty queue), and we define a system cycle as the minimal time until for the first time another empty queue state is reached after some busy period, i.e. after at least one person has been in the queue, then the system cycle is distributed as $E/\lambda + T_{10}$, where E is an exponential random variate, independent of T_{10}. The only M/M/1 queues of interest to us are those which have with probability one a finite value for T_{10}, i.e. those for which $\mu \geq \lambda$ (Heyman and Sobel, 1982, p. 91). The actual derivation of the distribution of T_{10} would lead us astray. What matters is that we can generate random variates distributed as T_{10} quite easily. This should of course not be done by generating all the arrivals and departures until an empty queue is reached, because the expected time of this method is not uniformly bounded over all values of $\lambda < \mu$. This is best seen by noting that $E(T_{10}) = 1/(\mu - \lambda)$.

The M/M/1 queue provides one of the few instances in which the distribution of the first passage times is analytically manageable. For example, $2\sqrt{\lambda\mu}\,T_{10}$ has density

$$f(x) = e^{-\frac{x}{2}(\xi + \frac{1}{\xi})} I_1(x) \frac{\xi}{x} \quad (x > 0) \, ,$$

where $\xi = \sqrt{\dfrac{\mu}{\lambda}}$ and I_1 is the Bessel function of the first kind with imaginary argument (see section IX.7.1 for a definition). Direct generation can be carried out based upon the following result.

Theorem 6.2.

When E is exponentially distributed, Y is a random variable with density

$$g(y) = c \frac{\sqrt{y(1-y)}}{\frac{1}{2}(\xi+\frac{1}{\xi})-1+2y} \qquad (0<y<1),$$

where $c = \frac{4\xi}{\pi}$ and $\xi = \sqrt{\frac{\mu}{\lambda}}$, and E, Y are independent, then $E/(\frac{1}{2}(\xi+\frac{1}{\xi})+2Y-1)$ has density f, and $E/(\mu+\lambda+2\sqrt{\mu\lambda}(2Y-1))$ is distributed as T_{10}.

Proof of Theorem 6.2.

This theorem illustrates once again the power of integral representations for densities. By an integral representation of I_1 (Magnus et al, 1966, p. 84),

$$f(x) = e^{-\frac{x}{2}(\xi+\frac{1}{\xi})} I_1(x) \frac{\xi}{x}$$

$$= e^{-\frac{x}{2}(\xi+\frac{1}{\xi})} \frac{\xi}{x} \frac{x}{\pi} \int_{-1}^{1} e^{-zx} \sqrt{1-z^2}\, dz$$

$$= \int_0^1 (\frac{1}{2}(\xi+\frac{1}{\xi})+2y-1) e^{-x(\frac{1}{2}(\xi+\frac{1}{\xi})+2y-1)} \frac{4\xi}{\pi} \frac{\sqrt{y(1-y)}}{\frac{1}{2}(\xi+\frac{1}{\xi})+2y-1}\, dy$$

$$= E((\frac{1}{2}(\xi+\frac{1}{\xi})+2Y-1) e^{-x(\frac{1}{2}(\xi+\frac{1}{\xi})+2Y-1)})$$

where Y has density g. Given Y, this is the density of $E/(\frac{1}{2}(\xi+\frac{1}{\xi})+2Y-1)$. ∎

Generation of Y can be taken care of very simply by rejection. Note that

$$g(y) \le \begin{cases} c \dfrac{\sqrt{y(1-y)}}{2y} \\ c \dfrac{\sqrt{y(1-y)}}{\frac{1}{2}(\xi+\frac{1}{\xi})-1} \end{cases},$$

where $c = \frac{4\xi}{\pi}$. The top upper bound, proportional to a beta $(\frac{1}{2},\frac{3}{2})$ density integrates to ξ. The bottom upper bound, proportional to a beta $(\frac{3}{2},\frac{3}{2})$ density, integrates to $(\xi/(\xi-1))^2$. One should always pick the bound which has the smallest integral. The cross-over point is at $\xi = \frac{1}{2}(3+\sqrt{5}) \approx 2.6$.

Generator for g

CASE

$\xi \leq \frac{3+\sqrt{5}}{2}$:

> REPEAT
>> Generate a uniform [0,1] random variate U.
>>
>> Generate a beta $(\frac{1}{2}, \frac{3}{2})$ random variate Y.
>
> UNTIL $\dfrac{U}{1-U} \leq \dfrac{2Y}{\frac{1}{2}(\xi+\frac{1}{\xi})-1}$

$\xi > \frac{3+\sqrt{5}}{2}$:

> REPEAT
>> Generate a uniform [0,1] random variate U.
>>
>> Generate a beta $(\frac{3}{2}, \frac{3}{2})$ random variate Y.
>
> UNTIL $\dfrac{U}{1-U} \leq \dfrac{\frac{1}{2}(\xi+\frac{1}{\xi})-1}{2Y}$
>
> RETURN Y

The expected number of iterations is $\min(\xi, (\frac{\xi}{\xi-1})^2)$. This is a unimodal function in ξ, taking the value 1 as $\xi \downarrow 1$ and $\xi \uparrow \infty$. The peak is at $\xi = (3+\sqrt{5})/2$. The algorithm is uniformly fast with respect to $\xi \geq 1$. In the case $\xi = 1$ the acceptance condition is automatically satisfied, and the combination of the g generator with the property of Theorem 6.2 is reduced to a generator already dealt with in Theorem IX.7.1.

6.4. Phase type distributions.

Phase type distributions (or simply **PH-distributions**) are the distributions of absorption times in absorbing Markov chains, which are useful in studying queues and reliability problems. We consider only discrete (or: discrete-time) Markov chains with a finite number of states. An absorption state is one which, when reached, does not allow escape. Even if there is at least one absorption state, it is not at all certain that it will ever be reached. Thus, phase type distributions can be degenerate.

Any state can also be "promoted" to absorption state to study the time needed until this state is reached. If we promote the starting state to absorption state immediately after we leave it, then this promotion mechanism can be used to simulate Markov chains by the shortcuts discussed in this section, at least if we can get a good handle on the times until absorption.

Discrete Markov chains can always be simulated by using a simple discrete random variate generator for every state transition (Neuts and Pagano, 1981). This generator is not uniformly fast over all Markov chains with m states and nondegenerate phase type distribution. In the search for uniformly fast generators, simple shortcuts are of little help.

For example, when we are in state i, we could generate the (geometrically distributed) time until we first leave i in constant expected time. The corresponding state can also be generated uniformly fast by a method such as Walker's, because we have a simple conditional discrete distribution with $m-1$ outcomes. This method can be used to eliminate the times spent idling in individual states. It cannot eliminate the times spent in cycles, such as in a Markov chain in which with high probability we stay in a cycle visiting states $i_1, i_2, \ldots, i_k$ in turn. Thus, it cannot possibly be uniformly fast over all Markov chains with m states.

It seems that in this problem, uniform speed does not come cheaply. Some preprocessing involving the transition matrix seems necessary.

6.5. Exercises.

1. Consider the rejection algorithm for the time $2X$ until the first return to the origin in a symmetric random walk given in the text. Show that when the time needed to compute p_{2X} is equal to X, then the expected time taken by the algorithm without squeeze steps is ∞.

2. A continuation of exercise 1. Show that when squeeze steps are added as in the text, then the algorithm halts in finite expected time.

3. **Discrete Markov chains.** Consider a discrete Markov chain with m states and initial state 1. You are allowed to preprocess at any cost, but just once. What sort of preprocessing would you do on the transition matrix so that you can design an algorithm for generating the state S_n at time n in expected time uniformly bounded over n. The expected time is however allowed to increase with m. Hint: can you decompose the transition matrix using a spectral representation so that the n-th power of it can be computed uniformly quickly over all n?

4. **The lost-games distribution.** Let X be the number of games lost before a player is ruined in the classical gambler's ruin problem, i.e. a gambler adds one to his fortune with probability p and loses one unit with probability $1-p$. He starts out with r units (dollars). The purpose of this exercise is to design an algorithm for generating X in expected time uniformly bounded in

r when $p < 1-p$ is fixed. Uniform speed in both r and p would be even better. Notice first that the restriction $p < 1-p$ is needed to insure that X is a proper random variable, i.e. to insure that the player is ruined with probability one.

A. Show that when $p < 1-p$, the player will eventually be ruined with probability one.

B. Show that X has discrete distribution given by

$$P(X=n) = \binom{2n-r}{n} p^{n-r}(1-p)^n \frac{r}{2n-r} \qquad (n=r, r+1,...)$$

(Kemp and Kemp, 1968).

C. Suppose that customers arrive at a queue according to a homogeneous Poisson process with parameter λ, that the service time is exponential with parameter $\mu < \lambda$, and that the queue has initially r customers. Show that the number of customers served until the queue first vanishes has the lost-games distribution with parameters r and $p = \lambda/(\lambda+\mu)$.

D. Using Stirling's approximation, determine the general dependence of $P(X=n)$ upon n, and use it to design a uniformly fast rejection algorithm.

For a survey of these and other waiting time mechanisms, see e.g. Patil and Boswell (1975).

7. THE GENERALIZATION OF A SAMPLE.

7.1. Problem statement.

As in section XIV.2, we will discuss an incompletely specified random variate generation problem. Assume that we are given a sample $X_1, \ldots, X_n$ of iid R^d-valued random vectors with common unknown density f, and that we are asked to generate a new independent sample $Y_1, \ldots, Y_m$ of independent random vectors with the same density f. Stated in this manner, the problem is obviously unsolvable, unless we are incredibly lucky.

What one can do is construct a **density estimate** $f_n(x) = f_n(x, X_1, \ldots, X_n)$ of $f(x)$, and then generate a sample of size m from f_n. This procedure has several drawbacks: first of all, f_n is typically not equal to f. Also, the new sample depends upon the original sample. Yet, we have very few options available to us. Ideally, we would like the new sample to appear to be distributed as the original sample. This will be called sample indistinguishability. This and other issues will be discussed in this section. The material appeared originally in Devroye and Gyorfi (1985, chapter 8).

7.2. Sample independence.

There is little that can be done about the dependence between $X_1, \ldots, X_n$ and $Y_1, \ldots, Y_m$ except to hope that for n large enough, some sort of asymptotic independence is obtained. In some applications, sample independence is not an issue at all.

Since the Y_i's are conditionally independent given $X_1, \ldots, X_n$, we need only consider the dependence between Y_1 and $X_1, \ldots, X_n$. A measure of the dependence is

$$D_n = \sup_{A,B} | P(Y \in A, X \in B) - P(Y \in A) P(X \in B) | ,$$

where the supremum is with respect to all Borel sets A of R^d and all Borel sets B of R^{nd}, and where $Y = Y_1$ and X is our shorthand notation for $(X_1, \ldots, X_n)$. We say that the samples are asymptotically independent when

$$\lim_{n \to \infty} D_n = 0 .$$

In situations in which $X_1, \ldots, X_n$ is used to design or build a system, and $Y_1, \ldots, Y_m$ is used to test it, the sample dependence often causes optimistic evaluations. Without the asymptotic independence, we can't even hope to diminish this optimistic bias by increasing n.

The inequality in Theorem 7.1 below provides us with a sufficient condition for asymptotic independence. First, we need the following Lemma.

Lemma 7.1. (Scheffe, 1947).

For all densities f and g on R^d,

$$\int | f - g | = 2 \sup_B | \int_B f - \int_B g | ,$$

where the supremum is with respect to all Borel sets B of R^d.

Proof of Lemma 7.1.

Let us take $B = \{f > g\}$, and let A be another Borel set of R^d. Because $\int (f - g) = 0$, we see that

$$\int | f - g | = 2 \int_B (f - g) .$$

Thus, we have shown that $\int | f - g |$ is at most twice the supremum over all Borel sets of $| \int_B (f - g) |$. To show the other half of the Lemma, note that if B' denotes the complement of B, then

$$| \int_A (f - g) | = | \int_{A \cap B} (f - g) + \int_{A \cap B'} (f - g) |$$

$$\leq \max(\int_{A \cap B} (f - g), \int_{A \cap B'} (f - g))$$

$$\leq \max(\int_{B}(f-g), \int_{B'}(g-f))$$

$$= \frac{1}{2}\int |f-g| \quad \text{(all } A\text{)}. \blacksquare$$

Scheffe's lemma tells us that if we assign probabilities to sets (events) using two different densities, then the maximal difference between the probabilities over all sets is equal to one half of the L_1 distance between the densities. From Lemma 7.1, we obtain

---

Theorem 7.1.

Let f_n be a density estimate, which itself is density. Then

$$D_n \leq E(\int |f_n-f|) .$$

Proof of Theorem 7.1.

Let $X_1, \ldots, X_{n+1}$ be iid. Then

$$D_n \leq \sup_{A,B} |P(Y \in A, X \in B)-P(X_{n+1} \in A, X \in B)|$$

$$+ \sup_{A,B} |P(X_{n+1} \in A, X \in B)-P(X_{n+1} \in A)P(X \in B)|$$

$$+ \sup_{A,B} |P(X_{n+1} \in A)P(X \in B)-P(Y \in A)P(X \in B)|$$

$$\leq \sup_{A,B} E(I_{X \in B} |\int_A f_n - \int_A f|) + 0 + \sup_A |P(X_{n+1} \in A)-P(Y \in A)|$$

$$\leq \sup_A E(|\int_A f_n - \int_A f|) + \sup_A |\int_A E(f_n)-\int_A f|$$

$$\leq E(\sup_A |\int_A f_n - \int_A f|) + \frac{1}{2}\int |E(f_n)-f|$$

$$= E(\frac{1}{2}\int |f_n-f|) + \frac{1}{2}\int |E(f_n)-f| . \blacksquare$$

We see that for the sake of asymptotic sample independence, it suffices that the expected L_1 distance between f_n and f tends to zero with n. This is also called **consistency**. Consistency does not imply asymptotic independence: just let f_n be the uniform density in all cases, and observe that $D_n \equiv 0$, yet

$\int |f_n - f|$ is a positive constant for all n and all nonuniform f.

7.3. Consistency of density estimates.

A density estimate f_n is **consistent** if for all densities f,

$$\lim_{n \to \infty} E(\int |f_n - f|) = 0.$$

Consistency guarantees that the expected value of the maximal error committed by replacing probabilities defined with f with probabilities defined with f_n tends to 0. Many estimates are consistent, see e.g. Devroye and Gyorfi (1985). Parametric estimates, i.e. estimates in which the form of f_n is fixed up to a finite number of parameters, which are estimated from the sample, cannot be consistent because f_n is required to converge to f for all f, not a small subclass. Perhaps the best known and most widely used consistent density estimate is the **kernel estimate**

$$f_n(x) = \frac{1}{nh^d} \sum_{i=1}^{n} K(\frac{x - X_i}{h}),$$

where K is a given density (or kernel), chosen by the user, and $h > 0$ is a smoothing parameter, which typically depends upon n or the data (Rosenblatt, 1956; Parzen, 1962). For consistency it is necessary and sufficient that $h \to 0$ and $nh^d \to \infty$ in probability as $n \to \infty$ (Devroye and Gyorfi, 1985). How one should choose h as a function of n or the data is the subject of a lot of controversy. Usually, the choice is made based upon the approximate minimization of an error criterion. Sample independence (Theorem 7.1) and sample indistinguishability (next section) suggest that we try to minimize

$$E(\int |f_n - f|).$$

But even after narrowing down the error criterion, there are several strategies. One could minimize the supremum of the criterion where the supremum is taken over a class of densities. This is called a **minimax strategy**. If f has compact support on the real line and a bounded continuous second derivative, then the best choices for individual f (i.e., not in the minimax sense) are

$$h = Cn^{-\frac{1}{5}},$$

$$K(x) = \frac{3}{4}(1 - x^2) \quad (|x| \le 1),$$

where C is a constant depending upon f only:

$$C = \left(\sqrt{\frac{15}{2\pi}} \frac{\int \sqrt{f}}{\int |f''|} \right)^{\frac{2}{5}}.$$

The optimal kernel coincides with the optimal kernel for L_2 criteria (Bartlett, 1963). The optimal formula for h, which depends upon the unknown density f, can be estimated from the data. Alternatively, one could compute the formula for a given parametric density, a rough guess of sorts, and then estimate the parameters from the data. For example, if this is done with the normal density as initial guess, we obtain the recommendation to take

$$h = \left(\frac{15e \sqrt{2\pi}}{8n} \right)^{\frac{1}{5}} \hat{\sigma} \, ,$$

where $\hat{\sigma}$ is a robust estimate of the standard deviation of the normal density (Devroye and Gyorfi, 1985). A typical robust estimate is the so-called quick-and-dirty estimate

$$\hat{\sigma} = \frac{X_{(np)} - X_{(nq)}}{x_p - x_q} \, ,$$

where x_p, x_q are the p-th and q-th quantiles of the standard normal density, and $X_{(np)}$ and $X_{(nq)}$ are the p-th and q-th quantiles in the data, i.e. the (np)-th and (nq)-th order statistics.

The construction given here with the kernel estimate is simple, and yields fast generators. Other constructions have been suggested in the literature with random variate generation in mind. Often, the explicit form of f_n is not given or needed. Constructions often start from an empirical distribution function based upon $X_1, \ldots, X_n$, and a smooth approximation of this distribution function (obtained by interpolation), which is directly useful in the inversion method. Guerra, Tapia and Thompson (1978) use Akima's (Akima, 1970) quasi-Hermite piecewise cubic interpolation to obtain a smooth monotone function coinciding with the empirical distribution function at the points X_i. Recall that the empirical distribution is the distribution which puts mass $\frac{1}{n}$ at point X_i. Hora (1983) gives another method for the same problem. Butler (1970) on the other hand uses Lagrange's quadratic interpolation on the inverse empirical distribution function to speed random variate generation up even further.

7.4. Sample indistinguishability.

In simulations, one important qualitative measure of the goodness of a method is the indistinguishability of $X_1, \ldots, X_m$ and $Y_1, \ldots, Y_m$ for the given sample size m. Note that we have forced both sample sizes to be the same, although for the construction of f_n we keep on using n points. The indistinguishability could be measured quantitatively by

$$S_{n,m} = \sup_A \ | E(N(A)) - E(M(A) | X_1, \ldots, X_n) |$$
$$= m \sup_A | \int_A f - \int_A f_n |$$

$$= \frac{m}{2} \int \mid f_n - f \mid \; .$$

Here, A is a Borel set of R^d, $N(A)$ is the cardinality of A for the original sample (the data, artificially inflated to size m), and $M(A)$ is the cardinality of A for the artificial Y_i sample. By cardinality of a set, we mean the number of data points falling in the set.

When $S_{n,m}$ is smaller than one, then the expected cardinality of a set A with a perfect sample of size m differs by at most one from the conditional expected cardinality of the generated sample of size m. We say that f_n is k-excellent for samples of size m when

$$E(S_{n,m}) \leq k \; .$$

This is equivalent to asking that the expected L_1 distance between f and f_n is at most $2k/m$. The notion of 1-excellence is very strong. For example, for most nonparametric estimates such as the kernel estimate 1-excellence forces us to use phenomenally large values of n for even moderate values of m. Devroye and Gyorfi (1985) have shown that for all kernel estimates (regardless of choice of K and h), and for all densities f, 1-excellence is not achievable for samples of size $m = 1000$ unless $n \geq 4,000,000$. For $m = 10,000$, we need $n \geq 1,300,000,000$. For the histogram estimate, the situation is even worse.

But even 1-excellence may not be good enough for one's application. For one thing, no assurances are given as to the discrepancy in moments between the generated sample and the original sample.

7.5. Moment matching.

Some statisticians attach a great deal of importance to the moments of the densities f_n and f. For $d = 1$, the i-th **moment mismatch** is defined as

$$M_{n,i} = \int x^i f_n - \int x^i f \qquad (i = 1,2,3,...) \; .$$

Clearly, $M_{n,i}$ is a random variable. Assume that we employ the kernel estimate with a zero mean finite variance (σ^2) kernel K. Then, we have

$$M_{n,1} = \frac{1}{n} \sum_{i=1}^{n} (X_i - E(X_i)) \; ,$$

$$M_{n,2} = \frac{1}{n} \sum_{i=1}^{n} (X_i^2 - E(X_i^2)) + h^2 \sigma^2 \; .$$

This follows from the fact that f_n is an equiprobable mixture of densities K shifted to the X_i's, each having variance $h^2\sigma^2$ and zero mean. It is interesting to note that the distribution of $M_{n,1}$ is not influenced by h or K. By the weak law of large numbers, $M_{n,1}$ tends to 0 in probability as $n \to \infty$ when f has a finite first moment. The story is different for the second moment mismatch.

Whereas $E(M_{n,1})=0$, we now have $E(M_{n,2})=h^2\sigma^2$, a positive bias. Fortunately, h is usually small enough so that this is not too big a bias. Note further that the variances of $M_{n,1}$, $M_{n,2}$ are equal to

$$\frac{Var(X_1)}{n}, \quad \frac{Var(X_1^2)}{n}$$

respectively. Thus, h and K only affect the bias of the second order mismatch. Making the bias very small is not recommended as it increases the expected L_1 error, and thus the sample dependence and distinguishability.

7.6. Generators for f_n.

For the kernel estimate, generators can be based upon the property that a random variate is distributed as an equiprobable mixture, as is seen from the following trivial algorithm.

Mixture method for kernel estimate

Generate Z, a random integer uniformly distributed on $\{1,2,\ldots,n\}$.
Generate a random variate W with density K.
RETURN $X_Z + hW$

For Bartlett's kernel $K(x)=\frac{3}{4}(1-x^2)_+$, we suggest either rejection or a method based upon properties of order statistics:

Generator based upon rejection for Bartlett's kernel

REPEAT
        Generate a uniform $[-1,1]$ random variate X and an independent uniform $[0,1]$ random variate U.
UNTIL $U \leq 1-X^2$
RETURN X

The order statistics method for Bartlett's kernel

Generate three iid uniform $[-1,1]$ random variates V_1, V_2, V_3.
IF $|V_3| > \max(|V_1|, |V_2|)$
      THEN RETURN $X \leftarrow V_2$
      ELSE RETURN $X \leftarrow V_3$

In the rejection method, X is accepted with probability 2/3, so that the algorithm requires on average three independent uniform random variates. However, we also need some multiplications. The order statistics method always uses precisely three independent uniform random variables, but the multiplications are replaced by a few absolute value operations.

7.7. Exercises.

1. **Monte Carlo integration.** To estimate $\int H(x) f(x) \, dx$, where H is a given function, and f is a density, the Monte Carlo method uses a sample of size n drawn from f (say, $X_1, \ldots, X_n$). The naive estimate is

$$\frac{1}{n} \sum_{i=1}^{n} H(X_i) \, .$$

When n is small, this estimate has a lot of built-in variance. Compute the variance and assume that it is finite. Then construct the **bootstrap estimate**

$$\frac{1}{m} \sum_{i=1}^{m} H(Y_i) \, ,$$

where the Y_i's are iid random variables with density f_n, the kernel estimate of f based upon $X_1, \ldots, X_n$. The sample size m can be taken as large as the user can afford. Thus, in the limit, one can expect the bootstrap estimate to provide a good estimate of $\int H(x) f_n(x) \, dx$.

A. Show that $|\int Hf - \int Hf_n| \le 2 (\sup H) \int |f - f_n|$ (Devroye and Gyorfi, 1985).

B. Compare the mean square errors of the naive Monte Carlo estimate and the estimate $\int Hf_n$ (the latter is a limit as $m \to \infty$ of the bootstrap estimate).

C. Compute the mean square error of the bootstrap estimate as a function of n and m, and compare with the naive Monte Carlo estimate. Also consider what happens when you let $m \to \infty$ in the expression for the mean square error.

2. The generators for the kernel estimate based upon Bartlett's kernel in the text use the mixture method. Still for Bartlett's kernel, derive the inversion method with all the details. Hint: note that the distribution function can be written as the sum of polynomials of degree three with compact support, and can therefore be considered as a cubic spline with at most $2n$ breakpoints when there are n data points (Devroye and Gyorfi, 1985).

3. Bratley, Fox and Schrage (1983) consider a density estimate f_n which provides fast generation by inversion. The X_i's are ordered, and f_n is constant on the intervals determined by the order statistics. In addition, in the intervals to the left of the minimum and to the right of the maximum exponential tails are added. The constant pieces and exponentail tails integrate to $1/(n+1)$ over their supports, i.e. all pieces are equally likely to be picked. Rederive their fast inversion algorithm for f_n. Is their estimate asymptotically independent? Show that it is not consistent for any density f. To cure the latter problem, Bratley, Fox and Schrage suggest coalescing breakpoints. Consider coalescing breakpoints by letting f_n be constant on the intervals determined by the k-th, $2k$-th, $3k$-th, $\cdots$ order statistics. How should one define the heights of f_n on these intervals, and how should k vary with n for consistency?

4. For the kernel estimate, show that for any density K, any f, and any sequence of numbers $h > 0$ with $h \to 0$, $nh^d \to \infty$, we have $E(\int |f - f_n|) \to 0$ as $n \to \infty$. Proceed as follows: first prove the statement for continuous f with compact support. Then, using the fact that any measurable function in L_1 can be approximated arbitrarily closely by continuous functions with compact support, wrap up the proof. In the first half of the proof, it is useful to split the integral by considering $|f - E(f_n)|$ separately. In the second half of the proof, you will need an embedding argument, in which you create a sample which with a few deletions can be considered as a sample drawn from f, and with a few different deletions can be considered as a sample drawn from the L_1 approximation of f.

Chapter Fifteen
THE RANDOM BIT MODEL

1. THE RANDOM BIT MODEL.

1.1. Introduction.

Chapters I-XIV are based on the premises that a perfect uniform [0,1] random variate generator is available and that real numbers can be manipulated and stored. Now we drop the first of these premises and instead assume a perfect bit generator (i.e., a source capable of generating iid $\{0,1\}$ random variates $B_1, B_2, ...$),while still assuming that real numbers can be manipulated and stored, as before: this is for example necessary when someone gives us the probabilities p_n for discrete random variate generation. The cost of an algorithm can be measured in terms of the number of bits required to generate a random variate. This model is due to Knuth and Yao (1976) who introduced a complexity theory for nonuniform random variate generation. We will report the main ideas of Knuth and Yao in this chapter.

If random bits are used to construct random variates from scratch, then there is no hope of constructing random variates with a density in a finite amount of time. If on the other hand we are to generate a discrete random variate, then it is possible to find finite-time algorithms. Thus, we will mainly be concerned with discrete random variate generation. For continuous random variate generation, it is possible to study the relationship between the number of input bits needed per n bits of output, and to develop a complexity theory based upon this relationship. This will not be done here. See however Knuth and Yao (1976).

1.2. Some examples.

Assume first that we wish to generate a binomial random variate with parameters $n = 1$ and $p \neq \dfrac{1}{2}$. This can be considered as the simulation of a biased coin flip, or the simulation of the occurrence of an event having probability p. If p were $\dfrac{1}{2}$, we could just exit with B_1. When p has binary expansion

$$p = 0.p_1 p_2 p_3 \cdots .$$

It suffices to generate random bits until for the first time $B_i \neq p_i$, and to return 1 if $B_i < p_i$ and 0 otherwise:

Binomial (1,p) generator

$i \leftarrow 0$

REPEAT

    $i \leftarrow i + 1$

    Generate a random bit B.

UNTIL $B \neq p_i$

RETURN $X \leftarrow I_{B < p_i}$

If we define the uniform [0,1] random variate

$$U = 0.B_1 B_2 B_3 \cdots ,$$

then it is easy to see that this simple algorithm returns

$$I_{U \leq p} .$$

Interestingly, the probability of exiting after i bits is 2^{-i}, so that the expected number of bits needed is precisely 2, independent of p. We recognize in this example the inversion method.

The rejection method too has a nice analog. Suppose that we want to generate a random integer X where $P(X = i) = p_i$, $1 \leq i \leq n$, and that all probabilities p_i are multiples of $\dfrac{1}{M}$, where $2^{k-1} < M \leq 2^k$ for some integer k. Then we can consider consecutive k-tuples in the sequence $B_1, B_2, \dots$ and set up a table with 2^k entries: M entries are used for storing integers between 1 and M, and the remaining entries are 0. If $p_i = l_i / M$, then the integer i should appear l_i times in the table. An integer 0 indicates a rejection. Now use

Rejection algorithm

REPEAT
   Generate k random bits, forming the number $Z \in \{0,1, \ldots, 2^k - 1\}$.
 UNTIL $Z < M$
 RETURN $X \leftarrow A[Z]$ (where A is the table of M integers)

In this algorithm, the expected number of bits required is k divided by the probability of immediate acceptance, i.e.

$$\frac{k}{\dfrac{M}{2^k}} \le 2k = 2 \left\lceil \log_2 M \right\rceil .$$

In both examples provided here, we can consider the complete unbounded binary tree in which we travel down by turning left when $B_i = 0$ and right when $B_i = 1$. In the rejection method, we have designated M nodes at the k-th level as terminal nodes. The remaining nodes at the k-th level are "rejection nodes", and are in turn roots of similar subtrees. Since these rejection nodes are identified with the overall root, we can superimpose them on the root, and form a pseudo-tree with some loopbacks from the k-th level to the root. But then, we have a finite directed graph, or a finite state machine.

In the inversion method, the expansion of p determines an unbounded path down the tree, and so does the expansion of U. Since we need only determine whether one path is to the left or the right of the other path, it suffices to travel down until the paths separate. With probability $\frac{1}{2}$, they separate right away. Otherwise, they separate with probability $\frac{1}{2}$ at the next level, and so forth.

What we will do in the sections that follow is

(i) Develop a lower bound for the expected number of bits in terms of $p_1, p_2, \ldots$, the probability vector of the discrete random variate.

(ii) Develop black box methods and study their expected complexity.

2. THE KNUTH-YAO LOWER BOUND.

2.1. DDG trees.

Suppose that we wish to generate a discrete random variate X with probability vector $p_1, p_2, \ldots$. The probability vector can be finite or infinite dimensional. Every algorithm based upon random bits can be represented as a binary tree (which is usually infinite), containing nodes of two types:

(i)   Branch nodes (or internal nodes), having two children. We can travel to the left child when a 0 bit is encountered, and to the right child otherwise.

(ii) Terminal nodes without children. These nodes are marked with an integer to be returned.

It is instructive to verify that this structure is present for the examples of the previous section. For example, for the binomial $(1,p)$ generator, consider the path for p, and assign terminal nodes marked 1 to all left children of nodes on the path that do not belong to the path themselves, and terminal nodes marked 0 to all right children of nodes on the path that do not belong to the path themselves.

Let us introduce the notation $t_i(k)$ for the number of terminal nodes on level k (the root is on level 0) which are marked i. Then we must have

$$\sum_{k \geq 0} \frac{t_i(k)}{2^k} = p_i \quad \text{(all } i \text{)} .$$

When these conditions are satisfied, we say that we have a DDG-tree (discrete distribution generating tree, terminology introduced by Knuth and Yao, 1976). The corresponding algorithms are called DDG-tree algorithms. DDG-tree algorithms halt with probability one because the sum of the probabilities of reaching the terminal nodes is

$$\sum_{i} \sum_{k \geq 0} \frac{t_i(k)}{2^k} = \sum_{i} p_i = 1 .$$

2.2. The lower bound.

Let us introduce the function $\chi(x) = x \bmod 1 = x - \lfloor x \rfloor$, the fractional part of x. Define furthermore

$$\nu(x) = \sum_{k \geq 0} \frac{\chi(2^k x)}{2^k} \quad (0 \leq x \leq 1) ,$$

and the entropy function

$$H(x) = x \log_2 \frac{1}{x} \quad (x > 0) .$$

Theorem 2.1

Let N be the number of random bits taken by a DDG-tree algorithm. Then:

A. $E(N) \geq \sum_{i} \nu(p_i)$.

B. Let $H(p_1, p_2, ...) = \sum_{i} H(p_i)$ be the entropy of the probability distribution $(p_1, p_2, ...)$. Then

$$H(p_1, p_2, ...) \leq \sum_{i} \nu(p_i) .$$

C. $\sum_{i} \nu(p_i) \leq H(p_1, p_2, ...) + 2$.

Proof of Theorem 2.1.

We begin with an expression for $E(N)$:

$$E(N) = \sum_{k \geq 0} P(N > k)$$
$$= \sum_{k \geq 0} \frac{b(k)}{2^k}$$

where $b(k)$ is the number of internal (or: branch) nodes at level k. We obtain the lower bound by finding a lower bound for $b(k)$. Let us use the notation $t(k)$ for the number of terminal nodes at level k. Thus,

$$b(0) + t(0) = 1 ,$$
$$b(k) + t(k) = 2b(k-1) \quad (k \geq 1) .$$

Using these relations, we can show that

$$b(k) = \sum_{j > k} \frac{t(j)}{2^{j-k}} .$$

(Note that this is true for $k = 0$, and use induction from there on.) But

$$\sum_{j > k} \frac{t_i(j)}{2^j} = p_i - \sum_{0 \leq j \leq k} \frac{t_i(j)}{2^j} \geq \frac{\chi(2^j p_i)}{2^j} .$$

This is true because the left-hand-sum is nonnegative, and the right-hand-sum is an integer multiple of 2^{-k}. Combining all of this yields

$$b(k) \geq \sum_{i} \chi(2^k p_i) .$$

This proves part A. Part B follows if we can show the following:

$$H(x) \leq \nu(x) \leq H(x) + 2x \quad (\text{all } x) .$$

Note that this is more than needed, but the second part of the inequality will be useful elsewhere. For a number $x \in [0,1)$, we will use the notation $x = 0.x_1 x_2 \cdots$ for the binary expansion. By definition of $\nu(x)$,

$$\nu(x) = \sum_{k \geq 0} 2^{-k} \sum_{j > k} \frac{x_j}{2^{j-k}}$$

$$= \sum_{j \geq 0} \sum_{0 \leq k \leq j} \frac{x_j}{2^j}$$

$$= \sum_{j \geq 0} \frac{j x_j}{2^j} \ .$$

Now, $\nu(0) = H(0) = 0$. Also, if $2^{-k} \leq x < 2^{1-k}$,

$$\nu(x) = \sum_{j \geq k} \frac{j x_j}{2^j}$$

$$\geq \sum_{j \geq k} \frac{\log_2(\frac{1}{x}) x_j}{2^j}$$

$$= H(x) \ .$$

Also, because $x_k = 1$,

$$H(x) + 2x - \nu(x) = \sum_{j \geq k} \frac{(\log_2(\frac{1}{x}) + 2 - j) x_j}{2^j}$$

$$> \sum_{j \geq k} \frac{(k + 1 - j) x_j}{2^j}$$

$$= 2^{-k} - \sum_{j \geq k+2} \frac{(j - k - 1) x_j}{2^j}$$

$$> 2^{-k} - \sum_{j \geq 1} \frac{j}{2^{j+k+1}}$$

$$= 0 \ . \blacksquare$$

The lower bound of Theorem 2.1 is related to the entropy of the probability vector. Let us briefly look at the entropy of some probability vectors: If $p_i = \frac{1}{n}$, $1 \leq i \leq n$, then

$$H(p_1, \ldots, p_n) = \log_2 n \ .$$

In fact, because H is invariant under permutations of its arguments, and is a concave   function,   it   is   true   that   for   probability   vectors

$(p_1, \ldots, p_n), (q_1, \ldots, q_n)$,

$$H(p_1, \ldots, p_n) \leq H(q_1, \ldots, q_n) ,$$

when the p_n vector is stochastically smaller than the q_n vector, i.e. if the p_i's and q_i's are in increasing order, then

$$p_1 \leq q_1 ;$$
$$p_1 + p_2 \leq q_1 + q_2 ;$$
$$\ldots$$
$$p_1 + p_2 + \cdots + p_n \leq q_1 + q_2 + \cdots + q_n .$$

This follows from the theory of Schur-convexity (Marshall and Olkin, 1979). In particular, for all probability vectors $(p_1, \ldots, p_n)$, we conclude that

$$0 \leq H(p_1, \ldots, p_n) \leq \log_2 n .$$

Both bounds are attainable. In a sense, entropy increases when the probability vector becomes smoother, more uniform. It is smallest when there is no randomness, i.e. all the probability mass is concentrated in one point. According to Theorem 2.1, we are tempted to conclude that uniform random variates are the costliest to produce. This is indeed the case if we compare optimal algorithms for distributions, and if the lower bounds can be attained for all distributions (this will be dealt with in the next sub-section). If we consider discrete distributions with n infinite, then it is possible to have $H(p_1, p_2, \ldots) = \infty$. To construct counterexamples very easily, we note that if the p_n's are $\downarrow$, then

$$H(p_1, \ldots) \geq E(\log(X))$$

where X is a random variate with the given probability vector. To see this, note that $p_n \leq \frac{1}{n}$, and thus that $-p_n \log(p_n) \geq p_n \log(n)$. Thus, whenever

$$p_n \sim \frac{c}{n \log^{1+\epsilon}(n)} ,$$

as $n \to \infty$, for some $\epsilon \in (0,1]$, we have infinite entropy. The constant c may be difficult to calculate except in special cases. The following example is due to Knuth and Yao (1976):

$$p_1 = 0 \; ; \; p_n = 2^{-\left\lfloor \log_2(n) \right\rfloor - 2 \left\lfloor \log_2(\log_2(n)) \right\rfloor - 1} \qquad (n \geq 2) .$$

Note that this corresponds to the case $\epsilon = 1$. Thus, we note that for any DDG-tree algorithm, $E(\log(X)) = \infty$ implies $E(N) = \infty$, regardless of whether the probability vector is monotone or not. The explanation is very simple: $E(\log_2(X))$ is the expected number of bits needed to store, or describe, X. If this is ∞, there is little hope of generating X requiring only $E(N) < \infty$ provided that the distribution of X is sufficiently spread out so that no bits are "redundant" (see exercises).

2.3. Exercises.

1. **The entropy.** This is about the entropy H of a probability vector $(p_1, p_2, ...)$. Show the following:

    A. There exists a probability vector such that $E(\log_2(X)) = \infty$, yet $E(N) < \infty$. Here X is a discrete random variate with the given probability vector. Hint: clearly, the counterexample is not monotone.

    B. Is it true that when the probability vector is monotone, then $E(\log_2(X)) < \infty$ implies $H(p_1, ...) < \infty$?

    C. Show that the p_n 's defined by

    $$p_1 = 0 \; ; \; p_n = 2^{-\left\lfloor \log_2(n) \right\rfloor - 2\left\lfloor \log_2(\log_2(n)) \right\rfloor - 1} \qquad (n \geq 2)$$

    form a probability vector, and that its entropy is ∞.

    D. Show that if one finite probability vector is stochastically larger than another probability vector, then its entropy is at most equal to the entropy of the second probability vector.

    E. Prove that when $x \in [0,1]$ is a power of 2, we have $\nu(x) = H(x)$, and that for any $x \in [0,1]$, $\nu(x) = 2^n \nu(\frac{x}{2^n}) - nx$.

3. OPTIMAL AND SUBOPTIMAL DDG-TREE ALGORITHMS.

3.1. Suboptimal DDG-tree algorithms.

We know now what we can expect at best from any DDG-tree algorithm in terms of the expected number of random bits. It is another matter altogether to construct feasible DDG-tree algorithms. Some algorithms require unwieldy set-up times and/or calculations which would overshadow the contribution to the total complexity from the random bit generator. In fact, most practical DDG-tree algorithms correspond to algorithms described in chapter III. Let us quickly check what kind of DDG-tree algorithms are hidden in that chapter.

In section III.2, we introduced inversion of a uniform [0,1] random variate U. In sequential inversion, we compared U with successive partial sums of p_n 's. This corresponds to the following infinite DDG-tree: consider all the paths for the partial sums, i.e. the path for p_1, for $p_1 + p_2$, etcetera. In case of a finite vector, we define the last cumulative sum by the binary expansion 0.111111111.... Then generate random bits until the path traveled by the random bits deviates for the first time from any of the p_n paths. If that path in question is for p_n, then return n if the last random bit was 0 (the corresponding bit on the path is 1), and return $n+1$ otherwise. This method has two problems: first, the set-up is impossible except in the following special case: all p_n 's have a finite binary expansion, and the probability vector is finite. In all other cases, the DDG-tree must be constructed as we go along.

The analysis for this DDG-tree algorithm is not very difficult. Construct (just for the analysis) the trie in which terminal nodes are put at the points where the paths for the p_n 's diverge for the first time. For example, for the probability vector

| $p_1 = 0.00101$ |
| $p_2 = 0.001001$ |
| $p_3 = 0.101101$ |

we have the cumulative probabilities 0.00101,0.010011,0.1111111111.... Thus, we can put terminal nodes at the positions 00, 01, and 1. It is easy to see that once the terminal nodes are reached, then on the average 2 more random bits are needed. Thus, $E(N)=2+$ expected depth of the terminal nodes in the trie defined above. In our example, this would yield $E(N)=2+\frac{1}{2}1+\frac{1}{2}2=\frac{7}{2}$. In another example, if all the p_n 's are equal to 2^{-k}, $1 \leq n \leq 2^k$, for some integer k, then $E(N)=2+k$, which grows as $\log_2 n$. In general, we have

$$E(N) \leq 2+\sum_{i=1}^{n} p_i \left\lceil \log_2(\frac{1}{p_i}) \right\rceil \leq 3+H(p_1, \ldots, p_n).$$

This follows from a simple argument. Consider the uniform [0,1] random variate U formed by the random bits of the random bit generator. Also mark the partial sums of p_i 's on [0,1], so that [0,1] is partitioned into n intervals. The expected depth of a terminal node in the trie is

$$\int_{0}^{1} D(x)\, dx$$

where $D(x)$ is the smallest nonnegative integer k such that the 2^k dyadic partition of [0,1] is such that only one of the partial sums (0 is also considered as a partial sum) falls in the same interval. The i-th partial sum "controls" an interval in which $D(x) \leq \left\lceil \log_2(\frac{1}{p_i}) \right\rceil$, and the size of the interval itself is a power of 2. Thus,

$$\int_{0}^{1} D(x)\, dx \leq \sum_{i=1}^{n} p_i \left\lceil \log_2(\frac{1}{p_i}) \right\rceil,$$

from which we derive the result shown above. We conclude that sequential search type DDG-tree algorithms are nearly optimal for all probability vectors (compare with Theorem 2.1).

The method of guide tables, and the Huffman-tree based methods are similar, with the sole exception that the probability vector is permuted in the Huffman tree case. All these methods can be translated into a DDG-tree algorithm of the type described for the sequential search method, and the performance bounds given above remain valid. In view of the lower bound of Knuth and Yao, we don't gain by using special truncation-based tricks, because truncation corresponds to search into a trie formed with equally-spaced points, and takes time proportional to $\log_2$ of the number of intervals.

Thus, it comes as no surprise that the alias method (section III.4) has an unimpressive DDG-tree analog. We can consider the following DDG-tree algorithm: first, generate a uniform $\{1, \ldots, n\}$ -valued random integer (this requires on the average $\geq \log_2 n$ and $\leq 1 + \log_2 n$ random bits, as we remarked above). Then, having picked a slab, we need to make one more comparison between a uniform random variate and a threshold, which takes on the average 2 comparisons by the binomial $(1, p)$ algorithm described in section XV.1. Thus,

$$2 + \log_2 n \ \leq \ E(N) \ \leq \ 3 + \log_2 n \ .$$

This performance grows with n, while for the optimal DDG-tree algorithms we will see that there are sequences of probability vectors for which $E(N)$ remain bounded as $n \rightarrow \infty$. In many cases, the alias algorithm does not even come close to the lower bound of Theorem 2.1.

The rejection method corresponds to the following DDG-tree: construct a DDG-tree in the obvious fashion with two types of terminal nodes, terminal nodes corresponding to a successful return (acceptance), and rejection nodes. Make the rejection nodes roots of isomorphic trees again, and continue at infinitum.

3.2. Optimal DDG-tree algorithms.

The notation of section XV.2 is inherited. We start with the following Theorem, due to Knuth and Yao (1976). It states that optimal algorithms achieving the lower bound do indeed exist.

Theorem 3.1.

Let $(p_1, p_2, \ldots, p_n)$ be a discrete probability vector (where n may be infinite). Assume first that $\sum_{i=1}^{n} \nu(p_i) < \infty$. Then there exists a DDG-tree algorithm for which

$$E(N) = \sum_{i=1}^{n} \nu(p_i) .$$

In fact, the following statements are equivalent for any DDG-tree algorithm:

(i) $P(N > k)$ is minimized for all $k \geq 0$ over all DDG-tree algorithms for the given distribution.

(ii)   For all $k \geq 0$ and all $1 \leq i \leq n$, there are exactly p_{ik} terminal nodes marked i on level k where p_{ik} denotes the coefficient of 2^{-k} in the binary expansion of p_i.

(iii) $E(N) = \sum_{i=1}^{n} \nu(p_i) .$

Assume next that $\sum_{i=1}^{n} \nu(p_i) = \infty$. Then, statements (i) and (ii) are equivalent.

Proof of Theorem 3.1.

We inherit the notation of the proof of Theorem 2.1. By inspecting that proof, we note that a DDG-tree algorithm attains the lower bound (if it is finite) if and only if for all i and k, we have equality in

$$\sum_{j > k} \frac{t_i(j)}{2^j} = p_i - \sum_{0 \leq j \leq k} \frac{t_i(j)}{2^j} \geq \frac{\chi(2^j p_i)}{2^j} .$$

This means that

$$\sum_{j=0}^{k} t_i(j) 2^{k-j} = \left\lfloor 2^k p_i \right\rfloor .$$

But this says simply that $t_i(k)$ is p_{ik} for all k. The number of terminal nodes at level k for integer i is 0 or 1 depending upon the value of the k-th bit in the binary expansion of p_i. To prove that such DDG-trees actually exist, define $t_i(k)$ and $t(k)$ by

$$t_i(k) = p_{ik}$$
$$t(k) = p_1(k) + \cdots + p_n(k) .$$

Thus, we certainly have

$$\sum_{k \geq 0} 2^{-k} t_i(k) = p_i$$
$$\sum_{k \geq 0} 2^{-k} t(k) = 1 .$$

A DDG-tree with these conditions exists if and only if the integers $b(k)$ defined by

$$b(0) + t(0) = 1 ,$$

$$b(k) + t(k) = 2b(k-1) \quad (k \geq 1)$$

are nonnegative. But the $b(k)$'s thus defined have a solution

$$b(k) = \sum_{j > k} \frac{t(j)}{2^{j-k}} .$$

Hence $b(k) \geq 0$, and such trees exist. This proves all the statements involving (iii). For the equivalence of (i) and (ii) in all cases, we note that in Theorem 2.1, we have obtained a lower bound for $b(k)$ for all k, and that the construction of the present theorem gives us a tree for which the lower bound is attained for all k. But $P(N > k) = \dfrac{b(k)}{2^k}$, and we are done. ∎

Let us give an example of the optimal construction.

Example 3.1. (Knuth and Yao, 1976)

Consider the transcendental probabilities

$p_1 = \dfrac{1}{\pi}$	$= 0.010100010111110...$
$p_2 = \dfrac{1}{e}$	$= 0.010111100010110...$
$p_3 = 1 - p_1 - p_2$	$= 0.010100000101010...$

The optimal tree is inherently infinite and cannot be obtained by a finite state machine (this is possible if and only if all probabilities are rational). The optimal tree has at each level between 0 and 3 terminal nodes, and can be constructed without too much trouble. Basically, all internal nodes have two children, and at each level, we put the terminal nodes to the right on that level. This usually gives an asymmetric left-heavy tree. Using the notation I for internal node, and 1,2,3 for terminal nodes for the integers 1,2,3 respectively, we can specify the optimal DDG-tree by specifying the nature of all the nodes on each level, from left to right. In the present example, this gives

Level	Nodes			
0	I			
1	I	I		
2	I	1	2	3
3	I	I		
4	I	1	2	3
5	I	2		
6	I	2		
7	I	2		
8	I	1		
9	I	I		
10	I	I	1	3
11	I	I	1	2
12	I	I	1	3
13	I	I	1	2
14	I	1	2	3
15	I	I		

■

3.3. Distribution-free inequalities for the performance of optimal DDG-tree algorithms.

We have seen that an optimal DDG-tree algorithm requires on the average

$$E(N) = \sum_{i=1}^{n} \nu(p_i)$$

random bits. By an inequality shown in Theorem 2.1, $H(x) \leq \nu(x) < H(x) + 2x$, $x \in [0,1]$, we see that for optimal algorithms,

$$\sum_{i=1}^{n} H(p_i) = H(p_1, \ldots, p_n)$$

$$\leq E(N) \leq H(p_1, \ldots, p_n) + 2 .$$

Thus, the performance is roughly speaking proportional to the entropy of the distribution. In general, this quantity is not known beforehand. Often one wants a priori guarantees about the performance of the algorithm. Thus, distribution-free bounds on $E(N)$ for the optimal algorithm can be very useful. We offer:

Theorem 3.2. (Knuth and Yao, 1976)

Let $p_1, \ldots, p_n$ be a finite probability vector. Then,

$$2 - 2^{2-n} \le \sum_{i=1}^{n} \nu(p_i) \le \left\lceil \log_2(n) \right\rceil + (n-1)2^{1-\left\lceil \log_2(n) \right\rceil}.$$

Proof of Theorem 3.2.

By definition of χ and ν,

$$\chi(2^k p_1) + \cdots + \chi(2^k p_n) \le \min(2^k, n-1)$$

for all $k \ge 0$. The $n-1$ upper bound follows by noting that the left hand side is less than n, and that it is integer valued because it can be written as

$$2^k - \left\lfloor 2^k p_1 \right\rfloor - \cdots - \left\lfloor 2^k p_n \right\rfloor.$$

Thus,

$$\sum_{i=1}^{n} \nu(p_i) = \sum_{k \ge 0} \sum_{i=1}^{n} 2^{-k} \chi(2^k p_i)$$

$$\le \sum_{k \ge 0} 2^{-k} \min(2^k, n-1)$$

$$= \sum_{0 \le k \le \left\lfloor \log_2(n-1) \right\rfloor} 1 + \sum_{k > \left\lfloor \log_2(n-1) \right\rfloor} \frac{n-1}{2^k}$$

$$= 1 + \left\lfloor \log_2(n-1) \right\rfloor + \frac{n-1}{2^{\left\lfloor \log_2(n-1) \right\rfloor}}.$$

The upper bound follows when we note that $\left\lfloor \log_2(n-1) \right\rfloor = \left\lceil \log_2(n) \right\rceil - 1$. Let us now turn to the lower bound. Using the notation of the proof of Theorem 2.1, an optimal DDG-tree always has

$$\sum_{i=1}^{n} \nu(p_i) = \sum_{k \ge 1} \frac{k t(k)}{2^k}$$

$$= \sum_{k \ge 1} \frac{k(2b(k-1) - b(k))}{2^k}$$

$$= \sum_{k \ge 0} \frac{b(k)}{2^k}.$$

Since $\sum_{k \ge 0} b(k) \ge n-1$ (there are $\ge n$ terminal nodes, and thus $\ge n-1$ internal nodes), and since conditional on the latter sum being equal to s, the minimum of $\sum_{k \ge 0} \frac{b(k)}{2^k}$ is reached for $b(0) = \cdots = b(s-1) = 1$, we see that

$$\sum_{i=1}^{n} \nu(p_i) \ge 2 - 2^{1-s} \ge 2 - 2^{2-n}. \quad \blacksquare$$

3.4. Exercises.

1. The bounds of Theorem 3.2 are best possible. By inspection of the proof, construct for each n a probability vector $p_1, \ldots, p_n$ for which the lower bound is attained. (Conclude that for this family of distributions, the expected performance of optimal DDG-tree algorithms is uniformly bounded in n.) Show that the upper bound of the theorem is attained for

$$
p_i = \begin{cases} 2^{-q}\left(\dfrac{2^n - 2^{i-1} - 1}{2^n - 1}\right) + 2^{-q} & ,1 \leq i \leq 2^q + 1 - n \ , \\[4mm] 2^{-q}\left(\dfrac{2^n - 2^{i-1} - 1}{2^n - 1}\right) & ,2^q + 1 - n < i \leq n \ , \end{cases}
$$

where $q = \left\lceil \log_2(n) \right\rceil$ (Knuth and Yao, 1976).

2. Describe an optimal DDG-tree algorithm of the shape described in Example 3.1, which requires storage of the probability vector only. In other words, the tree is constructed dynamically. You can assume of course that the p_n's can be manipulated in your computer.

3. **Finite state machines.** Show that there exists a finite state machine (edges correspond to random bits, nodes to internal nodes or terminal nodes) for generating a discrete random variate X taking values in $\{1, \ldots, n\}$ if and only if all probabilities involved are rational. Give a general procedure for constructing such finite state machines from (not necessarily optimal) DDG-trees by introducing rejection nodes and feedbacks to internal nodes. For simulating one die, find a finite state machine requiring on the average $\dfrac{11}{3}$ random bits. Is this optimal ? For simulating the sum of two dice, find a finite state machine which requires on the average $\dfrac{79}{18}$ random bits. For simulating two dice (NOT the sum), find a finite state machine which requires on the average $\dfrac{20}{3}$ random bits. Show that all of these numbers are optimal. Note that in the last case, we do better than just simulating one die twice with the first algorithm since this would have eaten up $\dfrac{22}{3}$ random bits on the average (Knuth and Yao, 1976).

4. Consider the following 5-state automaton: there is a START state, two terminal states, A and B, and two other states, S1 and S2. Transitions between states occur when bits are observed. In particular, we have:

$$
\begin{aligned}
&\text{START} + 0 \rightarrow \text{S1} \\
&\text{START} + 1 \rightarrow \text{S2} \\
&\text{S1} + 0 \rightarrow \text{A} \\
&\text{S1} + 1 \rightarrow \text{S2} \\
&\text{S2} + 0 \rightarrow \text{B} \\
&\text{S2} + 1 \rightarrow \text{START}
\end{aligned}
$$

If we start at START, and observe a perfect sequence of random bits, then what is $P(A)$, $P(B)$? Compute the expected number of bits before halting. Finally, construct the optimal DDG-tree algorithm for this problem and find a finite-state equivalent form requiring the same expected number of bits.

REFERENCES

M. Abramowitz and I.A. Stegun, *Handbook of Mathematical Tables,* Dover Publications, New York, N.Y., 1970.

A.G. Adams, "Algorithm 39. Areas under the normal curve," *Computer Journal,* vol. 12, pp. 197-198, 1969.

A.V. Aho, J.E. Hopcroft, and J.D. Ullman, *Data Structures and Algorithms,* Addison-Wesley, Reading, Mass., 1983.

J.H. Ahrens and U. Dieter, "Computer methods for sampling from the exponential and normal distributions," *Communications of the ACM,* vol. 15, pp. 873-882, 1972.

J.H. Ahrens and U. Dieter, "Extensions of Forsythe's method for random sampling from the normal distribution," *Mathematics of Computation,* vol. 27, pp. 927-937, 1973.

J.H. Ahrens and U. Dieter, "Computer methods for sampling from gamma, beta, Poisson and binomial distributions," *Computing,* vol. 12, pp. 223-246, 1974.

J.H. Ahrens and U. Dieter, "Sampling from binomial and Poisson distributions: a method with bounded computation times," *Computing,* vol. 25, pp. 193-208, 1980.

J.H. Ahrens and U. Dieter, "Generating gamma variates by comparison of probability densities," Institute of Statistics, Technical University Graz, Graz, Austria, 1981.

J.H. Ahrens and K.D. Kohrt, "Computer methods for efficient sampling from largely arbitrary statistical distributions," *Computing,* vol. 26, pp. 19-31, 1981.

J.H. Ahrens and U. Dieter, "Computer generation of Poisson deviates from modified normal distributions," *ACM Transactions on Mathematical Software,* vol. 8, pp. 163-179, 1982.

J.H. Ahrens and U. Dieter, "Generating gamma variates by a modified rejection technique," *Communications of the ACM,* vol. 25, pp. 47-54, 1982.

J.H. Ahrens, K.D. Kohrt, and U. Dieter, "Algorithm 599. Sampling from gamma and Poisson distributions," *ACM Transactions on Mathematical Software,* vol. 9, pp. 255-257, 1983.

J.H. Ahrens and U. Dieter, "Sequential random sampling," *ACM Transactions on Mathematical Software*, vol. 11, pp. 157-169, 1985.

J. Aitchison, "Inverse distributions and independent gamma-distributed products of random variables," *Biometrika*, vol. 50, pp. 505-508, 1963.

H. Akima, "A new method of interpolation and smooth curve fitting based on local procedures," *Journal of the ACM*, vol. 17, pp. 589-602, 1970.

B.J. Alder and T.E. Wainwright, "Phase transition in elastic disks," *Physical Review*, vol. 127, pp. 359-361, 1962.

M.M. Ali, N.N. Mikhail, and M.S. Haq, "A class of bivariate distributions including the bivariate logistic," *Journal of Multivariate Analysis*, vol. 8, pp. 405-412, 1978.

T.W. Anderson and D.A. Darling, "Asymptotic theory of certain goodness of fit criteria based on stochastic processes," *Annals of Mathematical Statistics*, vol. 23, pp. 193-213, 1952.

D.F. Andrews and C.L. Mallows, "Scale mixtures of normal distributions," *Journal of the Royal Statistical Society Series B*, vol. 36, pp. 99-102, 1974.

N.P. Archer, "The generation of piecewise linear approximations of probability distribution functions," *Journal of Statistical Computation and Simulation*, vol. 11, pp. 21-40, 1980.

G. Arfwedson, "A probability distribution connected with Stirling's second class numbers," *Skandinavisk Aktuarietidskrift*, vol. 34, pp. 121-132, 1951.

A.N. Arnason, "Simple exact, efficient methods for generating beta and Dirichlet variates," *Utilitas Mathematica*, vol. 1, pp. 249-290, 1972.

B.C. Arnold, "A note on multivariate distributions with specified marginals," *Journal of the American Statistical Association*, vol. 62, pp. 1460-1461, 1967.

B.C. Arnold and R.A. Groeneveld, "Some properties of the arcsine distribution," *Journal of the American Statistical Association*, vol. 75, pp. 173-175, 1980.

D.B. Arnold and M.R. Sleep, "Uniform random number generation of n balanced parenthesis strings," *ACM Transactions on Programming Languages and Systems*, vol. 2, pp. 122-128, 1980.

A.C. Atkinson and M.C. Pearce, "The computer generation of beta, gamma and normal random variables," *Journal of the Royal Statistical Society Series A*, vol. 139, pp. 431-461, 1976.

A.C. Atkinson and J. Whittaker, "A switching algorithm for the generation of beta random variables with at least one parameter less than one," *Journal of the Royal Statistical Society Series A*, vol. 139, pp. 462-467, 1976.

A.C. Atkinson, "An easily programmed algorithm for gewnerating gamma random variables," *Applied Statistics*, vol. 140, pp. 232-234, 1977.

A.C. Atkinson, "A family of switching algorithms for the computer generation of beta random variables," *Biometrika*, vol. 66, pp. 141-145, 1979.

A.C. Atkinson and J. Whittaker, "The generation of beta random variables with one parameter greater than one and one parameter less than one," *Applied Statistics*, vol. 28, pp. 90-93, 1979.

A.C. Atkinson, "The simulation of generalised inverse gaussian, generalised hyperbolic, gamma and related random variables," Research Report 52, Department of Theoretical Statistics, Aarhus University, Aarhus, Denmark, 1979.

A.C. Atkinson, "Recent developments in the computer generation of Poisson random variables," *Applied Statistics*, vol. 28, pp. 260-263, 1979.

A.C. Atkinson, "The computer generation of Poisson random variables," *Applied Statistics*, vol. 28, pp. 29-35, 1979.

A.C. Atkinson, "The simulation of generalized inverse gaussian and hyperbolic random variables," *SIAM Journal on Statistical Computation*, vol. 3, pp. 502-515, 1982.

S. Baase, *Computer Algorithms: Introduction to Design and Analysis,* Addison-Wesley, Reading, Mass., 1978.

M. Badel, "Generation de nombres aleatoires correles," *Mathematics and Computers in Simulation*, vol. 21, pp. 42-49, 1979.

B.J.R. Bailey, "Alternatives to Hastings' approximation to the inverse of the normal cumulative distribution function," *Applied Statistics*, vol. 30, pp. 275-276, 1981.

G. de Balbine, "Note on random permutations," *Mathematics of Computation*, vol. 21, pp. 710-712, 1967.

J. Banks and J.S. Carson, *Discrete Event Simulation,* Prentice-Hall, Englewood Cliffs, N.J., 1984.

G. Barbu, "On computer generation of a random variable by transformations of uniform variables," *Bull. Math. Soc. Sci. Math. Romanie*, vol. 26, pp. 129-139, 1982.

R.E. Barlow, A.W. Marshall, and F. Proschan, "Properties of probability distributions with monotone hazard rate," *Annals of Mathematical Statistics*, vol. 34, pp. 375-389, 1963.

R.E. Barlow and F. Proschan, *Mathematical Theory of Reliability,* John Wiley, New York, N.Y., 1965.

R.E. Barlow and F. Proschan, *Statistical Theory of Reliability and Life Testing,* Holt, Rinehart and Winston, New York, N.Y., 1975.

D.R. Barnard and M.N. Cawdery, "A note on a new method of histogram sampling," *Operations Research Quarterly*, vol. 25, pp. 319-320, 1974.

O. Barndorff-Nielsen, "Exponentially decreasing distributions for the logarithm of particle size," *Proceedings of the Royal Society of London Series A*, vol. 353, pp. 401-419, 1977.

O. Barndorff-Nielsen and C. Halgreen, "Infinite divisibility of the hyperbolic and generalized inverse gaussian distributions," *Zeitschrift fur Wahrscheinlichkeitstheorie und verwandte Gebiete*, vol. 38, pp. 309-311, 1977.

O. Barndorff-Nielsen, "Hyperbolic distributions and distributions on hyperbolae," *Scandinavian Journal of Statistics*, vol. 5, pp. 151-157, 1978.

O. Barndorff-Nielsen and P. Blaesild, "Hyperbolic Distributions," Encyclopedia of Statistical Sciences, John Wiley, New York, N.Y., 1980.

O. Barndorff-Nielsen, J. Kent, and M. Sorensen, "Normal variance-mean mixtures and z distributions," *International Statistical Review*, vol. 50, pp. 145-159, 1982.

V. Barnett, "Some bivariate uniform distributions," *Communications in Statistics*, vol. A9, pp. 453-461, 1980.

D.R. Barr and N.L. Slezak, "A comparison of multivariate normal generators," *Communications of the ACM*, vol. 15, pp. 1048-1049, 1972.

R. Bartels, "Generating non-normal stable variates using limit theorem properties," *Journal of Statistical Computation and Simulation*, vol. 7, pp. 199-212, 1978.

R. Bartels, "Truncation bounds for infinite expansions for the stable distributions," *Journal of Statistical Computation and Simulation*, vol. 12, pp. 293-302, 1981.

M.S. Bartlett, "Statistical estimation of density functions," *Sankhya Series A*, vol. 25, pp. 245-254, 1963.

D.E. Barton, "The matching distributions: Poisson limiting forms and derived methods of approximation," *Journal of the Royal Statistical Society, Series B*, vol. 20, pp. 73-92, 1958.

D.E. Barton and C.L. Mallows, "Some aspects of the random sequence," *Annals of Mathematical Statistics*, vol. 36, pp. 236-260, 1965.

D. Basu and R.C. Tiwari, "A note on the Dirichlet process," in *Statistics and Probability: Essays in Honor of C.R. Rao*, ed. G. Kallianpur, P.R. Krishnaiah, J.K. Ghosh, pp. 89-103, North-Holland, 1982.

J.D. Beasley and S.G. Springer, "The percentage points of the normal distribution," *Applied Statistics*, vol. 26, pp. 118-121, 1977.

J.R. Bell, "Algorithm 334. Normal random deviates," *Communications of the ACM*, vol. 11, p. 498, 1968.

R.B. Bendel and A.A. Afifi, "Comparison of stopping rules in forward stepwise regression," *Journal of the American Statistical Association*, vol. 72, pp. 46-53, 1977.

R.B. Bendel and M.R. Mickey, "Population correlation matrices for sampling experiments," *Communications in Statistics, Section Simulation and Computation*, vol. 7, pp. 163-182, 1978.

J.L. Bentley and J.B. Saxe, "Generating sorted lists of random numbers," *ACM Transactions of Mathematical Software*, vol. 6, pp. 359-364, 1980.

H. Bergstrom, "On some expansions of stable distributions," *Arkiv fur Mathematik, II*, vol. 18, pp. 375-378, 1952.

M.B. Berman, "Generating gamma distributed variates for computer simulation models," Technical Report R-641-PR, Rand Corporation, 1971.

D.J. Best, "Letter to the editor," *Applied Statistics*, vol. 27, p. 181, 1978.

D.J. Best, "A simple algorithm for the computer generation of random samples from a Student's t or symmetric beta distribution," in *COMPSTAT 1978: Proceedings in Computational Statistics*, ed. L.C.A. Corsten and J. Hermans, pp. 341-347, Physica Verlag, Wien, Austria, 1978.

D.J. Best, "Some easily programmed pseudo-random normal generators," *Australian Computer Journal*, vol. 11, pp. 60-62, 1979.

D.J. Best and N.I. Fisher, "Efficient simulation of the von Mises distribution," *Applied Statistics*, vol. 28, pp. 152-157, 1979.

D.J. Best, "A note on gamma variate generators with shape parameter less than unity," *Computing*, vol. 30, pp. 185-188, 1983.

W.H. Beyer, *Handbook of Tables for Probability and Statistics,* Cleveland: The Chemical Rubber Co., 1968.

G.P. Bhattacharjee, "Algorithm AS32. The incomplete gamma integral," *Applied Statistics*, vol. 19, pp. 285-287, 1970.

B.C. Bhattacharyya, "The use of McKay's Bessel function curves for graduating frequency distributions," *Sankhya Series A*, vol. 6, pp. 175-182, 1942.

A. Bignami and A. de Mattels, "A note on sampling from combinations of distributions," *Journal of the Institute of Mathematics and its Applications*, vol. 8, pp. 80-81, 1971.

P. Blaesild, "The shape of the generalized inverse gaussian and hyperbolic distributions," Research Report 37, Department of Theoretical Statistics, Aarhus University, Aarhus, Denmark, 1978.

M. Blum, R.W. Floyd, V. Pratt, R.L. Rivest, and R.E. Tarjan, "Time bounds for selection," *Journal of Computers and System Sciences*, vol. 7, pp. 448-461, 1973.

L.N. Bolshev, "On transformations of random variables," *Theory of Probability and its Applications*, vol. 4, pp. 136-149, 1959.

L.N. Bolshev, "Asymptotic Pearson transformations," *Theory of Probability and its Applications*, vol. 8, pp. 129-155, 1963.

L.N. Bolshev, "On a characterization of the Poisson distribution and its statistical applications," *Theory of Probability and its Applications*, vol. 10, pp. 446-456, 1965.

L. Bondesson, "On simulation from infinitely divisible distributions," *Advances in Applied Probability*, vol. 14, pp. 855-869, 1982.

G.E.P. Box and M.E. Muller, "A note on the generation of random normal deviates," *Annals of Mathematical Statistics*, vol. 29, pp. 610-611, 1958.

J.M. Boyett, "Random $R \times C$ tables with given row and column totals," *Applied Statistics*, vol. 28, pp. 329-332, 1979.

P. Bratley, B.L. Fox, and L.E. Schrage, *A Guide to Simulation,* Springer-Verlag, New York, N.Y., 1983.

R.P. Brent, "A gaussian pseudo-random number generator," *Communications of the ACM*, vol. 17, pp. 704-706, 1974.

G.W. Brown and J.W. Tukey, "Some distributions of sample means," *Annals of Mathematical Statistics*, vol. 17, pp. 1-12, 1946.

M.C. Bryson and M.E. Johnson, "Constructing and simulating multivariate distributions using Khintchine's theorem," Technical Report LA-UR 81-442, Los Alamos Scientific Laboratory, Los Alamos, New Mexico, 1981.

M.C. Bryson and M.E. Johnson, "Constructing and simulating multivariate distributions using Khintchine's theorem," *Journal of Statistical Computation and Simulation*, vol. 16, pp. 129-137, 1982.

I.W. Burr, "Cumulative frequency functions," *Annals of Mathematical Statistics*, vol. 13, pp. 215-232, 1942.

I.W. Burr and P.J. Cislak, "On a general system of distributions. I. The curve-shape characteristics. II. The sample median," *Journal of the American Statistical Association*, vol. 63, pp. 627-635, 1968.

I.W. Burr, "Parameters for a general system of distributions to match a grid of a3 and a4," *Communications in Statistics*, vol. 2, pp. 1-21, 1973.

J.C. Butcher, "Random sampling from the normal distribution," *Computer Journal*, vol. 3, pp. 251-253, 1961.

E.L. Butler, "Algorithm 370. General random number generator," *Communications of the ACM*, vol. 13, pp. 49-52, 1970.

T. Cacoullos, "A relation between t and F distributions," *Journal of the American Statistical Association*, vol. 60, pp. 528-531, p. 1249, 1965.

A.G. Carlton, "Estimating the parameters of a rectangular distribution," *Annals of Mathematical Statistics*, vol. 17, pp. 355-358, 1946.

C.P. Chalmers, "Generation of correlation matrices with given eigen-structure," *Journal of Statistical Computation and Simulation*, vol. 4, pp. 133-139, 1975.

J.M. Chambers, C.L. Mallows, and B.W. Stuck, "A method for simulating stable random variables," *Journal of the American Statistical Association*, vol. 71, pp. 340-344, 1976.

H.C. Chen and Y. Asau, "On generating random variates from an empirical distribution," *AIIE Transactions*, vol. 6, pp. 163-166, 1974.

R.C.H. Cheng, "The generation of gamma variables with non-integral shape parameter," *Applied Statistics*, vol. 26, pp. 71-75, 1977.

R.C.H. Cheng, "Generating beta variates with nonintegral shape parameters," *Communications of the ACM*, vol. 21, pp. 317-322, 1978.

R.C.H. Cheng and G.M. Feast, "Some simple gamma variate generators," *Applied Statistics*, vol. 28, pp. 290-295, 1979.

R.C.H. Cheng and G.M. Feast, "Gamma variate generators with increased shape parameter range," *Communications of the ACM*, vol. 23, pp. 389-393, 1980.

K.C. Cherian, "A bivariate correlated gamma-type distribution function," *Journal of the Indian Mathematical Society*, vol. 5, pp. 133-144, 1941.

M.A. Chmielewski, "Elliptically symmetric distributions: a review and bibliography," *International Statistical Review*, vol. 49, pp. 67-74, 1981.

Y.S. Chow, "Some convergence theorems for independent random variables," *Annals of Mathematical Statistics*, vol. 37, pp. 1482-1493, 1966.

Y.S. Chow and H. Teicher, *Probability Theory. Independence. Interchangeability. Martingales.*, Springer-Verlag, New York, N.Y., 1978.

E. Cinlar, *Introduction to Stochastic Processes*, Prentice-Hall, Englewood Cliffs, N.J., 1975.

J.M. Cook, "Rational formulae for the production of a spherically symmetric probability distribution," *Mathematics of Computation*, vol. 11, pp. 81-82, 1957.

R.D. Cook and M.E. Johnson, "A family of distributions for modeling non-elliptically symmetric multivariate data," *Journal of the Royal Statistical Society, Series B*, vol. 43, pp. 210-218, 1981.

B.E. Cooper, "Algorithm AS2. The normal integral," *Applied Statistics*, vol. 17, pp. 186-187, 1968.

D.R. Cox and P.A.W. Lewis, *The Statistical Analysis of Series of Events*, Methuen, London, 1966.

H. Cramer, *Mathematical Methods of Statistics*, Princeton University Press, Princeton, New Jersey, 1951.

J.S. Dagpunar, "Sampling of variates from a truncated gamma distribution," *Journal of Statistical Computation and Simulation*, vol. 8, pp. 59-64, 1978.

M. Darboux, "Sur les developpements en serie des fonctions d'une seule variable," *Journal de Mathematiques Series 3*, vol. 2, pp. 291-294, 1876.

D.A. Darling, "The Kolmogorov-Smirnov, Cramer-von Mises tests," *Annals of Mathematical Statistics*, vol. 26, pp. 1-20, 1955.

D. Davey and J.G. Vaucher, "Self-optimizing partition sequencing sets for discrete event simulation," *INFOR*, vol. 18, pp. 21-41, 1980.

F.N. David and E. Fix, "Rank correlation and regression in a non-normal surface," *Proceedings of the Fourth Berkeley Symposium*, vol. 1, pp. 177-197, 1961.

Ju.S. Davidovic, B.I. Korenbljum, and B.I. Hacet, "A property of logarithmically concave functions," *Dokl. Akad. Nauk SSR*, vol. 185, pp. 477-480, 1969.

P.J. Davis and P. Rabinowitz, *Methods of Numerical Integration*, Academic Press, New York, 1975.

I. Deak, "Comparison of methods for generating uniformly distributed random points in and on a hypersphere," *Problems of Control and Information Theory*, vol. 8, pp. 105-113, 1979.

I. Deak and B. Bene, "Random number generation: a bibliography," Working paper, Computer and Automation Institute, Hungarian Academy of Sciences, 1979.

I. Deak, "The ellipsoid method for generating normally distributed random vectors," *Zastosowania Matematyki*, vol. 17, pp. 95-107, 1979.

I. Deak, "Fast procedures for generating stationary normal vectors," *Journal of Statistical Computation and Simulation*, vol. 16, pp. 225-242, 1980.

I. Deak, "An economical method for random number generation and a normal generator," *Computing*, vol. 27, pp. 113-121, 1981.

I. Deak, "General methods for generating non-uniform random numbers," Technical Report DALTR-84-20, Department of Mathematics, Statistics and Computing Science, Dalhousie University, Halifax, NovaScotia, 1984.

E.V. Denardo and B.L. Fox, "Shortest-route methods:1. reaching, pruning, and buckets," *Operations Research*, vol. 27, pp. 161-186, 1979.

L. Devroye and A. Naderisamani, "A binomial random variate generator," Technical Report, School of Computer Science, McGill University, Montreal, 1980.

L. Devroye, "Generating the maximum of independent identically distributed random variables," *Computers and Mathematics with Applications*, vol. 6, pp. 305-315, 1980.

L. Devroye, "The computer generation of Poisson random variables," *Computing*, vol. 26, pp. 197-207, 1981.

L. Devroye and C. Yuen, "Inversion-with-correction for the computer generation of discrete random variables," Technical Report, School of Computer Science, McGill University, Montreal, 1981.

L. Devroye, "Programs for generating random variates with monotone densities," Technical report, School of Computer Science, McGill University, Montreal, Canada, 1981.

L. Devroye, "Recent results in non-uniform random variate generation," *Proceedings of the 1981 Winter Simulation Conference*, Atlanta, GA., 1981.

L. Devroye, "The computer generation of random variables with a given characteristic function," *Computers and Mathematics with Applications*, vol. 7, pp. 547-552, 1981.

L. Devroye, "The series method in random variate generation and its application to the Kolmogorov-Smirnov distribution," *American Journal of Mathematical and Management Sciences*, vol. 1, pp. 359-379, 1981.

L. Devroye and T. Klincsek, "Average time behavior of distributive sorting algorithms," *Computing*, vol. 26, pp. 1-7, 1981.

L. Devroye, "On the computer generation of random convex hulls," *Computers and Mathematics with Applications*, vol. 8, pp. 1-13, 1982.

L. Devroye, "A note on approximations in random variate generation," *Journal of Statistical Computation and Simulation*, vol. 14, pp. 149-158, 1982.

L. Devroye, "The equivalence of weak, strong and complete convergence in L_1 for kernel density estimates," *Annals of Statistics*, vol. 11, pp. 896-904, 1983.

L. Devroye, "Random variate generation for unimodal and monotone densities," *Computing*, vol. 32, pp. 43-68, 1984.

L. Devroye, "A simple algorithm for generating random variates with a log-concave density," *Computing*, vol. 33, pp. 247-257, 1984.

L. Devroye, "On the use of probability inequalities in random variate generation," *Journal of Statistical Computation and Simulation*, vol. 20, pp. 91-100, 1984.

L. Devroye, "Methods for generating random variates with Polya characteristic functions," *Statistics and Probability Letters*, vol. 2, pp. 257-261, 1984.

L. Devroye, "The analysis of some algorithms for generating random variates with a given hazard rate," *Naval Research Logistics Quarterly*, vol. 0, pp. 0-0, 1985.

L. Devroye and L. Gyorfi, *Nonparametric Density Estimation: The L_1 View*, John Wiley, New York, N.Y., 1985.

L. Devroye, "An automatic method for generating random variables with a given characteristic function," *SIAM Journal of Applied Mathematics*, vol. 0, pp. 0-0, 1986.

U. Dieter and J.H. Ahrens, "A combinatorial method for the generation of normally distributed random numbers," *Computing*, vol. 11, pp. 137-146, 1973.

P.J. Diggle, J.E. Besag, and J.T. Gleaves, "Statistical analysis of spatial point patterns by means of distance methods," *Biometrics*, vol. 32, pp. 659-667, 1976.

S.D. Dubey, "A compound Weibull distribution," *Naval Research Logistics Quarterly*, vol. 15, pp. 179-188, 1968.

D. Dugue and M. Girault, "Fonctions convexes de Polya," *Publications de l'Institut de Statistique des Universites de Paris*, vol. 4, pp. 3-10, 1955.

W.H. Dumouchel, "Stable distributions in statistical inference," Ph.D. Dissertation, Department of Statistics, Yale University, 1971.

R. Durstenfeld, "Random permutation," *Communications of the ACM*, vol. 7, p. 420, 1964.

A. Dvoretzky and H. Robbins, "On the parking problem," *Publications of the Mathematical Institute of the Hungarian Academy of Sciences*, vol. 9, pp. 209-225, 1964.

R. Englebrecht-Wiggans and W.L. Maxwell, "Analysis of the time indexed list procedure for synchronization of discrete event procedures," *Management Science*, vol. 24, pp. 1417-1427, 1978.

P. Erdos and M. Kac, "On certain limit theorems of the theory of probability," *Bulletin of the American Mathematical Society*, vol. 52, pp. 292-302, 1946.

P. Erdos and A. Renyi, "On random graphs," *Publ. Math. Debrecen*, vol. 6, pp. 290-297, 1959.

P. Erdos and A. Renyi, "On the evolution of random graphs," *Publications of the Mathematical Institute of the Hungarian Academy of Sciences*, vol. 5, pp. 17-61, 1960.

J. Ernvall and O. Nevalainen, "An algorithm for unbiased random sampling," *Computer Journal*, vol. 25, pp. 45-47, 1982.

V.N. Faddeeva, *Computational Methods of Linear Algebra,* Dover Publ., 1959.

R. Fagin and T.G. Price, "Efficient calculation of expected miss ratios in the independent reference model," *SIAM Journal on Computing*, vol. 7, pp. 288-297, 1978.

R. Fagin, J. Nievergelt, N. Pippenger, and H.R. Strong, "Extendible hashing - a fast access method for dynamic files," *ACM Transactions on Mathematical Software*, vol. 4, pp. 315-344, 1979.

E. Fama and R. Roll, "Some properties of symmetric stable distributions," *Journal of the American Statistical Association*, vol. 63, pp. 817-836, 1968.

C.T. Fan, M.E. Muller, and I. Rezucha, "Development of sampling plans by using sequential (item by item) selection techniques and digital computers," *Journal of the American Statistical Association*, vol. 57, pp. 387-402, 1962.

D.J.G. Farlie, "The performance of some correlation coefficients for a general bivariate distribution," *Biometrika*, vol. 47, pp. 307-323, 1960.

W. Feller, "On the Kolmogorov-Smirnov limit theorems for empirical distributions," *Annals of Mathematical Statistics*, vol. 19, pp. 177-189, 1948.

W. Feller, *An Introduction to Probability Theory and its Applications, Vol. 1*, John Wiley, New York, N.Y., 1965.

W. Feller, *An Introduction to Probability Theory and its Applications, Vol. 2*, John Wiley, New York, N.Y., 1971.

C. Ferreri, "A new frequency distribution for single variate analysis," *Statistica Bologna*, vol. 24, pp. 223-251, 1964.

G.S. Fishman, "Sampling from the gamma distribution on a computer," *Communications of the ACM*, vol. 19, pp. 407-409, 1975.

G.S. Fishman, "Sampling from the Poisson distribution on a computer," *Computing*, vol. 17, pp. 147-156, 1976.

G.S. Fishman, *Principles of Discrete Event Simulation,* John Wiley, New York, N.Y., 1978.

G.S. Fishman, "Sampling from the binomial distribution on a computer," *Journal of the American Statistical Association*, vol. 74, pp. 418-423, 1979.

A.I. Fleishman, "A method for simulating non-normal distributions," *Psychometrika*, vol. 43, pp. 521-532, 1978.

J.L. Folks and R.S. Chhikara, "The inverse gaussian distribution and its statistical application. A review," *Journal of the Royal Statistical Society, Series B*, vol. 40, pp. 263-289, 1978.

G.E. Forsythe, "von Neumann's comparison method for random sampling from the normal and other distributions," *Mathematics of Computation*, vol. 26, pp. 817-826, 1972.

B.L. Fox, "Generation of random samples from the beta and F distributions," *Technometrics*, vol. 5, pp. 269-270, 1963.

B.L. Fox, "Monotonicity, extremal correlations, and synchronization: implications for nonuniform random numbers," Technical Report, Universite de Montreal, 1980.

J.R. Fraker and D.V. Rippy, "A composite approach to generating autocorrelated sequences," *Simulation*, vol. 23, pp. 171-175, 1974.

J.N. Franklin, "Numerical simulation of stationary and nonstationary Gaussian random processes," *SIAM Review*, vol. 7, pp. 68-80, 1965.

W.R. Franta and K. Maly, "An efficient data structure for the simulation event set," *Communications of the ACM*, vol. 20, pp. 596-602, 1977.

W.R. Franta and K. Maly, "A comparison of heaps and TL structure for the simulation event set," *Communications of the ACM*, vol. 21, pp. 873-875, 1978.

M. Frechet, "Sur les tableaux de correlation dont les marges sont donnees," *Annales de l'Universite de Lyon, Sec. A, Series 3*, vol. 14, pp. 53-57, 1951.

M. Frechet, "Remarques au sujet de la note precedente," *Comptes Rendus de l'Academie des Sciences de Paris*, vol. 246, pp. 2719-2720, 1958.

E. Fredkin, "Trie memory," *Communications of the ACM*, vol. 3, pp. 490-499, 1960.

M.F. Freeman and J.W. Tukey, "Transformations related to the angular and square-root," *Annals of Mathematical Statistics*, vol. 21, pp. 607-611, 1950.

D.P. Gaver, "Analytical hazard representations for use in reliability, mortality and simulation studies," *Communications in Statistics, Section Simulation and Computation*, vol. B8, pp. 91-111, 1979.

H. Gebelein, "Das statistische Problem der Korrelation als Variations- und Eigenwertproblem und sein Zusammenhang mit der Ausgleichungsrechnung," *Zeitschrift fur Angewandte Mathematik und Mechanik*, vol. 21, pp. 364-379, 1941.

H. Gehrke, "Einfache sequentielle Stichprobenentnahme," Diplomarbeit, University of Kiel, West Germany, 1984.

E. Olusegun George and M.O. Ojo, "On a generalization of the logistic distribution," *Annales of the Institute of Statistical Mathematics*, vol. 32, pp. 161-169, 1980.

E. Olusegun George and G.S. Mudholkar, "Some relationships between the logistic and the exponential distributions," in *Statistical Distributions in Scientific Work*, ed. C. Taillie, G.P. Patil, B.A. Baldessari, vol. 4, pp. 401-409, D. Reidel Publ. Co., Dordrecht, Holland, 1981.

I. Gerontides and R.L. Smith, "Monte Carlo generation of order statistics from general distributions," *Applied Statistics*, vol. 31, pp. 238-243, 1982.

J.D. Gibbons, *Nonparametric Statistical Inference*, McGraw-Hill, New York, 1971.

M. Girault, "Les fonctions caracteristiques et leurs transformations," *Publications de l'Institut de Statistique des Universites de Paris*, vol. 4, pp. 223-299, 1954.

N. Glick, "Breaking records and breaking boards," *American Mathematical Monthly*, vol. 85, pp. 2-26, 1978.

H.J. Godwin, *Inequalities on Distribution Functions*, Charles Griffin, London, 1964.

G.H. Gonnet, "Heaps applied to event driven mechanisms," *Communications of the ACM*, vol. 19, pp. 417-418, 1976.

R.D. Gordon, "Values of Mills' ratio of area to bounding ordinate and of the normal probability integral for large values of the argument," *Annals of Mathematical Statistics*, vol. 12, pp. 364-366, 1941.

A. Grassia, "On a family of distributions with argument between 0 and 1 obtained by transformation of the gamma and derived compound distributions," *Australian Journal of Statistics*, vol. 19, pp. 108-114, 1977.

F.A. Graybill, *Introduction to Matrices with Applications in Statistics,* Wadsworth, Belmont, CA., 1969.

P.J. Green and R. Sibson, "Computing Dirichlet tesellations in the plane," *Computer Journal*, vol. 21, pp. 168-173, 1978.

A.J. Greenwood, "A fast generator for gamma-distributed random variables," in *COMPSTAT 1974 , Proceedings in Computational Statistics*, ed. G. Bruckmann, F. Ferschl, L. Schmetterer, pp. 19-27, Physica-Verlag, Vienna, Austria, 1974.

A.J. Greenwood, "Moments of the time to generate random variables by rejection," *Annals of the Institute of Statistical Mathematics*, vol. 28, pp. 399-401, 1976.

U. Grenander and M. Rosenblatt, "Statistical spectral analysis of time series arising from stationary stochastic processes," *Annals of Mathematical Statistics*, vol. 24, pp. 537-558, 1953.

V.O. Guerra, R.A. Tapia, and J.R. Thompson, "A random number generator for continuous random variables based on an interpolation procedure of Akima," *Proceedings of the 1978 Winter Simulation Conference*, pp. 228-230, 1978.

E.J. Gumbel, "Ranges and midranges," *Annals of Mathematical Statistics*, vol. 15, pp. 414-422, 1944.

E.J. Gumbel, *Statistics of Extremes,* Columbia University Press, New York, N.Y., 1958.

E.J. Gumbel, "Distributions a plusieurs variables dont les marges sont donnees (with remarks by M. Frechet)," *Comptes Rendus de l'Academie des Sciences de Paris*, vol. 246, pp. 2717-2720, 1958.

E.J. Gumbel, "Distributions des valeurs extremes en plusieurs dimensions," *Publications de l'Institut de Statistique des Universites de Paris*, vol. 9, pp. 171-173, 1960.

E.J. Gumbel, "Bivariate exponential distributions," *Journal of the American Statistical Association*, vol. 55, pp. 698-707, 1960.

E.J. Gumbel, "Sommes et differences de valeurs extremes independantes," *Comptes Rendus de l'Academie des Sciences de Paris*, vol. 253, pp. 2838-2839, 1961.

E.J. Gumbel, "Bivariate logistic distributions," *Journal of the American Statistical Association*, vol. 56, pp. 335-349, 1961.

G. Guralnik, C. Zemach, and T. Warnock, "An algorithm for uniform random sampling of points in and on a hypersphere," *Information Processing Letters*, vol.

21, pp. 17-21, 1985.

P. Hall, "The distribution of means for samples of size N drawn from a population in which the variate takes values between 0 and 1, all such values being equally probable," *Biometrika*, vol. 19, pp. 240-244, 1927.

J.M. Hammersley and D.C. Handscomb, *Monte Carlo Methods,* Methuen, London, 1964.

G.H. Hardy, J.E. Littlewood, and G. Polya, *Inequalities,* Cambridge University Press, London, U.K., 1952.

C.M. Harris, "The Pareto distribution as a queue service discipline," *Operations Research*, vol. 16, pp. 307-313, 1968.

J.F. Hart, *Computer Approximations,* John Wiley, New York, N.Y., 1968.

C. Hastings, *Approximations for Digital Computers,* Princeton University Press, Princeton, New Jersey, 1955.

R.M. Heiberger, "Generation of random orthogonal matrices," *Applied Statistics*, vol. 27, pp. 199-206, 1978.

J.O. Henriksen, "An improved events list algorithm," in *Proceedings of the 1977 Winter Simulation Conference*, pp. 554-557, 1977.

C. Heyde, "On a property of the lognormal distribution," *Journal of the Ryal Statistical Society, Series B*, vol. 25, pp. 392-393, 1963.

D.P. Heyman and M.J. Sobel, *Stochastic Models in Operations Research, Vol. 1,* McGraw-Hill, New York, 1982.

T. Hickey and J. Cohen, "Uniform random generation of strings in a context-free language," *SIAM Journal on Computing]*, vol. 12, pp. 645-655, 1983.

J.S. Hicks and R.F. Wheeling, "An efficient method for generating uniformly distributed points on the surface of an n-dimensional sphere," *Communications of the ACM*, vol. 2, pp. 17-19, 1959.

I.D. Hill, "Remark ASR2. A remark on algorithm AS2," *Applied Statistics*, vol. 18, pp. 299-300, 1969.

I.D. Hill, "Algorithm AS66. The normal integral," *Applied Statistics*, vol. 22, pp. 424-427, 1973.

T.W. Hill, "On determining a distribution function known only by its moments and/or moment generating function," Ph. D. Dissertation, Arizona State University, 1969.

D. Hitchin, "Remark ASR8. A remark on algorithms AS4 and AS5," *Applied Statistics*, vol. 22, p. 428, 1973.

W. Hoeffding, "Masstabinvariante Korrelationstheorie," *Schriften des Mathematischen Instituts und des Instituts fur Angewandte Mathematik der Universitat Berlin*, vol. 5, pp. 179-233, 1940.

R.G. Hoffman, "The simulation and analysis of correlated binary data," in *Proceedings of the Statistical Computing Section*, pp. 340-343, American Statistical Association, 1979.

S.C. Hora, "Estimation of the inverse function for random variate generation," *Communications of the ACM*, vol. 26, pp. 590-594, 1983.

F.C. Hsuan, "Generating uniform polygonal random pairs," *Applied Statistics*, vol. 28, pp. 170-172, 1979.

T.C. Hu and A.C. Tucker, "Optimal computer search trees and variable-length alphabetic codes," *SIAM Journal of Applied Mathematics*, vol. 21, pp. 514-532, 1971.

D. Huffman, "A method for the construction of minimum-redundancy codes," *Proceedings of the IRE*, vol. 40, pp. 1098-1101, 1952.

T.P. Hutchinson, "Compound gamma bivariate distributions," *Metrika*, vol. 28, pp. 263-271, 1981.

I.A. Ibragimov, "On the composition of unimodal distributions," *Theory of Probability and its Applications*, vol. 1, pp. 255-260, 1956.

I.A. Ibragimov and K.E. Chernin, "On the unimodality of stable laws," *Theory of Probability and its Applications*, vol. 4, pp. 417-419, 1959.

I.A. Ibragimov and Yu.V. Linnik, *Independent and Stationary Sequences of Random Variables,* Wolters-Noordhoff Publishing, Groningen, The Netherlands, 1971.

P. Inzevitov, "An estimate of the remainder of the asymptotic expansion in the local limit theorem for densities," *Lithuanian Mathematical Journal*, vol. 17, pp. 111-120, 1977.

P.A. Jacobs and P.A.W. Lewis, "A mixed autoregressive-moving average exponential sequence and point process, EARMA(1,1)," *Advances in Applied Probability*, vol. 9, pp. 87-104, 1977.

B. Jansson, *Random Number Generators,* pp. 189-191, V. Pettersons Bokindustri Aktiebolag, Stockholm, 1966.

M.D. Johnk, "Erzeugung von Betaverteilten und Gammaverteilten Zufallszahlen," *Metrika*, vol. 8, pp. 5-15, 1964.

D.G. Johnson and W.J. Welch, "The generation of pseudo-random correlation matrices," *Journal of Statistical Computation and Simulation*, vol. 11, pp. 55-69, 1980.

M.E. Johnson, "Models and Methods for Generating Dependent Random Vectors," Ph.D. Dissertation, University of Iowa, 1976.

M.E. Johnson and J.S. Ramberg, "Elliptically symmetric distributions: characterizations and random variate generation," *Proceedings of the American Statistical Association, Statistical Computing Section*, pp. 262-265, 1977.

M.E. Johnson and J.S. Ramberg, "A bivariate distribution system with specified marginals," Technical Report LA-6858-MS, Los Alamos Scientific Laboratory, Los Alamos, New Mexico, 1977.

M.E. Johnson and P.R. Tadikamalla, "Computer methods for sampling from the gamma distribution," In *Proceedings of the Winter Simulation Conference, Miami Beach, Florida*, pp. 131-134, 1978.

M.E. Johnson and M.M. Johnson, "A new probability distribution with applications in Monte Carlo studies," Technical Report LA-7095-MS, Los Alamos Scientific Laboratory, Los Alamos, New Mexico, 1978.

M.E. Johnson, "Computer generation of the exponential power distribution," *Journal of Statistical Computation and Simulation*, vol. 9, pp. 239-240, 1979.

M.E. Johnson and A. Tenenbein, "Bivariate distributions with given marginals and fixed measures of dependence," Technical Report LA-7700-MS, Los Alamos Scientific Laboratory, Los Alamos, New Mexico, 1979.

M.E. Johnson, G.L. Tietjen, and R.J. Beckman, "A new family of probability distributions with applications to Monte Carlo studies," *Journal of the American Statistical Association*, vol. 75, pp. 276-279, 1980.

M.E. Johnson and A. Tenenbein, "A bivariate distribution family with specified marginals," *Journal of the American Statistical Association*, vol. 76, pp. 198-201, 1981.

N.L. Johnson, "Systems of frequency curves generated by methods of translation," *Biometrika*, vol. 36, pp. 149-176, 1949.

N.L. Johnson and C.A. Rogers, "The moment problem for unimodal distributions," *Annals of Mathematical Statistics*, vol. 22, pp. 433-439, 1951.

N.L. Johnson, "Systems of frequency curves derived from the first law of Laplace," *Trabajos de Estadistica*, vol. 5, pp. 283-291, 1954.

N.L. Johnson and S. Kotz, *Distributions in Statistics: Discrete Distributions*, John Wiley, New York, N.Y., 1969.

N.L. Johnson and S. Kotz, *Distributions in Statistics: Continuous Univeriate Distributions - 1*, John Wiley, New York, N.Y., 1970.

N.L. Johnson and S. Kotz, *Distributions in Statistics: Continuous Univariate Distributions - 2*, John Wiley, New York, N.Y., 1970.

N.L. Johnson and S. Kotz, "Developments in discrete distributions," *International Statistical Review*, vol. 50, pp. 71-101, 1982.

A. Jonassen and O.J. Dahl, "Analysis of an algorithm for priority queue administration," *BIT*, vol. 15, pp. 409-422, 1975.

T.G. Jones, "A note on sampling from a tape-file," *Communications of the ACM*, vol. 5, p. 343, 1962.

B. Jorgensen, *Statistical Properties of the Generalized Inverse Gaussian Distribution*, Lecture Notes in Statistics 9, Springer-Verlag, Berlin, 1982.

V. Kachitvichyanukul, "Computer Generation of Poisson, Binomial, and Hypergeometric Random Variates," Ph.D. Dissertation, School of Industrial Engineering, Purdue University, 1982.

V. Kachitvichyanukul and B.W. Schmeiser, "Computer generation of hypergeometric random variates," *Journal of Statistical Computation and Simulation*, vol. 22, pp. 127-145, 1985.

F.C. Kaminsky and D.L. Rumpf, "Simulating nonstationary Poisson processes: a comparison of alternatives including the correct approach," *Simulation*, vol. 28, pp. 17-20, 1977.

M. Kanter, "Stable densities under change of scale and total variation inequalities," *Annals of Probability*, vol. 3, pp. 697-707, 1975.

J. Kawarasaki and M. Sibuya, "Random numbers for simple random sampling without replacement," *Keio Mathematical Seminar Reports*, vol. 7, pp. 1-9, 1982.

T. Kawata, *Fourier Analysis in Probability Theory*, Academic Press, New York, N.Y., 1972.

J. Keilson and F.W. Steutel, "Mixtures of distributions, moment inequalities and measures of exponentiality and normality," *Annals of Probability*, vol. 2, pp. 112-130, 1974.

D. Kelker, "Distribution theory of spherical distributions and a location-scale parameter generalization," *Sankhya Series A*, vol. 32, pp. 419-430, 1970.

D. Kelker, "Infinite divisibility and variance mixtures of the normal distribution," *Annals of Mathematical Statistics*, vol. 42, pp. 802-808, 1971.

F.P. Kelly and B.D. Ripley, "A note on Strauss' model for clustering," *Biometrika*, vol. 63, pp. 357-360, 1976.

A.W. Kemp and C.D. Kemp, "An alternative derivation of the Hermite distribution," *Biometrika*, vol. 53, pp. 627-628, 1966.

A.W. Kemp and C.D. Kemp, "On a distribution associated with certain stochastic processes," *Journal of the Royal Statistical Society, Series B*, vol. 30, pp. 160-163, 1968.

A.W. Kemp, "Efficient generation of logarithmically distributed pseudo-random variables," *Applied Statistics*, vol. 30, pp. 249-253, 1981.

A.W. Kemp, "Frugal methods of generating bivariate discrete random variables," in *Statistical Distributions in Scientific Work*, ed. C. Taillie et al., vol. 4, pp. 321-329, D. Reidel Publ. Co., Dordrecht, Holland, 1981.

A.W. Kemp, "Conditionality properties for the bivariate logarithmic distribution with an application to goodness of fit," in *Statistical Distributions in Scientific Work*, ed. C. Taillie et al., vol. 5, pp. 57-73, D. Reidel Publ. Co., Dordrecht, Holland, 1981.

C.D. Kemp and A.W. Kemp, "Some properties of the Hermite distribution," *Biometrika*, vol. 52, pp. 381-394, 1965.

C.D. Kemp and H. Papageorgiou, "Bivariate Hermite distributions," *Bulletin of the IMS*, vol. 5, p. 174, 1976.

C.D. Kemp and S. Loukas, "The computer generation of bivariate discrete random variables," *Journal of the Royal Statistical Society, Series A*, vol. 141, pp. 513-519, 1978.

C.D. Kemp and S. Loukas, "Fast methods for generating bivariate discrete random variables," in *Statistical Distributions in Scientific Work*, ed. C. Taillie et al., vol. 4, pp. 313-319, D. Reidel Publ. Co., Dordrecht, Holland, 1981.

D.G. Kendall, "On some modes of population growth leading to R.A. Fisher's logarithmic series distribution," *Biometrika*, vol. 35, pp. 6-15, 1948.

M. Kendall and A. Stuart, *The Advanced Theory of Statistics, Vol. 1,* Macmillan, New York, 1977.

W.J. Kennedy, Jr. and J.E. Gentle, *Statistical Computing,* Marcel Dekker, New York, 1980.

G. Kimeldorf and A. Sampson, "One-parameter families of bivariate distributions with fixed marginals," *Communications in Statistics,* vol. 4, pp. 293-301, 1975.

G. Kimeldorf and A. Sampson, "Uniform representations of bivariate distributions," *Communications in Statistics,* vol. 4, pp. 617-627, 1975.

G. Kimeldorf and A. Sampson, "Monotone dependence," *Annals of Statistics,* vol. 6, pp. 895-903, 1978.

A.J. Kinderman, J.F. Monahan, and J.G. Ramage, "Computer generation of random variables with normal and student's t distribution," *Proceedings of the American Statistical Association, Statistical Computing Section,* pp. 128-131, Washington, D.C., 1975.

A.J. Kinderman and J.G. Ramage, "The computer generation of normal random variables," *Journal of the American Statistical Association,* vol. 71, pp. 893-896, 1976.

A.J. Kinderman, J.F. Monahan, and J.G. Ramage, "Computer methods for sampling from Student's t-distribution," *Mathematics of Computation,* vol. 31, pp. 1009-1018, 1977.

A.J. Kinderman and J.F. Monahan, "Computer generation of random variables using the ratio of uniform deviates," *ACM Transactions on Mathematical Software,* vol. 3, pp. 257-260, 1977.

A.J. Kinderman and J.G. Ramage, "FORTRAN programs for generating normal random variables," Technical Report 23, Department of Statistics, The Wharton School, University of Pennsylvania, Philadelphia, PA., 1977.

A.J. Kinderman, J.F. Monahan, and J.G. Ramage, "FORTRAN programs for generating Student's t random variables," Technical Report No. 24, Department of Statistics, The Wharton School, University of Pennsylvania, 1977.

A.J. Kinderman and J.F. Monahan, "Recent developments in the computer generation of Student's t and gamma random variables," *Proceedings of the ASA Statistical Computing Section,* pp. 90-94, 1978.

A.J. Kinderman and J.F. Monahan, "New methods for generating student's t and gamma variables," Technical Report, Department of Management Science, California State University, Northridge, CA., 1979.

J.H. Kingston, "Analysis of Henriksen's algorithm for the simulation event list," Technical Report 232, Basser Department of Computer Science, University of Sydney, Australia, 1984.

J.H. Kingston, "Analysis of tree algorithms for the simulation event list," *Acta Informatica,* vol. 22, pp. 15-33, 1985.

P. Klingsberg, Doctoral Dissertation, University of Washington, Seattle, Washington, 1977.

K. Knopp, *Theorie und Anwendung der unendlichen Reihen,* Springer-Verlag, Berlin, 1964.

G.D. Knott, "A numbering system for binary trees," *Communications of the ACM,* vol. 20, pp. 113-115, 1977.

D.E. Knuth, *The Art of Computer Programming, Vol. 1. Fundamental Algorithms,* Addison-Wesley, Reading, Mass., 1968.

D.E. Knuth, *The Art of Computer Programming, Vol. 2: Seminumerical Algorithms,* Addison-Wesley, Reading, Mass., 1969.

D.E. Knuth, *The Art of Computer Programming, Vol. 3: Searching and Sorting,* Addison-Wesley, Reading, Mass., 1973.

D.E. Knuth and A.C. Yao, "The complexity of nonuniform random number generation," in *Algorithms and Complexity,* ed. J.E. Traub, pp. 357-428, Academic Press, New York, N.Y., 1976.

K.D. Kohrt, "Efficient sampling from non-uniform statistical distributions," Diploma Thesis, University Kiel, Kiel, West Germany, 1980.

A.N. Kolmogorov, "Sulla determinazione empirica di una legge di distribuzione," *Giorn. Inst. Ital. Actuari,* vol. 4, pp. 83-91, 1933.

S. Kotz and R. Srinivasan, "Distribution of product and quotient of Bessel function variates," *Annals of the Institute of Statistical Mathematics,* vol. 21, pp. 201-210, 1969.

C.J. Kowalski, "Non-normal bivariate distributions with normal marginals," *The American Statistician,* vol. 27, pp. 103-106, 1973.

R.A. Kronmal and A.V. Peterson, "On the alias method for generating random variables from a discrete distribution," *The American Statistician,* vol. 33, pp. 214-218, 1979.

R.A. Kronmal and A.V. Peterson, "The alias and alias-rejection-mixture methods for generating random variables from probability distributions," in *Proceedings of the 1979 Winter Simulation Conference,* pp. 269-280, 1979.

R.A. Kronmal and A.V. Peterson, "Programs for generating discrete random integers using Walker's alias method," Department of Biostatistics, University of Washington, 1979.

R.A. Kronmal and A.V. Peterson, "Generating normal random variables using the alias-rejection-mixture method," *Proceedings of the 1979 ASA Annual Meeting, Computer Section,* Washington, D.C., 1980.

R.A. Kronmal and A.V. Peterson, "A variant of the acceptance-rejection method for computer generation of random variables," *Journal of the American Statistical Association,* vol. 76, pp. 446-451, 1981.

R.A. Kronmal and A.V. Peterson, "An acceptance-complement analogue of the mixture-plus-acceptance-rejection method for generating random variables," *ACM Transactions on Mathematical Software,* vol. 10, pp. 271-281, 1984.

W.H. Kruskal, "Ordinal measures of association," *Journal of the American Statistical Association,* vol. 53, pp. 814-861, 1958.

N.H. Kuiper, "Tests concerning random points on a circle," *Koninklijke Nederlandse Akademie van Wetenschappen*, vol. A63, pp. 38-47, 1960.

S. Kullback, "An application of characteristic functions to the distribution problem of statistics," *Annals of Mathematical Statistics*, vol. 5, pp. 264-305, 1934.

R.G. Laha, "On some properties of Bessel function distributions," *Bulletin of the Calcutta Mathematical Society*, vol. 46, pp. 59-71, 1954.

R.G. Laha, "An example of a non-normal distribution where the quotient follows the Cauchy law," *Proceedings of the National Academy of Sciences*, vol. 44, pp. 222-223, 1958.

V.C. Lakhan, "Generating autocorrelated psuedo-random numbers with specific distributions," *Journal of Statistical Computation and Simulation*, vol. 12, pp. 303-309, 1981.

H.O. Lancaster, "Correlation and complete dependence of random variables," *Annals of Mathematical Statistics*, vol. 34, pp. 1315-1321, 1963.

A.M. Law and W.D. Kelton, *Simulation Modeling and Analysis,* McGraw-Hill, New York, 1982.

A.J. Lawrance and P.A.W. Lewis, "A moving average exponential point process, EMA(1)," *Journal of Applied Probability*, vol. 14, pp. 98-113, 1977.

A.J. Lawrance and P.A.W. Lewis, "The exponential autoregressive moving average EARMA (p,q) process," *Journal of the Royal Statistical Society Series B*, vol. 42, pp. 150-161, 1980.

A.J. Lawrance and P.A.W. Lewis, "A new autoregressive time series model in exponential variables," *Advances in Applied Probability*, vol. 13, pp. 826-845, 1981.

D.H. Lehmer, "The machine tool of combinatorics," in *Applied Combinatorial Mathematics*, ed. E. Beckenbach, pp. 19-23, John Wiley, New York, N.Y., 1964.

R. Leipnik, "The lognormal distribution and strong non-uniqueness of the moment problem," *Theory of Probability and its Applications*, vol. 26, pp. 850-852, 1981.

C.G. Lekkerkerker, "A property of log-concave functions I, II," *Indagationes Mathematicae*, vol. 15, pp. 505-521, 1953.

G. Letac, "On building random variables of a given distribution," *Annals of Probability*, vol. 3, pp. 298-306, 1975.

P. Levy, "Sur certains processus stochastiques homogenes," *Compositio Mathematica*, vol. 7, pp. 283-339, 1940.

P.A.W. Lewis and G.S. Shedler, "Simulation of nonhomogeneous Poisson point processes with log-linear rate function," *Biometrika*, vol. 63, pp. 501-505, 1976.

P.A.W. Lewis and G.S. Shedler, "Simulation of nonhomogeneous Poisson point processes with degree-two exponential polynomial rate function," *Operations Research*, vol. 27, pp. 1026-1040, 1979.

P.A.W. Lewis and G.S. Shedler, "Simulation of nonhomogeneous Poisson processes by thinning," *Naval Research Logistics Quarterly*, vol. 26, pp. 403-413, 1979.

S.T. Li and J.L. Hammond, "Generation of pseudorandom numbers with specified univariate distributions and correlation coefficients," *IEEE Transactions on Systems, Man and Cybernetics*, vol. 5, pp. 557-561, 1975.

J. Lieblein, "Properties of certain statistics involving the closest pair in a sample of three observations," *Journal of Research of the National Bureau of Standards*, vol. 48, pp. 255-268, 1952.

Yu.V. Linnik, "Linear forms and statistical criteria: I, II," *Selected Translations in Mathematical Statistics and Probability*, vol. 3, pp. 1-40, 41-90, 1962.

Yu.V. Linnik, "Linear forms and statistical criteria,I,II," *Selected Translations in Mathematical Statistics and Probability*, vol. 3, pp. 1-90, 1962.

M. Loeve, *Probability Theory*, Van Nostrand, Princeton, New Jersey, 1963.

H.W. Lotwick, "Simulation of some spatial hard core models and the complete packing problem," *Journal of Statistical Computation and Simulation*, vol. 15, pp. 295-314, 1982.

E. Lukacs, "A characterization of the gamma distribution," *Annals of Mathematical Statistics*, vol. 26, pp. 319-324, 1955.

E. Lukacs and R.G. Laha, *Applications of Characteristic Functions*, Griffin, London, 1964.

E. Lukacs, *Characteristic Functions*, Griffin, London, 1970.

D. Lurie and H.O. Hartley, "Machine-generation of order statistics for Monte Carlo computations," *The American Statistician*, vol. 26, pp. 26-27, 1972.

D. Lurie and R.L. Mason, "Empirical investigation of several techniques for computer generation of order statistics," *Communications in Statistics*, vol. 2, pp. 363-371, 1973.

E.J. Lusk and H. Wright, "Deriving the probability density for sums of uniform random variables," *The American Statistician*, vol. 36, pp. 128-130, 1982.

I. Lux, "A special method to sample some probability density functions," *Computing*, vol. 20, pp. 183-188, 1978.

I. Lux, "Another special method to sample probability density functions," *Computing*, vol. 21, pp. 359-364, 1979.

M.D. Maclaren, G. Marsaglia, and T.A. Bray, "A fast procedure for generating exponential random variables," *Communications of the ACM*, vol. 7, pp. 298-300, 1964.

M. Maejima, "The remainder term in the local limit theorem for independent random variables," *Tokyo Journal of Mathematics*, vol. 3, pp. 311-329, 1980.

W. Magnus, F. Oberhettinger, and R.P. Soni, *Formulas and Theorems for the Special Functions of Mathematical Physics*, Springer-Verlag, Berlin-Heidelberg, 1966.

S. Malmquist, "On a property of order statistics from a rectangular distribution," *Skandinavisk Aktuarietidskrift*, vol. 33, pp. 214-222, 1950.

B. Mandelbrot, "The variation of certain speculative prices," *Journal of Business*, vol. 36, pp. 349-419, 1963.

D. Mannion, "Random space-filling in one dimension," *Publications of the Mathematical Institute of the Hungarian Academy of Sciences,* vol. 9, pp. 143-153, 1964.

N. Mantel, "A characteristic function exercise," *The American Statistician,* vol. 27, p. 31, 1973.

K.V. Mardia, "Multivariate Pareto distributions," *Annals of Mathematical Statistics,* vol. 33, pp. 1008-1015, 1962.

K.V. Mardia, *Families of Bivariate Distributions,* Griffin, London, 1970.

K.V. Mardia, "A translation family of bivariate distributions and Frechet's bounds," *Sankhya Series A,* vol. 32, pp. 119-122, 1970.

K.V. Mardia, "Statistics of directional data," *Journal of the Royal Statistical Society Series B,* vol. 37, pp. 349-393, 1975.

G. Marsaglia, "Expressing a random variable in terms of uniform random variables," *Annals of Mathematical Statistics,* vol. 32, pp. 894-899, 1961.

G. Marsaglia, "Generating exponential random variables," *Annals of Mathematical Statistics,* vol. 32, pp. 899-902, 1961.

G. Marsaglia, "Generating discrete random variables in a computer," *Communications of the ACM,* vol. 6, pp. 37-38, 1963.

G. Marsaglia, M.D. Maclaren, and T.A. Bray, "A fast procedure for generating normal random variables," *Communications of the ACM,* vol. 7, pp. 4-10, 1964.

G. Marsaglia and T.A. Bray, "A convenient method for generating normal random variables," *SIAM Review,* vol. 6, pp. 260-264, 1964.

G. Marsaglia, "Choosing a point from the surface of a sphere," *Annals of Mathematical Statistics,* vol. 43, pp. 645-646, 1972.

G. Marsaglia, K. Ananthanarayanan, and N.J. Paul, "Improvements on fast methods for generating normal random variables," *Information Processing Letters,* vol. 5, pp. 27-30, 1976.

G. Marsaglia, "Random number generation," in *Encyclopedia of Computer Science,* ed. A. Ralston and C.L. Meek, pp. 1192-1197, Petrocelli/Charter, New York, N.Y., 1976.

G. Marsaglia, "The squeeze method for generating gamma variates," *Computers and Mathematics with Applications,* vol. 3, pp. 321-325, 1977.

G. Marsaglia, "Generating random variables with a t-distribution," *Mathematics of Computation,* vol. 34, pp. 235-236, 1979.

G. Marsaglia, "The exact-approximation method for generating random variables in a computer," *Journal of the American Statistical Association,* vol. 79, pp. 218-221, 1984.

G. Marsaglia and W.W. Tsang, "A fast, easily implemented method for sampling from decreasing or symmetric unimodal density functions," *SIAM Journal on Scientific and Statistical Computing,* vol. 5, pp. 349-359, 1984.

G. Marsaglia and I. Olkin, "Generating correlation matrices," *SIAM Journal on Scientific and Statistical Computations,* vol. 5, pp. 470-475, 1984.

A.W. Marshall and I. Olkin, "A multivariate exponential distribution," *Journal of the American Statistical Association*, vol. 62, pp. 30-44, 1967.

A.W. Marshall and I. Olkin, *Inequalities: Theory of Majorization and its Applications*, Academic Press, London, England, 1979.

A.W. Marshall and I. Olkin, "Domains of attraction of multivariate extreme value distributions," *Annals of Probability*, vol. 11, pp. 168-177, 1983.

J.H. May and R.L. Smith, "Random polytopes: their definition, generation and aggregate properties," *Mathematical Programming*, vol. 24, pp. 39-54, 1982.

W.M. McCormack and R.G. Sargent, "Analysis of future event set algorithms for discrete event simulation," *Communications of the ACM*, vol. 24, pp. 801-812, 1981.

E.J. McGrath and D.C. Irving, "Techniques for efficient Monte Carlo simulation: volume 2: random number generation for selected probability distributions," Technical Report SAI-72-590-LJ, Science Applications, Inc., La Jolla, CA., 1973.

A.T. McKay, "A Bessel function distribution," *Biometrika*, vol. 24, pp. 39-44, 1932.

J.K.S. McKay, "Algorithm 262. Number of restricted partitions of N," *Communications of the ACM*, vol. 8, p. 493, 1965.

J.K.S. McKay, "Algorithm 263. Partition generator," *Communications of the ACM*, vol. 8, p. 493, 1965.

J.K.S. McKay, "Algorithm 264. Map of partitions into integers," *Communications of the ACM*, vol. 8, p. 493, 1965.

P. McMullen, "The maximum number of faces of a convex polytope," *Mathematika*, vol. 17, pp. 179-184, 1970.

J.R. Michael, W.R. Schucany, and R.W. Haas, "Generating random variates using transformations with multiple roots," *The American Statistician*, vol. 30, pp. 88-90, 1976.

G.A. Mikhailov, "On modelling random variables for one class of distribution laws," *Theory of Probability and its Applications*, vol. 10, pp. 681-682, 1965.

G.A. Mikhailov, "On the reproduction method for simulating random vectors and stochastic processes," *Theory of Probability and its Applications*, vol. 19, pp. 839-843, 1974.

L. Mirsky, *Introduction to Linear Algebra*, Oxford University Press, 1955.

R. von Mises, "On the asymptotic distribution of differentiable statistical functions," *Annals of Mathematical Statistics*, vol. 18, pp. 309-348, 1947.

R. von Mises, *A Mathematical Theory of Probability and Statistics*, Academic Press, New York, N.Y., 1964.

R.L. Mitchell, "Table lookup methods for generating arbitrary random numbers," *IEEE Transactions on Computers*, vol. C-26, pp. 1006-1008, 1977.

S.S. Mitra, "Distribution of symmetric stable laws of index 2 to the power -n," *Annals of Probability*, vol. 9, pp. 710-711, 1981.

S.S. Mitra, "Stable laws of index 2^{-n}," *Annals of Probability*, vol. 10, pp. 857-859, 1982.

D.S. Mitrinovic, *Analytic Inequalities*, Springer-Verlag, Berlin, 1970.

J.F. Monahan, "Extensions of von Neumann's method for generating random variables," *Mathematics of Computation*, vol. 33, pp. 1065-1069, 1979.

W.J. Moonan, "Linear transformation to a set of stochastically dependent normal variables," *Journal of the American Statistical Association*, vol. 52, pp. 247-252, 1957.

P.A.P. Moran, "A characterization property of the Poisson distribution," *Proceedings of the Cambridge Philosophical Society*, vol. 48, pp. 206-207, 1951.

P.A.P. Moran, "Testing for correlation between non-negative variates," *Biometrika*, vol. 54, pp. 385-394, 1967.

B.J.T. Morgan, *Elements of Simulation*, Chapman and Hall, London, 1984.

D. Morgenstern, "Einfache Beispiele zweidimensionaler Verteilungen," *Mitteilungsblatt fur Mathematische Statistik*, vol. 8, pp. 234-235, 1956.

L.E. Moses and R.V. Oakford, *Tables of Random Permutations*, Allen and Unwin, London, 1963.

M.E. Muller, "An inverse method for the generation of random variables on large-scale computers," *Math. Tables Aids Comput.*, vol. 12, pp. 167-174, 1958.

M.E. Muller, "The use of computers in inspection procedures," *Communications of the ACM*, vol. 1, pp. 7-13, 1958.

M.E. Muller, "A comparison of methods for generating normal deviates on digital computers," *Journal of the ACM*, vol. 6, pp. 376-383, 1959.

V.N. Murty, "The distribution of the quotient of maximum values in samples from a rectangular distribution," *Journal of the American Statistical Association*, vol. 50, pp. 1136-1141, 1955.

M. Nagao and M. Kadoya, "Two-variate exponential distribution and its numerical table for engineering application," *Bulletin of the Disaster Prevention Research Institute*, vol. 20, pp. 183-215, 1971.

R.E. Nance and C. Overstreet, "A bibliography on random number generation," *Computing Reviews*, vol. 13, pp. 495-508, 1972.

S.C. Narula and F.S. Li, "Approximations to the chi-square distribution," *Journal of Statistical Computation and Simulation*, vol. 5, pp. 267-277, 1977.

A. Nataf, "Determination des distributions de probabilites dont les marges sont donnees," *Comptes Rendus de l'Academie des Sciences de Paris*, vol. 255, pp. 42-43, 1962.

J. von Neumann, "Various techniques used in connection with random digits," *Collected Works*, vol. 5, pp. 768-770, Pergamon Press, 1963. also: in Monte Carlo Method, National Bureau of Standards Series, Vol. 12, pp. 36-38, 1951

M.F. Neuts and M.E. Pagano, "Generating random variates from a distribution of phase type," Technical Report 73B, Applied Mathematics Institute, University of Delaware, Newark, Delaware 19711, 1981.

M.J. Newby, "The simulation of order statistics from life distributions," *Applied Statistics*, vol. 28, pp. 298-301, 1979.

T.G. Newman and P.L. Odell, *The Generation of Random Variates*, Griffin, London, 1971.

J. Neyman and E.S. Pearson, "On the use and interpretation of certain test criteria for purposes of statistical inference I," *Biometrika*, vol. 20A, pp. 175-240, 1928.

A. Nijenhuis and H.S. Wilf, *Combinatorial Algorithms*, Academic Press, New York, N.Y., 1975.

J.E. Norman and L.E. Cannon, "A computer program for the generation of random variables from any discrete distribution," *Journal of Statistical Computation and Simulation*, vol. 1, pp. 331-348, 1972.

R.M. Norton, "Moment properties of the arc sine law," *Sankhya Series A*, vol. 40, pp. 192-198, 1978.

R.E. Odeh and J.O. Evans, "The percentage points of the normal distribution," *Applied Statistics*, vol. 23, pp. 96-97, 1974.

E.G. Olds, "A note on the convolution of uniform distributions," *Annals of Mathematical Statistics*, vol. 23, pp. 282-285, 1952.

J.K. Ord, "The discrete Student's t distribution," *Annals of Mathematical Statistics*, vol. 39, pp. 1513-1516, 1968.

J.K. Ord, *Families of Frequency Distributions*, Griffin, London, 1972.

A.M. Ostrowski, *Solutions of Equations in Euclidean and Banach Spaces*, Academic Press, New York, N.Y., 1973.

W.J. Padgett, "Comment on inverse gaussian random number generation," *Journal of Statistical Computation and Simulation*, vol. 8, pp. 78-79, 1978.

E. Parzen, "On the estimation of a probability density function and the mode," *Annals of Mathematical Statistics*, vol. 33, pp. 1065-1076, 1962.

W.M. Patefield, "An efficient method of generating random $R \times C$ tables with given row and column totals," *Applied Statistics*, vol. 30, pp. 91-97, 1981.

I.D. Patel, "A generalized logarithmic series distribution," *Journal of the Indian Statistical Association*, vol. 19, pp. 129-132, 1981.

J.K. Patel, C.H. Kapadia, and D.B. Owen, *Handbook of Statistical Distributions*, Marcel Dekker, New York, N.Y., 1976.

G.P. Patil and V. Seshadri, "Characterization theorems for some univariate probability distributions," *Journal of the Royal Statistical Society, Series B*, vol. 26, pp. 286-292, 1964.

G.P. Patil and M.T. Boswell, "Chance mechanisms for discrete distributions in scientific modeling," in *Statistical Distributions in Scientific Work*, ed. G.P. Patil et al., vol. 2, pp. 11-24, D. Reidel Publ. Co., Dordrecht, Holland, 1975.

G.P. Patil, M.T. Boswell, and D.S. Friday, "Chance mechanisms in computer generation of random variables," in *Statistical Distributions in Scientific Work*, ed. G.P. Patil et al., vol. 2, pp. 37-50, D.Reidel Publ. Co., Dordrecht, Holland,

1975.

V. Paulauskas, "Convergence to stable laws and the modeling," *Lithuanian Mathematical Journal*, vol. 22, pp. 319-326, 1982.

E.S. Paulson, E.W. Holcomb, and R.A. Leitch, "The estimation of the parameters of the stable laws," *Biometrika*, vol. 62, pp. 163-170, 1975.

J.A. Payne, *Introduction to Simulation. Programming Techniques and Methods of Analysis*, McGraw-Hill, New York, 1982.

W.H. Payne, "Normal random numbers: using machine analysis to choose the best algorithm," *ACM Transactions on Mathematical Software*, vol. 3, pp. 346-358, 1977.

W.F. Perks, "On some experiments in the graduation of mortality statistics," *Journal of the Institute of Actuaries*, vol. 58, pp. 12-57, 1932.

A.V. Peterson and R.A. Kronmal, "On mixture methods for the computer generation of random variables," *The American Statistician*, vol. 36, pp. 184-191, 1982.

A.V. Peterson and R.A. Kronmal, "Analytic comparison of three general-purpose methods for the computer generation of discrete random variables," *Applied Statistics*, vol. 32, pp. 276-286, 1983.

V.V. Petrov, *Sums of Independent Random Variables*, Springer-Verlag, New York, 1975.

R.L. Plackett, "A class of bivariate distributions," *Journal of the American Statistical Association*, vol. 60, pp. 516-522, 1965.

R.L. Plackett, "Random permutations," *Journal of the Royal Statistical Society, Series B*, vol. 30, pp. 517-534, 1968.

R.J. Polge, E.M. Holliday, and B.K. Bhagavan, "Generation of a pseudo-random set with desired correlation and probability distribution," *Simulation*, vol. 20, pp. 153-158, 1973.

G. Polya, "Remarks on computing the probability integral in one and two dimensions," in *Proceedings of the First Berkeley Symposium on Mathematical Statistics and Probability*, ed. J. Neymann, pp. 63-78, University of California Press, 1949.

A. Prekopa, "On logarithmic concave measures and functions," *Acta Scientiarium Mathematicarum Hungarica*, vol. 34, pp. 335-343, 1973.

S.J. Press, "Stable distributions: probability, inference, and applications in finance - a survey, and a review of recent results," in *Statistical Distributions in Scientific Work*, ed. G.P. Patil et al., vol. 1, pp. 87-102, D. Reidel Publ. Co., Dordrecht, Holland, 1975.

B. Price, "Replicating sequences of Bernoulli trials in simulation modeling," *Computers and Operations Research*, vol. 3, pp. 357-361, 1976.

A.A.B. Pritsker, "Ongoing developments in GASP," in *Proceedings of the 1976 Winter Simulation Conference*, pp. 81-83, 1976.

M.H. Quenouille, "A relation between the logarithmic, Poisson and negative binomial series," *Biometrics*, vol. 5, pp. 162-164, 1949.

D.H. Raab and E.H. Green, "A cosine approximation to the normal distribution," *Psychometrika*, vol. 26, pp. 447-450, 1961.

M. Rabinowitz and M.L. Berenson, "A comparison of various methods of obtaining random order statistics for Monte Carlo computations," *The American Statistician*, vol. 28, pp. 27-29, 1974.

N.A. Rahman, "Some generalisations of the distributions of product statistics arising from rectangular populations," *Journal of the American Statistical Association*, vol. 59, pp. 557-563, 1964.

J.S. Ramberg and B.W. Schmeiser, "An approximate method for generating symmetric random variables," *Communications of the ACM*, vol. 15, pp. 987-990, 1972.

J.S. Ramberg and B.W. Schmeiser, "An approximate method for generating asymmetric random variables," *Communications of the ACM*, vol. 17, pp. 78-82, 1974.

J.S. Ramberg, "A probability distribution with applications to Monte Carlo simulation studies," in *Statistical Distributions in Scientific Work*, ed. G.P. Patil et al., vol. 2, pp. 51-64, D. Reidel Publ. Co., Dordrecht, Holland, 1975.

J.S. Ramberg and P.R. Tadikamalla, "On the generation of subsets of order statistics," *Journal of Statistical Computation and Simulation*, vol. 6, pp. 239-241, 1978.

J.S. Ramberg, P.R. Tadikamalla, E.J. Dudewicz, and E.F. Mykytka, "A probability distribution and its uses in fitting data," *Technometrics*, vol. 21, pp. 201-214, 1979.

C.M. Reeves, "Complexity analyses of event set algorithms," *Computer Journal*, vol. 27, pp. 72-79, 1984.

D.A. Relles, "A simple method for generating binomial random variables when n is large," *Journal of the American Statistical Association*, vol. 67, pp. 612-613, 1972.

A. Renyi, "On a one-dimensional problem concerning random space-filling," *Publications of the Mathematical Institute of the Hungarian Academy of Sciences*, vol. 3, pp. 109-127, 1958.

A. Renyi, "On measures of dependence," *Acta Mathematica Academia Scientifica Hungarica*, vol. 10, pp. 441-451, 1959.

P.R. Rider, "The distribution of the product of maximum values in samples from a rectangular distribution," *Journal of the American Statistical Association*, vol. 51, pp. 1142-1143, 1955.

B.D. Ripley, "Modelling spatial patterns," *Journal of the Royal Statistical Society, Series B*, vol. 39, pp. 172-212, 1977.

B.D. Ripley, "Simulating spatial patterns: dependent samples from a multivariate density," *Journal of the Royal statistical Society, series C*, vol. 28, pp. 109-112, 1979.

B.D. Ripley, "Computer generation of random variables: a tutorial," *International Statistical Review*, vol. 51, pp. 301-319, 1983.

S.A. Roach, "The frequency distribution of the sample mean where each member of the sample is drawn from a different rectangular distribution," *Biometrika*, vol. 50, pp. 508-513, 1963.

I. Robertson and L.A. Walls, "Random number generators for the normal and gamma distrubutions using the ratio of uniforms method," Technical Report AERE-R 10032, U.K. Atomic Energy Authority, Harwell, Oxfordshire, 1980.

C.L. Robinson, "Algorithm 317. Permutation.," *Communications of the ACM*, vol. 10, p. 729, 1967.

J.M. Robson, "Generation of random permutations," *Communications of the ACM*, vol. 12, pp. 634-635, 1969.

B.A. Rogozin, "On an estimate of the concentration function," *Theory of Probability and its Applications*, vol. 6, pp. 94-97, 1961.

G. Ronning, "A simple scheme for generating multivariate gamma distributions with non-negative covariance," *Technometrics*, vol. 19, pp. 179-183, 1977.

M. Rosenblatt, "Remarks on some nonparametric estimates of a density function," *Annals of Mathematical Statistics*, vol. 27, pp. 832-837, 1956.

H.L. Royden, "Bounds on a distribution function when its first n moments are given," *Annals of Mathematical Statistics*, vol. 24, pp. 361-376, 1953.

H.L. Royden, *Real Analysis,* Macmillan, London, U.K., 1968.

P.A. Rubin, "Generating random points in a polytope," *Communications in Statistics, Section Simulation and Computation*, vol. 13, pp. 375-396, 1984.

R.Y. Rubinstein, *Simulation and the Monte Carlo Method,* John Wiley, New York, 1981.

R.Y. Rubinstein, "Generating random vectors uniformly distributed inside and on the surface of different regions," *European Journal of Operations Research*, vol. 10, pp. 205-209, 1982.

F. Ruskey and T.C. Hu, "Generating binary trees lexicographically," *SIAM Journal on Computing*, vol. 6, pp. 745-758, 1977.

F. Ruskey, "Generating t-ary trees lexicographically," *SIAM Journal on Computing*, vol. 7, pp. 424-439, 1978.

T.P. Ryan, "A new method of generating correlation matrices," *Journal of Statistical Computation and Simulation*, vol. 11, pp. 79-85, 1980.

H. Sahai, "A supplement to Sowey's bibliography on random number generation and related topics," *Journal of Statistical Computation and Simulation*, vol. 10, pp. 31-52, 1979.

W. Sahler, "A survey of distribution-free statistics based on distances between distribution functions," *Metrika*, vol. 13, pp. 149-169, 1968.

H. Sakasegawa, "On a generation of pseudo-random numbers," *Annals of the Institute of Statistical Mathematics*, vol. 30, pp. 271-279, 1978.

O.V. Sarmanov, "The maximal correlation coefficient (nonsymmetric case)," *Selected Translations in Mathematical Statistics and Probability*, vol. 2, pp. 207-210, 1962.

O.V. Sarmanov, "The maximal correlation coefficient (symmetric case)," *Selected Translations in Mathematical Statistics and Probability*, vol. 4, pp. 271-275, 1963.

S.P. Satterthwaite and T.P. Hutchinson, "A generalization of Gumbel's bivariate logistic distribution," *Metrika*, vol. 25, pp. 163-170, 1978.

I.R. Savage, "Probability inequalities of the Tchebycheff type," *Journal of Research of the National Bureau of Standards*, vol. 65, pp. 211-222, 1961.

H. Scheffe, "A useful convergence theorem for probability distributions," *Annals of Mathematical Statistics*, vol. 18, pp. 434-458, 1947.

E.M. Scheuer and D.S. Stoller, "On the generation of normal random vectors," *Technometrics*, vol. 4, pp. 278-281, 1962.

B.W. Schmeiser, "Methods for modeling and generating probabilistic components in digital computer simulation when the standard distributions are not adequate: a survey," in *Winter Simulation Conference*, pp. 1-7, 1977.

B.W. Schmeiser, "Generation of the maximum (minimum) value in digital computer simulation," *Journal of Statistical Computation and Simulation*, vol. 8, pp. 103-115, 1978.

B.W. Schmeiser, "Random variate generation: a survey," in *Proceedings of the 1980 Winter Simulation Conference*, pp. 79-104, Orlando, Florida, 1980.

B.W. Schmeiser and R. Lal, "Squeeze methods for generating gamma variates," *Journal of the American Statistical Association*, vol. 75, pp. 679-682, 1980.

B.W. Schmeiser and A.J.G. Babu, "Beta variate generation via exponential majorizing functions," *Operations Research*, vol. 28, pp. 917-926, 1980.

B.W. Schmeiser, "Generation of variates from distribution tails," *Operations Research*, vol. 28, pp. 1012-1017, 1980.

B.W. Schmeiser, "Multivariate modeling in simulation: a survey," in *ASQC Technical Conference Transactions*, pp. 252-261, Atlanta, 1980.

B.W. Schmeiser and V. Kachitvichyanukul, "Poisson random variate generation," Research Memorandum 81-4, School of Industrial Engineering, Purdue University, West Lafayette, Indiana, 1981.

B.W. Schmeiser and R. Lal, "Bivariate gamma random vectors," *Operations Research*, vol. 30, pp. 355-374, 1982.

B.W. Schmeiser, "Recent advances in generating observations from discrete random variables," in *Computer Science and Statistics: The Interface*, ed. J.E. Gentle, pp. 154-160, North-Holland, 1983.

A. Schonhage, M. Paterson, and N. Pippenger, "Finding the median," *Journal of Computers and System Sciences*, vol. 13, pp. 184-199, 1976.

W.R. Schucany, "Order statistics in simulation," *Journal of Statistical Computation and Simulation*, vol. 1, pp. 281-286, 1972.

E.F. Schuster, "Generating normal variates by summing three uniforms," in *Proceedings of Computer Science and Statistics: 15th Symposium on the Interface*, Houston, Texas, 1983.

R. Seidel, "A convex hull algorithm optimal for point sets in even dimensions," Technical Report 81-14, Department of Computer Science, UBC, Vancouver, B.C., 1981.

J. Selgerstetter, "Pole-seeking Brownian motion and bird navigation," *Journal of the Royal Statistical Society Series B*, vol. 36, pp. 411-412, 1974.

J.G. Shanthikumar, "Discrete random variate generation using uniformization," Technical Report 83-002, Dept. of Systems and Industrial Engineering, University of Arizona, Tucson, AZ., 1983.

J.G. Shanthikumar, "Discrete random variate generation using uniformization," *European Journal of Operational Research*, vol. 21, pp. 387-398, 1985.

J.A. Shohat and J.D. Tamarkin, "The Problem of Moments," Mathematical Survey No. 1, American Mathematical Society, New York, 1943.

R.W. Shorrock, "On record values and record times," *Journal of Applied Probability*, vol. 9, pp. 316-326, 1972.

J. Shuster, "On the inverse gaussian distribution function," *Journal of the American Statistical Association*, vol. 63, pp. 1514-1516, 1968.

M. Sibuya, "On exponential and other random variable generators," *Annals of the Institute of Statistical Mathematics*, vol. 13, pp. 231-237, 1961.

M. Sibuya, "A method for generating uniformly distributed points on n-dimensional spheres," *Annals of the Institute of Statistical Mathematics*, vol. 14, pp. 81-85, 1962.

M. Sibuya, "Further consideration on normal random variable generator," *Annals of the Institute of Statistical Mathematics*, vol. 14, pp. 159-165, 1962.

M. Sibuya, "Generalized hypergeometric, digamma and trigamma distributions," *Annals of the Institute of Statistical Mathematics, Series A*, vol. 31, pp. 373-390, 1979.

H.A. Simon, "On a class of skew distribution functions," *Biometrika*, vol. 41, pp. 425-440, 1954.

H.A. Simon, "Some further notes on a class of skew distribution functions," *Information and Control*, vol. 3, pp. 90-98, 1960.

B.D. Sivazlian, "On a multivariate extension of the gamma and beta distributions," *SIAM Journal of Applied Mathematics*, vol. 41, pp. 205-209, 1981.

B.D. Sivazlian, "A class of multivariate distributions," *Australian Journal of Statistics*, vol. 23, pp. 251-255, 1981.

N.V. Smirnov, "On the distribution of the ω^2 criterion of von Mises," *Rec. Math.*, vol. 2, pp. 973-993, 1937.

N.V. Smirnov, "On the estimation of the discrepancy between empirical curves of distribution for two independent samples," *Bulletin Mathematique de l'Universite de Moscou*, vol. 2, 1939.

E.R. Sowey, "A chronological and classified bibliography on random number generation and testing," *International Statistical Review*, vol. 40, pp. 355-371, 1972.

E.R. Sowey, "A second classified bibliography on random number generation and testing," *International Statistical Review*, vol. 46, pp. 89-102, 1978.

M.D. Springer, *The Algebra of Random Variables,* John Wiley, New York, N.Y., 1979.

E.W. Stacy, "A generalization of the gamma distribution," *Annals of Mathematical Statistics*, vol. 33, pp. 1187-1192, 1962.

E. Stadlober, "Generating Student's t variates by a modified rejection method," *Proceedings of the 2nd Pannonian Symposium on Mathematical Statistics*, pp. 0-0, 1981.

T.A. Standish, *Data Structure Techniques,* Addison-Wesley, Reading, Mass., 1980.

J.F. Steffensen, "On certain inequalities between mean values, and their application to actuarial problems," *Skandinavisk Aktuarietidskrift*, vol. 1, pp. 82-97, 1918.

J.F. Steffensen, "On a generalization of certain inequalities of Tchebychef and Jensen," *Skandinavisk Aktuarietidskrift*, vol. 8, pp. 137-147, 1925.

A. Stuart, "Gamma-distributed products of independent random variables," *Biometrika*, vol. 49, pp. 564-565, 1962.

M.T. Subbotin, "On the law of frequency of errors," *Matematicheskii Sbornik*, vol. 31, pp. 296-301, 1923.

P.V. Sukhatme, "Tests of significance for samples of the chi square population with two degrees of freedom," *Ann. Eugen.*, vol. 8, pp. 52-56, 1937.

P. Survila, "A local limit theorem for densities," *Lithuanian Mathematical Journal*, vol. 4, pp. 535-540, 1964.

P.R. Tadikamalla and M.E. Johnson, "Rejection methods for sampling from the normal distribution," *Proceedings of the First International Conference on Mathematical Modeling*, pp. 573-578, St. Louis, Missouri, 1977.

P.R. Tadikamalla, "Computer generation of gamma random variables," *Communications of the ACM*, vol. 21, pp. 419-422, 1978.

P.R. Tadikamalla, "Computer generation of gamma random variables - II," *Communications of the ACM*, vol. 21, pp. 925-929, 1978.

P.R. Tadikamalla and M.E. Johnson, "A survey of computer methods for sampling from the gamma distribution," *Proceedings of the Winter Simulation Conference*, vol. 1, pp. 131-134, 1978.

P.R. Tadikamalla, "Random sampling from the generalized gamma distribution," *Computing*, vol. 23, pp. 199-203, 1979.

P.R. Tadikamalla, "A simple method for sampling from the Poisson distribution," Working Paper 365, Graduate School of Business, University of Pittsburgh, Pittsburgh, PA., 1979.

P.R. Tadikamalla and M.E. Johnson, "A survey of gamma variate generators," Technical Report LA-UR-3035, Los Alamos Scientific Laboratory, Los Alamos, New Mexico, 1980.

P.R. Tadikamalla, "A look at the Burr and related distributions," *International Statistical Review*, vol. 48, pp. 337-344, 1980.

P.R. Tadikamalla, "Random sampling from the exponential power distribution," *Journal of the American Statistical Association*, vol. 75, pp. 683-686, 1980.

P.R. Tadikamalla and M.E. Johnson, "A complete guide to gamma variate generation," *American Journal of Mathematical and Management Sciences*, vol. 1, pp. 213-236, 1981.

P.R. Tadikamalla, "On a family of distributions obtained by the transformation of the gamma distribution," *Journal of Statistical Computation and Simulation*, vol. 13, pp. 209-214, 1981.

K. Takahasi, "Note on the multivariate Burr's distribution," *Annals of the Institute of Statistical Mathematics*, vol. 17, pp. 257-260, 1965.

J. Talacko, "Perks' distributions and their role in the theory of Wiener's stochastic variables," *Trabajos de Estadistica*, vol. 7, pp. 159-174, 1956.

D.R.S. Talbot and J.R. Willis, "The effective sink strength of a random array of voids," *Proceedings of the Royal Society of London, Series A*, vol. 370, pp. 351-374, 1980.

J.C. Tanner, "A derivation of the Borel distribution," *Biometrika*, vol. 38, pp. 383-392, 1951.

M.A. Tanner and R.A. Thisted, "A remark on AS127. Generation of random orthogonal matrices," *Applied Statistics*, vol. 31, pp. 190-192, 1982.

Y. Tashiro, "On methods for generating uniform random points on the surface of a sphere," *Annals of the Institute of Statistical Mathematics*, vol. 29, pp. 295-300, 1977.

D. Teichroew, "The mixture of normal distributions with different variances," *Annals of Mathematical Statistics*, vol. 28, pp. 510-512, 1957.

J. Teuhola and O. Nevalainen, "Two efficient algorithms for random sampling without replacement," *International Journal of Computational Mathematics*, vol. 11, pp. 127-140, 1982.

G. Tinhofer, "On the generation of random graphs with given properties and known distribution," in *Graphs, Data Structures, Algorithms*, ed. M. Nagl, pp. 265-297, Carl Hanser Verlag, Munchen, West Germany, 1978.

G. Tinhofer, *Zufallsgraphen*, Carl Hansen Verlag, Munchen, West Germany, 1980.

G. Tinhofer and H. Schreck, "Linear time tree codes," *Computing*, vol. 33, pp. 211-225, 1984.

E.C. Titchmarsh, *The Theory of Riemann Zeta-Functions*, Oxford University Press, Oxford, 1951.

F.I. Toranzos, "An asymmetric bell-shaped frequency curve," *Annals of Mathematical Statistics*, vol. 23, pp. 467-469, 1952.

K.S. Trivedi, *Probability and Statistics, with Reliability, Queuing, and Computer Science Applications*, Prentice Hall, Englewood Cliffs, New Jersey, 1982.

REFERENCES

A.E. Trojanowski, "Ranking and listing algorithms for k-ary trees," *SIAM Journal on Computing*, vol. 7, pp. 492-509, 1978.

J.W. Tukey, "The practical relationship between the common transformations of percentages of counts and of amounts," Technical Report 36, Statistical Techniques Research Group, Princeton University, 1960.

E.G. Ulrich, "Event manipulation for discrete simulations requiring large numbers of events," *Communications of the ACM*, vol. 21, pp. 777-785, 1978.

G. Ulrich, "Computer generation of distributions on the m-sphere," *Applied Statistics*, vol. 33, pp. 158-163, 1984.

I. Vaduva, "On computer generation of gamma random variables by rejection and composition procedures," *Mathematische Operationsforschung und Statistik, Series Statistics*, vol. 8, pp. 545-576, 1977.

J.G. Vaucher and P. Duval, "A comparison of simulation event list algorithms," *Communications of the ACM*, vol. 18, pp. 223-230, 1975.

J.G. Vaucher, "On the distribution of event times for the notices in a simulation event list," *INFOR*, vol. 15, pp. 171-182, 1977.

J.S. Vitter, "Optimum algorithms for two random sampling problems," *Proceedings of the 24th IEEE Conference on FOCS*, pp. 65-75, 1983.

J.S. Vitter, "Faster methods for random sampling," *Communications of the ACM*, vol. 27, pp. 703-718, 1984.

J.S. Vitter, "Random sampling with a reservoir," *ACM Transactions on Mathematical Software*, vol. 11, pp. 37-57, 1985.

A.J. Walker, "New fast method for generating discrete random numbers with arbitrary frequency distributions," *Electronics Letters*, vol. 10, pp. 127-128, 1974.

A.J. Walker, "An efficient method for generating discrete random variables with general distributions," *ACM Transactions on Mathematical Software*, vol. 3, pp. 253-256, 1977.

C.S. Wallace, "Transformed rejection generators for gamma and normal pseudo-random variables," *Australian Computer journal*, vol. 8, pp. 103-105, 1976.

M.T. Wasan and L.K. Roy, "Tables of inverse gaussian probabilities," *Annals of Mathematical Statistics*, vol. 38, p. 299, 1967.

G.S. Watson, "Goodness-of-fit tests on a circle.I.," *Biometrika*, vol. 48, pp. 109-114, 1961.

G.S. Watson, "Goodness-of-fit tests on a circle.II.," *Biometrika*, vol. 49, pp. 57-63, 1962.

R.L. Wheeden and A. Zygmund, *Measure and Integral*, Marcel Dekker, New York, N.Y., 1977.

W. Whitt, "Bivariate distributions with given marginals," *Annals of Statistics*, vol. 4, pp. 1280-1289, 1976.

E.T. Whittaker and G.N. Watson, *A Course of Modern Analysis*, Cambridge University Press, Cambridge, 1927.

J. Whittaker, "Generating gamma and beta random variables with non-integral shape parameters," *Applied Statistics*, vol. 23, pp. 210-214, 1974.

W.A. Whitworth, *Choice and Chance,* Cambridge University Press, Cambridge, 1897.

D.V. Widder, *The Laplace Transform,* Princeton University Press, Princeton, N.J., 1941.

H.S. Wilf, "A unified setting for sequencing, ranking and random selection of combinatorial objects," *Advances in Mathematics*, vol. 24, pp. 281-291, 1977.

H.S. Wilf, "The uniform selection of free trees," *Journal of Algorithms*, vol. 2, pp. 204-207, 1981.

S.S. Wilks, *Mathematical Statistics,* Wiley, New York, 1962.

E.B. Wilson and M.M. Hilferty, "The distribution of chi-square," *Proceedings of the National Academy of Sciences*, vol. 17, pp. 684-688, 1931.

C.K. Wong and M.C. Easton, "An efficient method for weighted sampling without replacement," *SIAM Journal on Computing*, vol. 9, pp. 111-113, 1980.

N.C. Wormald, "Generating random regular graphs," *Journal of Algorithms*, vol. 5, pp. 247-280, 1984.

F.P. Wyman, "Improved event-scanning mechanisms for discrete event simulation," *Communications of the ACM*, vol. 18, pp. 350-353, 1975.

S.J. Yakowitz, *Computational Probability and Simulation,* Addison-Wesley, Reading, Mass., 1977.

C. Yuen, "The inversion method in random variate generation," M.Sc. Thesis, School of Computer Science, McGill University, Montreal, Canada, 1981.

M. Zelen and N.C. Severo, "Probability functions," in *Handbook of Mathematical Functions*, ed. M. Abramowitz and I.A. Stegun, pp. 925-995, Dover Publications, New York, N.Y., 1972.

K.S. Ziganglrov, "Expression for the Wald distribution in terms of normal distribution," *Radiotekhnika Electronika*, vol. 7, pp. 164-166, 1962.

S. Zimmerman, "An optimal search procedure," *American Mathematical Monthly*, vol. 66, pp. 690-693, 1959.

V.M. Zolotarev, "On analytic properties of stable distribution laws," *Selected Translations in Mathematical Statistics and Probability*, vol. 1, pp. 207-211, 1959.

V.M. Zolotarev, "On the representation of stable laws by integrals," *Selected Translations in Mathematical Statistics and Probability*, vol. 6, pp. 84-88, 1966.

INDEX